Development of Mathematics 1900 – 1950

Edited by
Jean-Paul Pier

Birkhäuser Verlag
Basel · Boston · Berlin

Editor

Jean-Paul Pier
Centre Universitaire de Luxembourg
Séminaire de Mathématique
162A, avenue de la Faïencerie
L-1511 Luxembourg

A CIP catalogue record for this book is available from the Library of Congress, Washington D.C., USA

Deutsche Bibliothek Cataloging-in-Publication Data

Development of mathematics 1900 – 1950 / ed. by Jean-Paul Pier. – Basel ; Boston ; Berlin : Birkhäuser, 1994

ISBN-13: 978-3-0348-9913-0 e-ISBN-13: 978-3-0348-9114-1
DOI: 10.1007/978-3-0348-9114-1

NE: Pier, Jean-Paul [Hrsg.]

Softcover reprint of the hardcover 1st edition 1994
Printed on acid-free paper produced from chlorine-free pulp
Revision and layout by *mathScreen Online*, CH-4123 Allschwil

9 8 7 6 5 4 3 2 1

Scientific Board

Local Committee

Table of Contents

Jean Dieudonné (1906–1992), Luxembourg 1981

Introduction

This volume grew out of the preparation and achievements of the Symposium organized by the Luxembourg Mathematical Society in June 1992, at Château Bourglinster (Luxembourg).

We express our best thanks to

- Monsieur Marc Fischbach, ministre de l'Education nationale, Luxembourg
- Monsieur René Steichen, ministre délégué aux Affaires culturelles, Luxembourg,
- Académie des sciences, Paris,
- Académie royale des sciences, des lettres et des beaux-arts de Belgique,
- Section des sciences de l'Institut grand-ducal, Luxembourg,
- and all lecturers and participants.

We hope that this contribution may ease a better understanding of the vitality in mathematics during the first half of the present century and provide clues for the explanation of its influence on later developments.

We say our gratitude to persons who assisted us with their advice and help for the technical realization of the volume, especially Mrs. Suzanne D'Addato-Muës, Mr. François Gaillard, Mr. Hubert Glesener. We are also grateful to Birkhäuser-Verlag for having accepted this publication.

Now the book should speak for itself . . .

April 1993 Jean-Paul Pier

Bourglinster 1992

List of participants

ARNAL DIDIER, Université de Metz F
AUSEJO ELENA, Facultad de Ciencias Zaragoza E

BAAYEN P. C., Centrum voor Wiskunde Amsterdam NL
BACKLUND ULF, University of Umea S
VAN BEELE JEROEN, J.Rijksuniversiteit Leiden NL
BERG CHRISTIAN, Københavns Universiteit DK
BINDER CHRISTA, Technische Universität Wien A
BISDORFF RAYMOND, Luxembourg L
BLIND BRUNO, Université de Nancy I F
BOHNKE GEORGES, Université de Nancy I F
BOSBACH CHRISTOF, Universität Trier D
BREMER JEAN-CLAUDE, Luxembourg L
BREYER MARCO, Luxembourg L
BRIGAGLIA ALDO, Università di Palermo I
BUEKENHOUT FRANCIS, Université Libre de Bruxelles B
BUREAU FLORENT, Université de Liège B

CHATTERJI SRISHTI D., Ecole polytechnique fédérale de Lausanne CH
CLEMENT ALBERT, Luxembourg L
COEN SALVATORE, Università degli studi di Bologna I
COLACITO FABRIZIO, Université de Mons-Hainaut B
CONTESSA MARIA, Università di Palermo I

DECKER EVA, Universität Karlsruhe D
DELAGARDELLE JEAN-CLAUDE, Luxembourg L
DI SIENO SIMONETTA, Università L. Bocconi Milano I
DOMBROWSKI PETER, Universität Köln D
DOOB JOSEPH L., University of Illinois USA
DOYEN JEAN, Université Libre de Bruxelles B

EYSCHEN JOSEPH, Luxembourg L

FÆLLSTRŒM ANDERS, University of Umea S
FICHERA GAETANO, Università di Roma I

GELFAND IZRAIL, Moscow Academy of Sciences / Rutgers University USA
GILAIN CHRISTIAN, Université Pierre-et-Marie-Curie Paris F
GLESENER HUBERT, Luxembourg L
GLODEN RAOUL, Euratom Ispra I
GLODT ROMAIN, Luxembourg L
GOLD BONNIE, Université Catholique de Louvain B
GREMLING MARIELOU, Luxembourg L
GRIMEE ETIENNE, Arlon B

GRIMEE JACQUELINE, Luxembourg L
GRIMEE-MORLOT VÉRONIQUE, Arlon B
GUERRAGIO ANGELO, Università L. Bocconi Milano I
GUILLAUME MARCEL, Université Blaise-Pascal Clermont-Ferrand F

HAYMAN WALTER, University of York UK
HEINTZ ROLAND, Luxembourg L
HINZ ANDREAS M., Universität München D
HORMIGON MARIANO, Facultad de Ciencias Zaragoza E
HOUZEL CHRISTIAN, Université Paris VII F

JACOBS KONRAD, Universität Erlangen-Nürnberg D
JUNG CLAUDE, Luxembourg L

KALF HUBERT, Universität München D
KASS GUY, Luxembourg L
KASS-HENGESCH DENISE, Luxembourg L
KIEFFER LUCIEN, Luxembourg L
KLEIN SUSANNE, St.Ingbert D
KLOPP RENÉ, Luxembourg L
KRIER RENÉ, Luxembourg L

LAMBERT DOMINIQUE, Fac. univ. N.-D. de la Paix Namur B
LELONG PIERRE, Université Pierre-et-Marie-Curie Paris F
LICHNEROWICZ ANDRÉ, Collège de France Paris F
LLOMBART JOSE, Lejona E
LORENTZ NICO, Luxembourg L
LUDWIG JEAN, Luxembourg L

MARQUE FRANÇOIS, Université de Nancy I F
MAURER STEFAN, Saarbrücken D
MAWHIN JEAN, Université catholique de Louvain B
MAY GERHARD, Universität Saarbrücken D
MEDINGER JEANNOT, Luxembourg L
METZ GEORGES, Luxembourg L
MOLITOR-BRAUN CARINE, Luxembourg L
MORTINI RAYMOND, Universität Karlsruhe D
MUELLER JÜRGEN, Universität Trier D
MULLER GUY, Luxembourg L
MULLER NIKI, Luxembourg L

NAVARRO LOIDI JUAN, Luxembourg L
NIELSEN, Scitech Holten S
NIRENBERG LOUIS, Courant Institute New York USA

PEIFFER JEANNE, C.N.R.S. Paris F
PEIFFER MONIQUE, Luxembourg L

PEIFFER-REUTER RENÉE, Schirmeck F
PESCATORE FRANÇOISE, Luxembourg L
PIER JEAN-PAUL, Luxembourg L
PIPER KARIN, Universität Saarbrücken D
PONCIN NORBERT, Luxembourg L
PONT JEAN-CLAUDE, Université de Genève CH

RAEDER ROBERT, Universität Saarbrücken D
RAUSSEN MARTIN, Aalborg Universitetscenter DK
VON RENTELN MICHAEL, Universität Karlsruhe D
RIBENBOIM PAULO, Queen's University Kingston Canada
RONVEAUX ANDRÉ, Fac. univ. N.-D. de la Paix Namur B
RUPP RUDOLF, Universität Karlsruhe D

SCHABO PIERRE, Luxembourg L
SCHMIDT KLAUS D., Universität Mannheim D
SCHMITT PETER, Universität Wien A
SCHROEDER PROSPER, Luxembourg L
SCHWARZ WOLFGANG, J. W. Goethe Universität Frankfurt D
SEILER CLAUS-MARIA, Universität Saarbrücken D
STAMMBACH URS, Eidgenössische Technische Hochschule Zurich CH
STEGEMAN JAN D., Rijksuniversiteit Utrecht NL
STRELCYN JEAN-MARIE, Université de Rouen F

THIBEAU MICHEL, Luxembourg L
TILLEUIL PHILIPPE, Université catholique de Louvain B
TITS JACQUES, Collège de France F

VAN FRANKENHUYSEN MACHIEL, Universiteit Nijmegen NL
VERGER-GAUGRY JEAN-LOUIS, Université Joseph-Fourier Grenoble F
WAGENER RAYMOND, Luxembourg L
WEBER EDMOND, Luxembourg L
WEILER DANIEL, Luxembourg L
WILLEM MICHEL, Université catholique de Louvain B

Titles of lectures

JEAN DIEUDONNÉ, *Académie des Sciences de Paris*
Une brève histoire de la topologie

JOSEPH L. DOOB, *University of Illinois*
The development of rigour in mathematical probability

GAETANO FICHERA, *Università di Roma*
Vito Volterra and the birth of Functional Analysis

IZRAIL GELFAND, *Moscow Academy of Sciences/Rutgers University*
The development of Functional Analysis from 1900 to 1950 and the comparison with the second part of the century

MARCEL GUILLAUME, *Université Blaise-Pascal Clermont-Ferrand*
La logique mathématique en sa jeunesse

WALTER HAYMAN, *University of York*
Function Theory 1900–1950

CHRISTIAN HOUZEL, *Université Paris–VII*
La préhistoire des conjectures de Weil

ANDRÉ LICHNEROWICZ, *Collège de France, Paris*
Géométrie différentielle et relativité

LOUIS NIRENBERG, *Courant Institute, New York*
On partial differential equations in the first half century

WOLFGANG SCHWARZ, *J.W. Goethe-Universität, Frankfurt*
History of prime number theorem 1896–1950, with some hints to recent developments

JACQUES TITS, *Collège de France, Paris*
La théorie des groupes de Lie semi-simples: L'oeuvre d'Elie Cartan et de Hermann Weyl (1900–1950)

Guidelines 1900–1950

Pierre Dugac, Beno Eckmann, Jean Mawhin
and Jean-Paul Pier[1])

1900

BACHELIER L., Mathematical formulation for Brownian motion.

FEJÉR L., (C,1)-summability of Fourier series.

FREDHOLM I., Fredholm integral equations.

FROBENIUS G., Young's schemes.

GOURSAT E., Cauchy integral theorem for a $\mathbb{C}$-differentiable function.

HILBERT D., Direct method of the calculus of variations.
The future of mathematics (Mathematical problems).
Axioms for elementary theory of real numbers.

LEVI-CIVITA T., RICCI M.G., Absolute differential calculus.

PAINLEVÉ P., Painlevé transcendants.

PEARSON K., χ^2-test.

POINCARÉ H., Combinatorial topology of cell complexes.

WEBER E. VON, *Vorlesungen über des Pfaff'sche Problem.*

1901

BENDIXSON I., Qualitative study of the curves defined by a differential equation.

BERNSTEIN F., *Untersuchungen aus der Mengenlehre.*

DEHN M., Equivalence of polyhedra (Solution of Hilbert third problem).

DICKSON L.E., *Linear Groups with an Exposition of the Galois Field Theory.*

FROBENIUS G., Solvable groups.

HADAMARD J., Invariant manifolds for an ordinary differential equation.

---

1) We want to acknowledge *L.H. Ahlfors, F. Aribaud, V.I. Arnold, M. Berger, F. van der Blij, C. Brezinski, B. Bru, F. Buekenhout, L. Carleson, S.S. Chern, G. Choquet, S.S. Demidov, J.M. Deshouillers, J. Dieudonné, J. Dixmier, P. Dombrowski, S. Eilenberg, P. Erdös, G. Faltings, C. Fefferman, J. Ferrand, G. Glaeser, M. Guillaume, W.K. Hayman, A.M. Hinz, J.P. Kahane, H. Kalf, U. Krengel, L. Le Cam, P. Lelong, P. Liberman, S. Mac Lane, M. Mendès France, A. Meyer, A.F. Monna, I. Netuka, G. Reeb, H. Reiter, M. von Renteln, W. Rudin, I.A. Rus, P. Samuel, K.D. Schmidt, L. Schwartz, W. Schwarz, U. Stammbach, J.D. Stegeman, D.J. Struik, E. Valabrega Gibellato, M. Waldschmidt, W. Walter, M. Willem, J. Wolfart* and *S. Zervos* for their remarks, critiques, corrections and additions to the first version of these guidelines.

HOLMGREN E., Uniqueness of regular solutions of analytic Cauchy problems.

LIAPUNOV A.M., Central limit theorem.

NETTO E., *Lehrbuch der Kombinatorik.*

POINCARÉ H., Rational points on curves of genus one.

SCHUR I., Matrix representation of groups.

WILSON E.B., *Vector Analysis Founded Upon the Lectures of J. Willard Gibbs.*

1902

BIANCHI L., Bianchi theorem.

CARTAN E., Infinite Lie groups.
Equivalence of differential systems.

ESCLANGON E., Quasi-periodic functions.

FURTWÄNGLER P., Higher reciprocity theorems.

GIBBS J.W., *Elementary Principles in Statistical Mechanics.*

GOURSAT E., *Cours d'analyse mathématique I.*

HENSEL K., LANDSBERG G., *Theorie der algebraischen Funktionen einer Variablen und ihre Anwendung auf algebraische Kurven und Abelsche Integrale.*

HILBERT D., Two-dimensional manifolds.
Completeness axiom for geometry.

LANDAU E., Prime ideals theorem.

LEBESGUE H., Lebesgue integral.
Lower semicontinuity in calculus of variations.

SCHOUTE, P.H., *Mehrdimensionale Geometrie.*

WHITTAKER E.T., *A Course of Modern Analysis.*

1903

CAMPBELL J.E., *Introductory Treatise on Lie's Theory of Finite Continuous Transformation Groups.*

FROBENIUS G., Linear representation of Molien-Cartan groups.
Generalization of Sylow theorem.

HADAMARD J., Uniqueness for nonlinear Cauchy problems.
Linear functionals on $C([0,1])$.

KRAZER A., *Lehrbuch der Thetafunktionen.*

LEBESGUE H., Riemann-Lebesgue theorem for Fourier series.

LERCH M., Uniqueness of the inverse of the Laplace transform.

LUNDBERG F., Collective risk theory.

POINCARÉ H., Notion of dimension.

RUSSELL B., *The Principles of Mathematics I.*

SCHMIDT E., Irregularities in the distribution of primes.

VESSIOT E., Continuous groups and integration of linear differential equations.

VORONOÏ G.F., Dirichlet's divisor problem.

1904

D'ADHÉMAR R.; HADAMARD J., Finite parts of integrals in hyperbolic equations.

BERNSTEIN S.N., Analyticity of smooth solutions of nonlinear elliptic equations.

BOHL P., Poincaré-Bohl theorem.

BOLZA O., *Lectures on the Calculus of Variations.*

BURNSIDE W., Solvability of groups of order $p^a q^b$ where p and q are prime.
Burnside problem.

FEJÉR L., Fejér's theorem on the convergence of Fourier series.

FRÉCHET M., Definition of a precompact set.

VON KOCH H., Geometric construction of a continuous curve without tangent.

LANDAU E.; SCHOTTKY F., Extension of Picard theorem on entire functions.

LEBESGUE H., *Leçons sur l'intégration et la recherche des fonctions primitives.*
First example of a non borelian function.

MINKOWSKI H., Geometry of numbers.

POINCARÉ H., The Poincaré conjecture.

PRANDTL L., Boundary layer theory.

SCHUR I., Schur multipliers.

VEBLEN O., Axiomatics of projective geometry.

ZERMELO E., Axiom of choice.
Every set can be well ordered.

1905

BAIRE R., *Leçons sur les fonctions discontinues.*

BURNSIDE W., Burnside theorem.

EINSTEIN A., Special theory of relativity.

FISCHER E., Minimax property of eigenvalues.

HAMEL G., Hamel basis.

HILBERT D., Invariant integral in the calculus of variations.

LASKER E., Concept of primary ideal.

LEVI E.E., Levi decomposition of Lie groups.

POINCARÉ H., Geodesics on convex surfaces.

SCHUR I., Frobenius-Schur representation theory of finite groups.

VITALI G., Non Lebesgue-mesurable set.

WEDDERBURN J.H.M., Finite fields are commutative.

1906

BOHL P., Quasi-periodic functions.

DICKSON L.E., Quasifields.

FATOU P., Fatou's theorem on complex functions.
Fatou's lemma.

FRÉCHET M., Abstract spaces.

GAMBIER P., Complete list of Painlevé transcendants.

HADAMARD J., Global implicit function theorem.

HARTOGS F., Analytic functions of several complex variables.
Subharmonic and superharmonic functions in potential theory.

HAUSDORFF F., Campbell-Hausdorff formula.

HILBERT D., Symmetric linear integral equations.
Weak convergence.

JENSEN J.L.W.V., Convex function of one variable.

LEBESGUE H., *Leçons sur les séries trigonométriques.*

LEVI B., Absolutely continuous functions of two variables.
Monotone convergence theorem.

LIAPUNOV A.M., Liapunov-Schmidt method for the equations of rotating fluids.

MARKOV A.A., Markov chains.

PICARD E., SIMART G., *Théorie des fonctions algébriques de deux variables indépendantes II.*

POINCARÉ H., Mathematical structure of the Lorentz group.

WILCZYNSKI E.J., *Projective Differential Geometry of Curves and Ruled Surfaces.*

1907

BROUWER L.E.J., Intuitionism.

FISCHER E., Weak convergence.

FISCHER E.; RIESZ F., Riesz-Fischer theorem : $l^2 \sim L^2$.

FRÉCHET M., Structure of continuous functionals on L^2.

FUBINI G., Fubini's theorem.

FURTWÄNGLER P., Proof of Hilbert theorems on class fields.

GOURSAT E., Goursat kernel in integral equation.

HENSEL K., p-adic numbers.

KOEBE P.; POINCARÉ H., Proof of the uniformization theorem.

LEBESGUE H., New results and methods in Dirichlet problem.

PERRON O., Perron-Frobenius theorem on positive matrices.

RIESZ F., Definition of Hilbert space.

1908

FEJÉR, L.; DE LA VALLÉE POUSSIN C.J.; LANDAU E., New proofs of Weierstrass approximation theorem.

FREDHOLM I., Elementary solutions of elliptic operators in $\mathbb{R}^3$.

FROBENIUS G., Perron-Frobenius theorem on positive matrices.

HADAMARD J., Elementary solutions of hyperbolic operators.

HENSEL K., *Theorie der algebraischen Zahlen I.*

MARKOV A.A., *Calculus of Probabilities* (Russian), (Second ed.).

MINKOWSKI H., Minkowski's space-time.

MOORE E.H., Abstract theory of linear operators.

PHRAGMÉN E., LINDELÖF E., Phragmén-Lindelöf theorem.

PLEMELJ J., Solution of the Riemann-Hilbert problem.

POINCARÉ H., The future of mathematics.

RITZ W., Approximation method for variational problems.

RUSSELL B., Type theory.

SCHMIDT E., Orthogonal projections in Hilbert spaces.
Liapunov-Schmidt method for nonlinear integral equations.

SCHOENFLIES A., *Die Entwicklung der Lehre von den Punktmannigfaltigkeiten II.*

STEINITZ E., Cartesian product of spaces.

TIETZE H., Orientable and non-orientable manifolds. Lens spaces.

VITALI G., Vitali's covering theorem.

WEBER H., *Lehrbuch der Algebra III* (Second ed.).

WEDDERBURN J.H., Wedderburn theorems on algebras over a field.

ZERMELO E., Axiomatic set theory.

1909

BOHL P., Uniform distribution of numbers modulo one.

BOREL E., 'Countable probability' on countable sets.

EISENHART L.P., *Differential Geometry of curves and surfaces.*

ERLANG A.K., Poisson process in telephone engineering.

HAMEL G., Foundations of mechanics.

HILBERT D., Solution of Waring's problem.
Grundlagen der Geometrie (Third ed.).

LANDAU E., *Handbuch der Lehre von der Verteilung der Primzahlen.*

LEVI E.E., Dirichlet problem for higher order elliptic equations.

RIESZ F., Representations of continuous functionals over $C([0,1])$.

THUE A., Finiteness of the set of entire solutions of homogeneous diophantine equations of two variables.

ZAREMBA S., Method of orthogonal projection in Dirichlet problem.
First example of domain for which Dirichlet problem is not solvable.

1910

BERNSTEIN S.N., A priori bounds for solutions of nonlinear elliptic equations.

BROUWER L.E.J., Simplicial approximation of continuous maps.

CARTAN E., Method of moving frame.

HADAMARD J., *Leçons sur le calcul des variations I.*
Topological consequences of Kronecker index and fixed point theory.

LEBESGUE H., Differentiation of multiple integrals.
Set functions.

LEVI E.E., Pseudo-convexity in functions of several complex variables.

MINKOWSKI H., *Geometrie der Zahlen.*

PLANCHEREL M., Plancherel formula.

RIESZ F., L^p spaces and continuous functionals over L^p.

SIERPINSKI W., Uniform distribution theory.

STEINITZ E., Algebraic theory of fields.

WEYL H., Singular boundary value problems for second order linear differential equations.

WHITEHEAD A.N., RUSSELL B., *Principia Mathematica I.*

YOUNG W.H., Functional approach to integration.

1911

BERNSTEIN S.N.; JACKSON D., Best approximation of continuous functions by polynomials (solution of de La Vallée Poussin problem).

BROUWER L.E.J., Invariance of dimension.
Extension of Jordan's separation theorem.

BURNSIDE W., Burnside conjecture.
Theory of Groups of Finite Order (Second ed.).

EGOROV D.F., Egorov theorem on sequences of measurable functions.

HERGLOTZ G.; RIESZ F., Representation theorem for holomorphic functions on the unit disc with positive real part.

LITTLEWOOD J.E., Tauberian theorem.

MINKOWSKI H., (posthumous publications), Mixed volumes.
Extremal points in convex sets.

MONTEL P., Normal families of functions.

1912

BACHELIER L., *Calcul des probabilités.*

BERNSTEIN S.N., Growth conditions in the equations of the calculus of variations.

BOREL E., Quasi-analytic functions.

BROUWER L.E.J., Topological degree of maps of manifolds.
Brouwer fixed point theorem.

DEHN M., Word problem for finitely presented groups.

DENJOY A., Totalisation.

DRACH J., Logical integration of ordinary differential equations.

FRÉCHET M., Fréchet differential.

HEAVISIDE O., *Electromagnetic Theory III.*

HILBERT D., *Grundzüge einer allgemeinen Theorie der linearen Integralgleichungen.*

LALESCO T., *Introduction à la théorie des équations intégrales.*

LANDAU E., General lattice point theorem.

LUZIN N.N., Luzin theorem on measurable functions.

PERRON O., *Die Lehre von den Kettenbrüchen.*

POINCARÉ H., *Calcul des probabilités* (Second ed.).
Poincaré-Birkhoff fixed point theorem.

POMPEIU D., Areolar derivative.

SEVERI F., Topological aspects of classical algebraic geometry.

SOMMERFELD A., Sommerfeld radiation condition.

STUDY E., Triality principle.

VASILEV N.A., Imaginary (non aristotelician) logic.

WEYL H., Asymptotic law for eigenvalues.

1913

BIRKHOFF G.D., Proof of Poincaré's last geometric theorem.
Riemann-Hilbert problem for a disc.

BOLZA O., Bolza problem in the calculus of variations.

CARATHÉODORY C., Geometric function theory.
Solution of Osgood problem.

CARTAN E., Theory of spinors.

GEVREY M., Method of potentials in parabolic equations.

HENSEL K., *Zahlentheorie.*

KÜRSCHAK J., Valuation of fields.

LEBESGUE H., Dirichlet problems without solution.

LUZIN N.N., Luzin's conjecture for Fourier series.

MACAULAY F.S., Distinction between 'immersed' and 'non-immersed' primary ideals.

RADON J., Radon measure.

RIESZ F., *Les systèmes d'équations linéaires à une infinité d'inconnues.*

SUNDMANN S., Convergent series for the solution of the three body problem.

VOLTERRA V., *Leçons sur les fonctions de ligne.*

WEYL H., *Die Idee der Riemannschen Fläche.*

YOUNG W.H., Young inequalities.

1914

CARATHÉODORY C., p-dimensional measure in $\mathbb{R}^n$.

CARTAN E., Classification of real semi-simple Lie algebras.

FRAENKEL A.A., Concept of abstract ring.

HARDY G.H., Infinitely many zeros on the critical line for the zeta function.

HAUSDORFF F., *Grundzüge der Mengenlehre.*

LIAPUNOV A.M., Figures of equilibrium of rotating fluids.

LITTLEWOOD J.E., Irregularities in the distribution of primes.

PERRON O., Perron's integral.

PÓLYA G., Entire functions with entire values.

1915

ALEXANDER J.W., Topological invariance of Betti numbers and torsion coefficients of a manifold.

BRUN V., Brun's sieve.

EINSTEIN A., General relativity.

GALERKIN B.G., Galerkin's approximation method.

HARDY G.H., H^p spaces. Lower estimate for the remainder in the circle problem.

LICHTENSTEIN L., Hilbert space techniques in the calculus of variations.

LUZIN N.N., *Integral and Trigonometric Series* (Russian).

MOORE E.H., Limit concept in general analysis.

PERRON O., Differential inequalities in the fundamental theory of ordinary differential equations.

SIERPINSKI W., Sierpinski triangle (gasket).

1916

BIEBERBACH L., Bieberbach conjecture.

BLASCHKE, W., Selection theorem for convex bodies.

LANDAU E., *Darstellung und Begründung einiger neuerer Ergebnisse der Funktionentheorie.*

DE LA VALLÉE POUSSIN C.J., *Intégrales de Lebesgue, fonctions d'ensemble, classes de Baire.*

LÖWENHEIM L., Existence of a countable model.

MACAULEY F.S., *The Algebraic Theory of Modular Systems.*
Polynomial ideals.

RAMANUJAN S., τ-function.

RIESZ F., RIESZ M., F. and M. Riesz theorem on analytic functions.

WEYL H., Uniform distribution of $(n\alpha)$ and $(P(n))$.

1917

BERNSTEIN S.N., Axiomatics of probability theory.

BIRKHOFF G.D., First minimax theorem in critical point theory.

HARDY G.H., RAMANUJAN S., Circle method.

HECKE E., Representation of prime ideals in the class of ideals.
Analytic continuation, functional equation and parts of Dedekind's Zeta function.
Generalized Dirichlet L-series for algebraic fields.

LEVI-CIVITA T., Parallel transport in Riemannian manifolds.

MIRIMANOFF D., Theory of cumulative types.

MORDELL L.J., Multiplicativity of Ramanujan τ-function.

OSTROWSKI A., Characterization of valuation on rationals.

RADON J., Radon transform.

SUSLIN M.YA., Analytic sets.

1918

BERNAYS P., Completeness and independence of axioms of the calculus of propositions, method of 'logical matrices'.

BROUWER L.E.J., Intuitionistic set theory.

CARATHÉODORY C., *Vorlesungen über reelle Funktionen.*

DANIELL P.J., Daniell's integral.

DUFFING G., Duffing's equation.

FINSLER P., *Ueber Kurven und Flächen in allgemeinen Räumen.*

GEVREY M., Gevrey classes.

HECKE E., L-functions with 'Grössencharakter'.
Decomposition of L-functions in Eulerian products.

HILBERT D., Problem of the axiomatization of mathematics.

JULIA G., Iteration of rational functions.

LEWIS C.I., *A Survey of Symbolic Logic.*

NOETHER E., Noether theorem for equivariant problems in the calculus of variations.

RAMANUJAN S., Trigonometric sums.

RIESZ F., Spectral theory of compact operators.

VEBLEN O., *Projective Geometry II.*

WEYL H., *Raum, Zeit, Materie. Vorlesungen über allgemeine Relativitätstheorie.*
Spaces with affine connexion.
Pure infinitesimal geometry.
Das Kontinuum.

1919

BRUN V., Eratosthènes' sieve.
Scarcity of prime twins.

FATOU P., Iteration of rational functions.

GATEAUX R., (posthumous publication) Integral on functional domains and functions of infinitely many variables.

HAUSDORFF F., Hausdorff measure and dimension.

VON MISES R., Foundations of the calculus of probability.

RAMANUJAN S., Rogers-Ramanujan identities.

1920

HARDY G.H., LITTLEWOOD J.E., Waring problem.

LAMSON K.W., Contraction mapping theorem in function spaces.

ŁUKASIEWICZ J., Three-valued logics.

SKOLEM T., Proof of the Löwenheim-Skolem theorem on the existence of a countable model.

TAKAGI T., Class field theory; abelian extensions of an algebraic number field.

VAN DER POL B., Van der Pol's equation for relaxation oscillations.

WIENER N., Normed vector spaces.

1921

ANTOINE L., Antoine sets in $\mathbb{R}^3$.

ARTIN E., Reciprocity law.
Function fields on a finite field.

BLASCHKE W., *Vorlesungen über Differentialgeometrie und geometrische Grundlagen von Einsteins Relativitätstheorie I.*

BOREL E., Game theory.

DANIELL P.J., Convolution of measures.

HELLY E., First example of nonreflexive Banach space.

NOETHER E., Noetherian rings.

NOETHER F., Singular integral equations.

POST E.L., Many-valued logics.

SEVERI F., *Vorlesungen über algebraische Geometrie.*

SIEGEL C.L., Approximation of algebraic numbers.

TONELLI L., *Fundamentali di calcolo delle variazioni I.*

WIENER N., Mathematical theory of Brownian motion.

1922

ALEXANDER J.W., Alexander duality for closed subsets of S^n.

BANACH S., Complete normed vector spaces.
Contraction mapping theorem in complete normed spaces.

BANACH S.; HAHN H., Uniform boundedness theorem.

BIRKHOFF G.D., KELLOGG O.D., Extension of Brouwer fixed point theorem to some function spaces.

CARTAN E., *Leçons sur les invariants intégraux.*

VAN DER CORPUT J.G., Estimates of sums of exponentials.

FRAENKEL A.A., Substitution axiom and weak form of the independence of the axiom of choice.

GOURSAT E., *Leçons sur le problème de Pfaff.*

HAMEL G., First existence proof for the Duffing's equation.

HILBERT D., Metamathematics and proof theory.

LÉVY P., *Leçons d'analyse fonctionnelle.*

MENGER K.; URYSOHN P., Dimension of metric spaces.

MOORE E.H., SMITH H.L., Moore-Smith convergence.

MORDELL L.J., Finite rank theorem.
Mordell conjecture.
Rational solutions of equations of degree smaller than four.

NEVANLINNA F., NEVANLINNA R., Behavior of analytic functions near a singular point or singular line.

RADEMACHER H., Theorem on almost sure convergence of orthogonal series.

SCHOUTEN J.A., Different types of connections.

SKOLEM T., Final form of Zermelo-Fraenkel axiomatics for set theory.

VEBLEN O., *Analysis Situs.*
Geometry of paths.

1923

BIRKHOFF G.D., *Relativity and Modern Physics.*

BLASCHKE W. *Vorlesungen über Differentialgeometrie* (Vol. II).
Affine Differentialgeometrie.

BOHR H., Real almost periodic functions.

BROUWER L.E.J., Intuitionistic theory of real functions of a real variable.

CARLEMAN T., *Sur les intégrales singulières à noyau réel et symétrique.*

DULAC H., Results on the number of limit cycles.

EDDINGTON A.S., *The Mathematical Theory of Relativity.*

HADAMARD J., *Lectures on Cauchy's Problem in Linear Partial Differential Equations.*

HASSE H., Quadratic forms on algebraic number fields.

HAUSDORFF F., Hausdorff-Young inequalities.

HECKE E. *Vorlesungen über die Theorie der algebraischen Zahlen.*

HILDEBRANDT T.H., First version of Banach-Steinhaus theorem.

KOLMOGOROV A.N., L^1-function with an a.e. divergent Fourier series.

KÜNNETH H., Künneth formula.

LEFSCHETZ S., Lefschetz fixed point formula.

PERRON O., Subharmonic functions in Dirichlet problem.

SKOLEM T., Recursive arithmetics.

STEINHAUS H., Probability as a measure.

STEKLOV V.A., *Main problems of mathematical physics* (Russian).

TRICOMI F., Partial differential equations of mixed type.

WEITZENBÖCK R., *Invariantentheorie.*

1924

ALEXANDROV P.S., Denjoy and Perron integrals are equivalent.
Locally compact spaces.

ALEXANDROV P.S., URYSOHN P.S., Concept of compact space.

ARTIN E., Artin's conjecture.

BANACH S., TARSKI A., Banach-Tarski-Hausdorff paradox.

COOLIDGE J.L., *The Geometry of the Complex Domain.*

COURANT R., HILBERT D., *Methoden der mathematischen Physik I.*

FUETER R., *Vorlesungen über die singulären Moduln und die komplexe Multiplikation der elliptischen Funktionen I.*

HARDY G.H., *Order of Infinity* (Second ed.).

KHINTCHIN A.YA., Law of iterated logarithm in Bernoulli case.

KNESER H., Topological equivalence of differential systems.

LEFSCHETZ S., *L'analysis situs et la géométrie algébrique.*

NEVANLINNA R., Estimates for meromorphic functions.

PÓLYA G., Seventeen wallpaper groups (Alhambra groups).

RIESZ M., Inequalities for conjugate functions.

SCHOUTEN J.A., *Ricci-Kalkül.*

TARSKI A., Finite sets.

WIENER N., Capacity in potential theory.

1925

ARTIN E., Theory of braids.

CARTAN E., *La géométrie des espaces de Riemann.*

FISCHER R.A., *Statistical Methods for Research Workers.*

HEYTING A., Axiomatics of intuitionistic mathematics.

KHINTCHIN A. YA., KOLMOGOROV A.N., Convergence of random series.

KLOOSTERMAN H.D., Kloosterman sums in number theory.

KOLMOGOROV A.N., Interpretability of classical mathematics in intuitionistic mathematics.

KRULL W., Groups with operators.

LAVRENTIEV M.A., Lavrentiev phenomenon.

LEJA F.; SCHREIER O., Topological groups.

LÉVY P., *Calcul des probabilités.*

MORSE M., Morse lemma and Morse inequalities.

VON NEUMANN J., Another axiomatics for set theory.

PÓLYA G., SZEGÖ G., *Aufgaben und Lehrsätze aus der Analysis.*

RADÓ T., Study of Riemann surfaces.

URYSOHN P.S., Normal spaces and extensions of bounded continuous functions.

WEYL H., Linear representation of semi-simple Lie groups.

1926

ALEXANDER J.W., Topological invariance of homology groups of polyhedra.

ARTIN E., Solution of the 17th Hilbert problem on rational functions.

ARTIN E., SCHREIER O., Formally real fields.

BERNAYS P., Consistency and independence of axioms of set theory.

CARLEMAN T., *Les fonctions quasi-analytiques.*

CARTAN E., Holonomy groups.

DIRAC P.A.M., Dirac 'function'.

EISENHART L.P., *Riemannian Geometry.*

FUBINI G., ČECH E., *Geometria proiettiva differenziale.*

GONSETH F., *Les fondements des mathématiques.*

HASSE H., *Höhere Algebra I.*

HOPF H., Mappings from S^n into S^n having the same degree are homotopic. Poincaré-Hopf theorem on singularities of vector fields on a manifold.

KOLMOGOROV A.N., An L^1-function with everywhere divergent Fourier series.

LEFSCHETZ S., Intersections in homology.

NOETHER E., Axiomatics of Dedekind rings.

RIESZ F., Subharmonic functions and potential theory.

SCHRÖDINGER E., Schrödinger equation.

VAN DER WAERDEN B.L., Precise determination of generic points of a variety.

WEYL H., Linear representations of semi-simple Lie groups (continued). Universal cover of a compact semi-simple Lie group.

WIENER N., Generalized Fourier transform and operational calculus.

1927

Artin E., Reciprocity law in class fields theory.

Banach S., Steinhaus H., Banach-Steinhaus theorem.

Birkhoff G.D., *Dynamical Systems.*

Bochner S., Generalized Fourier integrals.

Eisenhart L.P., *Non-Riemannian Geometry.*

Grell H., Rings of fractions.

Hahn H., Hahn-Banach theorem.
Dual space.

Hildebrandt T.H., Graves L.M., Implicit function theorem in Banach spaces.

Hopf E., Maximum principle for second order elliptic equations.

Lefschetz S., Relative homology.

Montel P., *Leçons sur les familles normales de fonctions analytiques et leurs applications.*

Peter F., Weyl H., Peter-Weyl theorem.

Riesz M., L^p-boundedness of the Hilbert transform.

Schauder J., Extension of Brouwer fixed point theorem to Banach spaces with bases.

Sidon S., Absolutely convergent lacunary Fourier series.

Vietoris L., Topologically defined homology groups.

van der Waerden B.L., van der Waerden theorem.

1928

Ackermann W., Ackermann functions.

Ackermann W., Hilbert D., *Grundzüge der theoretischen Logik.*

Alexander J.W., Torsion numbers and Alexander polynomials of classical knots.

Cartan E., *Leçons sur la géométrie des espaces de Riemann.*

Courant R., Friedrichs K.O., Lewy H., Difference methods for partial differential equations.

Einstein A., Absolute parallelism.

Fatou P., Averaging method for differential equations.

Fréchet M., *Les espaces abstraits.*

Grötzsch H., Quasi-conformal mappings.

Hellinger E., Toeplitz O., *Integralgleichungen und Gleichungen mit unendlichvielen Unbekannten.*

Herglotz G., Elementary solutions of partial differential equations.

Hopf H., Extension of Lefschetz fixed point formula.
Use of 'abstract' algebra in topology.

KRULL W., Galois theory of infinite algebraic extensions.

LIÉNARD A., Existence of limit cycles in second order differential equations.

MENGER K., *Dimensionstheorie.*

VON NEUMANN J., Minimax theorem in game theory.

RIESZ F., Ordered vector spaces.

SPERNER E., Sperner's lemma.

STOILOW S., Topological theory of functions.

VOLTERRA V., Lotka-Volterra systems in population dynamics.

WEIL A., Mordell-Weil theorem (the rank of a curve is finite for every genus).

WEYL H., *Gruppentheorie und Quantenmechanik.*

1929

AHLFORS L.V., Resolution of Denjoy's conjecture on the number of asymptotic values of an entire function of finite order.

ALEXANDROV P.S., Approximation of compact spaces by simplicial complexes.

ANDRONOV A.A., Self-oscillations in Van der Pol type equations are limit cycles.

BANACH S., Open mapping theorem.

BRAUER R., Brauer groups of central simple algebras over a field.

CARTAN E., Characterisation of compact Lie groups.

FANTAPPIÉ L., Analytical functionals.

FURTWÄNGLER P., Proof of the theorem on principal ideals of class fields.

GELFOND A.O., Transcendence problems in meromorphic functions.

KELLOGG O.D., *Foundations of Potential Theory.*

KNASTER B., KURATOWSKI K., MAZURKIEWICZ S., KKM lemma in fixed point theory.

KOLMOGOROV A.N., General law of the iterated logarithm.

LEFSCHETZ S., Poincaré-Lefschetz duality theorem.

VON NEUMANN J., Amenable groups.
General measure theory.
Hilbert spaces.

NEVANLINNA R., *Le théorème de Picard et la théorie des fonctions méromorphes.*

NOETHER E., Beginning of 'modern' algebra : groups, rings, ideals, representation theory.

PRESBURGER M., Completeness of the theory of addition on natural integers.

RAMSEY F.P., Ramsey theory.

SIEGEL C.L., Transcendental numbers and diophantine equations (finiteness on the number of integer points on curves of genus one).

STERNBERG W., Perron method for parabolic equations.

1930

AHLFORS L., Conformal mappings and entire functions.

BRELOT M., General theory of Dirichlet problem.

CARTAN E., *La théorie des groupes finis et continus et l'analysis situs.*

CHEVALLEY C.; HASSE H., Local class field theory.

CRAMÉR H., Risk theory.

CURRY H.B., Combinatorial logic.

DIRAC P.A.M., *The Principles of Quantum Mechanics.*

DOUGLAS J.; RADÓ T., Solution of Plateau's problem.

FATOU P., *Fonctions automorphes.*

FISCHER R.A., *Generical Theory of Natural Selection.*

GÖDEL K., Incompleteness and completeness theorems.

HAMMERSTEIN A., Nonlinear integral equations.

HARDY G.H., LITTLEWOOD J.E., Hardy-Littlewood maximal theorem.

HERBRAND J., *Recherches sur la théorie de la démonstration.*

HEYTING A., *Die formalen Regeln der intuitionistischen Logik.*

HOHEISEL G., Prime numbers in small intervals.

KARAMATA J., Tauberian theorem; slowly varying functions.

KOLMOGOROV A.N., The strong law of large numbers.

LANDAU E. *Grundlagen der Analysis.*

DE LA VALLÉE POUSSIN C.J., Generalized sweeping method in potential theory.

LEFSCHETZ S., *Topology.*

LUZIN N.N., *Leçons sur les ensembles analytiques et leurs applications.*

LYUSTERNIK L.A., SCHNIRELMANN L.G., *Topological Methods in Variational Problems* (Russian).

MAGNUS W., Freiheitssatz in group theory.

MAZUR S., Mazur theorem on the convex closure of compact sets.

MENGER K., NÖBELING G., Any separable metric space of dimension n is homeomorphic to a subspace of $\mathbb{R}^{2n+1}$.

VON NEUMANN J., Spectral theory of self-adjoint operators.

NIKODÝM O., Radon-Nikodým theorem.

PALEY R.E.A.C., ZYGMUND A., Series of random functions.

POLLACZEK F., G/G/1-queues.

RELLICH F., Compact injection of H^1 into L^2.

RIESZ F., Subharmonic functions and potential theory.
Riesz spaces.

STEINHAUS H., Probability for the convergence circule of an entire function to be a natural boundary.

STEINITZ E., *Algebraische Theorie der Körper.*

STONE M.H., One-to-one correspondence between self-adjoint operators and strongly continuous one-parameter unitary groups.

TARSKI A., Fundamental concepts of the methodology of deductive sciences.

TYCHONOFF A.N., Tychonoff theorem.

VAN DER WAERDEN B.L., Algebraic varieties are triangulable.
Moderne Algebra I (after E. ARTIN and E. NOETHER).

WIENER N., Generalized harmonic analysis.

ZORN M., Theory of alternative fields.

1931

BIRNBAUM Z.W., ORLICZ, W., Orlicz spaces.

BORSUK K., Theory of retracts.

CARTAN H., Convexity property of domains of holomorphy.

CASIMIR H.B.G., Casimir operators.

DAVENPORT H., Asymptotic formula for the distribution of quadratic residues.

FREUDENTHAL H., Ends of spaces and groups.

HOPF E., WIENER N. Solution of integral equations through factorization.
Wiener-Hopf integral equations.

HOPF H., Homotopically nontrivial maps from S^3 to S^2.
Hopf invariant.
Hopf fibration.

HOPF H., RINOW W., Hopf-Rinow theorem on complete Riemannian manifolds.

IKEHARA S., Tauberian theorem.

KOLMOGOROV A.N., Diffusion equation and stochastic processes.

LANGER R.E., Asymptotic solutions of linear differential equations.

VON NEUMANN J., Uniqueness of the operators in the canonical commutation relations.

DE RHAM G., The de Rham theorem.

TARSKI A., Decidability of elementary geometry.

1932

ASCOLI G., Linear manifolds of linear metric spaces.

BANACH S., *Théorie des opérations linéaires.*

BESICOVITCH A.S., *Almost Periodic Functions.*

BIRKHOFF G.D., Pointwise ergodic theorem.

BOCHNER S., *Vorlesungen über Fouriersche Integrale.*

BOGOLIUBOV N.N., KRYLOV, N.M., Invariant measures in dynamical systems.

BOHR H., *Fastperiodische Funktionen.*

BOULIGAND G., *Introduction à la géométrie infinitésimale directe.*

BRAUER R., Structure of central division algebras.

BRAUER R., HASSE H., NOETHER E., Structure of central simple algebras.

CARTAN H., THULLEN P., Characterisation of the domain of holomorphy by holomorphic convexity.

ČECH E., Čech homology.
Higher homotopy groups.

CHURCH A., λ-decidability (λ-calculus).

DENJOY A., Differential equations on a torus.

HOPF H., Hopf program in differential geometry.

KOLMOGOROV A.N., Computations of problems.

KRULL W., General concept of valuation.

MAHLER K., Classification of transcendental numbers.

VON NEUMANN J., *Mathematische Grundlagen der Quantenmechanik.*
Mean ergodic theorem.

NIKODÝM O., Dirichlet integral as a pseudo-norm.

PALEY R.E.A.C., WIENER N., ZYGMUND A., Random functions.

PETER R., Recursive functions.

SEIFERT H., Seifert fiber space.

STONE M.H., *Linear Transformations in Hilbert Space and their Application to Analysis.*

TARSKI A., Tarski theorem on the concept of truth.

VAN DANTZIG D., Topological rings and fields.

VEBLEN O., WHITEHEAD J.H.C., *The Foundations of Differential Geometry.*

WIENER N., Tauberian theorems.
Partition of unity.
Absolutely convergent Fourier series.

1933

ANDRONOV A.A., Cycle birth bifurcation in two-dimensional systems.

BEURLING A., Beurling inequality for holomorphic functions.

BIRKHOFF G., Lattice theory.

BOCHNER S., Bochner integral.

BORSUK K., Antipodal theorem.

CARLEMAN T., Estimates for the unique continuation of solutions of some partial differential equations.

GRAVES L.M., Lagrange problem in the calculus of variations.

HAAR A., Haar measure.

HALL P., p-groups.

HASSE H., Riemann conjecture for the congruence zeta-function.

HASSE H., SCHMIDT F.K., Characterization of locally compact nonarchimedian fields.

HODGE W.V.D., Hodge theory.

KÄHLER E., Kähler metric (Hermitian metric without torsion).

KHINTCHIN A.YA., *Asymptotische Gesetze der Wahrscheinlichkeitsrechnung.*

KOLMOGOROV A.N., *Grundbegriffe der Wahrscheinlichkeitsrechnung.*

KURATOWSKI K., *Topologie I.*

LEFSCHETZ S., Singular homology.

VON NEUMANN J., Contributions to Hilbert fifth problem for compact groups.

NEYMAN J., PEARSON E.S., Neyman-Pearson lemma, precise concept of statistical test.

PETROVSKY I.G., Theory of real algebraic varieties.

SAKS S., *Théorie de l'intégrale.*

SKOLEM T., Nonstandard models for integers.

VAN DER WAERDEN B.L., Foundations of algebraic geometry.

WIENER N., *The Fourier Integral and Certain of its Applications.*

YOUNG L.C., Generalized curves in the calculus of variations.

1934

BARBILIAN D., Ring geometry.

BEHNKE H., THULLEN P., *Theorie der Funktionen mehrerer komplexer Veränderlichen.*

BONNESEN T, FENCHEL W., *Theorie der konvexen Körper.*

CACCIOPPOLI R.; SCHAUDER J., Elliptic linear equations with Hölderian coefficients.

CAIRNS S.S., Triangulation of manifolds.

CARLEMAN T., Spectral resolution of Schrödinger operator.
Asymptotic relation for the eigenfunctions of a membrane.

EHRESMANN C., Topology of homogeneous spaces of compact Lie groups.

FRIEDRICHS K.O., Friedrichs extension of differential operators.

GELFOND A.O.; SCHNEIDER T., Solution of Hilbert seventh problem.

GÖDEL K., Concept of 'Herbrand-Gödel' recursivity.

HARDY G.H., LITTLEWOOD, J.E., PÓLYA, G., *Inequalities.*

HEILBRONN, H., There are at most ten imaginary-quadratic number fields with class number one.

HILBERT D., BERNAYS P., *Grundlagen der Mathematik I.*

JORDAN P., VON NEUMANN J., WIGNER E., Jordan's algebras.

KÄHLER E., *Einführung in die Theorie der Systeme von Differentialgleichungen.*

KHINTCHIN A. YA., Stationary processes.

KOLMOGOROV A.N., Normable topological vector spaces.

KÖTHE, G., TOEPLITZ O., Köthe spaces.

LERAY J., Existence of solutions of Navier-Stokes equations.
Generalized derivatives.
Mollifiers.

LERAY J., SCHAUDER, J., Degree theory in Banach spaces and method of a priori bounds.

LÉVY P., Lévy-Khintchin formula.
Absolute convergence of Fourier series.

MORSE M., *The Calculus of Variations in the Large.*

VON NEUMANN J., Almost periodic functions on an abstract group.

PALEY R.E.A.C., WIENER, N., *Fourier Transforms in the Complex Domain.*

PONTRYAGIN L.S., Pontryagin duality.

RIESZ F., Riesz theorem on the identification of a Hilbert space and its dual.

SCHNEIDER T., Solution of Hilbert seventh problem.

SEIFERT H., THRELFALL W., *Lehrbuch der Topologie.*

STEINITZ E., *Vorlesungen über die Theorie der Polyeder unter Einschluss der Elemente der Topologie ergänzt von H. Rademacher.*

STONE M., Representation theorem of Boolean algebras.

TCHETAEV N.G., Instability theorems for solutions of ordinary differential equations using Liapunov functions.

THOMAS T.Y., *Generalized Spaces.*

TURÁN P., Turán-Kubilius inequality.

VINOGRADOV I.M., Improvement of estimates in Waring problem.

WHITNEY H., Extension theorem of C^∞ functions on closed sets.

1935

AHLFORS L.V.; LAVRENTIEV M.A., Quasi-conformal mappings.

ALEXANDER J.W.; GORDON I.; KOLMOGOROV A.N., Beginnings of cohomology theory of groups and rings.

ALEXANDROV P.S., HOPF H., *Topologie I.*

BIRKHOFF G., Universal algebra.

BLASCHKE W., *Vorlesungen über Integralgeometrie I.*

BOURBAKI N., Bourbaki's first paper (on Carathéodory's measure in topological spaces).

BRAUER R., Modular characters.

BRAUER R., WEYL H., Clifford algebras and spinors.

CARATHÉODORY C., *Variationsrechnung und partielle Differentialgleichungen erster Ordnung.*

CARTAN H., Groups of analytic transformations in $\mathbb{C}^n$.

COXETER H.S.M., Classification of finite Coxeter groups.

DELSARTE J., Mean periodic functions.

FELLER W., Necessary and sufficient conditions for the central limit theorem.

FROSTMAN O., Radon integral in potential theory.

GENTZEN G., Hauptsatz.
Logic deduction.

HOPF H., Homotopically non-trivial maps $S^{4k-1} \to S^{2k}$.

HUREWICZ H., Homotopy type and homotopy groups.
Aspherical spaces and fundamental group.
Hurewicz homomorphism.

JESSEN B., MARCINKIEWICZ J., ZYGMUND A., Differentiation of multiple integrals on rectangular basis.

KACZMARZ S., STEINHAUS H., *Theorie der Orthogonalreihen.*

VAN KAMPEN E.R., Pontryagin-van Kampen duality.

KANTOROVITCH L.V., Partially ordered spaces.

KRULL W., *Idealtheorie.*

MANDELBROJT S., *Séries de Fourier et classes quasi-analytiques de fonctions.*

MYERS S.B., Riemannian manifolds in the large.

VON NEUMANN J., Locally convex topological vector spaces.

SCHAUDER J., Functional analytic approach to hyperbolic equations.

SIEGEL C.L., Analytic theory of quadratic forms.
Prime number theorem of Siegel-Walfisz.

SKOLEM T., New proof of Thue results by p-adic methods.

STONE M.H., Generalized Boolean algebras.

TITCHMARSH E.C., Computations on the validity of the Riemann conjecture.

WHITNEY H. Fiber spaces (sphere bundles).
Characteristic classes.

ZARISKI O., *Algebraic Surfaces.*

ZORN M., Zorn lemma.

ZYGMUND A., *Trigonometrical Series.*

1936

ADO I.D., Ado's theorem on Lie algebras.

BIRKHOFF G., VON NEUMANN J., Logic of quantum mechanics.

CARTAN E., *La topologie des groupes de Lie.*
Homology of compact Lie groups.

CESARI L., BV functions of two variables.

CHEVALLEY C., *L'arithmétique dans les algèbres de matrices.*
Concept of idele.

CHURCH A., Indecidability of the predicate calculus.

CHURCH A.; POST E.L.; TURING A.N., Effectively computable functions.

EHRESMANN C., Locally homogeneous spaces.

GENTZEN G., Proof of the relative non-contradiction of arithmetics through transfinite induction.

HALL P., Frobenius theorem for p-groups.

HASSE H., Proof of the Artin conjecture on Riemann hypothesis.

HILLE E., Semi-groups of linear operators.

KHINTCHIN A.YA., Infinitely divisible distributions as general limits in central limit problem.

KLEENE S.C., General theory of recursive functions.

KOLMOGOROV A.N., Markov chains with countable state space.
Reversibility of statistical laws.

KÖNIG D., *Theorie der endlichen und unendlichen Graphen.*
Kombinatorische Topologie der Streckenkomplexe.

LAPPO-DANILEVSKI I.A., Systems of linear differential equations.

MALCEV A.I., Extension of main theorems of model theory to the case of uncountable models.
Nonstandard models of analysis.

MARKOV A.A., Free equivalence of braids.

MAZUR S., SCHAUDER J., Minimization of functionals in reflexive spaces.

MIKHLIN S.G., Symbol of an integral operator.

MURRAY F.J., VON NEUMANN J., von Neumann algebras.

OKA K., Solution of the first Cousin problem.

ORE O., *L'algèbre abstraite* (structure concept).

POST E., Finite combinatory processes.

DE RHAM G., Currents.

SOBOLEV S.L., Sobolev spaces.
First results on distributions.

STIEFEL E., Stiefel-Whitney characteristic classes.

STONE M.H., Representation theorem for boolean lattice.

TURING A.N., Turing machines.

WHITNEY H., Immersions and embeddings of manifolds.

1937

ANDRONOV A.A., PONTRYAGIN L.S., Structural stability for differential systems.

ANDRONOV A.A., VITT A.A., KHAIKIN S.E., *Theory of Oscillators* (Russian).

BOCHNER S.; DIEUDONNÉ J., Partitions of unity.

CARTAN E., *La théorie des groupes finis et continus et la géométrie différentielle traitée par la méthode du repère mobile.*

CARTAN H., Filter theory.

ČECH E., Stone-Čech compactification.

CIMMINO G., Regularity of generalized solutions of smooth elliptic equations.

COURANT R., HILBERT D., *Methoden der mathematischen Physik II.*

CRAMÉR H., *Random Variables and Probability Distributions.*

DOEBLIN W., FORTET R., Chains with complete connections.

DOETSCH G., *Theorie und Anwendung der Laplace-Transformation.*

DOOB J.L., Stochastic processes in continuous time.

ERDÖS P., Conjecture on arithmetical progressions in sequences of positive densities.

FREUDENTHAL H., Homotopy suspension.

HECKE E., Hecke operators for modular forms.
Decomposition of L-functions in Eulerian products.

HOPF E., *Ergodentheorie.*

KOLMOGOROV A.N., PETROVSKY I.G., PISKUNOV N.S., Reaction-diffusion equations.

KRULL W. Lifting theorem for prime ideals in integral algebra.

KRYLOV N.M., BOGOLIUBOV N.N., Averaging method for nonlinear differential systems.

LEFSCHETZ S., *Lectures in Algebraic Geometry.*

LÉVY P., *Théorie de l'addition des variables aléatoires.*

NAGUMO M., Upper and lower solutions in boundary value problems.

PETROVSKY I.G., Cauchy problem for hyperbolic systems with variable coefficients.

RADÓ T., *Subharmonic Functions.*

RELLICH F., Perturbation theory of linear operators.

STONE M., $C(X)$ as a prototype of Banach algebra.
Stone-Weierstrass theorem.

VINOGRADOV I.M., Each large odd number is the sum of three primes.

WEIL A., *Sur les espaces à structure uniforme et sur la topologie générale.*
WITT E., Witt vectors, quadratic forms.

1938

AHLFORS L.V., Extension of Schwarz lemma.
BEURLING A., Formula for the spectral radius for the Fourier transform.
CARTAN E., Spinor theory.
CHOQUET G., An algorithm for minimal graph with given vertices.
DE FINETTI B., Theorem on exchangeable distributions.
DEHN M., Dehn braids.
DOEBLIN W., *Sur les propriétés asymptotiques des mouvements régis par certains types de chaînes simples.*
FRIEDRICHS K.O., Perturbation of operators with continuous spectrum.
GÖDEL K., Compatibility of consistence of the axiom of choice and continuum hypothesis.
HARDY G.H., WRIGHT E.M., *An Introduction to the Theory of Numbers.*
HEDLUND G.A., MORSE M., Symbolic dynamics.
KRASNER M., Abstract Galois theory.
KRULL W., Local rings.
MAZUR S., Normed rings.
MORREY C.B., Hölderian estimates for solutions of quasilinear elliptic equations.
PETROVSKY I.G., Wellposedness of Cauchy problem for hyperbolic equations.
PISOT CH., Pisot numbers.
PONTRYAGIN L.S., *Topological Groups* (Russian).
RIESZ M., Riesz operator for the solution of Cauchy problem for the wave equation.
SCHOENBERG I., Metric spaces and positive definite functions.
SKOLEM T., *Diophantische Gleichungen.*
SOBOLEV S.L., Mollifiers.
WHITEHEAD J.H.C., Homology equivalence and homotopy equivalence.
WHITNEY H., Tensor product of abelian groups.

1939

ANDRONOV A.A., LEONTOVICH E.A., Bifurcation of plane vector fields.
BOURBAKI N., *Théorie des ensembles. Fascicule de résultats.*
BRELOT M., Perron-Wiener-Brelot method in potential theory.
DOUGLAS J., General form of the Plateau problem.

ERDÖS P., KAC M., Central limit theorem in number theory.

GELFAND I.M., Normed algebras.

GNEDENKO B.V., Determination of the domain of attraction of all stable laws.

HOPF E., Ergodicity of geodesic flows on manifolds with negative curvature.

KANTOROVITCH L.V., General problem of linear programming.

KOLMOGOROV A.D., OBUCHOV A.M., MILLIONSCHIKOV M., Scaling theory and renormalization analysis of turbulence (law of $2/3, \ldots$).

LICHNEROWICZ A., *Sur certains problèmes globaux relatifs au système des équations d'Einstein.*

NAGUMO M., Upper and lower solutions for parabolic differential equations.

OKA K., Solution of the second Cousin problem.

SIEGEL C.L., Siegel's modular functions.

TEICHMÜLLER O., Teichmüller conjecture.

TURING A.N., Recursive reductibility.

VILLE J., *Etude critique de la notion de collectif.*

VAN DER WAERDEN B.L., *Einführung in die algebraische Geometrie.*

WEYL H., *The Classical Groups, their Invariants and Representations.*
Tube volume, starting point of characteristic classes.

WHITEHEAD J.H.C., A weak homotopy equivalence of CW-complexes is a homotopy equivalence.

WIENER N., Ergodic theory.

WIGNER E., Unitary representation in infinite dimension.

ZARISKI O., Resolution of singularities for algebraic surfaces.

1940

ALLENDOERFER C.B.; FENCHEL W., Generalization of Gauss-Bonnet formula to submanifolds of $\mathbb{R}^n$.

BIRKHOFF G., *Lattice Theory.*

CHEVALLEY C., Theory of idèles.

DOEBLIN W., General theory of Markov chains.
Set of powers of a law.
Universal laws.

DOOB J., Martingales, convergence theorem.

EILENBERG S., Obstruction theory.

GÖDEL K., *The Consistency of the Continuum Hypothesis.*

JACOBSON N., Galois theory for rings.

KOLMOGOROV A.N., Wiener spirals.

KREIN M.G., MILMAN D.P., Krein-Milman theorem.

LÉVY P., Planar Brownian motion.

LITTLEWOOD D.E., *The Theory of Group Characters and Matrix Representations of Groups.*

MARCINKIEWICZ J., Marcinkiewicz method for absolutely convergent Fourier series.

MONTGOMERY D., ZIPPIN L., Topological transformation groups.

MORREY C.B., Solution of Hilbert nineteenth problem.

RIESZ F., Ordered spaces.

SALEM R., *Essai sur les séries trigonométriques.*

SCHRÖDINGER E., One-dimensional eigenvalue problems by factorization of the operator.

STONE M.H., C^*-algebras.

WEIL A., *L'intégration dans les groupes topologiques et ses applications.*
Riemann hypothesis for algebraic curves on a finite field.

WEYL H., *Algebraic Theory of Numbers.*
Orthogonal projections in Dirichlet problem.

WHITNEY H., Product formula for the Whitney sum of sphere (or vector) bundles.

ZARISKI O., Local uniformization of algebraic varieties.

1941

BRAUER R., Brauer lifting in modular representation theory.

BRELOT M., Axiomatics of potential theory.

CARTAN H., Potential theory.

CHURCH A., *The Calculus of Lambda-conversion.*

DENJOY A., *Leçons sur le calcul des coefficients d'une série trigonométrique* (vol. I).

ECKMANN B.; EHRESMANN C., FELDBAU J.; HUREWICZ W., STEENROD N., Fiber spaces: homotopy lifting, exact homotopy sequence.

HODGE W.V.D., *The Theory and Applications of Harmonic Integrals.*

HOPF H., Homology of compact Lie groups. H-spaces. Hopf algebras.

HUREWICZ W., Exact sequences.

HUREWICZ W., WALLMAN, W., *Dimension Theory.*

KAKUTANI S., Kakutani's theorem.

KOLMOGOROV A.N., Interpolation and extrapolation of stationary sequences.

LINNIK YU.V., The large sieve.

MAL'TZEV A.I., Local theorems in algebra.

MARTIN R.S., Martin boundary.

SCHNEIDER T., Transcendence of the values of the Beta function.

SEGAL I.G., Group algebras.

TITCHMARSH E.C., Formula for the spectral function of Sturm-Liouville operators.

WHITNEY H., Beginning of differential topology.

1942

ARTIN E., *Galois Theory.*

BAER R., Homogeneity of projective planes.

BIRKHOFF G., Lattice-ordered groups.

DOOB J.L., Markov chains with continuous parameter.
Convergence theorem for discrete martingales.

ECKMANN B.; WHITEHEAD G.W., Non-parallelisability of spheres S^{4k+1}.

EILENBERG S., Extensions of groups and homology.

HAJOS G., Minkowski's theorem on linear forms.

HOPF H., Fundamental group and second homology group.

ITÔ K., Stochastic calculus.

LELONG J., Pluri-subharmonic functions.

LERAY J., Cohomology of topological spaces.

MANN H.B., Optimal lower estimate for the Schnirelmann density of sum sets.

OKA K., Solution of Levi's problem for functions of two variables.

SARD A., Sard theorem.

SIEGEL C.L., Iteration of analytic functions.

1943

ALLENDOERFER C.B., WEIL A., Gauss-Bonnet theorem for Riemannian polyhedra.

CHEVALLEY C., Semi-local rings.

EILENBERG S., MACLANE S., Homology of groups, Bar-resolution, Eilenberg-MacLane spaces.

GELFAND I.M., NAĬMARK M.A., Gelfand-Naĭmark theorem.

GELFAND I.M., RAĬKOV D.A., General theory of representation of a locally compact group.

HLAWKA E., Theorem of Minkowski-Hlawka.

HOPF E., Hopf bifurcation in higher dimension.

LEFSCHETZ S., Periodic solutions of forced Liénard equations.

LINNIK YU.V., Elementary solution of Waring's problem.

MACKEY G.W., Duality in locally convex spaces.

MONNA A.M., Banach spaces on a non Archimedian valued field.
PALM C., Point processes.
PREISSMANN A., Global properties of Riemannian spaces.
SALEM R., Characterization of uniqueness sets in terms of Pisot number.

1944

AMBROSE W., Spectrum of locally compact abelian group.
CAMERON R.N., MARTIN W.T., Cameron-Martin formula.
CARLEMAN T., *L'intégrale de Fourier et les questions qui s'y rattachent.*
CHERN S.S., Intrinsic Gauss-Bonnet formula for closed Riemannian manifolds.
DIEUDONNÉ J., Paracompact spaces.
EHRESMANN C., REEB G., Foliated structures.
EILENBERG S., Singular homology.
FRIEDRICHS K.O., Mollifiers and extensions of differential operators.
KAKUTANI S., Connection between Brownian motion and Dirichlet problem.
Brownian motion in an n-dimensional space.
LANDAU L.D., Transition to turbulence through torical bifurcations.
LEVINSON N., Dissipative dynamical systems.
VON NEUMANN J., MORGENSTERN O., *Theory of Games and Economic Behavior.*
ONSAGER L., Solution of two-dimensional Ising model.
POST E.L., Recursive degrees of unsolvability.
SALEM R., The set of Pisot numbers is closed.
On the theorem of Bohr and Pal.
WHITNEY H., Self-intersections of smooth manifolds.
ZARISKI O., Zariski topology.
Resolution of the singularities of a 3-dimensional algebraic variety.

1945

AMBROSE W., Systematization of Banach algebra theory.
BEURLING A., Spectral theorem.
BOGOLIUBOV N.N., *On some Statistical Methods in Mathematical Physics* (Russian).
BORG G., Inverse Sturm-Liouville problems.
BRELOT M., Fine potential theory, sweeping theory and modern form of the Dirichlet problem.
CARTAN H., Energy in potential theory.
CHEVALLEY C., Intersection multiplicities.

EILENBERG S., MACLANE S., Categories and functors.

EILENBERG S., STEENROD N.E., Axiomatic of homology theory of topological spaces.

HOPF H., Homology of groups : free resolution of modules over group algebras.

JACOBSON N., Theorem for simple associative and non-associative rings.

LICHNEROWICZ A., Solution of a fundamental global problem of Einstein and Pauli.

PETROVSKY I.G., Lacunas in hyperbolic equations.

SAMUEL P., Local rings.

SCHWARTZ L., Distributions.

1946

AIZERMAN M.A., Stability for automatic control systems.

BLANC-LAPIERRE A., FORTET R., Spectral analysis of random functions.

BOCHNER S., Vector fields and Ricci curvature.

CARTAN H., Sweeping in Newtonian potential.

CHERN S.S., Characteristic classes of Hermitian manifolds (Chern classes).

CHEVALLEY C., *Theory of Lie Groups I.*

COHEN I.S., Ideals of complete local rings.

GODEMENT R., Tauberian theorems on locally compact abelian groups.

ITÔ K., Stochastic integrals.

LERAY J., Sheaves.
Spectral sequences.

LEVINSON N., Asymptotic behavior of linear differential equations.

MAHLER K., Selection theorem for lattices.

VON NEUMANN J., 'Birth' of the computer.

DE RHAM G., Hodge-de Rham theory.

RICKART C.E., Non-commutative normed algebras.

WEIL A., *Foundations of Algebraic Geometry.*

ZARISKI O., Zariski rings.
Degeneracy theorem.

1947

CARTAN H., GODEMENT R., Duality and harmonic analysis on locally compact abelian groups.

CARTWRIGHT M.L., LITTLEWOOD J.E., Structure of solutions of forced Liénard equations.

CHOQUET G., Abstract 'contingent-paratingent' theorem.

DYNKIN E.B., Explicit computation of coefficients of Campbell-Hausdorff formula.

EHRESMANN C., Structures in fiber spaces.
Almost complex structures.

KELLEY J., PITCHER E., Exact sequences in homology theory.

LERAY J., Filtered homology.

MARKOV A.A., Markov algorithms.

MARKOV A.A.; POST E.L., Recursive non-solvability of the word problem for semi-groups.

SCHWARTZ L., Mean-periodic functions.

SELBERG A., Selberg sieve.

STEENROD E.N., Steenrod squares.

VINOGRADOV I.M., *The Method of Trigonometric Sums in Number theory* (Russian).

WAŻEWSKI T., Ważewski topological principle for ordinary differential equations.

1948

BOCHNER S.; LICHNEROWICZ A., Bochner-Lichnerowicz theorem for harmonic forms.

DE BRUIJN N.G., ERDÖS P., Fundamental inequality for finite linear spaces.

CHEVALLEY C., Classification of simple Lie algebras and their representations.

CHOQUET G., Convergences.

DANTZIG G.B., Simplex method in linear programming.

DOOB J.L., Ergodic theorem for Markov operators.

FEYNMAN R.P., Path integrals or Feynman integrals.

FRIEDRICHS K.O., Scattering theory.

GODEMENT R., Functions and measures of positive type.

HILLE E., *Functional Analysis and Semi-groups.*

HILLE E.; YOSIDA K., Complete characterization of the infinitesimal generator of a continuous semi-group.

KANTOROVITCH L.V., Newton method in functional spaces.

KÖTHE G., Stufenräume.

KREIN M.G., RUTMAN M.A., Krein-Rutman theorem for linear operators in cones.

LÉVY P., *Processus stochastiques et le mouvement brownien.*

MIKUSINSKI J.D., Generalized distributions.

SCHWARTZ L., Example of a non spectral synthesis set.

SEGRE B., *Lezioni di geometria moderna I.*

SHANNON C.E., Entropy in information theory.
Sampling theorem.

SZMIELEW W., Decidability of the theory of abelian groups.

THORIN G.O., Proof of M. Riesz convexity theorem.

VALIRON G., *Théorie des fonctions.*

WALD A., WOLFOWITZ J., Sequential analysis.

WALL H.S., *Analytic Theory of Continued Fractions.*

WEIL A., *Sur les courbes algébriques et les variétés qui s'en déduisent.*
Variétés abéliennes et courbes algébriques.
The future of mathematics.

WHITNEY H., Spectral synthesis of ideals of C^n functions.

WIENER N., *Cybernetics or Control and Communication in the Animal and the Machine.*

1949

BEURLING A., Invariant subspaces in Hilbert spaces.

BICADZE A.V., LAVRENTIEV M.A., Equations of mixed type.

CARTAN H., Sheaves in the theory of functions of several complex variables.

DEURING M., Class field of complex multiplication.

DIEUDONNÉ J., SCHWARTZ L., Extension of the properties of Banach spaces to the theory of distribution.

DOLPH C.L., Fredholm theory for nonlinear integral equations.

EBERLEIN W.F., Ergodic theorems for semi-groups.

ECKMANN B., GUGGENHEIMER H., Cohomology of Kählerian manifolds.

ENRIQUES F., *Le superficie algebriche.*

ERDÖS P.; SELBERG A., Elementary proofs of the prime number theorem.

FENCHEL W., Extension of Legendre transform to convex functions.

GELFOND A.O., Algebraic independence of transcendental numbers.

HASSE H., *Zahlentheorie.*

IWASAWA K., Toward a solution of Hilbert fifth problem.

KAC M., Probabilistic methods in analysis and number theory.

KODAIRA K., Harmonic fields on Riemannian manifolds (generalized potential theory).

LEVINSON N., Strange attractor of a forced piecewise linear van der Pol equation.

LICHNEROWICZ A., Gauss-Bonnet formula for Finsler spaces.

MAASS H., Non-holomorphic automorphic functions.

OLEINIK O.A., PETROVSKY I.G., Estimation of the Betti numbers of real algebraic varieties.

PONTRIAGIN L.S., Characteristic classes, $\pi_{n+1}(S^n)$.

SERRE J.P., Extensions of ordered fields.

SHAFAREVICH I.R., General reciprocity law.

SIEGEL C.L., *Analytic Functions of Several Complex Variables.*

TARSKI A., Concept of essential indecidability.

THOM R., Morse complexes.

TURÁN P., Method of sums of powers.

WEIL A., Weil conjectures.
Proof of Riemann's conjecture for the congruence zeta function.

WIENER N., *Extrapolation, Interpolation and Smoothing of Stationary Time Series.*

1950

AHLFORS L.V., BEURLING A., Conformal invariants.

BOREL A., SERRE J.P., Impossibility to fiber an Euclidian space by compact fibers.

BRAUER R. Modular representations of groups of finite order.

CARTAN H., Coherent sheaves; sheaf theory in holomorphic fiber spaces.
Transgression in Lie groups and principal fiber spaces.
Differential algebras.

CHOQUET G., Capacitability of the $K_{\sigma\delta}$.

DENY J., Schwartz distributions in potential theory.

DVORETZKI A., ERDÖS P., Double points of Brownian motion.

EHRESMANN C., Characterization of an infinitesimal connexion with a differential form.

FELLER W., *An Introduction to Probability theory and its Applications I.*

GÅRDING L., Generalized solution of the Dirichlet problem for elliptic equations with constant coefficients.

HENKIN L., Completude of the theory of types under admission of so-called non-standard models.

HESTENES M.R., First results on Pontryagin's maximum principle.

HOPF E., Galerkin method for nonstationnary Navier-Stokes equation.

JONES B.W., *The Arithmetic Theory of Quadratic Forms.*

KANTOROVITCH L.V., PINSKER A.G., VULIKH B.Z., *Functional Analysis in Partially Ordered Spaces* (Russian).

KODAIRA K., KREIN M.G., Boundary value problems for singular equations of higher order.

KODAIRA K., DE RHAM G., *Harmonic Integrals.*

KOSZUL J.L., Algebraic formulation of spectral sequences.
Hirsch-Koszul theory.
Homology and cohomology of a Lie algebra on a field of zero characteristic.

KREIN M.G., Parametric resonance in higher dimensional periodic Hamiltonian systems.

KUHN H.W., Extensive form of game theory.

KUHN H.W., TUCKER A.W., Kuhn-Tucker theorem in optimization.

MASSERA J.L., Bounded and periodic solutions of nonlinear differential equations.

OKA K., Oka's theorem on coherent analytic ideals.

SCHWARTZ L., *Théorie des distributions.*
Kernel theorem.

SOBOLEV S.L., *Applications of Functional Analysis in Mathematical Physics* (Russian).

WALD A., *Statistical Decision Functions.*

Une brève histoire de la topologie

Jean Dieudonné
Académie des Sciences de Paris

Introduction

Parmi les parties des mathématiques de notre temps, la Topologie est celle dont le développement me paraît le plus remarquable. Les Grecs étaient parvenus à faire correspondre à des concepts de la vie courante, comme les nombres et les figures géométriques simples, des *objets mathématiques* définis par des relations précises qui les relient. Tous les progrès en mathématiques, jusqu'à la fin du 18e siècle, ont reposé sur ces idées fondamentales, et la méthode des coordonnées a ramené la géométrie à des calculs sur des nombres. Même les nouveaux objets mathématiques qui, à partir de 1800, se font jour en Algèbre, Géométrie et Analyse, ne sont que des extensions hardies et fécondes des notions classiques.

La Topologie, à l'instar des Grecs, va devoir forger de *nouveaux* objets mathématiques, correspondant à d'autres notions communes, telles que la *position* des objets dans l'espace, l'un par rapport à l'autre, ou la *déformation* d'objets non rigides; notions dont l'étude scientifique n'avait guère commencé avant 1800. Ce sera long et difficile, et le nombre de nouveaux concepts qu'il faudra introduire est impressionnant. Mais à partir de la seconde moitié du 20e siècle, les succès remportés par les concepts topologiques dans toutes les parties des mathématiques, et même en physique théorique, révèleront la portée insoupçonnée de ces nouvelles notions issues de l'intuition spatiale.

Il faudra nous limiter aux très grandes lignes d'une histoire extraordinairement complexe, où s'entrecroisent et réagissent les unes sur les autres les idées venues des horizons mathématiques les plus variés. Il me semble que souvent les méthodes de démonstration sont plus intéressantes que les théorèmes qu'elles établissent, et j'ai essayé de les décrire en termes simples, ne supposant connues que les notions les plus élémentaires de la théorie des ensembles et de l'algèbre[1]). Ce n'est que dans les théories récentes que j'ai dû renoncer à ces descriptions en raison de leur technicité, et me borner à en énumérer les résultats les plus frappants.

---

1) Nous utilisons les symboles courants:
$\mathbb{N}$: ensemble des entiers «naturels» $0, 1, 2, \dots$
$\mathbb{Z}$: ensemble des entiers de signe quelconque.
$\mathbb{Q}$: ensemble des nombres rationnels m/n de signe quelconque.
$\mathbb{R}$: ensemble des nombres réels.
$\mathbb{C}$: ensemble des nombres complexes.
$\mathbb{H}$: ensemble des quaternions.
$\mathbb{F}_q$: corps fini à q éléments ($q = p^m$, puissance d'un nombre premier).

1 L'apport de Riemann

La Topologie ne deviendra une discipline mathématique (avec des preuves rigoureuses fondées sur des définitions précises) qu'au début du 20e siècle. Jusque là, les prétendus raisonnements de nature topologique se bornent à faire appel à l'«intuition», c'est-à-dire à la conception qu'on peut avoir, grâce aux sens de la vue et du toucher, d'objets situés dans le plan ou l'espace à 3 dimensions; ces raisonnements ne sont autre chose que des «expériences mentales», que le lecteur est prié de faire sur ces concepts, pour se convaincre de la véracité de l'énoncé qu'on lui propose. D'ailleurs, jusqu'au milieu du 19e siècle, ces objets ne sont autres que ceux de la géométrie euclidienne classique, bien qu'on se rende compte que les énoncés ne concernent pas les «grandeurs», mais ce qu'on appelle la «géométrie de situation». C'est le cas de la célèbre relation énoncée par Euler[2])

$$s - a + f = 2 \tag{1}$$

où s, a, f sont les nombres respectifs de sommets, d'arêtes et de faces d'un polyèdre. A partir de la fin du 18e siècle, cette formule va susciter de nombreux travaux, visant d'abord à suppléer à la démonstration fautive d'Euler dans le cas d'un polyèdre convexe, puis à examiner les «exceptions» que l'on rencontre quand on supprime l'hypothèse de convexité, sans qu'aucune définition *générale* de ce qu'il faut entendre par «polyèdre» soit formulée avant Poincaré.

Certaines notes de Gauss, non publiées de son vivant, montrent qu'il entrevoyait une «géométrie de situation» plus vaste. La plus remarquable contient une formule intégrale donnant le «nombre d'enlacement» de deux courbes lisses dans l'espace, dont il ne fournit pas de preuve.

C'est seulement vers le milieu du 19e siècle qu'apparaissent les idées qui donneront naissance à la Topologie. Elles sont essentiellement dues à Riemann, qui y est conduit par ses travaux d'Analyse et de Géometrie différentielle, dans plusieurs directions différentes.

1.1 Les «variétés»

L'idée d'une «géométrie à n dimensions» pour un entier $n > 3$ quelconque avait pris une forme précise vers 1845 avec les mémoires de Grassmann et de Cayley. En 1851, Schläfli étudia dans ce cadre des généralisations des polyèdres et des quadriques, dans le style de la géométrie différentielle de Monge et de son école (étude des surfaces plongées dans $\mathbb{R}^3$). Mais dans sa fameuse «leçon inaugurale» de 1854, Riemann, inspiré par les travaux de Gauss sur la «géométrie intrinsèque» des surfaces, conçoit l'idée de «multiplicités à n dimensions» (qui deviendront nos «variétés C^1») définies de façon intrinsèque sans partir d'un plongement dans un espace $\mathbb{R}^n$. La position d'un point dans une telle variété est déterminée par n nombres qui jouent le rôle des «coordonnées curvilignes» de Gauss, mais il

2) Dans un manuscrit de Descartes sur la géométrie sphérique, on trouve des relations qui impliquent la formule d'Euler.

n'y a pas encore chez Riemann de distinction nette entre coordonnées locales et coordonnées globales. Ces coordonnées possèdent des différentielles, permettant un calcul infinitésimal sur la variété, et notamment l'extension des notions «métriques» classiques (longueur, aire et volume). C'est cette extension qui est le but principal de Riemann, et son aspect local sera développé par ses successeurs de la fin du 19e siècle, notamment par l'introduction du calcul tensoriel et des formes différentielles. Mais il n'y aura pas d'étude des relations entre ces notions et les aspects globaux de la topologie de la variété avant les travaux de H. Weyl et E. Cartan vers 1930, qui créeront la Topologie différentielle.

1.2 Les surfaces de Riemann

Les idées entièrement neuves de Riemann sur les surfaces vont avoir un retentissement plus précoce. Il part de deux notions «intuitives» (qui le resteront jusque vers 1900): l'une est celle d'un arc de courbe joignant deux points, l'origine et l'extrémité, sur une surface (on dit que c'est un *chemin*, et sa définition précise sera l'image de l'intervalle $[0, 1]$ de $\mathbb{R}$ par une application continue α de cet intervalle dans la surface; on appellera *lacet* un chemin tel que $\alpha(0) = \alpha(1)$). L'autre est l'idée de «déformation» d'une courbe ou d'une surface, qui ne sera pas précisée avant 1910. Ces notions sont considérées comme comprises de tous, et Riemann ne fait pas d'effort pour les relier à des définitions mathématiques. Ce point de départ si banal lui fournit des outils pour un premier classement des surfaces dites «*fermées*» (ce sont celles qui plus tard seront définies comme compactes et sans bord). D'abord la notion de surface *simplement connexe* définie par la condition que tout lacet se déforme en un seul point; puis l'idée de décomposer la surface en réunion de morceaux simplement connexes, par des lacets issus du même point et qui en forment les frontières. Le plus petit nombre de morceaux obtenus par ce procédé est un nombre pair $2g$, et g est le nombre que Clebsch appellera le *genre* de la surface; on a $g = 0$ pour une surface simplement connexe comme la sphère, et $g = 1$ pour le tore; le fait essentiel est que g est invariant par déformation.

La conception la plus originale de Riemann est celle de surface formée de plusieurs «feuillets» plans, superposés mais reliés entre eux le long de lignes de passage. On sait que son idée géniale a été de lier cette conception à l'étude des intégrales abéliennes; elle donne le départ à la théorie des variétés holomorphes, qui englobera celle des fonctions analytiques de plusieurs variables complexes. Elle est aussi à l'origine de la notion de revêtement.

1.3 Les espaces fonctionnels

Depuis le 18e siècle, on avait été amené à considérer des ensembles d'objets autres que des points: courbes et surfaces en Calcul des variations, droites, plans, courbes algébriques en géométrie projective et en théorie des invariants. Dans son étude du «principe de Dirichlet», Riemann conçoit que des familles infinies de courbes ou de surfaces pourraient, elles aussi, être considérées comme des «multiplicités» de dimension infinie, mais où l'on pourrait parler de «variation continue», première ébauche de ce qui deviendra l'Analyse fonctionnelle.

A partir de Riemann, les travaux relatifs à ce que nous appelons à présent la Topologie vont se diviser en deux grands courants qui ne se rejoindront guère avant 1930:

1) La Topologie *générale* (ou «ensembliste»[3])), vouée à la définition précise et l'étude des notions topologiques les plus générales, d'abord dans $\mathbb{R}$ et les $\mathbb{R}^n$, puis dans les «espaces topologiques» quelconques.
2) La Topologie «combinatoire» (dite plus tard *algébrique*) qui développera et généralisera considérablement les idées de Riemann sur les surfaces.

2 Les notions topologiques dans les espaces $\mathbb{R}^n$

La notion de limite d'une suite de nombres, et celle de fonction réelle continue d'une variable réelle, qui s'en déduit, étaient acquises depuis les travaux de Cauchy et Abel. On avait aussi dû considérer des limites de suites de fonctions (notamment les séries entières et les séries trigonométriques), et après quelque flottement, on avait nettement distingué la convergence simple et la convergence uniforme d'une telle suite.

A partir de 1860, les cours de Weierstrass systématisent ces notions et les étendent aux fonctions de plusieurs variables réelles; cela implique la définition dans les ensembles $\mathbb{R}^n$ des notions topologiques élémentaires: voisinages, ensembles ouverts, ensembles fermés, points d'accumulation d'un ensemble infini. C'est par cette dernière notion que l'idée de *compacité* s'introduit sous la forme du «principe de Bolzano-Weierstrass» dans une partie bornée de $\mathbb{R}^n$ (son équivalence avec la définition par les recouvrements ouverts est due à Borel et Lebesgue à la fin du siècle). Avec ces notions, Weierstrass peut donner des preuves rigoureuses des propriétés élémentaires (jusque là admises sans démonstration) des fonctions réelles continues dans un intervalle compact de $\mathbb{R}$: théorème de la valeur intermédiaire et de l'existence d'un maximum; Heine montre d'autre part que dans un tel intervalle une fonction continue est uniformément continue. A partir de 1880, tout cela est devenu familier aux analystes.

Mais avec Cantor, à partir de 1872, commence une étude topologique beaucoup plus ambitieuse des sous-ensembles d'un espace $\mathbb{R}^n$. Il introduit dans $\mathbb{R}$ les ensembles dérivés successifs d'un sous-ensemble et les ensembles parfaits; il y décrit complètement les ensembles fermés, et découvre son fameux ensemble parfait totalement discontinu. La classification des sous-ensembles de $\mathbb{R}$ se poursuit après 1890 dans les travaux de R. Baire, E. Borel et H. Lebesgue; on prend conscience de leur complexité, qui ne fera que croître avec les recherches de Lusin et Suslin sur les ensembles projectifs, débouchant sur de difficiles questions de l'axiomatique de la théorie des ensembles (propositions «indécidables» dans un système d'axiomes).

---

3) Pendant longtemps, elle était considérée comme une partie de ce qu'on appelait «Théorie des ensembles».

Ce courant d'idées finit par s'écarter notablement de la plupart des problèmes d'Analyse. Par contre, on utilise de plus en plus dans ces problèmes les propriétés des fonctions semi-continues, introduites par Baire en 1899, et le théorème de Baire-Osgood sur l'intersection d'une famille dénombrable d'ensembles ouverts partout denses.

Mais pour $n \geq 2$, on s'aperçoit que, sans introduire de notions plus compliquées que celles de fonction continue et d'ensemble ouvert, on rencontre dans $\mathbb{R}^n$ des phénomènes qui vont à l'encontre de l'«intuition», tels que la courbe de Peano remplissant un carré (1889), et l'ensemble ouvert construit par Brouwer (1910) dans $\mathbb{R}^2$, qui a 3 composantes connexes ayant même frontière. Cela conduisit à une exigence accrue de «rigueur» même dans les démonstrations de propriétés qui paraissaient les plus «intuitives»; l'exemple classique est le fait que dans la sphère $\mathbb{S}_2$, une courbe C homéomorphe à un cercle a un complémentaire formé de deux ensembles ouverts simplement connexes; Jordan, en 1893, fut le premier à en donner une démonstration presque complète et Schönflies prouva que tout point de C est l'extrémité d'un arc de courbe entièrement contenu dans une de ces composantes (ce point excepté)[4]).

3 Espaces métriques et espaces topologiques

L'intuition de Riemann sur la possibilité d'étendre des notions topologiques à des ensembles d'objets qui ne sont plus des parties de $\mathbb{R}^n$, se prolonge chez les analystes de la fin du 19e siècle. Pour les fonctions, de nouvelles notions de convergence apparaissent en Calcul des variations: pour une suite de fonctions dérivables f_n dans un intervalle de $\mathbb{R}$, l'école de Weierstrass est amenée à distinguer la convergence uniforme de f_n vers une fonction dérivable f, d'une notion plus forte où l'on impose aussi la convergence uniforme de la dérivée f'_n vers f_n.

En 1887, V. Volterra forge le mot de «fonctionnelle» pour désigner, de façon plus générale que dans le Calcul des variations, une application dans $\mathbb{R}$ d'un ensemble de fonctions (définies dans un même ensemble). En 1903, Hadamard décrit les fonctionnelles $f \mapsto F(f)$ *linéaires* dans l'ensemble $C(0,1)$ des fonctions continues dans l'intervalle $0 \leq t \leq 1$, qui ont la propriété que $F(f_n)$ tend vers $F(f)$ lorsque f_n tend uniformément vers f dans $[0,1]$ (résultat que F. Riesz mettra en 1909 sous sa forme définitive à l'aide de l'intégrale de Stieltjes).

D'autre part, dans leurs recherches sur l'uniformisation des fonctions analytiques «multiformes», Klein et Poincaré n'hésitent pas à faire (sans beaucoup de justifications) des raisonnements «par continuité» dans des ensembles de fonctions.

---

4) $\mathbb{S}_n$ désigne le sous-espace fermé de $\mathbb{R}^{n+1}$ défini par l'équation

$$x_0^2 + x_1^2 + \ldots + x_n^2 = 1;$$

en particulier, $\mathbb{S}_0$ est l'ensemble des deux points $-1, +1$ dans $\mathbb{R}$. $\mathbb{D}_{n+1}$ est la boule fermée de $\mathbb{R}^{n+1}$ définie par $|x| \leq 1$; sa frontière dans $\mathbb{R}^{n+1}$ est $\mathbb{S}_n$.

Le pas décisif dans la création de la Topologie générale est franchi par M. Fréchet en 1906. Au lieu de définir des notions «topologiques» dans un ensemble de points, ou de courbes, ou de fonctions, il va d'emblée au maximum de généralité en cherchant à donner un sens à ces notions dans un ensemble *quelconque* d'objets mathématiques dont la nature n'est pas spécifiée. Son idée la plus féconde est de supposer définie, dans un ensemble E, une fonction $d : E \times E \longrightarrow \mathbb{R}$ dite «distance», qui a des valeurs $d(x, y) \geq 0$ et est seulement assujettie à 3 conditions inspirées de la géométrie élémentaire:

1) la relation $d(x, y) = 0$ équivaut à $x = y$;
2) on a $d(y, x) = d(x, y)$;
3) on a $d(x, z) \leq d(x, y) + d(y, z)$.

Il est remarquable qu'avec si peu d'hypothèses, *toutes* les notions topologiques introduites précédemment dans les espaces $\mathbb{R}^n$ s'étendent immédiatement à E muni d'une distance (on donne alors à E le nom d'*espace métrique*): limites, continuité, voisinages d'un point, ensembles ouverts, ensembles fermés, points d'accumulation, compacité, connexion, etc. Plusieurs distances (dites alors «équivalentes») peuvent d'ailleurs définir les mêmes notions topologiques sur un ensemble.

Toutefois, Fréchet constata que certaines propriétés des espaces $\mathbb{R}^n$ ne sont plus valables dans tous les espaces métriques: par exemple, la convergence des suites de Cauchy et le fait qu'un ensemble fermé borné (pour la distance choisie) est compact. Cela conduit à distinguer les espaces métriques *complets* (où le critère de Cauchy est valable) et les espaces métriques *localement compacts* (où tout point admet un voisinage compact).

Fréchet s'était aussi aperçu que certaines définitions de convergence pour une suite de fonctions ne pouvaient se définir à partir d'une distance sur l'espace de ces fonctions. En 1914, F. Hausdorff se rendit compte qu'en sacrifiant des propriétés de dénombrabilité non essentielles, on pouvait encore définir sur un ensemble quelconque les mêmes notions topologiques que dans un espace métrique, en ne faisant pas usage d'une distance. Une telle structure, qui finit par être appelée une «topologie», se définit par un système d'axiomes imposés à une famille de sous-ensembles, qualifiés d'ensembles ouverts, ou par un système équivalent imposé à une autre famille d'ensembles qualifiés de voisinages. Ces dénominations sont justifiées par le fait que, dans un espace métrique, les ensembles ouverts (resp. les voisinages) vérifient les axiomes en question. On constate alors qu'une bonne partie des théorèmes démontrés dans les espaces métriques demeurent valables totalement ou partiellement dans un espace topologique.

4 Homéomorphismes et dimension

Comme nous l'avons dit, l'idée de «déformation» est celle qui guide les mathématiciens du 19e siècle dans leurs tentatives de classification de nature «topologique»; pour mieux faire comprendre ce qu'ils entendent par là, beaucoup ajoutent qu'il faut opérer «sans déchirure» et invitent le lecteur à penser à des «surfaces en caoutchouc». Mais au 18e siècle, en géométrie euclidienne, on était arrivé, pour les objets rigides, à distinguer le mouvement du déplacement, ce dernier étant considéré, non plus comme le résultat d'un mouvement (comme chez les Grecs), mais comme une transformation de tout l'espace, conservant les longueurs. C'est sans doute Möbius, dans l'oeuvre de qui la notion de transformation est centrale, qui en 1858 dissocie le premier l'idée de déformation de ce qu'il appelle «corrélation élémentaire» entre deux parties de l'espace euclidien, et que depuis Poincaré nous qualifions d'*homéomorphisme* (Möbius cherche d'ailleurs à prouver qu'un homéomorphisme peut être considéré comme le résultat d'une déformation).

La notion de dimension des «grandeurs» spatiales, déjà conçue par les Grecs, se précisait depuis Gauss et Riemann par l'existence sur une variété de coordonnées locales, dont le nombre était la dimension de la variété. Möbius considère comme évident qu'un homéomorphisme conserve la dimension; mais la découverte par Cantor d'une bijection de $\mathbb{R}$ sur $\mathbb{R}^n$, puis la construction par Peano d'une courbe remplissant un carré, montraient la nécessité de donner une preuve rigoureuse de cette «évidence». Ce fut seulement en 1911 que Brouwer, utilisant les premiers résultats de Topologie algébrique, put prouver que $\mathbb{R}^m$ et $\mathbb{R}^n$ ne peuvent être homéomorphes si $m \neq n$ (voir plus loin).

Mais l'intervention des espaces $\mathbb{R}^n$ n'était possible que pour des variétés, et on se préoccupa assez tôt de définir une notion de dimension applicable à tous les espaces métriques et n'utilisant que la topologie de l'espace. En 1903, Poincaré esquissa une définition de ce genre, améliorée plus tard par Brouwer, Urysohn et Menger. On attribue à l'ensemble vide la dimension -1, puis on procède par induction: la dimension d'un espace non vide E est le plus petit entier $n \geq 0$ pour lequel tout point a un système fondamental de voisinages dont la frontière a une dimension $\leq n-1$; s'il n'existe aucun entier n ayant cette propriété, on dit que E a une dimension $+\infty$.

En 1911, H. Lebesgue proposa une autre définition. Pour un recouvrement ouvert fini R de E, l'ordre de R est défini comme le plus grand entier p tel qu'il existe $p+1$ ensembles distincts dans R dont l'intersection est non vide. Si $m(R)$ est le plus petit des ordres de tous les recouvrements ouverts finis de E, plus fins que R, la dimension de E est définie comme la borne supérieure des $m(R)$ pour tous les recouvrements ouverts finis de E. Pour les espaces métriques séparables (c'est-à-dire ceux dans lequel il existe un ensemble dénombrable partout dense), les deux définitions coïncident.

Il est clair que la dimension est invariante par homéomorphisme; pour $E = \mathbb{R}^n$, Brouwer prouva que la dimension ainsi définie est bien égale à n. Toute partie de $\mathbb{R}^n$ a une dimension $\leq n$; d'autre part, Menger et Nöbeling ont prouvé en

1930 que tout espace métrique séparable de dimension n est homéomorphe à un sous-espace de $\mathbb{R}^{2n+1}$. On peut donc dire que les espaces métriques séparables de dimension finie sont les sous-espaces des $\mathbb{R}^n$.

5 L'évolution de la Topologie générale

Après Hausdorff se détache tout un rameau de la Topologie générale, qui se limite surtout aux espaces de dimension finie, pour tenter de les classer et d'étudier leurs propriétés; on l'appelle à présent la *Topologie géométrique*, que nous aborderons plus loin.

Les autres travaux de Topologie générale traitent d'espaces de dimension quelconque. Les topologies définies par Hausdorff étaient *séparées*, c'est-à-dire que deux points distincts y admettent des voisinages sans point commun, ce qui assure qu'une suite de points ne peut avoir deux limites distinctes. Des «axiomes de séparation» plus restrictifs ont été étudiés, permettant notamment de disposer de propriétés utiles pour les fonctions réelles continues: par exemple, la possibilité de *prolonger*, en une fonction continue dans tout l'espace, une fonction définie seulement dans un sous-espace fermé[5]); il est utile aussi de pouvoir définir des partitions continues de l'unité, qui souvent permettent de passer de propriétés locales à des propriétés globales. Ces axiomes sont tous vérifiés dans les espaces métriques, mais n'entraînent pas que l'espace soit métrisable; on a découvert des «conditions de métrisabilité» supplémentaires, ne faisant intervenir que la topologie de l'espace.

La Topologie générale étudie aussi les espaces obtenus à partir d'autres espaces par les opérations usuelles sur les ensembles: sous-espaces, produits, quotients, images par des applications continues.

La plupart des espaces intervenant en Analyse sont *uniformisables*: cela signifie que leur topologie, même si elle n'est pas métrisable, peut néanmoins être définie par une famille infinie (d_α) de *pseudo-distances*; ce sont des fonctions vérifiant les axiomes 2) et 3) des distances, mais non l'axiome 1). On peut y généraliser les notions de suite de Cauchy et d'espace complet. Les espaces compacts séparés sont tous uniformisables.

Enfin, des espaces non séparés sont apparus naturellement dans des parties des mathématiques telles que la géométrie algébrique et la théorie spectrale des opérateurs.

5) Il y a des espaces séparés dénombrables, dans lesquels toute fonction réelle continue est nécessairement constante.

H. Poincaré at his home, rue Cl. Bernard, Paris (courtesy François Poincaré, grandson)

6 La préhistoire de la Topologie algébrique, de Riemann à Poincaré

Les idées de Riemann sur les surfaces et leur «déformation» étaient si nouvelles qu'elles suscitèrent chez les contemporains la nécessité de les clarifier et de les mettre sous une forme plus accessible (tout en restant comme lui tributaires de l'«intuition»). L'attention se concentra évidemment sur la difficile conception des «feuillets» étalés sur un plan et joints par des lignes de croisement; plusieurs mathématiciens s'y consacrèrent, et F. Klein affirma qu'une surface de Riemann de genre g est homéomorphe à une sphère munie de g «anses», forme popularisée par E. Picard, et qui acquerra une grande importance quand elle sera généralisée à toutes les dimensions après 1950.

C'est aussi Klein qui précise la notion d'*orientation* d'une surface (qui n'intervenait pas chez Riemann). Il considère en un point x un repère formé de deux vecteurs tangents v_1, v_2; après un mouvement continu de ce repère le long d'un lacet d'origine x, les vecteurs v_1, v_2 sont revenus en deux vecteurs tangents v'_1, v'_2; si le déterminant $\det(v'_1, v'_2)$ par rapport au repère (v_1, v_2) est toujours positif (pour tous les lacets d'origine x), on dit que la surface est orientable, et la classe d'équivalence des couples (v_1, v_2) se déduisant les uns des autres par une transformation linéaire de déterminant positif est appelée une orientation de la surface; les repères pour lesquels le déterminant est négatif forment l'orientation opposée. Si par contre il y a des lacets pour lesquels $\det(v'_1, v'_2)$ est négatif, on dit que la surface est non orientable. Klein et Schläfli indépendamment montrèrent que le plan projectif réel $\mathbb{P}_2(\mathbb{R})$ n'est pas orientable; Klein remarqua en outre qu'une surface non orientable plongée dans $\mathbb{R}^3$ n'a qu'un seul côté, phénomène dont Möbius et Listing indépendamment (et peut-être sur une suggestion de Gauss) avaient donné

Bourbaki meeting at Besse-en-Chandesse, Auvergne, 1935, H. Cartan, R. de Possel, J. Dieudonné, A. Weil, S. Mandelbrojt, C. Chevalley

un exemple (le fameux «ruban de Möbius») en 1858; mais leur surface avait un bord, et Klein en construisit une autre (dite «bouteille de Klein») sans bord[6]).

L'idée de «feuillets superposés» s'associe, chez Klein et Poincaré, à celle de «domaine fondamental» pour l'action d'un groupe sur un espace tel que le demi-plan $I(z) > 0$; c'est le fondement de la théorie des fonctions modulaires (dont le premier exemple est dû à Gauss) et des fonctions fuchsiennes qui les généralisent. On entend par là un ensemble ouvert dont la frontière est formée de droites non-euclidiennes (arcs de cercle de centre sur l'axe réel, ou droites perpendiculaires à cet axe); les transformées de son adhérence par les transformations du groupe recouvrent le demi-plan et n'ont aucun point intérieur commun. En «recollant» deux côtés d'un domaine fondamental qui sont transformés l'un de l'autre par une transformation du groupe, Poincaré obtient une surface dont le demi-plan est ce qui sera appelé un «revêtement» (ramifié ou non) qui joue vis-à-vis de cette surface le rôle des «feuillets» de Riemann. De là il s'élève, dès 1883, à la conception du «revêtement universel» simplement connexe d'une surface quelconque. Au début du 20e siècle, Boy observe de son côté qu'une surface non orientable a un revêtement orientable à 2 feuillets. Tout cela ne deviendra précis et rigoureux que dans le livre de H. Weyl (1913) sur les surfaces de Riemann.

Il en est de même de la classification des surfaces C^1 à homéomorphisme près. Après Riemann, elle a été poursuivie avec des succès divers par Möbius, Jordan, Klein et ses élèves. A la fin du siècle, on est arrivé à la conclusion qu'une surface compacte, orientable et sans bord, est déterminée à homéomorphisme près par son genre g. On sait construire explicitement une telle surface (dite «forme

6) Toutefois cette surface ne peut être plongée dans $\mathbb{R}^3$ sans acquérir des points singuliers; mais elle peut être plongée dans $\mathbb{R}^4$.

normale») en considérant un polygone régulier de $4g$ côtés; si on désigne par

$$a_1b_1a_1'b_1'a_2b_2a_2'b_2'\ldots a_gb_ga_g'b_g'$$

la suite des côtés lorsqu'on parcourt le polygone dans un certain sens, la surface est obtenue en «recollant» a_k (resp. b_k) parcouru dans ce sens, avec a_k' (resp. b_k') parcouru en sens inverse (pour $1 \leq k \leq g$). Il y a des «formes normales» analogues pour les surfaces compactes non orientables et les surfaces (orientables ou non) ayant un bord.

Il y a aussi quelques tentatives sans lendemain pour aborder l'étude des variétés C^1 de plus grande dimension. Von Dyck (élève de Klein) généralise aux variétés à n dimensions la notion d'orientation, en considérant des repères formés de n vecteurs tangents; il reconnaît que l'espace projectif $\mathbb{P}_m(\mathbb{R})$ est orientable si m est impair et non orientable si m est pair. Il distingue soigneusement la notion de variété non orientable de celle de variété à un seul côté, cette dernière notion n'ayant de sens que si la variété de dimension n est plongée dans une variété de dimension $n+1$. Par exemple, $\mathbb{P}_1(\mathbb{R})$ est orientable, mais n'a qu'un seul côté lorsqu'elle est plongée dans $\mathbb{P}_2(\mathbb{R})$.

A la fin de sa vie, Riemann avait songé à une étude de la structure topologique des variétés de dimension $n \geq 3$, sur le modèle de ses travaux sur les surfaces. Il ne put que laisser quelques notes relatives à ce problème, qui furent développées par son élève Betti. Généraliser la notion d'«ordre de connexion» d'une surface conduit à distinguer n «ordres de connexion» d'une variété de dimension n, en remplaçant les morceaux de surface limités par des courbes, par des morceaux de variétés de dimension $m < n$, limités par des sous-variétés de dimension $m-1$. Mais cela reste vague et confus chez Betti, sans la moindre tentative de démonstration des assertions qu'il énonce; si bien que le seul mérite de ses mémoires est d'avoir incité Poincaré à forger une théorie cohérente.

7 Les idées de Poincaré et l'intervention de l'algèbre

C'est en 1892 que H. Poincaré se décide à poser les fondements d'une théorie englobant les notions topologiques qu'il a rencontrées dans plusieurs de ses travaux d'Analyse, et qu'il pressent devoir devenir de plus en plus importantes en mathématiques. Tout en restant tributaire de l'«intuition» comme ses prédécesseurs, il s'en distingue en donnant toute une série de définitions et d'affirmations qui formeront la base de tous les travaux ultérieurs.

7.1 Le mémoire de 1895

L'apport essentiel de Poincaré est ce qui va devenir l'*homologie*. Dans un premier mémoire intitulé «Analysis Situs», il part des tentatives de Riemann et Betti. Au lieu de se borner, comme Betti, à faire le décompte du plus grand nombre de sous-variétés de dimension $m < n$ (dans une variété X de dimension n) qui «forment frontière» d'un ouvert connexe de dimension $m+1$ (dans une sous-variété de dimension $m+1$), il définit un *calcul* dont ces sous-variétés sont *elles-mêmes* les objets. Si, dans une sous-variété W de X, de dimension $m+1$, $V_1, V_2, \ldots, V_k$ sont des sous-variétés de dimension m, qui «forment frontière» dans W d'un ensemble ouvert connexe U, Poincaré exprime ce fait par une équation

$$V_1 + V_2 + \ldots + V_k \sim 0 \tag{2}$$

qu'il appelle une «homologie».

Il explique en outre[7]) qu'un voisinage ouvert Z de $\bar{U}$ dans W, ainsi que les V_j, doivent être munis d'une orientation; chaque V_j doit avoir deux côtés dans Z, et U est situé d'un seul de ces côtés. De plus, si $(u_1, \ldots, u_m)$ est un repère direct en un point d'un V_j et u_{m+1} un vecteur normal à V_j en ce point et contenu dans $\bar{U}$, $(u_1, \ldots, u_m, u_{m+1})$ doit être un repère direct dans Z; il dit alors que les orientations de V_j et de Z sont directement reliées, et sinon qu'elles sont inversement reliées.

Il ne se borne pas aux seules relations (2) et écrit aussi des «homologies» plus générales

$$r_1 V_1 + r_2 V_2 + \ldots + r_k V_k \sim 0 \tag{3}$$

où les entiers r_j sont positifs ou négatifs. Si $r_j > 1$, il se borne à expliquer qu'il faut remplacer V_j par r_j sous-variétés qui en sont «légèrement différentes»! Un exemple est la droite projective $\mathbb{P}_1(\mathbb{R})$ plongée dans le plan projectif $\mathbb{P}_2(\mathbb{R})$; on a $\mathbb{P}_1(\mathbb{R}) \not\sim 0$ car $\mathbb{P}_1(\mathbb{R})$ n'a qu'un seul côté; mais $2\mathbb{P}_1(\mathbb{R}) \sim 0$, car deux droites projectives «forment frontière» de l'angle dont elles sont les côtés. Quant à $-V_j$ c'est la sous-variété V_j dont on change l'orientation lorsqu'elle est inversement reliée à celle de Z.

Enfin Poincaré affirme sans justification que lorsque l'on a deux «homologies»

$$r_1 V_1 + r_2 V_2 + \ldots + r_p V_p \sim 0, \quad s_1 V_1' + s_2 V_2' + \ldots + s_q V_q' \sim 0$$

on a encore une «homologie» en les ajoutant:

$$r_1 V_1 + \ldots + r_p V_p + s_1 V_1' + \ldots + s_q V_q' \sim 0.$$

Ce qu'il appelle le *m-ème nombre de Betti* $p_m(X)$ est le plus grand entier k tel qu'il existe k sous-variétés V_j de dimension m qui ne sont pas liées par une «homologie» (3); une grande partie de son mémoire consiste à déterminer les nombres de Betti d'exemples divers de variétés. Malheureusement, il ne s'agit toujours dans tout cela que d'affirmations uniquement appuyées sur des «expériences mentales» qu'on demande de faire dans des espaces de dimension quelconque!

Toutefois il y a dans ce mémoire deux idées originales qui deviendront des théorèmes importants par la suite. La première est l'introduction du *groupe*

7) Pour la clarté de l'exposition, nous exprimons (lorsque c'est possible) les définitions de Poincaré dans le langage actuel de la Topologie.

fondamental, que Poincaré définit uniquement pour s'en servir dans le calcul d'«homologies» entre sous-variétés de dimension 1 (l'intérêt intrinsèque du groupe fondamental n'apparaîtra que plus tard). Dans un espace connexe par arcs X, il considère les lacets d'origine et extrémité un point x_0; on peut définir un tel lacet comme image de $[0, 1]$ par une application continue γ de $[0, 1]$ dans X telle que $\gamma(0) = \gamma(1) = x_0$. Poincaré les compose par *juxtaposition*, le lacet composé $\gamma = \gamma_1 \vee \gamma_2$ étant défini par

$$\gamma(t) = \gamma_1(2t) \text{ pour } 0 \leq t \leq \frac{1}{2}, \quad \gamma(t) = \gamma_2(2t-1) \text{ pour } \frac{1}{2} \leq t \leq 1.$$

Deux lacets sont rangés dans la même classe d'équivalence $[\gamma]$ si on peut «déformer» l'un dans l'autre en laissant fixe le point x_0. La composition de deux lacets définit alors la composition $[\gamma] = [\gamma_1][\gamma_2] = [\gamma_1 \vee \gamma_2]$ de leurs classes d'équivalence, et Poincaré montre que cette loi de composition est une loi de *groupe* (non commutatif en général) sur l'ensemble $\pi_1(X, x_0)$ de ces classes. Poincaré admet en outre que dans une variété X, un lacet qui est une sous-variété et qui de plus peut être «déformé» en un seul point (dont la classe est donc l'élément neutre de $\pi_1(X, x_0)$) «forme frontière» d'une surface connexe. Il en déduit l'affirmation qu'on obtient les «homologies» de dimension 1 en considérant toutes les relations

$$s_1^{m_1} s_2^{m_2} \dots s_k^{m_k} = 1$$

entre éléments $s_k = [\gamma_k]$ du groupe $\pi_1(X, x_0)$, puis en formant la relation correspondante

$$m_1\gamma_1 + m_2\gamma_2 + \dots + m_k\gamma_k \sim 0.$$

L'autre idée nouvelle est le fameux *théorème de dualité*: si X est une variété compacte orientable sans bord, de dimension n, alors on a la relation $p_r(X) = p_{n-r}(X)$ pour $0 \leq k \leq n$, entre ses nombres de Betti. Sa tentative de preuve repose sur une méthode très originale: pour deux sous-variétés V_1, V_2 de X, orientées, sans bord et de dimensions complémentaires k, $n-k$, en supposant qu'elles se coupent transversalement en un nombre fini de points M_j, il définit un *nombre d'intersection*, à la manière de Kronecker; c'est le nombre $N(V_1, V_2) = \sum_j S(M_j)$, où $S(M_j) = \pm 1$; le signe dépend des orientations, et est défini de sorte que $N(V_2, V_1) = (-1)^{k(n-k)} N(V_1, V_2)$. Poincaré affirme alors qu'une homologie (3) est équivalente au fait que $\sum_j r_j N(U, V_j) = 0$ pour toute sous-variété U orientée et de dimension complémentaire de celle des V_j, coupant transversalement toutes les V_j. Il en déduit que $p_k(X) \leq p_{n-k}(X)$, et l'inégalité inverse se prouve de même. Il va sans dire que ses raisonnements sont totalement incapables de convaincre de la vérité du théorème.

7.2 Les mémoires de 1899–1900

En 1898, un jeune mathématicien danois, P. Heegaard, publia un résultat qui paraissait contredire le théorème de dualité. Cela obligea Poincaré à revoir ses «démonstrations», et il se persuada qu'il faisait fausse route en essayant de suivre la voie de Riemann et Betti. Aucune technique, à la fin du 19e siècle, ne pouvait permettre de travailler sur un ensemble aussi vaste que celui de toutes les sous-variétés d'une variété donnée, ensemble dont personne ne pouvait à l'époque soupçonner l'extraordinaire complexité. C'est alors que Poincaré, dans deux «Compléments à l'Analysis Situs», publiés en 1899 et 1900, changea radicalement de méthode, en se limitant à certaines sous-variétés *fixées* et en faisant intervenir l'algèbre linéaire pour obtenir un algorithme régulier pour le calcul des groupes d'homologie; il posait ainsi les bases sur lesquelles allait reposer *toute* la Topologie algébrique.

Vers 1850, Schläfli avait généralisé la relation d'Euler (1) à des «polyèdres convexes» dans un espace $\mathbb{R}^n$ pour n quelconque; on peut définir un tel ensemble comme une intersection compacte ayant un intérieur non vide, d'un nombre fini de demi-espaces fermés. Les faces de dimension $n-1$ sont celles qui sont contenues dans un seul des hyperplans limitant ces demi-espaces; ce sont des polyèdres convexes de dimension $n-1$, et on définit de là, par récurrence descendante, les faces de dimension $n-2, \dots, 1, 0$. Si α_j $(0 \leq j \leq n-1)$ est le nombre des faces de dimension j, Schläfli propose une démonstration de la relation

$$\alpha_{n-1} - \alpha_{n-2} + \alpha_{n-3} - \dots + (-1)^{n-1}\alpha_0 = 1 - (-1)^n \tag{4}$$

par récurrence sur n. A la fin de son mémoire de 1895, Poincaré avait entrepris d'étendre cette formule à des polyèdres non convexes. Jordan avait déjà remarqué que la formule d'Euler est en fait un théorème de Topologie, et avait énoncé sans démonstration que pour un «polyèdre curviligne» qui est une surface compacte, orientable et de genre g, la formule doit être remplacée par

$$s - e + f = 2 - 2g. \tag{5}$$

Poincaré chercha aussi à étendre cette formule à toutes les dimensions, et cela lui donna d'abord l'occasion de définir pour la première fois, de façon générale, ce qu'il faut entendre par *polyèdre* (curviligne). C'est ce que l'on appellera plus tard une *triangulation* (différentiable) d'une *variété différentiable* compacte V, de dimension n, c'est-à-dire un ensemble fini T de «cellules» *disjointes* de diverses dimensions $\leq n$, telles que:

1) chaque cellule de dimension p ou *p-cellule* est l'image dans V d'une boule ouverte B de $\mathbb{R}^p$ par un difféomorphisme d'un voisinage ouvert de $\bar{B}$ sur une partie d'une sous-variété W de V de dimension p;
2) la frontière dans W d'une telle cellule de dimension p est réunion de cellules de dimension $\leq p-1$;
3) V est réunion des cellules de T.

Si V est connexe, orientable et sans bord, et si α_j est le nombre de cellules de T de dimension j ($0 \leq j \leq n$), Poincaré chercha à prouver la formule

$$\alpha_n - \alpha_{n-1} + \alpha_{n-2} - \ldots + (-1)^n \alpha_0 = 1 - p_{n-1} + p_{n-2} - \ldots + (-1)^{n-1} p_1 + (-1)^n \tag{6}$$

(dite maintenant «formule d'Euler-Poincaré»), où p_j est le j-ème nombre de Betti de V. Ses tentatives de preuve dans son mémoire de 1895 se bornent en fait aux cas $n = 2$ et $n = 3$, et avec sa définition de l'«homologie», il y rencontra des difficultés insurmontables.

Toutefois ce sont sans doute ces tentatives qui lui donnèrent l'idée, dans ses mémoires de 1899-1900, d'une nouvelle définition de l'«homologie» pour une variété compacte connexe V, où il suppose qu'*il existe une triangulation* T. Ayant fixé arbitrairement une orientation de chaque cellule de T, il définit, pour tout couple formé d'une p-cellule a_i^p et d'une $(p-1)$-cellule a_j^{p-1} un nombre ε_{ij}^p égal à 0 si a_j^{p-1} n'est pas contenu dans la frontière de a_i^p, à $+1$ (resp. -1) si elle est contenue dans cette frontière et si son orientation est directement (resp. inversement) reliée à celle de a_i^p. Pour chaque entier $p \geq 0$, il forme alors les combinaisons linéaires *formelles*

$$s = \sum_i \lambda_i a_i^p \tag{7}$$

des p-cellules de T, à coefficients $\lambda_i \in \mathbb{Z}$, qu'on appellera plus tard les *p-chaînes* de T, et qu'il ne faut pas confondre avec un sous-ensemble de V. Puis il associe à chacune des p-chaînes (7) la $(p-1)$-chaîne

$$b_p s = \sum_{i.j} \lambda_i \varepsilon_{ij} a_j^{p-1} \tag{8}$$

qu'on nomme maintenant le *bord* de la p-chaîne s. Mais cela fait, il n'y a plus aucun appel à la Topologie dans les définitions suivantes; tout est devenu purement *algébrique* (bien qu'évidemment inspiré par les idées du mémoire de 1895).

Le premier résultat est que ces définitions entraînent la relation

$$b_{p-1}(b_p s) = 0 \tag{9}$$

(correspondant à l'idée «intuitive» qu'un bord n'a pas de bord). Poincaré déduit de là les notions de «nombre de Betti» et de «coefficient de torsion» d'une triangulation, par un calcul de «matrices d'incidence» dont les éléments sont les coefficients des p-chaînes et de leurs bords. Vers 1925, E. Noether et L. Vietoris observèrent que ces définitions se présentent plus simplement dans le langage de l'algèbre moderne. On considère l'ensemble $C_p(T)$ des p-chaînes de T (pour $0 \leq p \leq n$); c'est un $\mathbb{Z}$-module ayant pour base (dite canonique) les a_i^p; la formule (8) définit une application $\mathbb{Z}$-linéaire

$$b_p : C_p(T) \longrightarrow C_{p-1}(T).$$

On considère alors le sous-$\mathbb{Z}$-module $Z_p(T) = \operatorname{Ker} b_p$ de $C_p(T)$, dont les éléments sont maintenant appelés les *p-cycles* de T, ainsi que le sous-$\mathbb{Z}$-module $B_p(T) = \operatorname{Im}(b_{p+1})$ dont les éléments sont les *p-bords* de T. Il résulte de (9) que l'on a $B_p(T) \subset Z_p(T)$, et le quotient

$$H_p(T) = Z_p(T)/B_p(T) \tag{10}$$

est le *p-ème groupe d'homologie* de T. On note[8])

$$H.(T) = \bigoplus_p H_p(T) \tag{11}$$

la somme directe des $H_p(T)$. Depuis H.S. Smith (1851), on dispose d'un algorithme régulier permettant, pour deux $\mathbb{Z}$-modules libres M, N de type fini, et pour une application $\mathbb{Z}$-linéaire $u : M \longrightarrow N$, de calculer de nouvelles bases de M et N, adaptées à u, et montrant que le $\mathbb{Z}$-module $N/u(M)$ est isomorphe à un groupe de la forme

$$\mathbb{Z}^q \oplus \mathbb{Z}/p_1^{t_1}\mathbb{Z} \oplus \mathbb{Z}/p_2^{t_2}\mathbb{Z} \oplus \ldots \oplus \mathbb{Z}/p_r^{t_r}\mathbb{Z}$$

où les p_k sont des nombres premiers (non nécessairement distincts); les entiers $q \geq 0$, $r \geq 0$ et $p_k^{t_k}$ ne dépendent que du groupe $N/u(M)$ et non des bases initialement données de M et N. Appliqué à l'application $b_{p+1} : C_{p+1}(T) \longrightarrow Z_p(T)$, q est le *p-ème nombre de Betti*, et Poincaré appelle les nombres $p_k^{t_k}$ *coefficients de torsion* (sans doute en raison de l'exemple de la bande de Möbius); ils peuvent exister même pour des variétés orientables comme les espaces projectifs de dimension impaire.

Le progrès considérable apporté par cette technique apparaît immédiatement sur les exemples considérés antérieurement par Betti et par Poincaré (dans son mémoire de 1895); il est facile d'y définir explicitement des triangulations simples et le calcul des groupes d'homologie est alors facile. En particulier, l'homologie des sphères est

$$H_0(\mathbb{S}_n) = \mathbb{Z}, \;\; H_p(\mathbb{S}_n) = 0 \text{ pour } 1 \leq p \leq n-1, \;\; H_n(\mathbb{S}_n) = \mathbb{Z} \tag{12}$$

et l'homologie du plan projectif réel

$$H_0(\mathbb{P}_2(\mathbb{R})) = \mathbb{Z}, \;\; H_1(\mathbb{P}_2(\mathbb{R})) = \mathbb{Z}/2\mathbb{Z}, \;\; H_2(\mathbb{P}_2(\mathbb{R})) = 0. \tag{13}$$

Il prouve aussi aisément (pour les nombres de Betti d'une triangulation) la formule (6). Le second membre de (6) est appelé la *caractéristique d'Euler-Poincaré* et noté $\chi(T)$ ou $\chi(V)$.

---

8) Une somme directe $\bigoplus_{p\in\mathbb{Z}} M_p$ de modules sur un anneau commutatif A est appelée un *A-module gradué*; on dit que les éléments de M_p sont *homogènes de degré* p et pour un tel élément $x \neq 0$, on pose $p = \deg x$.

Poincaré voulut ensuite obtenir de cette façon son théorème de dualité. Cela l'amena à inventer deux autres outils qui allaient aussi jouer un rôle essentiel chez ses successeurs: la *subdivision barycentrique* et la triangulation *duale* d'une triangulation donnée. Le *simplexe standard (ouvert) de dimension* p est l'ensemble des points $(x_0, x_1, \ldots, x_p)$ dans $\mathbb{R}^{p+1}$ vérifiant les relations

$$\Delta_p : x_j > 0 \text{ pour } 0 \leq j \leq p, \quad x_0 + x_1 + \ldots + x_p = 1. \tag{14}$$

Ses *faces* de dimension $q < p$ s'obtiennent en remplaçant dans les relations (14), $p-q$ des inégalités $x_j > 0$ par $x_j = 0$. Une triangulation T est dite *simpliciale* si l'adhérence de chaque p-cellule est l'image de $\bar{\Delta}_p$ par un homéomorphisme et si l'image par cet homéomorphisme d'une face de Δ_p de dimension q est une q-cellule de T. La subdivision barycentrique T' de T est une triangulation simpliciale qui se définit par récurrence sur la dimension des cellules de T. Si on suppose définies les cellules de T' contenues dans les q-cellules de T pour tout $q < p$ et si σ est une p-cellule de T, on considère un homéomorphisme transformant $\bar{\sigma}$ en un polyèdre convexe fermé K de dimension p dans $\mathbb{R}^p$; chaque face de K de dimension $q < p$ est réunion disjointe de simplexes c_i^r de dimension $r < p$, images de cellules de T'. On prend dans K son barycentre[9]) B et on considère tous les simplexes de dimension $r + 1 \leq p$ dont B est un sommet et la base opposée à B un des simplexes c_i^r de dimension r. En prenant dans σ les images de tous ces simplexes de dimension $r + 1$ et l'image de B comme 0-cellule, on a tous les simplexes de T' contenus dans σ, et on opère de même dans toutes les p-cellules de T.

Pour avoir la triangulation T^* *duale* de T, Poincaré regroupe autrement les cellules de T'. Si B_i est le barycentre de la cellule a_i^p de T, on considère les $(n - p)$-cellules de T' dont B_i est un sommet et dont les adhérences coupent a_i^p transversalement; la réunion $a_i^{*(n-p)}$ de ces cellules et des cellules de leurs frontières de dimension $< n - p$ ayant B_i pour sommet, est la cellule duale de a_i^p. La preuve du théorème de dualité esquissée par Poincaré repose alors sur deux propriétés: 1) le fait que pour une triangulation T et une subdivision barycentrique T', les nombres de Betti sont les mêmes; 2) le fait que T' est aussi une subdivision barycentrique de T^*. Il suffit alors de prouver que le nombre de Betti $p_m(T)$ est égal au nombre de Betti $p_{n-m}(T^*)$; pour cela Poincaré montre (en termes modernes) que l'application bord b_m pour T et l'application bord b_{n-m-1} pour T^* sont (au signe près) transposées l'une de l'autre.

---

9) Si $z_0, z_1, \ldots, z_p$ sont les sommets de K, le barycentre est le vecteur

$$B = \frac{1}{p+1}(z_0 + z_1 + \ldots + z_p).$$

8 Les débuts de l'homologie

Les successeurs de Poincaré dans les recherches de Topologie algébrique (dite alors «Topologie combinatoire») s'aperçurent assez vite que les mémoires de Poincaré, malgré le nombre des idées profondes et originales qui y foisonnaient, ne constituaient pas autre chose qu'un *programme de travail*. Les notions de base étaient bien là (en fait, aucune idée d'importance comparable n'apparut en Topologie algébrique avant 1920); mais il manquait les *outils* nécessaires pour fournir des démonstrations correctes, et Poincaré n'en disposait pas. Ce furent essentiellement Brouwer et Alexander qui vont les inventer et ils seront complétés par d'importantes contributions de Lefschetz.

Après Poincaré, le champ d'action de la Topologie algébrique s'élargit; on ne considère plus seulement les variétés C^1, mais les espaces compacts connexes munis d'une triangulation; cette dernière est uniquement soumise aux trois conditions énumérées dans 7.2, sauf qu'on remplace les difféomorphismes dans une variété par des homéomorphismes quelconques. Un tel espace muni de sa triangulation est appelé un *complexe cellulaire*. La notion de subdivision barycentrique est inchangée, et fournit les *complexes simpliciaux*. On montre aisément qu'un tel complexe est homéomorphe à un complexe simplicial contenu dans un espace $\mathbb{R}^N$ pour N assez grand, dont les cellules *sont* des simplexes rectilignes (chacun d'eux étant ouvert dans la variété affine qu'il engendre dans $\mathbb{R}^N$); on dira que ce sont des *complexes simpliciaux euclidiens*.

En 1908, H. Tietze a l'idée de remplacer dans la définition (6) des chaînes les coefficients entiers λ_i par des coefficients pris dans le corps $\mathbb{F}_2$ à 2 éléments («entiers modulo 2»), ce qui dispense de considérer des cellules orientées. Alexander remplace de même les coefficients par des entiers modulo m (quelconque) et Lefschetz par des nombres rationnels. Mais en fait les groupes d'homologie obtenus ainsi se déduisent des groupes d'homologie définis par Poincaré à l'aide de pures manipulations algébriques et ne constituent pas de nouveaux invariants. On note $H.(X;A)$ la somme directe des A-modules d'homologie $H_n(X;A)$ à coefficients dans un anneau commutatif A.

Le problème majeur est de prouver que ces groupes définis à l'aide d'une triangulation, ne dépendent en fait que de la topologie de l'espace. C'est le problème d'invariance, résolu seulement par Alexander en 1915.

8.1 Homotopie et applications simpliciales

C'est à Brouwer (1911) que l'on doit les *premières* démonstrations rigoureuses en Topologie algébrique. Il formule d'abord une définition qui donne (enfin!) un sens précis aux «déformations» chères à ses prédécesseurs et les généralise. C'est la notion générale d'*homotopie* qui deviendra une des plus fondamentales de la théorie. Deux applications continues f, g d'un espace topologique E dans un espace topologique F sont *homotopes* s'il existe une application continue H : $E \times [0,1] \longrightarrow F$ (qu'on appelle une *homotopie de f à g*) telle que $H(x,0) = f(x)$ et $H(x,1) = g(x)$ pour tout $x \in E$. Par exemple, l'identité id: $x \mapsto x$ de $\mathbb{R}^n$ est homotope à l'application constante $g : x \mapsto 0$ en prenant $H(x,t) = (1-t)x$. Ce qui correspond à une «déformation» c'est l'*isotopie*, où l'on suppose en outre que pour $0 \leq t \leq 1$, $x \mapsto H(x,t)$ est un homéomorphisme de E sur son image dans F.

L'apport essentiel de Brouwer est le concept d'*approximation simpliciale*. Etant donnés deux complexes simpliciaux (rectilignes) X, X' contenus dans un $\mathbb{R}^N$, dont T et T' sont les triangulations, une *application simpliciale* $f : X \longrightarrow X'$ est définie par la condition que pour tout entier $p \geq 0$, les images par f des sommets d'un p-simplexe σ de T sont des sommets (non nécessairement distincts) d'un q-simplexe de T' pour $q \leq p$ et la restriction de f à σ est *affine* de sorte que f est entièrement déterminée par ses valeurs aux sommets des simplexes de T. Ces applications sont donc les généralisations directes des «fonctions linéaires par morceaux» d'une variable réelle.

Soit maintenant $f : X \longrightarrow X'$ une application continue quelconque et ε un nombre > 0. Une *approximation simpliciale* de f à ε près est une application simpliciale $g : (X, T_i) \longrightarrow (X', T_i')$, où T_i (resp. T_i') est obtenue à partir de T (resp. T') par subdivisions barycentriques itérées, telles que

$$|f(x) - g(x)| \leq \varepsilon \text{ pour } x \in X.$$

L'existence d'une telle approximation a été prouvée par Brouwer dans un cas particulier, et dans le cas général par Alexander. On prend T_i' telle que le diamètre des simplexes de T_i' soit $\leq \varepsilon/3$. Pour tout sommet v' de T_i', on considère son *étoile* $\mathrm{St}(v')$ (définie déjà par Poincaré sous le nom d'«aster»): c'est la réunion de tous les simplexes de T_i' qui ont v' pour sommet, donc un ensemble ouvert dans X', et les $\mathrm{St}(v')$ forment un recouvrement de X'. Par l'uniforme continuité de f, il existe une subdivision barycentrique itérée T_i de T telle que, pour tout sommet v de T_i, $f(\mathrm{St}(v))$ soit contenu dans au moins une étoile de T_i'. Soient alors a_k ($1 \leq k \leq M$) les sommets distincts de T_i, b_j' ($1 \leq j \leq M'$) les sommets distincts de T_i'; pour tout k, soit $j(k)$ un des indices pour lesquels $f(\mathrm{St}(a_k)) \subset \mathrm{St}(b'_{j(k)})$. Si on considère $p+1$ des a_k qui sont les sommets d'un p-simplexe σ de T_i, les $b'_{j(k)}$ correspondants (qui ne sont pas nécessairement distincts) sont les sommets d'un q-simplexe τ' de T_i' avec $q \leq p$ par définition d'une étoile. On définit alors une application simpliciale $g : (X, T_i) \longrightarrow (X', T_i')$ en prenant $g(a_k) = b'_{j(k)}$; avec les notations précédentes, on a alors $g(\sigma) \subset \tau'$, donc $|f(x) - g(x)| \leq \varepsilon$ pour tout $x \in \sigma$. En outre, pour tout $x \in \sigma$, $f(x)$ est dans un simplexe de T_i' égal à $g(\sigma)$ ou dont $g(\sigma)$ est une face; donc le segment d'extrémités $f(x)$ et $g(x)$ est contenu dans X', et $(x,t) \mapsto (1-t)f(x) + tg(x)$ (pour $0 \leq t \leq 1$) est une *homotopie* de f à g.

Veblen et Alexander déduisirent de là un premier exemple de ce que l'on appellera plus tard «fonctorialité». Si $g : (X, T_i) \longrightarrow (X', T_i')$ est une application simpliciale, on en déduit, pour tout p, un *homomorphisme*

$$\tilde{g} : C_p(T_i) \longrightarrow C_p(T_i')$$

des $\mathbb{Z}$-modules de p-chaînes, en la définissant sur la base canonique des p-simplexes, par $\tilde{g}(\sigma) = g(\sigma)$ si les images par g des sommets de σ sont distinctes, et $\tilde{g}(\sigma) = 0$ sinon (l'image $g(\sigma)$ est appelée dans ce cas un p-simplexe dégénéré). On vérifie que $b_p(\tilde{g}(\sigma)) = \tilde{g}(b_p\sigma)$, d'où l'on déduit de $\tilde{g}$, par passage aux quotients, un homomorphisme de $\mathbb{Z}$-modules

$$g_* : H_p(T) \longrightarrow H_p(T'). \tag{15}$$

Si $h : (X', T_i') \longrightarrow (X'', T_i'')$ est une application simpliciale, on a

$$(h \circ g)_* = (h_*) \circ (g_*). \tag{16}$$

Veblen et Lefschetz prouvèrent en outre, en substance, une propriété fondamentale de l'homologie (admise sans preuve jusque là pour les «déformations»). Si f et g sont deux applications simpliciales de (X, T) dans (Y, U) qui sont *homotopes*, alors les homomorphismes correspondants f_* et g_* de $H.(T)$ dans $H.(U)$ sont *égaux*. Il s'agit de montrer que pout tout p-cycle z_p de T, $\tilde{f}(z_p) - \tilde{g}(z_p)$ est un bord de U. On se ramène au cas où l'homotopie

$$F : X \times [0, 1] \longrightarrow Y$$

de f à g est elle aussi une application simpliciale, pour une subdivision barycentrique itérée convenable de U, et une triangulation de $X \times [0, 1]$ définie comme suit.

Lorsque $a_0, a_1, \ldots, a_p$ sont les sommets d'un p-simplexe dans un $\mathbb{R}^N$, on écrit $a_0a_1 \ldots a_p$ ce p-simplexe muni de l'orientation pour laquelle les vecteurs joignant un point du simplexe à $a_0, a_1, \ldots, a_p$ forment un repère direct. En particulier, le simplexe standard Δ_p est orienté en le notant $e_0e_1 \ldots e_p$, les e_j étant les vecteurs unités sur les axes de $\mathbb{R}^{p+1}$. Le produit $\bar{\Delta}_p \times [0, 1]$ est alors réunion des adhérences des $(p + 1)$-simplexes

$$c_k = e_0e_1 \ldots e_k e_k' e_{k+1}' \ldots e_p' \quad (0 \leq k \leq p) \quad {}^{10)}$$

où e_k est identifié à $(e_k, 0)$ et $e_k' = (e_k, 1)$. On note $\Delta_p \times [0, 1]$ la $(p + 1)$-chaîne $\sum_{k=0}^{p} (-1)^k c_k$. On a alors les relations

$$b_{p+1}(\Delta_p \times [0, 1]) = \Delta_p \times \{1\} - \Delta_p \times \{0\} + \big((b_p\Delta_p) \times [0, 1]\big). \tag{17}$$

Par linéarité, on en déduit, pour toute p-chaîne z_p de T

$$\tilde{f}(z_p) - \tilde{g}(z_p) = b_{p+1}\big(\tilde{F}(z_p \times [0, 1])\big) - \tilde{F}\big((b_pz_p) \times [0, 1]\big) \tag{18}$$

---

10) Cette décomposition généralise la décomposition en 3 tétraèdres donnée par Euclide pour un prisme à base triangulaire.

et par suite, si $b_p z_p = 0$,

$$\tilde{f}(z_p) - \tilde{g}(z_p) = b_{p+1}\big(\tilde{F}(z_p \times [0,1])\big)$$

est bien un bord. La formule (18) est le premier exemple de ce qui deviendra plus tard un outil important, les «formules d'homotopie».

Si maintenant f est une application continue quelconque de X dans Y, la construction d'Alexander montre que toutes les approximations simpliciales assez voisines de f sont homotopes à f, donc homotopes entre elles, et tous les homomorphismes g_* associés à ces approximations sont donc égaux; cela définit l'homomorphisme $f_* : H.(X) \longrightarrow H.(Y)$ et ces homomorphismes vérifient encore la relation (16); en outre on a $f_* = g_*$ pour deux applications continues homotopes.

Une décomposition géométrique analogue en «prismes» permet de mettre sous une forme plus commode le fait que l'homologie $H.(T)$ pour une triangulation T est naturellement isomorphe à l'homologie $H.(T_1)$ de la subdivision barycentrique T_1 de T. Pour cela, on définit pour chaque $p \geq 0$, un homomorphisme canonique de p-chaînes

$$\mathrm{sd} : C_p(T) \longrightarrow C_p(T_1) \tag{19}$$

en définissant, pour tout p-simplexe σ de T, la p-chaîne $\mathrm{sd}(\sigma)$ comme la somme $\Sigma\tau$ des p-simplexes de T_1 contenus dans T, chacun étant muni de l'orientation induite par celle de σ. On montre que $\mathrm{sd}(b_p\sigma) = b_p(\mathrm{sd}(\sigma))$, et la décomposition «prismatique» de $\bar{\Delta}_p \times [0,1]$ à laquelle nous faisons allusion prouve que l'application $H_p(T) \longrightarrow H_p(T_1)$ déduite de sd est un isomorphisme.

8.2 La notion de degré

Il est très surprenant que Brouwer ne se soit jamais intéressé à la théorie de l'homologie et ne cite aucun des mémoires de Poincaré sur ce sujet. Tous les remarquables théorèmes qu'il prouva en 1911-12 sont ramenés, souvent de façon extrêmement détournée, à la notion découverte par lui en 1910, celle de *degré* d'une application continue $f : M \longrightarrow N$, où M et N sont des «polyèdres» au sens de Poincaré, connexes, orientés et de *même* dimension n.

L'idée de degré est très simple lorsqu'on suppose en outre que M et N sont des variétés C^∞ et f de classe C^∞. On montre alors qu'il y a des points $y \in N$ tels que $f^{-1}(y)$ soit un ensemble fini $\{x_1, x_2, \ldots, x_r\}$, et que pour chaque j, il y a un voisinage V_j de x_j dans M tel que la restriction $f|V_j$ soit un difféomorphisme sur un voisinage de y; si alors p (resp. q) est le nombre des x_j tels que $f|V_j$ préserve (resp. renverse) l'orientation, une formule intégrale qui remonte à Kronecker montre que le nombre $c = p - q$ est indépendant du choix de y vérifiant ces conditions; c'est le *degré* de f. Mais Brouwer voulait considérer des polyèdres quelconques M, N, et une application f que l'on suppose seulement continue. Or, même si $M = N = \mathbb{S}_1$, on peut aisément, à l'aide d'une courbe de Peano, donner des exemples où $f^{-1}(y)$ a la puissance du continu pour *tout* $y \in N$. L'idée de Brouwer fut de considérer des approximations simpliciales de f; si l'on peut définir le degré d'une application simpliciale et montrer que ce nombre est le même pour toutes les approximations simpliciales g dès que la distance de g à f est assez petite, ce nombre pourra être pris comme définition du degré de f.

Au début de ce programme, on considère donc une application simpliciale $g : (M,T) \longrightarrow (N,U)$; Brouwer montre qu'il y a un ouvert connexe Ω partout dense dans N, tel que, si $y \in \Omega$ n'est pas dans la réunion S des images par g des simplexes de dimension $n-1$, alors $g^{-1}(y)$ est un ensemble fini $\{x_1, x_2, \ldots, x_r\}$ dont chacun est dans un simplexe (ouvert) de T de dimension n; ces simplexes sont distincts, et la restriction de g à chacun d'eux σ est une bijection sur $g(\sigma)$; si p (resp. q) est le nombre des simplexes σ tels que $g|\sigma$ préserve (resp. renverse) l'orientation, Brouwer montre que le nombre $p-q$ est le même pour tous les points $y \in \Omega$ non dans S; pour cela il joint deux de ces points par une ligne brisée contenue dans Ω; il prouve que lorsque y se déplace sur cette ligne et traverse S, p et q augmentent ou diminuent simultanément d'un même nombre, donc $p-q$ ne change pas; cela tient à ce que tout $(n-1)$-simplexe de T est une face de deux n-simplexes de T exactement, ce qui est une condition imposée par Poincaré à ses «polyèdres», et est en fait vérifiée pour toute triangulation T dont les $(n-1)$-simplexes sont des sous-variétés de classe C^1.

Ce genre de raisonnement est typique, par son ingéniosité et sa complexité, des démonstrations de Brouwer. Pour arriver à la définition du degré et pour prouver ses propriétés, il doit encore développer plusieurs constructions du même genre. Il montre ainsi que le nombre $\deg(f)$ qu'il définit a les propriétés suivantes:

1) $\deg(f) = 0$ si f n'est pas surjective et $\deg(f) = \pm 1$ si f est un homéomorphisme;
2) $\deg(g \circ f) = \deg(g) \cdot \deg(f)$ pour deux applications continues $f : M \longrightarrow N$, $g : N \longrightarrow P$;
3) si $f_1 : M \longrightarrow N$ et $f_2 : M \longrightarrow N$ sont homotopes, alors $\deg(f_1) = \deg(f_2)$.

Nous reviendrons sur les remarquables théorèmes qu'il obtint en utilisant la notion de degré, et qui peuvent maintenant être prouvés plus simplement.

8.3 Invariance, dualité d'Alexander et homologie relative

J.W. Alexander a donné 3 preuves du théorème d'invariance établissant que les groupes d'homologie $H_p(X,T)$ d'un espace compact X muni d'une triangulation T, sont indépendants de la triangulation. La première preuve (1915) introduit de façon imparfaite l'idée de «simplexe singulier» que nous retrouverons plus loin. Entachée d'imprécisions, elle a été reprise par Veblen, Van der Waerden et Lefschetz, mais c'est seulement en 1938 que Tucker est parvenu à la rendre entièrement rigoureuse.

La seconde démonstration (1926) n'utilise plus les simplexes singuliers. Le théorème à prouver est le suivant: si X et X' sont deux sous-espaces d'un $\mathbb{R}^N$ munis de triangulations simpliciales (rectilignes) T, T', et si X et X' sont homéomorphes, alors, pour tout p, $H_p(T)$ est isomorphe à $H_p(T')$. On considère la suite (T_i) (resp. T'_j) des subdivisions barycentriques itérées de T (resp. T'); le diamètre maximum des simplexes de T_i (resp. T'_j) tend vers 0. Soient $f : X \longrightarrow X'$ un homéomorphisme, $g = f^{-1} : X' \longrightarrow X$ son inverse. La construction d'Alexander fournit un indice j assez grand et une approximation simpliciale $g_j : T'_j \longrightarrow T$ de g, puis un indice i assez grand et une approximation simpliciale $f_{ij} : T_i \longrightarrow T'_j$ de f. On en déduit une approximation simpliciale $h_i = g_j \circ f_{ij} : T_i \longrightarrow T$ de l'identité. Alexander montre que l'on a

$$\tilde{h}_i(\mathrm{sd}_i(\sigma_p)) = c \cdot \sigma_p \text{ pour } c \in \mathbb{Z}$$

pour tout p-simplexe de T. Il prouve ensuite par récurrence sur p que $c = 1$. C'est évident pour $p = 0$; d'autre part $\tilde{h}_i(\mathrm{sd}_i(b_p\sigma_p)) = c \cdot b_p\sigma_p$, et l'hypothèse de récurrence entraîne que $\tilde{h}_i(\mathrm{sd}_i(\sigma_{p-1})) = \sigma_{p-1}$ pour tout $(p-1)$-simplexe de T, d'où $c = 1$, et $(h_i)_*$ et est donc l'identité. Mais comme

$$(h_i)_* : H.(T_i) \xrightarrow{(f_{ij})_*} H.(T'_j) \xrightarrow{(g_j)_*} H.(T_i)$$

$(f_{ij})_*$ est injectif et $(g_j)_*$ surjectif. En échangeant les rôles de g et f et de T et T', on voit de même que $(f_{ij})_*$ est surjectif, donc il est bijectif.

Une fois prouvée l'invariance de l'homologie lorsque l'on change de triangulation, il devient possible de prouver que deux espaces compacts ne peuvent pas être homéomorphes. C'est le cas par exemple pour deux sphères $\mathbb{S}_m$ et $\mathbb{S}_n$ pour $m \neq n$.

La troisième preuve d'Alexander (1922) se rattache à un remarquable élargissement de l'homologie à des espaces non compacts et à la découverte d'une nouvelle propriété de *dualité*. Alors que jusque là il ne s'agissait que de complexes simpliciaux compacts, Alexander observa qu'on peut aussi définir une triangulation d'une partie ouverte U de $\mathbb{R}^n$. Les cellules sont définies comme précédemment, ainsi que les bords; on suppose que toute partie compacte de U ne rencontre qu'un nombre fini de cellules (on dit alors que le complexe cellulaire est *localement fini*). Moyennant quoi, la définition des groupes d'homologie reste la même. De plus, il est facile d'exhiber une triangulation pour *tout* ouvert non vide U de $\mathbb{R}^n$, à l'aide de cubes translatés des «cubes standard» $2^{-k} \cdot [0, 1]^n$, les cubes étant pris de plus en plus petits à mesure qu'on s'approche de la frontière de U (construction déjà utilisée par Runge pour $n = 2$). Cela étant, Alexander considère un complexe simplicial compact (curviligne) X contenu dans $\mathbb{S}_n$ et prouve, pour l'homologie à coefficients dans $\mathbb{F}_2$ les isomorphismes suivants:

$$H_p(X) \simeq H_{n-p-1}(\mathbb{S}_n - X) \text{ pour } 1 \leq p \leq n-2 \tag{20}$$

$$H_0(X) \simeq H_{n-1}(\mathbb{S}_n - X) \oplus \mathbb{F}_2, H_0(\mathbb{S}_n - X) \simeq H_{n-1}(X) \oplus \mathbb{F}_2. \tag{21}$$

En outre $H_{n-p-1}(\mathbb{S}_n - X) = 0$ si $p > \dim X$.

La démonstration est ingénieuse et assez compliquée. C'est sans doute le premier exemple d'une technique systématisée un peu plus tard par Mayer et Vietoris et très employée par la suite: elle consiste à écrire X comme réunion $Y \cup Z$ de deux sous-espaces fermés convenablement choisis, et déduire une propriété de l'homologie de X de la même propriété supposée connue pour Y, Z et $Y \cap Z$. Alexander applique cette méthode en 3 étapes:

1) Il suppose X homéomorphe au cube $[0, 1]^m$ et montre que, pour $m \leq n$, on a $H_p(\mathbb{S}_n - X) = 0$ pour $1 \leq p \leq n$ et $H_0(\mathbb{S}_n - X) \simeq \mathbb{F}_2$; il procède par contradiction en décomposant X en cubes de plus en plus petits.

2) Il suppose X est homéomorphe à une sphère $\mathbb{S}_m$ pour $m < n$; il décompose alors $\mathbb{S}_m$ en deux hémisphères fermés de frontière commune $\mathbb{S}_{m-1}$, chacun étant homéomorphe à un cube.

3) Dans le cas général, il raisonne par récurrence sur la dimension m de X, en décomposant X en réunion de l'adhérence d'un m-simplexe Y de la triangulation et de $X - Y$; puis il itère ce procédé jusqu'à ce qu'il ne reste qu'un complexe simplicial de dimension $m - 1$.

Le théorème d'invariance résulte de ce que les relations (20) et (21) ne dépendent pas du choix des triangulations de $\mathbb{S}_n$ et de X.

Il est intéressant de remarquer que les deux premières étapes de la preuve précédente entraînent comme cas particuliers deux célèbres théorèmes de Brouwer qu'il avait prouvés par des constructions très compliquées: 1) si X est homéomorphe à $[0, 1]^m$ avec $m < n$, $\mathbb{S}_n - X$ est connexe (pour $n = 2$, un arc de courbe homéomorphe à un intervalle fermé de $\mathbb{R}$ ne coupe pas le plan); 2) si X est homéomorphe à $\mathbb{S}_{n-1}$, $\mathbb{S}_n - X$ a deux composantes connexes (théorème dit «de Jordan-Brouwer»).

L'utilisation de la cohomologie a permis de simplifier et de généraliser la démonstration d'Alexander.

Le théorème de dualité d'Alexander incita Lefschetz en 1928 à introduire une nouvelle notion, l'*homologie relative*. Sous sa forme la plus simple, elle a trait à un complexe simplicial euclidien K et un sous-complexe L de K (ensemble nécessairement fermé puisqu'il contient les faces de tous ses simplexes). Pour une p-chaîne s de K, on définit son *bord relatif* $b_{L,p}s$ par rapport à L comme la somme des termes figurant dans l'expression du bord $b_p s$ qui sont des multiples de $(p - 1)$-simplexes n'appartenant pas à L. Un p-cycle (resp. p-bord) relatif à L est donc tel que $b_p s$ soit combinaison linéaire de $(p - 1)$-simplexes de L (resp. tel qu'il existe une $(p + 1)$-chaîne t de K telle que $b_{p+1}t - s$ soit combinaison linéaire de simplexes de L. On peut ainsi définir le *p-ème groupe d'homologie de K relatif à* L (ou *modulo* L), qu'on note $H_p(K, L)$. Lefschetz montra que ce groupe ne dépend que de l'espace $K - L$, et il étudia les relations entre $H.(K)$, $H.(L)$ et $H.(K, L)$. En particulier il prouva que si $K = \mathbb{S}_n$, $H_p(\mathbb{S}_n, L) \simeq H_{p-1}(L)$ pour $1 < p < n$, ce qui lui donna une autre preuve du théorème de dualité d'Alexander. Ses raisonnements sont les précurseurs de ce qui devait plus tard devenir la suite exacte d'homologie et le théorème d'excision (voir plus loin).

8.4 Homologie d'un produit

Au début de sa carrière, S. Lefschetz s'était consacré à la géométrie algébrique et avait appliqué les idées de Poincaré sur l'homologie à l'étude des surfaces algébriques et des variétés algébriques de toute dimension. Lorsqu'il se tourna vers la Topologie algébrique pour elle-même, à partir de 1920, il y transporta des techniques utilisées en géométrie algébrique, le produit cartésien $X \times Y$ de deux variétés et les intersections de sous-variétés. Si T (resp. U) est une triangulation d'un complexe cellulaire X (resp. Y), le produit $\sigma \times \tau$ d'une cellule de T et d'une cellule de U est encore une cellule; on définit ainsi $X \times Y$ comme complexe cellulaire; si σ est de dimension k et τ de dimension $m-k$, on a pour le bord la formule

$$b_m(\sigma \times \tau) = b_k\sigma \times \tau + (-1)^k \sigma \times b_{m-k}\tau. \tag{22}$$

A partir de là, il est facile de déterminer l'homologie de $X \times Y$ connaissant celles de X et de Y. Ce fut fait indépendamment en 1923 par Künneth et Lefschetz; en particulier, pour les nombres de Betti, on a la «formule de Künneth»

$$p_m(X \times Y) = \sum_{0 \leq k \leq m} p_k(X) p_{m-k}(Y) \tag{23}$$

et pour les caractéristiques d'Euler-Poincaré

$$\chi(X \times Y) = \chi(X)\chi(Y). \tag{24}$$

8.5 La notion de variété et les intersections

En 1934, S. Cairns prouva pour la première fois que si $r \geq 1$, une variété C^r est triangulable avec des cellules qui sont des sous-variétés; plusieurs autres preuves en furent données par la suite. Mais jusqu'en 1969 on ignora s'il existait des triangulations sur les variétés C^0, c'est-à-dire les espaces localement compacts séparables dont tout point a un voisinage homéomorphe à un $\mathbb{R}^n$.

Les successeurs de Poincaré durent combler les lacunes de sa preuve du théorème de dualité, et ils voulurent aussi étendre ce théorème à des espaces X aussi généraux que possible. Le point délicat se situe dans la construction de la triangulation T^* duale d'une triangulation T: il n'est pas évident que l'ensemble $a^{*(n-p)}$ défini par Poincaré soit effectivement une cellule lorsque X n'est pas une variété C^1; plusieurs mathématiciens aux environs de 1930 proposèrent des conditions pour qu'il en soit bien ainsi. Une des plus simples est que, pour tout $x_0 \in X$, on ait, pour l'homologie relative

$$H_q(X, X - \{x_0\}; \mathbb{Z}) = 0 \text{ pour } 0 \leq q \leq n-1 \text{ et } H_n(X, X - \{x_0\}; \mathbb{Z}) \simeq \mathbb{Z}. \tag{25}$$

Cette condition est vérifiée pour les variétés C^0, et on peut appeler «variété généralisée» un espace où elle est vérifiée. Lefschetz et E. Čech montrèrent en 1933 qu'elle entraîne la dualité de Poincaré. Il faut cependant noter que ces variétés généralisées peuvent avoir des propriétés très différentes de celles des variétés C^0; pour $n \geq 4$, il peut se faire que $V - \{x_0\}$ ne soit pas simplement connexe, quel que soit le voisinage assez petit V de x_0.

Une propriété importante d'une variété généralisée X de dimension n, compacte et connexe, est que si X est orientable, on a $H_n(X; \mathbb{Z}) \simeq \mathbb{Z}$; un générateur de ce groupe est la classe du n-cycle formé par la somme des n-cellules orientées par l'orientation de x; on dit que ce n-cycle est le *cycle fondamental* de X et sa classe est notée $[X]$. Si X n'est pas orientable, $H_n(X; \mathbb{Z}) = 0$. La notion de classe fondamentale fournit une définition simple du *degré* d'une application continue $f : X \longrightarrow X'$, où X et X' sont des variétés généralisées, compactes, connexes, orientées et de même dimension n; on a $f_*([X]) = c \cdot [X']$ pour $c \in \mathbb{Z}$, et $c = \deg(f)$.

L'idée de Lefschetz pour définir une «intersection» de deux sous-variétés dans une variété C^r (et plus généralement pour étendre cette notion aux variétés généralisées) est en fait d'obtenir une «intersection» de deux classes d'homologie. Partant de ce que dans une variété algébrique de dimension n, «en général» l'intersection d'une sous-variété de dimension p et d'une sous-variété de dimension q est vide si $p + q < n$, et une sous-variété de dimension $p + q - n$ sinon, on cherche à définir une application bilinéaire

$$(z_p, z_q) \mapsto z_p \cdot z_q$$

de $H_p(X) \times H_q(X)$ dans $H_{p+q-n}(X)$, satisfaisant aux relations

$$z_q \cdot z_p = (-1)^{(n-p)(n-q)} z_p \cdot z_q, (z_p \cdot z_q) \cdot z_r = z_p \cdot (z_q \cdot z_r)$$

de sorte que la somme directe $H.(X)$ est munie d'une structure d'anneau, appelé «anneau d'intersection». Lorsque $p+q = n$, on doit avoir $z_p \cdot z_{n-p} = \lambda z_0$, où z_0 est la classe d'homologie d'un point et λ un entier égal au «nombre d'intersection» défini par Poincaré et que Lefschetz note $(z_p \cdot z_{n-p})$.

Lorsque z_p et z_q sont les classes d'homologie de deux sous-variétés V_p, V_q, la classe d'homologie $z_p \cdot z_q$ devrait être la classe d'homologie de l'intersection $V_p \cap V_q$. Lefschetz chercha à étendre cette idée à des «simplexes singuliers» (dont la définition à cette époque manquait de clarté) qu'il approchait par des polyèdres (curvilignes) en «position générale». La cohomologie devait permettre ultérieurement une définition plus simple et plus générale.

9 Les premières applications de l'homologie

9.1 Points fixes d'une application et formule de Lefschetz

Un des plus fameux résultats de Brouwer (1911) est son théorème du point fixe: si f est une application du cube $[0, 1]^n$ de dimension n dans lui-même, dont on suppose uniquement qu'elle est continue, alors il existe au moins un point $x \in [0, 1]^n$ tel que $f(x) = x$. Une de ses preuves consiste à montrer d'abord que si $g : \mathbb{S}_n \longrightarrow \mathbb{S}_n$ est une application continue qui n'a pas de point fixe, alors $\deg(g) = (-1)^{n+1}$. Pour cela, en considérant le grand cercle (bien défini par hypothèse) qui passe par x et $g(x)$, il prouve que g est homotope à l'application $x \mapsto -x$, dont le degré est $(-1)^{n+1}$. Pour en déduire le théorème du point fixe, il remarque que l'on peut remplacer le cube $[0, 1]^n$ par l'hémisphère D_+ de $\mathbb{S}_n$ défini par l'inégalité $x_0 \geq 0$, qui lui est homéomorphe. On prolonge alors f en une application continue $g : \mathbb{S}_n \longrightarrow \mathbb{S}_n$, en posant $g(x) = f(s(x))$ si $x_0 < 0$, $s(x)$ étant le symétrique de x par rapport à l'hyperplan $x_0 = 0$. Comme $g(\mathbb{S}_n) \subset D_+$, g n'est pas surjective, donc $\deg(g) = 0$; mais cela contredit la relation $\deg(g) = (-1)^{n+1}$; g a donc nécessairement un point fixe, qui est aussi un point fixe de f.

Ce théorème permet de prouver l'existence de racines d'équations qui ne sont guère accessibles autrement, par exemple l'équation

$$z^n + h(z) = 0$$

dans $\mathbb{C}$, lorsque h est une fonction continue soumise seulement à l'inégalité $|h(z)| < |z|^n$ pour $|z|$ assez grand.

En 1922, G.D. Birkhoff et O. Kellogg ont montré qu'on peut généraliser le théorème du point fixe de Brouwer à certains espaces fonctionnels, et J. Schauder a étendu leurs résultats à tous les espaces de Banach: si K est un ensemble convexe et compact dans un tel espace, toute application continue $f : K \longrightarrow K$ admet un point fixe. L'idée est d'«approcher» K par un ensemble convexe compact K_n de dimension finie, et d'approcher f par une application continue $f_n : K_n \longrightarrow K_n$; on applique à f_n le théorème de Brouwer et on passe ensuite à la limite pour K_n et f_n. En 1934, J. Leray et J. Schauder ont même montré qu'on peut définir un *degré* pour une application $x \mapsto x + u(x)$ d'un espace de Banach dans lui-même, pourvu que u transforme tout ensemble borné en un ensemble relativement compact. Ces théorèmes ont permis de prouver l'existence de solutions d'équations aux dérivées partielles, inaccessibles par les méthodes antérieures.

En 1926, Lefschetz se proposa de relier à l'homologie de X le nombre de points fixes d'une application continue $f : X \longrightarrow X$, où X est une variété généralisée, compacte, sans bord, orientable et triangulable. Il considère plus généralement deux telles applications f, g et les «coïncidences» de f et g, c'est-à-dire les points $x \in X$ tels que $f(x) = g(x)$. Inspiré par la géométrie algébrique, il remarque que ces points sont les projections sur X de l'intersection $\Gamma_f \cap \Gamma_g$ des graphes de f et g dans $X \times X$. Toute une partie de la géométrie algébrique, débutant avec Chasles et développée notamment par Hurwitz et Severi, était consacrée à l'étude des «correspondances» sur une variété algébrique X de dimension n, qui ne sont autres que les sous-variétés de dimension n de $X \times X$, et il s'agissait en particulier d'étudier le nombre d'intersection $(C_1 \cdot C_2)$ de deux telles correspondances.

Utilisant sa théorie des intersections, Lefschetz considéra de même, sur une variété généralisée X de dimension n, le nombre d'intersection $(\Gamma_T \cdot \Gamma_{T'})$ des classes de deux n-cycles dans $X \times X$. S'inspirant toujours des constructions de la géométrie algébrique, il associa à Γ_T, pour tout entier p tel que $0 \leq p \leq n$, un endomorphisme $T_* : H_p(X;\mathbb{Q}) \longrightarrow H_p(X;\mathbb{Q})$, de la façon suivante: pour une classe d'homologie $\alpha \in H_p(X;\mathbb{Q})$, le produit $\alpha \times [X]$ de α par la classe fondamentale de X est une classe dans $H_{p+n}(X \times X;\mathbb{Q})$, et $T_*(\alpha)$ est l'image par la seconde projection de l'intersection $\Gamma_T \cdot (\alpha \times [X])$. Lefschetz put calculer le nombre $(\Gamma_T \cdot \Gamma_{T'})$ à l'aide des matrices des endomorphismes T_* et T'_*.

Le cas le plus important est celui où $\Gamma_{T'}$ est la classe de la diagonale Δ de $X \times X$ (qui est un n-cycle), et Γ_T est la classe Γ_f du graphe d'une application continue $f : X \longrightarrow X$, auquel cas T_* n'est autre que l'endomorphisme f_* défini par Veblen et Alexander. On a alors la célèbre formule de Lefschetz

$$(\Gamma_f \cdot \Delta) = \sum_{0 \leq m \leq n} (-1)^m \mathrm{Tr}((f_*)_m) \tag{26}$$

exprimant le nombre d'intersection à l'aide des traces des $(f_*)_m$. Après maints avatars, cette formule devait devenir le point de départ de la preuve des «conjectures de Weil».

Dans un de ses premiers travaux (1928), H. Hopf aborda autrement le problème. Pour *tout* complexe simplicial X de dimension n et toute application continue $f : X \longrightarrow X$, on peut définir un «nombre de Lefschetz»

$$\Lambda(f) = \sum_{0 \leq m \leq n} (-1)^m \mathrm{Tr}((f_*)_m). \tag{27}$$

Lorsque f est l'identité, ce nombre n'est autre que la caractéristique d'Euler-Poincaré $\chi(X)$. Par un raisonnement très simple (semblable à celui de Poincaré), Hopf montra que si $\Lambda(f) \neq 0$, f a nécessairement un point fixe. Si de plus l'ensemble $\mathrm{Fix}(f)$ de ces points est fini et si X est une variété généralisée, Hopf établit la formule

$$\Lambda(f) = \sum_{\alpha \in \mathrm{Fix}(f)} j_\alpha \tag{28}$$

où j_α est l'*indice* du point fixe α, qu'on définit comme suit. On peut supposer que α appartient à un n-simplexe σ de la triangulation. Il y a un homéomorphisme h de σ sur la boule ouverte $B : |x| < \rho$ de $\mathbb{R}^n$; alors $g = hfh^{-1}$ est définie dans B et est une application continue de B dans $\mathbb{R}^n$ avec un seul point fixe 0. Donc l'application $x \mapsto g(x)/|g(x)|$ est définie dans la sphère $S : |x| = \frac{1}{2}\rho$ et applique S dans $\mathbb{S}_{n-1}$; j_α est son degré.

Cette notion est apparentée à une notion voisine, introduite par Poincaré dès 1880 pour $n = 2$: l'indice i_α d'un champ de vecteurs tangents $x \longrightarrow v(x)$ sur une C^1-variété X de dimension n. On dit que $a \in X$ est un point singulier de ce champ si $v(a) = 0$. En supposant ce point isolé dans l'ensemble $\mathrm{Sing}(v)$ des points singuliers, on considère un homéomorphisme h d'un voisinage assez petit de a sur la boule $|y| < \rho$ de $\mathbb{R}^n$, et le champ de vecteurs $u = hvh^{-1}$ sur cette boule; i_a est encore le degré de l'application $y \mapsto u(y)/|u(y)|$ de $S : |y| = \frac{1}{2}\rho$ dans $\mathbb{S}_{n-1}$. Supposant que v est de classe C^1 et que $\mathrm{Sing}(v)$ est fini, on montre que si $(x,t) \mapsto F_v(x,t)$ est le flot du champ de vecteurs v (solution de l'équation différentielle $z'(t) = v(z(t))$ telle que $z(0) = x$), alors il y a un intervalle assez petit $|t| \leq \varepsilon$ tel que l'application $x \mapsto F_v(x,t)$ ait exactement pour points fixes les points singuliers de v pour $|t| \leq \varepsilon$, et i_a est exactement l'indice j_a défini par Hopf pour cette application; comme elle est homotope à l'identité, la formule (28) montre que

$$\sum_{a \in \mathrm{Sing}(v)} i_a = \chi(X) \tag{29}$$

formule qui généralise des résultats antérieurs de Poincaré et Brouwer pour $\mathbb{S}_n$.

9.2 Homologie des variétés algébriques

Dans ses travaux de géométrie algébrique, E. Picard avait cherché, dès 1888, à appliquer les notions introduites par Betti. Il fut un des premiers à utiliser les idées de Poincaré sur l'homologie, et ses méthodes et résultats furent développés par Poincaré lui-même, Severi et surtout Lefschetz.

Ce n'est qu'en 1930 que Van der Waerden donna une première démonstration du fait que toute variété algébrique projective quelconque (pouvant avoir des points singuliers) est triangulable; cette démonstration a ensuite été améliorée et étendue aux variétés analytiques par plusieurs mathématiciens.

Dans ses recherches de géométrie algébrique, qui datent des environs de 1920, Lefschetz supposait les variétés triangulables. Sa technique principale généralise une méthode utilisée par Picard pour l'étude de l'homologie d'une surface algébrique complexe S d'équation $P(x,y,z) = 0$ dans $\mathbb{C}^3$. Quand on fixe $y \in \mathbb{C}$, l'équation définit une courbe algébrique plane C_y; Picard montre que, par un changement de coordonnées, on peut supposer que C_y a le même genre g pour tous les $y \in \mathbb{C}$, à l'exception d'un nombre fini de points $a_k \in \mathbb{C}$, où C_{a_k} acquiert un point double supplémentaire à tangentes distinctes, de sorte que C_{a_k} est de genre $g-1$. Par des raisonnements «intuitifs», il montre que l'étude des 1-cycles sur S peut se ramener à celles des 1-cycles sur les courbes C_y (de dimension réelle 2). Partant d'un 1-cycle γ sur une courbe C_a (pour $a \in \mathbb{C}$ quelconque), il fait varier y le long d'un lacet, d'origine et extrémité a, et qui entoure un des points a_k; il montre que lorsque γ varie continûment avec y et que y est revenu en a, le cycle γ est devenu

$$\gamma' = \gamma + m \cdot \delta_k$$

où $m \in \mathbb{Z}$ et δ_k est un 1-cycle dont la classe ne dépend pas du lacet choisi, mais seulement de a et a_k; lorsque a tend vers a_k, Poincaré montra que δ_k tend vers 0, d'où le nom de cycles «évanescents» qu'il leur attribua.

Lefschetz généralisa cette méthode à une variété algébrique quelconque V de dimension n, plongée dans un espace projectif $\mathbb{P}_N(\mathbb{C})$. Il définit un «pinceau» de sous-variétés de dimension (complexe) $n-1$, obtenues par intersection de V

avec un hyperplan H_y, passant par une variété linéaire fixe Z de dimension $N-2$, convenablement choisie; H_y dépend d'un paramètre complexe y, et Lefschetz put définir encore des «cycles évanescents» dans ce cas général. Sous une forme très différente, cette technique est reparue à une époque récente, et forme un chaînon essentiel dans la preuve, par Grothendieck et Deligne, des «conjectures de Weil».

Un des principaux résultats de Lefschetz est que l'application

$$j_* : H_m(V \cap H_y) \longrightarrow H_m(V)$$

déduite de l'injection $j : V \cap H_y \longrightarrow V$, est (à l'exception d'un nombre fini de valeurs de y), bijective pour $0 \leq m \leq n-2$, et surjective pour $m = n-1$. Lefschetz en déduisit, pour les nombres de Betti de V, l'inégalité $p_m \geq p_{m-2}$ si $m \leq n$. Il montra aussi que p_{2m+1} est toujours un nombre pair (qui peut être 0) et que $p_{2m} > 0$ pour $0 \leq m \leq n$.

9.3 La théorie de Marston Morse

Soient X une variété C^∞ de dimension n, et $f : X \longrightarrow \mathbb{R}$ une fonction C^∞. Si en un point $x_0 \in X$, le développement de Taylor de $f(x) - f(x_0)$, en fonction de coordonnées locales en x_0, n'a pas de terme de degré 1, x_0 est un *point critique* de f. Si les termes de degré 2 constituent une forme quadratique non dégénérée φ, on dit que le point critique x_0 est *non dégénéré*, et l'indice de φ (dimension maximum d'un sous-espace vectoriel où $\varphi(u_0, u_1, \ldots, u_n) \leq 0$) est appelé l'*indice* de x_0. Les points critiques non dégénérés sont isolés. Un lemme prouvé par M. Morse montre l'existence de coordonnées locales $u_1, u_2, \ldots, u_n$ telles qu'au voisinage de x_0, on ait

$$f(x) = f(x_0) - u_1^2 - \ldots - u_k^2 + u_{k+1}^2 + \ldots + u_n^2$$

si x_0 est d'indice k.

Lorsque X est compacte, Morse montra que l'ensemble Φ des fonctions C^∞ dont *tous* les points critiques sont non dégénérés, est *dense* dans l'ensemble des fonctions C^∞ sur X, pour toute topologie C^r ($r \geq 0$). Soit alors C_k le nombre des points critiques d'indice k pour une fonction $f \in \Phi$ ($0 \leq k \leq n$). Généralisant des remarques de Poincaré et G.D. Birkhoff, Morse montra que si on pose $R_k = \dim H_k(X; \mathbb{F}_2)$, on a les relations

$$\begin{gathered} C_0 \geq R_0 \\ C_1 - C_0 \geq R_1 - R_0 \\ \cdots\cdots\cdots\cdots\cdots\cdots\cdots\cdots \\ C_{n-1} - C_{n-2} + \ldots + (-1)^{n-1} C_0 \geq R_{n-1} - R_{n-2} + \ldots + (-1)^{n-1} R_0 \\ C_n - C_{n-1} + \ldots + (-1)^n C_0 = R_n - R_{n-1} + \ldots + (-1)^n R_0. \end{gathered} \tag{30}$$

L'idée de la preuve (qui remonte à Birkhoff) est de considérer le sous-ensemble X_c de X formé des points vérifiant $f(x) < c$, et de voir comment varie l'homologie $H_0(X_c; \mathbb{F}_2)$ avec c, et notamment lorsque c franchit une valeur $f(x_0)$ pour un point critique x_0. Cette démonstration, assez compliquée, a progressivement été simplifiée par plusieurs mathématiciens, notamment grâce à la théorie des CW-complexes (voir plus loin).

Cette idée d'associer la topologie d'une variété au comportement d'une fonction définie sur la variété s'est révélée très féconde. Elle permit à Morse d'obtenir de grands progrès dans l'étude des géodésiques d'une variété riemannienne; son application à la théorie de l'homotopie a conduit depuis 1960 à des résultats spectaculaires.

10 La formation de l'armature algébrique

A partir de 1925 se manifestèrent des efforts divers pour élargir le cadre étroit des complexes simpliciaux et étendre la théorie de l'homologie à des espaces plus généraux. Ces généralisations s'accompagnèrent progressivement des techniques algébriques nouvelles; elles prenaient leur origine dans les raisonnements intervenant dans ces nouvelles théories homologiques, mais en éliminant l'origine topologique qui les utilisaient et en ne conservant que leur forme purement algébrique. Ces techniques se révélèrent rapidement adaptables à de nombreuses autres théories mathématiques, parfois très éloignées de la Topologie; elles furent exposées systématiquement pour la première fois dans le livre «Homological algebra» de H. Cartan et S. Eilenberg (1956) et ne cessent de s'accroître depuis lors.

Ce sont aussi ces techniques et leurs applications qui ont amené à mettre en évidence des concepts dépassant les notions usuelles de structures algébriques ou topologiques (groupes, modules, espaces topologiques, etc.). Eilenberg et MacLane, en 1942, ont codifié, sous le nom de «catégorie» et de «foncteur», ces concepts et les procédés de construction qui leur sont attachés; le langage qu'ils ont introduit, étendu ensuite par de nombreux mathématiciens, s'est montré très utile dans diverses parties des mathématiques.

Dans ce qui suit, nous limiterons strictement tous ces développements à ceux qui sont utilisés en Topologie algébrique; nous reporterons souvent l'histoire de leur introduction avec celle de leurs applications topologiques.

10.1 Catégories et foncteurs

La donnée d'une catégorie comporte d'une part une classe d'objets, et de l'autre une classe de «flèches» (ou «morphismes») reliant ces objets, et qui se «composent» entre elles: de deux flèches, $A \xrightarrow{u} B$ et $B \xrightarrow{v} C$ on déduit, par un procédé bien déterminé et dont la définition fait partie de celle de la catégorie considérée, une troisième flèche $A \xrightarrow{vu} C$, qui doit satisfaire à la condition $w(vu) = (wv)u$ pour des flèches dont la composition est définie. Les exemples les plus simples (auxquels il est commode de penser) sont les catégories dont les objets sont les ensembles munis d'un type de structure déterminé (algébrique, topologique, etc.) et les morphismes des applications satisfaisant à certaines conditions en relation avec le type de structure (homomorphismes, applications continues, etc.). Mais il y a d'autres types de catégories.

Un foncteur covariant (resp. contravariant) F d'une catégorie $\underline{\underline{C}}$ dans une catégorie $\underline{\underline{C}}'$ fait correspondre à tout objet M de $\underline{\underline{C}}$ un objet $F(M)$ de $\underline{\underline{C}}'$, et à toute flèche $M \xrightarrow{u} N$ de $\underline{\underline{C}}$ une flèche $F(M) \xrightarrow{F(u)} F(N)$ (resp. $F(N) \xrightarrow{F(u)} F(M)$) de $\underline{\underline{C}}'$, de sorte que l'on ait $F(vu) = F(v)F(u)$ (resp. $F(vu) = F(u)F(v)$).

Les foncteurs les plus utilisés en Algèbre homologique sont ceux où $\underline{\underline{C}} = \underline{\underline{C}}'$ est la catégorie dont les objets sont les A-modules et les morphismes les A-homomorphismes, A étant un anneau commutatif; le composé vu est l'application composée usuelle $x \mapsto v(u(x))$. Les deux foncteurs suivants interviennent fréquemment:

1) $\otimes$: Pour un A-module fixe G, le foncteur $F = \otimes$ est donné par

$$F(M) = M \otimes {}_A G, F(u) = u \otimes 1_G \text{ (homomorphisme } x \otimes g \mapsto u(x) \otimes g).$$

Ce foncteur covariant (dit «produit tensoriel») n'a été défini de façon générale que par Whitney en 1938; mais il était connu et utilisé antérieurement dans divers cas particuliers.

2) Hom: Pour un A-module fixe G, le foncteur contravariant $F = \text{Hom}$ est donné par

$$F(M) = \text{Hom}(M, G), F(u) = \text{Hom}(u, 1_G)$$

(homomorphisme $w \mapsto w \circ u$ de $\text{Hom}(N, G)$ dans $\text{Hom}(M, G)$ pour un homomorphisme $u : M \longrightarrow N$).

10.2 Chaînes et cochaînes

La notion de base, exprimée sans doute pour la première fois par W. Mayer (1929) est celle de *complexe de chaînes*. Si A est un anneau commutatif, on entend par là une suite (finie ou infinie) de A-modules et de A-homomorphismes

$$\begin{aligned} C. : 0 \leftrightarrow C_0 \xleftarrow{b_1} C_1 \xleftarrow{b_2} C_2 \longleftarrow \dots \\ \longleftarrow C_{n-1} \xleftarrow{b_n} C_n \xleftarrow{b_{n+1}} C_{n+1} \longleftarrow \dots \end{aligned} \tag{31}$$

soumis à la seule condition

$$b_n \circ b_{n+1} = 0 \text{ pour tout } n \geq 1. \tag{32}$$

Rien n'est spécifié sur l'origine des «chaînes» C_n et des «bords» b_n. On définit alors le sous-module $Z_n = \text{Ker } b_n$ des n-cyles, le sous-module $B_n = \text{Im } b_{n+1}$ des n-bords, et comme $B_n \subset Z_n$ par (32), on a un *n-ème module d'homologie* du complexe C.

$$H_n(C.) = Z_n/B_n. \tag{33}$$

Ces notions peuvent être généralisées, en considérant une famille $(C_n)_{n\in\mathbb{Z}}$ où les indices peuvent être négatifs (Eilenberg-Steenrod, 1945). Un complexe (31) est alors dit «positif» ou «borné inférieurement». On peut évidemment considérer aussi des complexes «négatifs» où $C_n = 0$ pour $n \geq 0$; mais ils se sont introduits en Topologie algébrique sous une autre terminologie: si l'on pose $C'_n = C_{-n}$ et $d_n = b_{-n}$ pour un tel complexe, on obtient une suite de A-modules et de A-homomorphismes

$$\begin{aligned} C'^{\cdot} : 0 \longrightarrow C'_0 \xrightarrow{d_0} C'_1 \xrightarrow{d_1} C'_2 \longrightarrow \dots \\ \longrightarrow C'_{n-1} \xrightarrow{d_{n-1}} C'_n \xrightarrow{d_n} C'_{n+1} \longrightarrow \dots \end{aligned} \tag{34}$$

avec $d_n \circ d_{n-1} = 0$ pour tout $n \geq 1$. On dit qu'une telle suite est un *complexe de cochaînes*, les d_n étant appelés «cobords». Les éléments de $Z'_n = \operatorname{Ker} d_n$ sont les n-cocycles, ceux de $B'_n = \operatorname{Im} d_{n-1}$ les n-cobords, et

$$H^n(C'^{\cdot}) = Z'_n / B'_n \tag{35}$$

est le *n-ème module de cohomologie* du complexe $C'^{\cdot}$.

Les complexes de chaînes sur un anneau commutatif A sont les objets d'une catégorie, lorsqu'on définit un *morphisme* $f. : E. \longrightarrow F.$ de complexes de chaînes comme une suite de A-homomorphismes $f_n = E_n \longrightarrow F_n$ tels que le diagramme

$$\begin{array}{ccccccccccccc}
0 & \longleftarrow & E_0 & \xleftarrow{b_1} & E_1 & \xleftarrow{b_2} & E_2 & \longleftarrow \cdots \longleftarrow & E_{n_1} & \xleftarrow{b_n} & E_n & \xleftarrow{b_{n+1}} & E_{n+1} \longleftarrow \cdots \\
 & & \downarrow{\scriptstyle f_0} & & \downarrow{\scriptstyle f_1} & & \downarrow{\scriptstyle f_2} & & \downarrow{\scriptstyle f_{n-1}} & & \downarrow{\scriptstyle f_n} & & \downarrow{\scriptstyle f_{n+1}} \\
0 & \longleftarrow & F_0 & \xleftarrow[b'_1]{} & F_1 & \xleftarrow[b'_2]{} & F_2 & \longleftarrow \cdots \longleftarrow & F_{n_1} & \xleftarrow[b'_n]{} & F_n & \xleftarrow[b'_{n+1}]{} & F_{n+1} \longleftarrow \cdots
\end{array}$$

soit *commutatif*, c'est-à-dire que $b'_{n+1} \circ f_{n+1} = f_n \circ b_{n+1}$ pour tout $n \geq 0$. On a alors $f_n(Z_n(E.)) \subset Z_n(F.)$ et $f_n(B_n(E.)) \subset B_n(F.)$; donc, par passage aux quotients, on a un homomorphisme

$$H_n(f.) : H_n(E.) \longrightarrow H_n(F.) \text{ pour tout } n \geq 0$$

qu'on écrit aussi $f_* = H.(f.) : H.(E.) \longrightarrow H.(F.)$. Pour deux morphismes de complexes de chaînes $f. : E. \longrightarrow F.$ et $g. : F. \longrightarrow G.$, on a

$$(g. \circ f.)_* = g_* \circ f_*. \tag{36}$$

Si $u.$ et $v.$ sont deux morphismes de chaînes $C. \longrightarrow C.'$, une *homotopie de chaînes* $h. : C. \longrightarrow C.'$ de $u.$ à $v.$ est une suite (h_j) d'homomorphismes $h_j : C_j \longrightarrow C'_{j+1}$ tels que pour tout $j \geq 1$ et tout $x \in C_j$, on ait

$$v_j(x) - u_j(x) = b'_{j+1}(h_j(x)) + h_{j-1}(b_j(x)) \tag{37}$$

(forme algébrique de la «formule d'homotopie» de Veblen et Lefschetz); on écrit $u. \sim v.$. Alors les homomorphismes u_* et v_* de $H.(C.)$ dans $H.(C.')$ sont *égaux*.

Une *équivalence de chaînes* consiste en deux morphismes $u. : C. \longrightarrow C.'$ et $v. : C.' \longrightarrow C.$ tels que $v. \circ u. \sim 1_{C.}$ et $u. \circ v. \sim 1_{C.'}$; cela entraîne que $H.(C.)$ est *isomorphe* à $H.(C.')$.

On a des résultats et définitions analogues pour les complexes de cochaînes en «renversant les flèches».

10.3 Suites exactes

L'étude des complexes de chaînes (ou cochaînes) et des groupes d'homotopie (voir plus loin) attira l'attention sur une propriété de deux homomorphismes de A-modules

$$M \xrightarrow{f} N \xrightarrow{g} P$$

à savoir la relation $\mathrm{Ker}(g) = \mathrm{Im}(f)$. On dit alors que ce couple d'homomorphismes est *exact* (ou «exact en N»). Une *suite exacte* est une suite (finie ou non) de A-modules et d'homomorphismes

$$\ldots \longrightarrow E_m \xrightarrow{f_m} E_{m+1} \xrightarrow{f_{m+1}} E_{m+2} \longrightarrow \ldots \longrightarrow E_n \xrightarrow{f_n} E_{n+1} \longrightarrow \ldots$$

qui est exacte en chacun des E_k. Dire qu'une suite de 5 termes

$$0 \longrightarrow M \xrightarrow{f} N \xrightarrow{g} P \longrightarrow 0$$

est exacte (on dit aussi que c'est une suite exacte courte) signifie donc que f est injectif, g surjectif et que $f(M) = g^{-1}(0)$, donc P est isomorphe à $N/f(M)$. On dit que cette suite courte est *scindée* si N est somme directe de $f(M)$ et d'un sous-module Q, ou, ce qui revient au même, s'il existe un homomorphisme $h : P \longrightarrow N$ tel que $g \circ h = \mathrm{id}_P$.

Dire qu'un complexe de chaînes (31) est exact en C_n signifie que $H_n(C.) = 0$. Si $C.$ est une suite exacte, tous les $H_n(C.)$ sont nuls; on dit que $C.$ est *acyclique*. Les modules d'homologie «mesurent» donc en quelque sorte le «défaut d'exactitude» d'un complexe de chaînes.

On peut définir de même une suite exacte de complexes de chaînes (ou de cochaînes). Dire qu'une suite courte de cochaînes

$$0 \longrightarrow M^{\cdot} \xrightarrow{f^{\cdot}} N^{\cdot} \xrightarrow{g^{\cdot}} P^{\cdot} \longrightarrow 0 \tag{38}$$

est exacte signifie, par définition, que chacune des suites courtes

$$0 \longrightarrow M^n \xrightarrow{f^n} N^n \xrightarrow{g^n} P^n \longrightarrow 0$$

est exacte.

Kelley et Pitcher (1947) (à la suite de résultats antérieurs concernant l'homologie, dus à H. Cartan et Eilenberg-Steenrod) montrèrent que, pour une suite exacte (38) de complexes de cochaînes, il existe pour chaque n un homomorphisme explicitement défini

$$\partial_n : H^n(P^\cdot) \longrightarrow H^{n+1}(M^\cdot)$$

tel que la suite

$$\ldots \longrightarrow H^n(M^\cdot) \xrightarrow{H^n(f^\cdot)} H^n(N^\cdot) \xrightarrow{H^n(g^\cdot)} H^n(P^\cdot) \xrightarrow{\partial_n} H^{n+1}(M^\cdot) \longrightarrow \ldots \tag{39}$$

soit exacte; on dit que c'est la *suite exacte de cohomologie* déduite de la suite exacte (38).

Dire que (38) est une suite exacte signifie que les diagrammes

$$\begin{array}{ccccccccc}
 & & \vdots & & \vdots & & \vdots & & \\
 & & \downarrow & & \downarrow & & \downarrow & & \\
0 & \longrightarrow & M^{n-1} & \xrightarrow{f^{n-1}} & N^{n-1} & \xrightarrow{g^{n-1}} & P^{n-1} & \longrightarrow & 0 \\
 & & d\downarrow & & d\downarrow & & d\downarrow & & \\
0 & \longrightarrow & M^{n} & \xrightarrow{f^{n}} & N^{n} & \xrightarrow{g^{n}} & P^{n} & \longrightarrow & 0 \\
 & & d\downarrow & & d\downarrow & & d\downarrow & & \\
0 & \longrightarrow & M^{n+1} & \xrightarrow{f^{n+1}} & N^{n+1} & \xrightarrow{g^{n+1}} & P^{n+1} & \longrightarrow & 0 \\
 & & d\downarrow & & d\downarrow & & d\downarrow & & \\
0 & \longrightarrow & M^{n+2} & \xrightarrow{f^{n+2}} & N^{n+2} & \xrightarrow{g^{n+2}} & P^{n+2} & \longrightarrow & 0 \\
 & & \downarrow & & \downarrow & & \downarrow & & \\
 & & \vdots & & \vdots & & \vdots & &
\end{array}$$

sont commutatifs, et les lignes exactes. L'homomorphisme ∂_n se définit comme suit. Soit z un n-cocycle dans P^n; il existe $y \in N^n$ tel que $z = g^n(y)$, et on a $0 = d(g^n(y)) = g^{n+1}(dy)$, donc il existe $x \in M^{n+1}$ tel que $dy = f^{n+1}(x)$. Montrons d'abord que x est un cocycle; en effet, on a

$$0 = d(dy) = d(f^{n+1}(x)) = f^{n+2}(dx)$$

donc $dx = 0$. En second lieu prouvons que la classe $\bar{x} \in H^{n+1}(M^\cdot)$ est bien déterminée par la classe $\bar{z} \in H^n(P^\cdot)$. En effet, y n'est déterminé qu'à l'addition près d'un élément $f^n(x')$ pour $x' \in M^n$; mais on a $d(f^n(x')) = f^{n+1}(dx')$, donc quand on remplace y par $y + f^n(x')$, x est remplacé par $x + dx'$ et $\bar{x}$ n'est pas changé. D'autre part, si $z \in P^n$ est un cobord, soit $z = dz'$ avec $z' \in P^{n-1}$, on a $g^n(y) = d(g^{n-1}(y'))$ pour un $y' \in N^{n-1}$, donc $g^n(y) = g^n(dy')$, donc $y = dy' + f^n(x')$ pour un $x' \in M^n$. Mais alors $dy = d(f^n(x')) = f^{n+1}(dx')$, donc $\bar{x} = 0$. Cela montre qu'on peut poser $\bar{x} = \partial_n(\bar{z})$ pour toute classe $\bar{z}$, sans ambiguïté.

Il faut ensuite prouver que la suite (39) est exacte en 3 termes consécutifs. Montrons-le par exemple pour le terme $H^n(P^\cdot)$. Il est immédiat que l'on a $\mathrm{Im}(H^n(g^\cdot)) \subset \mathrm{Ker}(\partial_n)$: en effet, avec les notations précédentes, si $z = g^n(y)$ où $y \in N^n$ est un cocycle, on a $f^{n+1}(x) = 0$, donc $x = 0$.

Montrons inversement que $\mathrm{Ker}(\partial_n) \subset \mathrm{Im}(H^n(g^\cdot))$. Si $\partial_n\bar{z} = 0$, x est un cobord, donc $x = dx'$ pour $x' \in M^n$ et $dy = f^{n+1}(dx') = d(f^n(x'))$; autrement dit $y - f^n(x')$ est un cocycle; donc $g^n(y) = g^n(y - f^n(x'))$ est l'image par g^n d'un cocycle, d'où $\bar{z} \in \mathrm{Im}(H^n(g^\cdot))$.

Les autres preuves d'exactitude sont analogues, ce qu'on appelle «la course dans le diagramme».

Pour une suite exacte de complexes de chaînes

$$0 \longrightarrow M. \xrightarrow{f.} N. \xrightarrow{g.} P. \longrightarrow 0 \tag{40}$$

on a de même une suite exacte d'homologie

$$\dots \longleftarrow H_{n-1}(M.) \xleftarrow{\partial_n} H_n(P.) \xleftarrow{H_n(g.)} H_n(N.) \xleftarrow{H_n(f.)} H_n(M.) \longleftarrow \dots \tag{41}$$

Un cas particulier important de complexes de chaînes $C.$ est celui où $C_n = 0$ sauf pour un nombre fini d'indices, et où pour ces derniers, C_n est un $\mathbb{Z}$-module de type fini. Si $\mathrm{rg}(C_n)$ («rang» de C_n) est le nombre d'éléments d'une base de C_n/T_n où T_n est le sous-module de torsion de C_n, W. Mayer (1929) avait introduit, sur le modèle de l'homologie, la caractéristique d'Euler-Poincaré

$$\chi(C.) = \sum_n (-1)^n \mathrm{rg}(C_n)$$

et il avait prouvé la relation

$$\chi(H.(C.)) = \chi(C.). \tag{42}$$

Kelley et Pitcher montrent que pour une suite exacte (40) de tels complexes de chaînes, on a

$$\chi(N.) = \chi(M.) + \chi(P.). \tag{43}$$

10.4 Les foncteurs Tor et Ext

Un foncteur covariant F de A-modules est exact si, pour toute suite exacte de A-modules

$$M \xrightarrow{u} N \xrightarrow{v} P$$

la suite

$$F(M) \xrightarrow{F(u)} F(N) \xrightarrow{F(v)} F(P)$$

est exacte (si F est contravariant, il faut remplacer cette suite par

$$F(P) \xrightarrow{F(v)} F(N) \xrightarrow{F(u)} F(M)).$$

Les foncteurs $\otimes$ et Hom ne sont pas exacts. Si la suite $M \xrightarrow{v} N \longrightarrow 0$ est exacte, il en est de même de $M \otimes G \xrightarrow{v\otimes 1} N \otimes G \longrightarrow 0$, car tout élément $v(m) \otimes g$ est image par $v \otimes 1$ de $m \otimes g$; mais si la suite $0 \longrightarrow M \xrightarrow{u} N$ est exacte, la suite $0 \longrightarrow M \otimes G \xrightarrow{u\otimes 1} N \otimes G$ ne l'est pas nécessairement; cela a été observé dans un cas particulier par Čech en 1935, puis par Whitney en 1938. Le «défaut d'exactitude» est mesuré par un nouveau foncteur $\mathrm{Tor}(N/M, G)$ défini par Čech dans un cas particulier, et par H. Cartan et Eilenberg de façon générale (1956); il ne dépend pas de u, mais seulement du quotient $N/u(M)$ (ou N/M si on identifie M et $u(M)$). On peut le définir en considérant une suite exacte

$$0 \longrightarrow Q \xrightarrow{u} P \longrightarrow N \longrightarrow 0 \tag{44}$$

où P est un A-module libre (il est immédiat de définir une telle suite). Alors $\mathrm{Tor}(M, N)$ est le noyau de l'homomorphisme $M \otimes Q \xrightarrow{1\otimes u} M \otimes P$; on montre qu'il ne dépend pas de la suite choisie (44).

Soit $C.$ un complexe de chaînes (31) et G un A-module; les A-modules $C_n \otimes {}_A G$ et les homomorphismes $b_n \otimes 1_G$ forment un complexe de chaînes

$$C. \otimes G : 0 \longleftarrow C_0 \otimes G \xleftarrow{b_1\otimes 1} C_1 \otimes G \xleftarrow{b_2\otimes 1} C_2 \otimes G \longleftarrow \dots \longleftarrow C_n \otimes G \longleftarrow \dots$$

et l'on dit que l'homologie $H.(C. \otimes {}_A G)$ est *l'homologie de $C.$ à coefficients dans G* (l'homologie d'un complexe simplicial à coefficients dans $\mathbb{Z}/m\mathbb{Z}$ ou $\mathbb{Q}$, mentionnée plus haut, en est un cas particulier). On montre que si A est un anneau principal et si les C_n sont des A-modules libres, on a des suites exactes scindées

$$0 \longrightarrow H_p(C.; A) \otimes {}_A G \longrightarrow H_p(C. \otimes {}_A G) \longrightarrow \mathrm{Tor}(H_{p-1}(C.; A), G) \longrightarrow 0. \tag{45}$$

C'est ce qu'on appelle la «formule des coefficients universels» pour l'homologie.

Si l'un des A-modules M, N est libre (et même plat), on a $\mathrm{Tor}(M, N) = 0$.

Etant donnés deux complexes de chaînes $C.'$, $C.''$, on définit un *produit tensoriel* $C. = C.' \otimes {}_A C.''$ comme formé des A-modules de chaînes

$$C_r = \bigoplus_{0 \le p \le r} C'_p \otimes {}_A C''_{r-p} \tag{46}$$

avec le bord

$$b_r(z'_p \otimes z''_{r-p}) = (b_p z'_p) \otimes z''_{r-p} + z'_p \otimes (b_{r-p} z''_{r-p}). \tag{47}$$

Si z'_p et z''_{r-p} sont des cycles, il en est donc de même de $z'_p \otimes z''_{r-p}$; si en outre z'_p ou z''_{r-p} est un bord, $z'_p \otimes z''_{r-p}$ est aussi un bord. On définit ainsi un homomorphisme canonique

$$H.(C.') \otimes {}_A H.(C.'') \longrightarrow H.(C.' \otimes {}_A C.''). \tag{48}$$

Lorsque A est un anneau principal et que les A-modules C'_n et C''_m sont tous libres, on démontre qu'il existe une suite exacte scindée

$$\begin{aligned} 0 \longrightarrow & \bigoplus_{0 \le p \le r} (H_p(C.') \otimes H_{r-p}(C.'') \longrightarrow H_r(C.' \otimes C.'') \\ \longrightarrow & \bigoplus_{0 \le q \le r-1} \operatorname{Tor}(H_q(C.'), H_{r-q-1}(C.'')) \longrightarrow 0 \end{aligned} \tag{49}$$

dite «formule de Künneth» pour les complexes de chaînes.

On a des propriétés analogues pour le foncteur Hom. Si la suite $N \xrightarrow{f} P \longrightarrow 0$ est exacte, il en est de même de $0 \longrightarrow \operatorname{Hom}(P, G) \xrightarrow{\operatorname{Hom}(f,1)} \operatorname{Hom}(N, G)$ (car si un homomorphisme $\varphi : P \longrightarrow G$ est tel que $\varphi \circ f = 0$, on a $\varphi = 0$). Par contre, si la suite $0 \longrightarrow M \xrightarrow{f} N$ est exacte, la suite $\operatorname{Hom}(N, G) \xrightarrow{\operatorname{Hom}(f,1)} \operatorname{Hom}(M, G) \longrightarrow 0$ ne l'est pas toujours (un homomorphisme $\psi : M \longrightarrow G$ n'est pas toujours la restriction à M d'un homomorphisme de N dans G). En 1942, Eilenberg et MacLane montrèrent qu'il y a un foncteur $\operatorname{Ext}(M, G)$ ne dépendant pas de N et tel que le conoyau $\operatorname{Coker}(\operatorname{Hom}(f, 1)) = \operatorname{Ext}(M, G)$. On peut encore le définir à partir d'une suite exacte $0 \longrightarrow Q \xrightarrow{v} P \longrightarrow M \longrightarrow 0$ où P est libre; alors $\operatorname{Ext}(M, G) = \operatorname{Coker}(\operatorname{Hom}(v, 1))$. Si M est libre, $\operatorname{Ext}(M, G) = 0$.

Pour tout complexe de chaînes $C.$, les $C^n = \operatorname{Hom}(C_n, G)$ et les homomorphismes $d_n = \operatorname{Hom}(b_{n+1}, 1)$ forment un complexe de cochaînes

$$C^{\cdot} : 0 \longrightarrow C^0 \xrightarrow{d_0} C^1 \xrightarrow{d_1} C^2 \longrightarrow \dots \longrightarrow C^n \xrightarrow{d_n} C^{n+1} \longrightarrow \dots \tag{50}$$

d'où l'on déduit des A-modules de cohomologie $H^n(C^{\cdot}; G)$. Si A est un anneau principal et les C_n des A-modules libres, on a des suites exactes scindées

$$0 \longrightarrow \operatorname{Ext}(H_{n-1}(C.), G) \longrightarrow H^n(C^{\cdot}; G) \longrightarrow \operatorname{Hom}(H_n(C.), G) \longrightarrow 0 \tag{51}$$

dite «formule des coefficients universels» pour la cohomologie.

10.5 Les opérateurs de Bockstein

Soit $C.$ un complexe de chaînes, où les C_n sont des A-modules libres; d'autre part, soit

$$0 \longrightarrow G' \longrightarrow G \longrightarrow G'' \longrightarrow 0$$

une suite exacte de A-modules. Alors la suite de complexes de chaînes

$$0 \longrightarrow C.\otimes G' \longrightarrow C.\otimes G \longrightarrow C.\otimes G'' \longrightarrow 0$$

est exacte, et on a donc une suite exacte d'homologie

$$\begin{aligned} \ldots \longrightarrow H_n(C.\otimes G') \longrightarrow H_n(C.\otimes G) \longrightarrow H_n(C.\otimes G'') \\ \xrightarrow{\beta} H_{n-1}(C.\otimes G') \longrightarrow \ldots \end{aligned} \tag{52}$$

Sous les mêmes conditions, on a, pour tout complexe de cochaînes $C^{\cdot}$, une suite exacte de cohomologie

$$\begin{aligned} \ldots \longrightarrow H^n(C^{\cdot}\otimes G') \longrightarrow H^n(C^{\cdot}\otimes G) \longrightarrow H^n(C^{\cdot}\otimes G'') \\ \xrightarrow{\beta} H^{n+1}(C^{\cdot}\otimes G') \longrightarrow \ldots \end{aligned} \tag{52bis}$$

Les homomorphismes β sont appelés opérateurs de Bockstein.

10.6 Suites spectrales

En 1946, J. Leray se proposa d'étudier, pour une application continue $f : X \longrightarrow Y$, les relations entre la cohomologie de X, celle de Y et celle des «fibres» $f^{-1}(y)$ pour $y \in Y$. Cela le conduisit à inventer une sorte de processus d'«approximation algébrique» pour la cohomologie d'un complexe de cochaînes. Un peu plus tard, J.-L. Koszul modifia cette définition, par une combinaison de techniques cohomologiques et de l'idée de *filtration*.

Au lieu de partir d'un complexe de cochaînes, il est plus simple de considérer d'abord un A-module M et un endomorphisme $d : M \longrightarrow M$ tel que $d \circ d = 0$. Une filtration (décroissante) sur M, compatible avec d, est une suite $(F^p(M))_{p\in\mathbb{Z}}$ de sous-modules de M, telle que $F^{p+1}(M) \subset F^p(M)$, et qu'on soumet à la seule condition

$$d(F^p(M)) \subset F^p(M) \text{ pour tout } p \in \mathbb{Z}.$$

Soit $Z^p = F^p(M) \cap \operatorname{Ker} d$ («cocycles»), $B^p = F^p(M) \cap \operatorname{Im} d$ («cobords»). L'image de Z^p dans $H(M) = \operatorname{Ker} d/\operatorname{Im} d$ est $Z^p/B^p = F^p(H(M))$; les $F^p(H(M))$ forment une filtration décroissante sur $H(M)$.

Dans les applications, on a souvent $\bigcap_{p\in\mathbb{Z}} F^p(M) = 0$, de sorte qu'on peut penser aux éléments de $F^p(M)$ comme «de plus en plus négligeables» quand p tend vers $+\infty$; un élément $z \in F^p(M)$ apparaît donc comme «presque un cocycle» lorsque $dz \in F^{p+r}(M)$ et que r devient de plus en plus grand. Autrement dit, le sous-module Z_r^p de ces éléments apparaît comme une «approximation» du sous-module Z^p. On a $Z_{r-1}^{p+1} \subset Z_r^p$, et si l'on pose $B_{r-1}^p = dZ_{r-1}^{p+1-r} \subset Z_r^p$, on peut former le quotient

$$E_r^p = Z_r^p/(Z_{r-1}^{p+1} + B_{r-1}^p). \tag{53}$$

On a les relations

$$\begin{aligned} 0 \subset B_0^p \subset B_1^p \subset \ldots \subset B_s^p \subset B_{s+1}^p \subset \ldots \subset B^p \subset Z^p \subset \ldots \\ \subset Z_{r+1}^p \subset Z_r^p \ldots \subset Z_0^p = F^p(M) \end{aligned}$$

donc les E_r^p peuvent être considérés comme «approchant» les sous-modules

$$E_\infty^p = Z^p/(Z^{p+1} + B^p) = F^p(H(M))/F^{p+1}(H(M)) \tag{54}$$

que l'on appelle l'«aboutissement» de la suite des E_r^p.

Le point essentiel est que $d(Z_r^p) \subset Z_r^{p+r}$ et $d(B_{r-1}^p + Z_{r-1}^{p+1}) \subset B_{r-1}^{p+r}$, d'où un homomorphisme

$$d_r : E_r^p \longrightarrow E_r^{p+r} \tag{55}$$

et comme $d \circ d = 0$, on a aussi $d_r \circ d_r = 0$. On a ainsi défini un complexe de cochaînes

$$E_r^{\cdot} : \ldots \longrightarrow E_r^p \xrightarrow{d_r} E_r^{p+r} \xrightarrow{d_r} E_r^{p+2r} \longrightarrow \ldots \tag{56}$$

et on voit facilement que

$$H^p(E_r^{\cdot}) = E_{r+1}^{\cdot} \text{ pour tout } r \in \mathbb{Z}. \tag{57}$$

Partons maintenant d'un complexe de cochaînes

$$M^{\cdot} : \ldots \longrightarrow M_q \xrightarrow{d} M_{q+1} \longrightarrow \ldots$$

et supposons que sur la somme directe $M = \bigoplus_{q\in\mathbb{Z}} M_q$, on ait une filtration *compatible* avec la graduation, c'est-à-dire telle que

$$F^p(M) = \bigoplus_{q\in\mathbb{Z}} (M_q \cap F^p(M)). \tag{58}$$

On applique ce qui précède à M; on a

$$Z_r^p = \bigoplus_{q\in\mathbb{Z}} Z_r^{pq}, B_r^p = \bigoplus_{q\in\mathbb{Z}} B_r^{pq}$$

avec

$$Z_r^{pq} = Z_r^p \cap M_{p+q}, B_r^{pq} = B_r^p \cap M_{p+q}.$$

Pour p et q fixés, on a les inclusions

$$\begin{aligned} 0 \subset B_0^{pq} \subset B_1^{pq} \subset \ldots \subset B_s^{pq} \subset B_{s+1}^{pq} \subset \ldots \\ \subset Z_{r+1}^{pq} \subset Z_r^{pq} \subset \ldots \subset Z_1^{pq} \subset Z_0^{pq} = F^p(M) \cap M_{p+q}. \end{aligned}$$

On pose

$$E_r^{pq} = Z_r^{pq}/(Z_{r-1}^{p+1,q-1} + B_{r-1}^{pq})$$

et on a

$$d_r(E_r^{pq}) \subset E_r^{p+r,q-r+1}, \tag{59}$$

et dans $H^{\cdot}(E_r^{\cdot}) = E_{r+1}^{\cdot}$, E_{r+1}^{pq} est formé des classes de cocycles appartenant à E_r^{pq}. Chaque $H^q(M^{\cdot})$ est filtré par $F^p(H^q(M^{\cdot})) = H^q(M^{\cdot}) \cap F^p(H^{\cdot}(M^{\cdot}))$, et les E_r^{pq} doivent être considérés comme approchant l'«aboutissement»

$$E_\infty^{pq} = F^p(H^{p+q}(M^{\cdot}))/F^{p+1}(H^{p+q}(M^{\cdot})).$$

Dans beaucoup d'applications, la connaissance des termes E_2^{pq} suffit à déterminer les E_∞^{pq} (d'où la détermination des $H^{p+q}(M^{\cdot})$ lorsque l'anneau des coefficients est un corps).

11 Les diverses théories homologiques

11.1 L'homologie singulière

L'idée de considérer, au lieu d'une triangulation d'un espace X, les images dans X de simplexes standard Δ_p par des applications continues, était apparue dès 1907 (Dehn-Heegaard). Mais les phénomènes tels que la courbe de Peano rendaient vaines les possibilités d'utiliser uniquement ces images pour définir des chaînes; on pensa alors à prendre comme objets les couples formés d'une application continue $f : \Delta_p \longrightarrow X$ et de son image; c'est ce qu'on appela un «simplexe singulier» $(f, f(\Delta_p))$. Si l'on définit convenablement le «bord» d'un tel simplexe singulier, on peut former des complexes de chaînes, combinaisons linéaires formelles de simplexes singuliers. L'homologie d'un tel complexe ne dépend évidemment que de la topologie de X, et Alexander entreprit de montrer que pour un espace triangulable cette homologie est isomorphe à celle déduite de la triangulation. La mise en oeuvre de cette idée, par Alexander et les mathématiciens qui s'y intéressaient, se heurta à des complications dans les définitions, qui ne furent pas surmontées avant 1938.

Mais, en 1943, Eilenberg dissipa d'un seul coup ces difficultés en montrant qu'il suffisait de considérer, comme «simplexes singuliers», les applications continues $s^p : \bar{\Lambda}_p \longrightarrow X$ *elles-mêmes*, sans s'occuper de leurs images. Une «p-chaîne singulière» est donc une combinaison linéaire *formelle* $\sum_i \lambda_j s_j^p$ de p-simplexes singuliers, à coefficients dans un anneau commutatif A; cela n'a plus aucune interprétation topologique. Les p-simplexes singuliers distincts s^p forment une base (en général infinie) d'un A-module libre $S_p(X)$ (ou $S_p(X; A)$). Le point essentiel est la définition du *bord*: on considère pour cela l'application affine («i-ème face») $\sigma_i : \bar{\Delta}_{p-1} \longrightarrow \bar{\Delta}_p$, définie par

$$\sigma_i(x_0, x_1, \dots, x_{p-1}) = (x_0, \dots, x_{i-1}, 0, x_{i+1}, \dots, x_{p-1}); \tag{60}$$

le bord est alors défini par

$$b_p s^p = \sum_{0 \le i \le p-i} (-1)^i s^p \circ \sigma_i \tag{61}$$

et on vérifie encore que $b_{p-1} \circ b_p = 0$, de sorte qu'on peut définir, pour tout $p \geq 0$ et tout espace topologique X, un *A-module $H_p(X; A)$ d'homologie singulière*.

On écrit $H.(X; A)$ la suite (ou la somme directe) des A-modules $H_p(X; A)$.

Si Y est un sous-espace quelconque de X, $S_p(Y; A)$ est un sous-module libre de $S_p(X; A)$, ayant pour base les applications continues s^p telles que $s^p(\bar{\Delta}_p) \subset Y$. En outre, ce A-module a un supplémentaire dans $S_p(X; A)$ qui est aussi un A-module libre, ayant pour base les s^p telles que $s^p(\bar{\Delta}_p) \not\subset Y$. On a donc une suite exacte scindée de complexes de chaînes

$$0 \longrightarrow S.(Y) \longrightarrow S.(X) \longrightarrow S.(X, Y) \longrightarrow 0 \tag{62}$$

où $S_p(X, Y) = S_p(X)/S_p(Y)$. Le bord $\bar{b}_p$ dans $S_p(X, Y)$ est tel que, pour toute classe $\bar{s}^p \in S_p(X)/S_p(Y)$, on ait $\bar{b}_p \bar{s}^p = \overline{b_p s^p}$. Le A-module $H_p(S.(X, Y); A)$ est le p-ème A-module d'homologie singulière *relative* de X par rapport à Y; il est noté $H_p(X, Y; A)$ ou $H_p(X, Y)$. On a $H_p(X, \phi) = H_p(X)$.

De la suite exacte (62) de complexes de chaînes, on déduit la suite exacte d'homologie singulière

$$\begin{aligned} 0 \longleftarrow H_0(X, Y) &\xleftarrow{H_0(j)} H_0(X) \xleftarrow{H_0(i)} H_0(Y) \xleftarrow{\partial_1} H_1(X, Y) \longleftarrow \dots \\ \dots \longleftarrow H_p(X, Y) &\xleftarrow{H_p(j)} H_p(X) \xleftarrow{H_p(i)} H_p(Y) \xleftarrow{\partial_{p+1}} H_{p+1}(X, Y) \longleftarrow \dots \end{aligned} \tag{63}$$

où $i(s^p) = s^p$ pour $s^p \in S_p(Y)$, et $j(s^p) = 0$ si $s^p \in S_p(Y)$, $j(s^p) = s^p$ sinon.

Pour toute application continue $u : X \longrightarrow X'$ et tout couple de sous-espaces $Y \subset X$, $Y' \subset X'$ tels que $u(Y) \subset Y'$, on définit un homomorphisme

$$S_p(u) : S_p(X, Y) \longrightarrow S_p(X', Y') \tag{64}$$

en posant $S_p(u)(\bar{s}^p) = \overline{u \circ s^p}$; ces applications commutent avec les opérateurs bords et forment donc un morphisme de chaînes, d'où l'on déduit un homomorphisme

$$H_p(u) : H_p(X, Y; A) \longrightarrow H_p(X', Y'; A)$$

aussi noté u_*; si $v : X' \longrightarrow X''$ est tel que $v(Y') \subset Y''$, on a

$$(v \circ u)_* = v_* \circ u_*.$$

Pour deux morphismes u_1, u_2 de (X, Y) dans (X', Y'), une homotopie de u_1 à u_2 est soumise à la restriction que pour l'application $F : X \times [0, 1] \longrightarrow X'$ qui la définit, on a $F(x, t) \in Y'$ pour $x \in Y$ et $t \in [0, 1]$. Alors on a $(u_1)_* = (u_2)_*$. La preuve est calquée sur celle de la propriété analogue pour l'homologie simpliciale à l'aide de la décomposition d'un «prisme» de Veblen et Lefschetz.

11.2 L'homologie de Vietoris et l'homologie de Čech

L'homologie singulière a l'inconvénient que la dualité d'Alexander n'est pas toujours valable quand un ensemble compact $X \subset \mathbb{S}_n$ n'est pas triangulable. Un exemple de P. Alexandroff (1925) est donné par la frontière X dans $\mathbb{R}^2$ de l'ensemble ouvert

$$U : 0 < x < \frac{2}{3\pi}, \quad \sin 1/x < y < 2.$$

Pour l'homologie singulière, on a en effet $H_0(\mathbb{R}^2 - X; \mathbb{F}_2) \simeq \mathbb{F}_2 \oplus \mathbb{F}_2$, $H_1(X; \mathbb{F}_2) = 0$.

A partir de 1926, plusieurs mathématiciens (P. Alexandroff, Vietoris, Lefschetz et Čech) proposèrent indépendamment de nouvelles méthodes de définition de modules d'homologie, d'abord pour un espace métrique compact, puis pour des espaces plus généraux. Leur idée commune est d'«approcher» l'espace, en un certain sens, par des ensembles finis, munis de structures de «complexes combinatoires» susceptibles de fournir des modules d'homologie.

Ce type de structure était apparu lorsqu'on s'était rendu compte que dans un complexe simplicial euclidien, la donnée d'un simplexe (rectiligne) orienté est équivalente à la donnée de la suite de ses sommets pris dans un certain ordre. La généralisation consiste à négliger entièrement la topologie; on part d'un ensemble quelconque V (fini ou non) totalement ordonné, et on suppose donné un ensemble $\mathfrak{S}(V)$ de parties finies $\{a_0, a_1, \ldots, a_n\}$ de V, avec $a_0 < a_1 < \ldots < a_n$, soumis à la seule condition que si $\{a_0, a_1, \ldots, a_n\} \in \mathfrak{S}(V)$, alors pour toute suite partielle d'indices $i_1 < i_2 < \ldots < i_k$, l'ensemble $\{a_{i_1}, a_{i_2}, \ldots, a_{i_k}\}$ appartient aussi à $\mathfrak{S}(V)$. Les ensembles à $n + 1$ éléments appartenant à $\mathfrak{S}(V)$ sont appelés «n-simplexes combinatoires». Les n-chaînes sont les combinaisons linéaires formelles des n-simplexes combinatoires, et le bord est donné par

$$b_n\{a_0, a_1, \ldots, a_n\} = \sum_{0 \leq k \leq n} (-1)^k \{a_0, \ldots, \hat{a}_k, \ldots, a_n\}$$

où l'accent circonflexe signifie que l'élément a_k est omis de la suite.

La méthode de Vietoris, pour un espace métrique compact X, est de considérer une suite (K_m) de complexes combinatoires formés de points de X et tels que le diamètre maximum de tous les simplexes combinatoires de K_m tende vers 0 avec

$1/m$. La méthode de Vietoris a été étendue à tous les espaces topologiques X en 1948 (Spanier, Dowker-Dugundji-Hurewicz) de la façon suivante. On considère un recouvrement ouvert quelconque $\mathcal{U} = (U_\alpha)$ de X, et le complexe combinatoire $\mathfrak{S}_\mathcal{U}$ dont les simplexes combinatoires sont des ensembles finis qui sont contenus dans un U_α (dépendant du simplexe combinatoire considéré); soit $C.(\mathfrak{S}_\mathcal{U})$ le complexe de chaînes correspondant. On dit qu'un recouvrement ouvert $\mathcal{V} = (V_\beta)$ est plus fin que $\mathcal{U}$ s'il existe une application φ de l'ensemble des indices de $\mathcal{V}$ dans celui de $\mathcal{U}$ telle que $V_\beta \subset U_{\varphi(\beta)}$ pour tout β. On en déduit une application

$$\phi : \mathfrak{S}_\mathcal{V} \longrightarrow \mathfrak{S}_\mathcal{U}$$

qui associe à tout simplexe combinatoire contenu dans un V_β le même simplexe contenu dans $U_{\varphi(\beta)}$. Cela définit un homomorphisme de A-modules

$$C.(\mathfrak{S}_\mathcal{V}) \longrightarrow C.(\mathfrak{S}_\mathcal{U})$$

qui commute avec le bord, et donne par passage à l'homologie un homomorphisme

$$H_{\mathcal{V},\mathcal{U}} : H.(\mathfrak{S}_\mathcal{V}) \longrightarrow H.(\mathfrak{S}_\mathcal{U})$$

qu'on vérifie être indépendant du choix de φ.

De plus, si $\mathcal{W}$ est un troisième recouvrement ouvert de X plus fin que $\mathcal{V}$, on a $h_{\mathcal{W},\mathcal{U}} = h_{\mathcal{W},\mathcal{V}} \circ h_{\mathcal{V},\mathcal{U}}$. Les recouvrements ouverts forment un ensemble ordonné filtrant pour la relation «plus fin», et $(H.(\mathfrak{S}_\mathcal{U}), h_{\mathcal{V},\mathcal{U}})$ est un système projectif[11]) sur cet ensemble. Sa *limite projective* est par définition l'homologie de Vietoris de X.

L'idée d'utiliser des recouvrements ouverts pour définir l'homologie remonte à P. Alexandroff (1925). Comme Vietoris, il se limitait aux espaces métriques compacts et aux recouvrements ouverts finis. Si $\mathcal{U} = (U_\alpha)$ est un tel recouvrement de X, Alexandroff lui associe son «nerf» $N(\mathcal{U})$ qui est un complexe combinatoire sur l'ensemble $\mathcal{F}(\mathcal{U})$ des parties finies de $\mathcal{U}$: un p-simplexe combinatoire de $N(\mathcal{U})$ est une partie finie $\{U_{\alpha_0}, U_{\alpha_1}, \ldots, U_{\alpha_p}\}$ telle que l'intersection $\bigcap_{0 \leq j \leq p} U_{\alpha_j}$ soit non vide.

11) Un ensemble ordonné par une relation $u \prec v$ est dit filtrant pour cette relation si, étant donnés deux éléments a, b de l'ensemble, il existe un élément c tel que $a \prec c$ et $b \prec c$. Un système projectif $(E_\alpha, \varphi_{\alpha\beta})$ dont les indices appartiennent à un ensemble ordonné filtrant I, est une famille $(E_\alpha)_{\alpha\in I}$ d'ensembles indexée par I, et $\varphi_{\alpha\beta}$, pour tout couple (α, β) tel que $\alpha \prec \beta$, une application définie dans E_β et à valeurs dans E_α; ces applications doivent en outre vérifier la relation $\varphi_{\alpha\beta} \circ \varphi_{\beta\gamma} = \varphi_{\alpha\gamma}$ lorsque $\alpha \prec \beta \prec \gamma$. La limite projective du système est la partie $\varprojlim(E_\alpha, \varphi_{\alpha\beta})$ (ou simplement $\varprojlim E_\alpha$) du produit $\prod_{\alpha\in I} E_\alpha$, formée des éléments (x_α) tels que $x_\alpha = \varphi_{\alpha\beta}(x_\beta)$ pour tout couple d'indices tels que $\alpha \prec \beta$.
Soient $(E_\alpha, \varphi_{\alpha\beta})$ et $(F_\alpha, \psi_{\alpha\beta})$ deux systèmes projectifs ayant même ensemble d'indices I. Une famille (f_α) d'applications $f_\alpha : E_\alpha \longrightarrow F_\alpha$ est un système projectif d'applications si l'on a $\psi_{\alpha\beta} \circ f_\beta = f_\alpha \circ \varphi_{\alpha\beta}$ lorsque $\alpha \prec \beta$. La limite projective $\varprojlim f_\alpha = f : \varprojlim E_\alpha \longrightarrow \varprojlim F$ est alors l'application vérifiant les relations $pr_\alpha \circ f = f \circ pr_\alpha$ pour tout $\alpha \in I$.

Čech (1933) a généralisé cette définition à tous les espaces topologiques, en ne considérant toujours que des recouvrements finis; mais C. Dowker (1950) a reconnu que cela conduisait à des propriétés pathologiques de l'homologie d'espaces aussi simples que la droite $\mathbb{R}$, et il est préférable d'utiliser des recouvrements ouverts $\mathcal{U}$ arbitraires. Si $\mathcal{V} = (V_\beta)$ est un recouvrement ouvert plus fin que $\mathcal{U}$, on a comme plus haut une application

$$\Phi : N(\mathcal{V}) \longrightarrow N(\mathcal{U})$$

en faisant correspondre à tout p-simplexe $\{V_{\beta_0}, V_{\beta_1}, \ldots, V_{\beta_p}\}$ le p-simplexe $\{U_{\varphi(\beta_0)}, U_{\varphi(\beta_1)}, \ldots, U_{\varphi(\beta_p)}\}$. On en déduit en homologie un homomorphisme

$$h_{\mathcal{V},\mathcal{U}} : H.(N(\mathcal{V})) \longrightarrow H.(N(\mathcal{U}))$$

qui ne dépend pas du choix de φ. Les $(H.(N(\mathcal{U})), h_{\mathcal{V},\mathcal{U}})$ forment encore un système projectif, et leur limite projective est l'homologie de Čech $\check{H}.(X; A)$ de X à coefficients dans A.

11.3 Les cohomologies

Dans son étude des intégrales abéliennes, Riemann avait considéré, pour une 1-forme différentielle ω sur une surface de Riemann S, telle que $d\omega = 0$, les «périodes» de ω, c'est-à-dire les valeurs des intégrales $\int_c \omega$ le long de courbes fermées c sur S. Il avait montré que l'ensemble de ces périodes est un $\mathbb{Z}$-module libre ayant une base de $2g$ éléments, valeurs de l'intégrale le long des 1-cycles dont les classes forment une base de l'homologie $H_1(S; \mathbb{Z})$. Betti énonça un résultat analogue pour les $(n-1)$-formes sur une variété de dimension n, puis Poincaré, au début de son mémoire de 1895, donna (sans démonstration) un énoncé général pour les p-formes $(1 \leq p \leq n)$.

Dans ses recherches sur les groupes de Lie compacts et leur topologie, à partir de 1927, E. Cartan, fidèle à ses méthodes reposant sur l'étude des formes différentielles, reprend une question qu'il avait déjà effleurée en 1922 dans son livre sur les «Invariants intégraux»: sur une variété lisse M, compacte, orientée, de dimension n, une p-forme ω qui est «fermée» (c'est-à-dire que $d\omega = 0$) est-elle «exacte» (c'est-à-dire de la forme $d\alpha$, où α est une $(p-1)$-forme)? Il suppose la variété triangulée; l'intégration définit une $\mathbb{Z}$-forme bilinéaire

$$(\omega, c) \mapsto \int_c \omega$$

sur les p-formes fermées ω et les p-cycles c de la triangulation T. La formule de Stokes montre que $\int_c \omega = 0$ si ω est exacte ou si c est un bord. E. Cartan conjectura en 1929 que:

1) si $\int_c \omega = 0$ pour tous les p-cycles c, ω est exacte;

2) si $\int_c \omega = 0$ pour toutes les p-formes fermées, c est un bord.

Par une ingénieuse construction de $(n-p)$-formes fermées associées à la triangulation T^* duale de T, G. de Rham prouva ces deux théorèmes dans sa thèse de 1931.

L'intégrale $\int_c \omega$ est définie pour toute p-forme lisse ω et toute p-chaîne de T, de sorte que $c \mapsto \int_c \omega$ peut être considérée comme une *forme linéaire* sur le $\mathbb{R}$-espace vectoriel $C_p(T;\mathbb{R})$ des p-chaînes de T. C'est une des remarques qui, en 1935, ont conduit indépendamment Alexander et Kolmogoroff, à considérer pour la première fois le complexe de cochaînes $C^\cdot = \mathrm{Hom}(C_., G)$ pour un complexe de chaînes $C_.$ sur un anneau A, et un A-module G. Ils se bornaient aux complexes de chaînes d'une triangulation; mais en 1944, Eilenberg appliqua la même idée au complexe des chaînes singulières $S_.(X)$ d'un espace topologique quelconque X, obtenant ainsi le complexe

$$S^\cdot(X;G) = \mathrm{Hom}(S_.(X), G)$$

des cochaînes singulières à valeurs dans G. On définit ainsi les groupes de cohomologie singulière $H^p(X;G)$, et on définit de même la cohomologie singulière relative $H^\cdot(X,Y;G)$. On a la suite exacte de cohomologie singulière

$$\ldots \longrightarrow H^{p-1}(Y) \longrightarrow H^p(X,Y) \longrightarrow H^p(X) \longrightarrow H^p(Y) \longrightarrow \ldots \tag{65}$$

(où l'on a supprimé la mention des coefficients). Toute application continue $u : X \longrightarrow X'$ définit un homomorphisme

$$S^p(u) : f \mapsto f \circ u$$

de $S^p(X';G)$ dans $S^p(X;G)$, commutant aux cobords, et donne par suite un homomorphisme

$$H^p(u) : H^p(X';G) \longrightarrow H^p(X;G),$$

aussi noté u^*; si $v : X' \longrightarrow X''$ est une application continue, on a $(v \circ u)^* = u^* \circ v^*$. On définit un homomorphisme analogue pour la cohomologie singulière relative.

Si X est un espace séparé, on dit qu'une p-cochaîne f a un *support compact* s'il existe une partie compacte $K \subset X$ telle que $f(s^p) = 0$ pour toute application continue $s^p : \bar{\Delta}_p \longrightarrow X$ telle que $s^p(\bar{\Delta}_p) \cap K = \phi$. Les p-cochaînes à support compact forment un sous-module $S_c^p(X;G)$ de $S^p(X;G)$, et on a $d_p(S_c^p(X;G)) \subset S_c^{p+1}(X;G)$, donc les $S_c^p(X;G)$ forment un complexe de cochaînes, d'où l'on déduit la définition du *p-ème groupe de cohomologie à supports compacts* $H_c^p(X;G)$. On a une définition analogue pour les groupes relatifs à un sous-espace, et il y a une suite exacte de cohomologie à supports compacts semblable à (65). Comme exemple de cohomologie à supports compacts, notons que $H_c^n(\mathbb{R}^n;\mathbb{Z}) \simeq \mathbb{Z}$, alors que $H^n(\mathbb{R}^n;\mathbb{Z}) = 0$.

La méthode d'Alexandroff et Čech utilisant des recouvrements ouverts fournit aussi des groupes de cohomologie, en définissant des complexes de cochaînes à partir des nerfs des recouvrements: $H^p(N(\mathcal{U});G)$ est le p-ème groupe de cohomologie du complexe de cochaînes $C^{\cdot} = \mathrm{Hom}(C_{\cdot}, G)$, où $C_{\cdot}$ est le complexe de chaînes du complexe combinatoire $N(\mathcal{U})$. Les $H^p(N(\mathcal{U});G)$ forment un système inductif[12]) sur l'ensemble ordonné filtrant des recouvrements ouverts, et le p-ème groupe de cohomologie de Čech à coefficients dans G est la limite inductive

$$H^p(X;G) = \varinjlim H^p(N(\mathcal{U});G).$$

On peut définir directement des théories de cohomologie en partant de complexes de cochaînes qui ne proviennent pas d'un complexe de chaînes par application du foncteur Hom. L'exemple le plus simple est le complexe

$$0 \longrightarrow \mathcal{E}^0 \xrightarrow{d} \mathcal{E}^1 \xrightarrow{d} \mathcal{E}^2 \longrightarrow \ldots \longrightarrow \mathcal{E}^n \xrightarrow{d} \mathcal{E}^{n+1} \longrightarrow \ldots \tag{66}$$

où $\mathcal{E}^n$ est le $\mathbb{R}$-espace vectoriel des n-formes différentielles lisses sur une variété lisse X, et d est la différentielle extérieure. La cohomologie correspondante est la *cohomologie de de Rham* $H^{\cdot}_{DR}(X)$. On a un autre complexe de cochaînes

$$0 \longrightarrow \mathcal{D}^0 \xrightarrow{d} \mathcal{D}^1 \xrightarrow{d} \mathcal{D}^2 \longrightarrow \ldots \longrightarrow \mathcal{D}^n \xrightarrow{d} \mathcal{D}^{n+1} \longrightarrow \ldots \tag{67}$$

où $\mathcal{D}^n$ est le $\mathbb{R}$-espace vectoriel des n-formes différentielles lisses à support compact, ce qui définit la cohomologie de de Rham à support compact $H^{\cdot}_{DR,c}(X)$.

Une autre méthode, applicable à tout espace topologique X, a été proposée en 1935 par Alexander, et développée par Spanier. Pour chaque entier $p \geq 0$, on considère le groupe commutatif $C^p(X;G)$ de toutes les applications (continues ou non) de X^{p+1} dans G, et le sous-groupe $C^p_0(X;G)$ formé de ces applications f qui sont nulles dans tout un voisinage (dépendant de f) de la diagonale Δ de X^{p+1} (image de X par l'application $x \mapsto \overbrace{(x, x, \ldots, x)}^{(p+1)\,\text{fois}}$ dans X^{p+1}). On définit sur $C^p(X;G)$ un cobord par la formule

$$(\delta_p f)(x_0, x_1, \ldots, x_{p+1}) = \sum_{0 \leq k \leq p+1} (-1)^k f(x_0, \ldots, \hat{x}_k, \ldots, x_{p+1}).$$

12) Un système inductif d'ensembles $(E_\alpha, \varphi_{\beta\alpha})$ est une famille $(E_\alpha)_{\alpha \in I}$ indexée par un ensemble ordonné filtrant I, pour laquelle on a une famille $(\varphi_{\beta\alpha})$ d'applications $\varphi_{\beta\alpha} : E_\alpha \longrightarrow E_\beta$ pour $\alpha \prec \beta$ telle que $\varphi_{\gamma\alpha} = \varphi_{\gamma\beta} \circ \varphi_{\beta\alpha}$ pour $\alpha \prec \beta \prec \gamma$. On définit la limite inductive $\varinjlim E_\alpha$ de cette famille de la façon suivante: on considère la réunion disjointe U des E_α, et la relation d'équivalence R dans U pour laquelle $x \in E_\alpha$ et $y \in E_\beta$ sont équivalents s'il existe $\gamma \in I$ avec $\alpha \prec \gamma$ et $\beta \prec \gamma$, et l'on a $\varphi_{\gamma\alpha}(x_\alpha) = \varphi_{\gamma\beta}(y_\beta)$; $E = \varinjlim E_\alpha$ est l'ensemble U/R de ces classes d'équivalence. Pour tout $\alpha \in I$, il existe une application canonique $\varphi_\alpha : E_\alpha \longrightarrow E$ qui fait correspondre à $x_\alpha \in E_\alpha$ sa classe d'équivalence dans E.
Si $(E_\alpha, \varphi_{\beta\alpha})$, $(F_\alpha, \psi_{\beta\alpha})$ sont deux systèmes inductifs d'ensembles ayant même ensemble d'indices, un système inductif d'applications $u_\alpha : E_\alpha \longrightarrow F_\alpha$ est tel que $\psi_{\beta\alpha} \circ u_\alpha = u_\beta \circ \varphi_{\beta\alpha}$ pour $\alpha \prec \beta$. La limite inductive $\varinjlim u_\alpha : \varinjlim E_\alpha \longrightarrow \varinjlim F_\alpha$ est l'application u telle que $u \circ \varphi_\alpha = \psi_\alpha \circ u_\alpha$ pour tout $\alpha \in I$.

On vérifie que $\delta_{p+1} \circ \delta_p = 0$ et que $\delta_p(C_0^p(X;G)) \subset C_0^{p+1}(X;G)$. Si on considère le groupe quotient $\bar{C}^p(X;G) = C^p(X;G)/C_0^p(X;G)$, on obtient donc à partir de δ_p un cobord

$$\bar{\delta}_p : \bar{C}^p(X;G) \longrightarrow \bar{C}^{p+1}(X;G).$$

Les $\bar{C}^p(X;G)$ et les $\bar{\delta}_p$ définissent donc un complexe de cochaînes, et ses groupes de cohomologie sont les *groupes de cohomologie d'Alexander-Spanier* de X à coefficients dans G, notés $\bar{H}^p(X;G)$.

11.4 Identifications et axiomatisation

Devant la multiplicité des définitions possibles de groupes d'homologie et de cohomologie, on fut amené à chercher si ces définitions fournissaient effectivement des groupes différents. En 1948, C. Dowker, par une ingénieuse méthode, prouva un résultat de théorie des ensembles qui entraîne simultanément l'existence d'un isomorphisme canonique de l'homologie de Vietoris sur l'homologie de Čech $\check{H}.(X;G)$, et d'un isomorphisme canonique de la cohomologie d'Alexander-Spanier $\bar{H}^{\cdot}(X;G)$ sur la cohomologie de Čech $\check{H}^{\cdot}(X;G)$. Ces résultats s'étendent aux homologies et cohomologies relatives à un sous-espace.

Pour les variétés C^∞, de Rham prouva en 1950 que les groupes de cohomologie $H_{DR}^p(X)$ du complexe de cochaînes (66) sont canoniquement isomorphes aux groupes $H^p(X;\mathbb{R})$ de cohomologie singulière à coefficients dans $\mathbb{R}$; de même $H_{DR,c}^p(X)$ est canoniquement isomorphe à $H_c^p(X;\mathbb{R})$.

En 1952, dans un livre resté classique (annoncé dès 1945), Eilenberg et Steenrod inaugurèrent une autre stratégie. Sans s'occuper de la façon particulière dont était définie chaque théorie, ils décrivent un système d'axiomes qui, à leur sens, fondent une théorie générale de l'homologie (ou de la cohomologie). Une telle théorie est définie par:

1) Une catégorie, dont les objets sont des couples (X,Y) d'espaces topologiques tels que $Y \subset X$ (avec la topologie induite), et les morphismes

$$f : (X,Y) \longrightarrow (X',Y')$$

sont des applications continues $X \longrightarrow X'$ telles que $f(Y) \subset Y'$. (Les espaces X, Y ne sont pas nécessairement des espaces topologiques quelconques, ils peuvent être soumis à diverses conditions supplémentaires; il en est de même pour les applications f.)

2) Un «foncteur» covariant de cette catégorie dans la catégorie des suites de A-modules (ou A-modules gradués), A étant un anneau commutatif: à l'objet (X,Y) il fait correspondre une suite $H.(X,Y;A) = (H_n(X,Y;A))_{n\geq 0}$, et à un morphisme $f : (X,Y) \longrightarrow (X',Y')$, une suite $H.(f) = (H_n(f))_{n\geq 0}$ (aussi notée f_*), où $H_n(f) : H_n(X,Y;A) \longrightarrow H_n(X',Y';A)$ est un homomorphisme de A-modules.

3) Pour chaque entier $n \geq 1$, un homomorphisme (de «décalage»)

$$\partial : H_n(X, Y; A) \longrightarrow H_{n-1}(Y; A).$$

Il y a 7 axiomes dans la théorie. Les 3 premiers expriment la «fonctorialité» de f_*: on a $(\mathrm{Id})_* = \mathrm{Id}$, $(g \circ f)_* = g_* \circ f_*$, et le diagramme

$$\begin{array}{ccc} H_n(X, Y; A) & \xrightarrow{f_*} & H_n(X', Y'; A) \\ \Big\downarrow \partial & & \Big\downarrow \partial \\ H_{n-1}(Y; A) & \xrightarrow[f_*]{} & H_{n-1}(Y'; A) \end{array}$$

est commutatif.

L'axiome 4 est l'existence de la suite exacte d'homologie (63), et l'axiome 5 exprime le fait que l'existence d'une homotopie de f_1 à f_2 entraîne $(f_1)_* = (f_2)_*$.

L'axiome 6 est le plus nouveau, mettant en valeur une notion, l'*excision*, qui (ainsi que la suite exacte d'homologie) était seulement en germe dans les travaux de Lefschetz sur l'homologie relative: si U est un ouvert non vide de X tel que Y soit un voisinage de $\bar{U}$, alors l'injection $j : (X - U, Y - U) \longrightarrow (X, Y)$ détermine un *isomorphisme*

$$j_* : H_n(X - U, Y - U) \xrightarrow{\sim} H_n(X, Y) \text{ pour } n \geq 0$$

(on peut «exciser» U).

Le dernier axiome impose que, pour l'espace P réduit à un point, on a $H_n(P) = 0$ pour $n \geq 1$. Si naturel qu'il paraisse, c'est le seul qui plus tard sera abandonné dans les théories d'«homologie généralisée» (voir plus loin).

On a une théorie axiomatique de la cohomologie en «renversant les flèches»; le foncteur, noté f^* ou $H^{\cdot}(f)$, est alors contravariant, associant à $f : (X, Y) \longrightarrow (X', Y')$ des homomorphismes $f^* : H^n(X', Y') \longrightarrow H^n(X, Y)$, et l'homomorphisme de décalage est $\partial : H^n(Y) \longrightarrow H^{n+1}(X, Y)$.

Eilenberg et Steenrod montrèrent que l'on peut déduire directement des axiomes certaines propriétés prouvées auparavant par des raisonnements particuliers aux théories homologiques envisagées. L'exemple le plus utile est une forme générale d'un résultat prouvé vers 1929 par W. Mayer et Vietoris pour certains complexes de chaînes: si X_1, X_2 sont deux sous-espaces de X tels que le morphisme injectif

$$(X_1, X_1 \cap X_2) \longrightarrow (X_1 \cup X_2, X_2)$$

donne pour les groupes d'homologie un *isomorphisme*

$$H_\cdot(X_1, X_1 \cap X_2) \xrightarrow{\sim} H_\cdot(X_1 \cup X_2, X_2)$$

alors la suite

$$\begin{aligned} \ldots \longrightarrow & H_p(X_1, X_1 \cap X_2) \longrightarrow H_p(X, X_2) \longrightarrow H_p(X, X_1 \cup X_2) \longrightarrow \\ & H_{p-1}(X_1, X_1 \cap X_2) \longrightarrow \ldots \end{aligned}$$

est exacte. Si en outre, on a $X = X_1 \cup X_2$, on obtient la *suite exacte de Mayer-Vietoris*

$$\begin{aligned} \ldots \longrightarrow & H_p(X_1 \cap X_2) \longrightarrow H_p(X_1) \oplus H_p(X_2) \longrightarrow H_p(X) \longrightarrow \\ \longrightarrow & H_{p-1}(X_1 \cap X_2) \longrightarrow \ldots \end{aligned} \tag{68}$$

très utile pour le calcul des groupes d'homologie. On a des résultats analogues pour la cohomologie.

Dans leur livre, Eilenberg et Steenrod examinent ensuite la validité de leurs axiomes pour les diverses théories homologiques connues en 1950. Leur résultat central est que pour les couples triangulables (X, Y) (où les simplexes de la triangulation de Y doivent être aussi des simplexes de la triangulation de X), toutes les théories homologiques qui satisfont aux axiomes fournissent nécessairement des groupes d'homologie (ou de cohomologie) *isomorphes*.

Il n'en est pas de même pour des espaces plus généraux. L'homologie et la cohomologie singulières satisfont à tous les axiomes, pour des couples (X, Y) d'espaces topologiques quelconques. Cela est encore vrai pour la cohomologie de Čech (ou d'Alexander-Spanier); mais par contre, il y a des couples (X, Y) pour lesquels X et Y sont compacts, mais dont les groupes d'homologie de Čech à coefficients dans $\mathbb{Z}$ ne vérifient pas l'axiome de la suite exacte; toutefois, cet axiome est alors vérifié lorsque l'anneau des coefficients est un corps.

La cohomologie d'Alexander-Spanier a des propriétés que ne possède pas la cohomologie singulière. Si X est paracompact, et Y un sous-espace fermé de X, les voisinages ouverts U de Y forment un ensemble ordonné filtrant pour la relation $U \supset V$, et on en déduit un système inductif $(\bar{H}^p(U; G))$ de groupes de cohomologie d'Alexander-Spanier. On a alors un *isomorphisme* canonique

$$\varinjlim \bar{H}^p(U; G) \xrightarrow{\sim} \bar{H}^p(Y; G). \tag{69}$$

On déduit de là que si X est compact et Y fermé dans X, on a

$$\bar{H}^p(X, Y; G) \xrightarrow{\sim} \bar{H}^p_c(X - Y; G)$$

d'où la suite exacte de cohomologie

$$\begin{aligned} \ldots \longrightarrow \bar{H}^p_c(X - Y; G) \xrightarrow{a} \bar{H}^p(X; G) \xrightarrow{j^*} \bar{H}^p(Y; G) \\ \longrightarrow \bar{H}^{p+1}_c(X - Y; G) \longrightarrow \ldots \end{aligned} \tag{70}$$

(où l'homomorphisme $\underline{a}$ associe à la classe d'un p-cocycle à support compact sur $X - Y$ la classe de ce même cocycle considéré comme un cocycle sur X ayant la valeur 0 sur le complémentaire de K^{p+1} pour un ensemble compact $K \subset X - Y$).

En outre, si deux couples (X, Y), (X', Y') sont formés d'espaces compacts et si $f : (X, Y) \longrightarrow (X', Y')$ est un morphisme tel que la restriction de f à $X - Y$ soit un *homéomorphisme* de $X - Y$ sur $X' - Y'$, alors

$$\bar{H}^{\cdot}(f) : \bar{H}^{\cdot}(X, Y) \longrightarrow \bar{H}^{\cdot}(X', Y')$$

est un isomorphisme (forme ultime des idées de Lefschetz sur l'homologie relative).

11.5 La cohomologie des faisceaux

Alors que tous les groupes d'homologie ou de cohomologie mentionnés jusqu'ici ont des «coefficients» dans un A-module commutatif fixe, Reidemeister en 1935 et Steenrod en 1942 furent conduits, au cours de leurs recherches, à considérer des «chaînes» $\sum_s g_s.s$, combinaisons linéaires de simplexes où le coefficient g_s du simplexe s appartient à un groupe G_s qui dépend de s. En fait, tous les groupes G_s sont isomorphes à un groupe fixe, mais par un isomorphisme qui dépend de s; ils forment ce qu'on a appelé un «système local de groupes».

Ces constructions furent englobées dans la théorie beaucoup plus générale de la *cohomologie des faisceaux*, fondée par J. Leray en 1946, et simplifiée plus tard dans des exposés de H. Cartan et R. Godement.

L'idée fondamentale dans la définition d'un *faisceau de A-modules* sur un espace topologique X est de se donner, non pas un A-module en chaque point de X, mais un A-module $\mathcal{F}(U)$ pour chaque *ouvert* U de X. Ces A-modules sont liés par la donnée d'homomorphismes («de restriction»)

$$\rho_{VU} : \mathcal{F}(U) \longrightarrow \mathcal{F}(V) \text{ pour deux ouverts } U \supset V$$

avec les conditions de compatibilité

$$\rho_{UU} = \mathrm{Id}_U, \rho_{WU} = \rho_{WV} \circ \rho_{VU} \text{ si } U \supset V \supset W.$$

Ces homomorphismes doivent en outre vérifier les deux axiomes suivants, qui expriment la possibilité du «passage du local au global». Lorsque U est réunion d'une famille quelconque (U_α) d'ouverts:

F1) Si s', s'' sont deux éléments de $\mathcal{F}(U)$ tels que $\rho_{UU_\alpha}(s') = \rho_{UU_\alpha}(s'')$ pour tout indice α, alors $s' = s''$.

F2) Si les $s_\alpha \in \mathcal{F}(U_\alpha)$ sont tels que, lorsque $U_\alpha \cap U_\beta \neq \phi$, on a

$$\rho_{U_\alpha, U_\alpha \cap U_\beta}(s_\alpha) = \rho_{U_\beta, U_\alpha \cap U_\beta}(s_\beta),$$

alors il existe $s \in \mathcal{F}(U)$ tel que $\rho_{UU_\alpha}(s) = s_\alpha$ pour tout α.

Les éléments de $\mathscr{F}(U)$ s'appellent les *sections* de $\mathscr{F}$ au-dessus de U, et on écrit aussi $\Gamma(U, \mathscr{F})$ au lieu de $\mathscr{F}(U)$.

Pour tout point $x \in X$, les voisinages ouverts U de x forment un ensemble ordonné filtrant pour la relation $\supset$, et les $\mathscr{F}(U)$ un système inductif de A-modules sur cet ensemble. Sa limite inductive $\mathscr{F}(x)$ est appelée la fibre de $\mathscr{F}$ au point x, et on dit que ses éléments sont les *germes* des sections de $\mathscr{F}$ au point x; deux sections ont même germe si et seulement si elles coïncident dans un voisinage de x.

Les notions de faisceau et de sections d'un faisceau concrétisent les idées de «localisation» et de «globalisation». Par exemple, soit X une variété analytique complexe connexe et compacte, et pour tout ouvert $U \subset X$, soit $\mathcal{O}(U)$ le groupe additif des applications holomorphes de U dans $\mathbb{C}$; ces groupes définissent le faisceau des germes d'applications holomorphes dans $\mathbb{C}$ mais $\Gamma(X, \mathcal{O})$ est réduit aux fonctions constantes, tandis que pour U assez petit, $\Gamma(U, \mathcal{O})$ est un espace vectoriel sur $\mathbb{C}$ de dimension infinie.

Il existe un faisceau $\mathscr{F}$ sur X tel que pour tout $x \in X$, la fibre $\mathscr{F}(x)$ soit isomorphe à G. On dit que $\mathscr{F}$ est un faisceau *constant*; mais $\Gamma(U, \mathscr{F})$ n'est formé d'applications constantes dans U que si U est connexe.

Si $\mathscr{F}$ et $\mathscr{F}'$ sont deux faisceaux de A-modules sur X, on appelle homomorphisme de $\mathscr{F}$ dans $\mathscr{F}'$ une famille $v_U : \mathscr{F}(U) \longrightarrow \mathscr{F}'(U)$ d'homomorphismes, telle que si $V \subset U$, le diagramme

$$\begin{array}{ccc} \mathscr{F}(U) & \xrightarrow{\;v_U\;} & \mathscr{F}'(U) \\ {\scriptstyle \rho_{VU}}\downarrow & & \downarrow{\scriptstyle \rho_{VU}} \\ \mathscr{F}(V) & \xrightarrow[\;v_V\;]{} & \mathscr{F}'(V) \end{array}$$

est commutatif. Par passage à la limite inductive, on en déduit un homomorphisme des fibres

$$v_x : \mathscr{F}(x) \longrightarrow \mathscr{F}'(x)$$

pour tout $x \in X$. On dit qu'une suite d'homomorphismes de faisceaux sur X

$$\mathscr{F}' \xrightarrow{u} \mathscr{F} \xrightarrow{v} \mathscr{F}''$$

est exacte si, pour tout $x \in X$, la suite d'homomorphismes de modules

$$\mathscr{F}'(x) \xrightarrow{u_x} \mathscr{F}(x) \xrightarrow{v_x} \mathscr{F}''(x)$$

est exacte.

Toutes les notions usuelles de la théorie des modules (sous-module, somme et intersection de sous-modules, module quotient, produit tensoriel, modules d'homomorphismes) se «localisent» en donnant des notions correspondantes pour les faisceaux.

La cohomologie d'un espace X à valeurs dans un faisceau $\mathcal{F}$ sur X est devenue un outil très efficace aussi bien pour l'étude de $\mathcal{F}$ (par exemple le faisceau $\mathcal{O}$ des germes de fonctions holomorphes sur une variété analytique complexe) que pour celle de la topologie de X. Sa définition la plus simple utilise une modification de la méthode d'Alexandroff-Čech. On part du même complexe de chaînes $C.(N(\mathcal{U}))$ du nerf d'un recouvrement ouvert quelconque $\mathcal{U} = (U_\alpha)$ de X. Mais au lieu d'associer à un p-simplexe combinatoire $s = \{U_{\alpha_0}, U_{\alpha_1}, \ldots, U_{\alpha_p}\}$ un élément de G, on lui associe un élément

$$f(s) \in \Gamma(M_s, \mathcal{F})$$

où $M_s = U_{\alpha_0} \cap U_{\alpha_1} \cap \ldots \cap U_{\alpha_p}$ (non vide par hypothèse). Les combinaisons linéaires à coefficients dans A de ces éléments forment un A-module $C^p(\mathcal{U}, \mathcal{F})$. Pour définir le cobord $d_p : C^p(\mathcal{U}, \mathcal{F}) \longrightarrow C^{p+1}(\mathcal{U}, \mathcal{F})$, on considère, pour tout $(p+1)$-simplexe combinatoire $t = \{U_{\alpha_0}, U_{\alpha_1}, \ldots, U_{\alpha_{p+1}}\}$, les p-simplexes $t_k = \{U_{\alpha_0}, \ldots, \hat{U}_{\alpha_k}, \ldots, U_{\alpha_{p+1}}\}$ $(0 \leq k \leq p+1)$, et on pose

$$(d_p f)(t) = \sum_{0 \leq k \leq p+1} (-1)^k \rho_{M_t M_{t_k}}(f(t_k)).$$

On vérifie qu'on obtient bien ainsi un complexe de cochaînes $C^\cdot(\mathcal{U}, \mathcal{F})$, dont on prend la cohomologie $H^\cdot(\mathcal{U}; \mathcal{F})$; on passe ensuite à la limite inductive suivant l'ensemble ordonné filtrant $\mathcal{U}$, ce qui donne la cohomologie $H^\cdot(X; \mathcal{F})$ à coefficients dans $\mathcal{F}$. Lorsque $\mathcal{F}$ est un faisceau constant de fibre G, le groupe $H^p(X; \mathcal{F})$ est isomorphe au groupe de cohomologie de Čech $\check{H}^\cdot(X; G)$.

Si

$$0 \longrightarrow \mathcal{F}' \longrightarrow \mathcal{F} \longrightarrow \mathcal{F}'' \longrightarrow 0$$

est une suite exacte de faisceaux sur X, la suite de complexes de cochaînes

$$0 \longrightarrow C^\cdot(\mathcal{U}, \mathcal{F}') \longrightarrow C^\cdot(\mathcal{U}, \mathcal{F}) \longrightarrow C^\cdot(\mathcal{U}, \mathcal{F}'') \longrightarrow 0$$

est exacte, et après passage à la limite inductive, on obtient la suite exacte de cohomologie à valeurs dans un faisceau

$$\ldots \longrightarrow H^p(X; \mathcal{F}') \longrightarrow H^p(X; \mathcal{F}) \longrightarrow H^p(X; \mathcal{F}'') \longrightarrow$$
$$\longrightarrow H^{p+1}(X; \mathcal{F}') \longrightarrow \ldots$$

En «localisant» les groupes de cohomologie, on obtient des *faisceaux de cohomologie* d'un faisceau de A-modules sur X: pour tout ouvert $U \subset X$, on considère le groupe $H^p(U; \mathcal{F})$; on montre qu'il existe un faisceau $\mathcal{H}^p(\mathcal{F})$ tel que

$$H^p(U; \mathcal{F}) = \Gamma(U, \mathcal{H}^p(\mathcal{F})).$$

Dès 1945, Leray s'était préoccupé de la détermination de la cohomologie d'un espace X par la connaissance de la cohomologie des ensembles d'un recouvrement $\mathcal{U} = (U_\alpha)$ (sans passer à la limite inductive). Il prouva un peu plus tard qu'il existe une suite spectrale dont les termes E_2 sont

$$E_2^{pq} = H^p(\mathcal{U}; \mathcal{H}^q(\mathcal{F}))$$

et l'aboutissement le module $H^\cdot(X; \mathcal{F})$ muni d'une filtration. En particulier si $\mathcal{H}^q(\mathcal{F}) = 0$ pour tout $q \geq 1$, on obtient un isomorphisme

$$H^\cdot(\mathcal{U}; \mathcal{F}) \xrightarrow{\sim} H^\cdot(X; \mathcal{F}).$$

Un résultat analogue (qui est à l'origine de l'idée de la suite spectrale) concerne les relations cohomologiques associées à une application continue $f : X \longrightarrow Y$. Si $\mathcal{F}$ est un faisceau sur X, il y a un faisceau $\mathcal{H}^q(f, \mathcal{F})$ sur Y tel que, pour tout ouvert U dans Y, $H^q(f^{-1}(U); \mathcal{F}) = \Gamma(U, \mathcal{H}^q(f, \mathcal{F}))$. Il y a aussi une suite spectrale (souvent citée comme «suite de Leray») dont les termes E_2 sont

$$E_2^{pq} = H^p(Y; \mathcal{H}^q(f, \mathcal{F}))$$

et dont l'aboutissement est le module $H^p(X; \mathcal{F})$ muni d'une filtration. Si X et Y sont localement compacts et f une application propre (i.e. telle que l'image réciproque d'un compact de Y soit un compact de X), la fibre $\mathcal{H}^q(f, \mathcal{F})(y)$ est isomorphe au groupe de cohomologie d'Alexander-Spanier $\bar{H}^q(f^{-1}(y); \mathcal{F})$. En particulier, si, pour $1 \leq q \leq m$, on a $\bar{H}^q(f^{-1}(y); A) = 0$, et $\bar{H}^0(f^{-1}(y); A) \simeq A$, alors on a un isomorphisme canonique $\bar{H}^j(X; A) \xrightarrow{\sim} \bar{H}^j(Y; A)$ pour $0 \leq j \leq m$; ceci est une vaste généralisation de résultats antérieurs de Vietoris et Begle pour les espaces compacts.

En 1959, par une utilisation assez compliquée de la cohomologie des faisceaux, A. Borel et J.C. Moore ont pu définir des groupes d'homologie $H_p(X; L)$ pour un espace X *localement compact* et des coefficients dans un anneau de Dedekind L; ces groupes n'ont pas les inconvénients des groupes d'homologie de Čech (avec lesquels ils se confondent lorsque X est compact et L un corps). Entre autres, ils permettent d'obtenir des versions plus générales des théorèmes de dualité (voir plus loin) et ils vérifient une suite exacte d'homologie

$$\ldots \longrightarrow H_p(F; L) \longrightarrow H_p(X; L) \longrightarrow H_p(U; L) \longrightarrow H_{p-1}(F; L) \longrightarrow \ldots \quad (71)$$

pour une partie fermée F de X et $U = X - F$.

12 Produits et coproduits

12.1 Les produits externes

La «formule de Künneth» de l'homologie simpliciale se généralise en homologie et cohomologie singulières. Cela est dû à un théorème fondamental d'Eilenberg et Zilber (1953): si $S.(X;A)$ et $S.(Y;A)$ sont les complexes de chaînes singulières à coefficients dans A pour deux espaces X, Y, alors il y a une équivalence de complexes de chaînes (non unique en général) entre les complexes de chaînes $S.(X\times Y;A)$ et $S.(X;A)\otimes_A S.(Y;A)$. La preuve consiste à construire des transformations de chaînes

$$u. : S.(X \times Y) \longrightarrow S.(X) \otimes S.(Y) \text{ et } v. : S.(X) \otimes S.(Y) \longrightarrow S.(X \times Y)$$

en les définissant par récurrence sur le degré des modules de chaînes, suivant un procédé qui a été plus tard généralisé sous le nom de «méthode des modèles acycliques».

On déduit de ce théorème un isomorphisme

$$H.(X \times Y;A) \simeq H.(S.(X;A) \otimes S.(Y;A))$$

qui ne dépend pas du choix des transformations de chaînes $u.$ et $v.$ précédentes. La «formule de Künneth» pour les complexes de chaînes (49) fournit d'autre part un homomorphisme canonique

$$H_p(S.(X;A)) \otimes_A H_q(S.(Y;A)) \longrightarrow H_{p+q}(S.(X;A) \otimes_A S.(Y;A))$$

d'où par composition des homomorphismes canoniques en homologie singulière

$$H_p(X;A) \otimes_A H_q(Y;A) \longrightarrow H_{p+q}(X \times Y;A). \tag{72}$$

Pour deux classes d'homologie $z'_p \in H_p(X;A)$, $z''_q \in H_q(Y;A)$, on note $z'_p \times z''_q$ (produit cartésien de classes d'homologie) l'image par (72) de $z'_p \otimes z''_q$.

Pour définir un produit analogue pour les classes de cohomologie, on remarque d'abord que pour deux complexes de chaînes $C.'$, $C.''$, et deux A-modules G', G'', on a une transformation de cochaînes

$$\varphi. : \mathrm{Hom}(C.', G') \otimes \mathrm{Hom}(C.'', G'') \longrightarrow \mathrm{Hom}(C.' \otimes C.'', G' \otimes G'') \tag{73}$$

définie par

$$(\varphi.(u \otimes v))(c' \otimes c'') = u(c') \otimes v(c'').$$

Par le théorème d'Eilenberg-Zilber, on a une équivalence de cochaînes

$$\mathrm{Hom}(S.(X) \otimes S.(Y), G' \otimes G'') \simeq \mathrm{Hom}(S.(X \times Y), G' \otimes G'')$$

d'où, par (73) une transformation de cochaînes

$$\mathrm{Hom}(S.(X), G') \otimes \mathrm{Hom}(S.(Y), G'') \longrightarrow \mathrm{Hom}(S.(X \times Y), G' \otimes G'')$$

et par suite un homomorphisme en cohomologie

$$H^p(X; G') \otimes H^q(Y; G'') \longrightarrow H^{p+q}(X \times Y; G' \otimes G''). \tag{74}$$

En particulier, si G' et G'' sont égaux au groupe additif de A, on obtient un homomorphisme

$$H^p(X; A) \otimes_A H^q(Y; A) \longrightarrow H^{p+q}(X \times Y; A); \tag{75}$$

on note alors $z'^p \times z''^q$ (produit cartésien de classes de cohomologie) l'image de $z'^p \times z''^q$ pour $z'^p \in H^p(X; A)$, $z''^q \in H^q(Y; A)$.

D'autres produits font intervenir à la fois l'homologie et la cohomologie. Avec les mêmes notations, on a deux morphismes de modules gradués

$$\alpha : C.' \otimes C.'' \otimes \mathrm{Hom}(G', G'') \longrightarrow \mathrm{Hom}(\mathrm{Hom}(C.', G'), C.'' \otimes G'')$$
$$\beta : \mathrm{Hom}(C.' \otimes C.'', \mathrm{Hom}(G', G'')) \longrightarrow \mathrm{Hom}(C.' \otimes G', \mathrm{Hom}(C.'', G''))$$

définis par les formules

$$\alpha(c' \otimes c'' \otimes h)(g') = c'' \otimes (h(g')(c'))$$
$$(\beta(v)(c' \otimes g'))(c'') = v(c' \otimes c'')(g')$$

pour tout $g' \in G'$. Compte tenu des définitions des bords et des cobords, on déduit de là des homomorphismes de modules gradués

$$\alpha_* : H.((C.' \otimes C.'') \otimes \mathrm{Hom}(G', G'')) \longrightarrow \mathrm{Hom}(H^{\cdot}(C'^{\cdot}; G'), H.(C.'' \otimes G''))$$
$$\beta^* : H^{\cdot}((C' \otimes C'')^{\cdot}; \mathrm{Hom}(G', G'')) \longrightarrow \mathrm{Hom}(H.(C.' \otimes G'), H^{\cdot}(C''^{\cdot}; G''))$$

où l'on a posé

$$C'^{\cdot} = \mathrm{Hom}(C.', G'), C''^{\cdot} = \mathrm{Hom}(C.'', G''),$$
$$(C' \otimes C'')^{\cdot} = \mathrm{Hom}(C.' \otimes C.'', \mathrm{Hom}(G', G'')).$$

Prenant $C.' = S.(X; A)$, $C.'' = S.(Y; A)$, on obtient, en tenant compte du théorème d'Eilenberg-Zilber, deux homomorphismes, dits «produits obliques» («slant products»)

$$/ H_n(X \times Y; \mathrm{Hom}(G', G'')) \longrightarrow \mathrm{Hom}(H^p(X; G'), H_{n-p}(Y; G''))$$

où l'on écrit $z_n / z'^p \in H_{n-p}(Y; G'')$ pour $z_n \in H_n(X \times Y; \mathrm{Hom}(G', G''))$, $z'^p \in H^p(X; G')$,

$$\backslash H^n(X \times Y; \mathrm{Hom}(G', G'')) \longrightarrow \mathrm{Hom}(H_p(X; G'), H^{n-p}(Y; G''))$$

où l'on écrit $z'^n \backslash z_p \in H^{n-p}(Y; G'')$ pour $z'^n \in H^n(X \times Y; \mathrm{Hom}(G', G''))$, $z_p \in H_p(X; G')$.

12.2 Cup-produit et algèbre de cohomologie

En 1942, Lefschetz a observé que la «fonctorialité» de la cohomologie à coefficients dans un anneau A, implique que l'application diagonale $\delta : x \mapsto (x,x)$ de X dans $X \times X$, pour un espace topologique quelconque X, définit un homomorphisme canonique de A-modules

$$\delta^* : H^{\cdot}(X \times X; A) \longrightarrow H^{\cdot}(X; A). \tag{76}$$

Mais comme on a des produits «cartésiens»

$$\times : H^p(X; A) \times H^q(X; A) \longrightarrow H^{p+q}(X \times X; A)$$

on obtient, par composition avec (76), des applications bilinéaires canoniques

$$\smile : H^p(X; A) \times H^q(X; A) \longrightarrow H^{p+q}(X; A) \tag{77}$$

qu'on appelle les *cup-produits*, et on note $z^p \smile z^q$ l'image par (77) de (z^p, z^q). Autrement dit, ces applications définissent sur le A-module gradué $H^{\cdot}(X; A) = \bigoplus_{p \geq 0} H^p(X; A)$ une structure de A-algèbre graduée, dite *algèbre de cohomologie singulière* de X à coefficients dans A.

On remarque en outre que l'on a des diagrammes commutatifs

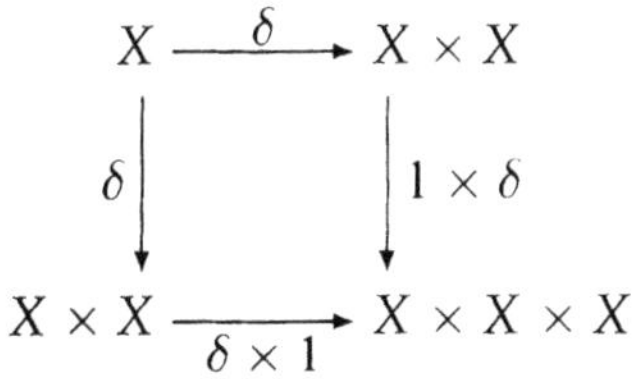

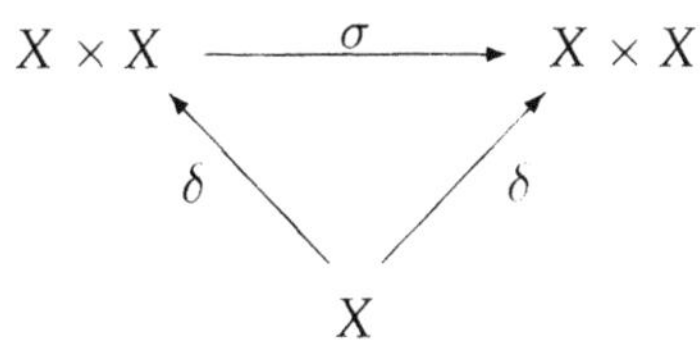

où $X \times X \times X$ est canoniquement identifié à $X \times (X \times X)$ et à $(X \times X) \times X$; on pose $(1 \times \delta)(x,y) = (x, \delta(y))$, $(\delta \times 1)(x,y) = (\delta(x), y)$ et $\sigma(x,y) = (y,x)$.

Cela entraîne par fonctorialité les propriétés d'*associativité* et d'*anticommutativité* pour l'algèbre graduée $H^{\cdot}(X; A)$

$$(z^p \smile z^q) \smile z^r = z^p \smile (z^q \smile z^r) \quad z^q \smile z^p = (-1)^{pq}(z^p \smile z^q).$$

En outre, si $f : X \longrightarrow Y$ est une application continue, on a un diagramme commutatif

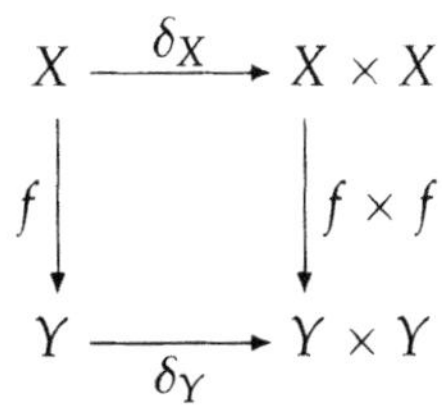

qui, en cohomologie, fournit la relation

$$f^*(z^p \smile z^q) = f^*(z^p) \smile f^*(z^q)$$

autrement dit, $f^* : H^\cdot(X;A) \longrightarrow H^\cdot(Y;A)$ est un homomorphisme d'*algèbres graduées*.

Le fait que l'on peut définir un produit canonique dans le module gradué de cohomologie d'un complexe simplicial *quelconque*, avait été découvert dès 1935 par Alexander et Kolmogoroff (indépendamment), et avait causé une vive surprise, allant à l'encontre de l'idée que l'on ne pouvait obtenir un tel produit que par dualité, à partir de l'anneau d'intersection. Leurs définitions partaient directement des cochaînes; elles furent améliorées par Čech et Whitney, et s'appliquent aussi bien à la cohomologie singulière. Pour $f \in S^p(X;A)$, $g \in S^q(X;A)$, on définit $(f \smile g)(s^{p+q})$ pour un simplexe singulier $s^{p+q} : \bar{\Delta}_{p+q} \longrightarrow X$ par la formule

$$(f \smile g)(s^{p+q}) = f(s^p)g(s^q)$$

où on a posé

$$\begin{aligned} &s^p = s^{p+q} \circ \sigma_p, s^q = s^{p+q} \circ \sigma_q \text{ avec } \sigma_p(e_i) = e_i \text{ pour } 0 \leq i \leq p, \\ &\sigma_q(e_j) = e_{p+j} \text{ pour } 0 \leq j \leq q. \end{aligned} \tag{78}$$

On vérifie que

$$d_{p+q}(f \smile g) = d_p f \smile g + (-1)^p f \smile d_q g$$

ce qui prouve que lorsque f et g sont des cocycles, $f \smile g$ est un cocycle dont la classe de cohomologie ne dépend que de celles de f et g; d'où la définition explicite du cup-produit.

12.3 Cap-produit et dualités

L'application diagonale $\delta : x \mapsto (x,x)$ donne en homologie singulière des homomorphismes

$$\delta_* : H.(X;A) \longrightarrow H.(X \times X;A)$$

ce qui, par composition avec le produit oblique

$$/ : H_{p+q}(X \times X;A) \longrightarrow \mathrm{Hom}(H^p(X;A), H_q(X;A))$$

définit une application linéaire

$$H_{p+q}(X;A) \longrightarrow \mathrm{Hom}(H^p(X;A), H_q(X;A)).$$

Pour $z \in H_{p+q}(X;A)$ et $w \in H^p(X;A)$, cela définit donc une classe

$$z \frown w \in H_q(X;A)$$

que l'on appelle le *cap-produit* de la classe d'homologie z et de la classe de cohomologie w.

La définition du cap-produit fut d'abord donnée indépendamment par Čech (1936) et Whitney (1938) par une méthode directe partant d'une application

$$\frown : S_{p+q}(X;A) \times S^p(X;A) \longrightarrow S_q(X;A);$$

pour $f \in S^p(X;A)$ et $s^{p+q} : \bar{\Delta}_{p+q} \longrightarrow X$, on prend, avec les notations de (78)

$$s^{p+q} \frown f = f(s^p)s^q$$

et le bord est donné par

$$b_p(z \frown f) = (-1)^p(b_{p+q}z \frown f) + (-1)^{p+1}(z \frown d_p f).$$

Cela montre que si z est un cycle et f un cocycle, $z \frown f$ est un cycle dont la classe d'homologie ne dépend que des classes de z et f, d'où la définition du cap-produit. On vérifie en outre que, pour $a \in H_{p+q}(X)$, $u \in H^p(X)$, $v \in H^q(X)$, et $c \in H_{p+q+r}(X)$, on a

$$\langle a \frown u, v\rangle = \langle a, u \smile v\rangle$$
$$c \frown (u \frown v) = (c \frown u) \frown v.$$

Si $q = 0$, et si X est connexe par arcs, $H_0(X;A)$ s'identifie à A et l'on écrit $\langle a, u\rangle \in A$ au lieu de $a \frown u$.

Enfin, pour toute application continue $g : X \longrightarrow Y$, on a

$$g_*(a \frown g^*(u)) = g_*(a) \frown u.$$

Soit X une variété C^0 de dimension n, compacte et orientée; on a encore alors un isomorphisme canonique $H_n(X;\mathbb{Z}) \simeq \mathbb{Z}$ et on appelle encore classe fondamentale un générateur $[X]$ de ce groupe d'homologie. Le cap-produit permet alors de définir la dualité de Poincaré pour X de façon générale, en homologie et cohomologie singulières: le cap-produit par la classe fondamentale

$$z^p \mapsto z^p \frown [X]$$

est un *isomorphisme*

$$j_p : H^p(X;A) \simeq H_{n-p}(X;A) \tag{79}$$

pour tout anneau de coefficients A. Lorsque A est un corps, $H^{\cdot}(X;A)$ et $H_{\cdot}(X;A)$ sont des espaces vectoriels de dimension finie, et $H^p(X;A)$ est le dual de l'espace vectoriel $H_p(X;A)$; donc (79) redonne la forme initiale de la dualité de Poincaré.

Si l'on prend $A = \mathbb{F}_2$, l'isomorphisme (79) est encore valable pour les variétés compactes non orientables. Pour les variétés non compactes, on a un isomorphisme généralisant (79)

$$H_c^p(X;A) \simeq H_{n-p}(X;A).$$

Utilisant (79), on définit sur le module $H.(X;A)$ une structure d'anneau gradué en posant

$$z_p \cdot z_q = j_{2n-p-q}(j_{n-p}^{-1}(z_p) \smile j_{n-q}^{-1}(z_q))$$

et Freudenthal a montré en 1937 que ce produit coïncide avec le produit d'intersection défini par Lefschetz, de sorte que pour ce produit, $H.(X;A)$ est une algèbre graduée isomorphe à $H^{\cdot}(X;A)$.

La dualité d'Alexander se généralise de la même manière à l'aide du cap-produit, et en utilisant la cohomologie d'Alexander-Spanier: pour toute partie fermée Y d'une variété orientée de dimension n, il y a un isomorphisme canonique

$$\bar{H}_c^p(Y;A) \simeq H_{n-p}(X, X-Y;A)$$

où le second membre est l'homologie singulière relative.

La cohomologie des faisceaux et l'homologie de Borel-Moore donnent des résultats plus généraux de dualité pour les variétés généralisées.

12.4 H-espaces et algèbres de Hopf

En 1935, R. Brauer avait déterminé l'homologie des groupes de Lie compacts *classiques*. E. Cartan observa que, pour chacun de ces groupes, l'homologie est la même que celle d'un produit de sphères

$$\mathbb{S}_{m_1} \times \mathbb{S}_{m_2} \times \ldots \times \mathbb{S}_{m_\ell}$$

où les dimensions m_k dépendent du groupe considéré, mais sont toujours des entiers impairs, et ℓ est égal au rang du groupe (dimension d'un tore maximal); il conjectura que ces coïncidences devaient provenir d'une propriété générale.

C'est cette propriété que découvrit H. Hopf en 1939, dans un mémoire qui introduisait des idées nouvelles très originales. Sa méthode (où il ne fait pas intervenir la cohomologie) peut être simplifiée en utilisant la «fonctorialité» de la cohomologie. On considère un espace topologique X sur lequel est définie une loi de composition continue $m : X \times X \longrightarrow X$. Cela entraîne l'existence d'un homomorphisme d'algèbres graduées de cohomologie singulière

$$m^* : H^{\cdot}(X;A) \longrightarrow H^{\cdot}(X \times X;A).$$

Hopf ne considère que le cas où l'anneau des coefficients est un corps K. Comme les K-modules sont alors libres, la «formule de Künneth» (49) pour les complexes de chaînes montre que l'algèbre de cohomologie $H^{\cdot}(X \times X;K)$ est isomorphe au «produit tensoriel gauche» d'algèbres graduées[13])

$$H^{\cdot}(X;K)^{g} \otimes H^{\cdot}(X;K).$$

On obtient donc un homomorphisme d'algèbres graduées

$$m^{*} : H^{\cdot}(X;K) \longrightarrow H^{\cdot}(X;K)^{g} \otimes H^{\cdot}(X;K). \tag{80}$$

De tels homomorphismes sont appelés *coproduits*.

Jusque là aucune restriction n'est imposée à la loi m. Hopf était naturellement intéressé au cas où il s'agit d'une loi de groupe; mais il s'aperçut que l'on pouvait obtenir la structure de $H^{\cdot}(X;K)$ à l'aide de l'existence d'un coproduit, en supposant que m possède des propriétés bien moins contraignantes que les axiomes des groupes, à savoir:

1) Il existe un élément $e \in X$ («unité homotopique») tel que $m(e,e) = e$.

2) Pour tout $x \in X$, les applications $x \mapsto m(e,x)$ et $x \mapsto m(x,e)$ sont homotopes à l'identité id_X.

Les espaces topologiques munis d'une telle loi de composition sont maintenant appelés *H-espaces*. On déduit des axiomes 1) et 2) que pour tout élément homogène z de $H^{\cdot}(X;K)$, on peut écrire

$$m^{*}(z) = 1 \otimes z + z \otimes 1 + \sum_{i,j} x_i \otimes y_j \tag{81}$$

où les x_i et les y_j sont des éléments homogènes (non uniquement déterminés par z) tels que $\deg(x_i) \geq 1$, $\deg(y_j) \geq 1$ et $\deg(x_i) + \deg(y_j) = \deg(z)$.

Le problème devient alors purement algébrique: quelle est la structure d'une algèbre graduée M sur un corps K, associative, anticommutative et ayant un élément unité (de degré 0), lorsqu'on suppose en outre qu'il existe un coproduit $c : M \longrightarrow M^{g} \otimes M$ pour lequel les relations (81) sont vérifiées? Ces algèbres s'appellent maintenant *algèbres de Hopf*, et jouent un grand rôle en Topologie algébrique et en théorie des groupes.

Lorsque K est un corps de caractéristique 0 et que M est de dimension finie sur K, Hopf prouva que M contient un espace vectoriel E ayant une base formée d'éléments homogènes $x_1, \ldots, x_\ell$ de degrés impairs $m_1, \ldots, m_\ell$, et tels que M soit isomorphe à l'algèbre extérieure $\wedge(E)$. En vertu de la formule de Künneth,

13) Si M et N sont deux algèbres graduées, le produit tensoriel gauche $M^{g} \otimes N$ a pour module sous-jacent le produit tensoriel $M \otimes N$, mais le produit de deux éléments $x_p \otimes y_q$, $x_r \otimes y_s$ où x_p, y_q, x_r, y_s sont homogènes de degrés respectifs p, q, r, s, est donné par

$$(x_p \otimes y_q) \cdot (x_r \otimes y_s) = (-1)^{qr}(x_p x_r) \otimes (y_q y_s).$$

cela explique déjà la coïncidence observée par E. Cartan, et la généralise à tous les H-espaces compacts.

Un peu plus tard, H. Samelson prouva que lorsque X est un groupe compact, il y a un choix canonique pour les x_j: on peut supposer que ce sont des éléments primitifs, c'est-à-dire ceux pour lesquels

$$c(z) = 1 \otimes z + z \otimes 1.$$

Utilisant ce résultat, Hopf donna une interprétation purement topologique du nombre ℓ de ces éléments x_j: pour deux applications continues u, v de X dans lui-même, on montre que l'application $w : x \mapsto u(x)v(x)$ est telle qu'en cohomologie on a $w^*(z) = u^*(z) + v^*(z)$ pour tout élément primitif z. Par suite, si p_k désigne l'application $x \mapsto x^k$ dans le groupe X, p_k^* est la multiplication par l'entier k^ℓ dans le groupe additif $H^n(X;K)$ (où $n = \dim X$); cela s'interprète en disant que k^ℓ est le *degré* de l'application p_k pour tout entier k.

Leray et A. Borel ont généralisé le théorème de Hopf au cas où K est un corps de caractéristique quelconque et où on suppose seulement que chacun des espaces vectoriels M_n est de dimension finie (mais il peut y avoir une infinité d'espaces M_n d'éléments homogènes, non réduits à 0).

13 Constructions topologiques

Vers 1935 débute une toute nouvelle partie de la Topologie algébrique, ayant pour base des notions qui s'introduisent pour la première fois, et dont les deux plus fondamentales sont la théorie algébrique de l'homotopie et la théorie des fibrations. Elles vont peu à peu occuper le devant de la scène, les théories homologiques leur cédant la place pour conserver un rôle indispensable d'outil algébrique dans ces nouvelles théories.

Mais en liaison avec cette expertise algébrique se réveille l'imagination géométrique, qui sort du cadre rigide des simplexes et de leurs combinaisons. La Topologie générale, parvenue à maturité, va être exploitée avec profit pour permettre, à partir des espaces que l'on étudie, de fabriquer de nouveaux espaces dont les propriétés se répercutent sur le problème initialement considéré. On commence ainsi à utiliser les espaces fonctionnels et leur topologie. Surtout, la notion d'espace quotient, qui acquiert sa définition précise après mainte tribulation, permet de justifier et de généraliser les «recollements» que s'autorisaient «intuitivement» à faire les pionniers du 19ᵉ siècle (qui se limitaient en général aux parties d'espaces $\mathbb{R}^n$).

13.1 Contraction et bouquet

L'exemple le plus simple d'espace quotient est ce qu'on obtient à partir d'un espace topologique X en «contractant» une partie Y de X en un seul point. Il suffit de considérer la partition de X formée de Y et des «singletons» $\{x\}$ pour $x \in X - Y$. On note souvent X/Y cette «contraction» de Y, correspondant à la partition précédente. Si Y est fermé dans X, l'application canonique $p : X \longrightarrow X/Y$ est un homéomorphisme de $X - Y$ sur l'ensemble ouvert de X/Y complémentaire du point $p(Y)$.

Un cas particulier est le «bouquet» d'une famille quelconque (X_α, x_α) d'espaces pointés. Dans l'espace «somme» des X_α, réunion disjointe des X_α, soit W l'ensemble des x_α. L'espace Z/W obtenu par contraction de W en un point est ce qu'on appelle le «bouquet» des espaces pointés (X_α, x_α). Si les espaces X_α sont séparés, il en est de même de Z/W; l'image $p(X_\alpha)$ de X_α par l'application canonique $p : Z \longrightarrow Z/W$ est alors un sous-espace fermé de Z/W et la restriction de p à X_α est un homéomorphisme de X_α sur $p(X_\alpha)$.

Lorsqu'il s'agit de deux espaces pointés (X, x_0), (Y, y_0), le bouquet de ces deux espaces est homéomorphe au sous-espace $(\{x_0\} \times Y) \cup (X \times \{y_0\})$ du produit $X \times Y$; on le note encore $X \vee Y$ («wedge»). La contraction $(X \times Y)/(X \vee Y)$ de $X \vee Y$ en un point se note $X \wedge Y$ et s'appelle le «produit écrasé» (smash product) de X et Y.

13.2 Jonction et suspension

Poincaré, dans le second «Complément» de 1900, avait déjà songé à définir la «jonction» de deux sous-espaces X, Y d'un $\mathbb{R}^n$, comme réunion des segments de droite joignant un point de X et un point de Y. On peut définir de manière générale la *jonction* («join») $X * Y$ de deux espaces topologiques quelconques, par la construction suivante. On considère le produit $Z = X \times Y \times [0, 1]$ et la partition de Z formée des singletons $\{(x, y, t)\}$ pour $x \in X$, $y \in Y$ et $0 < t < 1$, des ensembles $\{x\} \times Y \times \{0\}$, pour tout $x \in X$ et des ensembles $X \times \{y\} \times \{1\}$ pour tout $y \in Y$; $X * Y$ est l'espace quotient correspondant. Si $p : Z \longrightarrow X * Y$ est l'application canonique, l'image $p(X \times Y \times \{0\})$ (resp. $p(X \times Y \times \{1\})$ est un sous-espace fermé de $X * Y$ homéomorphe à X (resp. Y) avec lequel on l'identifie en général.

Dans le cas particulier où $Y = \mathbb{S}_0 = \{-1, 1\}$, $X * \mathbb{S}_0$ est appelé la *suspension* de X, introduite par H. Freudenthal en 1937 et notée SX; il joue un rôle important en théorie de l'homotopie. On montre qu'il est homéomorphe au produit écrasé $\mathbb{S}_1 \wedge X$. Pour $x_0 \in X$, le sous-espace $L = p(\{x_0\} \times \mathbb{S}_0 \times [0, 1])$ est homéomorphe à l'intervalle $[0, 1]$. La contraction SX/L de L en un point identifié à x_0 est appelée la *suspension réduite* de (X, x_0) et notée ΣX. On voit aisément que si $X = \mathbb{S}_n$, SX et ΣX sont homéomorphes à $\mathbb{S}_{n+1}$ pour tout $n \geq 0$.

13.3 Attachement d'un espace

La notion d'attachement d'un espace à un autre a été introduite par J.H.C. Whitehead en 1939 et utilisée par lui avec une grande virtuosité. Soient X, Y deux espaces topologiques, A une partie non vide de X et $f : A \longrightarrow Y$ une application continue. Dans l'espace «somme» Z de X et Y, on considère la partition formée des singletons $\{x\}$ pour $x \in X - A$ et les ensembles $\{y\} \cup f^{-1}(y)$ pour $y \in Y$ (ensemble réduit à y si $y \notin f(A)$). L'espace quotient correspondant est l'*attachement de X à Y le long de A* au moyen de f (dite *application d'attachement*); il est parfois noté $X \cup_f Y$. Si $p : Z \longrightarrow X \cup_f Y$ est l'application canonique, la restriction de p à Y est un homéomorphisme de Y sur $p(Y)$, avec lequel on identifie Y. Si A est fermé dans X, la restriction de p à $X - A$ est de même un homéomorphisme de $X - A$ sur $(X \cup_f Y) - p(Y)$, et on l'identifie aussi à $X - A$. Si X et Y sont des espaces normaux et si A est fermé dans X, $X \cup_f Y$ est aussi un espace normal.

13.4 Les CW-complexes

La définition des complexes cellulaires de Poincaré impose des conditions très strictes aux frontières des cellules, et on a cherché à utiliser des «décompositions cellulaires» moins contraignantes. Déjà Ehresmann en 1934 avait utilisé une telle décomposition dans sa détermination de l'homologie des grassmanniennes. C'est J.H.C. Whitehead qui, en 1941, introduisit, en vue de ses recherches sur l'homotopie, ce qu'il appela les CW-complexes, qui allient une grande flexibilité dans les applications à des propriétés très simples de leur homologie, et ont acquis une grande utilité dans toute la Topologie algébrique.

La définition d'un CW-complexe est très simple. C'est un espace séparé K, dans lequel on a défini une partition $(E_0, E_1, \dots, E_n, \dots)$, finie ou dénombrable. Chaque E_n non vide est lui-même réunion disjointe d'une famille $(e_j^n)_{j \in J_n}$ (où J_n est un ensemble quelconque); ce sont les *n-cellules* de K. Pour chaque e_j^n est donnée une application continue (dite *application d'attachement*) $f_j^n : \mathbb{D}_n \longrightarrow K$, et on impose à ces applications les conditions suivantes:

1) $f_j^n(\mathbb{D}_n) = \overline{e_j^n}$ (adhérence dans K, qui est donc compacte);
2) f_j^n, restreinte à la boule ouverte $\overset{o}{\mathbb{D}}_n$, est un homéomorphisme de $\overset{o}{\mathbb{D}}_n$ sur e_j^n ;
3) si on pose $K_{n-1} = \bigcup_{m \leq n-1} e_j^m$ (le $(n-1)$-*squelette*), alors $f_j^n(\mathbb{S}_{n-1}) \subset K_{n-1}$, et $f_j^n(\mathbb{S}_{n-1})$ ne rencontre qu'un nombre fini de m-cellules pour $m \leq n-1$.

Le n-squelette K_n est défini comme espace topologique obtenu par *attachement* de la réunion disjointe d'une famille $(\mathbb{D}_{n,j})_{j \in J}$ d'espaces homéomorphes à $\mathbb{D}_n$ au moyen des f_j^n, le long de la réunion des sphères $\mathbb{S}_{n,j}$.

Le CW-complexe K est dit *fini* s'il n'a qu'un nombre fini de cellules (donc est identitque à un K_n); sinon, la topologie de K est définie de la manière suivante:

4) Les ensembles fermés dans K sont les ensembles F tels que $F \cap \overline{e_j^n}$ soit fermé dans $\overline{e_j^n}$ pour toutes les cellules.

Dans cette définition, $E_0 = K_0$ est donc un espace discret quelconque dont les e_j^0 sont les points; mais on peut aussi prendre pour K_0 un espace séparé quelconque Y et définir ensuite les K_n pour $n \geq 1$ comme ci-dessus; on obtient ainsi un CW-complexe *relatif* à Y; on note parfois un tel espace (X, Y) et le n-squelette se note alors $(X, Y)^n$.

Comme exemples simples de CW-complexes, on a la sphère $\mathbb{S}_n$, où il y a une seule 0-cellule e^0 et une seule n-cellule e^n, la fonction d'attachement f^n étant l'application constante $\mathbb{S}_{n-1} \longrightarrow \{e^0\}$. Le plan projectif $\mathbb{P}_2(\mathbb{R})$ s'obtient par attachement d'une seule 2-cellule e^2 au cercle $\mathbb{S}_1$. La fonction d'attachement fait correspondre au point $e^{i\theta}$ de $\mathbb{S}_1$ le point $e^{2i\theta}$ pour $0 \leq \theta < 2\pi$.

14 L'aspect algébrique de l'homotopie

14.1 Les types d'homotopie

A partir de 1930, la notion d'homotopie est étudiée pour elle-même, alors que jusque là elle était utilisée comme outil auxiliaire permettant le calcul de l'homologie d'espaces particuliers.

La partie essentielle de la théorie se restreint à la catégorie des espaces pointés: les objets de cette catégorie sont les couples (X, x_0) formés d'un espace topologique non vide X et d'un point $x_0 \in X$; un morphisme $f : (X, x_0) \longrightarrow (Y, y_0)$ est une application continue f de X dans Y telle que $f(x_0) = y_0$. Une homotopie entre deux tels morphismes f, g est une application continue $F : (X, x_0) \times [0, 1] \longrightarrow (Y, y_0)$ telle que $F(x, 0) = f(x)$, $F(x, 1) = g(x)$ et $F(t, x_0) = y_0$ pour $0 \leq t \leq 1$. L'existence d'une telle application définit une relation d'équivalence dans l'ensemble $\mathscr{C}(X, x_0; Y, y_0)$ des morphismes, que l'on note $f \sim g$. L'objet principal d'études de la théorie est l'ensemble $[X, x_0; Y, y_0]$ (noté $[X; Y]$ si aucune confusion n'en résulte) des classes d'équivalence pour cette relation. Si f_1, f_2 appartiennent à $\mathscr{C}(X, x_0; Y, y_0)$, g_1, g_2 à $\mathscr{C}(Y, y_0; Z, z_0)$ (resp. à $\mathscr{C}(Z, z_0; X, x_0)$), alors les relations $f_1 \sim f_2$, $g_1 \sim g_2$ entraînent $g_1 \circ f_1 \sim g_2 \circ f_2$ (resp. $f_1 \circ g_1 \sim f_2 \circ g_2$).

En 1935, W. Hurewicz introduisit l'importante notion d'*équivalence d'homotopie*: c'est un morphisme $f : (X, x_0) \longrightarrow (Y, y_0)$ tel qu'il existe un morphisme $g : (Y, y_0) \longrightarrow (X, x_0)$ pour lequel $g \circ f \sim \mathrm{id}_X$ et $f \circ g \sim \mathrm{id}_Y$; g est alors une équivalence d'homotopie, inverse homotopique de f. De là découle la notion de *type d'homotopie*: deux espaces (X, x_0), (Y, y_0) ont même type d'homotopie s'il existe une équivalence d'homotopie $f : (X, x_0) \longrightarrow (Y, y_0)$; ils ont alors mêmes modules d'homologie et de cohomologie pour toutes les théories homologiques vérifiant les axiomes d'Eilenberg, mais non inversement, de sorte que la classification par type d'homotopie *raffine* la classification par l'homologie. Deux espaces homéomorphes ont même type d'homotopie, mais la réciproque n'est pas exacte. Un espace qui a le type d'homotopie d'un espace réduit à un seul point est appelé *rétractile*; il en est ainsi de tous les espaces $\mathbb{R}^n$.

Un exemple important d'équivalence d'homotopie est la *rétraction par déformation*, définie par Borsuk en 1933. C'est une application continue $r : X \longrightarrow Y$ d'un espace topologique X dans un sous-espace Y, pour laquelle il existe une application continue $h : X \times [0, 1] \longrightarrow X$ telle que $h(x, 0) = x$, $h(x, 1) = r(x)$ pour $x \in X$ et $h(y, t) = y$ pour $y \in Y$ et $0 \leq t \leq 1$. On dit alors que Y est rétracte par déformation de X; si $i : Y \longrightarrow X$ est l'inclusion, on a $r \circ i = \mathrm{id}_Y$ et $i \circ r \sim \mathrm{id}_X$ donc r est une équivalence d'homotopie. Pour tout espace Z tel que X soit un sous-espace de Z, on a alors $H_\cdot(Z, X) \simeq H_\cdot(Z, Y)$ et $H^\cdot(Z, X) \simeq H^\cdot(Z, Y)$.

Cette notion permet en particulier d'étudier l'homologie des CW-complexes. Soit M une partie du n-squelette K_n d'un CW-complexe K, telle que chaque intersection $M \cap e_j^n$ soit réduite à un seul point z_j (par exemple l'image par l'application d'attachement du centre 0 de $\mathbb{D}_n$), et que M soit la réunion des z_j. Alors K_{n-1} est rétracte par déformation de $K_n - M$: il suffit de noter que l'application continue $(x, t) \mapsto t\frac{x}{|x|} + (1 - t)x$ de $(\mathbb{D}_n - \{0\}) \times [0, 1]$ dans $\mathbb{D}_n$

rétracte $\mathbb{D}_n - \{0\}$ sur $\mathbb{S}_{n-1}$. On a donc $H.(K_n, K_n - M) \simeq H.(K_n, K_{n-1})$. Mais $E_n = K_n - K_{n-1}$ est ouvert et réunion disjointe des e_j^n ($j \in J_n$), donc on a, par excision, $H.(K_n, K_n - M) \simeq H.(E_n, E_n - M)$. Or, $H_p(e_j^n, e_j^n - \{z_j\}) = 0$ pour $p \neq n$, et $H_n(e_j^n, e_j^n - \{z_j\}) \simeq \mathbb{Z}$, d'où finalement

$$H_p(K_n, K_{n-1}) = 0 \text{ pour } p \neq n, H_n(K_n, K_{n-1}) \simeq \mathbb{Z}^{(J_n)} \text{ (somme directe).} \quad (82)$$

Par la suite exacte d'homologie, on tire de là que si K est un CW-complexe fini, et si m_p est le nombre de ses p-cellules, on a, pour tout corps F,

$$\dim_F(H_p(K;F)) \leq m_p \text{ et } \chi(K) = \sum_p (-1)^p m_p. \quad (83)$$

Un autre exemple très utile est fourni par la notion de *cylindre* d'une application continue $f : X \longrightarrow Y$ («mapping cylinder») introduite par J.H.C. Whitehead en 1939. C'est l'ensemble Z_f obtenu en attachant $X \times [0,1]$ à Y le long de $X \times \{1\}$, par l'application d'attachement $(x,1) \mapsto f(x)$; si $p : (X \times [0,1]) \cup Y \longrightarrow Z_f$ est l'application canonique, X (resp. Y) est identifié à un sous-espace fermé de Z_f par l'application $i : x \mapsto p(x,0)$ (resp. $y \mapsto p(y)$). On considère alors l'application continue $h : Z_f \times [0,1] \longrightarrow Z_f$ définie par $h(p(x,t),u) = p(x,(1-t)u+u)$ pour $x \in X$, $0 \leq t \leq 1$, $0 \leq u \leq 1$, $h(p(y),u) = p(y)$ pour $y \in Y$, $0 \leq u \leq 1$. Alors $r(z) = h(z,1)$ est une rétraction par déformation de Z_f sur Y, et on a une factorisation

$$f : X \xrightarrow{i} Z_f \xrightarrow{r} Y \quad (84)$$

en une inclusion i et une équivalence d'homotopie r.

14.2 Espaces de lacets et groupes d'homotopie

La définition du groupe fondamental $\pi_1(X, x_0)$, donnée par Poincaré, peut s'exprimer en disant que c'est l'ensemble $[\mathbb{S}_1, *; X, x_0]$ des classes d'homotopie des lacets d'origine x_0, muni de la loi de groupe déduite de la juxtaposition des lacets[14]). Il était naturel de chercher si une telle loi de groupe peut aussi se définir sur l'ensemble $[\mathbb{S}_n, *; X, x_0]$ pour $n > 1$, ce qui donnerait des groupes d'homotopie $\pi_n(X, x_0)$. Il n'est pas difficile de définir une telle loi en imitant le procédé de Poincaré, et cela fut effectivement fait par Dehn (qui ne le publia pas), et par Čech, qui se contenta d'une courte note en 1932; c'est que les groupes obtenus étaient *commutatifs*, et l'on pensait alors qu'une généralisation significative devait donner des groupes quelconques.

Sans tenir compte de cette objection, W. Hurewicz reprit le problème en 1935 d'une façon beaucoup plus féconde. Son idée est de partir de l'ensemble $\Omega(X, x_0)$ des lacets et *non* de leurs classes d'homotopie, et de munir cet ensemble d'une topologie naturelle. L'ensemble $\Omega(X, x_0)$ est une partie de l'ensemble $\mathscr{C}(\mathbb{S}_1; X)$

14) Pour toutes les sphères $\mathbb{S}_n$, on note $*$ le point de coordonnées $(1,0,\ldots,0)$ dans $\mathbb{R}^{n+1}$.

de toutes les applications continues de $\mathbb{S}_1$ dans X. Pour tout couple X, Y d'espaces topologiques, l'ensemble $\mathscr{C}(Y; X)$ des applications continues de Y dans X est muni de la topologie *compacte-ouverte*, dont les ouverts sont les réunions quelconques d'ensembles de la forme $T(K, U)$, où K est une partie compacte de Y, U une partie ouverte de X, et $T(K, U)$ est l'ensemble des applications continues $f : Y \longrightarrow X$ telles que $f(K) \subset U$.

Le point de départ de Hurewicz est une propriété fondamentale de cette topologie: si Y est localement compact et si Z est un espace séparé, on peut associer à toute application $f \in \mathscr{C}(Z \times Y; X)$ l'application $\tilde{f} \in \mathscr{C}(Z; \mathscr{C}(Y; X))$ définie par

$$\tilde{f}(z)(y) = f(z, y) \tag{85}$$

et on montre que l'application $f \mapsto \tilde{f}$ est alors un *homéomorphisme*.

On munit $\Omega(X, x_0) = \mathscr{C}(\mathbb{S}_1, *; X, x_0)$ de la topologie compacte-ouverte. Hurewicz définit alors par récurrence

$$\Omega^1(X, x_0) = \Omega(X, x_0), \Omega^n(X, x_0) = \Omega(\Omega^{n-1}(X, x_0), x_{n-1}) \text{ pour } n \geq 2 \tag{86}$$

où x_{n-1} est le point de $\Omega^{n-1}(X, x_0)$ défini par récurrence comme étant l'application constante $\mathbb{S}_1 \longrightarrow \{x_{n-2}\}$. Le n-ème groupe d'homotopie de (X, x_0) est alors défini par

$$\pi_n(X, x_0) = \pi_1(\Omega^{n-1}(X, x_0), x_{n-1}). \tag{87}$$

La propriété fondamentale de la topologie compacte-ouverte fournit une autre définition de ce groupe. Soit B la frontière dans $\mathbb{R}^n$ du cube $[0, 1]^n$, réunion des faces du cube. Alors l'espace $\Omega^n(X, x_0)$ des lacets itérés est canoniquement isomorphe au sous-espace de $\mathscr{C}([0, 1]^n; X)$ formé des applications f telles que $f(B) = \{x_0\}$. C'est évident par définition pour $n = 1$; la propriété fondamentale de la topologie compacte-ouverte définit un homéomorphisme $f \mapsto \tilde{f}$ de $\mathscr{C}([0, 1]^n; X)$ sur $\mathscr{C}([0, 1]; \mathscr{C}([0, 1]^{n-1}; X)$, où

$$\tilde{f}(t)(t_1, \ldots, t_{n-1}) = f(t, t_1, \ldots, t_{n-1}) \tag{88}$$

et si $f(B) = \{x_0\}$, on a $\tilde{f}(0) = \tilde{f}(1) = x_0$. On déduit de là des homéomorphismes canoniques

$$\Omega^{m+n}(X, x_0) \cong \mathscr{C}(\mathbb{S}_m, *; \Omega^n(X, x_0), x_n) \tag{89}$$

$$\begin{aligned} \pi_n(X, x_0) &\simeq \ldots \simeq \pi_{n-p}(\Omega^p(X, x_0), x_p) \simeq \ldots \\ &\simeq \pi_1(\Omega^{n-1}(X, x_0), x_{n-1}) \simeq \pi_0(\Omega^n(X, x_0)) \end{aligned} \tag{90}$$

où pour un espace topologique Z, $\pi_0(Z)$ désigne l'ensemble des composantes connexes par arcs de Z.

Si X est un H-espace, le groupe fondamental $\pi_1(X, e)$ (où e est l'«unité homotopique» de X) est *commutatif*. En effet, pour deux chemins $t \mapsto \alpha_1(t)$, $t \mapsto \alpha_2(t)$ dans X, notons $\alpha_1 * \alpha_2$ le chemin $t \mapsto m(\alpha_1(t), \alpha_2(t))$; si γ_1' (resp. γ_2')

est homotope à γ_1 (resp. γ_2), $\gamma_1' * \gamma_2'$ est homotope à $\gamma_1 * \gamma_2$. Si on identifie les points de X aux chemins constants, et si γ_1 est un chemin de e à a, γ_2 un chemin de e à b, $\gamma_1 * \gamma_2$ est homotope à chacune des juxtapositions $(\gamma_1 * e) \vee (a * \gamma_2)$ et $(e * \gamma_2) \vee (\gamma_1 * b)$. Pour $a = b = e$ tout lacet γ d'origine e peut être identifié à $\gamma * e$ et à $e * \gamma$; on en conclut que la loi de composition $([\gamma_1], [\gamma_2]) \mapsto [\gamma_1 \vee \gamma_2]$ est identique à la loi $([\gamma_1], [\gamma_2]) \mapsto [\gamma_1 * \gamma_2]$ et que cette loi est commutative.

D'autre part, il résulte aussitôt des définitions que pour tout espace pointé (X, x_0), l'espace des lacets $\Omega(X, x_0)$ est un H-espace pour la juxtaposition (l'«unité homotopique» e étant le lacet constant $[0,1] \mapsto \{x_0\}$). On conclut donc de la définition (87) que $\pi_n(X, x_0)$ est *commutatif* pour $n \geq 2$.

14.3 Homotopie relative et suite exacte d'homotopie

En 1938, Hurewicz généralisa la notion de groupe d'homotopie à celle de groupe d'homotopie relative. Pour un espace pointé (X, x_0) et une partie A de X telle que $x_0 \in A$, il considère les *chemins* $\alpha : [0,1] \longrightarrow X$ tels que $\alpha(0) = x_0$ et $\alpha(1) \in A$; leur ensemble $\Omega(X, A, x_0) \subset \mathscr{C}([0,1]; X)$ est encore muni de la topologie compacte-ouverte. Une homotopie *relative à* A de deux tels chemins α_1, α_2 est par définition une application continue $F : [0,1] \times [0,1] \longrightarrow X$ telle que

$$F(s,0) = \alpha_1(s), F(s,1) = \alpha_2(s), F(0,t) = x_0 \text{ et } F(1,t) \in A \text{ pour } 0 \leq t \leq 1.$$

On définit par récurrence

$$\Omega^1(X, A, x_0) = \Omega(X, A, x_0), \Omega^n(X, A, x_0) = \Omega(\Omega^{n-1}(X, A, x_0), x_{n-1}) \quad (91)$$

où x_n est défini par récurrence comme l'application constante $\mathbb{S}_1 \longrightarrow \{x_{n-1}\}$. On a $\Omega^n(X, \{x_0\}, x_0) = \Omega^n(X, x_0)$.

Le n-ème groupe d'homotopie de (X, x_0) relative à A est par définition

$$\pi_n(X, A, x_0) = \pi_1(\Omega^{n-1}(X, A, x_0), x_{n-1}) \text{ pour } n \geq 2. \quad (92)$$

Pour $n \geq 2$, $\Omega^n(X, A, x_0)$ est un H-espace, donc $\pi_n(X, A, x_0)$ est un groupe commutatif pour $n \geq 3$.

Si $u : (X, x_0) \longrightarrow (Y, y_0)$ est un morphisme d'espaces pointés, A (resp. B) une partie de X (resp. Y) contenant x_0 (resp. y_0), et si $u(A) \subset B$, $f \mapsto u \circ f$ est une application continue de $\Omega^n(X, A, x_0)$ dans $\Omega^n(Y, B, y_0)$, et on en déduit un homomorphisme de groupes

$$u_* : \pi_n(X, A, x_0) \longrightarrow \pi_n(Y, B, y_0). \quad (93)$$

On peut donc dire que π_n est un foncteur covariant de la catégorie des espaces pointés dans celle des groupes.

L'interprétation de $\Omega^n(X, x_0)$ comme sous-espaces de $\mathscr{C}([0,1]^n; X)$ se généralise à $\Omega^n(X, A, x_0)$ pour $n \geq 2$: si L est la partie de la frontière B de $[0,1]^n$ égale à $[0,1]^{n-1}\times\{1\}$ («face supérieure du cube»), on montre que $\Omega^n(X,A,x_0)$ est homéomorphe au sous-espace de $\mathscr{C}([0,1]^n;X)$ formé des applications continues u telles que $u(z)\in A$ pour $z\in L$ et $u(z)=x_0$ pour $z\in B-L$.

L'élément neutre de $\pi_n(X,A,x_0)$ correspond aux applications u telles que $u(z)\in A$ pour tout $z\in[0,1]^n$.

Il résulte de là que l'on a une application canonique

$$u\mapsto u|L:\Omega^n(X,A,x_0)\longrightarrow\Omega^{n-1}(A,x_0)$$

d'où un homomorphisme de groupes

$$\partial:\pi_n(X,A,x_0)\longrightarrow\pi_{n-1}(A,x_0)\text{ pour }n\geq 2. \tag{94}$$

En 1939, J.H.C. Whitehead a montré que si $i:(X,x_0)\longrightarrow(X,A)$ et $j:(A,x_0)\longrightarrow(X,x_0)$ sont les injections canoniques, la suite

$$\begin{aligned}\ldots\longrightarrow\pi_n(A,x_0)&\xrightarrow{j_*}\pi_n(X,x_0)\xrightarrow{i_*}\pi_n(X,A,x_0)\xrightarrow{\partial}\pi_{n-1}(A,x_0)\\ \longrightarrow\ldots\longrightarrow\pi_1(A,x_0)&\xrightarrow{j_*}\pi_1(X,x_0)\xrightarrow{i_*}\pi_1(X,A,x_0)\\ \xrightarrow{\partial}\pi_0(A)&\xrightarrow{j_*}\pi_0(X)\end{aligned} \tag{95}$$

est exacte[15]); c'est la *suite exacte d'homotopie relative*.

14.4 Suspension et espaces de lacets

L'isomorphisme canonique (89) s'écrit aussi

$$\mathscr{C}(\mathbb{S}_{n-1},*;\Omega(X,x_0),x_1)\simeq\mathscr{C}(\mathbb{S}_n,*;X,x_0).$$

C'est un cas particulier d'une propriété générale de la suspension réduite ΣX: pour deux espaces séparés pointés $(X,x_0),(Y,y_0)$, il y a un homéomorphisme canonique

$$\mathscr{C}(Y,y_0;\Omega(X,x_0),x_1)\simeq\mathscr{C}(\Sigma Y,y_0;X,x_0). \tag{96}$$

Cela résulte de la propriété générale de la topologie compacte-ouverte: l'espace $\mathscr{C}(Y,y_0;\Omega(X,x_0),x_1)$ est homéomorphe au sous-espace de $\mathscr{C}([0,1]\times Y;X)$ formé des applications $(t,y)\mapsto f(t,y)$ telles que $f(0,y)=f(1,y)=x_0$, $f(t,y_0)=x_0$ pour $0\leq t\leq 1$ et $y\in Y$. Mais une telle application peut, d'après la définition de la suspension réduite, s'écrire

$$[0,1]\times Y\xrightarrow{p}\mathscr{C}(\Sigma Y,y_0)\xrightarrow{g}(X,x_0)$$

15) Les trois derniers termes de cette suite ne sont pas des groupes, mais des ensembles munis d'un point privilégié: pour $\pi_1(X,A,x_0)$, c'est l'ensemble des classes des lacets d'origine x_0 et contenus dans A; pour $\pi_0(A)$ et $\pi_0(X)$, ce sont les composantes connexes par arcs de A et X respectivement, contenant x_0. L'exactitude signifie que les applications i_*, ∂ et j_* ont pour image le point privilégié.

où p est l'application canonique et g est continue. Passant aux classes d'homotopie, on a une bijection canonique

$$[Y, y_0; \Omega(X, x_0), x_1] \simeq [\Sigma Y, y_0; X, x_0]. \tag{97}$$

Du fait que $\Omega(X, x_0)$ est un H-espace, on déduit aisément que les deux membres de (97) sont des groupes; pour $p \geq 2$, $[Y, y_0; \Omega^p(X, x_0), x_p]$ est un groupe commutatif et tous les groupes

$$[\Sigma^k Y, y_0; \Omega^{p-k}(X, x_0), x_{p-k}]$$

sont isomorphes pour $k < p$.

Si l'on prend $(X, x_0) = (\Sigma Y, y_0)$, il correspond par (97), à $\mathrm{id}_{\Sigma Y}$ au second membre, une application canonique

$$s : (Y, y_0) \longrightarrow (\Omega\Sigma Y, y_1);$$

$s(y)$ n'est autre que le lacet $t \mapsto p(t, y)$ dans ΣY. Passant aux classes d'homotopie, on en déduit une application canonique $s_* : [Z, z_0; Y, y_0] \longrightarrow [Z, z_0; \Omega\Sigma Y, y_1]$, et finalement, compte tenu de la bijection canonique (97), on obtient une application canonique

$$E : [Y, y_0; X, x_0] \longrightarrow [\Sigma Y, y_0; \Sigma X, x_0] \tag{98}$$

introduite en 1937 par H. Freudenthal, et appelée *suspension homotopique*; pour toute application continue $f : (Y, y_0) \longrightarrow (X, x_0)$, $E([f])$ est la classe d'homotopie de l'application $E(f)$; $(\Sigma Y, y_0) \longrightarrow (\Sigma X, x_0)$ telle que $E(f)(p(t, y)) = p(t, f(y))$. Pour $(Y, y_0) = (\mathbb{S}_n, *)$, cela donne un homomorphisme canonique

$$E : \pi_n(X, x_0) \longrightarrow \pi_{n+1}(\Sigma X, x_0)$$

qui, en général, n'est ni injectif ni surjectif.

14.5 Produits de Whitehead

En 1941, pour étudier l'homotopie des CW-complexes, J.H.C. Whitehead introduisit une application canonique

$$\pi_m(X, x_0) \times \pi_n(X, x_0) \longrightarrow \pi_{m+n-1}(X, x_0). \tag{99}$$

Pour cela, posant $I = [0, 1]$, il part de deux applications continues $f : I^m \longrightarrow X$, $g : I^n \longrightarrow X$, telles que sur les frontières B_m de I^m et B_n de I^n, on ait $f(B_m) = \{x_0\}$, $g(B_n) = \{x_0\}$. La frontière du cube $I^{m+n} = I^m \times I^n$ est $B_{m+n} = (I^m \times B_n) \cup (B_m \times I^n)$, et on a $(I^m \times B_n) \cap (B_m \times I^n) = B_m \times B_n$. On définit alors $h : B_{m+n} \longrightarrow X$ par

$$h(s, t) = \begin{cases} f(s) & \text{pour } (s, t) \in I^m \times B_n \\ g(t) & \text{pour } (s, t) \in B_m \times I^n. \end{cases}$$

Cette définition est justifiée par le fait que $f(s) = g(t)$ pour $(s,t) \in B_m \times B_n$. Comme la frontière de I^{m+n} est homéomorphe à $\mathbb{S}_{m+n-1}$, la classe d'homotopie w de h appartient à $\pi_{m+n-1}(X, x_0)$ et ne dépend que des classes d'homotopie u, v de f et g; on la note $[u, v]$, ce qui définit l'application (99). On a

$$[u, v_1 + v_2] = [u, v_1] + [u, v_2] \text{ pour } n \geq 2$$
$$[v, u] = (-1)^{mn}[u, v] \text{ pour } m + n \geq 3$$

et pour tout morphisme $\varphi : (X, x_0) \longrightarrow (Y, y_0)$

$$\varphi_*([u, v]) = [\varphi_*(u), \varphi_*(v)].$$

Pour $m = n = 1$, on montre que $[u, v] = uvu^{-1}v^{-1}$, ce qui justifie la notation. Plusieurs mathématiciens ont montré ensuite que si $m+n+p \geq 4$ et $t \in \pi_p(X, x_0)$, on a une «identité de Jacobi»:

$$(-1)^{pn}\big[[t, u], v\big] + (-1)^{mp}\big[[u, v], t\big] + (-1)^{nm}\big[[v, t], u\big] = 0.$$

14.6 Changement de point base

On suppose que X est un espace connexe par arcs. Si x_0 et x_1 sont deux points de X, et $\alpha : [0, 1] \longrightarrow X$ un chemin d'origine x_0 et d'extrémité x_1, ce chemin détermine un homomorphisme

$$\sigma_n(\alpha) : \pi_n(X, x_0) \longrightarrow \pi_n(X, x_1) \tag{100}$$

de la façon suivante. Les éléments de $\pi_n(X, x_0)$ étant identifiés aux classes d'homotopie des applications continues $f : [0, 1]^n \longrightarrow X$ égales à x_0 sur la frontière B de ce cube, on voit aisément qu'il existe une application continue $F : [0, 1]^n \times [0, 1] \longrightarrow X$ telle que $F(x, t) = \alpha(t)$ dans $B \times [0, 1]$ et $F(x, 0) = f(x)$ dans $[0, 1]^n \times \{0\}$; on montre que la classe d'homotopie de $x \mapsto F(x, 1)$ ne dépend que de la classe $[f]$ de f, et $\sigma_n(\alpha)([f])$ est la classe de $x \mapsto F(x, 1)$ dans $\pi_n(X, x_1)$.

On montre aussi que si α et α' sont deux chemins de x_0 à x_1 qui sont homotopes, alors $\sigma_n(\alpha) = \sigma_n(\alpha')$; si β est un chemin de x_1 à un point x_2, l'homomorphisme composé

$$\pi_n(X, x_0) \xrightarrow{\sigma_n(\alpha)} \pi_n(X, x_1) \xrightarrow{\sigma_n(\beta)} \pi_n(X, x_2)$$

est égal à $\sigma_n(\alpha \vee \beta)$. En particulier, si $\beta(t) = \alpha(1 - t)$, de sorte que $\alpha \vee \beta$ est le chemin constant, on a $\sigma_n(\beta) \circ \sigma_n(\alpha) = \text{id}$; autrement dit, $\sigma_n(\alpha)$ est un *isomorphisme* pour tout chemin α.

Si γ est un lacet d'origine x_0, $\sigma_n(\gamma)$ est donc un automorphisme de $\pi_n(X, x_0)$ qui ne dépend que de la classe $[\gamma]$ dans le groupe fondamental $\pi_1(X, x_0)$. On a ainsi défini une *action* du groupe $\pi_1(X, x_0)$ sur $\pi_n(X, x_0)$

$$(u, v) \mapsto \sigma_n(u) \cdot v$$

et on montre que pour $n \geq 2$, cette action est donnée par

$$\sigma_n(u) \cdot v = [u, v] + v$$

où $[u, v]$ est le produit de Whitehead.

On dit que X est *n-simple* si $\sigma_n(u)$ est l'identité pour tous les éléments u de $\pi_1(X, x_0)$; l'isomorphisme $\sigma_n(\alpha)$ ne dépend alors que de l'extrémité de α.

Si $A \subset X$ est un sous-espace connexe par arcs, x_0 et x_1 deux points de A, on définit de même un isomorphisme $\sigma_n(\alpha)$ de $\pi_n(X, A, x_0)$ sur $\pi_n(X, A, x_1)$ pour tout chemin α de x_0 à x_1 contenu dans A; en particulier, on a une action du groupe fondamental $\pi_1(A, x_0)$ sur $\pi_n(X, A, x_0)$. On dit que le couple (X, A, x_0) est *n-simple* lorsque cette action est triviale.

14.7 Les premiers calculs

Les premiers résultats de la théorie algébrique de l'homotopie sont bien antérieurs aux Notes de Hurewicz de 1935 et sont dus à H. Hopf. En 1912, Brouwer avait conjecturé que deux applications continues de la sphère $\mathbb{S}_n$ dans elle-même, qui ont même degré, sont homotopes; il en donna pour $n = 2$ une esquisse de preuve, par une de ses constructions incroyablement compliquées et souvent obscures. En 1926, Hopf prouva la conjecture de Brouwer générale, en raisonnant par récurrence sur n, par une série d'ingénieuses réductions très longues, mais beaucoup plus claires que les raisonnements de Brouwer. Son résultat équivaut à un isomorphisme

$$\pi_n(\mathbb{S}_n) \simeq \mathbb{Z}$$

ce qui plus tard fut démontré beaucoup plus simplement.

En 1931, Hopf franchit une étape décisive dans l'étude de l'homotopie des sphères. Il est facile de voir, par approximation simpliciale, que $\pi_j(\mathbb{S}_n) = 0$ pour $j < n$; mais en 1930, personne ne savait s'il existait pour $m > n$ une application continue $f : \mathbb{S}_m \longrightarrow \mathbb{S}_n$ qui ne fût pas homotope à une application constante. Hopf prouva qu'il existe effectivement de telles applications $f : \mathbb{S}_3 \longrightarrow \mathbb{S}_2$ et que leur classe $[f]$ est entièrement déterminée par un invariant $H(f)$ qui peut prendre toutes les valeurs entières. Sa démonstration est encore longue et difficile, utilisant l'approximation simpliciale à la manière de Brouwer; l'idée de départ est de considérer pour $f : \mathbb{S}_3 \longrightarrow \mathbb{S}_2$ les fibres «génériques» $f^{-1}(y)$ pour $y \in \mathbb{S}_2$; si f est une application simpliciale ces fibres sont des 1-complexes simpliciaux et $H(f)$ est le «nombre d'enlacement» de deux d'entre elles. Par la suite, le résultat de Hopf s'interpréta comme un isomorphisme

$$\pi_3(\mathbb{S}_2) \simeq \mathbb{Z}.$$

En 1935, Hopf généralisa sa définition de $H(f)$ pour les applications continues $f : \mathbb{S}_{2r+1} \longrightarrow \mathbb{S}_{r+1}$; il prouva que si r est impair, $H(f)$ peut prendre toutes les valeurs paires; en outre, pour $r = 3$ et $r = 7$, $H(f)$ peut prendre toutes les valeurs entières, mais jusqu'en 1958, on ignora si cette propriété s'étend à d'autres valeurs de r.

L'étape suivante dans le calcul des $\pi_m(\mathbb{S}_n)$ est due à H. Freudenthal, qui, en 1937, montra que la suspension homotopique (définie par lui)

$$E : \pi_r(\mathbb{S}_n) \longrightarrow \pi_{r+1}(\mathbb{S}_{n+1})$$

est *bijective* si $1 \leq r < 2n - 1$ et *surjective* si $r = 2n - 1$. Sa méthode est voisine de celle de Hopf; elle lui donna l'isomorphisme

$$\pi_4(\mathbb{S}_3) \simeq \mathbb{Z}/2\mathbb{Z}.$$

Du théorème de Freudenthal on déduit que la suite

$$\pi_r(\mathbb{S}_n) \xrightarrow{E} \pi_{r+1}(\mathbb{S}_{n+1}) \xrightarrow{E} \dots \xrightarrow{E} \pi_{r+k}(\mathbb{S}_{n+k}) \longrightarrow \dots$$

est *stationnaire*: tous ses termes sont isomorphes dès que $k > r - 2n + 1$. Le groupe $\pi_{r+k}(\mathbb{S}_k)$ pour $k > r + 1$ est indépendant de k, à isomorphisme près; on dit que c'est le *r-ème groupe stable* d'homotopie des sphères.

C'est seulement après l'intervention des fibrations que de nouveaux progrès apparurent dans le calcul des groupes d'homotopie.

14.8 La suite des groupes d'homotopie d'un espace

En 1921, Veblen avait montré que, pour *tout* groupe G, il existe un espace topologique X pour lequel $\pi_1(X, x_0) \simeq G$. En 1949, J.H.C. Whitehead prouva, plus généralement, que pour *toute* suite infinie $G_1, G_2, \dots, G_n, \dots$, où on suppose seulement que G_n est commutatif pour $n \geq 2$, il existe un espace topologique Z tel que

$$\pi_n(Z, z_0) \simeq G_n \quad \text{pour } \textit{tout} \quad n \geq 1. \tag{101}$$

La preuve se fait en plusieurs étapes. On montre d'abord que si F est un groupe libre (resp. un groupe libre commutatif $\mathbb{Z}^{(I)}$ pour $n \geq 2$), il existe un CW-complexe (X, x_0) tel que

$$\pi_r(X, x_0) = 0 \text{ pour } r < n, \quad \pi_n(X, x_0) \simeq F. \tag{102}$$

Pour $n = 1$, X est un complexe simplicial de dimension 1 (un «graphe» connexe). Pour $n \geq 2$, (X, x_0) est un bouquet de sphères de dimension n.

En second lieu, on note que tout groupe G s'écrit $G = F/R$, où F est un groupe libre et R un sous-groupe normal de F. Le groupe R est aussi libre; soit $(r_\gamma)_{\gamma \in L}$ une famille libre de générateurs de R. Il correspond alors à chaque r_γ une application continue $h_\gamma \in \mathscr{C}(\mathbb{S}_n, *; X, x_0)$ telle que $[h_\gamma] = r_\gamma$. On attache à X une famille $(e_\gamma^{n+1})_{\gamma \in L}$ de sphères au moyen des h_γ. La suite exacte d'homotopie permet de montrer que le CW-complexe X_{n+1} obtenu ainsi est tel que

$$\pi_r(X_{n+1}, x_0) = 0 \text{ pour } r < n, \text{ et } \pi_n(X_{n+1}, x_0) \simeq G. \tag{103}$$

L'étape suivante est la plus importante: elle aboutit à la construction d'un CW-complexe $K(G,n)$ tel que $\pi_r(K(G,n)) = 0$ pour $r \neq n$ et $\pi_n(K(G,n)) \simeq G$. Tous les CW-complexes ayant ces propriétés ont même type d'homotopie, et sont appelés *espaces* (ou CW-complexes) *d'Eilenberg-MacLane*, ces mathématiciens ayant été les premiers à les définir et à montrer leur intérêt en 1943. La méthode de Whitehead pour construire ces espaces consiste à partir d'un CW-complexe X_{n+1} vérifiant (103), et de former par récurrence sur $m > n+1$ des CW-complexes X_m tels que $\pi_r(X_m, x_0) = 0$ pour $1 \leq r < m$ et $m \neq n$ et $\pi_n(X_m, x_0) \simeq G$. On passe de X_m à X_{m+1} par attachement de $(m+1)$-cellules qui «tuent» le groupe $\pi_m(X_m, x_0)$ sans modifier les $\pi_r(X_m, x_0)$ pour $r < m$. Le procédé est très voisin de celui de l'étape précédente, et $K(G,n)$ est alors le CW-complexe (en général de dimension infinie) dont les X_m sont les m-squelettes.

Enfin, pour obtenir un espace Z vérifiant (101), il suffit de prendre le produit $Z = \prod_{n \geq 1} K(G_n, n)$.

15 Premières relations entre homotopie et homologie

15.1 Les homomorphismes de Hurewicz

On a vu que Poincaré avait introduit le groupe fondamental $\pi_1(X, x_0)$ afin d'obtenir des relations entre classes d'homologie en «rendant commutatives» des relations entre éléments de $\pi_1(X, x_0)$. Le développement des théories homologiques a permis de donner à cette idée la forme précise suivante: pour tout espace pointé (X, x_0) connexe par arcs, il existe, pour l'homologie singulière, un homomorphisme canonique *surjectif*

$$h : \pi_1(X, x_0) \longrightarrow H_1(X; \mathbb{Z}) \tag{104}$$

dont le noyau est le *groupe des commutateurs* de $\pi_1(X, x_0)$. Pour définir h, on remarque que tout lacet γ d'origine x_0 est une application continue $\gamma = \bar{\Delta}_1 \longrightarrow X$ telle que $\gamma(e_0) = \gamma(e_1) = x_0$; c'est donc un 1-cycle singulier. En outre, si $F : [0,1] \times [0,1]$ est une homotopie entre deux lacets γ_1, γ_2, il y a un 2-simplexe singulier $s : \bar{\Delta}_2 \longrightarrow X$ tel que son bord soit donné par

$$b_2 s = \gamma_2 - \gamma_1$$

dans le groupe des 1-cycles singuliers. Il suffit pour cela de prendre $s = s_0 + c$, où c est le simplexe singulier $\bar{\Delta}_2 \longrightarrow \{x_0\}$, et s_0 est défini par

$$s_0(t(ue_2 + (1-u)e_0) + (1-t)(ue_1 + (1-u)e_0)) = F(t,u).$$

En 1935, Hurewicz a défini, pout tout $n > 1$, un homomorphisme analogue

$$h_n : \pi_n(X, x_0) \longrightarrow H_n(X; \mathbb{Z}).$$

Pour cela, il remarque que l'adhérence $\bar{\Delta}_n$ du simplexe standard Δ_n est homéomorphe au cube $[0,1]^n$, et l'on peut donc considérer $\Omega^n(X,x_0)$ comme l'espace des applications continues $f : \bar{\Delta}_n \longrightarrow X$ telles que $f(\bar{\Delta}_n - \Delta_n) = \{x_0\}$. Or une telle application définit par fonctorialité un homomorphisme en homologie singulière

$$H_n(f) : H_n(\bar{\Delta}_n, \bar{\Delta}_n - \Delta_n; \mathbb{Z}) \longrightarrow H_n(X, \{x_0\}; \mathbb{Z}).$$

Mais on a $H_n(X, \{x_0\}; \mathbb{Z}) = H_n(X; \mathbb{Z})$, et la suite exacte d'homologie singulière fournit un isomorphisme

$$\partial : H_n(\bar{\Delta}_n, \bar{\Delta}_n - \Delta_n; \mathbb{Z}) \simeq H_{n-1}(\mathbb{S}_{n-1}; \mathbb{Z})(\cong \mathbb{Z}).$$

Soit ε_n l'image réciproque de la classe fondamentale $[\mathbb{S}_{n-1}]$ par cet isomorphisme. L'axiome d'homotopie pour l'homologie singulière montre que $H_n(f)(\varepsilon_n)$ ne dépend que de la classe d'homotopie $[f]$ dans $\pi_n(X, x_0)$. On a donc défini une application canonique

$$h_n : [f] \longrightarrow H_n(f)(\varepsilon_n) \tag{105}$$

et on prouve que c'est un homomorphisme de groupes. Pour $n \geq 2$, des exemples simples montrent que h_n n'est en général ni injective ni surjective.

Mais Hurewicz découvrit que si l'on suppose que $\pi_j(X) = 0$ pour $1 \leq j \leq n-1$ (ce qu'on exprime en disant que X est $(n-1)$-*connexe*), alors h_n est *bijectif* (théorème d'isomorphisme «absolu» de Hurewicz).

En 1938, Hurewicz (dans un cours non publié) généralisa la définition de l'homomorphisme h_n aux groupes d'homotopie relatifs. On suppose que le sous-espace A de X est connexe par arcs, et avec les mêmes notations que ci-dessus, on suppose en outre que $f(\bar{\Delta}_n - \Delta_n) \subset A$; on a cette fois un homomorphisme

$$h_n(f) : H_n(\bar{\Delta}_n, \bar{\Delta}_n - \Delta_n; \mathbb{Z}) \longrightarrow H_n(X, A; \mathbb{Z})$$

et la même définition (105) donne ici un homomorphisme de groupes

$$h_n : \pi_n(X, A, x_0) \longrightarrow H_n(X, A; \mathbb{Z}) \text{ pour } n \geq 2. \tag{106}$$

Le théorème d'isomorphisme «relatif» de Hurewicz est ici que si $\pi_k(X, A, x_0) = 0$ pour $1 \leq k \leq n-1$ et pour tout $x_0 \in A$ (ce qu'on exprime en disant que le couple (X, A) est $(n-1)$-*connexe*) et si en plus (X, A) est *n-simple*, alors h_n est un isomorphisme.

Hurewicz a donné en 1935 une démonstration du théorème d'isomorphisme «absolu» pour les complexes simpliciaux, par récurrence sur la dimension; mais il ne publia aucune preuve du théorème «relatif». Des démonstrations des deux théorèmes furent publiées par R. Fox en 1943; il prouva en outre que, sous les mêmes hypothèses sur (X, A), l'homomorphisme h_{n+1} est surjectif. D'autres démonstrations et des généralisations de ces théorèmes ont été obtenues par les techniques plus puissantes des fibrations et des suites spectrales.

Le théorème de Hurewicz fournit un isomorphisme canonique

$$h_n : \Pi \xrightarrow{\sim} H_n(K(\Pi, n); \Pi)$$

lorsque Π est commutatif; en vertu du théorème des coefficients universels, l'application $H^n(K(\Pi, n); \Pi) \longrightarrow \mathrm{Hom}(H_n(K(\Pi, n); \Pi), \Pi)$ est bijective, l'isomorphisme h_n^{-1} peut donc être considéré comme un élément u de $H^n(K(\Pi, n); \Pi)$, appelé la *classe fondamentale* de cohomologie de $K(\Pi, n)$. Si maintenant X est un CW-complexe quelconque, on a une bijection

$$[X; K(\Pi, n)] \xrightarrow{\sim} H^n(X; \Pi)$$

donnée par $[f] \mapsto f^*(u)$.

15.2 L'équivalence d'homotopie faible

J.H.C. Whitehead a déduit du théorème d'isomorphisme relatif de Hurewicz une remarquable relation entre les homomorphismes $H_r(f) : H_r(X; \mathbb{Z}) \longrightarrow H_r(Y; \mathbb{Z})$ en homologie singulière, et $f_* : \pi_r(X, x_0) \longrightarrow \pi_r(Y, y_0)$ pour un morphisme $f : (X, x_0) \longrightarrow (Y, y_0)$ d'espaces pointés connexes par arcs. Il montre que:

1) Si f_* est bijectif pour $1 \leq r < n$ et surjectif pour $r = n$, alors $H_r(f)$ est bijectif pour $1 \leq r < n$ et surjectif pour $r = n$.

2) Si de plus X et Y sont simplement connexes, la réciproque est vraie.

L'utilisation du cylindre Z_f de f permet de se ramener au cas où f est l'injection canonique d'un sous-espace X de Y. Le théorème d'isomorphisme relatif et les suites exactes d'homotopie et d'homologie singulière prouvent alors 1). Si $\pi_1(X, x_0) = 0$ et $\pi_1(Y, y_0) = 0$, on a aussi $H_1(X; \mathbb{Z}) = 0$ et $H_1(Y; \mathbb{Z}) = 0$; donc les hypothèses de 2) entraînent $H_q(X, Y; \mathbb{Z}) = 0$ pour $1 \leq q \leq n - 1$ par la suite exacte d'homologie; on conclut de nouveau par le théorème d'isomorphisme relatif et la suite exacte d'homotopie (en tenant compte de ce que Z_f est simplement connexe si X et Y le sont).

L'assertion 2) n'est plus exacte si on ne suppose pas que X et Y sont simplement connexes.

Il faut prendre garde que ce théorème ne signifie pas que deux espaces X, Y ayant les mêmes groupes d'homotopie ont aussi les mêmes groupes d'homologie (ni inversement lorsqu'ils sont simplement connexes). Il faut pour l'appliquer disposer d'une application continue d'où proviennent les isomorphismes.

On dit qu'une application continue $f : (X, x_0) \longrightarrow (Y, y_0)$ est une *équivalence d'homotopie faible* si les homomorphismes $f_* : \pi_r(X, x_0) \longrightarrow \pi_r(Y, y_0)$ sont bijectifs pour *tout* $r \geq 1$. En 1939, J.H.C. Whitehead montra que, si X et Y sont des CW-complexes, non seulement les $H_n(f)$ sont alors des isomorphismes, mais f est une *équivalence d'homotopie*. La preuve se fait par récurrence sur n, en montrant que si P est un CW-complexe de dimension n, l'application $f^* : [P; X] \longrightarrow [P; Y]$ est bijective; on en déduit que les applications

$$f^*_{(X)} : [X; X] \longrightarrow [X; Y] \text{ et } f^*_{(Y)} : [Y; X] \longrightarrow [Y; Y]$$

sont bijectives. Il y a donc une application $g : Y \longrightarrow X$ telle que $f^*_{(Y)}([g]) = [\mathrm{id}_Y]$; on en conclut que g est inverse homotopique de f, car

$$f^*_{(X)}([g \circ f]) = [f \circ g \circ f] = [\mathrm{id}_Y \circ f] = [f \circ \mathrm{id}_X] = f^*_{(X)}([\mathrm{id}_X])$$

d'où $[g \circ f] = [\mathrm{id}_X]$.

15.3 Obstructions

L'existence d'une homotopie entre deux applications continues f, g de X dans Y est équivalente à l'existence d'un *prolongement* continu d'une application continue: on considère en effet, dans l'espace $X \times [0, 1]$ le sous-espace $C = (X \times \{0\}) \cup (X \times \{1\}) \cup (\{x_0\} \times [0, 1])$ et l'application continue $G : C \longrightarrow Y$ définie par

$$G(x, 0) = f(x), \ G(x, 1) = g(x), \ G(x_0, t) = y_0$$

et il s'agit de prolonger cette application en une application continue $F : X \times [0, 1] \longrightarrow Y$.

En 1940, Eilenberg montra qu'on peut donner des critères pour l'existence d'un prolongement (ou d'une homotopie) à l'aide de la cohomologie et des groupes d'homotopie. Le cas le plus simple est celui d'une application continue $f : \mathbb{S}_n \longrightarrow Y$; pour qu'elle se prolonge en une application continue de la boule $\mathbb{D}_{n+1}$ dans Y, il faut et il suffit que la classe $[f]$ de f dans le groupe $\pi_n(Y)$ soit nulle.

Soit alors X un complexe simplicial, A un sous-complexe de X et f une application continue de A dans un espace Y, qui est supposé n-simple pour tout n. Eilenberg cherche à prolonger f en la prolongeant, par récurrence sur n, à tout sous-complexe $X_n \cup A$, où X_n est le n-squelette de X. Si f est définie dans $X_n \cup A$, pour la prolonger à $X_{n+1} \cup A$, il faut la prolonger à tout $(n + 1)$-simplexe τ, ce qui n'est possible que si la restriction de f à la frontière Fr τ, a une classe nulle dans $\pi_n(Y)$ (Fr τ étant homéomorphe à $\mathbb{S}_n$). On est donc amené à associer à f l'application

$$s_{n+1}(f) : C_{n+1}(X; \mathbb{Z}) \longrightarrow \pi_n(Y)$$

qui, à τ, fait correspondre l'élement $[f|\mathrm{Fr}\,\tau]$ de $\pi_n(Y)$; $s_{n+1}(f)$ est donc une $(n+1)$-cochaîne à valeurs dans $\pi_n(Y)$. Pour $n \geq 2$, Eilenberg prouva que $s_{n+1}(f)$ est un $(n + 1)$-cocycle, et définit donc une classe de cohomologie

$$\beta_{n+1}(f) \in H^{n+1}(X, A; \pi_n(Y)).$$

Cette méthode fournit entre autres à Eilenberg le théorème suivant: si on suppose que $\pi_j(Y) = 0$ pour $1 \leq j \leq n - 1$, et que $X - A$ est de dimension $\leq n + 1$, alors il existe un élément bien déterminé $\beta_{n+1}(f) \in H^{n+1}(X, A; \pi_n(Y))$ tel que la condition $\beta_{n+1}(f) = 0$ soit nécessaire et suffisante pour qu'il existe un prolongement continu de f à X.

On a des théorèmes analogues pour l'existence d'une homotopie entre deux prolongements. On en déduit en particulier une bijection canonique

$$[X, x_0; Y, y_0] \simeq H^n(X; \pi_n(Y))$$

lorsque X est de dimension n et que Y est $(n-1)$-connexe; cette propriété avait été établie auparavant par Hopf, Hurewicz et Whitney par d'autres méthodes. En particulier, on a une bijection

$$[X, x_0; \mathbb{S}_n, *] \simeq H^n(X; \mathbb{Z}).$$

15.4 Groupes d'homotopie et homologie des groupes

Si X est un complexe simplicial connexe par arcs, l'application de Hurewicz

$$h_2 : \pi_2(X, x_0) \longrightarrow H_2(X; \mathbb{Z})$$

n'est pas surjective en général. Mais en 1941, H. Hopf découvrit le fait remarquable que le quotient $H_2(X; \mathbb{Z})/h_2(\pi_2(X, x_0))$ est entièrement déterminé par le groupe fondamental $\pi_1(X, x_0)$. De façon plus précise, il montra qu'on obtient ce quotient par un procédé purement algébrique applicable à *tout* groupe G. On écrit le groupe $G = F/R$ comme quotient d'un groupe libre et on forme le quotient $G_1 = (R \cap [F, F])/[F, R]$, qui ne dépend pas du choix de F et R; alors, si $G = \pi_1(X, x_0)$, on a $G_1 \simeq H_2(X; \mathbb{Z})/h_2(\pi_2(X, x_0))$.

Ce résultat n'est plus valable pour les groupes d'homotopie de dimension ≥ 3. Toutefois, Hopf observa en 1942 que si l'on suppose que $\pi_j(X, x_0) = 0$ pour $2 \leq j \leq n-1$, le groupe $Q_n(X) = H_n(X; \mathbb{Z})/h_n(\pi_n(X, x_0))$ est encore déterminé par le groupe fondamental $\pi_1(X, x_0)$. Il chercha alors un procédé de pure théorie des groupes qui donnerait $Q_n(X)$ à partir du groupe $\Pi = \pi_1(X, x_0)$. Indépendament, Eilenberg et MacLane étudiaient le même problème. La solution qu'ils obtinrent est équivalente à celle de Hopf, mais elle est plus simple et elle a créé l'*homologie des groupes*, qui est devenue un outil important en algèbre.

Pour un groupe quelconque Π (commutatif ou non) on considère le complexe de chaînes

$$\begin{aligned} K.(\Pi) : 0 \longleftarrow K_0(\Pi) \xleftarrow{b_1} K_1(\Pi) \longleftarrow \ldots \longleftarrow K_{n-1}(\Pi) \\ \xleftarrow{b_n} K_n(\Pi) \longleftarrow \ldots \end{aligned}$$

où $K_n(\Pi) = \mathbb{Z}^{(\Pi^n)}$ est le module libre ayant pour base tous les n-uples $(x_1, \ldots, x_n)$ d'éléments de Π, avec pour bord

$$\begin{aligned} b_n(x_1, \ldots, x_n) &= (x_2, \ldots, x_n) \\ &+ \sum_{1 \leq k \leq n-1} (-1)^k (x_1, \ldots, x_{k-1}, x_k x_{k+1}, x_{k+2}, \ldots, x_n) \\ &+ (-1)^n (x_1, \ldots, x_{n-1}). \end{aligned}$$

Pour tout groupe commutatif G, on considère comme d'ordinaire le complexe de chaînes $K.(\Pi) \otimes G$ et le complexe de cochaînes $K^{\cdot}(\Pi, G) = \mathrm{Hom}(K.(\Pi), G)$. Les groupes d'homologie et de cohomologie du groupe Π à coefficients dans G sont alors par définition

$$H_n(\Pi; G) = H_n(K.(\Pi) \otimes G)$$
$$H^n(\Pi; G) = H^n(K^{\cdot}(\Pi; G))$$

et les résultats d'Eilenberg-MacLane concernant l'homologie de l'espace X sous l'hypothèse que $\pi_j(X, x_0) = 0$ pour $2 \leq j \leq r - 1$ sont les suivants:

$$\left\{ \begin{array}{l} H_n(X; G) \simeq H_n(\pi_1(X); G) \\ H^n(X; G) \simeq H^n(\pi_1(X); G) \end{array} \right\} \text{ pour } n < r \tag{107}$$

$$H_r(X; G)/h_r(\pi_r(X, x_0)) \simeq H_r(\pi_1(X); G).$$

La construction des groupes $H^n(\Pi; G)$ est le premier exemple de l'intervention de l'algèbre homologique dans une théorie autre que la Topologie. Depuis lors, ces interventions se sont multipliées dans les théories les plus diverses: homologie des algèbres (Hochschild), des algèbres de Lie (Chevalley-Eilenberg), des groupes finis (Tate), longuement exposées dans le livre de Cartan-Eilenberg. Puis, plus récemment, l'homologie cyclique dans les C^*-algèbres (Connes), l'homologie d'intersection dans les variétés stratifiées, etc.

16 Fibrations

16.1 Espaces fibrés

La notion d'espace fibré trouve son origine dans des conceptions telles que les «champs de vecteurs» et les «repères mobiles». Il s'agit d'associer, de façon continue, à tout point x d'une variété X, un objet variable avec x, qui n'est plus un nombre, mais un ensemble tel qu'un groupe ou un espace vectoriel (idée très voisine de celle qui conduit aux faisceaux, qui ont d'ailleurs des liens étroits avec les espaces fibrés).

La première définition générale a été donnée par Whitney en 1935, et développée par lui-même, Steenrod et Ehresmann aux environs de 1940. Un *espace fibré* est un quadruplet (E, B, F, p), où E (l'«espace total»), B (l'«espace de base»), et F (la «fibre») sont des espaces topologiques, et $p : E \longrightarrow B$ une application continue (la «projection»). Ces données sont assujetties à la condition suivante:

(EF) Pour tout $y \in B$, il existe un voisinage ouvert U de y et un homéomorphisme

$$\varphi : U \times F \longrightarrow p^{-1}(U) \tag{108}$$

tel que

$$p(\varphi(y, t)) = y \text{ pour } (y, t) \in U \times F.$$

L'application p est alors ouverte[16]), B est homéomorphe à l'espace quotient E/R, où R est la relation d'équivalence dont les classes sont les fibres $p^{-1}(y)$ pour $y \in B$. L'espace fibré se note souvent (E, B, p), F étant déterminé par ces données à homéomorphisme près. On écrit $p^{-1}(y) = E_y$.

Un espace produit $E = B \times F$ est un espace fibré (dit *trivial*) pour la première projection pr_1 et l'homéomorphisme $\varphi = \mathrm{id}_E$. On peut considérer un espace fibré comme «localement» un produit, ou «localement trivial».

Les espaces fibrés sont les objets d'une catégorie, dont les morphismes $(E, B, p) \longrightarrow (E', B', p')$ sont les couples (f, g) où $f : E \longrightarrow E'$, $g : B \longrightarrow B'$ sont des applications continues, telles que le diagramme

$$\begin{array}{ccc} E & \xrightarrow{f} & E' \\ {\scriptstyle p}\downarrow & & \downarrow{\scriptstyle p'} \\ B & \xrightarrow[g]{} & B' \end{array}$$

soit commutatif. La restriction f_y de f à la fibre E_y est une application continue dans la fibre $E'_{g(y)}$. Si g est l'identité, on dit que (f, id_B) est un B-morphisme. S'il existe un B-isomorphisme $(E, B, F, p) \xrightarrow{\sim} (B \times F, B, F, \mathrm{pr}_1)$ sur un fibré trivial, on dit que (E, B, F, p) est trivialisable et que ce B-isomorphisme en est une trivialisation.

Il n'est pas toujours aisé de montrer qu'une application $p : E \longrightarrow B$ définit un espace fibré. Un théorème général a été prouvé par Ehresmann: si E et B sont des variétés C^∞, p une application C^∞ qui est une submersion (l'application linéaire tangente est surjective en tout point de E) et propre ($p^{-1}(K)$ est compact pour toute partie compacte K de B), alors, si B est connexe, (E, B, p) définit un espace fibré.

16.2 Revêtements

L'exemple le plus simple d'espace fibré (E, B, F, p) est celui où la fibre F est un espace *discret*; E est alors appelé un *revêtement* (non ramifié) de B, ce qui est la forme générale de la notion utilisée par Poincaré, Klein et H. Weyl (1913) pour les surfaces. On se borne aux espaces B connexes et localement connexes par arcs.

La propriété fondamentale des revêtements est l'existence d'un *relèvement unique* d'un chemin quand l'origine du relèvement est donnée (propriété énoncée par H. Weyl, mais sans doute utilisée auparavant): pour tout chemin $\alpha : [0, 1] \longrightarrow B$ tel que $\alpha(0) = a$, et pour tout point $y \in p^{-1}(a)$, il existe un chemin et un seul $\tilde{\alpha} : [0, 1] \longrightarrow E$ tel que $\tilde{\alpha}(0) = y$ et $\alpha = p \circ \tilde{\alpha}$. En outre, si φ est une homotopie entre deux chemins α_1, α_2 d'origine a, avec $\varphi(0, t) = s$ pour $0 \leq t \leq 1$, il y a

16) On rappelle qu'une application $f : X \longrightarrow Y$ où X, Y sont des espaces topologiques, est dite *ouverte* (resp. *fermée*) si l'image par f de tout ouvert (resp. fermé) de X est un ouvert (resp. fermé) de Y.

une unique homotopie $\tilde{\varphi}$ entre $\tilde{\alpha}_1$ et $\tilde{\alpha}_2$ telle que $\tilde{\varphi}(0,t) = y$ pour $0 \leq t \leq 1$ et $\varphi = p \circ \tilde{\varphi}$.

Cela entraîne une *action* du groupe fondamental $\pi_1(B,a)$ sur la fibre E_a. Pour tout point $y \in E_a$, un lacet γ d'origine a dans B se relève en un chemin unique $\tilde{\gamma}$ d'origine y dans E, et son extrémité, qui appartient à E_a, ne dépend que de la classe d'homotopie u de γ dans $\pi_1(B,a)$; on la note $y \cdot u$, et on a $y \cdot e = y$ pour l'élément neutre e de $\pi_1(B,a)$, et $(y \cdot u) \cdot v = y \cdot (uv)$ pour deux éléments u, v de $\pi_1(B,a)$.

Avec les hypothèses faites sur B, toute composante connexe par arcs Y de E est encore un revêtement de B. Pour que E soit connexe par arcs, il faut et il suffit que $\pi_1(B,a)$ opère *transitivement* dans E_a. Alors, pour tout $y \in E_a$, l'homomorphisme $p_* : \pi_1(E,y) \longrightarrow \pi_1(B,a)$ est injectif, et son image est le stabilisateur de y (sous-groupe des $u \in \pi_1(B,a)$ tels que $y \cdot u = y$). Si (E,B,p) et (E',B,p') sont deux revêtements de B connexes par arcs, pour qu'il existe un B-isomorphisme $f : E \longrightarrow E'$, il faut et il suffit que les sous-groupes $p_*(\pi_1(E,y))$ et $p'_*(\pi_1(E',y'))$ soient conjugués dans $\pi_1(B,a)$.

Il n'existe pas toujours de revêtement (E,B,p) de B qui soit *simplement connexe*. C'est cependant le cas s'il existe un recouvrement de B par des ouverts simplement connexes (ce qui est le cas des C^0 variétés et des complexes simpliciaux). Le revêtement simplement connexe Z de B est alors appelé son *revêtement universel*. Il est unique à un B-isomorphisme près, et on le construit par la méthode de Poincaré, comme ensemble de classes d'homotopie de chemins d'origine a. Alors, pour tout sous-groupe H de $\pi_1(B,a)$, il existe un revêtement (Y,B,p) connexe par arcs, tel que pour tout point $y \in p^{-1}(a)$, $\pi_1(E,y)$ soit isomorphe à H; tous les revêtements connexes par arcs de B s'obtiennent de cette façon.

Pour tout revêtement connexe par arcs (E,B,p) et tout point $a \in B$, il y a un voisinage ouvert U de a dans B tel que $p^{-1}(U)$ soit réunion disjointe d'ouverts V_α de E; le groupe G des B-automorphismes de E opère transitivement sur les V_α et est isomorphe à $\pi_1(E,y)$ pour tout $y \in p^{-1}(a)$.

Inversement, on dit qu'un groupe G opère de façon *proprement discontinue* sur un espace connexe et localement connexe par arcs Z si, pour tout point $z \in Z$, il existe un voisinage V de z dans Z tel que $V \cap s \cdot V = \phi$ pour tout $s \neq e$ dans G. Alors l'espace des orbites de G, noté Z/G (espace quotient correspondant à la partition de Z formée des orbites de G) est connexe et localement connexe par arcs et $(Z, Z/G, p)$ (où p est l'application canonique) est un revêtement de Z/G; le groupe des (Z/G)-automorphismes de ce revêtement est isomorphe à G. Si Z est simplement connexe, c'est un revêtement universel de Z/G.

16.3 Exemples de revêtements

1) Toute variété lisse X non orientable possède un revêtement X' qui est une variété lisse orientable connexe, dont la fibre est formée de 2 points (revêtement «à deux feuillets»).

2) Un groupe de Lie connexe G est orientable et admet un revêtement universel $\tilde{G}$ sur lequel on peut définir une structure de groupe de Lie et une seule pour laquelle la projection $p : \tilde{G} \longrightarrow G$ soit un homomorphisme de groupes; la fibre $p^{-1}(e)$, isomorphe au groupe fondamental $\pi_1(G, e)$, est un sous-groupe discret du centre de $\tilde{G}$.

3) Dans ses premiers travaux, Poincaré avait émis la conjecture qu'une variété compacte orientée de dimension 3, qui a la même homologie que $\mathbb{S}_3$, est homéomorphe à $\mathbb{S}_3$. En 1905, il s'aperçut qu'on pouvait au contraire définir une telle «sphère d'homologie» non simplement connexe. Sa définition est très compliquée, et la théorie des revêtements en fournit une autre beaucoup plus simple. La sphère $\mathbb{S}_3$ s'identifie au groupe multiplicatif des quaternions de norme 1, et il existe un sous-groupe fini I d'ordre 120, qui est égal à son groupe des commutateurs, et opère par translations à gauche sur $\mathbb{S}_3$; alors $X = \mathbb{S}_3/I$ est une variété orientable telle que $\pi_1(X) \simeq I$, donc $H_1(X;\mathbb{Z}) = H_2(X;\mathbb{Z}) = 0$ et $H_0(X;\mathbb{Z}) \simeq H_3(X;\mathbb{Z}) \simeq \mathbb{Z}$.

Poincaré modifia en conséquence sa conjecture, qui s'exprime en disant que toute variété compacte simplement connexe de dimension 3 est homéomorphe à $\mathbb{S}_3$; à l'heure actuelle, on ne sait ni démontrer ni infirmer cette «conjecture de Poincaré».

4) Un exemple analogue est obtenu en considérant $\mathbb{S}_3$ comme définie dans $\mathbb{C}^2$ par l'équation $|z_1|^2 + |z_2|^2 = 1$. Pour deux entiers p, q sans facteur commun et tels que $1 \leq q < p$, le groupe cyclique $G = \mathbb{Z}/p\mathbb{Z}$ opère sur $\mathbb{S}_3$ par

$$(k, (z_1, z_2)) \mapsto (\omega^k z_1, \omega^{qk} z_2) \text{ où } \omega = e^{2\pi i/p}.$$

Alors $L(p,q) = \mathbb{S}_3/G$ est une variété connexe orientable, pour laquelle $\pi_1(L(p,q)) \simeq G$. On définit aisément une triangulation de $L(p,q)$ qui fournit l'homologie

$$H_0(L(p,q);\mathbb{Z}) \simeq H_3(L(p,q);\mathbb{Z}) \simeq \mathbb{Z}, H_1(L(p,q);\mathbb{Z}) \simeq G, H_2(L(p,q);\mathbb{Z}) = 0.$$

Ces espaces ont été définis en 1908 par H. Tietze et sont appelés les *espaces lenticulaires*. Leur homologie et leur groupe fondamental ne dépendent pas de q, et cependant Tietze soupçonnait que $L(5,1)$ et $L(5,2)$ ne sont pas homéomorphes. Cela fut effectivement prouvé par Alexander en 1919; de Rham en donna une démonstration plus simple en 1931, en montrant qu'une condition nécessaire pour que $L(p,q)$ et $L(p,q')$ soient homéomorphes est que

$$q' = \pm\nu^2 q (\text{mod} \cdot p)$$

pour un entier ν, lorsque p est premier. En 1940, J.H.C. Whitehead prouva que cette condition est nécessaire et suffisante pour que $L(p,q)$ et $L(p,q')$ aient même type d'homotopie; il fait usage d'un invariant combinatoire, la *torsion* d'une triangulation, introduite d'abord par Reidemeister, et que Whitehead généralisa dans sa théorie du *type simple d'homotopie*. De Rham, en 1960, démontra directement le critère de Whitehead, n'utilisant pas cette théorie mais seulement la théorie des obstructions.

16.4 Relèvement des homotopies et notion de fibration

Dans un espace fibré (E, B, F, p) où la fibre n'est pas discrète, il n'y a plus unicité du relèvement d'un chemin dans E quand on se donne son origine. Mais il subsiste une importante propriété, le *relèvement des homotopies*.

Cette propriété a un sens pour tout morphisme $p : (X, x_0) \longrightarrow (Y, y_0)$; étant donnés un morphisme $\tilde{f} : (Z, z_0) \longrightarrow (X, x_0)$, sa projection $f = p \circ \tilde{f} : (Z, z_0) \longrightarrow (Y, y_0)$, et une homotopie $F : (Z, z_0) \times [0, 1] \longrightarrow (Y, y_0)$ entre f et $g : (Z, z_0) \longrightarrow (Y, y_0)$, il doit exister une homotopie $\tilde{F} : (Z, z_0) \times [0, 1] \longrightarrow (X, x_0)$ telle que $F = p \circ \tilde{F}$ («relèvement» de F) telle que $\tilde{F}(z, 0) = \tilde{f}(z)$ et $g(z) = p(\tilde{F}(z, 1))$ («relèvement» de g). On dit alors que (Z, X, Y, p) a la propriété de relèvement des homotopies.

Cette propriété avait été notée en 1935 par Hurewicz dans un cas très particulier; puis, en 1940, son intérêt dans l'étude de l'homotopie fut remarqué indépendamment par Hurewicz et Steenrod, Eckmann, et Ehresmann et Feldbau. En fait, dès 1937, Borsuk avait démontré la propriété de relèvement des homotopies pour (Z, X, Y, p), où Z est quelconque, $X = \mathscr{C}(U, V)$, $Y = \mathscr{C}(W, V)$, où W est un sous-espace fermé de U, et p est l'application de restriction $f \mapsto f|W$; les espaces U, V, W sont soumis à des conditions assez générales, vérifiées lorsque ce sont des variétés ou des complexes simpliciaux.

C'est un cas particulier du théorème de Borsuk qui attira l'attention, à l'occasion des travaux de J.-P. Serre sur l'homotopie: Y est un espace connexe par arcs, $X = \Omega(Y, X, \{y_0\})$ l'espace de *tous* les chemins dans Y d'origine y_0, et p l'application $\alpha \mapsto \alpha(1)$.

Cela conduisit à appeler *fibration* un triplet (X, Y, p) de deux espaces X, Y connexes par arcs et d'une application continue $p : X \longrightarrow Y$, lorsque pour tout espace topologique Z, (Z, X, Y, p) a la propriété de relèvement des homotopies. C'est le cas lorsque (X, Y, p) est un espace fibré; mais l'exemple de l'espace des chemins montre qu'il y a d'autres exemples. Les «fibres» $p^{-1}(y)$ d'une fibration ne sont plus toujours homéomorphes, mais elles ont même type d'homotopie.

Une *section* d'une fibration est une application continue $s : Y \longrightarrow X$ telle que $p \circ s = \mathrm{id}_Y$. Lorsque X et Y sont des variétés C^r et p une application de classe C^r, on définit de même les sections de classe C^r.

16.5 La suite exacte d'homotopie des fibrations

En 1940, Hurewicz et Steenrod ont montré que pour une fibration (X, Y, p) et toute fibre $F = p^{-1}(y_0)$ et tout point $x_0 \in F$, l'application

$$p_* : \pi_n(X, F, x_0) \longrightarrow \pi_n(Y, y_0) \tag{109}$$

est un isomorphisme pour tout $n \geq 1$.

Pour voir que p_* est injectif, on considère un morphisme $g : (\mathbb{S}_n, *) \longrightarrow (Y, y_0)$ qui est homotope à l'application constante $\mathbb{S}_n \longrightarrow \{y_0\}$. En relevant l'homotopie on obtient un morphisme $f : (\mathbb{S}_n, *) \longrightarrow (X, x_0)$ homotope à un morphisme $(\mathbb{S}_n, *) \longrightarrow (F, x_0)$, donc on a par définition $[f] = 0$ dans $\pi_n(X, F, x_0)$, ce qui prouve l'injectivité puisque $p_*([f]) = [g] = 0$.

Pour montrer que p_* est surjectif, on identifie $\Omega^n(Y, y_0)$ au sous-espace de $\mathscr{C}([0,1]^n, Y)$ formé des applications continues $h : [0,1]^n \longrightarrow Y$ égales à y_0 dans la frontière B_n du cube $[0,1]^n$. On peut considérer h comme une homotopie

$$h : [0,1]^{n-1} \times [0,1] \longrightarrow Y$$

entre deux applications constantes $[0,1]^{n-1} \longrightarrow \{y_0\}$, avec $h(z,t) = y_0$ pour tout $z \in B_{n-1}$, frontière de $[0,1]^{n-1}$. Il y a donc une homotopie

$$H : [0,1]^{n-1} \times [0,1] \longrightarrow (X, x_0)$$

qui relève h; donc $H(z) \in F$ pour $z \in L = [0,1]^{n-1} \times \{1\}$ (face supérieure du cube) et par ailleurs $H(z,t) = x_0$ dans le complémentaire $B_n - L$. Cela signifie que $H \in \Omega^n(X, F, x_0)$, et l'on a $p_*([H]) = [h]$.

La suite exacte d'homotopie relative (95) et l'isomorphisme (109) établissent donc que la suite

$$\begin{aligned} \ldots \longrightarrow \pi_n(F, x_0) \xrightarrow{j_*} \pi_n(X, x_0) \xrightarrow{p_*} \pi_n(Y, y_0) \xrightarrow{\partial} \pi_{n-1}(F, x_0) \\ \longrightarrow \ldots \longrightarrow \pi_0(F) \xrightarrow{j_*} \pi_0(X) \longrightarrow \pi_0(Y) \longrightarrow 0 \end{aligned} \tag{110}$$

est exacte: c'est la *suite exacte d'homotopie des fibrations*. Plusieurs cas particuliers sont importants:

1) S'il existe une section $s : (Y, y_0) \longrightarrow (X, x_0)$, on a $p_* \circ s_* = \mathrm{id.}$ d'où

$$\pi_n(X, x_0) \simeq \pi_n(Y, y_0) \oplus \pi_n(F, x_0).$$

2) Si l'injection $j : (F, x_0) \longrightarrow (X, x_0)$ est homotope à l'application constante $F \longrightarrow \{x_0\}$, on a

$$\pi_n(Y, y_0) \simeq \pi_n(X, x_0) \oplus \pi_{n-1}(F, x_0) \text{ pour } n \geq 2.$$

3) Si X est rétractile, on a $\pi_n(X, x_0) = 0$ pour tout $n \geq 1$, donc

$$\pi_n(Y, y_0) \simeq \pi_{n-1}(F, x_0) \text{ pour } n \geq 2.$$

16.6 Espaces fibrés principaux et espaces fibrés vectoriels

Un espace fibré *principal* (P, B, G, p) est un espace dont la fibre générique G est un *groupe* topologique, appelé *groupe structural* du fibré principal. Le groupe G *opère* sur P, par la définition suivante: pour tout homéomorphisme φ dans l'axiome (EF), on a

$$\varphi(y,t) \cdot s = \varphi(y, ts) \text{ pour tout } s \in G. \tag{111}$$

Les orbites de G pour cette action sont les fibres P_a, et G opère de façon simplement transitive dans chacune de ces fibres.

Les espaces fibrés principaux sont les objets d'une sous-catégorie de la catégorie des espaces fibrés, mais la définition des morphismes y est plus restrictive: pour deux espaces fibrés principaux (P, B, G, p), (P', B', G', p'), un morphisme (f, g) doit être tel en outre qu'il existe un homomorphisme de groupes topologiques $\rho : G \longrightarrow G'$, pour lequel on a

$$f(x \cdot s) = f(x) \cdot \rho(s) \text{ pour } x \in P, s \in G. \tag{112}$$

Lorsqu'il existe une section continue d'un espace fibré principal, ce fibré est trivialisable. C'est toujours le cas lorsque B est paracompact et que G est homéomorphe à un $\mathbb{R}^n$.

Un exemple important de fibré principal est celui où $P = \mathbb{S}_{2n+1}$, considéré comme sous-espace de $\mathbb{C}^{n+1}$; les fibres sont les orbites de l'action

$$(\zeta, (z_1, \ldots, z_n)) \mapsto (\zeta z_1, \zeta z_2, \ldots, \zeta z_n)$$

du groupe $\mathbb{S}_1$ des nombres complexes de valeur absolue $|\zeta| = 1$; l'espace de base est l'espace projectif complexe $\mathbb{P}_n(\mathbb{C})$, et p est l'application canonique. Lorsque $n = 1$, $\mathbb{P}_1(\mathbb{C})$ s'identifie à la sphère «de Riemann» $\mathbb{S}_2$, et on a donc un fibré principal $(\mathbb{S}_3, \mathbb{S}_2, \mathbb{S}_1, p)$; l'application $p : \mathbb{S}_3 \longrightarrow \mathbb{S}_2$ est celle utilisée par H. Hopf pour donner un exemple d'application dont l'invariant de Hopf est égal à 1.

L'exemple le plus important de fibré principal est celui où l'espace total P est un groupe de Lie, G un sous-groupe fermé, et l'espace de base B est l'espace homogène P/G des classes à droite xG pour tout $x \in P$; G opère sur P par translation à droite $(x, s) \mapsto xs$.

Un *espace fibré vectoriel* («vector bundle») réel (resp. complexe) est un espace fibré $\xi = (E, B, F, p)$, où la fibre générique F est un espace vectoriel réel (resp. complexe) de dimension finie m (appelée le *rang* du fibré); l'axiome (EF) est ici remplacé par l'axiome plus restrictif

(EFV). Pour tout $y \in B$, il existe un voisinage ouvert U de y et un homéomorphisme

$$\varphi : U \times F \longrightarrow p^{-1}(U)$$

tel que $p(\varphi(y, t)) = y$ pour $(y, t) \in U \times F$, et en outre que l'application $t \mapsto \varphi(y, t)$ soit une application $\mathbb{R}$-linéaire (resp. $\mathbb{C}$-linéaire) de F dans $p^{-1}(y)$ pour tout $y \in U$.

Ici encore, les fibrés vectoriels sont les objets d'une sous-catégorie de la catégorie des espaces fibrés, avec des morphismes plus restreints: un morphisme $(f, g) : (E, B, F, p) \longrightarrow (E', B', F', p')$ doit être tel que, pour tout $y \in B$, $f_y : E_y \longrightarrow E'_{g(y)}$ soit une application $\mathbb{R}$-linéaire (resp. $\mathbb{C}$-linéaire).

Pour les fibrés vectoriels de même base, on peut définir toutes les notions algébriques relatives aux espaces vectoriels: sous-espace, quotient, somme directe, produit tensoriel, espace d'applications linéaires, dual, etc. Par exemple, la somme directe $\xi' \oplus \xi''$ (dite aussi «somme de Whitney») de deux fibrés vectoriels $\xi' = (E', B, F', p')$, $\xi'' = (E'', B, F'', p'')$ a pour fibre générique $F' \oplus F''$, et les restrictions des homéomorphismes $\varphi : U \times (F' \oplus F'') \longrightarrow p^{-1}(U)$ à $U \times F'$ (resp. $U \times F''$) sont les homéomorphismes correspondants de ξ' et ξ''.

L'exemple le plus important de fibré vectoriel est le *fibré tangent* $T(M)$ d'une variété lisse M, dont la fibre en chaque point $y \in M$ est l'espace tangent $T_y(M)$ à M en ce point. Son dual $T^*(M)$ est appelé le *fibré cotangent* de M; les sections de $T(M)$ (resp. $T^*(M)$) sont les champs de vecteurs sur M (resp. les 1-formes différentielles).

Soit $\lambda : (P, B, G, \pi)$ un fibré principal localement compact, et supposons qu'il existe une action continue $(s, y) \mapsto s \cdot y$ de G sur un espace localement compact F; alors G opère à droite sur le produit $P \times F$ par

$$(x, y) \cdot s = (x \cdot s, s^{-1} \cdot y).$$

L'espace des orbites pour cette action s'écrit $P \times {}^G F$; il est localement compact. Soit p l'application canonique $P \times F \longrightarrow P \times {}^G F$; alors $\pi_F(p(x, y)) = \pi(x) \in B$ et si $U \subset B$ est un ouvert tel qu'il y ait une section $\sigma : U \longrightarrow \pi^{-1}(U)$ dans P, l'application $(b, y) \mapsto p(\sigma(b), y)$ est un homéomorphisme de $U \times F$ sur $\pi_F^{-1}(U)$ tel que $\pi_F(p(\sigma(b), y) = b$. Donc $(P \times {}^G F, B, F, \pi_F)$ est un fibré qu'on dit *associé* à λ et à l'action de G sur F. En particulier, tout espace fibré vectoriel réel (resp. complexe) de rang m peut être considéré comme associé à un fibré principal de groupe structural le groupe orthogonal $O(m)$ (resp. le groupe unitaire $U(m)$) et à l'action canonique de ce groupe sur $\mathbb{R}^m$ (resp. $\mathbb{C}^m$).

16.7 Images réciproques de fibrés et classification des fibrés principaux

Etant donnés un espace fibré $\lambda = (E, B, F, p)$ et une application continue $g : B' \longrightarrow B$ Whitney a défini en 1935 l'*image réciproque* («pull-back») $g^*(\lambda)$ de λ par g, de la façon suivante: on considère le «produit fibré» $E' = E \times {}_B B'$, sous-espace du produit $E \times B'$ formé des couples (x, y') tels que $p(x) = g(y')$; la projection $p' : E' \longrightarrow B'$ est l'application $(x, y') \mapsto y'$. Si $\varphi : U \times F \longrightarrow p^{-1}(U)$ est un homéomorphisme vérifiant (EF) pour λ, on définit un homéomorphisme $\varphi' : g^{-1}(U) \times F \longrightarrow p'^{-1}(g^{-1}(U))$ en prenant $\varphi'(y', t) = (y', \varphi(g(y'), t))$. L'image réciproque $g^*(\lambda) = (E', B', F, p')$.

Si $f : E' \longrightarrow E$ est l'application $(x, y') \mapsto x$, le couple (f, g) est un morphisme de $g^*(\lambda)$ dans λ, tel que la restriction de f à toute fibre $E'_{y'}$ est un homéomorphisme de $E'_{y'}$ sur $E_{g(y')}$ (les fibres se transportent). Lorsque λ est un fibré principal (resp. un fibré vectoriel), il en est de même de $g^*(\lambda)$.

Un cas particulier est celui de l'injection canonique $j : B' \longrightarrow B$ d'un sous-espace B' de B; on dit alors que $j^*(\lambda)$ est le fibré induit sur B' par λ. Par exemple, si M est une variété lisse, X un sous-espace de M et $j : X \longrightarrow M$ l'injection canonique, $j^*(T(M))$ est un fibré de base X; si X est une sous-variété lisse de M, $T(X)$ est un sous-fibré de $j^*(T(M))$ et on dit que le quotient $j^*(T(M))/T(X)$ est le *fibré normal* de X dans M.

Vers 1940, Whitney et Steenrod se posèrent le problème de trouver une *classification* des fibrés principaux de base B et de groupe structural G donnés, à B-isomorphisme près. L'idée de Whitney a été de les obtenir par image réciproque d'un fibré principal *fixe* (E', B', G, p') de même groupe structural. Le résultat de leurs recherches est que l'on doit supposer que l'espace total P' est *rétractile*. Milnor a montré en 1956 que pour tout groupe topologique G, on peut construire un fibré principal $\lambda_G = (P_G, B_G, G, p_G)$ tel que P_G soit rétractile et que tout fibré principal $\lambda = (E, B, G, p)$ soit B-isomorphe à une image réciproque $g^*(\lambda_G)$ par une application continue, lorsque λ satisfait à la condition suivante: il y a une suite d'applications continues $u_n : B \longrightarrow [0, 1]$ telle que les ensembles ouverts $u_n^{-1}(]0, 1])$ forment un recouvrement de B et que le fibré induit par λ sur chacun de ces ensembles soit trivialisable (fibrés «numérables»). Pour que deux applications continues g_1, g_2 de B dans B_G donnent des images réciproques B-isomorphes, il faut et il suffit que g_1 et g_2 soient homotopes; l'ensemble des classes de fibrés principaux B-isomorphes, de groupe structural G, est donc en correspondance bijective avec l'ensemble $[B; B_G]$. On dit que λ_G est le fibré *universel* et B_G l'espace *classifiant* pour les fibrés principaux de groupe structural G.

Lorsque B est un CW-complexe de dimension $\leq n$, on peut obtenir une classification des fibrés principaux de base B sans utiliser la construction de Milnor: il suffit de considérer un fibré principal (P', B', G, p') tel que $\pi_j(P') = 0$ pour $1 \leq j \leq n$; on dit qu'un tel fibré est *n-universel* et sa base *n-classifiante*. Lorsque G est le groupe orthogonal $0(r)$, on peut prendre pour P' la *variété de Stiefel* réelle $\mathbb{S}_{N,r}(\mathbb{R})$ où $N > \max(n, r)$; c'est le sous-espace de $\mathbb{R}^{Nr}$ formé des systèmes $(x_1, x_2, \ldots, x_r)$ de r vecteurs orthogonaux x_j dans $\mathbb{R}^N$. L'espace de base est la *grassmannienne* réelle $\mathbb{G}_{N,r}(\mathbb{R})$ dont les éléments sont les sous-espaces vectoriels de $\mathbb{R}^N$ de dimension r, et p' est l'application qui, au système de vecteurs $(x_1, \ldots, x_r)$, fait correspondre le sous-espace vectoriel qu'ils engendrent. On a un fibré principal n-universel pour le groupe unitaire $U(r)$ en remplaçant le produit scalaire euclidien dans $\mathbb{R}^N$ par le produit scalaire hermitien dans $\mathbb{C}^N$, le fibré ayant pour espace total la variété de Stiefel complexe $\mathbb{S}_{N,r}(\mathbb{C})$ et pour base la grassmannienne complexe $\mathbb{C}_{N,r}(\mathbb{C})$.

La classification des fibrés principaux de base B et de groupe G entraîne aussitôt celle des fibrés associés à une action de G sur un espace F. En particulier, on obtient les fibrés vectoriels réels (resp. complexes) de rang r comme images réciproques du fibré vectoriel $(U_{N,r}(\mathbb{R}), \mathbb{G}_{N,r}(\mathbb{R}), \mathbb{R}^r, p)$ (resp. $(U_{N,r}(\mathbb{C}), \mathbb{G}_{N,r}(\mathbb{C}), \mathbb{C}^r, p)$ où $U_{N,r}(\mathbb{R})$ (resp. $U_{N,r}(\mathbb{C})$) est appelé le *fibré tautologique*, la fibre au point V, espace vectoriel de dimension r, étant l'ensemble des vecteurs appartenant à V.

16.8 Classes caractéristiques

Dans leurs premiers travaux sur les fibrés vectoriels, Stiefel et Whitney, en 1935, cherchaient des conditions pour qu'un fibré vectoriel réel $\xi = (E, B, F, p)$, où B est un complexe simplicial, admette une section $s : B \longrightarrow E$ telle que $s(y) \neq 0$ pour tout $y \in B$. Ils trouvèrent qu'une condition nécessaire est l'annulation d'une certaine classe de cohomologie dans $H^{\cdot}(B; \mathbb{F}_2)$. En 1942, Pontrjagin observa que la classification de Whitney-Steenrod, pour les fibrés principaux de base B, détermine en particulier, pour un fibré vectoriel $\xi = (E, B, \mathbb{R}^m, p)$ une sous-algèbre de l'algèbre de cohomologie $H^{\cdot}(B; \mathbb{Z})$, à savoir l'image

$$g^*(H^{\cdot}(\mathbb{G}_{N+m,m}(\mathbb{R})); \mathbb{Z})$$

où $g : B \longrightarrow \mathbb{G}_{N+m,m}(\mathbb{R})$ définit le fibré ξ par image et est déterminé à homotopie près. La décomposition cellulaire canonique (les «cellules de Schubert») qu'avait donnée Ehresmann en 1937 définit des générateurs bien déterminés de cette sous-algèbre; leurs images par g^* sont des éléments $p_k(\xi) \in H^{4k}(B; \mathbb{Z})$ qu'on appelle maintenant les *classes de Pontrjagin* de ξ. Pontrjagin a aussi considéré le fibré principal universel correspondant au groupe orthogonal unimodulaire $\mathrm{SO}(m)$. L'espace classifiant correspondant est la grassmannienne $\mathbb{G}'_{N+m,m}(\mathbb{R})$ des sous-espaces vectoriels *orientés* de dimension m, et les images réciproques correspondantes sont les fibrés vectoriels orientés. Pour un tel fibré de rang pair $m = 2m'$, il y a en outre un générateur $e(\xi) \in H^m(B; \mathbb{Z})$ appelé *classe d'Euler* et tel que $e(\xi) \smile e(\xi) = p_{m'}(\xi)$. Les *classes de Stiefel-Whitney* $w_k(\xi) \in H^k(B; \mathbb{F}_2)$ se définissent de même à partir de l'algèbre $H^{\cdot}(\mathbb{G}_{N+m,m}(\mathbb{R}); \mathbb{F}_2)$.

En 1945, S. Chern a défini de même des classes caractéristiques pour les fibrés vectoriels complexes, en remplaçant les grassmanniennes réelles par les grassmanniennes complexes; ce sont cette fois des éléments $c_k(\xi) \in H^{2k}(B; \mathbb{Z})$ qu'on appelle *classes de Chern* du fibré vectoriel. Les classes de Pontrjagin d'un fibré vectoriel réel peuvent se définir à partir des classes de Chern du fibré complexifié $\xi \otimes \mathbb{C}$ du fibré vectoriel réel ξ.

Indépendamment, Pontrjagin et Chern ont montré que, pour le fibré tangent d'une variété riemannienne ou hermitienne M, les classes caractéristiques à valeurs dans $\mathbb{R}$, considérées comme appartenant à la cohomologie de de Rham, s'expriment comme classes de cohomologie de formes différentielles explicitement données par la structure riemannienne ou hermitienne.

Les classes de Chern peuvent s'interpréter comme les obstructions à l'existence de sections de la variété de Stiefel complexe.

16.9 La suite spectrale des espaces fibrés

Un des buts de Leray en définissant les suites spectrales, et notamment celle qu'on appelle «suite de Leray», était d'appliquer cette dernière à la cohomologie des espaces fibrés $\lambda = (E, B, F, p)$. Il suppose E, B, F localement compacts et utilise la cohomologie d'Alexander-Spanier à coefficients dans un A-module M. Lorsque $\pi_1(B)$ opère trivialement sur la cohomologie d'une fibre $H^{\cdot}(E_a; M)$, la suite spectrale a pour termes E_2

$$E_2^{pq} = H^p(B; H^q(F; M)).$$

Lorsque M est un corps, ou que $M = \mathbb{Z}$ et que B ou F n'ont pas de torsion, ce terme est isomorphe à $H^p(B; M) \otimes H^q(F; M)$. Ces expressions lui permirent de prouver de nombreux résultats sur les relations entre les cohomologies de B, F et E. Si deux des trois modules $H^{\cdot}(B;M)$, $H^{\cdot}(F;M)$, $H^{\cdot}(E;M)$ sont de type fini, il en est de même du troisième. Lorsque M est un corps, les nombres de Betti correspondants $b_k(X) = \dim_M(H_k(X;M))$ sont tels que

$$b_k(E) \leq b_k(B \times F), \quad \chi(E) = \chi(B)\chi(F).$$

Supposons que l'espace B possède un revêtement universel; soit X un revêtement de B correspondant à un sous-groupe normal Π de $\pi_1(B,a)$. Leray et H. Cartan ont montré qu'il existe, pour tout anneau A de coefficients, une suite spectrale dont le terme E_2 est formé de groupes de cohomologie du groupe Π : $E_2^{pq} = H^p(\Pi; H^q(X;A))$ et dont l'aboutissement est $H^{\cdot}(B;a)$ muni d'une filtration. On en déduit diverses relations entre les cohomologies de B et de X, et la cohomologie du groupe Π.

Dans ses travaux sur les groupes d'homotopie (voir plus loin), J.-P. Serre utilisa des suites spectrales déduites d'une fibration générale, dont les fibres ne sont plus localement compactes; il dut donc utiliser la cohomologie singulière au lieu de la cohomologie d'Alexander-Spanier, et adapter des raisonnements de Leray à ce cas; il eut aussi besoin de définir une suite spectrale en homologie singulière, où la filtration est croissante.

Des suites spectrales de Leray et Serre, il est possible de déduire d'importantes suites exactes, obtenues antérieurement par d'autres méthodes. Si $\xi = (E,B;\mathbb{R}^m,p)$ est un fibré vectoriel réel orienté de rang m, soit E^0 l'ouvert de l'espace total E, complémentaire de la section nulle (identifiée à B); on a alors la *suite exacte de Gysin*

$$\begin{aligned} \ldots \longrightarrow H^{r-m}(B;\mathbb{Z}) \xrightarrow{g} H^r(B;\mathbb{Z}) \xrightarrow{p_0^*} H^r(E^0;\mathbb{Z}) \\ \longrightarrow H^{r-m+1}(B;\mathbb{Z}) \longrightarrow \ldots \end{aligned} \tag{113}$$

où p_0 est la restriction de p à E^0, et g le cup-produit $c \mapsto c \smile e(\xi)$ par la classe d'Euler de ξ. Il y a des suites exactes analogues pour l'homologie et pour les fibrés vectoriels non orientés, lorsque l'anneau des coefficients est $\mathbb{F}_2$.

Si $\lambda = (E, \mathbb{S}_n, F, p)$ est un espace fibré dont la base est une sphère $\mathbb{S}_n$ et dont la fibre F est un complexe simplicial, on a la *suite exacte de Wang* en homologie singulière

$$\ldots \longrightarrow H_{r-n+1}(F) \longrightarrow H_r(F) \longrightarrow H_r(E) \longrightarrow H_{r-n}(F) \longrightarrow \ldots \tag{114}$$

et il y a une suite exacte analogue en cohomologie singulière.

17 Applications des fibrations

17.1 La tour de Postnikov

Le théorème de Hopf montrant que le quotient $H_2(X)/h_2(\pi_2(X))$ ne dépend que du groupe $\pi_1(X)$ souleva aussitôt la question de savoir si le groupe $h_2(\pi_2(X))$ ne dépend aussi que des groupes $\pi_1(X), \pi_2(X)$ et de l'action du premier sur le second. En 1946, Eilenberg et MacLane montrèrent qu'il n'en est rien, et en outre définirent un élément explicite $k^3 \in H^3(\pi_1(X); \pi_2(X))$ qui, joint aux autres données, détermine entièrement le groupe $H_2(X; \mathbb{Z})$.

Ce résultat a été considérablement généralisé par M. Postnikov en 1957. Il a réussi à décrire, pour un CW-complexe X, sa classe d'équivalence d'homotopie faible, à l'aide d'un système d'invariants généralisant l'invariant k^3 d'Eilenberg-MacLane. Il forme une suite

$$Z_0 \xleftarrow{p_1} Z_1 \xleftarrow{p_2} Z_2 \longleftarrow \dots \xleftarrow{p_n} Z_n \longleftarrow \dots$$

où $(Z_{n+1}, Z_n, F_n, p_{n+1})$ est une fibration dont la fibre F_n est l'espace d'Eilenberg-MacLane $K(\pi_{n+1}(X), n+1)$, et la projection p_{n+1} est complètement déterminée par une classe de cohomologie

$$k^{n+2} \in H^{n+2}(Z_n; \pi_{n+1}(X)).$$

Il y a en outre un système projectif d'applications continues $f_n : X \longrightarrow Z_n$, et la limite projective

$$f = \varprojlim f_n : X \longrightarrow Z = \varprojlim Z_n$$

est une équivalence d'homotopie faible. En vertu du théorème de Whitehead, l'homologie

$$H.(X; \mathbb{Z}) \simeq H.(Z; \mathbb{Z})$$

est donc entièrement déterminée par les groupes d'homotopie $\pi_n(X)$ et les invariants de Postnikov.

17.2 Les groupes d'homotopie des sphères

En 1951, J.-P. Serre fit faire de grands progrès à la détermination des groupes d'homotopie $\pi_m(\mathbb{S}_n)$ par l'utilisation de la fibration définie par l'espace des chemins d'un espace, et de la suite spectrale des fibrations. Son plus remarquable théorème est que si n est impair, ou si n est pair et $m \neq 2n-1$, le groupe $\pi_m(\mathbb{S}_n)$ (où $m > n$) est *fini*; pour n pair et $m = 2n-1$, $\pi_{2n-1}(\mathbb{S}_n)$ est somme directe de $\mathbb{Z}$ et d'un groupe fini.

Lorsque n est impair, Serre divise la preuve en deux parties:

1) $\pi_m(\mathbb{S}_n)$ est un groupe de type fini.

2) Pour un corps K de caractéristique 0, $\pi_m(\mathbb{S}_n) \otimes K = 0$.

Les deux parties suivent une même méthode générale. Partant d'un CW-complexe fini X, on forme la suite d'espaces

$$X_0 = X, T_1, X_1, T_2, X_2, \dots$$

où T_j est le revêtement universel de X_{j-1} et $X_j = \Omega(T_j)$ l'espace des lacets de T_j; on a alors $\pi_{n+1}(X) = H_1(X_n;\mathbb{Z})$. Sous la seule hypothèse que les groupes $H_j(X;\mathbb{Z})$ sont de type fini, la suite spectrale des espaces de chemins et la suite spectrale des revêtements montrent que les $H_j(X_m;\mathbb{Z})$ sont de type fini pour *tout* m. Cela implique l'assertion 1) pour $X = \mathbb{S}_n$.

Pour prouver l'assertion 2), on note d'abord que pour $X = \mathbb{S}_n$ et $m \leq n-1$, la suite exacte de Wang en cohomologie montre que $H^{\cdot}(X_m;K)$ est alternativement; pour m impair, une K-algèbre de polynômes engendrée par un unique élément homogène de degré $n - m$; pour m pair, une algèbre extérieure engendrée par un unique élément homogène de degré $n - m$. La suite spectrale des revêtements prouve que $H_i(T_n;K) = 0$ pour $i > 0$. On en déduit que $\pi_i(T_n) \otimes K = 0$ pour tout $i > 0$, et la conclusion résulte de ce que $\pi_i(T_n) = \pi_{i+n-1}(\mathbb{S}_n)$. Les détails sont compliqués et reposent essentiellement sur l'usage des suites spectrales.

17.3 La topologie des groupes de Lie

L'application des suites spectrales à la fibration $(G, G/H, H, p)$ définie par un sous-groupe fermé H d'un groupe de Lie G, a été commencée dès 1946 par Leray, tout d'abord lorsque G est compact et H un sous-groupe contenant un tore maximal T de G; il prouva une formule conjecturée par G. Hirsch et qui donne les nombres de Betti de G/H lorsque ceux de G et ceux de H sont connus.

Plusieurs mathématiciens ont ensuite étudié la cohomologie des espaces fibrés principaux dont le groupe structural est un groupe de Lie. Les résultats les plus complets ont été obtenus par A. Borel en 1953, à l'aide d'une étude algébrique originale des suites spectrales correspondantes. Leur principale application concerne les relations entre la cohomologie d'un groupe de Lie compact G et celle de son espace classifiant B_G. Le fait essentiel est que si K est un corps (de caractéristique quelconque) et si $H^{\cdot}(G;K)$ est une algèbre extérieure $\Lambda(x_1,\ldots,x_m)$ sur K engendrée par des éléments homogènes x_j de degrés impairs (condition vérifiée lorsque K est de caractéristique 0, par le théorème de Hopf), alors

$$H^{\cdot}(B_G;K) \simeq K[y_1,\ldots,y_m]$$

algèbre de polynômes sur K, où chaque y_j est homogène de degré

$$\deg y_j = 1 + \deg x_j.$$

La preuve est basée sur le fait que la suite spectrale du fibré principal universel (P_G, B_G, G) a pour terme E_2.

$$K[y_1,\ldots,y_m] \otimes \Lambda(x_1,\ldots,x_m)$$

une algèbre sur K introduite peu de temps auparavant par A. Weil pour l'étude des algèbres de Lie lorsque $K = \mathbb{R}$. Les résultats de Borel ont non seulement permis de calculer les nombres de Betti de G et des espaces homogènes G/H, mais aussi leurs coefficients de torsion.

Un autre problème abordé après 1950 a été la détermination des groupes d'homotopie $\pi_i(G)$ pour un groupe de Lie compact G. Pour les groupes de Lie classiques $U(n, K)$ avec $K = \mathbb{R}, \mathbb{C}$ ou $\mathbb{H}$[17]), plusieurs méthodes, réduisant le problème au calcul des groupes d'homotopie de sphères, avaient permis de calculer les $\pi_i(G)$ pour $i \leq 12$. On constata, comme pour les sphères, un phénomène de stabilité de ces groupes: pour i donné, tous les groupes $\pi_i(U(n))$ sont isomorphes dès que $n > i/2$. Mais un nouveau phénomène se présenta: pour n fixé et $2 \leq i < 2n$, les groupes $\pi_i(U(n))$ sont alternativement 0 et $\mathbb{Z}$, tandis qu'ils deviennent très irréguliers lorsque $i \geq 2n$. En 1956, R. Bott montra, par des méthodes entièrement nouvelles, que cette périodicité a lieu pour tous les couples (n, i) tels que $i < 2n$; il établit des périodicités analogues mais de période 8, pour les groupes $SO(n)$ et $U(n, \mathbb{H})$.

La méthode repose sur la théorie de Morse. L'espace d'orbites $U(2m)/U(m)$ est la variété de Stiefel complexe $\mathbb{S}_{2m,m}(\mathbb{C})$, et l'espace d'orbites $\mathbb{S}_{2m,m}(\mathbb{C})/U(m)$ la grassmannienne complexe $\mathbb{G}_{2m,m}(\mathbb{C})$; par la suite exacte d'homotopie des fibrés, on a $\pi_{i-1}(U(m)) \simeq \pi_i(\mathbb{G}_{2m,m}(\mathbb{C}))$ pour $i < 2m$; on a d'autre part $\pi_j(SU(n)) \simeq \pi_j(U(n))$ pour $j \geq 2$ et tout n. Le point essentiel est de montrer que

$$\pi_{i+1}(SU(2m)) \simeq \pi_i(\mathbb{G}_{2m,m}(\mathbb{C})) \text{ pour } i < 2m,$$

et par définition des groupes d'homotopie, cela équivaut à

$$\pi_i(\Omega(SU(2m), \mathbb{I}_{2m})) \simeq \pi_i(\mathbb{G}_{2m,m}(\mathbb{C})).$$

L'espace des lacets $\Omega(SU(2m), \mathbb{I}_{2m})$ d'origine l'élément neutre $\mathbb{I}_{2m}$, a même type d'homotopie que l'espace P des chemins joignant $\mathbb{I}_{2m}$ et $-\mathbb{I}_{2m}$. Bott montre d'abord que pour $i < 2m$

$$\pi_i(P) \simeq \pi_i(P_{\min}) \tag{115}$$

où $P_{\min}$ est le sous-espace de P formé des géodésiques minimales joignant $\mathbb{I}_{2m}$ et $-\mathbb{I}_{2m}$ pour la structure riemannienne canonique du groupe de Lie $SU(2m)$. Ces géodésiques étant connues explicitement par la théorie des groupes de Lie, Bott put montrer que le groupe (115) est bien isomorphe à $\pi_i(\mathbb{G}_{2m,m}(\mathbb{C}))$, par des raisonnements faisant ample usage de la théorie de Morse.

---

17) $U(n, K)$ est le groupe des automorphismes $u \in GL_n(K)$ qui laissent invariante la forme

$$x_1\bar{x}_1 + x_2\bar{x}_2 + \ldots + x_n\bar{x}_n$$

où $\bar{x} = x$ si $K = \mathbb{R}$, $\bar{x} = a - bi$ si $x = a + bi$ dans $\mathbb{C}$ et $\bar{x} = a - bi - cj - dk$ si $x = a + bi + cj + dk$ dans $\mathbb{H}$. Le groupe $U(n, \mathbb{R})$ se note $0(n)$ et est appelé le groupe orthogonal; le groupe $U(n, \mathbb{C})$ est appelé groupe unitaire et noté $U(n)$; enfin le groupe $U(n, \mathbb{H})$ est le groupe unitaire quaternionique et se note aussi $Sp(n)$.
On note $SO(n)$ le groupe orthogonal unimodulaire, dont les éléments ont pour déterminant +1; de même $SU(n)$ est le groupe unitaire unimodulaire, dont les éléments ont pour déterminant +1.

Il est commode d'interpréter le théorème de Bott en introduisant le «groupe unitaire infini» $U(\infty) = \varinjlim U(m)$. Le théorème s'exprime alors en disant que l'on a des équivalences d'homotopie faible

$$U(\infty) \simeq \Omega(B_{U(\infty)}) \text{ et } B_{U(\infty)} \times \mathbb{Z} \simeq \Omega(U(\infty))$$

pour l'espace classifiant de $U(\infty)$. D'autres preuves de ces équivalences n'utilisant pas la théorie de Morse ont été obtenues depuis lors.

* * *

Les développements de la Topologie algébrique depuis 1950 ont présenté un foisonnement tellement ample qu'il est tout à fait impossible de mentionner tous les résultats importants en restant dans des limites raisonnables; je devrai donc me limiter à un nombre restreint de résultats, choisis parmi les plus spectaculaires ou les plus surprenants. Les méthodes qui y conduisent sont toujours d'une complexité et d'une technicité qui excluent toute tentative de description d'allure élémentaire et je devrai les passer sous silence.

* * *

17.4 L'algèbre de Steenrod

En 1947, N. Steenrod prouva l'existence, en cohomologie des CW-complexes finis à coefficients dans $\mathbb{F}_2$, d'applications fonctorielles, les «carrés de Steenrod»

$$Sq^i : H^m(X; \mathbb{F}_2) \longrightarrow H^{m+i}(X; \mathbb{F}_2) \text{ pour tout } i > 0.$$

La définition en est compliquée; pour $i = m$, $Sq^m(u)$ est le «cup-carré» $u \smile u$. En 1950, il put définir des applications analogues pour tout nombre premier $p > 2$

$$P_n^k : H^m(X; \mathbb{F}_p) \longrightarrow H^{m+2k(p-1)}(X; \mathbb{F}_p);$$

sa définition, tout aussi compliquée, utilise le produit oblique et l'homologie des groupes.

Les divers carrés de Steenrod Sq^i ne sont pas indépendants. En 1952, J. Adem explicita certaines relations entre ces applications, par exemple, pour $a < 2b$,

$$Sq^a Sq^b = \sum_{0 \leq c \leq a/2} \binom{b-c-1}{a-2c} Sq^{a+b-c} Sq^c.$$

En 1953, il obtint des relations analogues entre les P_p^k. En 1954, en liaison avec sa détermination de l'homologie des espaces d'Eilenberg-MacLane, H. Cartan considéra l'algèbre $\mathcal{A}_p$ engendrée par les opérateurs P_p^k pour tous les indices k, et par l'opérateur β de Bockstein correspondant à l'homomorphisme $z \mapsto pz$ de $\mathbb{Z}/p^2\mathbb{Z}$ sur $\mathbb{Z}/p\mathbb{Z}$; il détermina une base de l'espace vectoriel $\mathcal{A}_p$ sur le corps $\mathbb{F}_p$. Puis Milnor, en 1958, prouva que $\mathcal{A}_p$ est une algèbre de Hopf, dont il détermina explicitement la multiplication et la comultiplication. Il n'est pas possible de résumer les méthodes employées par tous ces auteurs.

Les algèbres de Steenrod $\mathcal{A}_p$ sont rapidement devenues des outils très performants en Topologie algébrique. Steenrod avait découvert les Sq^i en étudiant le problème suivant, posé par Pontrjagin en 1938. Lorsque K est un complexe simplicial de dimension n, l'ensemble $[K; \mathbb{S}_n]$ est déterminé par la relation $[K; \mathbb{S}_n] \simeq H^n(K; \mathbb{Z})$ prouvée par Hopf, Hurewicz et Whitney, en considérant l'application $[f] \mapsto f^*(s_n)$, où s_n est la classe fondamentale de $\mathbb{S}_n$. Il s'agissait de déterminer $[K; \mathbb{S}_n]$ lorsque la dimension de K est $n+1$, et Pontrjagin résolut le problème pour $n = 2$ en 1941. Steenrod en obtint une solution pour tout n en procédant par récurrence sur la dimension des squelettes de K et utilisant la théorie des obstructions: pour tout $c \in H^n(K; \mathbb{Z})$, l'ensemble des classes d'homotopie des applications continues $f : K \longrightarrow \mathbb{S}_n$ telles que $f^*(s_n) = c$ est en bijection canonique avec le quotient

$$H^{n+1}(K; \mathbb{F}_2)/Sq^2(H^{n-1}(K; \mathbb{F}_2)).$$

Une autre application des carrés de Steenrod a été faite par Serre, dès 1951, à la détermination de la cohomologie des espaces d'Eilenberg-MacLane $K(\mathbb{Z}/2\mathbb{Z}, q)$, à coefficients dans $\mathbb{F}_2$. Utilisant les résultats de Borel sur les fibrés principaux, il montra que $H^{\cdot}(K(\mathbb{Z}/2\mathbb{Z}, q); \mathbb{F}_2)$ est formé de polynômes par rapport à des produits d'éléments $Sq^j(u_q)$ explicitement déterminés, où u_q est l'unique élément $\neq 0$ de $H^q(K(\mathbb{Z}/2\mathbb{Z}, q); \mathbb{F}_2) \simeq \mathbb{F}_2$.

Les carrés de Steenrod ont aussi de remarquables relations avec les classes caractéristiques des espaces fibrés. Soit $\xi = (E, B, F, p)$ un fibré vectoriel réel de rang m. Si E^0 est l'ouvert de l'espace total E, complémentaire de la section nulle (identifiée à B), et u l'unique élément $\neq 0$ de $H^m(E_a, E_a \cap E^0; \mathbb{F}_2)$, un théorème général de Leray-Hirsch montre que l'homomorphisme

$$\Phi_j : c \mapsto p^*(c) \smile u$$

de $H^j(B; \mathbb{F}_2)$ dans $H^{j+m}(E, E^0; \mathbb{F}_2)$ est bijectif; on dit que c'est l'*isomorphisme de Thom-Gysin*. Thom a montré que les classes de Stiefel-Whitney $w_j(\xi)$ sont données par

$$w_j(\xi) = \Phi_0^{-1}(Sq^j(\Phi_0(1))).$$

Sa démonstration utilise une triangulation de B et la définition de $w_j(\xi)$ comme obstruction à l'existence de sections de ξ; d'autres preuves ont été données depuis lors.

Thom s'intéressait particulièrement au cas où ξ est le fibré tangent $\tau = T(M)$ à une variété lisse; il découvrit que les classes $w_j(\tau)$ ne dépendent que de la topologie de M et non de sa structure différentielle. Cela résulte d'une interprétation de ces classes due à Wu Wen Tsün. Si M est compacte, connexe et de dimension n, on a $H^n(M;\mathbb{F}_2) \simeq \mathbb{F}_2$; si u est l'unique élément $\neq 0$ de ce groupe, on a par dualité de Poincaré

$$x_k \cup y_{n-k} = \langle x_k, y_{n-k}\rangle u$$

pour $x_k \in H^k(M;\mathbb{F}_2)$, $y_{n-k} \in H^{n-k}(M;\mathbb{F}_2)$. Il y a alors, pour $0 \leq k \leq n$, une classe unique $v_k \in H^k(M;\mathbb{F}_2)$ telle que

$$v_k \smile x = Sq^k(x) \text{ pour } x \in H^{n-k}(M;\mathbb{F}_2).$$

Ces classes (appelées maintenant classes de Wu) sont telles que

$$w_k(\tau) = \sum_{i+j=k} Sq^i(v_j)$$

ce qui prouve le théorème de Thom.

Par contre, J. Milnor a montré en 1964 que les classes de Pontrjagin de $\tau = T(M)$ dépendent de la structure différentielle de M.

17.5 La cohomologie des espaces d'Eilenberg-MacLane

L'utilisation des espaces d'Eilenberg-MacLane dans de nombreuses questions exige qu'on connaisse leur homologie. A l'aide de la suite spectrale des chemins dans un espace $K(\Pi, q)$ Serre, en 1951, a obtenu l'expression de la cohomologie $H^{\cdot}(K(\Pi, n); k)$ lorsque k est un corps et Π un groupe commutatif de type fini. En 1950, Eilenberg et MacLane ont attaqué le problème de façon purement algébrique, en construisant un complexe de chaînes et en itérant cette construction (dite «bar construction»), de sorte que l'homologie de ces complexes soit celle des espaces $K(\Pi, n)$. Ils purent ainsi déterminer explicitement les groupes $H_{n+q}(K(\mathbb{Z}, n); \mathbb{Z})$ pour $n > q$ et $q \leq 10$.

H. Cartan, en 1954, a défini une suite de «constructions» algébriques plus générales que celles d'Eilenberg-MacLane. Elles font intervenir des triplets d'algèbres graduées qui imitent, en algèbre, les complexes de chaînes de l'espace total, de l'espace de base et de la fibre d'une fibration. L'exposition détaillée est très longue et très complexe, mais conduit en principe à la détermination explicite des groupes $H_n(K(\Pi, m); A)$ pour tout anneau commutatif A, lorsque Π, m et n sont donnés. Comme le montraient les premiers résultats de Serre, les algèbres que l'on obtient sont bâties à partir d'algèbres de polynômes et d'algèbres extérieures.

17.6 Calculs de groupes d'homotopie et localisations

Jusqu'en 1950, on ne connaissait les groupes d'homotopie $\pi_{n+k}(\mathbb{S}_n)$ que pour $k = 1$; Pontrjagin et G.W. Whitehead prouvèrent les premiers que $\pi_{n+2}(\mathbb{S}_n) \simeq \mathbb{Z}/2\mathbb{Z}$ pour tout n. En 1951-52, il fallut les efforts combinés de plusieurs mathématiciens pour voir que $\pi_6(\mathbb{S}_3) \simeq \mathbb{Z}/12\mathbb{Z}$. A l'aide des carrés de Steenrod, Serre put ensuite déterminer les $\pi_{n+k}(\mathbb{S}_n)$ pour $3 \leq k \leq 8$. Après 1953, H. Toda et ses élèves se sont consacrés au calcul des $\pi_{n+k}(\mathbb{S}_n)$ en utilisant de nombreuses méthodes spéciales; chaque année, le nombre des valeurs de k pour lesquelles les groupes sont connus augmente peu à peu. Mais même les spécialistes sont découragés devant l'extrême irrégularité des résultats, en l'absence d'une idée nouvelle qui pourrait éclairer le problème.

Quelques résultats généraux ont été obtenus, notamment grâce à l'étude par de nombreux mathématiciens des espaces $\Omega\Sigma X$ et des itérés $\Omega^m\Sigma^n X$. I.M. James a découvert une suite exacte infinie valable pour n impair

$$\ldots \longrightarrow \pi_q(\mathbb{S}_n) \xrightarrow{E} \pi_{q+1}(\mathbb{S}_{n+1}) \xrightarrow{H} \pi_{q+1}(\mathbb{S}_{2n+1}) \longrightarrow \pi_{q-1}(\mathbb{S}_n) \longrightarrow \ldots \tag{116}$$

où E est la suspension homotopique et H une généralisation de l'invariant de Hopf. Pour tout n, la suite (116) est encore exacte à partir de $\pi_{3n-2}(\mathbb{S}_n)$.

G.W. Whitehead utilise l'application canonique $\mathbb{S}_n \longrightarrow \mathbb{S}_n \vee \mathbb{S}_n$, où $\mathbb{S}_n \vee \mathbb{S}_n$ est considéré comme la contraction $\mathbb{S}_n/\mathbb{S}_{n-1}$ de l'équateur en un point.

P. Hilton a pu déterminer explicitement les groupes d'homotopie des bouquets de sphères

$$W = \mathbb{S}_{n_1} \vee \mathbb{S}_{n_2} \vee \ldots \vee \mathbb{S}_{n_k};$$

à l'aide de produits de Whitehead itérés, il montre par une preuve longue et difficile que $\pi_n(W)$ est somme directe de groupes $\pi_n(\mathbb{S}_{m(a)})$ où a est un indice qui parcourt un ensemble de «produits de Whitehead basiques» à chacun desquels on attribue un «poids» $m(a)$. L'outil essentiel est toujours l'utilisation de l'espace des lacets de W et des sphères. Appliqué au cas $\mathbb{S}_n \vee \mathbb{S}_n$, cela entraîne l'existence d'homomorphismes canoniques $H_0, H_1, \ldots$, qui généralisent l'invariant de Hopf.

Faute de lois apparentes dans l'expression des $\pi_{n+k}(\mathbb{S}_n)$, Serre a été le premier à chercher si l'on peut obtenir des régularités pour les *composants p-primaires* de ces groupes, tout au moins pour certaines valeurs du nombre premier p. Il étudie pour cela les groupes $\pi_i(\mathbb{S}_n) \otimes \mathbb{F}_p$, par les mêmes méthodes que celles qu'il avait appliquées au cas où l'on tensorise par un corps de caractéristique 0. Par exemple, il montre que si n est impair, on a $\pi_i(\mathbb{S}_n) \otimes \mathbb{F}_p = 0$ pour $n < i < n + 2p - 3$.

Cette idée a débouché sur une nouvelle idée géométrique, la localisation d'un CW-complexe X en un nombre premier p; c'est un CW-complexe $X_{(p)}$ qu'on forme par attachement de cellules à X, et qui a la propriété que les groupes d'homologie et d'homotopie de $X_{(p)}$ s'obtiennent en tensorisant ceux de X par l'anneau local $\mathbb{Z}_{(p)}$ (anneau des nombres rationnels r/s où s n'est pas multiple de p). Beaucoup de travaux sur ces espaces ont été faits depuis 30 ans avec des applications aux p-composantes des $\pi_i(\mathbb{S}_n)$.

18 L'homologie et la cohomologie généralisées

18.1 Les groupes de Grothendieck

Au cours de ses premières recherches en géométrie algébrique (1957), en vue de généraliser le théorème de Riemann-Roch, Grothendieck se trouva en présence du problème suivant: si l'on considère les classes de B-isomorphisme des fibrés vectoriels complexes (de rang quelconque) de base donnée B, peut-on donner un sens à une «combinaison linéaire» de ces classes à coefficients entiers? Pour deux fibrés E_1, E_2, la classe de B-isomorphisme de la somme directe $E_1 \oplus E_2$ ne dépend que des classes de E_1 et E_2; mais les combinaisons à coefficients négatifs ne semblent pas avoir de sens. Le problème se pose de façon générale: étant donné un monoïde M (ensemble muni d'une loi additive $(m_1, m_2) \mapsto m_1 + m_2$ associative, commutative et ayant un élément neutre 0), lui associer un groupe commutatif $K(M)$ tel que, pour tout homomorphisme de monoïdes $u : M \longrightarrow G$, où G est un groupe commutatif, u se *factorise* de façon unique en $u : M \xrightarrow{\varphi} K(M) \xrightarrow{v} G$ où φ est un homomorphisme canonique de monoïdes et v un *homomorphisme de groupes*.

Grothendieck résolut ce problème par la méthode analogue à celle suivie par Whitney en 1938 pour définir le produit tensoriel, et devenue usuelle dans tous les problèmes «universels». On forme le $\mathbb{Z}$-module $\mathbb{Z}[M]$ des combinaisons linéaires formelles $\sum_j \lambda_j \{m_j\}$ des «singletons» de M, à coefficients dans $\mathbb{Z}$; $K(M)$ est le quotient de ce groupe additif par le sous-groupe engendré par les éléments $\{m + m'\} - \{m\} - \{m'\}$, et φ applique m sur la classe de $\{m\}$ dans $K(M)$; on dit maintenant que $K(M)$ est le *groupe de Grothendieck* de M.

En 1959, Atiyah et Hirzebruch appliquèrent cette définition pour définir un foncteur contravariant K, de la catégorie **CW** dont les objets sont les CW-complexes finis et les morphismes les applications continues, dans la catégorie **Ann** des anneaux commutatifs. Pour tout fibré vectoriel complexe E sur un CW-complexe X, soit $\gamma(E)$ la classe dans le monoïde M des classes de X-isomorphismes. L'objet $K(X)$ est le groupe de Grothendieck $K(M)$, muni en outre de la multiplication définie par $\gamma(E_1) \cdot \gamma(E_2) = \gamma(E_1 \otimes E_2)$; pour un morphisme $g : X \longrightarrow Y$, $K(g) : K(Y) \longrightarrow K(X)$ est l'application $\gamma(E) \mapsto \gamma(g^*(E))$.

Pour un espace P réduit à un point, on a $K(P) \simeq \mathbb{Z}$. Pour tout CW-complexe fini X l'application $P \longrightarrow \{x_0\}$ pour un $x_0 \in X$ définit un homomorphisme d'anneaux $K(X) \longrightarrow \mathbb{Z}$, qui est surjectif; on note $\tilde{K}(X)$ l'idéal de $K(X)$, noyau de cet homomorphisme.

Le théorème de classification des fibrés vectoriels montre que $K(X)$ peut être identifié au groupe $[X, B_{U(\infty)}]$ des classes d'homotopie dans le classifiant du groupe unitaire $U(\infty)$. Comme le théorème de Bott définit une équivalence d'homotopie faible $B_{U(\infty)} \simeq \Omega^2(B_{U(\infty)})$, Atiyah et Hirzebruch obtinrent le résultat principal de leurs recherches: l'application canonique

$$\beta : K(X) \otimes_{\mathbb{Z}} K(\mathbb{S}_2) \longrightarrow K(X \times \mathbb{S}_2) \tag{117}$$

définie par

$$\beta(\gamma(E)\otimes\gamma(E'))=\gamma(\mathrm{pr}_1^*E)\otimes\gamma(\mathrm{pr}_2^*E')$$

est un isomorphisme d'anneaux. Cela leur permit, pour deux variétés lisses X, Y compactes, orientées et soumises à quelques conditions restrictives, d'obtenir pour une application continue $f : X \longrightarrow Y$ un théorème analogue au théorème prouvé par Grothendieck pour les variétés algébriques projectives complexes.

18.2 La K-théorie comme cohomologie généralisée

Poursuivant leurs travaux sur le groupe $K(X)$ pour un CW-complexe fini X, Atiyah et Hirzebruch définirent des groupes $K^{-n}(X, Y)$, pour Y sous-CW-complexe de X, par

$$K^{-n}(X,Y)=K(\Sigma^n(X/Y)), K^{-n}(X)=K^{-n}(X,\phi) \text{ pour } n\geq 0$$

et prouvèrent l'existence d'une suite exacte

$$\begin{aligned}&\ldots\longrightarrow K^{-n-1}(Y)\xrightarrow{\partial}K^{-n}(X,Y)\longrightarrow K^{-n}(X)\longrightarrow K^{-n}(Y)\longrightarrow\\&\ldots\longrightarrow K^0(X)\longrightarrow K^0(Y)\end{aligned}\qquad(118)$$

et une suite exacte analogue pour $\tilde{K}(X)$. En outre, le fait que l'homomorphisme (117) soit bijectif leur fournit un isomorphisme explicite

$$K^{-n}(X,Y)\simeq K^{-n-2}(X,Y)$$

si bien que la suite (118) ne contient en fait que 6 groupes différents et peut donc s'étendre en une suite doublement infinie

$$\begin{aligned}&\ldots\longrightarrow K^n(X,Y)\longrightarrow K^n(X)\longrightarrow K^n(Y)\xrightarrow{\partial}K^{n+1}(X,Y)\\&\longrightarrow\ldots \text{ pour } n\in\mathbb{Z}.\end{aligned}\qquad(119)$$

Autrement dit, le foncteur K vérifie un des axiomes d'Eilenberg-Steenrod pour la cohomologie; Atiyah et Hirzebruch montrèrent que les autres axiomes sont aussi vérifiés, à l'exception de l'axiome de dimension: on a en effet $K^n(P)\simeq\mathbb{Z}$ si n est pair et $K^n(P)=0$ si n est impair. Ils considérèrent que la «K-théorie» qu'ils avaient ainsi construite était une théorie de «*cohomologie généralisée*». Ils montrèrent qu'il y a une relation entre la K-théorie et la cohomologie ordinaire: ils filtrent chaque $K^n(X)$ par les sous-groupes

$$K_p^n(X)=\mathrm{Ker}(K^n(X)\longrightarrow K^n(X^{p-1}))$$

où on note X^p le p-squelette de X; ils en déduisirent une suite spectrale dont les termes E_2 et E_∞ sont

$$E_2^{pq}=H^p(X;K^q(P)),\ E_\infty^{pq}=K_p^{p+q}(X)/K_{p+1}^{p+q}(X).$$

La K-théorie a permis à J.F. Adams de résoudre un problème posé depuis longtemps: trouver le nombre maximum k de champs de vecteurs linéairement indépendants sur une sphère $\mathbb{S}_n$. On écrit le nombre $n+1$ sous la forme

$$(2a+1)2^{c+4d}, a, c, d \text{ entiers } \geq 0, \quad 0 \leq c \leq 3.$$

Alors $k = 2^c + 8d - 1$; en particulier on n'a $k = n$ (cas où on dit que $\mathbb{S}_n$ est *parallélisable*) que si $n = 1, 3$ ou 8. Un autre résultat d'Adams a été la détermination des applications continues $f : \mathbb{S}_{2n+1} \longrightarrow \mathbb{S}_{n+1}$ dont l'invariant de Hopf $H(f) = \pm 1$. Les valeurs possibles de n avaient progressivement été de plus en plus restreintes par plusieurs mathématiciens. Adams parvint à prouver que les seules valeurs possibles étaient celles déjà connues de Hopf, $n = 1, 3$ et 7. Steenrod avait montré qu'on a toujours $H(f) \neq \pm 1$ si l'application

$$Sq^n : H^n(X; \mathbb{F}_2) \longrightarrow H^{2n}(X; \mathbb{F}_2) \tag{120}$$

est 0 pour tous les espaces X tels que $H^j(X; \mathbb{F}_2) = 0$ pour $n < j < 2n$. La première démonstration d'Adams utilisait des relations très compliquées entre les carrés de Steenrod, obtenues par des «opérations cohomologiques secondaires». En 1966, Adams et Atiyah montrèrent qu'une preuve beaucoup plus simple s'obtenait à l'aide de la K-théorie. La caractérisation des valeurs de n pour lesquelles $H(f) = \pm 1$ établit aussi que ce sont les seules valeurs pour lesquelles $\mathbb{S}_n$ peut être munie d'une structure de H-espace.

Plusieurs mathématiciens ont remarqué vers 1960 que la catégorie des fibrés vectoriels sur un CW-complexe fini X est équivalente à la catégorie des modules projectifs (sommandes directes de modules libres) sur l'anneau $\mathscr{C}(X)$ des fonctions réelles continues dans X. Cela a donné l'idée de définir, pour un anneau commutatif quelconque A, le groupe de Grothendieck $K_0(A)$ des classes d'isomorphisme des A-modules projectifs (qui forment un monoïde commutatif). Puis, sur le modèle de la théorie d'Atiyah et Hirzebruch, on a cherché à définir des groupes $K_n(A)$ pour tout $n \geq 0$. Ces efforts ont abouti à une théorie très générale de Quillen (1970) qui permet de définir des groupes $K_n(M)$ pour des objets de catégories très variées. Cette «K-théorie algébrique» a été fertile en applications en théorie des groupes, en géométrie algébrique, en théorie des nombres, en théorie des C^*-algèbres, etc.

18.3 Spectres d'espaces et cohomologie généralisée

D'autres généralisations des théories homologiques apparurent en 1958. Spanier et J.H.C. Whitehead avaient forgé une théorie générale fondée sur la considération de la limite inductive $\varinjlim[\Sigma^k X, \Sigma^k Y]$ des classes d'homotopie de suspensions itérées. Pour X et Y sous-CW-complexes d'une sphère, cela leur donnait une intéressante généralisation de la dualité d'Alexander. En 1958, E. Lima voulut élargir cette théorie au cas où X et Y ne sont plus des CW-complexes. Au lieu de prendre la suite $\Sigma^k X$ des suspensions itérées d'un même espace, il étudia ce qu'il appela un *spectre d'espaces*; c'est une suite (X_n) d'espaces munis d'une suite d'applications continues

$$\rho_n : \Sigma X_n \longrightarrow X_{n+1}$$

où ρ_n n'est plus nécessairement l'identité. Un exemple est donné par les équivalences d'homotopie faibles

$$K(\Pi, n) \longrightarrow \Omega K(\Pi, n+1)$$

qui peuvent aussi s'écrire

$$\rho_n : \Sigma K(\Pi, n) \longrightarrow K(\Pi, n+1)$$

et définissent donc ce qu'on appelle le «spectre d'Eilenberg-MacLane». Le théorème de Bott permet de définir le «spectre d'Atiyah-Hirzebruch», avec $E_{2n} = \mathbb{Z} \times B_{U(\infty)}$, $E_{2n+1} = U(\infty)$. Depuis 1960, beaucoup d'autres spectres ont été définis.

Pour les espaces $K(\Pi, n)$, la relation $[X; K(\Pi, n)] \simeq H^n(X; \Pi)$ pour Π commutatif peut s'écrire

$$H^q(X; \Pi) = \varinjlim_n [\Sigma^n X; K(\Pi, q+n)].$$

Cela conduit, pour un spectre quelconque $E = (E_n, \rho_n)$ d'espaces, et pour un CW-complexe X, à considérer la limite inductive

$$H_E^q(X) = \varinjlim_n [\Sigma^n X; E_{q+n}]$$

comme q-ème *groupe de cohomologie généralisée* de X, relatif au spectre E (le cas du spectre d'Eilenberg-MacLane redonne donc la cohomologie usuelle). Pour un sous-CW-complexe Y de X, on pose $H_E^q(X, Y) = H_E^q(X/Y)$. Pour cette nouvelle cohomologie, on vérifie les axiomes d'Eilenberg-Steenrod à l'exception de l'axiome de dimension; on a cette fois

$$H_E^q(P) = \varinjlim_n \pi_n(E_{n+q}).$$

En 1954, G.W. Whitehead, en étudiant l'homotopie d'un produit écrasé $X \wedge Y$ de CW-complexes, avait prouvé, à l'aide d'une suite spectrale déduite de la filtration croissante $(X^p \wedge Y)$ de $X \wedge Y$, que l'on a, pour Π commutatif,

$$H_q(X; \Pi) = \varinjlim_n \pi_{n+q}(X \wedge K(\Pi, n)).$$

En 1962, cela lui donna l'idée de définir, pour un spectre quelconque E d'espaces, les *groupes d'homologie généralisée* de X relatifs à E

$$H_q^E(X) = \varinjlim_n \pi_{n+q}(X \wedge E_n).$$

Il fit une étude approfondie de ces groupes, où de nouveau on vérifie les axiomes d'Eilenberg-Steenrod, l'axiome de dimension étant cette fois remplacé par

$$H_n^E(P) = \varinjlim_k \pi_{n+k}(E_k)$$

qui sont appelés *groupes d'homotopie stables* du spectre E. Ces groupes, ainsi que les divers spectres d'espaces qui sont apparus dans la théorie du cobordisme (voir plus loin) font l'objet depuis 30 ans de très nombreuses recherches. Une suite spectrale découverte par J.F. Adams, qui relie la cohomologie de l'algèbre de Steenrod à l'homotopie stable, joue un grand rôle dans ces travaux (on a fait un volume entier sur la structure des termes de cette suite spectrale).

19 La topologie géométrique des variétés lisses

Jusque vers 1950, il n'était guère question, en Topologie algébrique, que des variétés lisses «sans bord». Au contraire, depuis cette date, les «variétés-à-bord» lisses et leurs relations avec leurs «bords» sont au centre des recherches. Une «variété-à-bord» lisse de dimension n est un espace paracompact X muni d'une famille de cartes $\varphi_\alpha : U_\alpha \longrightarrow H_n$, où les U_α forment un recouvrement ouvert de X, H_n est le demi-espace fermé de $\mathbb{R}^n$ formé des points $(x_1, x_2, \ldots, x_n)$ tels que $x_1 \geq 0$, et chaque φ_α est un homéomorphisme de U_α sur un ensemble ouvert *de l'espace* H_n; en outre, si $U_\alpha \cap U_\beta \neq \phi$; l'homéomorphisme de transition

$$\varphi_\beta \circ \varphi_\alpha^{-1} : \varphi_\alpha(U_\alpha \cap U_\beta) \longrightarrow \varphi_\beta(U_\alpha \cap U_\beta)$$

doit être un difféomorphisme[18]). Les points $x \in X$ tels qu'il existe une carte $\varphi_\alpha : U_\alpha \longrightarrow H_n$ tels que $x \in U_\alpha$ et que $\varphi_\alpha(x)$ appartienne à l'hyperplan $x_1 = 0$, forment le *bord* de X, noté ∂X; c'est une variété lisse de dimension $n-1$, connexe ou non et sans bord. Il y a un difféomorphisme h de $\partial X \times [0, 1[$ sur un voisinage ouvert de ∂X dans X (un «faux-col» qui transforme $\partial X \times \{0\}$ en ∂X). L'espace tangent $T_x(\partial X)$ au point $x \in \partial X$ est un hyperplan dans l'espace tangent $T_x(X)$; les vecteurs de $T_x(X)$ qui sont images par l'application linéaire tangente $T_x(h)$ des vecteurs $v = (t, u_2, \ldots, u_n)$ avec $t > 0$ sont dits «dirigés vers l'intérieur» de X ($-v$ est alors «dirigé vers l'extérieur»). Le fibré normal $j^*(T(X))/T(\partial X)$ ($j : \partial X \longrightarrow X$ étant l'inclusion) est trivialisable.

---

18) Un difféomorphisme h d'un voisinage V de x dans H_n sur un voisinage W de $y = h(x)$ dans H_n, lorsque x appartient à l'hyperplan $x_1 = 0$ (auquel cas y doit aussi appartenir à cet hyperplan), est par définition la restriction à V d'un difféomorphisme d'un voisinage V' de x dans $\mathbb{R}^n$ tel que $V' \cap H_n = V$, sur un voisinage W' de y dans $\mathbb{R}^n$ tel que $W' \cap H_n = W$.

19.1 Les travaux de R. Thom

Un problème naturel (formulé par Steenrod) est de se demander quand une variété lisse de dimension $n-1$ peut être bord d'une variété lisse de dimension n. Pour résoudre ce problème, Thom a créé la théorie du *cobordisme* (non orienté), qui a donné le départ à toutes les études sur les variétés qu'on peut qualifier de «Topologie géométrique», vu la prépondérance qu'y détiennent les constructions géométriques; mais bien entendu il y est fait usage de tous les acquits antérieurs de la Topologie algébrique.

Comme dans les premières recherches de Poincaré, il est question de variétés lisses compactes sans bord. Mais elles sont considérées indépendamment de tout plongement dans un $\mathbb{R}^N$, ce qui supprime toute difficulté dans la définition de l'«addition» de deux variétés M_1, M_2 de même dimension: c'est tout simplement leur réunion disjointe $M_1 \coprod M_2$. On dit que M_1 et M_2 sont *cobordantes* si $M_1 \coprod M_2$ est le bord d'une variété lisse compacte de dimension $n+1$. C'est là une relation d'équivalence, et les classes d'équivalence forment (pour la loi déduite de $\coprod$) un groupe commutatif $\mathfrak{N}_n$, dont tout élément est d'ordre 2. La somme directe $\mathfrak{N}_. = \bigoplus_{n \geq 0} \mathfrak{N}_n$ est un anneau commutatif gradué pour le produit de classes déduit du produit cartésien $M_1 \times M_2$ des variétés.

Thom a obtenu le remarquable résultat que l'anneau gradué $\mathfrak{N}_0$ est isomorphe à l'algèbre de polynômes

$$\mathbb{F}_2[v_2, v_4, v_5, v_6, v_8, v_9, \ldots]$$

où chaque v_j est une indéterminée de degré j, et seuls interviennent les entiers j qui ne sont pas de la forme $2^k - 1$. Sa méthode, qui développe une idée antérieure de Pontrjagin, se déroule en plusieurs étapes:

1. Pour tout fibré vectoriel $\xi = (E, B, \mathbb{R}^k)$, où B est un CW-complexe fini, on considère le CW-complexe fini $T(\xi)$ (appelé maintenant l'*espace de Thom* de ξ) obtenu en compactifiant E par un point à l'infini ∞.

2. Lorsque B est une variété compacte lisse sans bord, on définit, pour tout $m > k$, un homomorphisme canonique de groupes

$$\lambda_\xi : \pi_m(T(\xi)) \longrightarrow \mathfrak{N}_{m-k}$$

en associant à tout morphisme d'espaces pointés $f : (\mathbb{S}_m, *) \longrightarrow (T(\xi), \infty)$ une application lisse g homotope à f, pour laquelle $g^{-1}(B)$ est une sous-variété lisse de $\mathbb{S}_m$ de dimension $m-k$; on montre que deux quelconques de ces variétés sont cobordantes.

3. On applique le résultat de 2) au fibré vectoriel «tautologique» $\gamma(m,k) = (U_{m,k}, \mathbb{G}_{m,k}(\mathbb{R}), \mathbb{R}^k)$. Thom prouve que pour m et k assez grands, l'homomorphisme

$$\lambda_{\gamma(m,k)} : \pi_{n+k}(T(\gamma(m,k))) \longrightarrow \mathfrak{N}_n$$

est bijectif (nouveau phénomène de stabilité).

Il utilise pour cela des constructions géométriques ingénieuses et compliquées, analogues à celles de Steenrod pour la classification des fibrés principaux.

4. A l'aide des carrés de Steenrod, Thom construisit un produit d'espaces d'Eilenberg-MacLane dont les groupes d'homotopie sont les mêmes que les $\pi_{n+k}(T(\gamma(m,k)))$ pour $n < k$ et m assez grand. Cela lui donna aussi la loi de multiplication dans $\mathfrak{N}_.$.

Une autre partie des travaux de Thom concerne le cobordisme *orienté*. Il s'agit cette fois de variétés lisses, compactes et orientées; pour une telle variété X, $-X$ est la variété munie de l'orientation opposée. La relation de cobordisme est ici que X_1 et X_2 sont cobordantes si elles ont même dimension n, et s'il existe une variété-à-bord W lisse, compacte et orientée, dont le bord, muni de l'orientation directement reliée à celle de W, est $X_1 \coprod(-X_2)$. Thom définit alors un anneau gradué $\Omega_. = \bigoplus_{n\geq 0} \Omega_n$ des classes de cobordisme orienté; mais il ne put déterminer explicitement les Ω_n que pour $n \leq 7$. Il prouva toutefois que le produit tensoriel $\Omega_. \otimes_{\mathbb{Z}} \mathbb{Q}$ est isomorphe à l'algèbre graduée de polynômes sur $\mathbb{Q}$

$$\mathbb{Q}[w_4, w_8, \ldots, w_{4j}, \ldots]$$

où w_{4j} est la classe de cobordisme orienté de l'espace projectif complexe $\mathbb{P}_{2j}(\mathbb{C})$. La structure assez compliquée de l'algèbre $\Omega_.$ elle-même, a été ensuite obtenue en combinant des contributions de Milnor, Averbuch, Rokhlin et C. Wall, qui acheva la description de cette algèbre.

Ces travaux donnent une réponse explicite au problème de Steenrod. Soit $\tau = T(M)$ le fibré tangent d'une variété compacte lisse M, et soient $w_j(\tau)$ ses classes de Stiefel-Whitney, éléments de l'algèbre $H^\cdot(M; \mathbb{F}_2)$; les éléments

$$\langle w_1(\tau)^{r_1} w_2(\tau)^{r_2} \ldots w_n(\tau)^{r_n}, [M]\rangle \in \mathbb{F}_2$$

avec $r_1 + 2r_2 + \ldots + nr_n = n$, sont appelés les nombres de Stiefel-Whitney de M. Pour les variétés non orientées, Pontrjagin avait montré qu'une condition nécessaire pour que M soit un bord, est que tous ses nombres de Stiefel-Whitney soient nuls; Thom prouva que cette condition est suffisante. Pour le même problème concernant les variétés orientées, on forme les «nombres de Pontrjagin»

$$\langle p_1^{a_1} p_2^{a_2} \ldots p_r^{a_r}, [M]\rangle \in \mathbb{Z}$$

où les $p_k \in H^{4k}(M; \mathbb{Z})$ sont les classes de Pontrjagin de τ et $n = 4r$, $r = a_1 + \ldots + a_r$; Pontrjagin avait montré que si M est le bord d'une variété orientée, tous les nombres de Pontrjagin sont nuls; C. Wall prouva inversement que si les nombres de Stiefel-Whitney et les nombres de Pontrjagin sont nuls, alors M est le bord d'une variété orientée.

En 1952, Thom avait étudié un invariant topologique d'un CW-complexe (connexe ou non) de dimension paire $2m$: l'application

$$(x, y) \mapsto \langle x \smile y, [M]\rangle$$

est une forme bilinéaire sur l'espace vectoriel $H^m(M; \mathbb{Q})$; elle est symétrique pour m pair, antisymétrique pour m impair. Pour $m = 2k$, la signature de cette forme symétrique est donc un invariant $\sigma(M)$ du type d'homotopie de M, appelé la *signature* de M. Thom remarqua que $\sigma(M)$, pour une variété orientée, ne dépend que de la classe de M dans l'anneau $\Omega_{\cdot}$. Comme les classes de Pontrjagin d'une variété orientée M sont les mêmes pour toute variété orientée cobordante à M, on devait s'attendre à des relations entre $\sigma(M)$ et les classes de Pontrjagin de M: Thom prouva effectivement que pour $\dim M = 4$, on a

$$\sigma(M) = \frac{1}{3}\langle p_1, [M]\rangle \tag{121}$$

et en 1953, Hirzebruch obtint l'expression explicite de polynômes $L_k(p_1, p_2, \ldots, p_k)$ tels que $\sigma(M) = \langle L_k(p_1, \ldots, p_k), [M]\rangle$ pour $\dim M = 4k$; on a par exemple

$$L_2(p_1, p_2) = \frac{1}{45}(7p_2 - p_1^3), L_3(p_1, p_2, p_3) = \frac{1}{945}(62p_3 - 13p_1p_2 + 2p_1^3).$$

Thom avait remarqué que la définition de l'espace $T(\xi)$, qui dépend du fibré universel de groupe structural $O(k)$ pour le cobordisme non orienté, et du fibré universel de groupe structural $SO(k)$ pour le cobordisme orienté, devait s'étendre à d'autres fibrés principaux dont le groupe structural est un groupe de Lie. Cela a été fait et on a maintenant étudié beaucoup de «cobordismes».

19.2 Les présentations par anses et le h-cobordisme

Indépendamment, Thom en 1949 et E. Pitcher en 1950, observèrent que la théorie développée par M. Morse pour l'étude, sur une variété compacte lisse M, d'une fonction f de classe C^∞ dont tous les points critiques sont non dégénérés, peut s'interpréter de la façon suivante: soit c une valeur critique de f, telle que $f^{-1}(c)$ soit réduit à un seul point; alors la variété $M_{c+\varepsilon}$ des $x \in M$ tels que $f(x) < c + \varepsilon$ est homéomorphe à la variété obtenue par *attachement* à $M_{c-\varepsilon}$ d'une n-cellule dont la fonction d'attachement dépend de l'indice de f au point c. Les travaux de S. Smale à partir de 1960 allaient montrer que cette conception conduit à des résultats inattendus.

Le point de départ consiste, dans la définition de l'attachement de la boule $\mathbb{D}_n$ par une k-*anse*: on désigne ainsi le produit $\mathbb{D}_k \times \mathbb{D}_{n-k}$, dont la frontière est $(\mathbb{S}_{k-1} \times \mathbb{D}_{n-k}) \cup (\mathbb{D}_k \times \mathbb{S}_{n-k-1})$; l'attachement se fait par un difféomorphisme de $\mathbb{S}_{k-1} \times \mathbb{D}_{n-k}$ sur une partie du bord ∂M (pour $n = 3$, $k = 1$, on retrouve bien les «anses» attachées à la boule $\mathbb{D}_3$, décrites par Klein). Il est possible de décrire une structure de variété lisse sur le CW-complexe relatif obtenu par ces attachements, et la théorie de Morse montre que la variété à bord définie par $f(x) \geq a$ pour a voisin du minimum de f, s'obtient à homéomorphisme près, à partir de $\mathbb{D}_n$, par attachements successifs de k-anses, chacune d'elles correspondant à un point critique d'indice k.

La nouveauté dans l'approche de Smale est que si V_1 résulte de l'attachement à V d'une k-anse, et V_2 résulte de l'attachement à V_1 d'une $(k+1)$-anse, alors, sous des conditions convenables d'intersection des anses, il y a une isotopie transformant V_2 en V (autrement dit, on a «tué» la k-anse). Inversement, une isotopie permet de faire «naître» simultanément une k-anse et une $(k+1)$-anse sur V. On peut aussi procéder en sens inverse, en «tuant» une k-anse par une $(k-1)$-anse (ce qui correspond au remplacement de la fonction de Morse f par $-f$). Smale a su avec virtuosité manier ces isotopies en faisant «glisser» les anses les unes sur les autres, et il a obtenu ainsi des résultats qui paraissaient à l'époque inaccessibles, puisqu'ils affirmaient l'existence d'*homéomorphismes*, là où les méthodes de Topologie algébrique ne donnaient que des équivalences d'homotopie:

1) Toute variété lisse, compacte, sans bord, de dimension $n \geq 5$, qui a le type d'homotopie de $\mathbb{S}_n$, est en fait homéomorphe à $\mathbb{S}_n$ (extension pour $n \geq 5$ de la «conjecture de Poincaré» pour $n = 3$).

2) Soit M une variété compacte lisse de dimension $n \geq 6$, ayant un bord ∂M connexe et simplement connexe. Alors M est difféomorphe à la boule $\mathbb{D}_n$.

3) On dit que deux variétés V_1, V_2 lisses, compactes, orientées et de même dimension n, sont *h-cobordantes* s'il existe une variété W lisse, compacte, orientée, de dimension $n+1$ telle que le bord ∂W soit réunion disjointe de V_1 et $-V_2$, et en outre telle que V_1 et V_2 soient des rétractes par déformation de W. Si $n \geq 5$ et si V_1 est simplement connexe, V_1 et V_2 sont alors *difféomorphes.*

19.3 Le problème des immersions

Si X et Y sont deux variétés lisses paracompactes, une *immersion* $f : X \longrightarrow Y$ est une application de classe C^∞ dont l'application linéaire tangente $T_x(f)$ est injective en tout point $x \in X$; une immersion injective est appelée un *plongement* et est alors un difféomorphisme sur son image si f est propre. Une immersion n'est pas nécessairement injective: par exemple, il y a des immersions de la bouteille de Klein dans $\mathbb{R}^3$, mais pas de plongement. Si un point $y \in f(X)$ est tel que $f^{-1}(y)$ comporte plus d'un point, on dit que y est une *auto-intersection* de $f(X)$.

Un des premiers théorèmes prouvés par H. Whitney en 1935 est que, pour toute variété lisse paracompacte X de dimension n, il existe une immersion de X dans $\mathbb{R}^{2n}$ et un plongement propre de X dans $\mathbb{R}^{2n+1}$; il utilisait le théorème de Baire mais aucun résultat de Topologie algébrique. En 1944, il améliora ce théorème en montrant qu'en fait il existe une immersion de X dans $\mathbb{R}^{2n-1}$ et un plongement propre dans $\mathbb{R}^{2n}$. La preuve est beaucoup plus difficile et consiste en des isotopies qui éliminent les auto-intersections; une d'elles, dite «déformation de Whitney» («Whitney trick») a été fort utilisée par la suite.

Le problème des immersions se formule de la façon suivante: quel est le plus petit entier k telle que toute variété lisse, compacte, sans bord et de dimension n, admette une immersion dans un espace $\mathbb{R}^{n+k}$?

Les travaux de Smale et M. Hirsch (1959) ont ramené le problème à une question sur les fibrés vectoriels. Deux fibrés vectoriels ξ_1, ξ_2 sur un CW-complexe X sont dits stablement équivalents s'il existe deux fibrés vectoriels triviaux θ_1, θ_2 de base X tels que $\xi_1 \oplus \theta_1$ et $\xi_2 \oplus \theta_2$ soient X-isomorphes. Si X est une variété lisse de dimension n, et ν_X le fibré normal de X pour une immersion dans $\mathbb{R}^{2n}$, il y a une immersion de X dans $\mathbb{R}^{n+k}$ si ν_X est stablement équivalent à un fibré de rang k.

Une condition nécessaire pour cela est que les classes de Stiefel-Whitney $w_i(\nu_X)$ soient nulles pour $i > k$. En 1960, W. Massey a prouvé que cette propriété dépend de l'écriture dyadique de l'entier n: si

$$n = 2^{m_1} + 2^{m_2} + \ldots + 2^{m_r} \quad \text{avec} \quad 0 \leq m_1 < m_2 < \ldots < m_r$$

on pose $\alpha(n) = r$; le plus petit nombre $k'(n)$ tel que $w_i(\nu_X) = 0$ pour $i > k'(n)$ pour toutes les variétés X, est exactement $n-\alpha(n)$. L'exemple du produit d'espaces projectifs

$$X = \mathbb{P}_{2^{m_1}}(\mathbb{R}) \times \mathbb{P}_{2^{m_2}}(\mathbb{R}) \times \ldots \times \mathbb{P}_{2^{m_r}}(\mathbb{R})$$

montre qu'on a effectivement $w_{n-\alpha(n)}(\nu_X) \neq 0$ pour cette variété. On conjectura dès lors que le plus petit nombre k pour lequel toutes les variétés de dimension n admettent une immersion dans $\mathbb{R}^{n+k}$ est $n - \alpha(n)$.

A partir de 1970, le problème a fait l'objet de recherches de plusieurs mathématiciens: R.L. Brown, E.H. Brown, R. Gitler, M. Mahowald, F. Peterson; finalement, R.L. Cohen, en 1982, est parvenu à prouver la conjecture en complétant tout ce qui avait été acquis par les travaux antérieurs. La technicité des moyens employés est impressionnante: espaces de Thom, spectres d'espaces, localisation d'espaces et de spectres, tour de Postnikov, etc.

20 La théorie générale des variétés

20.1 Les diverses catégories de variétés

La catégorie des variétés lisses (avec ou sans bord) et des applications de classe C^∞ se note souvent DIFF. Pour les variétés C^0, la notion de bord se définit comme pour les variétés lisses, en n'imposant aucune condition de différentiabilité aux homéomorphismes de transition $\varphi_\beta \circ \varphi_\alpha^{-1}$ entre ouverts de H_n; on note TOP la catégorie de ces variétés et des applications continues. On introduit aussi la catégorie PL (linéaire par morceaux) où les homéomorphismes de transition $\varphi_\beta \circ \varphi_\alpha^{-1}$ sont des applications simpliciales pour des triangulations des ouverts de H_n considérés. Une variété PL est triangulable.

A partir de 1960 environ, on a étudié de façon systématique le passage d'une de ces catégories à une autre: une variété TOP admet-elle une structure PL, et si oui, combien y en a-t-il? De même pour le passage de PL à DIFF.

L'idée générale qui a inspiré les recherches sur ces problèmes est de travailler sur le fibré tangent (en catégorie DIFF) et sur des notions analogues (les «microfibrés») en catégories PL et TOP. Cela introduit les groupes structuraux des fibrés principaux correspondants, et les espaces classifiants de ces groupes: pour la catégorie DIFF, le groupe structural est le groupe $\mathrm{O}(\infty)$, limite inductive des groupes orthogonaux $\mathrm{O}(m)$; on note BO son espace classifiant. On a des espaces classifiants analogues BPL et BTOP pour les deux autres catégories, avec des fibrations

$$\mathrm{BO} \longrightarrow \mathrm{BPL} \longrightarrow \mathrm{BTOP}$$

dont on désigne par PL/O et TOP/PL les fibrés génériques. A une structure DIFF (resp. PL, TOP) sur M correspond donc une application continue f de M dans BO (resp. BPL, BTOP). Pour qu'il y ait une structure DIFF sur une variété PL, il faut et il suffit que l'application $f : M \longrightarrow \mathrm{BPL}$ se factorise en $M \longrightarrow \mathrm{BO} \longrightarrow \mathrm{BPL}$, et les classes de structures DIFF possibles sont en correspondance bijective canonique avec l'ensemble $[M; \mathrm{PL/O}]$ des classes d'homotopie.

Pendant longtemps on avait espéré que ces études aboutiraient à ce que suggéraient les propriétés des variétés de dimension ≤ 3 (voir plus loin), à savoir qu'il n'y a pas lieu de distinguer les trois catégories de variétés. Le premier surprenant résultat fut le célèbre exemple de J. Milnor (1956) montrant qu'il y a, sur la sphère $\mathbb{S}_7$, 28 structures différentielles compatibles avec la topologie et deux à deux non difféomorphes. Un peu plus tard, Milnor et Kervaire montrèrent que l'ensemble $[\mathbb{S}_n, \mathrm{PL/O}]$ a une structure naturelle de groupe commutatif fini, dont l'ordre est lié de façon inattendue aux nombres de Bernoulli; par exemple, pour $n = 11$, ce groupe est cyclique d'ordre 992. Depuis lors, on s'est aperçu que ces sphères «exotiques» interviennent naturellement dans des problèmes de géométrie différentielle: E. Brieskorn a montré que l'intersection de la sphère $\mathbb{S}_{4m+1}$ définie par l'équation $\sum_{j=0}^{2m} z_j\bar{z}_j = 1$ dans $\mathbb{C}^{2m+1}$, et de la variété algébrique d'équation $z_0^3 + z_1^{6k-1} + z_2^2 + \ldots + z_{2m}^2 = 0$ est une sphère «exotique» pour la structure différentielle induite par la structure usuelle de $\mathbb{R}^{4m+2}$.

Kervaire a découvert d'autre part un exemple de variété PL de dimension 10 qui n'admet pas de structure DIFF.

En 1969, des résultats décisifs sur le passage de TOP à PL furent obtenus par R. Kirby et L. Siebenmann, mettant le point final à toute une série de résultats préliminaires obtenus par plusieurs mathématiciens. La propriété centrale est que l'espace TOP/PL a le type d'homotopie faible de l'espace d'Eilenberg-MacLane $K(\mathbb{Z}/2\mathbb{Z}, 3)$. On en déduit que si M est une variété TOP de dimension ≥ 5, il y a une «obstruction» à l'existence d'une structure PL sur M, qui est une classe de cohomologie dans $H^4(M; \mathbb{F}_2)$; si cette classe est nulle, les classes d'isomorphie des structures PL possibles sont en correspondance bijective canonique avec $H^3(M; \mathbb{F}_2)$. On a pu donner des exemples explicites de variétés TOP de dimension 5 n'admettant pas de structures PL, et d'autres ayant plusieurs structures PL non isomorphes. Cela donne une réponse négative à la conjecture qu'une structure PL sur une variété est unique à isomorphisme près, conjecture qui, sous le nom de «Hauptvermutung» remontait au début de la Topologie algébrique. Des exemples de CW-complexes qui ne sont pas des variétés et qui ne vérifient pas cette conjecture avaient d'ailleurs été donnés un peu auparavant par Milnor.

Ces résultats sont les plus spectaculaires de toute une théorie de Topologie géométrique qui en comporte une foule d'autres. Outre les techniques de la Topologie algébrique, il y est fait grand usage du cobordisme et de la théorie des immersions, ainsi que de ce qu'on appelle la «chirurgie», qui consiste en gros à des remplacements de k-anses par des $(n-k)$-anses dans une présentation par anses d'une variété de dimension n. Ces travaux, qui témoignent d'une rare originalité, sont dispersés dans des volumes entiers et de nombreux mémoires, qui font sans cesse référence des uns aux autres. En dépit de nombreux dessins destinés à aider les spécialistes dans le déroulement des démonstrations, il est à peu près impossible à un mathématicien non spécialiste de s'en faire une idée d'ensemble de façon précise, faute d'un ouvrage qui décrirait les méthodes de Topologie géométrique avec le même détail que ceux consacrés à l'algèbre homologique.

Si on ne se préoccupe pas des passages d'une catégorie à une autre, un des résultats les plus frappants est que, pour les dimensions ≥ 5, les principaux théorèmes pour les variétés DIFF restent valables pour les variétés TOP: conjecture de Poincaré (une variété TOP ayant même type d'homotopie que $\mathbb{S}_n$ est homéomorphe à $\mathbb{S}_n$), présentation par anses, cobordisme, h-cobordisme, etc.

20.2 La Topologie géométrique au-delà des variétés

Avant 1950, les recherches dans ce qu'on peut appeler «Topologie géométrique» tendaient plutôt à s'écarter de l'étude des variétés, en évitant autant que possible de faire intervenir les espaces $\mathbb{R}^n$ et leur définition algébrique, et en n'utilisant au contraire que des propriétés et des techniques où seules sont invoquées les notions de Topologie générale; on se préoccupe même de caractériser $\mathbb{R}$ et les $\mathbb{R}^n$ en termes purement topologiques, où l'algèbre n'intervient pas.

Nous nous limiterons à quelques définitions et résultats qui ont trouvé plus tard d'importantes applications.

Un rétracte d'un espace X est un sous-espace Y de X pour lequel il existe une rétraction de X sur Y, c'est-à-dire une application continue $r : X \longrightarrow Y$ telle que $r(y) = y$ pour $y \in Y$. Borsuk a défini en 1933 la notion de *rétracte absolu de voisinage* (ANR en abrégé): c'est un espace métrique séparable Y tel que pour tout homéomorphisme de Y sur un sous-espace Z d'un espace métrique X, Z soit un rétracte d'un voisinage de Z dans X. Il revient au même de dire que pour toute partie fermée A d'un espace métrique X, toute application continue $f : A \longrightarrow Y$ a un prolongement continu $U \longrightarrow Y$ dans un voisinage U de A dans X. Les ANR ont d'utiles propriétés de permanence: tout rétracte d'un ANR est un ANR, tout ouvert d'un ANR est un ANR, tout produit d'un nombre fini d'ANRs est un ANR. Toute variété C^0 compacte est un ANR. Tout ANR est localement rétractile, c'est-à-dire que tout point admet un voisinage qui est un espace rétractile; inversement, tout espace métrique compact, de dimension finie et localement rétractile, est un ANR.

Les phénomènes tels que la courbe de Peano ont très tôt incité à étudier les images, par des applications continues, d'espaces tels que les variétés. La courbe de Peano est un cas particulier du théorème général de Hahn-Mazurkiewicz: les images de $[0, 1]$ par une application continue dans un espace séparé sont exactement tous les espaces compacts, connexes et localement connexes.

A l'opposé, il s'est révélé utile d'étudier les espaces quotients E/R qui sont homéomorphes à E; on se borne au cas où toute classe d'équivalence est compacte, et l'application canonique $p : E \longrightarrow E/R$ est fermée («upper semi-continuous decomposition»). Les premiers travaux sur cette question remontent à 1925; R.L. Moore a prouvé que si $E = \mathbb{R}^2$ et si aucune fibre $p^{-1}(y)$ ne sépare le plan, alors E/R est homéomorphe à E.

A partir de 1950, les recherches s'élargirent à des espaces E plus généraux. Le théorème central est le «critère de rétrécissement» de R. Bing (1957). Dans le cas le plus simple, E est un espace métrique compact; alors E/R est métrisable, et si d est une distance sur E/R, le critère s'énonce comme suit: pour tout $\varepsilon > 0$, il existe un homéomorphisme h de E sur lui-même tel que: 1) l'image par h de toute fibre $p^{-1}(y)$ a un diamètre $\leq \varepsilon$; 2) on a $d(p(h(x)), p(x)) \leq \varepsilon$ (l'homéomorphisme h rend les fibres très petites et déplace peu leurs images dans E/R). Alors, ces conditions entraînent, non seulement que E/R est homéomorphe à E, mais que p est limite uniforme d'homéomorphismes de E sur E/R; réciproquement, ces deux propriétés entraînent le critère de rétrécissement.

Ce résultat suscita la construction, par Bing et son école, d'une nuée d'exemples et contre-exemples d'espaces aux propriétés plus ou moins étranges. Il faut signaler que le premier en date, et qui servit de modèle à beaucoup d'autres, avait été trouvé par L. Antoine en 1921. Pour le construire, on part d'un tore C_0 dans $\mathbb{R}^3$, et on définit un ensemble $C_1 \subset C_0$ formé de 4 tores égaux, de diamètre (des méridiens) moitié de celui de C_0, et qui s'enlacent à la façon des anneaux d'une chaîne; puis l'ensemble C_2 est réunion de 4 ensembles dont chacun est formé de 4 tores contenus dans un des 4 tores de C_1, de diamètre moitié et qui s'enlacent comme ceux de C_1; puis on recommence indéfiniment la construction, obtenant la réunion de 4^k tores à la k-ème étape (ce genre de constructions par itération remontait aux exemples de fonctions sans dérivée du début du siècle, et elles sont souvent qualifiées de «fractales» depuis quelque temps). L'ensemble A, intersection des C_n (et appelé le «collier d'Antoine») a la propriété d'être homéomorphe à l'ensemble totalement discontinu de Cantor, mais $\mathbb{R}^3 - A$ n'est pas simplement connexe. Un des exemples de Bing (le «dogbone») est particulièrement surprenant; on prend $E = \mathbb{R}^3$, E/R n'est pas homéomorphe à $\mathbb{R}^3$, mais le produit $(E/R) \times \mathbb{R}$ est homéomorphe à $\mathbb{R}^4$.

A partir de 1970, ces recherches se sont combinées avec la Topologie géométrique des variétés, et deux nouvelles idées ont fait leur apparition.

En premier lieu, les quotients «celluliques»: on entend par là un quotient E/R où chaque fibre $p^{-1}(y)$ admet un système fondamental de voisinages U_n dans E tels que dans U_n l'injection canonique $p^{-1}(y) \longrightarrow U_n$ soit homotope à une application constante. Parmi les propriétés de ces quotients, la plus importante est que si E est un ANR et si E/R satisfait au critère de rétrécissement, alors le quotient E/R est cellulique.

Pour obtenir une réciproque, il faut faire intervenir une autre propriété: celle de «disjonction des disques» dans un espace métrique X. Cela signifie que pour deux applications continues quelconques $f_1 : \mathbb{D}_2 \longrightarrow X$, $f_2 : \mathbb{D}_2 \longrightarrow X$, et pour tout $\varepsilon > 0$, il existe deux applications $g_1 : \mathbb{D}_2 \longrightarrow X$, $g_2 : \mathbb{D}_2 \longrightarrow X$, respectivement homotopes à f_1 et f_2, telles que $d(f_1, g_1) \leq \varepsilon$ et $d(f_2, g_2) \leq \varepsilon$ et enfin ayant la propriété que $g_1(\mathbb{D}_2) \cap g_2(\mathbb{D}_2) = \phi$. R.D. Edwards a prouvé en 1977 que si E est une variété TOP de dimension ≥ 5 et sans bord, et si E/R est un quotient cellulique de E qui possède la propriété de disjonction des disques, alors $p : E \longrightarrow E/R$ est limite d'homéomorphismes.

Ce théorème lui a ensuite permis de résoudre un problème ouvert depuis 1960: si V est une variété TOP compacte qui a la même homologie que $\mathbb{S}_n$, alors la double suspension non réduite S^2V est homéomorphe à $\mathbb{S}_{n+2}$.

21 Les variétés de dimension infinie

21.1 Les variétés hilbertiennes

Il est facile de définir une variété de dimension infinie X «modelée» sur un espace de Banach: il suffit, dans la définition des cartes φ_α, de les définir comme des homéomorphismes $\varphi_\alpha : U_\alpha \longrightarrow E$, (U_α) étant un recouvrement ouvert de X. A partir de 1960 environ plusieurs mathématiciens étudièrent en particulier les variétés hilbertiennes, où E est l'espace ℓ^2 de Hilbert. On s'aperçut vite que de tels espaces ont des propriétés très différentes de celles des espaces de dimension finie: par exemple, si V est un voisinage ouvert V d'un point $x \in \ell^2$, il existe un difféomorphisme de ℓ^2 sur $\ell^2 - \{x\}$ qui transforme V en $V - \{x\}$ et est l'identité en dehors de V. On eut donc tendance à penser qu'on ne pourrait pas obtenir de théorie topologique intéressante de ces variétés dont les propriétés paraissaient «pathologiques». Mais à partir de 1967, toute une série de travaux de R.D. Anderson, R. Miller, J. West, T. Chapman, R.D. Edwards et H. Toruńczyk, ont dégagé de remarquables relations entre les variétés hilbertiennes et les ANR.

On peut en effet identifier les ANR à des sous-espaces fermés de ℓ^2. Toruńczyk a caractérisé ces sous-espaces de la façon suivante: ce sont ceux tels que $Y \times \ell^2$ soit une variété hilbertienne. En particulier, un complexe simplicial localement fini K est tel que $K \times \ell^2$ soit une variété hilbertienne et inversement, *toute* variété hilbertienne est de cette forme (autrement dit, elle est *triangulable*). Plus précisément, si K_1, K_2 sont deux complexes simpliciaux localement finis et si une application continue

$$f : K_1 \times \ell^2 \longrightarrow K_2 \times \ell^2$$

est homotope à un homéomorphisme, alors l'application composée

$$K_1 \longrightarrow K_1 \times \{0\} \longrightarrow K_1 \times \ell^2 \xrightarrow{f} K_2 \times \ell^2 \xrightarrow{pr_1} K_2 \qquad (122)$$

est une équivalence d'homotopie. En outre, deux homéomorphismes $V_1 \longrightarrow V_2$ de variétés hilbertiennes qui sont homotopes sont aussi isotopes.

21.2 Les $\mathbf{I}^\infty$-variétés et le type simple d'homotopie
On note I^∞ le produit $[0, 1]^{\mathbb{N}}$ d'une infinité dénombrable d'intervalles compacts de $\mathbb{R}$ («cube infini»). Dans la définition donnée ci-dessus des variétés hilbertiennes, on peut remplacer ℓ^2 par I^∞; cela définit les I^∞-variétés. Leur théorie est très voisine de celle des variétés hilbertiennes; Torúnczyk a prouvé que, pour que le produit $X \times I^\infty$ soit une I^∞-variété, il faut et il suffit que X soit un ANR localement compact; toute I^∞-variété est homéomorphe à $K \times I^\infty$, où K est un complexe simplicial localement fini. Mais l'homéomorphie, à homotopie près, de deux triangulations par une application $f : K_1 \times I^\infty \longrightarrow K_2 \times I^\infty$ introduit un résultat inattendu, découvert par Chapman en 1972: si on définit une application $K_1 \longrightarrow K_2$ par la composition (122) où ℓ^2 est remplacé par I^∞, cette application n'est plus seulement une équivalence d'homotopie, mais bien une équivalence d'homotopie *simple*.

Cette notion, essentiellement combinatoire, avait été introduite par J.H.C. Whitehead vers 1940, pour généraliser la recherche de nouveaux invariants faite par Reidemeister en 1939, afin de distinguer les espaces lenticulaires à homéomorphisme près. Whitehead travaille sur les CW-complexes finis; sur un tel espace X, il définit des «déformations élémentaires»; cela consiste, soit en un attachement, d'un type assez spécial, de deux cellules de dimensions consécutives $k - 1$ et k, soit en la suppression d'un tel attachement. Si X, Y sont deux CW-complexes, on dit qu'une application continue $f : X \longrightarrow Y$ est *cellulaire* si $f(X^n) \subset Y^n$ pour tout n-squelette X^n de X. Si alors $f : X \longrightarrow Y$ est une application cellulaire qui est aussi une équivalence d'homotopie, Whitehead dit que f est une équivalence d'homotopie *simple* si elle s'obtient par une suite finie de «déformations élémentaires».

Pour caractériser ces équivalences d'homotopie simples, Whitehead reprit une idée de Reidemeister: faire intervenir le groupe fondamental d'un CW-complexe fini X: on «relève» de façon canonique les cellules de X dans le revêtement $\tilde{X}$ correspondant au groupe $\pi = \pi_1(X)$; l'algèbre $\mathbb{Z}[\pi]$ du groupe π opère alors canoniquement sur les cellules de $\tilde{X}$. Une application cellulaire $f : X \longrightarrow Y$ qui est une équivalence d'homotopie se relève en un homomorphisme $\tilde{f} : C.(\tilde{X}) \longrightarrow C.(\tilde{Y})$ de complexes de chaînes, qui est un morphisme de $\mathbb{Z}[\pi]$-complexes. Pour tout groupe π (commutatif ou non), Whitehead définit un groupe commutatif $\mathrm{Wh}(\pi)$ (appelé maintenant *groupe de Whitehead*) de la façon suivante: on considère le groupe $\mathrm{GL}_\infty(\mathbb{Z}[\pi])$, limite inductive des groupes linéaires $\mathrm{GL}_n(\mathbb{Z}[\pi])$; $\mathrm{Wh}(\pi)$ est le quotient de $\mathrm{GL}_\infty(\mathbb{Z}[\pi])$ par le groupe engendré par le groupe des commutateurs et par les images dans $\mathrm{GL}_\infty(\mathbb{Z}[\pi])$ des éléments de π et l'élément -1.

Whitehead montre qu'on peut attacher à l'homomorphisme $\tilde{f}$ un élément $\tau(f) \in \mathrm{Wh}(\pi)$, qu'il appela la *torsion* de f; si $g : X \longrightarrow Y$ est une autre application cellulaire qui est une équivalence d'homotopie, et qui est homotope à f, alors

$\tau(g) = \tau(f)$. Le résultat fondamental prouvé par Whitehead est que f est une équivalence d'homotopie *simple* si et seulement si $\tau(f) = 0$.

Malheureusement, ce travail d'une remarquable originalité restait, comme celui de Reidemeister, attaché à une décomposition particulière de X et Y en cellules; Whitehead lui-même n'espérait plus que la notion d'équivalence d'homotopie simple puisse avoir une signification topologique. C'est ce progrès décisif qui découle du théorème de Chapman: un homéomorphisme est une équivalence d'homotopie simple.

Si X est un CW-complexe fini, Y un sous-CW-complexe, et si l'inclusion $j : Y \longrightarrow X$ est une équivalence d'homotopie, on définit $\tau(Y, X)$ comme égal à la torsion $\tau(j)$. Cette définition permet de généraliser la notion de h-cobordisme: les données étant les mêmes que dans le théorème de Smale, on suppose, non plus que V_1 et W sont simplement connexes, mais seulement que $\tau(V_1, W) = 0$; on dit alors que cela définit un *s-cobordisme*. Le théorème du s-cobordisme est alors que, pour $\dim V_1 \geq 5$, W est difféomorphe à $V_1 \times I$ si et seulement si $\tau(V_1, W) = 0$.

22 Les variétés de petite dimension

22.1 Les variétés de dimension ≤ 3

Par contraste avec les dimensions ≥ 5, les distinctions entre les catégories TOP, PL et DIFF disparaissent pour les dimensions ≤ 3; en d'autres termes, toute C^0 variété M de dimension 2 ou 3 peut être munie d'une structure lisse, donc est triangulable; deux structures DIFF (resp. PL) sur M sont isomorphes. Cela a été prouvé pour $n = 2$ par T. Rado en 1935, et pour $n = 3$ par E. Moise et C. Papakyriakopoulos en 1952.

Les problèmes pour ces variétés sont donc la classification des variétés à homéomorphisme (ou difféomorphisme) près, et la classification des couples (X, Y) formés d'une variété X de dimension ≤ 3 et d'une sous-variété Y.

Pour $n = 2$, la classification des surfaces est donnée par leur «forme normale», mais il y a eu de nombreux travaux sur les automorphismes des surfaces, notamment ceux de Nielsen vers 1940, et récemment ceux de W. Thurston, en liaison avec l'étude des courbes lisses tracées sur la surface.

La classification des variétés de dimension 3 est beaucoup moins avancée. Pour deux variétés M_1, M_2 de dimension n quelconque, connexes et orientées, on définit leur *somme connexe* $M_1 \# M_2$ de la façon suivante. On considère deux ouverts assez petits $U_1 \subset M_1$, $U_2 \subset M_2$, dont l'adhérence est homéomorphe à $\mathbb{D}_n$; si S_1, S_2 sont les frontières de U_1, U_2, on prend un homéomorphisme $h : S_1 \to S_2$ renversant l'orientation; puis on «recolle» S_1 et S_2 par h, c'est-à-dire qu'on prend le quotient de $(M_1 - U_1) \coprod (M_2 - U_2)$ par la relation d'équivalence dont les classes sont les singletons n'appartenant ni à S_1 ni à S_2, et les ensembles à 2 éléments $\{x, h(x)\}$ pour $x \in S_1$. La variété $M_1 \# M_2$ ne dépend pas (à homéomorphisme près) des choix de U_1, U_2 et h; la somme connexe $M_1 \# \mathbb{S}_n$ est homéomorphe à

M_1. Pour $n = 3$, Milnor a montré que toute variété connexe, orientée et sans bord, est somme connexe $M_1 \# M_2 \# \ldots \# M_k$, où les M_j ne sont plus décomposables en somme connexe de variétés autres que $\mathbb{S}_3$; on dit que ce sont des variétés *primaires*, et elles sont bien déterminées à l'ordre près et à homéomorphisme près.

Une variété M de dimension 3 est dite irréductible si pour tout plongement f de $\mathbb{S}_2$ dans M, il existe un plongement de la boule $\mathbb{D}_3$ dans M dont $f(\mathbb{S}_2)$ est le bord; on a alors nécessairement $\pi_2(M) = 0$. Inversement, si $\pi_2(M) \neq 0$, Papakyriakopoulos a prouvé en 1957 qu'il y a un plongement de $\mathbb{S}_2$ dans M qui n'est pas homotope à une application constante. Papakyriakopoulos et Homma ont aussi démontré indépendamment un lemme énoncé par M. Dehn avec une preuve insuffisante: s'il existe une immersion du disque $\mathbb{D}_2$ dans M qui est injective au voisinage de $\mathbb{S}_1$, alors il y a un plongement PL de $\mathbb{D}_2$ dans M, qui transforme $\mathbb{S}_1$ en $f(\mathbb{S}_1)$ (autrement dit, on peut supprimer les auto-intersections); ce lemme joue un rôle important dans l'étude des variétés de dimension 3.

Les sous-variétés compactes sans bord et de dimension 1 dans $\mathbb{S}_3$ sont appelées des *entrelacs* («links»); un entrelacs est réunion disjointe d'un nombre fini de courbes difféomorphes à un cercle $\mathbb{S}_1$; un entrelacs connexe est un *noeud*. Les noeuds et les entrelacs font l'objet depuis 150 ans d'une énorme littérature. Le problème central est de classer les entrelacs pour une relation d'équivalence, deux entrelacs étant équivalents s'il existe une isotopie de $\mathbb{S}_3$ sur elle-même transformant l'un en l'autre. Un invariant évident pour la classe d'équivalence d'un entrelacs L est le groupe fondamental $\pi_1(\mathbb{S}_3 - L)$; mais jusqu'ici on ne connaît pas de système fini d'invariants qui caractériserait un entrelacs à équivalence près.

E. Artin a introduit en 1927 une méthode d'étude des entrelacs utilisant ce qu'il a appelé les *groupes de tresses*. Pour tout $n \geq 1$, soit Y_n l'ensemble des parties de $\mathbb{C}$ ayant n éléments; on peut le munir d'une topologie de la façon suivante. On considère le sous-espace X_n des vecteurs $(z_j)_{1 \leq j \leq n}$ de $\mathbb{C}^n$ dont toutes les composantes sont distinctes. Le groupe symétrique $\mathfrak{S}_n$ opère par permutation des coordonnées sur X_n, et Y_n s'identifie à l'espace quotient $X_n/\mathfrak{S}_n$. Le groupe de tresses B_n est le groupe fondamental $\pi_1(Y_n)$. Un lacet de $\Omega(Y_n)$ peut être représenté par n chemins d'équations $t \mapsto (\gamma_j(t), t)$ dans $\mathbb{C} \times [0, 1]$, d'origine le point $(j, 1)$ et d'extrémité $(\sigma(j), 0)$ pour une permutation $\sigma \in \mathfrak{S}_n$, chaque γ_j étant strictement décroissante et les n points $\gamma_j(t)$ pour $1 \leq j \leq n$ étant toujours distincts; d'où le nom de «tresses». A chaque tresse correspond un entrelacs obtenu en «raccordant» les extrémités aux origines par des chemins d'origine $(\sigma(j), 0)$ et d'extrémité $(\sigma(j), 1)$. On montre que tout entrelacs peut être obtenu de cette façon. E. Artin a montré que le groupe B_n est engendré par $n - 1$ éléments $g_1, \ldots, g_{n-1}$ liés par les relations

$$g_i g_j = g_j g_i \text{ si } |i - j| > 1, g_{i+1} g_i g_{i+1} = g_i g_{i+1} g_i \text{ pour } 1 \leq i \leq n - 2. \qquad (123)$$

En 1983, V. Jones a remarqué que les algèbres de Hecke $H_n(q)$, qui interviennent dans la théorie du groupe linéaire $\mathrm{SL}_n(\mathbb{F}_q)$, ont des systèmes de générateurs qui satisfont aux relations (123); il en a déduit, grâce aux propriétés des algèbres

$H_n(q)$, la formation d'un polynôme qui ne dépend que de la classe d'isotopie d'un entrelacs. Un autre polynôme invariant avait été défini par Alexander en 1928 à partir de l'homologie de $\mathbb{S}_3 - L$. Les polynômes de Jones et d'Alexander sont différents, mais depuis 1985 ils ont été déduits par des changements de variables convenables, d'un même polynôme à 3 variables qui est aussi attaché de façon invariante à une classe d'entrelacs.

22.2 Les variétés de dimension 4

En 1975, les progrès considérables réalisés en Topologie géométrique avaient apporté une moisson de résultats remarquables sur les variétés, à l'exception de celles de dimension 4. Pour une telle variété X, compacte, orientée, sans bord et simplement connexe, le cup-produit définit une forme bilinéaire symétrique non dégénérée

$$\Phi : H^2(X;\mathbb{Z}) \times H^2(X;\mathbb{Z}) \longrightarrow \mathbb{Z}$$

(dite «forme d'intersection» par abus de langage). On s'était rendu compte que tous les invariants topologiques de X étaient entièrement déterminés par cette forme (y compris les classes caractéristiques du fibré tangent à X lorsque X est lisse); pour deux telles variétés X, Y, Milnor avait montré en 1958 qu'elles ont même type d'homotopie si elles ont même forme d'intersection, et Wall ajouta en 1964 qu'elles sont même h-cobordantes. Il manquait malheureusement un théorème de h-cobordisme; les méthodes de preuve utilisées en dimension ≥ 5 échouaient en dimension 4, parce qu'on ne pouvait plus utiliser la déformation de Whitney. Mais vers 1973 apparurent de nouvelles méthodes, et en 10 ans un véritable feu d'artifice d'idées originales dissipa les ténèbres où gisait la dimension 4, en révélant des propriétés tout à fait insoupçonnées et spéciales à cette dimension.

En ce qui concerne les variétés TOP (compactes, sans bord et simplement connexes) les résultats décisifs ont été obtenus par M. Freedman en 1982. Si une telle variété a la même homologie que $\mathbb{S}_4$, elle est homéomorphe à $\mathbb{S}_4$ («conjecture de Poincaré en dimension 4»). Le théorème de h-cobordisme est vrai pour les variétés TOP de dimension 4, et ces variétés sont entièrement classifiées à homéomorphisme près par leur forme d'intersection; à toute forme bilinéaire symétrique non dégénérée sur $\mathbb{Z}$, il correspond une structure TOP et une seule à homéomorphisme près.

Ces résultats ont été rendus accessibles par l'introduction, à partir de 1976, par A. Casson, d'objets nouveaux (d'allure «fractale»), appelés «anses ouvertes de Casson» et qui permettent une variante de la «chirurgie». M. Freedman se rendit compte que l'usage de ces «anses» donnerait le résultat cherché si l'on pouvait prouver qu'une anse ouverte de Casson, malgré sa définition compliquée, est en fait homéomorphe à $\mathbb{D}_2 \times \mathbb{R}^2$. C'est ce qu'il parvint finalement à prouver, en montrant qu'un certain espace quotient de $\mathbb{D}_2 \times \mathbb{R}^2$ est homéomorphe à une anse de Casson, puis en utilisant le critère de rétrécissement de Bing pour montrer qu'en fait cet espace quotient est homéomorphe à $\mathbb{D}_2 \times \mathbb{R}^2$. Malgré diverses améliorations, l'accumulation des techniques variées nécessaires pour les démonstrations en interdit l'accès à d'autres que les «connaisseurs».

Le théorème de Freedman ne donnait pas de renseignement sur les variétés DIFF de dimension 4. La surprise fut donc très grande lorsqu'en 1983, S. Donaldson prouva qu'une variété X de dimension 4 (compacte, orientée, sans bord et simplement connexe) ne peut être munie d'une structure DIFF que sous des conditions très restrictives sur sa forme d'intersection: si elle est définie négative, $H^2(X;\mathbb{Z})$ admet une base sur $\mathbb{Z}$ pour laquelle la forme d'intersection a pour matrice $-\mathbb{I}_m$.

Ce qui est encore plus remarquable est la preuve de ce théorème: il s'agit d'établir un cobordisme orienté entre la variété lisse X et la somme disjointe de m variétés difféomorphes à $\mathbb{P}_2(\mathbb{C})$ muni de son orientation canonique. Ce qui est tout à fait inattendu, c'est la façon dont Donaldson définit ce cobordisme W, qui doit être une variété lisse, orientée, de dimension 5: elle a pour points les classes de X-isomorphisme (des «modules» au sens de Riemann) de certaines connexions sur un fibré principal de groupe structural SU(2) et de base X; lorsque X est muni d'une structure riemannienne, ces connexions sont les solutions d'une équation aux dérivées partielles dite «équation de Yang-Mills» d'après son origine dans la physique quantique. La preuve est un tour de force qui nécessite des moyens techniques considérables tirés de la géométrie différentielle (notamment la formule d'Atiyah-Singer), et de théorèmes difficiles d'existence et d'approximation de solutions d'équations aux dérivées partielles non linéaires.

Il résulte des théorèmes de Freedman et Donaldson qu'il y a très peu de variétés TOP de dimension 4 qui puissent être munies d'une structure lisse, et que le théorème du h-cobordisme n'est pas valable pour la catégorie DIFF en dimension 4. Une conséquence surprenante est qu'il existe sur $\mathbb{R}^4$ une infinité de structures lisses non difféomorphes deux à deux, alors que tous les autres espaces $\mathbb{R}^n$ n'admettent qu'une seule structure lisse à difféomorphisme près.

Index

Seminar Alexandrov, Moscow, 1975

A.N. Tihonov, P.S. Alexandrov

I.G. Petrovsky, P.S. Alexandrov

The Development of Rigor in Mathematical Probability, (1900–1950)

Joseph L. Doob
University of Illinois

1 Introduction

This paper is a brief informal outline of the history of the introduction of rigour into mathematical probability in the first half of this century. Specific results are mentioned only in so far as they are important in the history of the logical development of mathematical probability.

The development of science is not a simple progression from one advance to the next. Judged by hindsight, the development is slow, proceeds in a zigzag course, with many wrong turns and blind alleys, and frequently moves in directions condemned by leading scientists. In the 1930's Banach spaces were sneered at as absurdly abstract, later it was the turn of locally convex spaces, and now it is the turn of nonstandard analysis. Mathematicians are no more eager than other humans to embrace new ideas, and full acceptance of mathematical probability was not realized until the second half of the century. In particular, many statisticians and probabilists resented the mathematization of probability by measure theory, and some still place mathematical probability outside analysis. The following quotations (in translation) are relevant.

Planck: *A new scientific truth does not triumph by convincing its opponents and making them see the light, but rather because its opponents eventually die, and a new generation grows up with it.*

Poincaré: *Formerly when one invented a new function, it was to further some practical purpose; today one invents them in order to make incorrect the reasoning of our fathers, and nothing more will ever be accomplished by these inventions.*

Hermite: (in a letter to Stieltjes) *I recoil with dismay and horror at this lamentable plague of functions which do not have derivatives.*

Probability theory began, and remained for a long time, an idealization and analysis of certain real life phenomena outside mathematics, but gradually, in the first half of this century, mathematical probability became a normal part of mathematics. The mathematization of probability required new ideas, and in particular required a new approach to the idea of acceptability of a function. In view of the above quotations it is not surprising that acceptance of this mathematization was slow and faced resistance. In fact even now some probabilists fear that mathematization has removed the intrinsic charm from their subject. And they are right in the sense that the charm of the old vague probability-mathematics, based on nonmathematical definitions, has split into two quite different charms: those of real world probability and of mathematical precision. But it must be stressed that many of the most essential results of mathematical probability have been suggested by the nonmathematical context of real world probability, which has never even had a universally acceptable definition. In fact the relation between real world probability and mathematical probability has been simultaneously the bane of and inspiration for the development of mathematical probability.

2 What is the real world (nonmathematical) problem?

What is usually called (real world) probability arises in many contexts. Besides the obvious contexts of gambling games, of insurance, of statistical physics, there are such simple contexts as the following. Suppose an individual rides his bicycle to work. The rider would be surprised if, when the bicycle is parked, the valve on the front tire appeared in the upper half of the tire circle 10 successive days, just as surprised as if 10 successive tosses of a coin all gave heads. However it is clear that (tire context) if the ride is very short, or (coin context) if the coin starts close to the coin landing place and the initial rotational velocity of the coin is low, the surprise would decrease and the probability context would become suspect. The moral is that the specific context must be examined closely before any probabilistic statement is made. If philosophy is relevant, an arguable question, it must be augmented by an examination of the physical context.

3 The law of large numbers

In a repetitive scheme of independent trials, such as coin tossing, what strikes one at once is what has been christened the *law of large numbers*. In the simple context of coin tossing it states that in some sense the number of heads in n tosses divided by n has limit $1/2$ as the number of tosses increases. The key words here are *in some sense*. If the law of large numbers is a mathematical theorem, that is, if there is a mathematical model for coin tossing, in which the law of large numbers is formulated as a mathematical theorem, either the theorem is true in one of the various mathematical limit concepts or it is not. On the other hand, if the law of large numbers is to be stated in a real world nonmathematical context, it is not at all clear that the limit concept can be formulated in a reasonable way. The most obvious difficulty is that in the real world only finitely many experiments can be performed in finite time. Anyone who tries to explain to students what happens when a coin is tossed mumbles words like *in the long run, tends, seems to cluster near*, and so on, in a desperate attempt to give form to a cloudy concept. Yet the fact is that anyone tossing a coin observes that for a modest number of coin tosses the number of heads in n tosses divided by n seems to be getting closer to $1/2$ as

J.L. Doob, Luxembourg 1992

n increases. The simplest solution, adopted by a prominent Bayesian statistician, is the vacuous one: never discuss what happens when a coin is tossed. A more common equally satisfactory solution is to leave fuzzy the question of whether the context under discussion is or is not mathematics. Perhaps the fact that the assertion is called a *law* is an example of this fuzziness. The following statements have been made about this law (my emphasis):

Laplace: (1814) *This theorem, implied by common sense, was difficult to prove by analysis.*

Ville: (1939) *One sees no reason for this proposition to be true; but as it is impossible to prove experimentally that it is false, one can at least safely state it.*

Bauer: (translated from the context of dice to that of coins) *It is an experimentally established fact that the quotient ... exhibits a deviation from* $1/2$ *which approaches 0 for large* n.

These statements illustrate the enduring charm of discussions of real world probability. Mathematicians, unfortunately, have felt forced to think about the following question, or at least to write about it.

4 What is probability?

Here are some attempts to answer this question and to discuss the teaching of the subject.

Poincaré: (1912) *One can scarcely give a satisfactory definition of probability.*

Mazurkiewicz: (1915) *The theory of probability is not an independent element of mathematical instruction; nevertheless it is very desirable that a mathematician knows its general principles. Its fundamental concepts are incompletely determined. They contain many unsolved difficulties.*

v. Mises: (1919) *In fact one can scarcely characterize the present state other than that probability is not a mathematical discipline.* (He proceeded to make it into a mathematical discipline by basing mathematical probability on a sequence of observations («Beobachtungen») with properties that cannot be satisfied by a mathematical well defined sequence. In a lighter mood he is said to have defined probability as a number between 0 and 1 about which nothing else is known.)

E. Pearson: (1935) (Oral communication) *Probability is so linked with statistics that, although it is possible to teach the two separately, such a project would be just a tour de force.*

Uspensky: (1937) In a useful textbook, he gave the following once common textbook definition. *If, consistent with condition S, there are n mutually exclusive, and equally likely cases, and m are favorable to the event A, then the mathematical probability of A is defined as m/n.*

The foregoing should make obvious the advisability of separating mathematical probability theory from its real world applications. Note however that noone doubts the real world applicability of mathematical probability. Gambling, genetics, insurance and statistical physics are here to stay.

Only *mathematical* probability will be discussed below, except for the following remark on coin tossing. Newtonian mechanics provides a partial mathematical model for coin tossing. In coin tossing, a solid body falls under the influence of gravity. Its motion is determined in Newton's model by his laws, and any discussion of what the coin does cannot be complete unless these laws are applied. Only these laws, rather than philosophical remarks, can explain the quantitative influence and importance of the initial and final conditions of the coin motion in order to justify allusions to equal likelihood of heads and tails. Of course these laws can at best reduce the analysis to considerations of the initial and final conditions of the toss, but these conditions can show what the «equal likelihood» depends on and thereby give it a plausible interpretation.

5 Mathematical probability before the era of precise definitions

There were many important advances in mathematical probability before 1900, but the subject was not yet mathematics. Although nonmathematical probabilistic contexts suggested problems in combinatorics, difference equations and differential equations, there was a minimum of attention paid to the mathematical basis of the contexts, a maximum of attention to the pure mathematics problems they suggested. This unequal treatment was inevitable, because measure theory, needed for mathematical modeling of real world probabilistic contexts, had not yet been invented.

It was always clear that, however classical mathematical probability was to be developed, the concept of additivity of probability as applied to incompatible real world events was fundamental. Additive functions of sets were of course familiar to mathematicians from concepts of volume, mass and so on, long before 1900. It was realized that contexts involving averages led to probability. It was frequently clear how to use the contexts to suggest problems in analysis, but it was not clear how to formulate an overall mathematical context, that is, how to define a mathematical structure in which the various contexts could be placed.

A weaker condition than additivity was less familiar but turned out to be essential later. The standard loose language will be used here. If $x_1, x_2, \ldots$ are numbers obtained by chance, and if A is a set of numbers, consider the probability that at least one of the members of this sequence lies in A, or, in more colorful language, consider the probability that an orbit of this motion through points of a line hits A. The usual calculation (ignoring here all notions of rigour), defines a function $A \to \phi(A)$ which in general is not additive. In fact ϕ satisfies the inequality

$$\phi(A) + \phi(B) - \phi(A \cup B) \geq \phi(A \cap B). \tag{5.1}$$

whereas additivity of ϕ would imply equality in (5.1). The point is that the left side of (5.1) is the probability that the sequence $x\bullet$ hits both A and B, a probability at least equal to and in general greater than $\phi(A \cap B)$, the probability that the sequence hits $A \cap B$. The inequality (5.1), the *strong subadditivity* inequality, is satisfied also by the electrostatic capacity of a body in $\mathbf{R}^3$, and this fact hints at the close connection between potential theory and probability, developed in great detail in the second half of the century with the help of Choquet's theory of mathematical capacity.

6 The development of measure theory

Recall that a *Borel field* (= σ algebra) of subsets of a space is a collection of subsets which is closed under the operations of complementation and the formation of countable unions and intersections. The class of *Borel sets* of a metric space is the smallest set σ algebra containing the open sets of the space. A *measurable space* is a pair, $(S, \mathbb{S})$, where S is a space and $\mathbb{S}$ is a σ algebra of subsets of S. The sets of $\mathbb{S}$ are the *measurable sets* of the space. In the following, if S is metric, the coupled σ algebra making it into a measurable space will always be the σ algebra of its Borel sets. In particular $(\mathbf{R}^N, \mathbb{R}^N)$ denotes N dimensional Euclidean space coupled with its Borel sets. The superscript will be omitted when $N = 1$. A measurable function from a measurable space $(S_1, \mathbb{S}_1)$ into a measurable space $(S_2, \mathbb{S}_2)$ is a function from S_1 into S_2 with the property that the inverse image of a set in $\mathbb{S}_2$ is a set in $\mathbb{S}_1$.

Measure theory started with Lebesgue's thesis (1902), which extended the definition of volume in $\mathbf{R}^N$ to the Borel sets. Radon (1913) made the further step to general measures of Borel sets of $\mathbf{R}^N$ (finite on compact sets). These measures are usually extended to slightly larger classes than the class of Borel sets, by *completion*. Finally Fréchet (1915), 13 years after Lebesgue's thesis, pointed out that all that the usual definitions and operations of measure theory require is a σ algebra of subsets of an abstract space on which a measure, that is, a positive countably additive set function, is defined. At each step of this progression not necessarily positive countably additive set functions — signed measures — were incorporated into the theory. As noted below, the Radon-Nikodym theorem (1930), which gives conditions necessary and sufficient that a countably additive function of sets can be expressed as an integral over the sets, turned out to be the final essential result needed to formulate the basic mathematical probability definitions. Thus it was 28 years before Lebesgue's theory was extended far enough to be adequate for the mathematical basis of probability. This extension was not developed in order to provide a basis for probability, however. Measure theory was developed as a part of classical analysis, and applications in analysis were immediate, for example to the (Lebesgue measure) almost everywhere derivability of a monotone function.

There has been criticism of the fact that mathematical probability is usually prescribed not only to be additive but even to be countably additive. The question whether real world probability is countably additive, if the question is to be meaningful, asks whether a mathematical model of real world probabilistic phenomena *necessarily* always involves *countably* additive set functions. In fact there may well be real world contexts for which the appropriate mathematical model is based on finitely but not countably additive set functions. But there have been very few applications of such set functions in either mathematical or nonmathematical contexts, and such set functions will not be discussed further here.

7 Early applications of explicit measure theory to probability

Some probabilistic slang will be needed, enduring relics of the historical background of probability theory. A probability space is a triple $(S, \mathbb{S}, \mathrm{P})$, where $(S, \mathbb{S})$ is a measurable space and P is a measure on $\mathbb{S}$ with $\mathrm{P}(S) = 1$. A measure with this normalization is a *probability measure*. A *random variable* is a measurable function from a probability space $(S, \mathbb{S}, \mathrm{P})$, into a measurable space $(S', \mathbb{S}')$. The space S', or, when one writes carefully, $(S', \mathbb{S}')$, is the *state space* of the random variable. Mutual independence of random variables is defined in the classical way. The *distribution* of a random variable x is the measure P_x on $\mathbb{S}'$ defined by setting

$$P_x(A') = P\{s \in S : x(s) \in A'\}.$$

The joint distribution of finitely many random variables defined on the same probability space is obtained by making x into a vector and specifying $\mathbb{S}'$ and S' correspondingly. A stochastic process is a family of random variables $\{x(t, \bullet), t \in I\}$ from some probability space $(S, \mathbb{S}, P)$, into a state space $(S', \mathbb{S}')$. The set I is the *index set* of the process. Thus a stochastic process defines a function of two variables, $(t, s) \rightarrow x(t, s)$, from $I \times S$ into the state space. The function $x(t, \bullet)$ from S into S' is the tth random variable of the process; the function $x(\bullet, s)$ from I into S' is the sth sample function, or sample path, or sample sequence if I is a sequence.

Borel (1909) pointed out that in the dyadic representation $x = .x_1x_2\cdots$ of a number x between 0 and 1, in which each digit x_j is either 0 or 1, these digits are functions of x, and if the interval [0,1] is provided with Lebesgue measure, a probability measure on this interval, these functions miraculously become random variables which have exactly the distributions used in calculating coin tossing probabilities. That is, 2^{-n} is the probability assigned to the event that, in a tossing experiment, the first n tosses yield a specified sequence of heads and tails, and 2^{-n} is also the total length (= Lebesgue measure) of the finite set of intervals whose points x have dyadic representations with a specified sequence of 0's and 1's in the first n places. Thus a mathematical version of the law of large numbers in the coin tossing context is the existence in some sense of a limit of the sequence of function averages $\{(x_1 + \cdots + x_n)/n,\ n \geq 1\}$. Classical elementary probability calculations imply that this sequence of averages converges in measure to 1/2, but a stronger mathematical version of the law of large numbers was the fact deduced by Borel — in an unmendably faulty proof — that this sequence of averages converges to 1/2 for (Lebesgue measure) almost every value of x. A correct proof was given a year later by Faber, and much simpler proofs have been given since. [Fréchet remarked tactfully: «Borel's proof is excessively short. It omits several intermediate arguments and assumes certain results without proof.»] This theorem was an important step, an example of a new kind of convergence theorem in probability. Observe that (fortunately) pure mathematicians need not interpret this theorem in the real world of real people tossing real coins. Some of the quotations given above indicate that they not only need not but should not.

Daniell (1918) used a deep approach to measure theory in which integrals are defined before measures to get a (rather clumsy) approach to infinite sequences of random variables by way of measures in infinite dimensional Euclidean space.

The Brownian motion stochastic process in $\mathbf{R}^3$ is the mathematical model of Brownian motion, the motion of a microscopic particle in a fluid as the particle is hit by the molecules of the fluid. The process is normalized by supposing it starts at the origin of a cartesian coordinate system in $\mathbf{R}^3$, and a (normalized) Brownian motion process in $\mathbf{R}$ is the process of a coordinate function of a normalized process in $\mathbf{R}^3$, vanishing initially. A (normalized) Brownian motion process in $\mathbf{R}^N$ is a process defined by N mutually independent Brownian motion processes in $\mathbf{R}$. It was well known what the joint distributions of the random variables of a Brownian motion process should be, and it had been taken for granted that in a proper mathematical model the class of continuous paths would have probability 1. By 1900, Bachelier had even derived various important distributions related to the Brownian motion process in $\mathbf{R}$, such as that of the maximum change during a time interval, by finding corresponding distributions for a certain discrete random walk and then going to the limit as the walk steps tended to 0. More precisely, what Bachelier derived were distributions valid for a Brownian motion process if in fact there was such a thing as a Brownian motion process, and if it was approximable by his random walks. Observe that there was no question about the existence of Brownian motion; Brownian motion is observable under a microscope. But there was as yet no proof of the existence of a stochastic process, a mathematical construct, with the desired properties. Wiener (1923) constructed the desired Brownian motion process, now sometimes called the *Wiener process*, by applying the Daniell approach to measure theory to obtain a measure with the desired properties on a space S of continuous functions: if $x(t, \bullet)$ is the random variable defined by the value at time t of a function in S, the stochastic process of these random variables is a stochastic process with sample functions the members of S, and with the joint random variable distributions those prescribed for the Brownian motion process.

Bachelier's results remained unnoticed for years, and in fact were rediscovered several times. Wiener's work, like his fundamental work in potential theory, had little immediate influence because it was published in a journal which was not widely distributed. It was an aspect of his genius that he carried out his Brownian motion research then and later without knowledge of the slang and some of the useful elementary mathematical techniques of probability theory.

Steinhaus (1930) demonstrated that classical arguments to derive standard probability theorems could be placed in a rigorous context by taking Lebesgue measure on a linear interval of length 1 as the basic probability measure, interpreting random variables as Lebesgue measurable functions on this interval, and expectations of random variables as their integrals. No new proofs were required; all that was required was a proper translation of the classical terminology into his context. If this were all mathematization of probability by measure theory had to offer, the scorn of rigorous mathematics expressed by some nonmathematicians would be justified.

8 Kolmogorov's 1933 monograph

Kolmogorov (1933) constructed the following mathematical basis for probability theory.

(a) The context of mathematical probability is a probability space $(S, \mathbb{S}, \mathrm{P})$. The sets in $\mathbb{S}$ are the mathematical counterparts of real world events; the points of S are counterparts of elementary events, that is of individual (possible) real world observations.

(b) Random variables on $(S, \mathbb{S}, \mathrm{P})$, are the counterparts of functions of real world observations. Suppose $\{x(t, \bullet), t \in I\}$ is a stochastic process on a probability space $(S, \mathbb{S}, \mathrm{P})$, with state space S'. A set of n of the process random variables has a probability distribution on S'^n. Such finite dimensional distributions are mutually compatible in the sense that if $1 \leq m < n$, the joint distribution of $x(t_1, \bullet), \ldots, x(t_m, \bullet)$ on S'^m is the m-dimensional distribution induced by the n-dimensional distribution of $x(t_1, \bullet), \ldots, x(t_n, \bullet)$ on S'^n.

(c) Conversely, Kolmogorov proved that given an arbitrary index set I, and a suitably restricted measurable space $(S'', , \mathbb{S}')$, (for example the measurable space can be a complete separable metric space together with the σ algebra of its Borel sets) and a mutually compatible set of distributions on S'^n, for integers $n \geq 1$, indexed by the finite subsets of I, there is a probability space and a stochastic process $\{x(t, \bullet), t \in I\}$ defined on it, with state space S', with the assigned joint random variable distributions. To prove this result he constructed a probability measure on a σ algebra of subsets of the product space S'^I, the space of all functions from I into S', and obtained the required random variables as the coordinate functions of S'^I.

(e) The expectation of a numerically valued integrable random variable is its integral with respect to the given probability measure.

(f) The classical definition of the conditional probability of an event (measurable set) A, given an event B of strictly positive probability, is $\mathrm{P}(A \cap B)/P(B)$. In this way, for fixed B, new probabilities are obtained, and expectations of random variables for given B are computed in terms of these new *conditional* probabilities. More generally, given an arbitrary collection of random variables, conditional probabilities and expectations relative to given values of those random variables are needed, functions of the values assigned to the conditioning random variables. If $(S, \mathbb{S}, \mathrm{P})$ is a probability space, and if a collection of random variables is given, let $\mathbb{F}$ be the smallest sub σ algebra of $\mathbb{S}$ relative to which all the given random variables are measurable. This σ algebra is the *σ algebra generated by conditions imposed on the given random variables*. A reasonable interpretation of a measurable real valued function of the given collection of random variables is a measurable function from $(S, \mathbb{F})$ into **R**. The Kolmogorov *conditional expectation* of a real valued integrable random variable x on $(S, \mathbb{S}, \mathrm{P})$, relative to a σ algebra $\mathbb{C}$ of measurable sets, is a random variable which is measurable relative to $\mathbb{C}$ and has the same integral as x on every set in $\mathbb{C}$. The existence of such a random variable, and its uniqueness up to P-null sets, is assured by the Radon-Nikodym theorem. The conditional expectation of x relative to a collection of random variables is defined as the conditional expectation of x relative to the σ algebra generated by conditions on the random variables. A conditional probability of a measurable set A is defined as the conditional expectation of the random variable which is 1 on A and 0 elsewhere.

Kolmogorov's 1933 exposition paints a discouraging picture of mathematical progress. In the first pages of his monograph he states explicitly that real valued random variables are measurable functions and expectations are their integrals. Even as late as 1933, however, he must have thought that mathematicians were not familiar with measure theory. In fact in the body of his monograph, when he comes to the definition of a real valued random variable, he does not simply refer back to the first pages of the monograph and say that a random variable is a measurable function. Instead he actually defines measurability of a real valued function, and similarly when he defines the expected value of a random variable he does not simply state that it is the integral of the random variable with respect to the given probability measure, but he actually defines the integral. Later in the monograph, when he needs Lebesgue's theorem allowing taking limits of convergent function sequences under the sign of integration, he does not simply refer to Lebesgue but gives a detailed proof of what he needs. As confirmation of Kolmogorov's caution in invoking measure theory, the author recalls his student experience in 1932 when there were professorial disapproving remarks on the extreme generality of a seminar lecture given by Saks on what is now called the Vitali-Hahn-Saks theorem, a theorem which has since become an important tool in probability theory. [He also recalls that he did not understand the point of Kolmogorov's measure on function space until long after he had read the monograph.]

It was some time before Kolmogorov's basis was accepted by probabilists. The idea that a (mathematical) random variable is simply a function, with no romantic connotation, seemed rather humiliating to some probabilists. A prominent statistician in 1935 wondered whether two orthogonal real valued random variables with zero means (integrals) are necessarily independent, as they are under the added hypothesis that they have a bivariate Gaussian distribution. He was rather surprised by the example of the sine and cosine functions on the interval $[0, 2\pi]$, with probability measure defined as Lebesgue measure divided by 2π. These two functions, orthogonal and with zero means but not independent, are not the kind of random variables probabilists were used to. Some analysts may be gratified, some humiliated, to learn that in discussing Fourier series they can be accused of discussing probabilities and expectations.

9 Expansion backwards of the Kolmogorov basis

Kolmogorov's basis for mathematical probability can be expanded, and should be expanded in the view of some probabilists, who want to start with some not necessarily numerical mathematical version of the confidence of observers that certain events will occur, and to proceed postulationally to numerical evaluations of this confidence, and finally to additivity. Such an analysis may be enlightening in discussing the appropriateness of mathematical probability as a model for real world phenomena, but any approach to the subject which ends with a justification of the classical calculations and is mathematically usable, will end with Kolmogorov's basis, however phrased, because all the measure basis to probability does is to give a formal precise mathematical framework for the classical calculations and their present refinements. This framework has made it possible to apply mathematical probability in many other mathematical fields, for example to potential theory and partial differential equations. Although such applications were made in the past before the acceptance of measure theory as the basis of probability the probabilistic context served only to suggest mathematics and was not an integral part of the mathematics. The meaning of solutions as probabilities and expectations could not be formulated and exploited.

10 Uncountable index sets

If the index set I of a stochastic process $\{x(t, \bullet), t \in I\}$ is an interval of the line, and if the state space of the random variables is $\mathbf{R}$, the class of continuous sample functions may not be measurable. This difficulty arises in the processes derived by the Kolmogorov construction of a measure on a function space, for example, whatever the choice of joint distributions of the process random variables. To understand the difficulty, observe that if the index set I of a stochastic process with state space $\mathbf{R}$ is an interval, and if J is a subset of I, the function $s \rightarrow \sup_{t \in J} x(t, s)$ is measurable if J is countable, but not necessarily if J is uncountable. If boundedness and continuity of sample functions are to be discussed, some modification of the probability relations of the random variables of a stochastic process should be devised to make such suprema measurable functions. A clumsy approach was proposed by Doob (1937) but a more usable one was not devised until after 1950.

11 Reluctance to accept measure theory by probabilists

There was considerable resistance to the acceptance and exploitation of measure theory by probabilists, both in Kolmogorov's day and later. The following quotation is an example of the reluctance of some mathematicians to separate the mathematics from the context that inspired it.

Kaç (1959) *How much fuss over measure theory is necessary for probability theory is a matter of taste. Personally I prefer as little fuss as possible because I firmly believe that probability theory is more closely related to analysis, physics and statistics than to measure theory as such.*

12 New relations between functions made possible by the mathematization of probability

Probability theory suggested new relations between functions. For example consider the sequence $x_1, x_2, \ldots$ of real valued integrable random variables on a probability space $(S, \mathbb{S}, \mathrm{P})$ and suppose that the conditional expectation of x_n given $x_1, \ldots, x_{n-1}$ vanishes P almost everywhere, for $n > 1$, that is, the integral of x_n over any set determined by conditions on the preceding random variables vanishes. If these random variables are square integrable this condition is equivalent to the condition, much stronger than mutual orthogonality, that x_n is orthogonal to every square integrable function of $x_1, \ldots, x_n$. Bernstein (1927) seems to have been the first to treat such sequences systematically. This condition on a sequence of functions means that in a reasonable sense the sequence of partial sums of the given sequence is the counterpart of a fair game. In fact the partial sums $y_1, y_2, \ldots$ are characterized by the property that the expectation of y_n relative to $y_1, \ldots, y_{n-1}$ is equal to y_{n-1} almost everywhere on the probability space. Processes with this property, called *martingales*, first used explicitly by Ville (1939), have had many applications, for example to partial differential equations, to derivation and to potential theory. Another important class of sequences of random variables is the class of sequences with the Markov property. These sequences are characterized by the fact that when $n \geq 1$ the conditional probabilities for x_n relative to $x_1, \ldots, x_{n-1}$ are equal almost everywhere to those for x_n relative to x_{n-1}. Roughly speaking, the influence of the present, given the past, depends only on the immediate past. The Markov property, introduced in a very special case by Markov in 1906, (named in his honor by others) has proved very fruitful, for example leading in the second half of the century to a probabilistic potential theory, generalizing and including classical potential theory.

13 What is the place of probability theory in measure theory, and more generally in analysis?

It is considered by some mathematicians that if one deals with analytic properties of probabilities and expectations then the subject is part of analysis, but that if one deals with sample sequences and sample functions then the subject is probability but not analysis. These authors are in the interesting situation that in considering a function of two variables, $(t, s) \to x(t, s)$ — as in considering stochastic processes

— they call it analysis if the family of functions $x(t, \bullet)$ as t varies is studied, but call it probability and definitely not analysis if the family of functions $x(\bullet, s)$ as s varies is studied. More precisely, they regard discussions of distributions and associated questions as analysis, but not discussions in terms of sample functions. This point of view is expressed in the following quotation.

Protter *By developing his integral in 1944 with stochastic processes as integrands, Itô was able to study multidimensional diffusions with purely probabilistic techniques, an improvement over the analytic methods of Feller.*

The following remark on the convergence of a sum of orthogonal functions illustrate the difficulty in separating (mathematical) probability from the rest of analysis. The measure space is a probability space, but with trivial changes the discussion is valid for any finite measure space.

If $x\bullet$ is an orthogonal sequence of functions, on a probability measure space, and if x_n^2 has integral σ_n^2, then (Riesz-Fischer), Σx_n converges in the mean if

$$\Sigma\sigma_n^2 < +\infty. \tag{13.1}$$

The orthogonal series converges almost everywhere if either (Menšov-Rademacher) (13.1) is strengthened to

$$\Sigma\sigma_n^2 \log^2 n < +\infty, \tag{13.2}$$

or Lévy (1937) the condition (13.1) is kept but the orthogonality condition is strengthened to the condition in Section 10.

The reader should judge which of these results is measure theoretic and which is probabilistic, whether there is any point in evicting mathematical probability from analysis, and if so whether measure theory should also be evicted.

JOSEPH L. DOOB
208, West High Street
Urbana Ill 61801 USA

A. Kolmogorov

Vito Volterra and the birth of functional analysis

Gaetano Fichera
Università di Roma

The argument I shall deal with in this paper concerns the birth of functional analysis which can be traced back to the end of the last century and the beginning of the current one, even though it is difficult to pin-point the birth of a new branch of Science. It often happens that several concepts and results exist before the official constitution of a theory and it is only later on that they are perceived as belonging to the same theory. This is particularly true for functional analysis. One of its important aspects, the calculus of variations, had already appeared in the second half of the seventeenth century in the work of Isaac Newton (1642–1727) (problems of the body of revolution of minimum resistance) and, successively, in the work of Jacques (1645–1705) and Jean (1667–1748) Bernoulli (brachistochrone and isoperimetric problems). It acquired an even wider breadth with the immortal contributions of Leonard Euler (1701–1783), first, and Giuseppe Luigi Lagrange (1736–1813), after. But, if it is worthwhile to make the source of a theory coincide with the work of one who shows that he fully understands that, with his studies, he his opening up a new field of Science, then there is no doubt that Vito Volterra (1860–1940) must be considered as the initiator of functional analysis. In a Lincei Note of 1887 [1], he introduces *the functions which depend on other functions*. Subsequently, considering the graph of the *independent variable* function, rather than the function itself, he uses the expression functions of lines. He later on definitely adopts the term *functional* proposed by Jacques Hadamard (1865–1963).

Volterra's ideas on the new concepts that he introduced met with immediate success among the mathematicians of the times. Jacques Hadamard, in particular, appreciated and justly valued their enormous potential. He not only used the new ideas in his important research on the variation of Green's function in dependence of the relative domain, but he prompted his students, men like Maurice Fréchet (1878–1973), René Gateaux (1880–1914), Paul Lévy (1886–1960) to promote the new theory initiated by Volterra.

In Jean Dieudonné's *History of Functional Analysis* [2] we read:

> *«We must finally mention the first attempt at «Functional Analysis» of the young Volterra in 1887, to which, under the influence of Hadamard, has been attributed an exaggerated historical importance».*

In order to explain this severe judgement of Volterra's pioneer work by an eminent French Colleague, it is necessary to clarify two concepts, which, even though they are quite distinct from one another, are often confused by historical scholars and, in particular, by those studying the History of Science. They are the concepts of *Revisitation* and *History*. Revisitation consists in re-examining with today's mentality, with the experience and complex of new acquisitions that the passage of time has made available to us, those themes which belong to a past historical period. This approach corresponds more to an *historical review* rather than to *history* itself. *History*, instead, is something more subtle and more difficult. In fact, the true historian must make the effort to shed himself of today's way of thinking and of all the experience he has acquired in the time interval between the moment in which he is working and the epoch he is examining. He must try to reconstruct the typical mentality of the period to which his historical research is directed, discarding all the superstructures which have arisen during the passage of the years. Only in this way can he give an unbiased judgement of a given historical figure and, in the case of a scientist, correctly evaluate his work. Ideas and concepts, which today appear evident or even commonplace, often belong to categories which, in past times, were quite inconceivable. Let us consider, for example, the concept of function, today of common dominion; anyone who reads a newspaper or watches television and sees a graph drawn can understand it. However, the conviction that I have heard expressed by various mathematicians, according to which infinitesimal calculus would have been born seventeen centuries earlier if Archimedes had not been so barbarously killed by the Roman soldier, is incorrect. Archimedes, even with his sovereign mind, was far from possessing the idea of function, so abstract a conceptual formation that seventeen centuries would be needed before it could be, even though in a rather particular sense, understood by mankind. Only after this abstract concept had been conquered was it possible for Newton and Leibniz to create infinitesimal calculus... The same can be said of functional analysis: none of the distinguished scholars of calculus of variations had, before Volterra, conceived the integrals which are studied *as functions of lines*! It is certainly true that the approach indicated by Volterra did not immediately lead to the heart of the great problems of functional analysis. But the demonstration of this fact is not an easy consequence of simple axiomatic affirmations. It requires a careful and committed analysis, which is what we propose to do in this presentation. On the other side we shall show, when examining, even though briefly, Luigi Fantappiès (1901–1956) work, that, although we remain in a strictly «Volterrian» ambit, the ideas introduced by the great Italian pioneer were susceptible to considerable developments.

Many were the problems deriving from both pure and applied mathematics which led Volterra to introduce the concept of *function of lines*. But among these, those which particularly justify the advent of this idea, more as a «discovery» than as an «invention», were the problems of elastic or electromagnetic inheritance, which he set out in his Monograph of 1913, entitled «*Fonctions de lignes*» [3], originating from a course of lectures held at the Sorbonne in 1912.

Let Ω be a 3-dimensional elastic solid in its natural configuration. For any deformation originated by body- and surface-forces (either directly applied or reactions of constraints) it is assumed, in the context of classical elasticity, that the *Hooke law* holds, i.e.

$$\sigma_{ij} = a_{ijhk}\varepsilon_{hk}; \tag{1}$$

$\sigma \equiv \{\sigma_{ij}\}(\sigma_{ij} = \sigma_{ji})$ is the *stress tensor* and $\varepsilon \equiv \{\varepsilon_{hk}\}(\varepsilon_{hk} = \varepsilon_{kh})$ the *strain tensor*, both referred to the same system of rectangular cartesian coordinates; a_{ijhk} is the *elastic tensor* of Ω, which enjoyes the symmetry properties: $a_{ijhk} = a_{hkij} = a_{jihk}$. The function $a_{ijhk}(x)$ of the point x of Ω is constant if Ω is a homogeneous body. It is assumed that the quadratic form $a_{ijhk}(x)\varepsilon_{ij}\varepsilon_{hk}$ is positive definite for any $x \in \Omega$.

Eq. (1) states that the *actual* stress σ depends on the *actual* strain ε. Volterra, following Boltzmann (1844–1906) [4], observed that in several elastic bodies the *past history* of the body cannot be neglected and in consequence the actual stress must depend not only on the actual strain but also on the *whole* of the deformations under which the solid was subjected in the past. Hence Eq. (1) should be modified as follows:

$$\sigma_{ij}(t) = a_{ijhk}\varepsilon_{hk}(t) + F_{ij}[\varepsilon\overset{t}{\underset{\gamma}{(}}\tau)]; \tag{2}$$

γ is the instant when the phenomenon starts to be considered, t the actual instant, F_{ij} a *functional* which depends on all the values that the strain tensor assumes in the instants τ between γ and t. Volterra, remaining in the context of linear elasticity, proposes the following analytical representation for F_{ij}:

$$F_{ij}[\varepsilon\overset{t}{\underset{\gamma}{(}}\tau)] = \int_\gamma^t \psi_{ijhk}(t,\tau)\varepsilon_{hk}(\tau)d\tau; \tag{3}$$

$\psi_{ijhk}(t,\tau)$ is the memory which the solid has at the instant t of the deformation $\varepsilon(\tau)$ it had in the past instant τ. The *memory* or *hereditary coefficients* ψ_{ijhk} are subjected to the following conditions: $\psi_{ijhk} = \psi_{jihk} = \psi_{ijkh}$ besides some qualitative properties which also depend on the fact that one assumes either $\gamma = -\infty$ or $\gamma > -\infty$. This last case is the one almost exclusively considered by Volterra.

In the classical case (1) the problem consisting in representing ε by means of σ is an elementary algebraic problem. In the *hereditary* case (2), (3) for getting $\varepsilon(t)$ in terms of $\sigma(t)$ one has to solve a system of integral equations, which nowadays are denoted as *Volterra integral equations*. It is easily seen that (2), (3) can be written as follows:

$$f(t) = u(t) + \int_\gamma^t K(t,\tau)u(\tau)d\tau, \tag{4}$$

where f is the *known term* i.e. a given n-vector [in the case (2), (3) $f(t) = \sigma(t); n = 6$], the n-vector u is the *unknown function* and $K(t,\tau)$ is a given $n \times n$ matrix. Eq. (4) is the Volterra integral equation of *the second kind*. For it Volterra under rather general hypotheses on $f(t)$ and $K(t,\tau)$ proved existence and uniqueness when $\gamma > -\infty$.

The problems connected with equilibrium or motion of Ω are more complicated. In the classical case (1) by imposing on σ to satisfy either to equilibrium or to motion equations and representing ε_{hk} as $\varepsilon_{hk} = 2^{-1}(u_{h/k} + u_{k/h})$ (u is the *displacement vector*) one gets the classical equation of Elasticity either for equilibrium or for motion.

The same procedure applied to (2), (3) leads to *integro-differential* equations which, at the time of Volterra, were of a completely new kind.

In the case $\gamma = -\infty$ even today a complete theory is not known.

Volterra conceived functionals as functions depending on an infinity of variables: all the values of the *independent variable* function $u(t)$. The method he used to affront functional related problems or, more generally, to lay the basis of *Functional Analysis*, which would extend the concepts and results of common functions, object of *Calculus*, to functionals, was what he called *the passage from finiteness to infinity*. This method could be viewed as a sort of passage to the limit from a finite set of variables to an infinite set, but, in substance, it consisted not so much in a process of passage to the limit, as in a *principle of analogy*, which, through opportune transpositions, transferred operations and symbols of a finite scheme to an infinite one. Volterra never gave a definition of *the passage from finiteness to infinity*, but he used it as a heuristic method in transferring concepts and results of Calculus to Functional Analysis [1], [5].

Let us briefly expound this method of Volterra. To this end we consider a particularly significant example: existence of a *potential functional*.

One of the main problems of Infinitesimal Calculus is the one consisting in

i) giving necessary and sufficient conditions for a given differential form $\varphi_1(u_1, \ldots, u_n)du_1 + \ldots + \varphi_n(u_1, \ldots, u_n)du_n$ defined in a domain A of $\mathbb{R}^n$ to be the exact differential of a scalar function $f(u_1, \ldots, u_n)$;

ii) constructing the *potential function* $f(u)$.

It is well known that assuming $\varphi_\alpha(u)[u \equiv (u_1, \ldots, u_n); \alpha = 1, \ldots, n]$ continuous in A, the potential function f exists when and only when for each regular closed curve Γ contained in A of parametric equations $u_\alpha = u_\alpha(s)$, $0 \leq s \leq L$, $u_\alpha(0) = u_\alpha(L)$, one has

$$\int_\Gamma \sum_\alpha \varphi_\alpha[u(s)]\frac{du_\alpha}{ds} = 0. \tag{5}$$

Regularity of Γ means that each $u_\alpha(s)$ is continuous in $[0, L]$ with its first derivative and, moreover, for $s \in [0, L]$

$$\sum_\alpha [u'_\alpha(s)]^2 > 0. \tag{6}$$

If (5) is satisfied, every solution of the problem is given by

$$f(u) = f(u^0) + \int_{\Lambda(u^0,u)} \sum_\alpha \varphi_\alpha[v(s)]\frac{dv_\alpha}{ds}ds, \tag{7}$$

where u^0 is arbitrarily fixed in A, $\Lambda(u^0, u)[v_\alpha = v_\alpha(s),\ 0 \le s \le \ell,\ v_\alpha(0) = u^0_\alpha,\ v_\alpha(\ell) = u_\alpha]$ is a regular curve joining u^0 with u and the constant $f(u^0)$ is arbitrarily chosen.

To say that $f(u)$ is a solution of the problem means that for each $u \in A$ and each η such that $u + \eta \in A$, one has

$$f(u+\eta) - f(u) = \sum_{\alpha=1}^{n} \varphi_\alpha(u)\eta_\alpha + o(\delta) \tag{8}$$

where $\delta = \max_\alpha |\eta_\alpha|$.

With his *passage from finiteness to infinity* Volterra considers α like a point of a (bounded) continuum C of $\mathbb{R}^m$ and for $\varphi_\alpha(u)$ substitutes the *mixed functional* $\varphi[u;\alpha]$. This means that, for each α of C, φ is a functional of u defined in a domain $\mathbb{A}$ of the space of the continuous functions in C. He considers the following problem:

i′) to give necessary and sufficient conditions for the existence of a *potential functional* $f(u)$ of $\varphi[u;\alpha]$ i.e. of a functional $f(u)$ such that for each $u \in \mathbb{A}$ and $\eta, u+\eta \in \mathbb{A}$, one has

$$f(u+\eta) - f(u) = \int_C \varphi[u;\alpha]\eta(\alpha)d\alpha + o(\delta) \tag{9}$$

where $\delta = \max_{\alpha \in C} |\eta(\alpha)|$;[1])

ii′) to construct the potential function $f(u)$.

Of course one assumes that, for each $u \in \mathbb{A}$, $\varphi(u;\alpha)$ is a continuous function of α in C. This permits to give a meaning to the integral at the right hand side of (9).

It is obvious how (9) is obtained by analogy from (8) by *passage from finiteness to infinity.*

The problem is solved by transposing the *finite* scheme into the *infinite* one.

Let us consider, in fact, a *regular closed curve* Γ in the domain $\mathbb{A}$. Γ is determined by the *parametric equation* $u = u(\alpha, s)$, where $u(\alpha.s)$ for each $s \in [0, L]$ belongs to $\mathbb{A}$; $u(\alpha, 0) \equiv u(\alpha, L)$. Moreover $u(\alpha, s)$ is a continuous function of s in $[0, L]$ with the derivative $\frac{\partial u(\alpha,s)}{\partial s}$ and, in analogy with (6),

$$\int_C \left[\frac{\partial u(\alpha,s)}{\partial s}\right]^2 d\alpha > 0, \quad s \in [0, L].$$

1) Actually the relationship between f and φ in the work of Volterra is slightly different from the one we have considered here. In fact Volterra assumes that $\varphi[u;\alpha]$ is the *differential* of f in a more general sense than the one we have considered here. The Volterra definition of differential corresponds to a concept which is now called *Gateaux differential*, while we have considered the more particular concept of *Fréchet differential* (see [9]).

The necessary and sufficient condition for the existence of $f(u)$ is in analogy with (5) (supposing $\varphi[u(x,s);\alpha] \in C^0\{C \times [0,L]\}$)

$$\int_\Gamma ds \int_C \varphi[u(x,s);\alpha]\frac{\partial u(\alpha,s)}{\partial s}d\alpha = 0 \tag{10}$$

for every closed Γ. If (10) is satisfied one has

$$f(u) = f(u^0) + \int_{\Lambda(u^0,u)} ds \int_C \varphi[v(x,s);\alpha]\frac{\partial v(\alpha,s)}{\partial s}d\alpha = 0, \tag{11}$$

where the symbols used in (11) are easily explained by passage from finiteness to infinity from (7).

Eq. (10) and (11) besides their conceptual value are of noteworthy interest in several problems of Analysis and Mathematical Physics.

Volterra's method, suggestive and rich in remarkable results though it appears, has, however very precise confines, which limit the range of action and prevent from entering into the heart of the more important problems of functional analysis. In fact, to obtain results of functional analysis from the results of classical Calculus by means of the passage from finiteness to infinity tacitly presumes the existence of the following principle: *Nothing can occur in the spaces of infinite dimension that does not already occur in those of finite dimension.*

This statement is quite far from reality! Actually, the more significant and profound concepts and results, relative to the spaces of infinite dimension, as function spaces usually are, have no significant comparison to those of the finite dimension, in so much as they both flatten themselves on concepts and results which, passing from a finite to an infinite dimension, admit at least two possibilities of generalization; of these, Volterra's method leads us to consider the more obvious and, often, the less important one. Let us take into consideration the concepts, fundamental in functional analysis, of compactness, of total continuity, of weak convergence, etc., that coincide in the finite dimension with those of boundedness and closure, of continuity, of ordinary convergence, etc.

Let us now show how the Volterra method is inadequate for investigating some basic problems of Analysis. We refer, for instance, to integral equation, with fixed limits, of the second kind

$$u(s) + \lambda \int_0^1 K(s,t)u(t)dt = f(s) \quad (0 \le s \le 1). \tag{12}$$

We assume $K(s,t) \in L^2[(0,1)\times(0,1)]$, $f(s) \in L^2(0,1)$ and seek the unknown function $u(s)$ in $L^2(0,1)$. We suppose complex-valued all the considered functions; λ is a complex parameter.

We briefly set

$$Ku = \int_0^1 K(s,t)u(t)dt, \quad \langle u, v\rangle = \int_0^1 u(s)v(s)ds. \tag{13}$$

Eq. (12) can be considered as obtained by a passage from finiteness to infinity from the algebraic system

$$u_s + \lambda \sum_{t=0}^{m} K_{st} u_t = f_s \quad (s = 0, 1, \ldots, m); \tag{14}$$

$u \equiv (u_0, \ldots, u_m)$ and $f \equiv (f_0, \ldots, f_m)$ are now (complex) $(m+1)$-vectors of $\mathbb{C}^{m+1}$ and K_{st} is a complex $(m+1) \times (m+1)$-matrix.

By analogy with (13) we set

$$Ku = \sum_{t=0}^{m} K_{st} u_t, \quad \langle u, v\rangle = \sum_{s=0}^{m} u_s v_s. \tag{15}$$

Eq.

$$u + \lambda K u = f \tag{16}$$

in both cases (12) and (14) has the formal solution

$$u = f - \lambda K f + \lambda^2 K^2 f + \ldots + (-1)^n \lambda^n K^n f + \ldots \tag{17}$$

which is an actual solution of (16) if $|\lambda| < \rho$, $\rho = \lim_{n\to\infty} ||K^n||^{-\frac{1}{n}}$, where

$$||K^n|| = \sup \langle K^n u, \overline{K^n u}\rangle^{\frac{1}{2}} \text{ for } \langle u, \bar{u}\rangle = 1$$

($\bar{a}$ denotes the conjugate of a).

If $\rho = +\infty$, i.e. $\lim_{n\to\infty} ||K^n||^{\frac{1}{n}} = 0$, (17) gives the only solution of (16) for any λ. This is the case when one has in (12), $K(s,t) \equiv 0$ for $s \leq t$ and, accordingly in (14), $K_{st} = 0$ for $s \leq t$. In both cases this hypothesis is called *triangular hypothesis*. If triangular hypothesis holds, Eq. (12) can be written

$$u(s) + \lambda \int_0^s K(s,t)u(t)dt = f(s).$$

This is the classical *Volterra's integral equation of the second kind*, we have already met.

So far there is a perfect analogy between (14) and (12) and the principle of the passage from finiteness to infinity works beautifully.

The situation is much more complicated if $|\lambda| \geq \rho$. It is obvious that in this case the triangular hypothesis is not satisfied. Concerning (14) we know from elementary algebra that two cases are possible:

1) λ is not an eigenvalue of the homogeneous system associated to (14);

2) λ is an eigenvalue for such system.

If 1) holds, (14) has one and only one solution for any given f. In the case 2) the range V_λ of the linear transformation $I + \lambda K$ (I = identity) is characterized by the equations $\langle v, w_h \rangle = 0$, $h = 1, \ldots, p$ where $w_1, \ldots, w_p$ are p independent vectors of $\mathbb{C}^{m+1}$. These vectors are a complete system of eigenvectors of the *transposed homogeneous equation*

$$w + \lambda K^* w = 0,$$

where K^* is the transposed matrix of K. Hence equation (14) has a solution if and only if $\langle f, w_h \rangle = 0$ ($h = 1, \ldots, p$).

Extension of these results to Eq. (12) *cannot be obtained by the Volterra principle of the passage from finiteness to infinity*. Actually if we interpret $K^* w$ as follows

$$K^* w = \int_0^1 K(t, s) w(t) dt,$$

it would be possible to extend the results of the finite dimensional case to $L^2(0, 1)$ if we had the additional information that the range V_λ of $I + \lambda K$ is a *closed* manifold of $L^2(0, 1)$ and has a *finite codimension*.

The Volterra method is not able to supply this information: every linear manifold of $\mathbb{C}^{m+1}$ is closed and has finite codimension!

All the efforts of Volterra to get a complete theory for Eq. (12) were in vain. Later Fredric Riesz (1880–1956) and Stefan Banach (1892–1945) proved that the above properties for V_λ hold. However for getting this result they used a property of the linear transformation K which Volterra's method is far from suggesting, i.e. the *total continuity* (or *compactness*) of the transformation K. This means that if $\{u_n\}$ is any sequence of $L^2(0, 1)$ *weakly converging* to u (for any $w \in L^2(0, 1)$, $\lim_{n \to \infty} \langle u_n, w \rangle = \langle u, w \rangle$) then $K u_n$ *converges strongly* to Ku ($\lim_{n \to \infty} ||K_n u - Ku|| = 0$).

It must be observed that in a finite dimensional normed space *weak convergence* and *strong convergence* are equivalent. This is not true in $L^2(0, 1)$. Hence in $L^2(0, 1)$ *total continuity* of a transformation K does not coincide with *continuity*.

Still before F. Riesz and S. Banach the Swedish mathematician Ivar Fredholm (1866–1927) succeeded in 1903 to achieve, in a really ingenious way, a complete theory of Eq. (12), without any use of concepts of Functional Analysis (total continuity, weak convergence, etc.). He followed a purely algorithmical way [7]. He starts from observing that for $|\lambda|$ small ($|\lambda| < \rho$) solution of (12) given by

(17) can be put under the following form

$$u(s) = f(s) + \int_0^1 R(s, t; \lambda) f(t) dt$$

where $R(s, t; \lambda)$ is a function of (s, t) belonging to $L^2[(0, 1) \times (0, 1)]$. Moreover $R(s, t; \lambda)$ for any fixed (s, t) is a function of λ holomorphic for $|\lambda| < \rho$. Fredholm succeeded in proving that $R(s, t; \lambda)$ can be continued in a function of λ meromorphic in the complex plane. Moreover, with incredible analytical skill, he constructed the two entire functions of λ such that $R(s, t; \lambda)$ is expressed as their ratio. From this analytical representation of $R(s, t; \lambda)$ he deduced all the results concerning resolution of Eq. (12). Volterra, after Fredholm's discovery, he himself used possibility of expressing solution of (12) like a meromorphic function of λ. This reduces to a rational function in the finite dimensional case (14). He could in this way recover, somehow, his principle of the passage from finiteness to infinity. But it was too late. On the other hand, the Volterra method is just *misleading* in the analysis of the *integral equation of the first kind*

$$Ku = f. \tag{18}$$

If (18) is considered in the finite dimensional case, i.e. K given by (15), the range of K is, obviously, a *closed* linear manifold of $\mathbb{C}^{m+1}$.

In the case (13) such range is not, in general, a *closed* linear manifold of $L^2(0, 1)$. This circumstance, which the method of the passage from finiteness to infinity is unable to foresee, makes so that, for technical reasons we cannot report here, problem (18) is *ill posed*. This means that a solution of (18) (when it exists!) even if it is unique, *does not depend continuously on the datum* f.

The great and merited influence that Volterra had on the Italian mathematicians saw to it that the trends cultivated abroad by mathematicians like David Hilbert, Maurice Fréchet, by the above mentioned F. Riesz, Banach and others, were not cultivated in Italy, and in this manner the entry of Italian analysts into the main stream of functional analysis was delayed by at least twenty years. Neither did help the work of Salvatore Pincherle (1853–1936), author of a dissertation on linear transformations which, maintained on an essentially formal plane, met with the severe, but unfortunately justified, criticism of eminent foreign analysts, first among whom was Edmund Landau (1877–1938).

But apart from the limitations to which Volterra's approach to functional analysis was subject, his pioneer work was outstanding, especially when it is considered from the point of view of the philosophy which inspired it: to tackle the great problems of applications with profoundly new ideas and methods [6].

His research on hereditary phenomena and his theory on elastic dislocations developed between 1905 and 1907, represent milestones in classical mathematical physics. And even in pure mathematics some of his pioneeristic ideas later proved to be productive. The intention of extending to the functionals, dependent on a line or, more in general, on a variety, the ordinary concept of harmonic function, induced Volterra to define *harmonic functionals* which, taken up again by William Hodge (1903–1975), were to lead to what can be considered one of the major mathematical results of this century: *Hodge's theorem of existence of harmonic integrals on a Riemannian manifold of dimension* n; in itself a wonderful theorem and full of important consequences for algebraic geometry and for the continuous group theory.

In the period between the two wars Volterra mostly devoted himself to the applications of mathematical analysis to biological problems. He was attracted to this research by the noteworthy biologist Umberto d'Ancona (1896–1964), who was also his son-in-law and who collaborated closely with him. In particular, Volterra studied the problems that arise from the cohabitation of different species of animals, among which the *prey* and the *predators*, and eventually formulated a mathematical theory on the *struggle for life*. The mathematical analysis used is usually not particularly complicated and requires the study of systems of differential or, at the most, integrodifferential equations of Volterra's type. But quite suggestive is the biological interpretation of the analytical results achieved, some of the more simple of which had already been attained by other authors. This is the case of the now famous «Lotka-Volterra equations», which the American scholar Alfred James Lotka (1880–1949) had discovered, even though under particular assumptions, some years before Volterra.

Volterra's attempt to deal with *population dynamics* in the same way as with the *dynamical systems* of classical analytical mechanics appears rather fascinating. To this aim, in problems relating to *biological associations*, he introduces variational principles similar to those of Hamilton (1805–1869) and of Maupertuis (1698–1759) (*minimum action*) of classical dynamics, developing a parallel biological theory. To what degree this mathematical model can reproduce biological reality is, today, still an uncertain question and, perhaps, unascertainable, given the relative simplicity of the first with respect to the enormous complexity of the second [6].

When examining one of the fundamental concepts introduced by Volterra, that is the derivative of a functional, Dieudonné maintains the inadequacy of Volterra's theory in the absence of the use of topology in the considered function space. He writes «*With our experience of 50 years of Functional Analysis we cannot help feeling that without even the barest notion of general topology, these ad hoc definitions were decidedly premature*».

It is quite true that Volterra's approach and his fundamental principle of the passage from finiteness to infinity show their limits when, introducing topology in space, we realize that, in infinite dimensional space, situations, which the above mentioned principle cannot foresee, are created. However one cannot drastically affirm that Volterra's ideas, in order to evolve, necessarily require a topology in function space. Instead, one should keep in mind the possibility that the introduction of a topology can be substituted with other opportune hypotheses which would allow the development of the theory. In fact if, for example, one chooses to operate in spaces of analytic functions, one can do without topology in developing an even rather thorough functional analysis. This was very well demonstrated by Luigi Fantappiè, a student of Volterra and the creator of the *Theory of Analytic Functionals* [8].

Fantappiè did not make use of the geometry of function spaces where his functionals are defined, but he was able to construct his theory by purely algorithmic means, exploiting only the fact that each of them is considered as defined on classes of analytic functions of the z complex variable and that, besides, if the independent variable function depends analytically, other than on z, on a complex parameter α, then the functional is, by definition, an analytic function of α. Only later on was it noted that a structure of topological space T_0 could be given to the function classes where Fantappiè's functionals are defined.

This theory was particularly important in the case of linear functionals. For these Fantappiè supplied a theorem of representation by means of which he then managed to build a rigorous *symbolic calculus* for a wide class of linear operators, representing, under opportune hypotheses, the operator as an analytic functional of its *characteristic function*. He then partially extended the theory to analytic functionals dependent on analytic functions of several complex variables for which he managed to establish certain results interesting in themselves.

Let K be a closed set of the complex sphere Σ and let $\mathbb{A}$ be the function class defined by the following conditions: i) each $f \in \mathbb{A}$ is holomorphic in an open set A containing K; ii) if $\infty \in A$ then $f(\infty) = 0$.

It is obvious that $\mathbb{A}$ is a complex linear space. Let $F(f)$ be a functional defined on $\mathbb{A}$ such that $F(f_1) = F(f_2)$ if, being f_k defined on A_k $(k = 1, 2)$, one has $f_1 \equiv f_2$ in $A_1 \cap A_2$. If $A \supset K$ and $f(z, \alpha)$ is defined for $z \in A$ and $\alpha \in \Omega$ ($\Omega =$ open set of the complex plane) and is a holomorphic function of (z, α) in $A \times \Omega$, then $F[f(z, \alpha)]$ is a holomorphic function of α in Ω. F is said an *analytic functional defined in* $\mathbb{A}$. If $F(af_1 + bf_2) = aF(f_1) + bF(f_2)$ (a, b complex constants), F is a *linear analytic functional*. Let $K \subset A$ (A open set) and let D be an open set such that $K \subset D$, $\bar{D} \subset A$. Let the boundary C of D be made by one or several closed smooth contours. For any $\alpha \in C$ the function $(\alpha - z)^{-1}$ as a function of z is holomorphic in D. Set $u(\alpha) = F[(\alpha - z)^{-1}]$. For every $F \in \mathbb{A}$ the following representation theorem holds:

$$F(f) = \frac{1}{2\pi i} \int_{+C} u(\alpha) f(\alpha) d\alpha. \tag{19}$$

The function $u(\alpha)$ is called the *indicatrix* of F.

The proof of (19) is obtained by observing that in D one has

$$f(z) = \frac{1}{2\pi i} \int_{+C} \frac{f(\alpha)}{\alpha - z} d\alpha.$$

Hence, after proving that F can be interchanged with $\int_{+C}$

$$F(f) = \frac{1}{2\pi i} F\left[\int_{+C} \frac{f(\alpha)}{\alpha - z} d\alpha\right] = \frac{1}{2\pi i} \int_{+C} F(\frac{1}{\alpha - z}) f(\alpha) d\alpha.$$

Let us remark that this proof is conceptually analogous to the one by F. Riesz concerning the representation of a linear continuous functional F in $C^0[a,b]$. F. Riesz, after continuing F in the class of piece-wise continuous functions, considers the function

$$\varphi(\alpha, x) \begin{cases} = 0 & \alpha \leq x, \\ = 1 & \alpha > x \end{cases}$$

and proves that for $f \in C^0[a,b]$ one has

$$F[f(x)] = F[\int_a^b f(\alpha) d\varphi(\alpha, x)] = \int_a^b f(\alpha) du(\alpha);$$

$u(\alpha)$ is the function of bounded variation such that $u(\alpha) = F[\varphi(\alpha, x)]$.

Let T be a linear operator which maps a function linear space S into itself. If $f(\lambda)$ is an analytic function of λ, Fantappiè intends to give a meaning to the operator $f(T)$. Assuming that $f(T)$ operates on $v(x) \in S$, he considers, for x and v fixed, $f(T)[v(x)]$ like a linear analytical functional of $f(\lambda)$. Under suitable hypotheses for applying his representation theorem, he supposes that the indicatrix $u(\alpha, x)$ of this functional is the only solution of the equation

$$\alpha u(\alpha, x) - Tu(\alpha, x) = v(x).$$

This permits him to determine $u(\alpha, x)$ and to define $f(T)[v(x)]$ in the following way:

$$f(T)[v(x)] = \frac{1}{2\pi i} \int_{+C} u(\alpha, x) f(\alpha) d\alpha$$

with an obvious meaning for C.

Starting from this result, he constructs a *symbolic calculus* which has a very attractive feature and leads him to discover the *resolvent formulas* of several classic problems of Analysis, considered in the complex domain.

His approach to the theory of analytic functions of a $n \times n$ matrix is particularly elegant.

Fantappiè's work is interwoven with, and often precedes, the work of other authors who, between the two wars, studied the theory of the analytic functions of an operator.

It is a rather suggestive demonstration of how far the deeply new ideas first introduced by Vito Volterra could lead.

Bibliography

[1] V. VOLTERRA, *Sopra le funzioni che dipendono da altre funzioni*, Rend. R. Accad. dei Lincei, 4, 3, 97–105, 141–146, 153–158, 1887.

[2] J. DIEUDONNÉ, *History of Functional Analysis*, Mathematics Studies 49, North Holland, Amsterdam-New York-Oxford, 1981.

[3] V. VOLTERRA, *Leçons sur les fonctions de lignes*, Collect. de Monograph. sur la théorie des fonctions, Gauthier-Villars, Paris, 1913.

[4] L. BOLTZMANN, *Zur Theorie der elastischen Nachwirkung*, Sitzber. Kaiserl. Akad. Wiss. Wien, Math.-Natur. Kl. 70, II, 275–300, 1874.

[5] V. VOLTERRA, *Theory of Functionals and of Integral and Integro-Differential Equations*, Dover Publ. Inc. New York, 1959.

[6] V. VOLTERRA, *Opere matematiche*, 5 vol. Accad. Naz. dei Lincei, 1954–1962.

[7] I. FREDHOLM, *Sur une classe d'équations fonctionnelles*, Acta Math. 27, 365–390, 1903.

[8] L. FANTAPPIÈ, *Opere scelte*, vol. I & vol. II, Unione Matematica Italiana, 1973.

[9] P. LÉVY, *Problèmes concrets d'analyse fonctionnelle*, Collect. de Monograph. sur la Théorie des fonctions, Gauthier-Villars, Paris, 1931.

GAETANO FICHERA
7, Via Pietro Mascagni
I-00199 Roma

La Logique Mathématique en sa jeunesse

Marcel Guillaume

Essai sur l'Histoire de la Logique durant la première moitié du vingtième siècle

Avant-Propos

Le texte ci-après n'est qu'un Essai, écrit sur l'aimable invitation des organisateurs du Colloque sur le développement des mathématiques de 1900 à 1950, que je remercie encore de m'avoir donné l'occasion de l'écrire. Ni moi, ni eux, je pense, ne soupçonnaient à quel point l'objet à l'étude, même restreint à une seule discipline, allait se révéler fractal.

J'ai entamé cet ouvrage en visant à donner aux autres mathématiciens, qui dans leur ensemble ne savent pas grand'chose de la logique, et aux informaticiens, qui ne la connaissent que dans certains aspects de son être récent, une idée de son histoire, et de ce que les logiciens ont fait; faute à n'écrire que pour le cercle étroit des logiciens eux-mêmes, il n'était donc pas possible d'user des termes qu'ils utilisent couramment sans fournir quelques explications sur leur signification, ni de raconter comment les idées ont cheminé, sans reproduire de temps en temps quelques raisonnements typiques.

Certes, ce parti-pris est pour une bonne part responsable de la longueur de ce travail. Mais il m'est apparu, chemin faisant, qu'une histoire totalement fidèle de l'évolution de la logique de 1900 à 1950 demanderait probablement des centaines de pages. Je n'ai ni cité tous les hommes qui auraient, à des titres divers, mérité de l'être; ni mentionné tous les travaux comportant des contributions notables de ceux dont je parle; ni analysé dans toute leur portée, tous ceux sur lesquels je dis quelques mots. Je n'ai pas, non plus, complètement respecté, encore que je m'y sois efforcé au maximum de mes possibilités, les règles de la méthodologie de l'histoire. Les choix que j'ai faits de suivre ou non tel ou tel cheminement d'idées, de traiter ou non tel ou tel développement, ou d'aller jusqu'à tel ou tel point dans son exposé; ou encore, de mentionner ou non, après la premières, les démonstrations parfois sans cesse faites et refaites du même théorème, avec des nuances plus ou moins subtiles, ou les multiples définitions d'une même notion, ou les interminables variations des résultats acquis pas à pas, par petites touches successives; ces choix ont été souvent dictés par le souci d'être compris, si possible, de mes putatifs interlocuteurs mathématiciens, d'autres fois par les plus ou moins grandes facilités d'accès à telle ou telle documentation qui ont été les miennes, c'est-à-dire, par de l'arbitraire, du point de vue de l'histoire.

Ce texte est d'ailleurs plein de questions, le plus souvent tacites, de rapprochements qui ne constatent peut-être que le fait du hasard, ou de l'air du temps, et qu'il faudrait une équipe de je ne sais combien de personnes pour tous vérifier.

J'ai, néanmoins, disposé d'une documentation assez riche; sur certains points, on peut maintenant consulter des travaux d'historiens spécialisés; en outre, les logiciens sont souvent très attentifs à l'histoire, et les notes, parfois inexactes sur des points de détail, qui parsèment de nombreux ouvrages consacrés avant tout à la discipline, par les recoupements qu'elles permettent, constituent, prises ensembles, un guide assez sûr dans ses grandes lignes. J'ai par ailleurs consulté beaucoup d'ouvrages originaux — les remaniements des rééditions réservent parfois des surprises. En réponse à certaines méconnaissances courantes de ce qu'ont réellement été les conceptions de certains des acteurs de l'hisoire, il m'arrive parfois de les citer dans le texte, sans traduire, puisqu'il s'est agi d'un colloque européen. Au total, j'espère, malgré tout, être parvenu au but visé: donner à un lecteur que je suppose doté d'une bonne culture mathématique, ou informatique, une idée d'ensemble correcte des démarches les plus importantes, ou les plus caractéristiques, ou les plus informatives sur la vie de la discipline et sur ses rapports avec celle des autres, de ce qu'a été, dans ses grandes lignes, la grande aventure intellectuelle des logiciens durant ce demi-siècle.

Afin de mieux suivre les cheminements d'idées, les développements ultérieurs d'un même thème, tant qu'il n'est pas traité d'interférences avec quelque autre, sont en principe traités aussitôt après l'exposé de ses débuts, ceux-ci seuls étant rangés à peu près selon l'ordre chronologique. Il se superpose à cela, dans la dernière partie, un classement par sous-disciplines, dans la mesure où celles-ci apparaissent; de certains domaines sur lesquels il y a peu à dire, il est traité parfois incidemment, à l'occasion d'une interférence avec un autre domaine. Les sous-titres insérés dans le texte, et ceux des subdivisions de la dernière partie, sont destinés à aider le lecteur à s'y reconnaître dans les dédales des connexions rationnelles et chronologiques, qui n'en finissent pas de se ramifier, entre les thèmes traités.

L'aide matérielle que m'ont fournie le Département de Mathématiques de l'Université Blaise Pascal, et ses secrétaires, particulièrement Mesdames Bernard, Bressoulaly et Rebord, mes collègues du Laboratoire de Logique, Algorithmique et Informatique Clermontois, de l'Université d'Auvergne, et sa secrétaire Madame Jallat, m'a été précieuse: je les en remercie tous chaleureusement, ainsi que Jean-Paul Pier qui s'est donné la peine de corriger une première frappe. Ces remerciements s'adressent tout particulièrement à François Gaillard, sans l'aide matérielle et morale de qui la réalisation de ce travail dans des délais assez brefs n'eût tout simplement pas été possible.

Table des Matières

Les temps optimistes

Des *Principia Mathematica* comme objet de travaux.

Efforts vers la réduction du nombre des primitives; origines de la logique combinatoire.

De la théorie des types simple.

De cheminements d'idées chez des philosophes, issus des *Principia Mathematica* et de travaux postérieurs, et finissant par avoir un effet en retour sur la logique, jusqu'en 1930.

De l'école de logique de Varsovie: ŁUKASÍEWÍCZ et ses logiques multivalentes.

De l'école de logique de Varsovie: aperçu sur des conceptions de LEŚNIEWSKI ayant influencé le développement de la logique.

De la théorie des ensembles arrangée en théorie des types cumulative: axiome de fondement.

De quelques étapes importantes dans l'histoire de l'axiome de choix, jusqu'en 1930.

Du stade atteint par HILBERT dans le développement de ses idées, en 1917.

Des contributions de BERNAYS et de POST à l'étude du calcul des propositions.

Des théorèmes de complétude et de compacité pour la logique élémentaire.

Des mathématiques et de la logique intuitionnistes, de 1918 à 1925.

Des formalisations de la logique intuitionniste et des interprétations intuitionnistes de fragments de la théorie des nombres classique, jusque vers 1933.

Des résultats partiels concernant le problème de la décision, avant 1933.

Des affinements du programme de HILBERT en 1922.

De l'arithmétique primitive récursive de SKOLEM.

Des principales contributions à la théorie des ensembles axiomatique de FRAENKEL, SKOLEM, VON NEUMANN et BERNAYS.

Du concept de métamathématique chez TARSKI.

Des achèvements et questionnements de l'école de HILBERT, de 1922 à 1930.

Les premieres maturations de l'ère de la connaissance limitée
Du «séisme» de 1930 et de ses suites durant les vingt années suivantes.

Des théorèmes d'incomplétude de GÖDEL, et du théorème de non-définissabilité arithmétique des énoncés vrais dans leur application aux nombres entiers naturels, de TARSKI.

Sur les réactions aux découvertes de GÖDEL.

Sur les tendances générales de l'évolution de la logique, de 1930 à 1950.

Du développement de la théorie de la démonstration, après 1930.

Des calculs de déduction naturelle et des calculs de «séquences».

Des origines des conceptions de GENTZEN dans les travaux de théorie de la démonstration de Paul HERTZ.

Des preuves intuitionnistes de non-contradiction de l'arithmétique de PEANO du premier ordre classique.

De la théorie des ensembles, de 1930 à la veille de la fin de la première moitié du vingtième siècle.

Des «grands cardinaux» durant la première moitié du vingtième siècle.
Des progrès relatifs aux axiomatiques de la théorie des ensembles reposant sur la «limitation de taille», entre 1930 et 1950.
De nouvelles axiomatiques de théorie des ensembles, ne reposant pas sur la «limitation de taille», proposées par QUINE.

De la notion de modèle, jusqu'aux temps de la sémantique.
Des origines lointaines, dans l'«Algèbre symbolique», de la notion de modèle en logique.
Des origines de la notion de modèle en logique: la tradition géométrique et axiomatique.
Les origines de la notion de modèle en logique: la tradition de l'algèbre de la logique.
Des transpositions formalistes de la notion d'interprétation, avec des termes sans variable libre pour individus, aux alentours de 1930.
Des origines de la notion de modèle en logique: du rôle de l'émergence de l'algèbre universelle.
De quelques emplois du terme de «modèle» entre 1925 et 1935.
De l'évolution vers le terme de modèle dans les écrits de TARSKI, GENTZEN et MOSTOWSKI.
Des difficultés d'appréhension des notions de système formel et de modèle, jusqu'aux alentours de 1945.
De l'introduction, par TARSKI et CARNAP, de la définition mathématique rigoureuse de la notion de satisfaction.
Des axiomatisation, formalisation, et arithmétisation des métadisciplines selon TARSKI.
De la notion de modèle dans l'*Introduction to Mathematical Logic* de CHURCH.
Du mémoire de KEMENY, *Models of logical systems*, de 1948.

De la formation de la province de la récursivité, jusqu'en 1950.
Des travaux de Rôsza PÉTER sur les récurrences des niveaux supérieurs.
Des calculs de λ-conversion en 1931.
Des fonctions récursives générales de HERBRAND-GÖDEL.
De l'établissement de la coextensivité de trois concepts: λ-définissabilité, récursivité générale, et μ-récursivité générale.
De la «Thèse de CHURCH».
Des notion de fonction calculable par une machine de TURING et de fonction calculable au sens de POST.
De la nouvelle allure prise par les problèmes de décision en raison de la thèse de CHURCH.
Des résultats d'indécidabilité obtenus en combinant interprétabilité et indécidabilité essentielle.
Des fonctions partiellement récursives.
Des fonctions récursives élémentaires de KALMÁR.

Du développement de la logique combinatoire et du calcul de λ-conversion jusqu'aux alentours de 1950.
Du théorème de CHURCH-ROSSER.
De l'indécidabilité récursive du calcul de λ-conversion et du calcul des prédicats du premier ordre.
Des preuves d'indécidabilité par réduction.
Des ordinaux constructibles de CHURCH et KLEENE.
Des notions de classe récursive et de classe arithmétique, chez GÖDEL en 1931.
Des ensembles récursivement énumérables considérés dans leurs rapports avec les ensembles récursifs.
Des travaux de TARSKI concernant la géométrie, les nombres réels, et les algèbres de BOOLE de parties.
De la définissabilité et de la définissabilité arithmétique dans la structure de la droite réelle chez TARSKI, en 1930.
De la théorie de la récursivité: de la hiérarchie arithmétique.
Des degrés de non résolubilité récursive.
Point sur divers cas spéciaux du problème de la décision de formules du premier ordre, vers 1950.
Des débuts de l'«analyse calculable».

Des progrès et controverses, postérieurs à 1930, touchant mathématiques et logique intuitionnistes.

Des avancées, après 1930, au carrefour entre logique propositionnelle, algèbre et topologie

Des rapports entre algèbres de BOOLE, systèmes déductifs, calcul des propositions classique et calcul des propositions intuitionniste, dégagés par TARSKI en 1935.
Des rapports entre topologie, algèbre de BOOLE et algèbre linéaire, dégagés par STONE aux alentours de 1935.
Des rapports entre métamathématique générale, algèbre de BOOLE et topologie, dégagés par MOSTOWSKI et TARSKI, en 1938.
Des rapports entre calcul des propositions intuitionniste, topologie, algèbres de BOOLE et calcul des propositions classique, dégagés par TARSKI en 1938.
Des logiques modales de LEWIS, et des premiers rapports notés entre calcul des propositions intuitionniste, calcul des propositions de la logique S4, et topologie.
Des rapports entre calcul des propositions des logiques classique, intuitionniste, S4, algèbres de BOOLE, algèbres de BROUWER et de HEYTING, et «closure algebras», dégagés par TARSKI et MCKINSEY.
Des avancées de l'algèbre universelle entre 1935 et 1950.
Des axiomatiques du calcul des relations des alentours de 1940 et des problèmes relatifs aux algèbres de relations.
Des théorèmes et méthodes généraux issus des débuts de la théorie des modèles.
Des débuts de la théorie des modèles: premiers théorèmes et méthodes.
Des succès de la méthode d'élimination des quantificateurs après 1930.
Du théorème de LÖWENHEIM-SKOLEM «ascendant».

Introduction

Absence de théorie des modèles

Une discipline *est* un système dynamique; en étudier l'évolution entre deux dates requiert un mot sur ses conditions initiales, sans pour autant en négliger l'état final. Sur celui-ci, je me bornerai, pour l'instant, afin de justifier mon titre, à citer une appréciation portée par Andrzej MOSTOWSKI en 1964 [*Mtw*14][1]) sur le *théorème de complétude pour la théorie des types* de Léon HENKIN, publié en 1950 [*Hnk*3]: «This result was first obtained by HENKIN whose paper has essentially contribued to a better understanding of what higher-order logic really is». Ceci n'est pas le fait d'une discipline parvenue à complète maturité. En fait une branche manquait encore à la logique peu avant 1950: si la notion de *modèle*, certes depuis longtemps utilisée, venait, après un temps d'élaboration, d'être théorisée en tenant compte des formalisations, les logiciens en étaient à peine à entamer le défrichement du nouveau champ d'investigations qu'elle leur offrait, la *théorie des modèles*, à savoir l'étude systématique des rapports entre les théories et les structures susceptibles d'en être des modèles, entre les divers modèles d'une théorie donnée, entre ces modèles et les extensions de ces théories, entre les théories des structures sur un même support, etc ...

Racines de la logique au dix-neuvième siècle: l'algèbre de la logique

Quant à ses origines, fille, comme sa soeur l'*algèbre universelle*, de l'*algèbre symbolique* de George PEACOCK [*Pck*1, 2], la *logique symbolique* est née au milieu du dix-neuvième siècle, dans le cerveau et sous la plume de George BOOLE [*Boo*]; les ouvrages antérieurs n'ont pas eu le retentissement des siens. L'oeuvre, elle aussi fameuse en ce temps, d'Augustus DE MORGAN [*Mrg*], dont le grand mérite fut l'introduction de la notion de relation, n'avait pas le même caractère algébrique. Grand philosophe et logicien, Charles Sanders PEIRCE, entre autres contributions, développa ([*Prc*1, 2]) *l'algèbre de la logique* de BOOLE, en y incorporant une algèbre des relations, et les quantificateurs modernes. Ernst SCHRÖDER recueillit l'héritage de ces travaux, les systématisa, les développa, surtout en ce qui concerne la résolution des équations booléiennes — une histoire qui commence avec BOOLE lui-même. S'il y a lieu d'en parler, c'est parce que la publication du bilan que SCHRÖDER a fait des travaux de ses prédécesseurs et des siens, celle des trois tomes des *Vorlesungen über die Algebra der Logik* [*Sdr*1], qui s'étend de 1890 à 1905, suivie, en 1909 et 1910, de celle d'un *Abriß* [*Sdr*2] publié par un élève, mord sur le début du vingtième siècle; et parce que ce traité, qui influença fortement les achèvements de Bertrand RUSSELL, David HILBERT, Thoralf SKOLEM et Alfred TARSKI, était encore, en 1950, l'un des ouvrages de référence proposé aux jeunes gens désireux d'apprendre la logique.

Il y a dans ces travaux beaucoup plus que le grand public — même mathématicien — connaît communément de «l'algèbre de BOOLE», car le *problème de*

---

1) Voir la bibliographie alphabétique à la fin du texte.

la résolution d'une équation comportant des «coefficients», voire des «inconnues» *relations*, est, sous un déguisement devenu inhabituel à nos yeux, très voisin (en encore plus difficile) de ce qui sera plus tard dénommé le *problème de la décision* pour le calcul des prédicats, reconnu récursivement insoluble en 1936: donc, si l'ouvrage de SCHRÖDER semble se perdre en une multitude de recettes assez semblables à ce que sont les répertoires de méthodes de résolution d'équations différentielles, on ne devrait pas lui en faire un reproche, et ne pas manquer de célébrer, en 1995, le centenaire de la sortie de l'un des volumes de l'*Algebra der Logik*, dont on a déjà injustement laissé passer, en 1990 et 1991, deux occasions de célébrer les mérites.

De l'intérêt des mathématiciens pour les questions de fondements, à la fin du dix-neuvième siècle

Outre la réussite de l'algébrisation de la logique entreprise par BOOLE, trois autres facteurs peuvent expliquer l'intérêt accordé à la logique par les plus portés sur la philosophie parmi les mathématiciens du dernier quart du dix-neuvième siècle, et, notamment, par les deux plus grands d'entre eux, David HILBERT et Henri POINCARÉ: d'une part le questionnement général des mathématiciens sur les fondements, particulièrement quant à la géométrie, et sur l'objet de leur connaissance, parcourt tout le dix-neuvième siècle; les progrès de la géométrie, ceux de l'analyse, et la perte de confiance dans le recours à l'intuition qu'ils induisent, ravivent les discussions que suscite l'axiomatique de la géométrie, et font rechercher un recours dans une rigueur sans cesse accrue des formulations et des raisonnements; à partir de l'article fameux d'Eugenio BELTRAMI [*Blt*] en 1868, l'indépendance de l'axiome des parallèles établie, ces progrès amènent à théoriser la situation ainsi créée dans les rapports entre des géométries différentes, et tournent l'attention vers la problématique générale de l'axiomatique. D'un autre côté, c'est aussi tout au long du dix-neuvième siècle que les mathématiques entrent dans «l'age de l'abstraction»; des opérations «abstraites», puis des structures «abstraites», sont introduites peu à peu, groupes, espaces vectoriels, anneaux, réseaux, etc ... Enfin, à partir de 1840, sont publiés des textes inédits de LEIBNIZ sur ses conceptions mathématiques [*Lbz*1, 2], textes qui, dans cette atmosphère, devenaient plus aisément compréhensibles qu'ils n'avaient pu l'être au siècle précédent; — Gottlob FREGE [*Frg*1] et Guiseppe PEANO [*Pno*5], pour ne citer qu'eux, ont attesté de l'influence exercée par LEIBNIZ sur leurs conceptions.

La promotion progressive, tout au long de cette évolution, de méthodes *non constructives* de raisonnement, culmine, intempérée, dans la théorie des ensembles, dont on sait quelles sont, entre les mains de Georg CANTOR et de Richard DEDEKIND, les racines au dix-neuvième siècle. Par son acquiescement à l'infini actuel, et l'usage qu'elle en fait, elle brise un tabou millénaire, et déclenche, en réaction, le débat primordial en philosophie contemporaine des mathématiques, ouvert par Leopold KRONECKER, entre les partisans d'un recours exclusif aux méthodes constructives et les adeptes de la liberté d'user de toutes les armes de la pensée pour assurer les progrès de la connaissance.

Racines de la logique au dix-neuvième siècle: les innovations de FREGE
Bien timide, et trop peu rigoureuse, apparaît cependant l'algèbre de la logique, du moins telle qu'elle se présentait en 1879, en comparaison avec le second départ pris par la logique mathématique à ce moment-là, avec le projet qu'eut FREGE [*Frg*1] d'étendre les fondements des mathématiques, en son temps considérés comme fermement établis en ce qui concerne l'analyse, à l'arithmétique. Une réflexion approfondie le persuada que l'arithmétique ne pouvait être qu'un développement de la logique. Il entreprit donc d'établir des fondements axiomatiques pour la logique, ce qu'il fit très bien, d'emblée, dès 1879, puisque, de nos jours encore, l'axiomatique de son premier ouvrage [*Frg*1] conviendrait parfaitement à une première initiation à la logique. Et il le fit, non sans apporter en outre à la logique une innovation qui en renouvelait complètement l'approche formalisante, et qu'on ne trouve pas chez DE MORGAN: il fut le premier à reconnaître, et à dire explicitement, que l'analyse de la pensée, et du discours qui en rend compte, en sujet, prédicat et copule, avec quantification de la proposition, n'était pas pertinente, et il y substitua l'analyse moderne, en «sujets multiples» et prédicats à sujets multiples (pour parler le langage employé par David HILBERT et Paul BERNAYS en 1934 dans les premières pages de leurs *Grundlagen der Mathematik* [*HbB*]), et avec des quantificateurs portant sur un quelconque de tels termes. Influencé par LEIBNIZ, dont les *Mathematische Schriften* [*Lbz*2] venaient d'être publiés, et par les travaux des géomètres de la génération précédente, qui cherchaient à analyser de plus en plus finement leurs déductions, FREGE avait reconnu qu'un développement *sans lacune* de son projet requérait l'explicitation de chacune des inférences d'un raisonnement, donc l'explicitation des règles d'inférence admises, et, pour mieux maîtriser la complexité des énoncés à traiter ainsi, l'introduction d'un «langage de formules» — analogue à celui de la chimie par exemple — d'un langage formel, dirions-nous; et il s'en était forgé un. Au moment de la publication de son *chef d'oeuvre* — au sens que ce mot avait au Moyen-Age — où il présentait le traitement qu'il avait envisagé, les *Grundgesetze der Arithmetik, Begriffsschriftlich abgeleitet* [*Frg*2], parus en deux tomes, l'un en 1893, l'autre en 1903, donc au début de vingtième siècle aussi, il avait aussi reconnu, de plus en plus, au fil des années, que la bonne réalisation de son projet exigeait une rigueur d'analyse et une précision de langage, de loin inégalées dans les travaux qui avaient précédé les siens.

Racines de la logique au dix-neuvième siècle:
les travaux de PEANO et de ses collaborateurs
De son côté, à partir de 1889 [*Pno*1], partant de l'algèbre de la logique, PEANO s'était fixé, parmi d'autres, l'objectif de proposer des notations autorisant une formalisation intégrale, non seulement de la logique pure, mais encore des autres branches des mathématiques, en introduisant des symboles pour les relations, primitives ou définies, répondant, mieux que celles de FREGE, aux canons de transparence de signification des caractères qui avaient été ceux de LEIBNIZ. La publi-

cation de son *Formulaire de Mathématiques* [*Pno*5], qui s'étale de 1895 à 1908, concerne, elle aussi, le début du vingtième siècle.

Pour insuffisantes qu'elles aient pu être sur bien des points (dont quelques-uns sont aussitôt relevés par FREGE [*Frg*3, 4]), les analyses effectuées par PEANO en vue de cette avancée vers la formalisation; et, qu'elles subsistent ou non de nos jours, ses notations, bien choisies, en vue desquelles ces analyses ont été effectuées et auxquelles elles ont mené, ont les unes et les autres contribué au progrès de la logique, en attirant, plus que ne l'ont fait les travaux d'autres logiciens, l'attention des mathématiciens. Corrélativement, les axiomatisations proposées par PEANO et par ses collaborateurs, dans le domaine de l'arithmétique — où l'axiomatique de PEANO [*Pno*2] est restée en vue — et dans celui de la géométrie, ont occupé une place importante dans le mouvement de recherches d'axiomatisations de diverses branches des mathématiques, qui s'accélère à la fin du dix-neuvième siècle.

Des vicissitudes de la théorie des ensembles durant l'ultime décennie du dix-neuvième siècle

Jusqu'en 1890, la théorie des ensembles s'était développée sans souci d'axiomatisation. C'est encore à PEANO [*Pno*3], qui, pourtant, incorporait la théorie des ensembles à ses conceptions, et en faisait, dans ses travaux, un généreux usage, que l'on doit d'avoir détecté, cette année-là, la possible intervention d'une infinité de *choix* arbitraires *simultanés* dans un raisonnement mathématique (le premier usage d'un terme faisant référence à un *choix*, celui de «legge di scelta», remonte à Rodolfo BETTAZZI [*Btz*]), et d'avoir ouvert la voie à ce qui allait devenir au début du vingtième siècle une ardente polémique, en rejetant et en évitant *explicitement* une telle intervention. La découverte des «paradoxes» modernes — le premier d'entre eux, dit «de BURALI-FORTI», et issu de remarques [*BuF* 1, 2] faites par celui-ci en 1897, est celui du type d'ordre de l'ensemble des types de bons ordres, qui est un type de bon ordre isomorphe à l'un de ses segments propres, contredisant une propriété des types de bons ordres — oblige à réfléchir aux principes premiers de la théorie des ensembles, ainsi que CANTOR, qui s'était rendu compte de ce genre de difficulté dès 1895, le fit en 1899 (cf. [*NCv*]), mais sans rien publier officiellement.

De certaines racines, au dix-neuvième siècle, de l'orientation particulière à l'école de logique polonaise

La genèse et le développement de l'école de logique polonaise, qui éclôt vers 1915, soutenue, au moins jusque vers 1930, par les mathématiciens de ce pays, emmenés par Wacław SIERPINSKI, doivent beaucoup à l'intérêt accordé à la logique mathématique par «le père de la philosophie polonaise», Kasimierz TWARDOWSKI, dont les fondateurs de cette école, Jan ŁUKASÍEWÍCZ et Stanisław LEŚNIEWSKI, ont été, dès le tout début du vingtième siècle, les élèves. TWARDOWSKI lui-même avait hérité l'exigence de précision à laquelle il appelait la réflexion philosophique, et l'intérêt qu'il accordait, par voie de conséquence, à la connaissance des développements scientifiques, de son propre maître, l'empiriste viennois Franz BRENTANO. D'un autre côté, quand, après avoir étudié les mathématiques pendant six semestres sous les férules de Leopold KRONECKER et de Karl WEIERSTRASS, Edmund HUSSERL décida de se tourner vers la philosophie, c'est auprès de BRENTANO qu'il décida de la faire — et il en fut influencé dans le même sens. Or nous retrouvons HUSSERL à une conférence présentée par HILBERT en 1901 à la Société Mathématique de Göttingen (il y prend de fort intéressantes notes sur ce que dit HILBERT; sans doute n'est-il pas inintéressant non plus de noter qu'il est nommé cette année-là «professeur extraordinaire» à l'Université de Göttingen, où il enseignera jusqu'en 1916), et nous verrons Alfred TARSKI attester, dans le mémoire fondateur de la sémantique logique [*Trk*9], que l'une des conceptions fondamentales sur lesquelles repose l'édification du système de logique élaboré par LEŚNIEWSKI, est issue d'une idée de HUSSERL.

De la périodisation de l'histoire de la logique mathématique durant la première moitié du vingtième siècle

Si, ainsi qu'il a déjà été dit, il y a, un peu avant 1950, un tournant dans l'histoire de la logique, débouchant sur des temps nouveaux, et durant lequel se multiplient mises en forme de notions auparavant manipulées sur le seul plan intuitif et notions nouvelles, telles celle de modèle et celle de modèle non standard, et premiers travaux d'une génération bien plus nombreuse de logiciens mathématiciens, il n'en reste pas moins que ce n'est pas une date marquée par des ruptures aussi nettes que celles dont je vais dire un mot ci-après. J'émets la *conjecture* que l'importance de cette date pour l'ensemble des mathématiques, et en particulier pour la logique, tient en grande partie à ce que c'est à ce moment-là que commencent à se manifester les effets de la reprise des échanges scientifiques internationaux, après leur relative interruption durant la seconde guerre mondiale, reprise qui suscite un rattrapage dans la diffusion de travaux effectués entre temps, et un foisonnement d'idées nouvelles.

Par contre, il y a, dans l'histoire de la logique mathématique durant la première moitié du vingtième siècle, deux dates marquantes, parce que s'y sont produits des ébranlements des idées reçues, immédiatement perçus comme tels par la communauté mathématique: en 1903, avec la révélation [*Rsl*3] du paradoxe de RUSSELL (auquel celui-ci est parvenu de décembre 1900 à juin 1901 [*Anl*]); et 1930, avec les théorèmes d'*incomplétude* de GÖDEL [*Gdl*1]. La première de ces dates, conjointe avec la parution, en 1899, de la première édition des *Grundlagen der Geometrie* de HILBERT [*Hbt*1], suivie, dès 1901 et 1903, de leur traduction française [*Hbt*4] et de leur deuxième édition [*Hbt*5]; compte tenu, aussi, des échanges intervenus aux deux Congrès internationaux, de mathématiques et de philosophie, qui se sont tenus à Paris en 1900; puis, à partir de 1905, de l'éclatement d'une série de controverses, situe au début du siècle le commencement d'une nouvelle période dans l'histoire de la logique. La seconde de ces dates sépare en deux intervalles presque égaux la première moitié du vingtième siècle.

Une troisième date, 1915, où Leopold LÖWENHEIM découvrit l'un des théorèmes fondamentaux de la logique moderne [*Lwh*], sépare en deux la première de ces périodes, traversée d'essais souvent immatures et de polémiques. La découverte de LÖWENHEIM *aurait dû*, si elle avait été aussitôt comprise (et peut-être l'eût-elle été plus vite, si cela n'avait été durant la première guerre mondiale), provoquer un ébranlement analogue, mais les logiciens eux-mêmes, à peu d'exceptions près, n'en comprirent que peu à peu la portée.

Les premiers fruits du dix-neuvième siècle

De l'algèbre de la logique à l'aube du vingtième siècle: le mémoire de LÖWENHEIM

Durant la première des périodes que je viens de mentionner (1900–1915), seuls les travaux ressortissant à l'algèbre de la logique vont se poursuivre sans encore interagir avec les préoccupations des courants qui viennent d'être mentionnés. LÖWENHEIM étudie les méthodes de résolution des équations à coefficients et inconnues relations. L'article qu'il publie en 1915 dans les *Mathematische Annalen* [*Lwh*] est le premier travail dont l'objectif affiché se restreint à la notion d'expression dénuée de quantificateurs portant sur les relations, c'est-à-dire, d'expression «du premier ordre» au sens d'aujourd'hui; il nomme *Zählausdruck* une telle expression. Il prouve que si toutes les relations que comporte une telle expression n'ont qu'un argument (à l'exception, éventuelle, de l'égalité), la classe des nombres d'éléments (finis ou non) des domaines où elle «vaut» forme, ou une partie finie de l'ensemble des nombres entiers naturels, ou la complémentaire d'une telle partie; il donne des bornes pour ces parties finies, résolvant ainsi, avant la lettre, un cas particulier du *problème de la décision*; mais surtout, il fournit une preuve de ce qu'une Zählausdruck, valide dans tout domaine fini, mais non valide dans quelque domaine infini, est déjà non valide dans un domaine infini dénombrable. De ce théorème-là, il faudra au moins vingt années pour tirer les premières conséquences, et on n'a pas encore fini, aujourd'hui, de faire le tour de celles qui concernent la théorie des ensembles. Après une mise sous forme normale quelque peu hasardeuse, LÖWENHEIM utilise pour sa preuve une argumentation qui sera réutilisée par Thoralf SKOLEM, Jacques HERBRAND, et Kurt GÖDEL durant les quinze années qui suivront.

Bref aperçu sur la situation avant les Congrès de Paris, en 1900

Auparavant, ce qui a d'abord tenu le devant de la scène, aux alentours du début du siècle, dans les études touchant les fondements, ce sont les travaux de CANTOR sur l'arithmétique transfinie; les axiomatisations, parfois multiples, de diverses disciplines; les formalisations proposées à cet effet par PEANO et ses collaborateurs, et leur souci méthodologique quant à l'indépendance d'un système d'axiomes (en 1891 [*Pno*4], PEANO prouve celle de ses axiomes de l'arithmétique); et quelques réflexions éparses de HILBERT et de POINCARÉ: il est significatif que ce dernier collabore, dès le début de sa parution, en 1893, à la *Revue de Métaphysique et de Morale*.

K. Gödel (from: Greenberg, *Euclidean and Non-Euclidean Geometry*. Copyright © 1993 by W.H. Freeman and Company. Reprinted with permission.)

Il n'est pas indifférent non plus, au regard de l'histoire, qu'en 1900 nous le trouvions achevant, déjà, une polémique avec RUSSELL ([*Pnc*1], [*Rsl*2], [*Pnc*3]), mais partie, celle-là, de ce que, rendant compte [*Pnc*1] de l'*Essay on the Foundations of Geometry* [*Rsl*1] de celui-ci, paru en 1897, POINCARÉ en conteste la thèse, encore retenue là par RUSSELL, d'un certain degré de vérifiabilité expérimentale de la géométrie.

Au congrès des mathématiciens qui se tient à Paris cette année-là, POINCARÉ donne, lui aussi, une conférence, *Du Rôle de l'Intuition et de la Logique en Mathématiques* [*Pnc*2]. S'il concède que «les logiciens, comme les intuitifs, ont fait de grandes choses que les autres n'auraient pu faire», il n'en parle pas moins déjà en

D. Hilbert

procureur contre la logique et en défenseur de l'intuition, qui «doit conserver son rôle comme complément, j'allais dire comme contrepoids ou comme contrepoison de la logique».

Au moment du Congrès, la première édition des *Grundlagen der Geometrie* [*Hbt*1] vient de paraître, l'année précédente; entre temps, HILBERT a donné sa conférence *Über den Zahlbegriff* [*Hbt*2]. Il s'agit des nombres réels, à la compatibilité de la théorie desquels les *Grundlagen der Geometrie* ont ramené la compatibilité de la géométrie (en usant, pour la géométrie des figures sur la droite, dans le plan, ou dans l'espace, de ce que l'on pourrait appeler le *modèle cartésien*, constitué par la géométrie analytique). A la méthode *génétique*, procédant par extensions successives, et selon laquelle, partant des nombres entiers naturels, les diverses espèces de nombres ont été progressivement introduites, HILBERT oppose la méthode *axiomatique*, selon laquelle on peut introduire les nombres réels comme éléments d'un «domaine fondamental» sur lequel sont définies des opérations et des relations premières, répondant à un système d'axiomes, qu'il produit. Ces axiomes comportent deux postulats dits «de continuité» (pour des raisons qui tiennent à l'histoire de la géométrie au dix-neuvième siècle): l'un est l'axiome d'ARCHIMÈDE, et l'autre est un axiome nouveau, probablement introduit pour répondre à une critique qui reprochait à l'axiomatique des *Grundlagen der Geometrie* de ne pas réellement *caractériser*, à isomorphie près, le domaine fondamental de la géométrie; ce nouvel axiome, dit «de complétude», stipule qu'à l'inverse du système des nombres rationnels, celui des nombres réels ne peut être étendu. En 1902, une transposition de cet axiome à la géométrie sera incorporée à la traduction française des *Grundlagen* [*Hbt*4], et on la retrouvera dans les éditions en allemand [*Hbt*5] qui suivront.

Des travaux des théoriciens des postulats américains, et de diverses autres conceptions de la complétude

Dans la même conférence, cependant, rappelant le système des axiomes de la géométrie, HILBERT en définit *d'une autre façon* la complétude: ils doivent «suffire à la preuve de toutes les propositions de la géométrie». Par la suite, diverses autres notions de complétude d'une axiomatique seront introduites encore. Il sera montré, en 1928, par BALDUS [*Bld*], que l'axiome rajouté par HILBERT à l'axiomatique initiale des *Grundlagen der Geometrie* ne remplit pas son objet: à un même système d'axiomes, il peut être répondu par des interprétations différentes, de domaines fondamentaux non isomorphes; tel est le cas, par exemple, de la géométrie hyperbolique.

Les premières autres conceptions de la complétude introduites dans les premières années du vingtième siècle, et dont certaines ne seront pas retenues comme telles, apparaissent dans les travaux de mathématiciens américains qui, reprenant en somme certaines des préoccupations de PEANO et de ses collaborateurs, s'attachent à l'axiomatisation de diverses branches des mathématiques, en portant attention à soumettre leurs axiomatiques à des canons qu'ils rattachent à une aspiration à des fondements rigoureux, dont le type de traitement des *Grundlagen der Geometrie* est l'archétype, même s'ils en révisent l'axiomatique. La théorisation, à laquelle ils ont ainsi été conduits, des conditions auxquelles doit répondre une axiomatique, les a fait qualifier de *postulate theorists*.

Il en est deux dont l'activité concerne plus particulièrement l'histoire de la logique: Edward Vermilye HUNTINGTON, parce qu'il propose des axiomes pour «l'algèbre de la logique» — en fait, sur le modèle des *Grundlagen der Geometrie*, pour la structure d'algèbre de BOOLE — en 1904 [*Hun*2], d'abord, puis à nouveau, en 1932 [*Hun*5] et 1933 [*Hun*6] — et aussi, pour une raison sur laquelle je reviendrai tout à l'heure; et Oswald VEBLEN, parce que, la jugeant indispensable, il fera démarrer, en 1925, la recherche en logique à Princeton. VEBLEN est avant tout un géomètre; mais, suivant un cheminement analogue, sur ce point, à celui de HILBERT, son attachement aux questions de fondements et de logique dérive de son intérêt pour la géométrie, où ses travaux s'étendent, de 1903 à 1932, de l'établissement d'une axiomatique concurrente de celle des *Grundlagen der Geometrie* [*Vbl*1] à celui de fondements pour la géométrie différentielle [*VWh*1,2], en passant par le corps des réels [*Vbl*3,5], la géométrie projective [*VbY*1,2] — il est coauteur des deux volumes de la *Projective Geometry* de VEBLEN et YOUNG

[*VbY*2], parus en 1910 et 1918 — et la topologie [*Vbl*2, 4, 7], où il est crédité de la première preuve rigoureuse, dès 1905, du théorème de JORDAN, qu'il traitait à l'aide de l'«addition modulo 2» (alias, «différence symétrique») [*Vbl*4].

Après *Über den Zahlbegriff*, l'unicité d'interprétation est, dans un premier temps, juste rebaptisée: HUNTINGTON, en 1902, discutant une axiomatique de la théorie des «grandeurs» [*Hun*1], qualifie de «suffisante» une axiomatique assurant l'isomorphisme de ses modèles; en 1904 [*Vbl*2], VEBLEN, faisant de même avec une nouvelle axiomatique de la géométrie dont il peut montrer l'indépendance, préfèrera pour la même propriété le qualificatif de *catégorique*, suggéré par John DEWEY ([*Mor*2]). VEBLEN envisage que la catégoricité puisse déterminer, pour tout énoncé, lequel, de cet énoncé et de sa négation, est valide (ce qui est une de nos notions de complétude d'aujourd'hui); HUNTINGTON encore, en 1905 [*Hun*3], envisage que la catégoricité puisse déterminer, pour tout énoncé, lequel, de cet énoncé et de sa négation, est déductible (ce qui est une autre de nos notions de complétude). Et, à côté de la compatibilité, de l'indépendance, et de la nette distinction entre axiomes et définitions, la catégoricité sera une des conditions requises d'une axiomatique pour répondre aux canons des théoriciens des postulats.

Il est significatif que ce soit dans le paragraphe consacré à la «postulate theory», dans la version développée [*Chu*13] de son *Introduction to mathematical logic*, parue en 1956, qu'Alonzo CHURCH range les diverses versions, dont la plupart ne se sont introduites qu'après la période qui nous occupe maintenant, des concepts de conséquence, de compatibilité, d'indépendance, de complétude, de catégoricité, et de modèle, tous relatifs aux systèmes d'axiomes; l'école polonaise, elle, entre les deux guerres mondiales, les range dans ce qu'elle appelle la «méthodologie des sciences déductives». Avant la première guerre mondiale, les versions précises de ces concepts qui s'appliquent aux axiomatiques formalisées ne sont pas plus disponibles que les axiomatiques formalisées elles-mêmes, et l'on ne dispose que de versions maintenant qualifiées d'«informelles», encore très vagues en ce qui concerne la complétude. De ces versions, CHURCH mentionne le traitement qu'en donnent VEBLEN et YOUNG dans l'introduction de leur *Projective Geometry* [*VbY*2]; pour les bien plus nombreuses versions disponibles vers 1930, Rudolf CARNAP et Alfred TARSKI renvoient à un paragraphe de la troisième édition de l'*Einleitung in die Mengenlehre* [*Fkl*5] de Abraham Adolf FRAENKEL, parue en 1928.

De l'attitude philosophique de HILBERT dans sa conférence sur les problèmes mathématiques, en 1900

L'adresse de HILBERT au Congrès de 1900, on le sait, s'achève par sa fameuse liste de problèmes [*Hbt*3]. On sait moins que HILBERT y fait état de son «Überzeugung von der Lösbarkeit eines jeden mathematischen Problems» — «durch eine *endliche* Anzahl von Schlüssen», insiste-t-il à *trois* reprises, «auf Grund einer *endlichen* Anzahl von Voraussetzungen» [c'est moi qui souligne]. Mais n'a-t-on pas complètement oublié qu'il y renvoie *deux fois plus souvent* à l'expérience, à l'intuition, au «wiederholenden und wechselnden Spiel zwischen Denken und Erfahrung», et qu'il y demande ce que deviendraient les mathématiques, privées de la géométrie et de la physique mathématique? Le HILBERT développeur de l'abord formaliste des fondements des mathématiques, et celui dont on pourrait rappeler quelles furent, à la même époque, les importantes contributions à la physique mathématique, ne font qu'un seul et même homme.

Des problèmes de HILBERT attaqués par les logiciens avant le milieu du vingtième siècle

La liste de problèmes débute par le problème du continu: y a-t-il, ou non, des puissances intermédiaires entre la sienne et celle du dénombrable? Peut-on en exhiber un bon ordre, qui peut-être aiderait à trouver la réponse à la première question? Le second demande une preuve *directe*, c'est-à-dire ne procédant pas par une réduction à une autre théorie, de la non-contradiction de l'arithmétique [ce sont là les deux problèmes dont HILBERT va s'occuper lui-même]. Le dixième problème de la liste se rapporte, dans l'optique de l'époque, à la théorie des nombres: trouver une méthode par laquelle, en un nombre fini d'opérations, on pourra distinguer si une équation diophantienne est résoluble en nombres entiers rationnels. Il apparaîtra, quelque quarante années plus tard, qu'il s'agit là d'un problème de logique. En 1950, en dépit de résultats partiels, mis à part le second problème, auquel les théorèmes de GÖDEL [*Gdl*1] seront venus, à la fin 1930, apporter une réponse d'insolubilité, inattendue pour HILBERT, aucune de ces questions n'aura été vraiment été tranchée, car celle de la possibilité d'une bonne ordination ayant très vite été ramenée à l'axiome de choix, a aussitôt été transférée à la recherche d'une preuve de celui-ci.

De la «méthode de PADOA», du Congrès de Philosophie de Paris en 1900, au Théorème de BETH

Typique des contributions, aux alentours du début du siècle, des collaborateurs de PEANO, et d'influence profonde, à terme, sur la maturation de la logique, est l'adresse présentée, quelques jours auparavant, au Congrès de philosophie, par l'un d'entre eux, Alessandro PADOA [*Pad*]. On retrouve presque mot pour mot, chez TARSKI, en 1936 [*Trk*17], sa conception, inspirée de celle de Moritz PASCH [*Pch*], d'une théorie déductive partant d'un système de symboles non définis (un langage formel, pour nous), pensables, souligne-t-il, *complètement dénués de signification* — dans l'esprit de ce que demandera HILBERT vingt ans après — et d'un système de propositions non démontrées, constitutifs d'une théorie *générique* réalisée dans chacune des théories *particulières* obtenues en restituant des contenus de sens aux symboles non définis, ses *interprétations* (ses *réalisations*, selon la troisième édition, de 1928, de l'*Einleitung in die Mengenlehre* de FRAENKEL). Entre temps, Cassius KEYSER élabore, en 1918 [*Ksr*3], une conception, adaptée aux systèmes formalisés, des axiomes, fonctions propositionnelles des notions primitives, et de leurs interprétations, les propositions que ces fonctions prennent pour valeurs après substitution, aux notions primitives, de constantes qui leur sont attribuables pour valeurs, analogue et source très probable des développements «sémantiques» de TARSKI entre les deux guerres mondiales.

En 1900, PADOA indique une méthode pour prouver qu'une notion primitive d'une théorie déductive est indépendante du système des autres notions primitives de cette théorie, déterminée par les principes premiers que sont supposés vérifier toutes ces notions: elle consiste à chercher deux interprétations de ces notions, qui coïncident pour ce qui est des notions primitives différentes de la concernée, qui diffèrent pour ce qui est de la concernée, et qui satisfont l'une et l'autre aux principes premiers. TARSKI sut, en 1926 ([*Trk*11, 34]), adapter cette idée aux théories formalisées dans la théorie des types, et pourvoir à une preuve rigoureuse du théorème de non-définissabilité correspondant. Cette preuve use de quantifications sur les prédicats; elle ne s'applique donc pas à la logique élémentaire, celle du premier ordre, où l'on ne dispose pas de ces quantifications. Un théorème analogue pour la logique du premier ordre (le théorème de BETH [*Bth*2]) ne devait intervenir qu'en 1953.

Les grands traits de l'oeuvre de Bertrand RUSSELL: un nouveau visage de la logique

Ce fut aussi à ce Congrès qu'en 1900, Bertrand RUSSELL découvrit les travaux de PEANO et la logique symbolique ([*Rsl*8]). Durant les mois suivants (cf [*Anl*]), il réfléchit au théorème de CANTOR selon lequel il n'y a pas de plus grand cardinal, conséquence de cet autre théorème de CANTOR, énonçant qu'un ensemble est de puissance strictement inférieure à l'ensemble de ses parties. En reprenant la démonstration du premier de ces théorèmes, RUSSELL découvrit le paradoxe, auquel on a donné son nom, du prédicat d'être un prédicat qui ne s'applique pas à lui-même (Indépendamment de RUSSELL, ZERMELO avait, à peu près à la même époque, et peut-être même avant RUSSELL, fait la même découverte [*RgT*]). Publiée en 1903, dans le gros livre des *Principles of Mathematics* [*Rsl*3] de RUSSELL, la contradiction qui en résulte a eu un retentissement considérable, car elle heurte la conviction naïve, et extrêmement répandue jusque là parmi les logiciens et les mathématiciens, qu'à chaque concept est naturellement associée une extension déterminée; ou, dit en termes plus modernes, que toute fonction propositionnelle est collectivisante.

Cette contradiction ruinait l'axiomatique de la logique proposée dans un ouvrage que RUSSELL avait entre temps entrepris de lire, les *Grundgesetze* [*Frg*2] de FREGE, car elle se dérive de cette axiomatique. FREGE entendait seulement prouver que l'arithmétique n'est qu'une branche de la logique. Or, cette idée avait enthousiasmé RUSSELL, que les formalisations de PEANO avaient fait bondir de là à l'idée, tout aussi enthousiasmante et qu'il soutient dans ses *Principles of Mathematics* [*Rsl*3], que «all Mathematics is Symbolic Logic» — «one of the greatest discoveries of our age». Afin de sauver cette philosophie, qu'il détaillait dans cet ouvrage, il y introduisit un *Appendice* où il esquissait une théorie qu'il espérait susceptible de permettre, sans paradoxe, la formalisation de la totalité des mathématiques, *la théorie des types logiques*. Selon celle-ci, il est attribué à chaque variable un *domaine de signification*, son *type logique*; ces types sont numérotés à partir de 1, et une proposition n'est signifiante que si et seulement si le type de chacun de ses termes est strictement supérieur à ceux des termes auxquels ce terme s'applique, ce qui interdira les raisonnements qui mènent aux paradoxes; dans le cas du paradoxe de RUSSELL, ce sera parce que la notion de prédicat qui s'applique à lui-même est ainsi déclarée non signifiante. Des «individus» constituent le type 1; le type 2 se constitue des *énoncés* (dénués de variables libres), dits aussi *du premier ordre*, où les quantificateurs, s'il y en a, portent au plus sur des individus; le type 3 se constitue des énoncés, dits aussi *du second ordre*, où le type maximum des variables quantifiées est 2 (ce qui suppose l'emploi de *variables d'énoncés*); etc ... Le type d'une *fonction propositionnelle* (c'est-à-dire d'une phrase dépendant de termes indéterminés, qui se représente, selon une formalisation donnée, par une *formule* comportant des variables libres, c'est-à-dire non quantifiées, représentant ces termes) se détermine, en considérant comme des constantes les variables (libres) dont elle dépend (nous laissons de côté les règles de détermination de l'ordre d'une fonction propositionnelle). Enfin, une fonction propositionnelle n'est collectivisante que si et seulement si chacun de ses termes est d'un type supérieur, juste d'une unité, au maximum des types des termes auxquels ce terme s'applique.

Las, il fallut bientôt déchanter: il apparut vite que le développement de l'arithmétique exige la présence d'une infinité d'individus (axiome de l'infini) et l'usage de classes définies par des fonctions propositionnelles qui ne répondent pas à ces exigences.

A la fin de 1905, dans un article *On some difficulties in the theory of transfinite numbers and order types* [*Rsl*4], paru en 1907, RUSSELL mentionne, pour éliminer les paradoxes, trois options possibles, dont aucune n'est la théorie des types: 1) la théorie «zig-zag», où seules seraient collectivisantes (RUSSELL dit «predicatives», mais ce terme va très vite perdre ce sens) les fonctions propositionnelles «suffisamment simples»; il rejette cette théorie, au motif de la complication et du manque de plausibilité d'une définition du qualificatif de «suffisamment simple»; 2) la théorie de la *limitation de taille*, qui consiste à n'accorder comme collectivisantes que les fonctions propositionnelles dont l'extension ne s'avèrerait pas excessive, à l'encontre, par exemple, de la notion de nombre ordinal de CANTOR; ce qui arrête RUSSELL, c'est qu'il ne voit pas de critère délimitant où commence l'excès de taille: bientôt, la théorie axiomatique des ensembles, que ZERMELO présentera en 1908, pourra, en un sens, être reçue comme une mise en œuvre de cette idée; 3) la théorie «sans aucune classe», dans laquelle, dit RUSSELL, les théorèmes d'existence sont difficiles à démontrer, et les théorèmes d'arithmétique, difficiles à énoncer.

Il apparaîtra finalement à RUSSELL que la moins mauvaise solution est de revenir à la théorie des types et à ses axiomes subreptices dont il est tout à fait discutable qu'ils soient d'un caractère purement logique: l'axiome de l'infini, et, en vue de rétablir les développements, avant tout, de l'arithmétique, des *axiomes de réductibilité* assurant que toute fonction propositionnelle signifiante est équivalente à une fonction propositionnelle du type immédiatement supérieur au maximum des types de ses variables libres, et donc collectivisante.

RUSSELL expose ces principes dans un article *Mathematical theories as based on the theory of types* [*Rsl6*], paru en 1908. Avec l'aide de Alfred North WHITEHEAD, il rédige et publie, complètement formalisé dans des notations inspirées de celles de PEANO, un développement complet de la théorie des types, et des éléments de toutes les branches des mathématiques, les *Principia Mathematica*, un monument de près de 2000 pages en trois grands volumes [*WhR*], parus en 1910, 1912 et 1913. Devenus aussitôt le principal ouvrage de référence pour les logiciens, ils le resteront au moins jusqu'en 1930.

Des conceptions de HILBERT, de 1901 à l'esquisse de programme présentée au Congrès des mathématiciens de 1904

Les notes prises par HUSSERL, en 1901 ([*Hsl2*]), lors de la conférence prononcée par HILBERT à Göttingen, ci-dessus mentionnée, font apparaître un questionnement, quant à la complétude, au départ: comment est-on en droit d'affirmer que toute proposition ne concernant que les nombres entiers positifs doit être, ou vraie, ou fausse, en se fondant sur les axiomes concernant les nombres entiers positifs? puis, quant à ce qui s'appellera plus tard le *problème de la décision*: de quelles branches peut-on utiliser les connaissances, pour affirmer qu'une proposition est décidée, en se fondant sur les axiomes d'un domaine? quant à la logique enfin, qui recouvre tous les *théorèmes* libres de toute particularité relative à un domaine de connaissances particulier, et qui valent indépendamment de tout *axiome spécifique*: n'arrivons-nous pas à une impasse, dans les domaines du calcul logique, des cardinaux finis, des ordinaux, voire de la théorie générale des ensembles?

En 1904, aux yeux de HILBERT, le paradoxe découvert par RUSSELL et par ZERMELO créait une situation rendant plus urgentes encore les recherches sur les fondements de l'arithmétique, jusqu'à la refonte des premiers principes de la logique même. Ceci le détermina à présenter, au Congrès des Mathématiciens qui s'est tenu cette année-là à Heidelberg, une communication *Über die Grundlagen der Logik und der Arithmetik* [*Hbt6*]. Là, il a tranché: constatant d'abord que «... bei der hergebrachten Darstellung der Gesetze der Logik gewisse arithmetische Grundbegriffe, zum Beispiel der Begriff der Menge, zum Teil auch der Begriff der Zahl, insbesondere als Anzahl, bereits zur Verwendung kommen. Wir geraten so in eine Zwickmühle, und zur Vermeidung von Paradoxien ist daher eine teilweise gleichzeitige Entwicklung der Gesetze der Logik und der Arithmetik erforderlich», il présente une première ébauche d'un plan de travail comportant déjà, sous une forme plus ou moins explicite, une grande partie des idées directrices de sa future *théorie de la démonstration*:

- traduction, dans une langue de formules, c'est-à-dire, formalisation, des termes, énoncés et déductions mathématiques, afin de transformer la recherche d'une preuve de non-contradiction en un problème de caractère arithmétique élémentaire;
- manipulation simultanée d'un nombre fini d'objets seulement, procédures récurrentes pour évaluer leur identité ou différence;

- introduction progressive des axiomes, en veillant à chaque étape à préserver la compatibilité;
- introduction par étapes des expressions, avec à chaque étape restriction des quantificateurs aux termes antérieurement introduits, et réexamen de la validité des axiomes.

Une méthode est proposée pour reconnaître la compatibilité, qui consiste à répartir les propositions en deux classes, dont l'une est celle des propositions auxquelles est attribuée «l'existence» (nous dirions la valeur Vrai), l'autre celle des propositions auxquelles est attribuée «l'inexistence» (nous dirions la valeur Faux), «l'existence» respectant, bien sûr, ce que nous appellerions les tables de vérité, et en s'arrangeant pour qu'aux axiomes, ce soit «l'existence» qui soit attribuée.

Ces idées sont mises en oeuvre, à titre d'exemple, sur des théories très élémentaires, dont il est montré, par récurrence sur des constructions d'expressions ou sur des preuves formelles, que les propositions admettent une telle valuation; dans le langage de maintenant, cela revient à suggérer la construction d'un modèle constitué de termes.

HILBERT mis à part, un autre participant au Congrès de 1904, Gyula KÖNIG, reprendra, dans ses *Neue Grundlagen der Logik, Arithmetik und Mengenlehre* posthumes [*GKg*2], parus en 1914, l'idée de preuves de non-contradiction utilisant des valuations, ainsi que le note Johann VON NEUMANN lorsqu'il emploie la même méthode, en 1927, dans un article *Zur Hilbertschen Beweistheorie* [*vNm*2].

De la communication de G. KÖNIG au Congrès des mathématiciens de 1904

KÖNIG avait d'ailleurs fait sensation à ce Congrès, en y présentant une preuve de la non-existence d'un bon ordre du continu [*GKg*1], qui avait beaucoup affecté CANTOR ([*Snf*]). Cette preuve reposait sur une formule de Felix BERNSTEIN,

$$\aleph_\alpha^{\aleph_\beta} = 2^{\aleph_\beta}.\aleph_\alpha$$

dont on croyait, à tort, qu'il l'avait prouvée dans sa Thèse, en 1901, pour tout ordinal α. Dans les jours qui suivirent le Congrès, il fut établi par Felix HAUSDORFF [*Hdf*1] que cette formule ne vaut en fait que pour α fini ou pour $\alpha \leq \beta$, et KÖNIG retira sa conclusion. Il nous reste, de sa communication, établie sans l'axiome de choix, l'*inégalité de* KÖNIG pour deux suites transfinies de cardinaux (puissances d'ensembles bien ordonnés) de même type, $(\mathfrak{m}_\xi)_{\xi<\lambda}$ et $(\mathfrak{o}_\xi)_{\xi<\lambda}$ telles que l'on ait $\mathfrak{m}_\xi < \mathfrak{o}_\xi$ pour tout $\xi < \lambda$:

$$\sum_{\xi<\lambda} \mathfrak{m}_\xi < \prod_{\xi<\lambda} \mathfrak{o}_\xi;$$

Elle a été étendue par ZERMELO au cas d'un ensemble d'indices non nécessairement bien ordonné, toujours sans recourir à l'axiome de choix, en 1908 [*Zml*3].

Aperçus sur le développement «naïf» de la théorie des ensembles, et sur les controverses qui l'accompagnent, jusque vers 1915

Le 24 septembre 1904, ZERMELO écrit à HILBERT, qui va s'empresser de la faire publier dans les *Mathematische Annalen*, une lettre pour lui communiquer une *Beweis, daß jede Menge wohlgeordnet werden kann* [*Zml*1], qu'il vient de découvrir lors de conversations avec Erhard SCHMIDT. ZERMELO mentionne que celui-ci lui a suggéré d'utiliser, à cet effet, un principe «logique»: une famille d'ensembles non vides admet toujours une application qui associe, à chacun de ces ensembles, un de ses éléments. On sait que Beppo LEVI a fait état de cet énoncé comme d'un postulat dans une conversation avec CANTOR et BERNSTEIN vers 1901 ([*FBH*], p. 48); en 1902, LEVI le mentionne dans un écrit [*Lvi*] où il le rejette.

La théorie des ensembles en est encore, à cette époque, à un stade préaxiomatique. Les mathématiciens qui en font l'objet de leurs recherches, exclusif ou non, sont principalement des jeunes. Il en va de même de ceux qui utilisent la théorie des ensembles dans leurs recherches. En dehors, d'une part, de travaux qui concernent les relations et les fonctions, dont les logiciens s'occupent aussi, et qui seront, durant la période qui nous occupe maintenant, transférés en théorie des ensembles, d'une façon maladroite, en utilisant la notion de *paire* (non ordonnée), ou en cherchant à utiliser des inclusions, des appartenances et des opérations booléiennes, finies ou non, par ZERMELO (cf. [*Lel*]) et son élève Gerhard HESSENBERG [*Hbg*1, 2, 3]; en dehors, aussi, de l'étude de notions qui seront plus tard reclassées dans ce que nous appelons la topologie, la grande affaire en théorie des ensembles est alors la théorie des nombres transfinis, ordinaux, puissances, et cardinaux, entendus comme puissances des ensembles bien ordonnés; leur arithmétique, la structure de ce que nous appelons aujourd'hui la classe des ordinaux, et des fonctions de cette classe vers elle-même. Je n'en signalerai, dans la suite, quelques étapes, que dans la mesure où elles relèvent, à cette époque, ou bien où elles relèveront plus tard, de la problématique de l'axiomatique de la théorie des ensembles.

C'est ainsi qu'en 1902 WHITEHEAD (dans un article [*Whd*] écrit avec la participation de RUSSELL) définit les sommes infinies et les produits infinis de cardinaux; en 1907 HESSENBERG [*Hbg*] établit que la somme et le produit de deux cardinaux *infinis* sont égaux au plus grand des deux; en 1908, VEBLEN [*Vbl*6] étudie les fonctions des ordinaux vers les ordinaux qui sont strictement croissantes et continues, c'est-à-dire, associent, à un ordinal limite, la limite des images de ses prédécesseurs (en d'autres termes, celles que l'on qualifie maintenant de *normales*: c'est la propriété essentielle de l'application qui, à un ordinal α, associe le cardinal $\aleph_\alpha$), leurs itérées (transfinies !) et leurs «points fixes», ainsi que les *classes fermées cofinales* d'ordinaux — c'est-à-dire, celles qui ont des éléments au-delà de tout ordinal fixé à l'avance, et qui contiennent tous les ordinaux limites vers lesquels tendent les suites transfinies d'ordinaux qui leur appartiennent — et les «dérivées» (d'ordres transfinis) de ces classes.

La même année, en termes de types d'ordre d'ensembles bien ordonnés, Hausdorff [*Hdf*2] énonce l'*hypothèse du continu généralisée*: $2^{\aleph_\alpha} = \aleph_{\alpha+1}$, à la preuve de laquelle on étend le premier problème de Hilbert, et introduit la définition de la *cofinalité* d'un ordinal: plus petit des types d'ordres d'une suite d'ordinaux précédant cet ordinal et dont cet ordinal est la limite, ainsi que les notions d'ordinal *régulier*: égal à sa cofinalité; et *singulier*: non régulier; et d'ordinal (ou cardinal) *faiblement inaccessible*: de la forme $\aleph_\lambda$ avec λ limite, régulier, non dénombrable (quelques Auteurs n'en excluent pas $\aleph_0$ qui remplit les autres conditions).

En matière d'inaccessiblité, c'est Paul Mahlo qui, entre 1911 et 1913, aura les concepts les plus forts de la première moitié du vingtième siècle [*Mhl*1, 3, 4] — mais ce ne sera que vers 1960 que seront introduits, en théorie des ensembles, des «principes d'infinitude forte», indépendants des axiomes usuels, et dont on s'apercevra qu'ils sont équivalents à l'existence des «π_0-Zahlen» et «ρ_0-Zahlen» — maintenant «cardinaux de Mahlo»: les premiers sont les ordinaux faiblement inaccessibles admettant, dans toute classe fermée cofinale, un prédécesseur régulier; les seconds sont de plus «limites fortes», c'est-à-dire tels que l'ensemble de leurs prédécesseur est stable par passage de λ à 2^λ.

Du grand débat sur l'axiome de choix, et des controverses sur la théorie des ensembles qui l'accompagnent

Il est bien connu que le nouveau «principe» proposé par Zermelo a aussitôt suscité un débat, dont l'expression la plus frappante a sans doute résidé dans la publication, dans les *Mathematische Annalen* de 1905, d'une première réaction de Emile Borel [*Brl*1], rédigée dès le début de décembre 1904, et dans le *Bulletin de la Société Mathématique de France* de 1905, des fameuses *Cinq lettres sur la théorie des ensembles* [*HBLB*], échangées quelques temps auparavant entre Jacques Hadamard, Emile Borel, René Baire et Henri Lebesgue. Les réactions vont de l'hostilité d'un certain nombre de mathématiciens, dont quelques-uns sont revenus peu à peu sur leurs préventions premières, à l'approbation de quelques autres, en passant par quelques attitudes embarrassées, ou plus nuancées.

Telle est, par exemple, celle de Borel, dans la Note de décembre 1904, parue l'année suivante aux *Mathematische Annalen* [*Brl*1]: il refuse «tout raisonnement où l'on suppose un *choix arbitraire* fait une infinité non-dénombrable de fois»; et s'il accepte le choix *dénombrable*, ce n'est pas sans demander qu'on lui indique à l'avance une marche à suivre pour l'effectuer. Ernest William Hobson fait montre d'un souci d'effectivité analogue [*Hbs*1], mais finira ([*Hbs*2, 3]) par revenir peu à peu sur un refus complet initial (cf. [*CsG*]). Baire, qui refuse aussi d'admettre l'ensemble des parties d'un ensemble [*HBLB*], et, avec prescience, attire l'attention sur le fait qu'on ne sait pas quelles parties l'on y met; Alfred Cardew Dixon [*Dxn*]; Gyula König [*GKg*3, 4]; et Peano [*Pno*6], maintiendront leur refus. Du côté de ceux qui acceptent le nouveau principe, on trouve Godfrey Harold Hardy [*Hdy*]; Jacques Hadamard [*HBLB*], [*Hdm*1, 2, 3, 4, 5]; Felix Hausdorff [*Hdf*2, 3, 4, 5, 6]; Georg Hamel [*Hml*], qui l'applique, aussitôt après la publication de la lettre de ZERMELO, pour établir l'existence de solutions non continues de l'équation fonctionnelle $f(x+y) = f(x) + f(y)$; Constantin CARATHÉODORY [*Ctr*1, 2]; Gerhard HESSENBERG [*Hbg*1, 2, 3]; et, tacitement, nous disent Jean CASSINET et Michel GUILLEMOT [*CsG*], Maurice FRÉCHET, Arnaud DENJOY, Giulio VIVANTI, et Guiseppe VITALI, qui édite à frais d'auteur, en 1905, une plaquette *Sul problema della misura dei gruppi di punti di una retta* [*Vit*], où il prouve que si le continu peut être bien ordonné, il doit nécessairement admettre des parties non mesurables pour la mesure de LEBESGUE — un résultat auquel les logiciens s'intéresseront beaucoup, quelque cinquante ans plus tard.

LEBESGUE, «plutôt enclin à partager le point de vue de BOREL» [*HBLB*], [*Lbg*3], mais dans le doute, semble avoir cherché à prouver la non-validité du principe, en en étudiant des conséquences, comme en attestent les commentaires qu'il apporte à la suggestion d'une preuve analogue à celle de HAMEL, qu'il fait à Conado SEGRE en 1907 [*Lbg*1] — «ce dernier résultat étant démontré en prenant les définitions dans un sens idéaliste, sera légitimé aux yeux des Idéalistes; mais comme c'est un résultat négatif, il sera démontré, a fortiori, pour les Empiristes» — et une *Contribution à l'étude des correspondances de M. Zermelo*, (c'est-à-dire, des fonctions de choix) [*Lbg*2] publiée la même année, et que LEBESGUE conclut: «l'on ne pourra jamais définir une telle correspondance par des procédés analytiques, et cela, même si l'on se borne à la considération de certains sous-ensembles très particuliers».

Il y a un moment, entre 1904 et 1906, où RUSSELL se rend compte, sans disposer pour autant d'arguments pour ou contre, qu'il va lui falloir, pour développer l'arithmétique des cardinaux transfinis dans les *Principia Mathematica*, utiliser la non-vacuité d'un produit (infini) d'ensembles non vides — énoncé où ZERMELO, dans sa lettre à HILBERT [*Zml*1], voit curieusement «l'expression formelle» de son principe. En raison de cette référence au produit, dans l'entourage de RUSSELL, on appelle cet énoncé l'«axiome multiplicatif». Dans sa communication *On some difficulties in the theory of transfinite numbers and order types* [*Rsl*4], du 14 décembre 1905, RUSSELL, réagissant à une communication antérieure de HOBSON [*Hbs*1], dit, de l'«axiome de ZERMELO»: «It may turn out to be capable of disproof by *reductio ad absurdum*. It may also, of course, be capable of proof, but that is far less probable». Il ne voit pas là de difficulté *fondamentale*, c'est-à-dire jetant la suspicion sur la totalitéé des raisonnements concernant classes et relations; il pourrait se faire qu'il y ait des classes «zermeliennes» et d'autres qui ne le soient pas: à l'avenir de le dire.

En 1906 [*Pnc*5], POINCARÉ, commentant à son tour la citation précédente, taxe d'illusoire l'espoir qu'elle exprime. «Chacun n'en croira que son intuition». Se disant «plutôt disposé à admettre l'axiome de ZERMELO», mais spéculant sur sur son indémontrabilité à partir «des autres postulats», il y voit un «jugement synthétique a priori» (c'est-à-dire qui n'est susceptible ni d'une démonstration analytique, ni d'une preuve logique, ni d'un établissement expérimental).

A cette date, ZERMELO est déjà en possession de son axiomatique de 1908, et POINCARÉ, faute de temps pour en prendre connaissance, lui écrit qu'il en attendra la parution pour la commenter (cf. [*Hzm*2], [*CsG*]). Auparavant, ZERMELO, dans une *Neuer Beweis für die Möglichkeit einer Wohlordnung* [*Zml*2], parue cette année-là, va répondre aux objections qui lui ont été faites. Et le principal de sa réponse est que tout le monde, ou guère s'en faut, et partout, ou guère s'en faut, se sert de ce qu'il nomme maintenant «l'axiome de choix».

L'attitude de la communauté mathématique finira par rejoindre celle de RUSSELL et de LEBESGUE, à ceci près que, dans l'affaire, elle aura pris conscience du véritable rôle de l'axiome de choix, et qu'elle aura bientôt de bonnes raisons de cesser de croire à sa réfutabilité, en raison de la preuve de compatibilité relative apportée par FRAENKEL [*Fkl*2, 3] en 1922. RUSSELL et WHITEHEAD, dans les *Principia Mathematica*, indiqueront scrupuleusement les démonstrations et propositions dépendant de cet axiome, et l'habitude se prendra de les imiter.

Cette même année 1908, BOREL défend à nouveau son attitude devant le Congrès des mathématiciens, qui se tient à Rome. Dans sa communication *Sur les principes de la théorie des ensembles* [*Brl*2], il introduit la notion d'ensemble *effectivement énumérable*, et remarque qu'il n'est pas nécessaire de recourir à l'axiome de choix pour démontrer qu'une réunion dénombrable d'ensembles *effectivement énumérables* est dénombrable.

En 1914 paraissent les *Grundzüge der Mengenlehre* [*Hdf*6] de HAUSDORFF, qui vont constituer le premier traité de référence en matière de théorie des ensembles. Quant aux thèmes délimités plus haut, la notion de couple ordonné y est utilisée pour traiter des relations binaires en les assimilant à leur graphe, et le «principe de la chaîne maximale» (tout ensemble ordonné admet une partie totalement ordonnée maximale) y est montré équivalent à l'axiome de choix (c'est dans cet ouvrage qu'apparaît d'ailleurs la notion d'ensemble partiellement ordonné). Il y est aussi établi la conséquence «étrange» de l'axiome de choix connue sous le nom de «paradoxe de HAUSDORFF-BANACH-TARSKI»: on peut, à un ensemble dénombrable près, trouver une partition de la sphère en trois parties «équivalentes par découpage fini» entre elles et avec la réunion de deux d'entre elles (les termes de même rang des partitions se correspondent par des déplacements). On en déduit qu'une sphère est équivalente par découpage fini à deux sphères de même rayon. En 1924, Stephan BANACH et Alfred TARSKI [*BnT*] ont étendu le résultat à tout couple de parties bornées de l'espace, d'intérieur non vide.

En 1915, Friedrich HARTOG'S prouve [*Hrt*] que la «loi de trichotomie» — qui énonce que deux puissances sont toujours comparables — implique que tout ensemble peut être bien ordonné, et donc que les deux propositions sont équivalentes à l'axiome de choix, la réciproque étant connue depuis les *Beiträge* [*Cnt*1] de CANTOR, de 1897.

De la théorie axiomatique des ensembles de ZERMELO

En 1908, ZERMELO présente la première axiomatisation de la théorie des ensembles [*Zml*3], qui est bien connue. Il est assez curieux que, commençant, comme le font, et HILBERT, dans les *Grundlagen der Geometrie* [*Hbt*1], SCHRÖDER [*Sdr*1], et RUSSELL, dans sa théorie des types [*Rsl*3], par introduire un *domaine d'individus*, dont certains sont des ensembles, et une relation sur ce domaine, l'appartenance, ZERMELO traite plutôt de ce qu'est un *modèle* de la théorie des ensembles que de la notion d'ensemble proprement dite.

A l'énoncé près de certains d'entre eux, les axiomes sont ceux que l'on enseigne couramment de nos jours, à l'exception de «l'axiome de fondement», absent, et de «l'axiome de remplacement», remplacé par un «principe de sélection» disant que «les éléments d'un ensemble qui satisfont une propriété *définie* forment un ensemble». Selon ZERMELO, une assertion est *définie* «si les relations fondamentales du domaine, par les axiomes et les lois universellement valides de la logique, déterminent sans arbitraire si elle vaut ou non»; une «Klassenaussage» comportant une variable x est définie si elle est «définie pour chaque individu de la classe où x prend ses valeurs».

Du paradoxe de RICHARD et des polémiques soutenues par POINCARÉ: le concept d'imprédicativité

Au milieu de l'année 1905, un nouveau type de paradoxes apparaît, dont un seul sera mentionné ici, celui de Jules RICHARD [*Rch*], en raison des répercussions, plus importantes que celles des autres, que lui ont valu son intervention dans une série de polémiques qui vont s'engager, entre POINCARÉ d'une part, et entre RUSSELL, PEANO, HILBERT et ZERMELO d'une autre.

Expliqué dans une terminologie d'aujourd'hui, le «paradoxe» s'atteint en partant du rangement, selon la longueur, et, à longueur égale, selon l'ordre alphabétique, de toutes les suites finies de lettres de l'alphabet. Il y apparaît les définitions, à l'aide d'un nombre fini de mots, de tous les nombres (réels) définissables en un nombre fini de mots. Rangés dans l'ordre de leurs définitions, ceux-ci forment une suite infinie, que l'on suppose avoir formée en parcourant, dans l'ordre, toutes les suites finies de lettres, et en barrant au fur et à mesure toutes celles qui ne définissent pas un terme de la suite. Le nombre, «diagonal» par rapport à cette suite, de partie entière 0, et dont, pour chaque n, la n^e décimale est égale modulo 10 au successeur de la n^e décimale du terme de rang n de cette suite (supposée commencer au rang 1), quoique ainsi défini en un nombre fini de mots, ne peut figurer dans cette suite.

Avec le «paradoxe», RICHARD fournit une solution: la définition du nombre «diagonal» a été biffée au moment où elle a été atteinte, car, faute de disposer à ce moment-là de la totalité de la suite définie, elle n'a pu être reconnue comme définissante au moment où l'on est arrivé à elle.

Or, depuis le début du vingtième siècle, Louis COUTURAT diffuse et défend, en France, les travaux de PEANO ([*Ctr*1]), les idées de RUSSELL ([*Ctr*2, 5, 6, 8]), l'algèbre universelle ([*Ctr*3]), et l'algèbre de la logique ([*Ctr*9]); en 1904 et 1905, il publie dans la *Revue de Métaphysique et de Morale*, à laquelle POINCARÉ collaborait régulièrement, une série d'articles sur *Les principes des mathématiques, avec un appendice sur la philosophie des mathématiques de* KANT [*Ctr*6, 7, 8]; il y conteste le caractère synthétique de l'arithmétique, s'en prenant aux opinions émises à ce sujet dans des articles antérieurs de Poincaré; et va jusqu'à critiquer la philosophie kantienne [*Ctr*4, 7, 8].

Poincaré va donc répondre, et la polémique, que Mooij [*Moj*] et Gerhard Heinzmann [*Hzm*1] ont fort bien étudiée par ailleurs, va se poursuivre jusqu'en 1910. Seuls en seront évoqués ici les éléments où s'est manifestée quelque prémonition (comme plus haut concernant l'axiome de choix) ou qui ont infléchi le cours de l'histoire.

En 1905 et 1906 paraît, en deux parties, dans la *Revue de Métaphysique et de Morale*, un premier article de Poincaré, intitulé *Les mathématiques et la logique* [*Pnc*4]. A Couturat et à Russell, qui lui disent avoir ramené à la logique le principe d'induction complète, Poincaré objecte qu'il leur faut une preuve de non-contradiction de leurs axiomes, qu'elle ne peut se faire *par l'exemple*, et qu'elle doit donc se faire par induction complète sur les démonstrations; il faut donc que l'induction complète soit justifiée par une intuition, c'est-à-dire, qu'elle soit un jugement synthétique a priori. Il adresse le même avertissement à Hilbert, tout en se félicitant de ce que celui-ci ait expliqué, dans sa conférence de 1904, à l'encontre de Russell, que pour étudier les fondements de la logique et de l'arithmétique, il faille recourir à un développement simultané de quelques rudiments d'arithmétique.

Là-dessus vient, dans l'article de Russell *On some difficulties in the theory of transfinite numbers and order types* [*Rsl*4], la définition d'une fonction propositionnelle comme *prédicative* si elle est collectivisante, et l'interrogation, qui parcourt l'article, sur la délimitation du domaine des fonctions propositionnelles prédicatives, en vue de laquelle trois solutions sont envisagées. Dans un second article, intitulé, comme le précédent, *Les mathématiques et la logique* [*Pnc*5], Poincaré revient sur les trois *«antinomies cantoriennes»*, au nombre desquelles, celle de Richard [*Rch*], dont il endosse la solution: l'ensemble E des nombres considérés avant l'introduction du nombre «diagonal» est celui «de *tous* les nombres que l'on peut définir par un nombre fini de mots, *sans introduire la notion de l'ensemble E lui-même*. Sans quoi la définition de E contiendrait un cercle vicieux; on ne peut pas définir E par l'ensemble E lui-même.» Et il propose que «les définitions qui doivent être regardées comme non prédicatives sont celles qui contiennent un cercle vicieux.»

Dans une *Additione* à un article en latin [*Pno*6], paru en 1906, Peano ripostera à Poincaré en montrant que la définition du nombre «diagonal» qu'il veut proscrire n'est qu'une transposition de définitions d'usage courant, et que le «puncto debile» ne gît pas dans le cercle vicieux que Poincaré croit y voir. Mais Russell, ripostant au même moment à Poincaré par un article en français intitulé *Les paradoxes de la logique* [*Rsl*5], acceptera la suggestion. En l'élaborant à partir d'une remarque de Poincaré: «le mot *tous* a un sens bien net quand il s'agit d'un nombre fini d'objets», il la fait déboucher sur l'énoncé: «Tout ce qui contient une variable apparente (nous disons aujourd'hui: liée) ne doit pas être une des valeurs possibles de cette variable» (traduire: une fonction propositionnelle, pour être collectivisante, doit ne comporter que des quantificateurs portant sur un type auquel la classe qu'elle définit n'appartient pas). Il élabore la théorie des types de façon à respecter ce principe et de façon à proscrire la transposition formelle de raisonnements de l'espèce de celui de l'antinomie de RICHARD.

Avec un certain flou, d'un auteur à un autre, voire d'un article à un autre, provenant des différentes versions s'étalant de 1906 à 1912, suscitées par les réactions de PEANO et de ZERMELO, et que divers travaux s'efforcent de préciser, tels sont de nos jours les sens des notions de *prédicativité* et d'*imprédicativité* d'une définition, qui ont traversé jusqu'à nous l'histoire de la logique — le legs de POINCARÉ à la logique.

Le programme intuitionniste de BROUWER

Plus radical encore dans sa défense du rôle de l'intuition et dans sa critique des attitudes classiques est le programme avancé par Luitzen Egbertus Jan BROUWER dans sa thèse [*Brw*1], soutenue en 1907. Ce programme s'enracine lui aussi au dix-neuvième siècle: KRONECKER requérait des preuves effectives et voulait bannir des mathématiques l'infini actuel, que dans sa conférence-programme de 1904, HILBERT entendait explicitement défendre contre ses attaques (POINCARÉ [*Pnc*5] imputait lui aussi, à l'infini actuel le surgissement des paradoxes; PEANO [*Pno*6] et RUSSELL [*Rsl*5] lui avaient répondu là-dessus).

BROUWER conçoit une activité mathématique a priori, qu'il prétend indépendante de l'expérience, et dont il prétend l'expérience indépendante. Préalable même à l'exercice du langage, qui n'est que grossièrement adapté à en rendre compte, cette activité s'exerce selon lui dans l'intuition fondamentale de la suite des nombres entiers naturels, fournissant la justification indispensable à la position de celle-ci en tant que processus susceptible d'une continuation indéfinie, et à son obéissance au principe de récurrence. Au-delà, la mathématique ne connaît d'existence que de ce qu'elle justifie en le construisant elle-même; l'existence d'un objet mathématique doit être prouvée, et ne peut l'être que par l'appréhension d'un procédé de «construction» de cet objet (au même moment, POINCARÉ soutient toujours qu'«en mathématiques, le mot exister ne peut avoir qu'un sens, il signifie exempt de contradiction» [*Pnc*6]); les successions temporelles d'affections distinctes de la pensée par ses objets constituent un mécanisme fondamental de l'intellect, qui pourvoit celui-ci de l'intuition originelle du procédé de construction de toute la suite des nombres naturels, construction en quoi consiste l'induction complète (en 1904 [*Ksr*1], KEYSER en appelait déjà à l'infinitude des répétitions dont est susceptible un acte mental éxécutable une fois, pour justifier l'«axiome de l'infini»). La logique n'est qu'une application, au langage du raisonnement, de la mathématique du tout à la partie; l'activité mathématique en est indépendante.

BROUWER distingue dans la pratique de l'activité mathématique une hiérarchie de niveaux:

- celui de la construction de systèmes intuitifs d'entités mathématiques;

- celui du langage mathématique, reflet discursif de la pensée mathématique;
- celui de l'analyse mathématique de ce langage, activité qui mène à la découverte de constructions discursives obéissant aux principes de la logique;
- celui que l'on atteint en faisant abstraction du sens des mots, parvenant de la sorte à des systèmes abstraits «du second ordre», les systèmes formels mathématiques étudiés par la logique formelle;
- celui que l'on atteint en introduisant le langage de la logique symbolique associé aux contructions de la logique (niveau atteint, par exemple, dans les travaux de PEANO et de RUSSELL);
- celui de l'analyse mathématique du langage des logiciens (niveau atteint par HILBERT, et auquel PEANO et RUSSELL ne sont pas parvenus);
- celui que l'on atteint en faisant abstraction, etc ...

Entre autres conséquences, une propriété ne détermine pas toujours une classe des objets qui la satisfont: on n'a pas toujours un moyen de construire cette classe; et tant qu'elle n'a pas été construite, on ne peut se poser la question d'en construire une autre dont elle soit — ou ne soit pas — un élément: ainsi échappe-t-on aux paradoxes. De même, une démonstration de non-contradiction qui ne décrirait pas la construction d'un modèle ne serait pas concluante. Dès sa thèse, BROUWER critique la méthode axiomatique, la logique symbolique selon PEANO et RUSSELL, dénuée d'enseignements touchant les mathématiques, et la théorie des ensembles de CANTOR (dont il accepte cependant certains ordinaux transfinis); il rejette, par exemple, les preuves des théorèmes de CANTOR-BERNSTEIN et du bon ordre, et l'axiome de choix, en les qualifiant de purs édifices verbaux, dénués de sens.

L'année suivante, BROUWER va conclure, de ces principes, au rejet du tiers exclu, ainsi que du principe de résolvabilité de tous les problèmes mathématiques [*Brw*2], et en 1912, il réagira à la publication des *Principia Mathematica* en ouvrant une polémique ([*Brw*3, 4]) contre les tenants de l'opinion que tout le contenu des mathématiques réside dans ce qui s'en formalise.

Les temps optimistes

Des *Principia Mathematica* comme objet de travaux

RUSSELL et WHITEHEAD concevaient les *Principia Mathematica* à la fois comme un système déductif, où l'on déduit des théorèmes d'un système d'axiomes, et comme un *calcul formel*, où toutes les propositions sont formalisées — une terminologie qui renvoie à la fois aux origines booléiennes de l'algèbre de la logique et à la conception frégéienne de l'objet de la théorie, les fonctions propositionnelles. De PEIRCE et de SCHRÖDER, ils avaient conservé la division de la logique en trois branches, le calcul des propositions, le calcul des classes, et le calcul des relations. Il ne faut pas s'en étonner: qu'était-ce après tout que la logique symbolique au début du siècle, si ce n'est l'algèbre de la logique et les contributions partielles récentes de FREGE, et de PEANO et de ses collaborateurs? Ces contributions reprises, et surpassées, par les développements des *Principia*, ce fut l'algèbre de la logique et les *Principia*.

Cependant, RUSSELL et WHITEHEAD estimaient traiter d'une réalité plus fondamentale que celles dont s'occupent les autres branches des mathématiques, et qui rend compte de vérités absolues et indiscutables, dans un langage certes formalisé, mais où les formules ne fonctionnent que comme substituts des mots qu'elles traduisent, et dont la signification leur est déléguée. Il n'était donc pas question, même simplement d'en concevoir des interprétations divergentes, et les questions de compatibilité et de complétude des axiomes ne leur effleuraient même pas l'esprit. Ils ne faisaient pas non plus de différence entre les propriétés du système formel, que nous rangeons aujourd'hui dans sa syntaxe, et les propriétés exprimées par lui, du fait du sens attribué aux symboles.

Sur ce point, le manuel *A Survey of Symbolic Logic* [*Lws*], écrit par Clarence Irving LEWIS et paru en 1918, ouvre une voie qu'il qualifie d'«hétérodoxe»: d'une part, il réintroduit quelques-unes des notations employées par les algébristes de la logique — dont celles des quantificateurs; d'autre part, et notamment pour expliquer ces derniers, qu'il conçoit sans plus comme des conjonctions ou disjonctions infinies, il traite le système en formalisme érigé en objet d'étude, indépendant de la signification des formules, et dont on met à jour les mécanismes formels associés à l'emploi des divers symboles: «A mathematical system is any set of strings of recognizable marks in which some strings are taken initially and the remainder derived from these by operations performed according to rules which are independent of any meaning assigned to the marks» (cité par [*Gdl*18] **I**, p. 45).

Mais sur le moment, dans leur grande majorité, les lecteurs des *Principia* adhérèrent à l'attitude philosophique de leurs auteurs quant à ce rapport entre le système et une réalité, celle des propriétés indépendantes de toute particularité des objets concernés, et que les formules exprimeraient. Il y eut néanmoins, parmi ceux-là, des lecteurs attentifs pour être gênés par certains défauts des *Principia*, qui les amenèrent à en rechercher des perfectionnements ou à proposer de les remplacer par un système, à trouver, qui remplisse mieux les objectifs mêmes que RUSSELL et WHITEHEAD s'étaient assignés; trouver **le** système formalisé supposé inclus, et à découvrir, dans les mathématiques, fut même jusque vers la seconde guerre mondiale un objectif de recherches reconnu, dont l'entreprise de BOURBAKI elle-même porte les traces, dans la partie consacrée à son fondement. La double hiérarchie des types et des ordres (qui se différencie sur les fonctions propositionnelles), introduite pour éviter la dérivabilité d'antinomies telles que celle de RICHARD, et qui justifie le nom de théorie des types *ramifiée* sous lequel on a désigné plus tard le système des *Principia*, en fait d'ailleurs un objet compliqué; par surcroît, cette hiérarchie était expliquée de façon très embrouillée.

Efforts vers la réduction du nombre des primitives; origines de la logique combinatoire

Il n'est donc pas surprenant que, comme on le faisait à la même époque pour d'autres disciplines, on ait cherché à réduire, et le nombre des notions primitives —

la redécouverte, par SHEFFER, en 1913, de la définissabilité des autres connecteurs à partir de l'incompatibilité [Shf], que PEIRCE avait déjà trouvée, et présentée dans une publication peu connue ([$Prc3$],t.4, pp. 13–18), avait impressionné les logiciens — et le nombre des axiomes: la trouvaille, que fit Jean NICOD en 1917 [Ncd], d'un axiome, écrit à l'aide de la «barre de SHEFFER», et dont se déduit tout le calcul des propositions, fit également sensation.

«Une curiosité», dit d'abord HILBERT de cet axiome. Il n'en reste pas moins que l'essentiel de la thèse de TARSKI [$Trk1$], selon TARSKI lui-même [$Trk34$], résida dans une preuve de ce que l'implication réciproque pouvait, à l'aide d'un seul axiome, servir de primitive unique (associée, toutefois, aux quantifications) pour axiomatiser le calcul propositionnel étendu par des variables de fonctions propositionnelles et des quantifications portant sur toutes les variables.

Le 7 décembre 1920 voit Moses SCHÖNFINKEL présenter, à la Société Mathématique de Göttingen, une communication [Sfk], qui finira par être publiée dans les *Mathematische Annalen* en 1924, et où il pousse à l'extrême cette idée de réduire le nombre des notions primitives de la logique, en s'appuyant sur quelques autres idées directrices: en premier lieu, ne considérer que des fonctions; en second lieu, dans ce but, revenir aux conceptions frégéiennes selon lesquelles les prédicats sont des fonctions, non pas propositionnelles, mais ayant les «valeurs de vérité» (le «Vrai», le «Faux») pour valeurs; en troisième lieu, n'admettre comme primitives, en sus d'un minimum de fonctions génératrices données, que la seule opération d'*application*, savoir, l'opération qui, à un opérateur et à un opérande, associe la valeur associée par cet opérateur à cet opérande (notée selon l'un des procédés habituels). Or, en convenant de comprendre «$F(x,y)$ est la valeur en (x,y) de la fonction F» comme voulant dire «$F(x,y)$ est la valeur en y de la fonction valeur en x de la fonction F», on ramène tout ce que l'on exprime avec des fonctions de plusieurs variables à ce que l'on exprime avec des fonctions d'une seule variable; en convenant d'expliciter toutes les quantifications implicites dans l'usage des variables libres (non quantifiées), on n'altère pas le pouvoir d'expression des écritures, mais on n'a plus affaire qu'à une seule sorte de variables, les variables liées (par des quantifications); si l'on convient d'écrire $(f(x) \mid^x g(x))$ au lieu de $(x)(f(x) \mid g(x))$ (où la barre verticale est la «barre de SHEFFER», qui représente la négation de la conjonction de ses termes (NAND), et où l'écriture (x) représente, comme chezRUSSELL, la quantification universelle sur la variable x), on chasse des écritures explicites, au profit de cette «barre de SHEFFER généralisée», les quantificateurs et les autres connecteurs; en posant $Ufg = f(x) \mid^x g(x)$, on en élimine la variable liée x; postérieurement, les variables provenant du système initial, comme f et g, s'élimineront au profit de U encore, mais, pour tenir compte de la multiplicité des quantifications et des enchevêtrements de leurs champs d'action, deux autres *combinateurs* auront été nécessaires: C tel que $Cxy = x$ pour tous x et y (mais ici, x et y ne sont plus des variables provenant du système initial), et S tel que $S\varphi\chi x = (\varphi x)(\chi x)$, pour parvenir, finalement à ce que toutes les formules s'écrivent comme combinaisons, à l'aide de *l'application*, des seuls C, S et U, c'est-à-dire comme suites finies de ces trois signes et de parenthèses, dont on peut se dispenser de diverses manières: l'idée de la *logique combinatoire* vient de naître; il faudra, en fait, encore beaucoup l'élaborer (le combinateur *U*, dans un contexte aussi mal délimité, est source de difficultés).

La santé de SCHÖNFINKEL ne lui permettra pas de le faire, mais le flambeau sera repris par Haskell Brooks CURRY. En 1927, sans connaître le travail de SCHÖNFINKEL ([*CyF*], p. 184), CURRY était parvenu à une conception analogue, à partir de l'idée d'amender les *Principia Mathematica* sur le point suivant: si certaines des règles d'inférence utilisées étaient simples — comme les règles explicitement posées, la règle dite *modus ponens*, accommodée à la FREGE (elle autorise ainsi l'inférence de A et de $A \to B$ à B), et la règle de généralisation — par contre, une règle de *substitution* (d'une expression signifiante à une variable), oubliée, elle, dans l'axiomatique, est si compliquée à formuler qu'il a fallu attendre 1934 (ou 1929, selon Eugene LUSCHEI [*Lus*], en ce qui concerne le système de LEŚNIEWSKI [*Lnw*3]), pour qu'en paraisse, dans le premier volume des *Grundlagen der Mathematik* de HILBERT et BERNAYS [*HbB*], la première formulation correcte — et 1950, pour que CHURCH, qui avait d'abord jugé cette formulation incorrecte ([*Chu*10] p. 63), reconnaisse [*Chu*12] qu'en définitive elle ne l'est pas! L'idée était donc de décomposer cette règle en règles plus simples, à l'aide de combinateurs analogues à ceux de SCHÖNFINKEL, dont CURRY reconnut la parenté avec les siens quand, quelques mois après, il prit connaissance du travail de celui-ci. CURRY vint donc à Göttingen, et dans sa thèse [*Cur*1, 2], soutenue en 1929, introduisit et étudia les propriétés du premier système, encore partiel, de logique combinatoire: la partie qui correspond au calcul des propositions; il en prouvait, en particulier, la non-contradiction.

De la théorie des types simple

Norbert WIENER [*Wnr*] a introduit, en 1914, une représentation, dans la théorie des types, de la notion de couple (ordonné), conçue de façon à permettre de traiter une relation à au moins deux arguments en classe de couples; la traduction en termes de théorie des ensembles de cette représentation diffère de celle que nous utilisons aujourd'hui. Elle apportait à la théorie des types une simplification bien plus conséquente que celle qui résultait de l'introduction de la «barre de SHEFFER», et ouvrait la voie à la théorie des types *simple*, épurée de la hiérarchie des ordres à la RUSSELL. Préconisée dès 1921 par Leon CHWISTEK [*Chw*1, 2], d'un côté, et d'un autre, par LEŚNIEWSKI, qui devait bientôt l'abandonner ([*Lus*], p. 34), celle-ci fut à nouveau proposée en 1925 par Frank Plumpton RAMSEY [*Rms*1]. La hiérarchie des ordres à la RUSSELL avait été introduite pour parer à un raisonnement analogue à celui de RICHARD. Or, fait observer RAMSEY, il y a deux groupes de paradoxes: ceux qui, comme celui de RUSSELL, ne font intervenir que des notions de mathématiques et de logique; ceux-là peuvent se retrouver formalisés; et les paradoxes qui, comme celui de RICHARD, renvoient en outre à des notions empiriques imprécises, qui n'appartiennent ni aux mathématiques, ni à la logique, telles celles de signification et de définition (dans le langage courant, celle-ci peut aller jusqu'à désigner les articles d'un dictionnaire); PEANO avait remarqué cela dès 1906: «exemplo de RICHARD non pertine ad Mathematica, sed pertine ad Linguistica» [*Pno6*]. Ces notions-là ne se formalisent pas quand on ne formalise que les notions de mathématiques et de logique; il n'y a donc pas de risque que l'on puisse dériver avec elles une contradiction dans le système; les complications de la hiérarchie des ordres des *Principia* pour les fonctions propositionnelles sont donc inutiles. Par la suite, diverses variantes de la théorie des types simple, avec des hiérarchies en «ordres» simplifiées, ont été proposées pour constituer un support logique au développement des théories déductives, par exemple par CARNAP en 1929 [*Cnp1*], TARSKI en 1930 [*Trk7*], CHURCH en 1940 [*Chu7*], ...

De cheminements d'idées chez des philosophes,
issus des *Principia Mathematica* et de travaux postérieurs,
et finissant par avoir un effet en retour sur la logique, jusqu'en 1930

Il a plus haut été noté le souci de précision dans l'expression qui était celui de philosophes comme TWARDOWSKI, ŁUKASÍEWÍCZ et LEŚNIEWSKI (cf. [*Lus*]), à l'époque aux prises, de surcroît, avec l'élucidation de ce qui pêche dans l'argumentation du paradoxe de RUSSELL, et du paradoxe apparenté, le paradoxe du Menteur, issu de l'Antiquité grecque, remis à l'ordre du jour par la similitude des problèmes qu'ils soulèvent. Des solutions ont pu être cherchées en scrutant les ressources du langage, et le rapport du signifiant au signifié, ou, pour HILBERT (dans le texte, publié en 1922, d'une conférence [*Hbt9*] sur laquelle je reviendrai), de la forme (d'un terme mathématique) à son contenu (de sens). Par analogie avec la pratique des linguistes — à quoi, en somme, avait renvoyé PEANO [*Pno6*] — ce type d'études a été rattaché à la discipline dont l'objet est l'étude de la signification, la *sémantique*.

L'avènement de systèmes formels, tel celui des *Principia Mathematica*, d'ailleurs pensé pour expliquer et éliminer de tels paradoxes — et tel le sytème de FREGE [*Frg2*], auquel le précédent renvoie — était pour ces recherches le bienvenu, et il semble que l'idée germe très vite que ces sytèmes sont les prototypes de systèmes offrant à la science un langage doté d'une précision toute mathématique, dans lequel les axiomatiques peuvent être couchées, étudiées pour elles-mêmes, et comparées les unes avec les autres, avec une précision égale à celle des mathématiques.

Il semble aussi que vers 1920 des philosophes d'horizons divers en arrivent, de là, à l'idée de la distinction à faire entre le langage décrivant les faits de connaissance et les discours destinés à en élucider les règles — entre ce qui est dit *dans* le langage étudié, et ce qui se dit dans l'étude *à son propos* (la distinction est déjà très claire chez FREGE quand, dans le second volume des *Grundgesetze*, il explique ce que devrait être le traitement d'une arithmétique formelle, qu'il rejette). C'est l'une des idées développées par Ludwig WITTGENSTEIN, un élève de RUSSELL, originaire de Vienne, dans le *Tractatus Logico-Philosophicus* [*Wtg*], écrit entre 1913 et 1918, et qu'il publie en 1921. Et selon le témoignage de TARSKI [*Trk9*], des éléments, au moins, de cette différenciation, déjà en germe dans la distinction faite par FREGE [$Frg2$] entre *l'usage* et *la mention* d'un symbole, étaient présents, dès la fin de l'année 1919, dans les enseignements de LEŚNIEWSKI. Le système élaboré par celui-ci, aux alentours de 1922, et dont il n'a publié les grandes lignes qu'en 1927 ([Lus]), se fondait sur le concept de *Bedeutungskategorie* «que nous devons à HUSSERL, et qui fut introduit par LEŚNIEWSKI dans les recherches sur les fondements des sciences déductives», écrit TARSKI en 1933 [$Trk9$] (en fait, HUSSERL, dans la deuxième édition, parue en 1913, de ses *Logische Untersuchungen* [$Hsl1$], n'avait fait qu'évoquer la notion de catégorie sémantique).

Si donc la distinction entre deux niveaux de langages, l'un étant en principe voué à l'étude de l'autre, procède de la différence promue par HILBERT dans ses conférences de 1922 [$Hbt9, 10$] entre les mathématiques formalisées et la *métamathématique* qui en étudie les propriétés, ce ne peut être que par des voies indirectes, ayant leur origine dans des conversations antérieures; comme écrit antérieur, où soient distingués, relativement aux mathématiques effectivement pratiquées, différents niveaux de langage, il semble bien n'y avoir que la thèse de BROUWER [$Brw1$].

Le *Tractatus*, dont WITTGENSTEIN avait remis un exemplaire à RUSSELL et à FREGE, a exercé une grande influence sur le Cercle de Vienne, un groupe de philosophes et de scientifiques qui s'est formé dans cette ville en 1923 avec l'intention de développer une philosophie *exacte* de la science, prétendant exclure tout présupposé métaphysique, et qui tenait des réunions hebdomadaires. La rigueur entendue était attendue de la reprise — d'une façon que WITTGENSTEIN lui-même n'approuvait pas ([Cvr]) — de l'autre grande idée du *Tractatus*, de voir, dans des formalismes du type de celui des *Principia*, la forme incontournable de toute expression de connaissance, et par conséquent, d'évaluer la véracité des (propositions exprimant les) «faits» à partir de celle des propositions atomiques (celles qui affirment l'établissement d'un rapport déterminé entre des objets déterminés). Logique formalisée, formalismes dans lesquels coucher les théories scientifiques, et travaux concernant ces formalismes, étaient donc de la plus haute importance pour le Cercle, dont un membre, élève de FREGE, Rudolf CARNAP, publie en 1929 un *Abriß der Logistik*; le Cercle fonde, la même année, une revue, *Erkenntnis*; et les mathématiciens du Cercle dirigent, vers la logique, l'attention et les travaux d'un étudiant du nom de Kurt GÖDEL.

De l'école de logique de Varsovie: ŁUKASÍEWÍCZ et ses logiques multivalentes
A partir de 1915, sous l'impulsion des philosophes ŁUKASÍEWÍCZ et LEŚNIEWSKI, éclôt à Varsovie une école de logique où se côtoient philosophes versés dans les mathématiques et mathématiciens, et qui partagera le rayonnement des mathématiques polonaises jusqu'à la seconde guerre mondiale. De nos jours, les centres mathématiques et logiques polonais, quoique très lourdement touchés pendant la guerre, contribuent toujours de façon très active au progrès de ces disciplines.

ŁUKASÍEWÍCZ était l'aîné. Il avait milité, dès 1912, pour la fondation d'un «Institut de Méthodologie» [$Łuk1$]. Il était devenu titulaire d'un chaire de *philosophie* à Varsovie en 1915. Il fut, de 1920 à 1928, membre du comité de Rédac-

J. Łukasíewícz

tion de la revue de *mathématiques Fundamenta Mathematicae*, et, en 1928, à l'occasion d'une réorganisation de l'Université de Varsovie, il opta pour la Faculté des *Sciences*. Son influence sur les mathématiciens et les logiciens, qui fut grande, est reconnue (cf. [*BkS*], [*Mtw*13], [*Slz*]).

Sa grande préoccupation personnelle a été l'étude des textes écrits par les logiciens de l'Antiquité, dont il a déchiffré ([*Łuk*6, 7]) l'exposition du calcul des propositions. Dans le domaine qui nous occupe ici, il est surtout connu pour sa proposition, en 1920 [*Łuk*3], mais annoncée en 1918 [*Łuk*2], d'étudier des logiques à plus de deux valeurs de vérité — il est allé dès 1922 (cf. [*Łuk*9], p. 140) jusqu'à envisager des logiques à une infinité de valeurs [*Łuk*5, 8]. Ces logiques cherchent à résoudre un problème laissé ouvert par ARISTOTE, celui du statut logique des *modalités* du Possible et du Nécessaire (ou de variantes de celles-ci, telles celles du Permis et de l'Obligé) et de leur incorporation à un traitement formel. Ainsi, dans le cas de la logique trivalente, ŁUKASÍEWÍCZ introduit une «valeur de vérité» supplémentaire, et, à ses yeux, intermédiaire entre le «Vrai» et le «Faux», et pouvant être vue comme celle du «Possible», ou, si l'on veut raffiner, du «Probable à 50%». A partir de «tables de vérité» étendues à ces valeurs, élaborées par ŁUKASÍEWÍCZ à partir de considérations logico-philosophiques et mathématiques, on tente d'écrire des axiomes et des règles d'inférence à l'aide desquels on puisse démontrer les propositions prenant la valeur «Vrai» pour tout système de valeurs attribuées aux variables dont elles dépendent, et rien que ces propositions.

Durant la période qui nous occupe en ce moment, ce thème de recherches fut, dès son introduction, et de bout en bout, considéré comme prometteur. En fait, les difficultés qui devaient par la suite s'y présenter n'étaient pas soupçonnées. On peut dire qu'elles n'ont pas été surmontées: rien de nouveau n'a été annoncé à ce propos depuis la publication, en 1952, du manuel *Many-Valued Logics* [*RrT*] de ROSSER et TURQUETTE, où ceux-ci présentent les axiomatisations effectuées par eux depuis 1948, qui s'étendent à des calculs des prédicats.

Mais, entre 1920 et 1930, l'exploration du domaine, encore inconnu, des calculs des propositions de ces logiques multivalentes, a joué un rôle important. Dès le début, l'époque était à la *méthodologie des sciences déductives*, discipline à laquelle Kazimierz AJDUKIEWICZ a consacré une monographie [*Ajd*], parue en 1921, à Łwôw; TARSKI raconte ([*Trk*34]) que c'est une discussion figurant dans cette monographie qui l'a mené à établir, en 1921, le *théorème de la déduction* pour le formalisme des *Principia Mathematica* (ce théorème dit que, d'une démonstration formelle d'un énoncé — une formule sans variable libre — établie à partir d'un certain nombre d'hypothèses, l'on peut tirer une démonstration formelle de l'implication, par l'une de ces hypothèses, de cet énoncé, à partir des hypothèses restantes — il sert à guider, en la rationalisant quelque peu, la recherche de démonstrations formelles; trouvé indépendamment, plus tard, par Jacques HERBRAND, qui le prouve dans sa thèse, en 1930, ce n'est que la même année qu'il en paraît un énoncé dans une publication de TARSKI [*Trk*5]). Les logiques multivalentes de ŁUKASÍEWÍCZ constituaient des sciences déductives, dont l'axiomatisation et la formalisation, par exemple à la manière des *Principia*, apportait un gage de traitement rigoureux. Ainsi ŁUKASÍEWÍCZ et son entourage furent-ils conduits à développer, parallèlement à, et en contact avec l'entourage de HILBERT, l'étude comparative de toute une pléiade de calculs propositionnels; peu à peu, s'élabora une méthodologie de ce type d'études, jusqu'à la constitution d'une théorie générale des théorèmes généraux concernant les théories déductives formalisées (tel celui qui dit que le théorème de la déduction s'applique à toutes celles qui obéissent à des axiomes et à des règles d'inférence appropriés) — c'est-à-dire, les connaissances de cette nouvelle discipline, et les méthodes d'investigation qui lui sont propres, s'organisèrent en une théorie, dont une part fondamentale fut axiomatisée par TARSKI [*Trk*5, 6], vers la fin de la décennie: on verra plus loin que par là se sont préparés des progrès parmi les plus importants de la période qui suit. En 1930, ŁUKASÍEWÍCZ et TARSKI ont utilisé ce cadre théorique, lors de la publication, sous le titre *Untersuchungen über den Aussagenkalkül* [*ŁkT*], d'une récapitulation de tout ce qu'eux-mêmes et leurs collaborateurs ont fait dans ce domaine, dont beaucoup n'avait pas été publié.

De l'école de logique de Varsovie: aperçu sur des conceptions de LEŚNIEWSKI ayant influencé le développement de la logique

LEŚNIEWSKI a rejoint ŁUKASÍEWÍCZ à Varsovie, où il occupa aussi une chaire de philosophie, en 1919. Extrêmement soucieux de rigueur et de clarté, perfectionniste dans l'analyse des concepts et des raisonnements qu'il poussait jusqu'à leurs composants primitifs les plus extrêmes, il avait été fasciné quand ŁUKASÍEWÍCZ lui avait fait découvrir la logique, mais n'en était pas moins très critique à l'égard des ouvrages de l'époque ([*Lus*]). C'est ainsi qu'en 1915, il avait entrepris de développer une théorie formelle générale des «multiplicités», inspirée de l'algèbre de la logique, et dont la notion primitive est celle de partie [*Lnw*1]; il devait l'incorporer plus tard, sous le nom de «méréologie» [*Lnw*2], à son système, inspiré de même de celui des *Principia Mathematica*, mais selon une hiérarchie conçue de façon toute différente. Convaincu en 1920 par CHWISTEK de tirer avantage de traiter dans un système formel les développements qu'il avait traité auparavant en langue courante, il avait abordé le projet de ce système en 1922, désireux d'abandonner la théorie des types qu'il jugeait inadéquate, ce qu'il fit effectivement peu après, après avoir confié une «étude de faisabilité» à un élève: Alfred TARSKI, dont ce fut le sujet de thèse.

LEŚNIEWSKI eut ainsi toute une cohorte d'élèves, qui participèrent aussi aux séminaires de ŁUKASÍEWÍCZ et de TARSKI et furent actifs dans des domaines intermédiaires entre les mathématiques et la logique; après TARSKI, le plus brillant d'entre eux fut Adolf LINDENBAUM, disparu pendant la guerre après son arrestation par la Gestapo.

Aux yeux de LEŚNIEWSKI, la formalisation ne représentait qu'un but second, un instrument de précision pour atteindre le maximum de rigueur et de clarté. «Je ne connais pas», disait-il ([*Lnw*3], §11), «d'autre méthode effective pour expliquer mes «intuitions logiques» à mes lecteurs que d'en «formaliser» la théorie déductive». Gardant à l'esprit les applications à d'autres disciplines de la logique et des mathématiques, il tenait que «des mathématiques sans intuition ne peuvent remédier efficacement aux maladies de l'intuition». Le but premier est de communiquer un sens. Il approuvait RUSSELL d'avoir écrit «Le sens des réalités [...] doit être préservé jusque dans les études les plus abstraites. La logique [...] concerne le monde réel, tout autant que la zoologie». D'un autre côté, il croyait à la réalité d'un système de logique parfait, de validité universelle, constituant le fondement des mathématiques, et lui imposait des conditions esthétiques assez contradictoires avec les convictions ci-dessus affichées (une seule notion primitive, un seul axiome par branche — et de forme autre que celle d'une conjonction, ...); elles l'ont fait différer jusqu'en 1927 la publication, par morceaux, de son système, hiérarchisé en catégories sémantiques, et déjà prêt dans ses grandes lignes en 1922; publication qu'il ne put achever en raison d'un différent avec l'éditeur, puis de sa mort prématurée juste avant la guerre, enfin de la destruction de ses manuscrits durant celle-ci, de telle sorte que l'on ne connaîtra jamais dans son intégralité ce système, encore en voie d'élaboration au moment de sa mort.

RUSSELL avait déjà expliqué que sa théorie des types supposait que les propositions ne sont pas des individus ([*Rsl6*]). Les catégories sémantiques des expressions signifiantes du système de LEŚNIEWSKI sont associées à des classes d'équivalence sur ces expressions; ces catégories se déterminent de proche en proche par récurrence sur la formation de ces expressions, à partir des catégories des expressions dénuées d'arguments; ces expressions-là se répartissent à leur tour entre deux catégories génératrices, celle des noms et celle des propositions; quant aux catégories des expressions comportant des arguments (qui se répartissent aussi entre catégories de formateurs de noms et catégories de formateurs de propositions), elles se déterminent selon la règle (maintenant bien connue en informatique): une catégorie (de valeurs à prendre) et une suite de catégories (d'arguments) déterminent la catégorie de toute représentation d'une fonction à valeurs appartenant à la catégorie donnée d'abord, et définie sur les suites d'expressions dont les catégories reproduisent, dans l'ordre où ces expressions se présentent, la suite de catégories donnée ensuite (cf. [*Lus*]).

Mais ce ne sont pas les détails techniques du système de LEŚNIEWSKI qui nous importent ici, mais le fait et le contenu de l'influence exercée, par les idées se rattachant aux siennes, sur les recherches entreprises en logique à Varsovie à partir de 1920. Cette influence sera évoquée à plusieurs reprises plus loin, dans la genèse de la notion de modèle, et dans le développement, après 1930, de certains aspects de la *Beweistheorie* de HILBERT (dont on parlera plus loin).

Il y a eu ainsi, depuis les *Principia Mathematica*, de nombreux systèmes formels concurrents du leur, proposés par leurs auteurs pour remplir la même fonction, et présentant une structure hiérarchisée inspirée de la sienne. Il serait d'autant plus oiseux de les passer tous en revue qu'ils n'ont pas été associés à une aussi grande influence sur le cours de l'histoire que celui de LEŚNIEWSKI.

De la théorie des ensembles arrangée en théorie des types cumulative: axiome de fondement

En 1917, Dimitri MIRIMANOFF [*Mmf*] avait remarqué que l'axiomatique de ZERMELO n'exclut pas la présence d'ensembles «extraordinaires», qui appartiennent à eux-mêmes, ou plus généralement à l'aide desquels on peut constituer une suite infinie d'ensembles, descendante pour l'appartenance (chaque terme étant élément du précédent, sauf le premier). Les autres ensembles sont «bien fondés», et l'on peut assigner à chacun un ordinal, son *rang*, en convenant que le rang de l'ensemble vide est zéro, et que les autres ensembles bien fondés sont du rang immédiatement supérieur à tous ceux de leurs éléments. Ainsi, les ensembles de rang ordinal $\alpha+1$ constituent l'ensemble des parties de l'ensemble des ensembles de rang α, et si λ

est un ordinal limite, les ensembles de rang λ constituent la réunion des ensembles des rangs strictements inférieurs à λ. En conséquence, partant de l'ensemble vide, et itérant — transfiniment — le passage à l'ensemble des parties, en prenant la réunion dans les passages à la limite, on obtient tous les ensembles bien fondés. Selon une remarque faite par GÖDEL en 1944 [*Gdl*15] (et probablement bien plus tôt dans ses enseignements, compte tenu de ce que MOSTOWSKI, qui a suivi ses cours à Vienne en 1937, dit à ce propos dans sa thèse de 1938 [*Mtw*2]), les «rangs» (si l'on peut se permettre d'appeler ainsi chacun des ensembles constitués des ensembles d'un rang donné) se comportent ainsi comme des *types cumulatifs* (c'est-à-dire les types d'une théorie des types où les ensembles d'un type α sont aussi de tous les types $\beta \geq \alpha$), une partie des liens qu'entretient la théorie des ensembles avec la théorie des types s'explicite comme suit (MOSTOWSKI dit seulement que «la seule théorie de types logiques en théorie des ensembles est due à MIRIMANOFF»): s'ils étaient vus de cette façon, les étapes de la génération des types de RUSSELL et WHITEHEAD consisteraient à passer d'un type donné au suivant, alors constitué des seules parties du précédent *définissables* comme extensions des fonctions propositionnelles s'appliquant à ses éléments. GÖDEL fait aussi ressortir par cette remarque le caractère «relativement prédicatif» de la théorie des ensembles qui s'ensuit: chaque ensemble est défini postérieurement à ses éléments — mais l'engendrement des ordinaux est imprédicatif à partir d'un certain rang.

MIRIMANOFF avait aussi proposé, en 1917, de compléter l'axiomatique de ZERMELO par un axiome excluant les ensembles extraordinaires; cette conception a été formalisée au moyen d'un axiome par von Neumann en 1925 [*vNm*1], sous une forme peu élaborée, puis en 1929 [*vNm*3], sous une forme apparentée à sa forme actuelle; en 1930, ZERMELO [*Zml*4] l'a appelé *Axiom der Fundierung*, comme on le fait aujourd'hui.

De quelques étapes importantes dans l'histoire de l'axiome de choix, jusqu'en 1930

On sait quelle attention a été accordée à la délimitation du rôle de l'axiome de choix par l'école mathématique polonaise. L'élan a été donné, en 1918, par un mémoire de SIERPINSKI, *L'axiome de M. Zermelo et son rôle dans la Théorie des Ensembles et l'Analyse* [*Srp*1]. SIERPINSKI y énumère un grand nombre de propositions de divers domaines, qu'on ne sait alors démontrer qu'avec l'axiome de choix. Il préconise de toujours attirer l'attention sur l'emploi de celui-ci, et de l'éviter si possible, y compris pour des théorèmes déjà établis avec son aide. Il mentionne l'opinion de Nicolas Nicolaïevitch LUSIN, avec lequel il a beaucoup travaillé durant son exil à Moscou, selon laquelle il est inutile de chercher à infirmer une proposition déjà établie avec l'aide de l'axiome de choix, appréciation que viendra confirmer, en 1938, la démonstration de compatibilité de GÖDEL [*Gdl*13]. Il introduit la notion de *choix dépendants* — à quoi HADAMARD fait allusion dans la première de ses lettres à BOREL, en 1905 ([*HBLB*]).

TARSKI s'est intéressé à ce type de questions; il a, dès 1924 [*Trk*2], produit plusieurs énoncés d'arithmétique des puissances infinies, équivalents à l'axiome de choix, et montré, dans un article bien connu [$Trk3$], que l'axiome de choix semble indispensable, dans l'état des connaissances de l'époque, à la preuve d'équivalence de diverses définitions de la notion d'ensemble fini. En 1926 [$LdT1$], LINDENBAUM et TARSKI avaient affirmé que l'hypothèse du continu généralisée entraîne l'axiome de choix; SIERPINSKI le rappelle quand, en 1947, il en donne une preuve [$Srp2$]. Dans le même ordre d'idées, c'est en 1926 [DKg] que Dénes KÖNIG a établi le «lemme de KÖNIG» sur l'existence d'une branche infinie dans un arbre infini admettant une ramification finie en chacun de ses noeuds.

Du stade atteint par HILBERT dans le développement de ses idées, en 1917

Depuis le Congrès de 1904, HILBERT n'avait pas abandonné son projet, et il avait, à quatre reprises, dispensé des cours sur les principes ou fondements de la logique et des mathématiques ([$Mor2$]). Le 11 septembre 1917, il prononce à Zürich une conférence intitulée *Axiomatisches Denken* [$Hbt8$], à l'issue de laquelle il propose à Paul BERNAYS de devenir son Assistant pour développer des recherches en logique, ce que BERNAYS accepte. Ainsi débute l'activité d'un autre centre de recherches en logique, à Göttingen.

Dans sa conférence, HILBERT dit que l'on pourrait, dans l'axiomatisation de la logique selon les *Principia Mathematica*, «die Krönung des Werkes der Axiomatisierung überhaupt erblicken», mais aussi, aussitôt, que l'achèvement de l'entreprise demande «noch neuere und vielseitigere Arbeit.» Et il énumère une série de «schwierigster erkenntnistheoretischer Fragen» qui accompagnent celle de la non-contradiction de l'arithmétique et de la théorie des ensembles; entre autres, «das Problem der prinzipiellen *Lösbarkeit einer jeden mathematischen Frage*, [...], die Frage nach dem Verhältnis zwischen *Inhaltlichkeit und Formalismus* in Mathematik und Logik und endlich das Problem der *Entscheidbarkeit* einer mathematischen Frage durch eine endliche Zahl von Operationen».

Peu après, il commençait un cours sur les *Prinzipien der Mathematik und Logik*, dont les *Grundzüge der theoretischen Logik* [HbA] de HILBERT et ACKERMANN, parus en 1928, reproduisent, après révision, de larges parties. A la manière des *Principia*, HILBERT y traitait à part, dès 1917, le calcul des propositions, mais en en donnant une preuve de non contradiction; c'était aussi la première fois que, sous le nom de «calcul fonctionnel», la logique du premier ordre était traitée en *branche* de la logique.

HILBERT et ACKERMANN en fournissent dans les *Grundzüge* une preuve de non-contradiction, qui repose sur l'idée suivante: si l'axiomatique du «calcul fonctionnel» ne se transforme pas en une axiomatique dont on puisse dériver formellement une contradiction par adjonction de l'hypothèse de présence d'un seul individu, on ne peut pas non plus, sans cette hypothèse, en dériver formellement une contradiction. Or, sous cette hypothèse, on peut déduire l'implication réciproque des formules $\forall x P(x), \exists x P(x)$ et $P(x)$. A partir de là, on constate que les axiomes du «calcul fonctionnel» deviennent déductivement équivalents à des tautologies du calcul des propositions; et que cette propriété se transmet des prémisses d'une règle d'inférence à sa conclusion, et donc, à toute conséquence déductive des axiomes du «calcul fonctionnel». Or, elle ne se transmet pas d'une formule à sa négation. En 1936 [*Gtz*4], Gerhard GENTZEN utilise la même idée pour prouver la non-contradiction de la théorie des types simple; mais TARSKI avait déjà prouvé un résultat plus fort, en 1933 [*Trk*10].

Des contributions de BERNAYS et de POST à l'étude du calcul des propositions En 1918, dans son *Habilitationschrift* [*Brn*1], publiée en 1926, BERNAYS faisait le tour des propriétés de complétude et d'*indépendance* des axiomes auxquelles satisfait le calcul des propositions. De ce calcul, il prouvait ce qu'on appelle aujourd'hui la complétude fonctionnelle: l'ajout d'une formule dénuée de démonstration formelle reposant sur les axiomes en détruit la compatibilité; la complétude déductive: on ne peut faire reposer sur les axiomes des démonstrations formelles d'une formule et de sa négation; et la complétude sémantique: une formule admet une démontration formelle reposant sur les axiomes, si et seulement si elle est valide — en l'occurrence, prenant la valeur «Vrai» pour tout système de valeurs attribuées aux variables dont elles dépend; l'élaboration de la notion de modèle procède aussi de cette source-là. Et de cette autre: BERNAYS prouvait une foule de résultats d'indépendance par rapport à des sous-sytèmes du système des axiomes — d'une façon qui fait irrésistiblement penser aux preuves analogues dans les *Grundlagen der Geometrie* — en usant de ce que ŁUKASÍEWÍCZ et TARSKI allaient appeler, en 1930 [*ŁkT*], la *méthode des matrices*. Cette méthode, détectée et décrite *en tant que méthode* à champ d'action général par TARSKI entre temps, consistait alors à se donner un ensemble fini (ou, à la rigueur, dénombrable) et des lois de composition pour interpréter les connecteurs ET, OU, NON, ..., de façon à ce que, selon toute assignation de valeurs prises dans cet ensemble aux variables, les axiomes acceptés et leurs conséquences prennent la valeur «désignée», faisant office du «Vrai» (ou une de plusieurs valeurs «désignées»), mais non l'axiome dont on veut prouver l'indépendance. Ce sont les formules qui prennent la valeur «Vrai» selon toute assignation de valeurs aux variables dans une «matrice» à trois éléments particulière, qui constituaient la logique trivalente introduite par ŁUKASÍEWÍCZ en 1920 [*Łuk*3]; le problème étant de trouver une axiomatique dont ces formules, et elles seules, se dérivent.

En 1920 aussi, à Columbia, un élève de SHEFFER, Emil Leon POST, utilisait la méthode des matrices [*Pst*1, 2, 3] pour établir la compatibilité, la complétude sémantique, et la complétude fonctionnelle du calcul des propositions. POST faisait explicitement la différence entre les formules formellement démontrables *du* système, et les théorèmes *concernant* le système, objets de son étude, et renvoyait au *Survey* de LEWIS. Il introduisait aussi, à l'aide de matrices, des calculs des propositions multivalents, différents de ceux de ŁUKASÍEWÍCZ, et dont les rapports, assez étroits, avec ces derniers, ont été élucidés en 1939 par Jerzy SŁUPECKI [*Słp*1, 2]; ce qui a été dit plus haut des systèmes de Łukasíewícz s'applique tout aussi bien à ces systèmes de POST, y compris en ce qui concerne leur axiomatisation par ROSSER et TURQUETTE [*RrT*] (cela s'applique également aux tentatives de développer une *logique quantique*, où feraient défaut les lois de distributivité de la conjonction par rapport à la disjonction, et de la disjonction par rapport à la conjonction, qui ont fait suite à une suggestion faite en 1936 par Garret BIRKHOFF et John VON NEUMANN [*BvN*]).

POST usait d'un système d'axiomes non indépendants, aménagé de façon à permettre une conversion d'un test positif d'égalité des tables de vérité de deux formules du calcul des propositions en preuve formelle de leur implication réciproque. En tentant d'étendre un tel algorithme au reste du noyau logique des *Principia* et à l'arithmétique, il eut, vers la fin de 1921, l'intuition d'un «argument diagonal» analogue à celui par lequel GÖDEL devait prouver, non loin de dix ans plus tard, l'existence de propositions arithmétiques indécidables. Il en conclut, à bon droit, à la non résolubilité algorithmique du «calcul fonctionnel» (cf. [*Pst*10]), mais une santé déficiente, et les difficultés de la vie, lui firent laisser de côté ces travaux (restés ignorés jusqu'en 1941) — et en particulier, l'élaboration d'une preuve détaillée — durant une quinzaine d'années.

Des théorèmes de complétude et de compacité pour la logique élémentaire

C'est la publication des *Grundzüge* qui semble avoir suscité la position du problème de la complétude sémantique du «calcul fonctionnel» sous la forme: une formule admet-elle une démonstration formelle reposant sur les axiomes, si et seulement si elle est juste dans tous les Individuenbereichen? Sa résolution, par GÖDEL, dans sa thèse [*Gdl*1], en 1929, allait très vite suivre. Du même coup, GÖDEL établissait ce que nous appelons aujourd'hui le *théorème de compacité*: si tout sous-système fini d'un système d'axiomes est satisfaisable (dans un modèle), il en va de même de ce système.

Des mathématiques et de la logique intuitionnistes, de 1918 à 1925

1918 est aussi l'année où Hermann WEYL, au départ élève de HILBERT, manifeste, dans son livre *Das Kontinuum* [*Wyl*2], qui paraît cette année-là, son évolution vers une forme d'intuitionnisme. WEYL avait eu, en 1910 [*Wyl*1], pour préciser la notion de «propriété définie», invoquée et laissée dans le vague par ZERMELO dans son axiomatique de 1908, l'idée proposée par SKOLEM, indépendamment, en 1922 [*Skm*4]. Il l'avait abandonnée parce que l'intervention apparente de la notion de nombre entier naturel dans le décompte des arguments d'une expression lui y avait fait voir un cercle vicieux. Dans son opuscule de 1918, partant de domaines incluant la suite des nombres entiers naturels et constituant un premier niveau d'objets, il leur superpose un second niveau, constitué des formules où les arguments représentent des objets du premier niveau, et des ensembles que ces formules définissent; il laisse de côté la possibilité, qu'il trouve illusoire, de poursuivre à des niveaux plus élevés. Faisant le tour des définitions imprédicatives qui interviennent en analyse, il explore ce que l'on peut reconstruire de mathématiques sans les définitions imprédicatives. Il parvient à définir une partie dénombrable des nombres réels irrationnels et à reconstituer une bonne part de l'analyse, mais l'ouvrage montre que l'on est arrêté dans cette voie par l'absence de preuve de ce qu'un ensemble borné non vide de nombres réels admet nécessairement une borne supérieure.

Hermann WEYL, par la suite, n'a pas continué dans cette voie, s'étant rallié à la conception brouwérienne du continu, apparue la même année.

A ce moment-là commence en effet la publication, qui se poursuivra l'année suivante, du premier article de BROUWER sur les mathématiques intuitionnistes, *Begründung der Mengenlehre unabhängig vom logischen Satz vom ausgeschlossenen Dritten* [$Brw5, 7$].

Rebaptisés plus tard «déploiements» (cf. [$Htg1, 3, 4$]), les ensembles dont il s'agit sont des ensembles de nombres réels, ou plutôt, de suites de nombres rationnels génératrices de nombres réels. Une suite génératrice se construit pas à pas, une fois donnée une règle de déploiement qui trie, à chaque adjonction d'un terme, les suites finies de nombres entiers naturels à admettre, et une règle d'association, à chaque suite finie de nombres entiers naturels admise, d'un nombre rationnel. BROUWER montre ([$Brw5$]) que pour qu'un ensemble en ce sens soit un intervalle fermé, il faut et il suffit que ce soit un «éventail» (*fan* en anglais), c'est-à-dire, que sa loi de déploiement soit telle que, pour chaque suite finie retenue, il ne soit retenu qu'un nombre fini de prolongements par adjonction d'un terme supplémentaire. Il apparaît que les concepts concernant les nombres réels doivent être revus: par exemple, si deux générateurs de réels ne sont pas «écartés» (c'est-à-dire, si l'on ne peut assigner une minoration uniforme strictement positive à la différence de leurs termes de mêmes rangs), on ne peut pas nécessairement affirmer qu'ils sont égaux. Il en va de même pour les théorèmes, et BROUWER entreprend de reconnaître dans quelles circonstances le théorème de BOLZANO-WEIERSTRASS subsiste dans l'analyse intuitionniste [$Brw5, 11$].

Entre 1919 et 1923, BROUWER développe une théorie de la mesure et amorce une théorie des fonctions ([$Brw7, 9$]). En 1923 («seulement», note Wim RUITENBURG [Rtg], y voyant une rançon de sa répugnance au formel), il remarque [$Brw11$] qu'en logique intuitionniste, trois négations équivalent à une seule. Dans un autre article paru la même année [$Brw12$], il pointe, afin de montrer les effets (dévastateurs) du rejet du tiers exclu sur l'analyse, quelques théorèmes auxquels ce rejet fait obstruction, tels l'existence d'un maximum atteint par une fonction d'une variable réelle continue sur un intervalle fermé et le théorème de HEINE-BOREL, et il prouve le *«fan theorem»*: si f est une fonction définie en tout point d'un éventail et à valeurs entières naturelles, on peut calculer, à partir de la définition de f, un nombre entier naturel N tel que la valeur de f en un point de cet éventail est déterminée par les N premiers termes d'une suite génératrice de ce point.

Un exemple de corollaire «intuitionniste» de ce théorème est le suivant: une fonction réelle d'une variable réelle, définie sur un intervalle fermé, y est uniformément continue.

En 1925, la recherche en mathématiques intuitionnistes se diversifie; un théorème de D'ALEMBERT intuitionniste est établi par BROUWER et DE LOOR ([BdL]); Arend HEYTING soutient une thèse [$Htg1$] sur l'axiomatisation de la géométrie projective intuitionniste.

Des formalisations de la logique intuitionniste et des interprétations intuitionnistes de fragments de la théorie des nombres classique, jusque vers 1933

C'est alors que paraît un mémoire [*Kmg*1] de Andrei Nikolaievitch KOLMOGOROFF sur le tiers exclu, dont les innovations les plus porteuses d'avenir résident d'abord dans une confrontation de la logique propositionnelle intuitionniste et de la logique propositionnelle classique, à cet effet formalisées l'une et l'autre comme HILBERT le fait dans ses conférences de 1922 [*Hbt*9, 10]; en second lieu, dans le fait de suggérer une axiomatisation de la logique intuitionniste, explicitée complètement pour le calcul des propositions; et en troisième lieu, dans la proposition d'une «lecture intuitionniste» des mathématiques classiques, c'est-à-dire, en termes techniques, dans la production d'un algorithme transformant chaque formule (entendue représenter une assertion relevant des mathématiques classiques) en une autre formule (entendue représenter une assertion relevant des mathématiques intuitionnistes), dite (ici) interpréter la première.

Les seuls connecteurs primitifs étant l'implication et la négation, d'un noyau d'axiomes propositionnels commun, on passe à la version donnée du calcul intuitionniste en lui adjoignant l'axiome (dans une notation, ici, qui n'est pas celle de KOLMOGOROFF)

$$(A \rightarrow B) \rightarrow \big((A \rightarrow \text{NON}B) \rightarrow \text{NON}A\big),$$

exprimant le rejet d'une hypothèse impliquant contradiction, et à la version donnée du calcul classique, en adjoignant à la place l'axiome

$$\text{NONNON}A \rightarrow A.$$

La transformation qui, à une formule, associe la formule qui va l'interpréter, consiste à y intercaler des doubles négations, à raison d'une préfixée à chaque champ d'action d'une implication (qui a deux tels champs: un antérieur, et un postérieur), d'une négation ou d'une quantification — les champs d'action des négations intercalées n'étant bien entendu pas pris en compte — puis à préfixer encore pour finir une double négation à la formule obtenue. KOLMOGOROFF prouve que si, dans son système de logique propositionnelle classique, une formule se déduit formellement d'hypothèses, alors, dans son système de logique propositionnelle intuitionniste, l'interprétation de cette formule se déduit formellement des interprétations de ces hypothèses. Après avoir noté que tous les axiomes des mathématiques classiques — sans préciser ce qu'il entend par là — ont des interprétations vraies (du point de vue intuitionniste), il induit de là qu'il en va de même pour tous les théorèmes des mathématiques classiques.

Que le lecteur y prenne garde: cette conclusion n'est pas fondée, non parce que l'axiomatique donnée par KOLMOGOROFF pour le calcul propositionnel classique n'est pas complète — cela pourrait se corriger sans incidence sur les conclusions ici en cause — mais en raison de ce que sa preuve ne concerne explicitement que le calcul des propositions, alors que l'une des règles d'inférence propres au calcul des prédicats est susceptible de manquer à transmettre, de sa prémisse à sa conclusion, l'acceptabilité intuitionniste de l'interprétation des formules selon cette version. Cependant, comme GÖDEL le découvre en 1932 [*Gdl*8], selon une version un peu différente, les axiomes usuels de l'arithmétique des nombres entiers naturels permettent de rétablir la situation, en ce qui concerne ce domaine limité; mais on n'a pas pu étendre ce résultat depuis.

La Wiskundig Genootschap réagit en 1927 par l'institution d'un prix devant récompenser, entre autres, la production d'une formalisation des mathématiques intuitionnistes ([*Rtg*]). HEYTING le gagne, l'année suivante, pour la production, entre autres, d'une axiomatique de la logique du premier ordre avec égalité intuitionniste. Elaborée en triant, dans les théorèmes et déductions des *Principia Mathematica*, ceux que les intuitionnistes pouvaient accepter, et formalisée à la HILBERT, cette axiomatique bien connue est publiée en 1930 dans un article intitulé *Die formalen Regeln der intuitionistischen Logik* [*Htg*2].

Le hasard des délais de publication fait que l'extension, par Valère GLIVENKO, aux calculs *des propositions* de HILBERT et de HEYTING, d'une preuve analogue à celle de KOLMOGOROFF, relative à l'interprétation d'une formule par sa double négation ([*Gvk*]), paraît en 1929. En 1932, GÖDEL [*Gdl*8], et, indépendamment, GENTZEN et BERNAYS [*Gtz*3], obtiennent des interprétations conduisant ainsi à des théorèmes analogues; ceux dont le champ d'application est le plus vaste concernent l'arithmétique (de PEANO, du premier ordre) et l'arithmétique intuitionniste — un résultat important dans le contexte de l'époque, puisqu'il constitue une preuve de compatibilité de l'arithmétique classique relativement à l'arithmétique intuitionniste.

Des résultats partiels concernant le problème de la décision, avant 1933

En ce qui concerne le problème de la décision, les premiers résultats partiels obtenus avant 1930, après ceux de LÖWENHEIM [*Lwh*], ont été obtenus par SKOLEM en 1919 [*Skm*2], puis par Heinrich BEHMANN en 1922 [*Bmn*]. SKOLEM a produit un algorithme de décision pour le Klassenkalkül — soit, en termes d'aujourd'hui, pour la théorie du premier ordre des algèbres de BOOLE de toutes les parties d'un ensemble (attention, en l'occurrence, ce sont les *parties* de cet ensemble qui sont prises comme individus, les éléments de l'ensemble n'interviennent que par le truchement des parties à un élément), ou encore, sous un autre angle de vue, pour le calcul des prédicats à un argument *du second ordre* (sans l'égalité — l'application au calcul des prédicats à un argument, *avec l'égalité*, suppose que l'on reste au premier ordre); BEHMANN, indépendamment, semble-t-il ([*Chu*13]), produit des algorithmes de décision pour les mêmes trois calculs des prédicats — c'est de cet article-là que proviennent les dénominations de «forme normale disjonctive» et de «forme normale conjonctive».

L'algorithme de SKOLEM reposait sur la mise en oeuvre, avec soin et rigueur, d'une méthode juste esquissée par LÖWENHEIM dans son mémoire de 1915 pour établir la décidabilité du calcul des prédicats à un argument du premier ordre. Dégagée des particularités de ses premiers emplois, cette méthode consiste à adjoindre, aux moyens d'expression d'une théorie, des formules particulières, de signification intuitive claire, dont le choix bien sûr résulte d'une analyse perspicace des structures dont parle la théorie, et qui soit de nature à permettre de montrer que, des axiomes de la théorie, il résulte formellement que toute formule de la théorie est équivalente à une formule sans quantificateur — dans laquelle peuvent apparaître, certes, des notions primitives adjointes dans les formules ajoutées, mais qu'il ne reste plus qu'à «vérifier» pour compléter un procédé de décision, permettant de dire si la formule testée à l'origine était ou non un théorème. Dans le cas du calcul des classes, l'une de ces formules particulières est celle qui dit d'une classe qu'elle inclut un atome, c'est-à-dire une classe qui ne peut être partagée qu'en elle-même et en la classe vide.

Cette idée attira l'attention de Cooper Harold LANGFORD, qui l'adapta de façon à obtenir, en 1927, des procédures de décision pour les théories élémentaires (c'est à dire, du premier ordre) de l'ordre dense sans premier ni dernier élément, de l'ordre dense avec premier élément sans dernier élément, de l'ordre dense avec premier et dernier élément ([*Lfd*1]), et de l'ordre discret avec premier élément et sans dernier élément ([*Lfd*2]), avec à la clef des preuves de complétude déductive. TARSKI mit l'étude de cette méthode au programme de son séminaire, de 1927 à 1929, la qualifiant, ce faisant, de méthode d'*élimination des quantificateurs*.

Des procédures de décision furent activement recherchées, et trouvées, notamment pour la théorie élémentaire de l'ordre discret sans premier ni dernier élément, et pour la théorie élémentaire de l'ordre discret avec premier et dernier élément; de surcroît, des descriptions de toutes leurs extensions complètes furent obtenues. TARSKI établit aussi (dans sa terminologie de l'époque) que les algèbres de BOOLE atomiques, c'est-à-dire, celles dont tous les éléments majorent un atome, ont la même théorie élémentaire que les algèbres de BOOLE réalisant le Klassenkalkül; donc, la théorie élémentaire des algèbres de BOOLE atomiques est décidable, un résultat qu'il ne publie qu'en 1935 [*Trk*13]; ceux de son étude de la théorie de l'ordre dense et de ses extensions [*Trk*12] paraissent l'année suivante.

En 1929, en appliquant la même méthode, Mojżesz PRESBURGER a établi [*Pbg*] un algorithme de décision pour l'arithmétique du premier ordre restreinte à l'addition. Indépendamment, SKOLEM [*Skm*8] en établissait un autre peu après, assorti d'un algorithme de décision pour la théorie de l'ordre et de la multiplication des nombres entiers naturels. TARSKI lui-même a annoncé en 1931 [*Trk*8] avoir obtenu grâce à cette méthode un procédé de décision, sur lequel je reviendrai plus loin, pour la théorie du premier ordre de l'algèbre élémentaire, soit, vue d'une autre façon, celle du corps ordonné des nombres réels.

L'algèbre de la logique a légué un algorithme qui transforme n'importe quelle formule en une formule qui lui est logiquement équivalente mais est sous forme *prénexe*, c'est-à-dire, commençant par un *préfixe* constitué d'une suite de quantificateurs suivie d'une formule où il n'y a plus de quantificateurs du tout (la «matrice» de la forme prénexe). Il y a des informaticiens qui sont surpris d'apprendre que les *fonctions* de SKOLEM, qui servent à la réduction des problèmes de satisfaisabilité de formules à celui de formules sous *forme normale* de SKOLEM, ont été introduites par celui-ci [*Skm*3] en 1920 (en fait, après des anticipations de LÖWENHEIM [*Lwh*], et même de SCHRÖDER [*Sdr*1], en 1895), en vue de donner une preuve sans lacune du théorème de LÖWENHEIM, étendu là par SKOLEM, en utilisant l'axiome de choix, à des expressions pouvant comporter des conjonctions et des disjonctions infinies, ou bien, une infinité dénombrable de quantifications.

Pour ce qui est de la satisfaisabilité et/ou de la validité d'un *schéma* de formules, Paul BERNAYS et Moses SCHÖNFINKEL, en 1928 [*BnS*], ont décrit, pour un langage du premier ordre sans l'égalité, un procédé de décision qui couvre le cas où le préfixe se compose d'une suite de quantificateurs universels *suivie* d'une suite de quantificateurs existentiels, l'une ou l'autre de ces deux suites pouvant être vide (et l'égalité, alors, présente). Cette année-là, RAMSEY a découvert [*Rms*2] les théorèmes de combinatoire qui portent son nom pour obtenir un procédé de décision de l'existence d'un modèle infini, dans le cas où le préfixe ne comporte que des quantificateurs universels. Le résultat le plus fort avant 1930 a été obtenu par Wilhelm ACKERMANN en 1928 [*Ack*3]: il concerne un préfixe où tous les quantificateurs sont universels, sauf un. L'analogue pour deux quantificateurs existentiels *consécutifs* et en l'absence de l'égalité, constitue la limite de ce que l'on peut obtenir sans imposer d'autres types de restrictions à la formule considérée; il a été obtenu entre 1932 et 1934, indépendamment, par Kurt GÖDEL [*Gdl*4, 10], Lázsló KALMÁR [*Kmr*1], et Kurt SCHÜTTE [*Sct*1, 2]. Un résultat typique de ce qui peut être obtenu sous des conditions de nature différente, et relatives à la seule matrice, est dû à HERBRAND en 1931 [*Hbd*1]; il couvre le cas où cette formule est équivalente à une disjonction de formules «premières», c'est-à-dire dénuées de symbole logique, ou une formule équivalente à une telle disjonction.

Des affinements du programme de HILBERT en 1922

Dans une conférence donnée à Copenhague au printemps de 1922 et répétée l'été suivant à Hambourg [*Hbt*9], faisant reproche à WEYL et à BROUWER de vouloir «eine Verbotsdiktatur à la KRONECKER errichten», HILBERT leur avait opposé un programme: il s'agit de rendre «die Axiome, Formeln und Beweise der mathematischen Theorie selbst Gegenstand einer inhaltlichen Untersuchung. Dazu müssen aber zunächst die üblichen inhaltlichen Überlegungen der mathematischen Theorie durch Formeln und Regeln ersetzt bzw. durch Formalismen nachgebildet werden, d.h. es muß eine strenge Formalisierung der ganzen mathematischen Theorien einschließlich ihrer Beweise durchgeführt werden, so daß die mathematischen Schlüsse und Begriffsbildungen — nach dem Muster des Logikkalküls — in das Gebäude der Mathematik als formale Bestandsteile einbezogen sind.» C'est pour cela, comme il l'a dit auparavant, qu'il faut déprendre les signes de leur signification; «[...] dadurch werden die inhaltlichen Überlegungen, die selbstverständlich niemals völlig entbehrt oder ausgeschaltet werden können, an eine andere Stelle, gewisssermaßen auf ein höheres Niveau verlegt [...]». «[...] wir werden so zu einer Art *Beweistheorie* gedrängt, die von dem Operieren mit den Beweisen selbst handelt.» En résumé: «*Erstens:* Alles, was bisher die eigentliche Mathematik ausmacht, wird nunmehr streng formalisiert, so daß *die eigentliche Mathematik* oder die Mathematik in engerem Sinne zu einem Bestande an beweisbaren Formeln wird. [...] *Zweitens:* Zu dieser eigentlichen Mathematik kommt eine gewissermaßen neue Mathematik, eine *Metamathematik*, hinzu, die zur Sicherung jener dient, indem sie sie vor dem Terror der unnötigen Verbote sowie der Not der Paradoxien schützt.»

En septembre 1922, HILBERT, dans une autre conférence, intitulée *Die logischen Grundlagen der Mathematik* [$Hbt10$], décrit dans ses grandes lignes une preuve de non-contradiction *finitiste* de la théorie de deux fonctions, dont l'une ne prend pas la valeur 0 (qu'il écrit addition de 1), et dont l'autre est la «réciproque à gauche» de la première (qu'il pense être le prédecesseur), avec adjonction d'une fonction «transfinie» (nous dirions une «fonction logique de choix». HILBERT la qualifie de *transfinie* parce qu'elle dit quelques chose d'un domaine, fût-il au-delà du fini, et il appelle l'axiome qui en régit le comportement, *axiome du transfini*). Sans doute a-t-il entre temps décidé de convaincre ses adversaires, et pense-t-il aussi, par là, lever les objections de POINCARÉ à l'égard de ses suggestions de 1904, car il souligne parvenir à sa preuve «mittels *finiter Logik* durch rein anschauliche Überlegungen, wozu die Rekursion und die anschauliche Induktion für vorliegende endliche Gesamtheiten gehört.»

De l'arithmétique primitive récursive de SKOLEM

En 1923 paraît, sous le titre *Begründung der elementaren Arithmetik durch die rekurrierende Denkweise ohne Anwendung scheinbarer Veränderlichen mit unendlichem Ausdehnungsbereich*, un mémoire de SKOLEM [$Skm5$], écrit, sauf ses brèves remarques finales, en 1919, après étude des *Principia Mathematica*. Sans formalisation, SKOLEM y délimite ce que l'on appelle aujourd'hui *l'arithmétique primitive récursive*. On ne s'y autorise des définitions et des raisonnements par récurrence que dans la mesure où il n'intervient, dans le définissant, ou dans la proposition à démontrer, que des quantifications *bornées*, c'est-à-dire, portant sur une variable dont il est exprimé que le domaine de variation se compose des nombres entiers naturels ne dépassant pas un nombre donné (pouvant dépendre de paramètres).

SKOLEM développe cette théorie jusqu'au théorème d'unique factorisation en facteurs premiers et jusqu'au principe du plus petit nombre pour une fonction propositionnelle répondant aux conditions ci-dessus. Il note dans une des remarques finales ajoutées au moment de la publication le strict point de vue *finitiste* qu'il maintient de façon cohérente dans son ouvrage. HILBERT et ses collaborateurs devaient en être d'accord: une bonne place est consacrée à cette théorie dans le premier volume, paru en 1934, des *Grundlagen der Mathematik* [HbB] de HILBERT et BERNAYS, qui notent que les théorème de cette arithmétique sont «vérifiables», ce qui en prouve la compatibilité. Une première formalisation de cette théorie a été proposée par CURRY en 1940 [$Cur8$], pour aider à l'étude des fonctions récursives, et d'autres encore depuis. Aujourd'hui, des recherches liées à l'informatique théorique continuent d'être consacrées à cette théorie, et l'on considère que la dernière formalisation qui en a été donnée est bel et bien celle de la notion de «raisonnement finitiste» de l'école de HILBERT.

Des principales contributions à la théorie des ensembles axiomatique de FRAENKEL, SKOLEM, VON NEUMANN **et** BERNAYS

En 1921 et 1922, deux importantes contributions à la théorie axiomatique des ensembles sont le fait de FRAENKEL et de SKOLEM (dont les textes paraîtront respectivement en 1922 et 1923).

FRAENKEL portait attention à l'amélioration de l'axiomatique de la théorie des ensembles et aux propriétés métamathématiques de cette axiomatique; on lui doit de nombreuses contributions dans ce domaine, s'étalant sur près de vingt années; il a notamment essayé d'assurer la catégoricité de la théorie en adjoignant à ses axiomes un *axiome de restriction* limitant au maximum la prolifération des ensembles dont les autres axiomes assurent l'engendrabilité. Il a tenu à jour, d'abord seul, puis en collaboration, des sommes récapitulant les travaux et les essais effectués dans ce domaine; durant la période qui nous occupe, ce sont les trois éditions, très enrichies de chacune à la suivante, de son *Einleitung in die Mengenlehre*, parues en 1919 [$Fkl1$], 1923 [$Fkl4$] et 1928 [$Fkl5$], et dont la dernière a longtemps été un ouvrage de référence.

En 1921 [$Fkl2, 3$], il introduit, sous une forme qui n'est pas restée, un axiome de remplacement, et il démontre l'indépendance de l'axiome de choix relativement à une théorie où il y a, et c'est sa faiblesse, un ensemble dénombrable de paires d'individus qui ne sont pas des ensembles. Il montre qu'alors, tout ensemble est invariant sous l'action de toutes les transpositions des éléments d'une de ces paires de non ensembles, sauf un nombre fini d'entre elles (dépendant de cet ensemble), remerciant au passage ZERMELO de son insistance sur la notion d'invariance. Or, l'ensemble des valeurs d'une fonction de choix pour l'ensemble des paires d'individus ne peut être tel. En 1950 encore, on n'aura pas mieux, concernant l'indépendance de l'axiome de choix.

En 1928 [$Fkl6$], selon une méthode analogue, où interviennent les permutations d'un ensemble infini dénombrable de non ensembles, FRAENKEL démontre l'indépendance du «principe d'ordination», selon lequel tout ensemble porte un(e) (relation d') ordre.

SKOLEM, qui se trouvait à Göttingen au moment de la sortie de l'article de LÖWENHEIM [Lwh], en avait immédiatement perçu les conséquences, contrairement à ce qu'il en a été de l'ensemble des mathématiciens. Il s'en était alors immédiatement ouvert à BERNSTEIN (cf. [$Skm4$]), qui ne semble pas avoir eu de réaction notable.

Au congrès des mathématiciens scandinaves qui se tient à Helsinki en 1922, il fait d'abord remarquer [$Skm4$] la circularité présentée par l'usage d'un «domaine» pour expliquer la notion d'«ensemble». Puis, il propose, pour préciser la notion de condition «définie» de ZERMELO, d'y voir la notion de condition exprimable par une *Zählaussage* (la thèse [$Skm1$] soutenue par SKOLEM en 1913 portait sur l'algèbre de la logique). Il propose, sous une forme voisine (au langage près) de celle que nous utilisons aujourd'hui, un axiome (notre «axiome de remplacement») auquel il ne donne pas de nom (rappelons que WEYL [$Wyl1$] avait anticipé cette idée); il établira plus tard, en 1929 [$Skm7$], que cet axiome recouvre l'axiome de remplacement de FRAENKEL. Il rappelle les objections de POINCARÉ contre les définitions non prédicatives, et contre l'incohérence de l'emploi de l'induction complète sur les formules pour justifier l'induction complète sur les nombres du domaine. Puis il note que les axiomes ne déterminent pas celui-ci, puisque les ensembles bien fondés en forment une partie qui satisfait aux axiomes — et qu'en conséquence, il est douteux que l'on puisse prouver l'hypothèse du continu. Il rappelle aussi que l'axiome de choix n'a pas reçu un assentiment universel. Mais tout ceci n'est rien, à côté de ce qui résulte du théorème de LÖWENHEIM, qu'il redémontre, mais cette fois-ci, sans utiliser l'axiome de choix, dans le cas d'un ensemble dénombrable d'énoncés: s'il existe un modèle de ces énoncés, il en existe un sur les nombres entiers naturels. Or, cela s'applique à la théorie des ensembles! Tel est l'énoncé de ce qui s'appelle toujours «le paradoxe de SKOLEM». L'explication qu'en fournit SKOLEM, et que les travaux de COHEN, en 1963 [$Coh1, 2$], ont pleinement confirmée, est qu'un ensemble pris dans un domaine peut y être non dénombrable *relativement à ce domaine*, alors qu'un autre domaine admet peut-être une bijection de cet ensemble sur l'ensemble des nombres entiers naturels, qui n'appartient pas au premier domaine. Il peut en aller de même de parties d'un ensemble; en conséquence, même les notions de «fini», d'«infini», et de «suite infinie» ... ne sont que *relatives*.

Sur le moment, les remarques de SKOLEM n'eurent pas beaucoup d'écho. Elles devaient cependant trouver un relais dans un article de 1925 [$vNm1$] où VON NEUMANN a proposé, pour la théorie des ensembles, une autre axiomatique, qui devait être appelée à jouer un rôle important, et qui repose sur des idées directrices divergeant un peu plus de celles de ZERMELO. Une de ces idées est de réintroduire les «multiplicités inconsistantes» dont CANTOR dit quelques mots dans une lettre adressée à DEDEKIND durant l'année 1899 (cf. [NCv]), tout en appliquant à celles-ci, seules, le traitement imaginé par RUSSELL, c'est-à-dire en refusant un sens à toute expression exprimant que l'une d'elles est élément — ceci est d'ailleurs conforme au sentiment de CANTOR à propos de ces multiplicités, pour lesquelles, écrit-il en 1899, l'hypothèse d'un *Zusammensein* conduit à contradiction.

Il serait trop long de détailler par le menu les axiomes de VON NEUMANN, dont par ailleurs, d'ailleurs, le contenu ne s'écarte pas de celui des axiomes de ZERMELO, FRAENKEL et SKOLEM. A la manière de HILBERT [$Hbt1$], il les répartit par groupes; plus tard, BERNAYS [$Brn2$] et GÖDEL [$Gdl14$], dans les variantes qu'ils donneront de son système, feront de même. Ce système de VON NEUMANN présente les originalités suivantes:

- comme chez FREGE, les valeurs de vérité et la notion de fonction sont présentées comme primitives, et l'appartenance est définie (on écrit $p(x)$ = Vrai au lieu de $x \in p$);

- il y a deux domaines, celui des objets arguments et celui des objets fonctions; la plupart ont les deux qualités, mais certains n'en n'ont qu'une: par exemple, les valeurs de vérité ne sont pas des fonctions; une classe étant définie comme une fonction dont les valeurs sont toutes des valeurs de vérité, la classe universelle (c'est-à-dire, la constante prenant la valeur Vrai) n'est pas un argument; un ensemble est une classe argument, à laquelle il est donc permis d'être élément;
- le caractère fonctionnel d'une classe de couples s'exprime avec les moyens ainsi autorisés; on peut donc parler de (graphe de) bijection entre deux classes; les classes admises comme ensembles sont celles qui ne sont pas trop trop grandes, c'est-à-dire ne sont pas en bijection avec la classe universelle, composée de tous les objets arguments: on en déduit le schéma de remplacement pour les ensembles et l'existence d'un bon ordre sur la classe universelle, le paradoxe de Burali-Forti prouvant que la classe des ordinaux n'est pas un ensemble, et donc que la classe universelle est image de la classe des ordinaux par une bijection qui en transporte le bon ordre.

Mentionnant leurs travaux, Von Neumann donne une preuve du théorème de Löwenheim-Skolem pour son sytème d'axiomes, en décrivant un procédé pour extraire, d'un supposé «système» de deux domaines satisfaisant à ses axiomes, un «sous-système» dénombrable, satisfaisant encore aux axiomes; il n'emploiera le terme de *modèle* que plus loin. Il rejoint, en les précisant, les conclusions de Skolem: les cardinalités infinies d'un «système» sont spécieuses, puisqu'on peut lui superposer un «système» où il y aurait des bijections n'appartenant pas au premier. Les définitions connues de la finitude font qu'il en va peut-être de même pour elle. Et pour le bon ordre, il n'est pas exclu qu'une partie d'un ensemble bien ordonné dans le premier système ne possède, dans le second, une partie inaperçue du premier, et dénuée de premier élément dans le second. Le système d'axiomes n'est visiblement pas catégorique: l'approche de Hilbert sera donc impuissante [puisque l'objection ne concerne pas la compatibilité, mais la catégoricité].

Bernays, dans une série de mémoires [*Brn*2] dont la publication, débutant en 1937, a été longtemps différée, restitue les traits essentiels du système de Von Neumann dans un langage plus coutumier, en abandonnant la notion de fonction en tant que notion primitive, et en lui substituant les notions d'ensemble et de classe; il introduit deux notions primitives d'appartenance, l'une d'ensemble à ensemble, l'autre d'ensemble à classe, et dit d'un ensemble qu'il *représente* une classe s'ils ont les mêmes éléments. Il superpose une axiomatique pour les classes (gouvernant l'inclusion d'ensemble à classe) et une axiomatique assurant la représentabilité des classes nécessaires à un comportement des ensembles conforme aux axiomes de Zermelo et Fraenkel, abandonnant la caractérisation des ensembles par leur non-équipotence à la classe universelle. Gödel, dans son exposé de 1940 [*Gdl*14], réidentifie les ensembles aux classes qu'il représentent, définit la notion d'ensemble comme celle de classe qui admet une classe dont il est élément, et ne modifie les axiomes que de façon inessentielle.

Du concept de métamathématique chez TARSKI

En 1930, TARSKI consacre, à *«certains concepts fondamentaux de métamathématique»*, une Note mettant au net une esquisse exposée lors d'une conférence en 1928 ([*Trk*5]), et d'entrée de jeu, explique qu'il s'agit de «concepts appartenant à la *méthodologie des sciences déductives*, qu'en suivant HILBERT, on a coutume d'appeler *métamathématique*»; dans le texte du mémoire développé [*Trk*6] qui suit bientôt, intitulé, lui, *«Concepts fondamentaux de la méthodologie des sciences déductives»*, cette assimilation est maintenue. Par là, TARSKI opère une nette extension du sens du terme «métamathématique». Une autre résulte du passage à un degré supérieur d'abstraction, auquel certaines des données issues de la formalisation (des formules, en l'occurrence) sont prises comme des entités abstraites, non analysées, sauf particularisation momentanée, ce qui revient à ignorer les processus effectifs de construction de ces données. Une troisième, enfin, va résulter de ce que TARSKI va, dans l'étude de ces données issues de la formalisation, dédaigner toute restriction de type finitiste.

De plus, le préfixe «méta» va aussitôt se voir adopté par l'école polonaise, puis, au fil des années, universellement, pour désigner tout ce qui touche à l'étude des objets formels d'un système formel, ou, par extension, à l'étude des règles grammaticales d'une langue: «métalangage», pour désigner le langage dans lequel s'effectue cette étude; «métalogique», pour qualifier la science des systèmes formels de logique, et les notions auxquelles elle fait appel; «métathéorème», pour désigner les théorèmes de métamathématique; et quand TARSKI introduit, en 1933 [*Trk*9], une formalisation de la métamathématique, le terme de «métamétamathématique» vient aussitôt; et ainsi de suite.

Dès 1928 [*Trk*5], appelant *ensemble des «énoncés»* un ensemble dénombrable dont les éléments sont appelés des *«énoncés»*, TARSKI propose une axiomatisation et une première étude d'une notion abstraite de *passage aux conséquences*, traité en opérateur à un argument sur les ensembles d'«énoncés» (savoir, en application de l'ensemble des parties de l'ensemble des «énoncés» vers lui-même), particularisé, dans certains passages, à titre de paradigme, mais aussi, de cas particulier intéressant à examiner de plus près, et au prix d'axiomes supplémentaires, en passage aux conséquences *déductives* (ce qui suppose des opérateurs de formation, et des règles d'inférence).

Ceci justifie l'appellation de *système déductif* proposée pour un ensemble d'«énoncés» égal à celui de ses conséquences. Pour les ensembles d'«énoncés» sont proposées des notions abstraites d'«équivalence logique» (identité des conséquences, mais, attention ! il s'agit de paires d'ensembles d'énoncés), de compatibilité (existence d'un «énoncé» hors les conséquences), de complétude (absence d'extension compatible non logiquement équivalente), d'indépendance (absence de partie propre logiquement équivalente), et d'axiomatisabilité finie (existence d'une partie finie logiquement équivalente). TARSKI énonce et prouve le théorème de LINDENBAUM disant que tout ensemble d'«énoncés» compatible admet au moins une extension compatible et complète. On comprendra, *plus tard*, que *«si le calcul des propositions est présupposé»* (stabilité par formation de négations et d'implications, répondant à des axiomes adéquats), ce théorème revient au théorème sur les idéaux maximaux dans une algèbre de Boole; en 1930, Tarski annonce que, sous les présupposés ci-dessus, tout ensemble d'«énoncés» [dénombrable, là !] est celui des conséquences d'un ensemble indépendant [et, sans doute, encore sous l'hypothèse tacite qu'il s'agit d'axiomes *spécifiques*]. En 1933 [$Trk9, 10$], il restreindra les notions ci-dessus aux systèmes déductifs, et, au prix d'une modification mineure des présupposés ci-dessus, il appliquera sa théorie aux *formules*, qualifiées de *fonctions propositionnelles*, savoir, de ce que les formules représentent, au lieu d'énoncés: voilà pour la raison pour laquelle j'ai mis des guillemets aux «énoncés» dont il est question depuis un moment.

Des achèvements et questionnements de l'école de Hilbert, de 1922 à 1930
Depuis longtemps, le plan et la méthode de travail de Hilbert visaient à traiter d'abord des cas simples: présence du seul successeur, formules ouvertes, ... et à réintroduire pas à pas les notions, les formules, ... laissées de côté, fût-ce au prix d'avoir à reprendre à chaque fois les preuves.

Celles-ci s'effectuaient en soumettant à des transformations formelles les démonstrations formalisées, et faisaient usage de vérifications, effectives ou supposées telles, s'effectuant par des calculs, effectifs ou supposés tels. On en trouve de multiples variétés dans les deux volumes des *Grundlagen der Mathematik* [HbB] de Hilbert et Bernays, parus en 1934 et 1939.

Les fonctions définies par des applications répétées de substitutions d'expressions à des variables (du schéma de composition, si l'on préfère, ou de spécialisations des variables en les expressions substituées) et de schémas de récurrence répondaient bien à la condition d'être effectivement calculables selon un procédé uniforme, indépendant des valeurs affectées aux variables, et ne requérant qu'un jeu fini de calculs. Hilbert pensait même que le domaine des fonctions définissables par applications répétées de substitutions d'expressions préalablement obtenues à des variables et de schémas de récurrence de plus en plus compliqués, recouvre la totalité du domaine des fonctions à arguments et valeurs nombres entiers naturels (sinon, se pose la question de la délimitation du concept de fonction calculable). Cependant, de toutes façons, la question se posait, de prouver que l'adjonction de schémas de récurrence de complications croissantes permettait bien de définir des fonctions qui n'étaient pas définissables sans eux. Skolem avait montré, dans son mémoire de 1923 [$Skm5$], que des schémas de récurrence en apparence bien plus compliqués que le bon vieux schéma de récurrence de Dedekind [$Ded1$] (notre schéma de récurrence primitive), par exemple le schéma de récurrence sur le «cours des valeurs», pouvaient se ramener au précédent, et ne permettaient donc pas de définir des fonctions que celui-ci ne permettait pas de définir.

Aux alentours du début de 1925, presque simultanément, et indépendamment (cf. [ClM]), Gabriel Sudan [Sud] et Wilhelm Ackermann [$Ack2$] découvrent un schéma de récurrence double, non réductible au schéma de récurrence primitive, et une fonction définissable selon ce nouveau schéma, mais non récursive primitive, comme le montre, par exemple, la preuve d'ACKERMANN, parue en 1928. Cette fonction est un peu différente de celle qu'on appelle aujourd'hui de son nom, mais le principe de génération en est le même; et son procédé de détermination à partir d'une suite de fonctions récursives primitives se généralise d'ailleurs visiblement à toute classe, dont elle permet de sortir, de fonctions définissables par applications répétées de spécialisations de variables et de schémas de récurrence donnés.

Tout cela fournit aux intuitions de HILBERT un appui qui lui paraît assez solide pour le décider à présenter, en juin 1925, à une journée d'hommage à WEIERSTRASS, son plan d'attaque d'une preuve de l'hypothèse du continu. De son adresse [*Hbt*11], intitulée *Über das Unendliche*, les auditeurs sortirent, selon le témoignage de ZERMELO (cf. [*PLv*]), sans l'avoir comprise. Il vaut la peine, cependant, d'examiner ce texte, qui éclaire les motifs d'importants travaux ultérieurs. HILBERT y introduit une hiérarchie de types (de fonctionnelles à un argument), issue d'un type initial dit de niveau 0 — et composé, là, des nombres entiers naturels — par applications répétées (transfiniment) de deux processus: le processus de passage de la donnée d'une suite infinie (ordinaire) de types de fonctionnelles à même type de valeurs, au type des fonctionnelles définies sur les suites infinies de fonctionnelles dont la suite des types reproduit la suite donnée, et à valeurs dans le type commun à leurs valeurs — le niveau du type ainsi formé est la limite ordinale des niveaux de la suite de types donnée; et le processus ordinaire de la formation du type des fonctionnelles définies sur les fonctionnelles d'un type donné, et à valeurs dans un autre, de niveau dépassant de 1 le maximum de leurs niveaux.

HILBERT remarque alors ceci: premièrement, il correspond ainsi, à chaque ordinal dénombrable [car les limites prises sont celles de suites ordinaires], des types dont le niveau est égal à cet ordinal; deuxièmement (ce qui demande un peu de technique, dont il esquisse le cheminement), grâce à l'usage de ces fonctionnelles, tous les schémas de récurrence utilisés à la définition de fonctions à arguments et valeurs nombres entiers naturels peuvent s'obtenir, par spécialisation de variables en expressions préalablement obtenues, et dépendant de telles fonctionnelles, à partir d'un unique schéma de schémas d'itération [un par système de types attribués aux variables] — c'est le cas, par exemple, du schéma de récurrence double, non réductible au schéma de récurrence primitive, et qui sert à la définition de la fonction d'ACKERMANN ([*Ack*2]); troisièmement, les fonctions (à arguments et valeurs nombres entiers naturels) définissables par applications répétées de spécialisations de variables et de schémas de récurrence, se rangent en classes, selon le plus petit des niveaux des types des variables qu'il faut utiliser dans les applications du schéma d'itération unique pour en obtenir une définition; quatrièmement, si l'on borne strictement [nécessairement par un ordinal dénombrable] les niveaux des types des expressions ainsi utilisées dans les définitions de fonctions, non seulement on en obtient au plus une infinité dénombrable, mais encore on peut, en en formalisant le processus de définition par applications répétées de spécialisations de variables et de schémas de récurrence, en obtenir une énumération — c'est-à-dire une définition uniforme, chacune de ces fonctions étant obtenue par fixation, à une valeur convenable, d'un paramètre entier naturel dans l'expression $\varphi(a, n)$ d'une fonction de deux variables, elle aussi définissable par applications répétées de spécialisations de variables et de schémas de récurrence; et cinquièmement, en appliquant le procédé de la diagonale à cette expression, on obtient toujours une expression $\varphi(a, a) + 1$ qui représente une fonction échappant à l'énumération précédente, et requérant donc, pour quelque définition que l'on puisse en donner par applications répétées de spécialisations de variables et de schémas de récurrences issus du schéma d'itération, l'intervention, dans celui-ci, d'une variable d'un type dont le niveau atteint la borne fixée.

Pariant alors: premièrement, qu'il y ait, pour toute fonction à arguments et valeurs nombres entiers naturels, un schéma de définition par récurrence d'un niveau approprié, grâce auquel on puisse définir cette fonction par applications répétées de spécialisations de variables et de schémas de récurrence; deuxièmement, que l'usage de récurrences transfinies en vue de définir les niveaux de types eux-mêmes, c'est-à-dire, les ordinaux dénombrables, n'est pas indispensable, en d'autres termes, que tous ces ordinaux peuvent être définis en usant de récurrences ordinaires (sur les nombres entiers naturels), HILBERT en conclut que les $2^{\aleph_0}$ fonctions se répartissent en $\aleph_1$ classes dénombrables et dont il indique un bon ordre canonique associé à celui des processus de définition des fonctions qui les composent.

En ces temps-là, HERBRAND était venu à Göttingen, pour la préparation de sa thèse [*Hbd*1], qu'il acheva en 1929, et soutint en 1930. Il y donne des versions finitistes, savoir, répondant aux critères de la «finiter Logik» de HILBERT, du théorème de LÖWENHEIM et du théorème de complétude de GÖDEL, que BERNAYS va incorporer au second volume des *Grundlagen der Mathematik* [*HbB*], paru en 1939. HERBRAND utilise pour cela des lemmes sur les fonctions calculables (il s'agit pour lui d'une notion intuitive, dont il n'a pas de délimitation mathématique précise), et un théorème qui porte son nom: ce théorème décrit un procédé algorithmique, qui, appliqué à une formule du calcul des prédicats du premier ordre, permet, une fois cette formule écrite sous forme normale prénexe de SKOLEM, de construire une suite infinie de disjonctions du calcul des propositions, telle que la formule initiale est démontrable dans le calcul des prédicats, si et seulement si l'une de ces disjonctions est démontrable dans le calcul des propositions. La «méthode de résolution» utilisée en informatique de nos jours, dans le moteur d'inférence du langage PROLOG, se fonde sur ce théorème.

HILBERT et son entourage se croyaient à ce moment-là près de réussir: si ACKERMANN lui-même, au moment de publier, en 1924 [*Ack*1], une preuve de compatibilité de l'arithmétique du premier ordre, y avait décelé une faiblesse, dont la rectification réduisait à un tout petit peu plus qu'à l'arithmétique récursive primitive la non-contradiction établie; et si, après avoir critiqué d'autres défauts de cette preuve, VON NEUMANN en avait produit une autre [*vNm*2], mais sans améliorer le résultat, ACKERMANN venait d'écrire à BERNAYS pour lui indiquer encore une autre preuve, qui cette fois avait été jugée bonne. En 1928, dans une adresse sur les *Probleme der Grundlegung der Mathematik* [*Hbt*12], HILBERT avait annoncé au Congrès international des mathématiciens de Bologne qu'ACKERMANN et VON NEUMANN avaient mené à bien une preuve finitiste de la compatiblité de l'arithmétique, et énuméré quatre problèmes à attaquer: fournir une preuve de la compatibilité de l'analyse — c'est-à-dire, du calcul fonctionnel du second ordre; étendre une telle preuve à des calculs fonctionnels d'ordres plus élevés; fournir des preuves de complétude déductive pour la théorie des nombres et pour l'analyse; montrer la complétude du système des règles de la logique sous leur forme générale. Ce problème de complétude sémantique, précisé dans les *Grundzüge*, GÖDEL [*Gdl*1] venait de le résoudre pour le calcul fonctionnel du premier ordre.

C'est alors qu'intervinrent ses théorèmes d'incomplétude [*Gdl*2, 3].

Les premières maturations de l'ère de la connaissance limitée

Du «séisme» de 1930 et de ses suites durant les vingt années suivantes.

Des théorèmes d'incomplétude de GÖDEL, et du théorème de non-définissabilité arithmétique des énoncés vrais dans leur application aux nombres entiers naturels, de TARSKI

Durant l'été 1930, GÖDEL prouve [*Gdl*2] que, dans tout système formel répondant à des conditions minimales de compatibilité et dont les axiomes sont ceux de PEANO, même renforcés à l'aide des axiomes de l'un quelconque des systèmes formels antérieurement proposés pour les mathématiques classiques, celui des *Principia Mathematica* ou ceux de ZERMELO-FRAENKEL ou de VON NEUMANN, il y a des énoncés du langage de l'arithmétique, de la forme $(Ex)F(x)$, intuitivement vrais, tels que $F(x)$ est décidable, alors que ni $(Ex)F(x)$, ni sa négation n'y sont formellement démontrables: c'est là ce que HILBERT et BERNAYS, dans le second volume de leurs *Grundlagen der Mathematik* [*HbB*], paru en 1939, appelleront le premier théorème de «non-déductibilité», et que l'on appelle le premier théorème *d'incomplétude*, de GÖDEL [dans la forme que l'on donne *aujourd'hui* à l'énoncé de ce théorème, les «conditions minimales de compatibilité» ci-dessus se réduisent à la non-contradiction pure et simple, hypothèse sous laquelle il a été montré en 1936 [*Rsr*2], par John Barkley ROSSER, que l'on peut établir le théorème, en modifiant l'énoncé $F(x)$ et la preuve; dans la version produite à l'origine par GÖDEL, il s'agissait d'ω-compatibilité, c'est-à-dire de ce que, pour toute formule φ, l'on ne puisse prouver à la fois $\varphi(1)$, $\varphi(2)$, ... $\varphi(n)$, ... ad infinitum, et $(Ex)\text{NON}\varphi(x)$].

Après avoir montré, quoiqu'au prix d'une ultime étape non finitiste, la complétude sémantique du calcul des prédicats du premier ordre [*Gdl*1], GÖDEL s'était attaqué (cf. [*Wng*3]), après l'annonce [*Hbt*12], par HILBERT, à Bologne, d'une preuve de non-contradiction de l'arithmétique, et suivant les requêtes de la même communication, à la recherche d'une preuve de non-contradiction de l'analyse.

Mais au lieu d'envisager une preuve finitiste opérant directement sur les démonstrations formelles de l'analyse, son idée était de décomposer la preuve en deux étapes: une fois l'arithmétique «sécurisée» par une preuve de non-contradiction finitiste, employer toutes les ressources des propriétés des nombres entiers naturels — c'est-à-dire, des propositions vraies dans leur application aux nombres entiers naturels — pour fournir une preuve de non-contradiction de l'analyse.

Faute de réponse, d'ailleurs requise à Bologne, à la question, pendante à l'époque, de savoir s'il existe une axiomatique formalisée sémantiquement complète de théorie des nombres du premier ordre, GÖDEL fut amené à rechercher une définition de la notion d'«énoncé du premier ordre vrai dans son application aux nombres entiers naturels» — et à s'apercevoir que, si une telle définition est possible en termes arithmétiques du premier ordre, et si on la formalise, le vieux raisonnement du paradoxe du Menteur peut se transposer en arithmétique. Or, la complétude d'une telle axiomatique entraînerait la coïncidence des notions d'«énoncé du premier ordre vrai dans son application aux nombres entiers naturels» et de théorème formel, et *celle-ci se définit en termes arithmétiques du premier ordre*; en effet, formules, démonstrations formelles, ... peuvent se «lire», en imaginant que les variables représentent l'écriture de leur rang alphabétique, comme écritures de nombres dans un système de base convenable; on se rend tout de suite compte de ce que leurs définitions sont récursives (primitives); et GÖDEL [*Gdl*3] montre — ce qui est déjà une contribution substantielle — que des définitions par récurrence, comme les leurs, se ramènent à des définitions explicites. Aucune axiomatique formalisée de théorie des nombres du premier ordre ne saurait donc être complète sans être contradictoire. Il restait à trouver un énoncé *indécidable,* c'est-à-dire, tel que ni lui, ni sa négation, n'admette de démonstration formelle. Et c'est en suivant l'idée naturelle de reproduire le raisonnement du paradoxe du Menteur en substituant la notion de démontrabilité formelle à celle de vérité que GÖDEL, après avoir résolu quelques difficultés techniques, est finalement parvenu à son théorème.

A la fin d'août 1930, il s'en ouvre à son entourage viennois (cf. [*Dws*]). Ses interlocuteurs d'alors ne semblent pas en avoir bien perçu le ressort de la preuve, ni la portée; mais GÖDEL est invité à présenter une communication sur ce travail, une quinzaine de jours plus tard, à la Deuxième Conférence sur l'Epistémologie des Sciences Exactes, qui se tient à Königsberg. De ses auditeurs, le seul à réagir vivement est VON NEUMANN, qui va, dans les deux mois qui suivent, en tirer tout seul le corollaire que GÖDEL en aura déjà tiré lui-même à ce moment-là, et que l'on appelle le deuxième théorème *d'incomplétude* de GÖDEL: aucune théorie formelle, répondant aux hypothèses du théorème précédent, ne peut admettre de preuve formelle d'un énoncé, quel qu'il soit, exprimant la non-contradiction de cette théorie.

Ces deux résultats sont publiés en janvier 1931 [*Gdl*3]; là, leurs preuves reposent sur un *codage*, en nombres entiers naturels (ou aussi, fugitivement, sur la suggestion d'une «lecture», comme nombres entiers naturels), des variables, termes, formules, démonstrations formelles, ..., et sur un développement théorique concernant les fonctions et prédicats que GÖDEL appelle récursifs, et que l'on appelle aujourd'hui *récursifs primitifs* (et dont font partie, selon cette «lecture», variables, termes, formules, ..., et surtout la relation d'un énoncé à une démonstration formelle de celui-ci). Ce développement théorique est le premier de son genre (le mémoire de SKOLEM [*Skm*5], paru en 1923, se bornait à établir une liste de fonctions et de prédicats définissables à l'aide de récurrences sur des formules sans variable liée). En raison du climat intellectuel de l'époque, GÖDEL n'y fait aucune allusion à la conclusion qui l'a mis sur la voie et qui est connue sous le nom de «théorème de TARSKI», pour avoir été publiée [*Trk*9] par celui-ci en 1933 (en polonais) et 1936 (en allemand), sa dette à GÖDEL dûment signalée: l'ensemble (des codes) des énoncés d'arithmétique vrais de l'ensemble des nombres entiers naturels n'est l'extension d'aucune propriété s'exprimant dans le langage de l'arithmétique.

Sur les réactions aux découvertes de GÖDEL

De nos jours, on n'a pas encore fini de prendre la mesure de l'impact des théorèmes d'incomplétude de GÖDEL sur les mathématiques et sur la philosophie. Jusqu'aux alentours de 1950, pas mal de philosophes, de mathématiciens, voire même de logiciens, n'en comprennent pas la totalité des ressorts ou des nuances de l'énoncé. Il en résulte, jusqu'au-delà de la période que nous avons en vue, des polémiques, fondées sur des erreurs de compréhension et conduisant à des rectifications et à des mises au point; le seul intérêt, du point de vue auquel on se place ici, de ce type de polémiques, réside dans leur existence. Mais, parmi les chercheurs alors activement attelés au développement de la logique, ces difficultés s'évanouirent très vite; pour une part, certes, en raison du soin apporté par GÖDEL à la rédaction de son mémoire; mais comment les réactions immédiatement totalement compréhensives et enthousiastes de mathématiciens aussi influents que VON NEUMANN et TARSKI (cf. [*Trk*34], pp. 247 et 279) pourraient-elles ne pas y avoir été aussi pour beaucoup? — et la découverte de GÖDEL fut pratiquement immédiatement acceptée et intégrée au savoir.

L'effet produit sur les esprits a été comparé par Karl POPPER [*Ppr*] à celui d'un tremblement de terre. C'est que cette découverte remettait en cause les *perspectives* de tout le travail des logiciens, non seulement depuis le programme de HILBERT, auquel elle opposait ce que GÖDEL appelait «eine Schwierigkeit», non seulement depuis l'entreprise de FREGE, mais depuis l'oeuvre de LEIBNIZ lui-même, qui rêvait de pouvoir résoudre paisiblement, après formalisation, par des raisonnements formels, toutes les controverses.

Que vont faire les logiciens mathématiciens, face à cette situation? Désormais dotés d'instruments, les théories axiomatiques formalisées, sur lesquels effectuer des travaux mathématiques, auxquels ils viennent de se mettre, et qui s'avèrent intéressants en eux-mêmes, même si les mathématiques appropriées diffèrent passablement dans leur style de celles des analystes et des géomètres, ils vont s'éloigner des préoccupations philosophiques — il est d'ailleurs dans l'esprit du temps d'éluder les prises de position métaphysiques — et, tout simplement, tout en faisant le bilan des achèvements de la décennie passée, continuer l'exploration des domaines qui viennent d'être ouverts, et principalement, de:

- celui de la décidabilité d'énoncés, dont les premiers résultats d'indécidabilité rehaussent l'intérêt de parvenir à cerner les limites;
- celui de nouveaux principes d'axiomatisation des théories déductives, lié au développement recherché de la théorie de la preuve;
- celui des notions et problèmes de théorie des ensembles, et plus particulièrement de ceux qui touchent à l'indépendance de l'axiome du choix, à l'hypothèse du continu, et à la définition de mesures (sur des ensembles) en cause en physique mathématique;
- celui de la théorisation des rapports entre dénotant et dénoté;
- celui des formalisations plus affinées de la logique combinatoire;
- celui de la théorie des opérations définies de proche en proche, sur les nombres entiers naturels, à partir des opérations usuelles de l'arithmétique, par applications répétées de spécialisations de variables et de schémas de récurrence de complexités croissantes, et qui, en 1931, ne se confond pas encore avec cet autre:
- celui de la recherche d'une définition mathématique précise des concepts de fonction calculable, a priori plus large que le précédent, et de moyens de preuve finitistes, laissant subsister l'espoir, explicitement non exclu par GÖDEL [*Gdl*3], qu'il existe de tels moyens qui aient échappé aux formalisations des mathématiques connues en 1931;
- pour tenter de concrétiser cet espoir, celui de logiques se départant de façons diverses de la logique sous-jacente à ces formalisations, qu'il s'agisse de logiques classiques enrichies, par exemple de nouvelles règles de déduction, ou de logiques non classiques, multivalentes, modales, intuitionniste, ..., les tenants de cette dernière proclamant, d'ailleurs, les résultats d'incomplétude conformes à leurs conceptions;
- celui, enfin, des rapports de ces logiques entre elles, et avec les mathématiques, et, plus généralement, des rapports entre théories formalisées, avec un retour à la recherche de preuves de compatibilité *relative*, où la non-contradiction d'un système est établie à l'aide d'un autre, supposé non-contradictoire, telle celle de la géométrie relativement à la théorie des nombres réels, que l'on trouve dans les *Grundlagen der Geometrie*.

Au chapitre des bilans, HEYTING décrit, dans un livre [*Htg*3] intitulé *Mathematische Grundlagenforschung. Intuitionismus. Beweistheorie*, paru en 1934, les conceptions et les résultats de l'école intuitionniste; dans le millier de pages de leurs *Grundlagen der Mathematik* [*HbB*], le premier traité développé consacré à la théorie de la preuve, dont le premier volume paraît en 1934 et le second en 1939, HILBERT et BERNAYS rapportent le meilleur des contributions de l'école de

Göttingen à cette théorie — non sans tenir compte, surtout dans le second volume, de certains des apports intervenus entre temps.

Sur les tendances générales de l'évolution de la logique, de 1930 à 1950
Désormais, aucun des divers courants de philosophie des mathématiques pratiqués par les mathématiciens et qui orientaient les recherches avant 1930 n'est plus en mesure de justifier son point de vue par des résultats exempts de grosses difficultés: du fait des théorèmes d'incomplétude, le programme de HILBERT est en échec, sauf à bénéficier de fort nébuleuses possibilités de sauvetage, et bien qu'il en subsiste de passionnantes questions de mathématiques à étudier — mais justement, en restant dans le champ mathématique; les mathématiques intuitionnistes s'avèrent trop compliquées, et cherchant par trop à se calquer sur les mathématiques classiques, pour être vraiment convaincantes [il faudra attendre l'ère de l'informatique, et celle de la théorie des catégories, pour que l'on puisse, avec quelque chance de recueillir quelque écho favorable au-delà du petit monde des logiciens, plaider, en faveur de leur admission au sein des mathématiques, à titre de chapitre digne d'intérêt, comme Gehrard GENTZEN [*Gtz*5]le fait dès 1938; mais en 1950, l'informatique en sera encore à se débattre avec les difficultés de fonctionnement de ses prototypes ...]; et on voit de moins en moins comment ramener les mathématiques à la logique — ce dernier programme dégénère en celui de ramener les mathématiques à la théorie des ensembles, et en 1950, il n'aura pas encore fini de susciter, chez les mathématiciens, le ralliement quasi universel dont il profite de nos jours.

Ainsi, l'état d'esprit dans lequel les logiciens développent et défendent leurs conceptions durant les deux dernières décennies de la première moitié du vingtième siècle diffère beaucoup de ce qu'il était auparavant: les positions philosophiques que certains affichent encore sont moins tranchées, à très peu d'exceptions près, et elles ne les amènent plus, désormais, à opposer d'obstacle absolu à la prise en compte de travaux mathématiques sous-tendus par une autre position, d'autant plus que, formalisées, au moins pour une grande part, la logique et les mathématiques intuitionnistes deviennent aussi perméables aux logiciens classiques que la logique et les mathématiques classiques l'étaient auparavant aux intuitionnistes.

Cette attitude est théorisée sur le plan philosophique par CARNAP, qui énonce en 1934 ([*Cnp*2]) un *principe de tolérance* selon lequel, en logique, il n'y a pas de morale: le logicien serait libre de construire sa logique, sur sa propre forme de langage, comme il le veut; tout ce qu'on lui demande, s'il veut en discuter, est d'énoncer ses méthodes clairement, et de donner des règles de syntaxe et non des arguments philosophiques (FREGE, dans l'introduction des *Grundgesetze* [*Frg*2], notait déjà qu'il allait plus loin qu'EUCLIDE, en spécifiant les règles d'inférence qu'il *allait* suivre). Les commentaires que CARNAP en fait montrent qu'il vise les interdictions, qu'il voit remplacées par des différenciations selon les définitions — de façon à faire avec les différentes logiques comme l'on fait en géométrie avec les différentes géométries. Bien entendu, cette attitude ne se répandra que peu à peu, et il restera toujours des gens qui creuseront le sillon de philosophies déterminées. Mais s'il subsiste des polémiques, elles seront désormais souriantes, et ne constitueront plus le moteur de travaux mathématiques.

Ainsi, le développement de la logique obéit après 1930 à d'autres ressorts. La *théorie de la preuve* va connaître, jusqu'à la seconde guerre mondiale, des développements importants, dont celle-ci freinera l'élan, appelé à reprendre seulement après 1950. Il va s'organiser une *théorie de la récursivité*, qui sera d'abord une théorie des fonctions récursives, mais qui, par la suite, venant à englober les théories de la calculabilité et de la décidabilité, se développera très vite bien au-delà, interagissant peu à peu avec toutes les branches de la logique, jusqu'à tendre à occuper, au sein de celle-ci, une place dominante. Dans le cadre d'une branche naissante vers 1930, la *sémantique logique*, ainsi dénommée sous l'influence d'interactions avec la linguistique, et qui finira par se diluer dans la théorie des modèles au cours du troisième quart du siècle, le contenu de la notion de *modèle*, abstrait de diverses notions d'interprétation, et de la prise en compte de la notion de structure algébrique, s'élabore et se structure peu à peu; vers la fin de la première moitié du vingtième siècle, elle achève de se constituer telle qu'elle est de nos jours. Au même moment apparaissent les premiers éléments de la *théorie des modèles*, au sens entendu dans l'Introduction de ce travail. En rapport étroit, dès les travaux de Gödel et de Tarski des alentours de 1930, avec la théorie de la récursivité et avec la sémantique logique, les problèmes de *définissabilité* vont entrer en jeu dans des mémoires importants. Diverses contributions vont être apportées à l'affinement de l'axiomatique de la *théorie des ensembles*; les statuts de l'axiome de choix et de l'hypothèse du continu vont être mieux déterminés, du fait de la preuve de leur compatibilité relative à celle des autres axiomes, apportée par Gödel en 1938 [*Gdl*13], et ce succès sera suivi d'une période de quasi-stagnation de vingt-cinq années; d'un autre côté, les nouveautés les plus importantes de la théorie durant les vingt années qui séparent 1930 de 1950 dernières décennies de la première moitié du vingtième siècle viennent d'un retour à des travaux sur la théorie de la mesure, induit par les interactions avec la physique théorique; ces travaux auront même une incidence sur l'axiomatique, dont les effets ne se feront sentir que durant la seconde moitié du vingtième siècle. Enfin, parallèlement aux progrès accomplis dans l'étude des calculs des propositions *non-classiques*, des progrès seront faits dans l'étude des modèles qu'en offrent les mathématiques classiques; aux mathématiques *intuitionnistes*, seront lentement adjoints de nouveaux chapitres, inspirés de chapitres correspondants des mathématiques classiques; et la logique intuitionniste se verra proposer de nouvelles interprétations, qui connaîtront des fortunes diverses.

Durant ces vingt années, en dehors du cercle étroit des spécialistes, le rayonnement de ces travaux est obéré. Les philosophes ont du mal à suivre leur complexité mathématique croissante, source de fréquentes méprises. La différence de style évoquée plus haut en écarte beaucoup de mathématiciens; la plupart d'entre eux sont heurtés par l'insertion, dans les mathématiques, des notations mathématiques. L'absence de livres réunissant et mettant en ordre les connaissances acquises durant ces deux décennies, qui paraîtront *tous* après le milieu du siècle, n'arrange rien. Mais ce qui joue surtout, ce sont les limites des rapports apparents des résultats obtenus avec le reste des mathématiques, laissant à quelques rares esprits informés la perception de l'applicabilité potentielle des travaux des logiciens; celle-ci n'est pas encore en évidence, et les seuls résultats de logique qui aient fait sensation parmi les mathématiciens, et que l'ensemble d'entre eux connaissent, sont les théorèmes d'incomplétude de GÖDEL [*Gdl*3], qui font intervenir des énoncés indécidables, mais paraissant tout à fait construits ad hoc, hors des préoccupations mathématiques ordinaires, et son théorème sur la compatibilité de l'axiome de choix et de l'hypothèse du continu.

Ce relatif isolement des logiciens, rendant plus difficile, en tout cas aux Etats-Unis, la publication de leurs travaux ([*Asp*]); l'accroissement de leur nombre, du fait du recrutement des diverses écoles; un besoin de solidarité, propre à l'époque; un début de rapprochement des points de vue, évoqué plus haut, font sans doute partie des raisons qui suscitent un développement institutionnel: 1935 voit se créer l'*Association for Symbolic Logic*, qui assure, à partir de 1936, la publication, dans des conditions souvent précaires, d'un journal consacré exclusivement à la logique, le *Journal of Symbolic Logic*, dans le premier volume duquel paraît, établie par CHURCH, une bibliographie, non loin d'être exhaustive, des publications touchant à la logique depuis LEIBNIZ.

Un effort de diffusion au niveau des classes terminales des Lycées ou des propédeutiques universitaires, dont le succès a été exceptionnel, a été entrepris par TARSKI, dont le manuel élémentaire en polonais *O logice Matematycznej i metodzie dedukcyjnej* [*Trk*14] de 1936 a été traduit et édité en allemand l'année d'après sous le titre *Einführung in die mathematische Logik und die Methodologie der Mathematik*, puis, après les premiers d'une série d'ajouts et de révisions, en anglais, en 1941, sous le titre *Introduction to logic and to the methodology of deductive sciences*. Des vingt-neuf éditions postérieures, partielles ou complètes, de celle-ci ou de traductions en une douzaine de langues, deux, dont une en russe, appartiennent à la décennie qui précède 1950.

D'un autre côté, de 1935 à 1950, conscience va être prise du rôle de la logique dans divers mécanismes naturels ou artificiels, et dont le développement va ouvrir à la logique d'autres domaines d'application. Le recours à la logique dans l'ingénierie des circuits à commutateurs, préconisé dès 1910 par le physicien russe EHRÉNFÉST [*Erf*], fait l'objet de mémoires de V.I. CHESTAKOV, en 1935, en Union Soviétique — qui cependant n'est publié qu'en 1941 (cf. [*Ank*]) — de Akira NAKASIMA et Masao HANZAWA [*NkH*], en 1936, au Japon, et de Claude SHANNON [*Shn*], en 1938, aux Etats-Unis, suivis, dans ces pays, par de nombreuses publications sur ce sujet. Le mémoire de Walter PITTS et Warren Sturgis MCCULLOCH, *A logical calculus of the ideas immanent in nervous activity* [*MCP*], de 1943, retient immédiatement l'attention de VON NEUMANN (cf. [*Gls*]). Dès 1928, l'astronome Leslie John COMRIE détourne, des usages statistiques et actuariels auxquels elles étaient dédiées, les machines électromécaniques à cartes perforées fonctionnant, depuis la fin du dix-neuvième siècle, selon le système HOLLERITH, pour les utiliser à des calculs de mécanique céleste (cf. [*Gls*]). En 1937, tandis que Howard Hathaway AIKEN met à l'étude ([*Aik*]) les modifications à apporter à l'architecture des machines électromécaniques pour en faire des calculateurs aptes aux calculs scientifiques, John Vincent ATANASOFF conçoit un calculateur électronique dédié à la résolution de grands sytèmes d'équations linéaires — et réalise un prototype, qui fonctionne quelques temps en 1940 ([*Gls*]). Conçus, aux Etats-Unis et en Allemagne, aux alentours de 1940, les prototypes d'ordinateurs électromécaniques à relais, mis en service aux alentours de 1944, sont presque aussitôt dépassés par les protopypes d'ordinateurs électroniques conçus entre temps et mille fois plus rapides à l'origine, l'ENIAC en service en 1946, rejoint en 1950 par l'EDVAC, d'architecture conçue par VON NEUMANN ([*vNm6*]; cf.[*BGN*]), et qui sert de modèle aux machines construites pendant les décennies qui suivent ([*Gls*]).

Du développement de la théorie de la démonstration, après 1930

Des calculs de déduction naturelle et des calculs de «séquences»

En 1933, à l'âge de 23 ans, Gerhard GENTZEN, élève et assistant de Hermann WEYL, soutient à Göttingen une thèse intitulée *Untersuchungen über das logische Schliessen* [*Gtz2*], et publiée en 1934 et 1935 à la *Mathematische Zeitschrift.* Il y introduit de nouvelles axiomatiques des logiques classique et intuitionniste du premier ordre, et établit les théorèmes fondamentaux de leur théorie de la preuve (incidemment, il semble bien que ce soit là que le symbole ∀ pour le quantificateur universel soit employé pour la première fois). Ces nouvelles axiomatiques sont de deux sortes: les «calculs de déduction naturelle» et les «calculs de séquences», termes qui vont être expliqués plus loin.

Destinés à produire des démonstrations formelles serrant au plus près les raisonnements effectivement pratiqués, ces systèmes ont en commun d'exprimer le sens des connecteurs et des quantificateurs par des règles de déduction plutôt que par des axiomes, et de ne comporter qu'un tout petit nombre (de schémas) d'axiomes. Ils présentent aussi la particularité que la différence entre leurs versions classiques et leurs versions intuitionnistes repose juste sur la présence éventuelle d'un axiome (le tiers exclu) ou sur celle d'une restriction sur le nombre de formules en cause dans l'un des composants des objets considérés.

Les systèmes de *déduction naturelle* formalisent, en la complétant de façon cohérente, une *pratique* d'écriture commentée, ligne par ligne, de déductions formelles, dont LEŚNIEWSKI employait dès 1916 des formes embryonnaires (cf. [*Lus*]), et dont on trouve la trace dans les premiers écrits de TARSKI [*Trk*1]. En gros, elle consiste à introduire de temps en temps, et pour un temps, des hypothèses, dont on peut inférer ensuite des conséquences, jusqu'à ce qu'au bout d'un certain temps, on s'en «décharge», par exemple en appliquant le théorème de la déduction. GENTZEN donne ainsi des règles pour introduire ou éliminer une conjonction, une disjonction, une implication, une négation ... En 1934 même est publiée par

JAŚKOWSKI [$Jkk1$] une autre version d'un système de «déduction naturelle», issue, indépendamment de GENTZEN, de suggestions émises par ŁUKASÍEWÍCZ en 1926. Cependant, pour la logique classique, les règles de GENTZEN relatives à la négation n'entrent pas dans un schéma général permettant de procéder constante logique par constante logique, soit qu'il faille traiter simultanément deux négations consécutives, soit qu'il faille introduire le seul tiers exclu comme schéma d'axiomes: GENTZEN [$Gtz2$] met ces inconvénients en balance avec l'avantage que représente un traitement formel des raisonnements plus proche de la pratique que le recours, à côté de la règle de généralisation, au seul *modus ponens*.

Dans la période qui nous occupe, les calculs de déduction naturelle ont plus été utilisés, par d'autres que GENTZEN, comme outil de travaux menés à d'autres fins, que développés ou étudiés pour eux-mêmes. Il en est allé autremment des calculs de séquences, en raison de l'usage que GENTZEN en a fait par la suite, et d'une parenté formelle entre les théorèmes les concernant et des propriétés analogues de déductions formelles établies par HERBRAND [$Hbd1$].

Sur le plan purement formel, une *«Sequenz»* est un couple de suites d'énoncés, composé d'un *antécédent* et d'un *succédent*, dont la vacuité éventuelle de l'un des deux, ou des deux, n'est pas exclue. La «signification attendue» d'une séquence est la vérité logique de l'implication, par la conjonction des énoncés de l'antécédent, de la disjonction des énoncés du succédent (si l'antécédent est vide, la vérité logique de la disjonction des énoncés du succédent; si le succédent est vide, la fausseté logique de la conjonction des énoncés de l'antécédent).

Les démonstrations formelles d'un calcul de séquences se composent de séquences, et procèdent par transformation de prémisses en conclusions selon des règles d'inférence, dont les unes sont des règles d'introduction et d'élimination de connecteurs et de quantificateurs, analogues à celles des calculs de déduction naturelle, et les autres, des règles dites «structurelles», autorisant l'élimination de répétitions, les permutations et les adjonctions de formules, effectuées sur un antécédent ou sur un succédent. Une règle dite de *coupure* autorise, lorsqu'un même énoncé apparaît dans l'antécédent d'une séquence et dans le succédent d'une autre, à conclure de ces prémisses à une séquence dont l'antécédent *fusionne* en une suite les *autres* énoncés entrant dans les antécédents des prémisses, et le succédent, les *autres* énoncés entrant dans les succédents des prémisses. Sont axiomes les séquences dont antécédent et succédent sont réduits à un même énoncé — et, en présence de l'égalité, les séquences d'antécédent vide et de succédent de la forme $a = a$. Enfin, en raison de la persistance de la disposition en colonnes, et de la présence de règles à deux prémisses, les démonstrations formelles sont disposées en arbres, dont la racine est la séquence à démontrer, et dont les feuilles sont des axiomes.

Le sens en lequel les calculs de séquences sont équivalents aux calculs fonctionnels de HILBERT et de HEYTING est le suivant: une séquence d'antécédent vide et de succédent réduit à un énoncé est formellement démontrable dans un calcul de séquences, si et seulement si cet énoncé est formellement démontrable dans le calcul fonctionnel qui lui correspond ([$Gtz2$]). Les avantages des calculs de séquences tiennent à ce qu'ils formalisent, implicitement, le comportement de la notion de conséquence logique: compte tenu du théorème de *complétude* de GÖDEL [$Gdl1$] et de ce qui a été dit plus haut, une séquence vaut si et seulement si la disjonction des énoncés de son succédent est conséquence logique de la conjonction des énoncés de son antécédent. Il en résulte que la recherche d'une démonstration formelle peut en principe procéder à l'envers, par application quasi mécanique des règles d'élimination, moyennant, cependant, des précautions à prendre avec les quantifications, qui peuvent entraîner l'introduction d'une suite indéfinie de noms d'individus. Il en découle des propriétés importantes: toute séquence démontrable possède une preuve où n'interviennent que des *sous-énoncés* des énoncés entrant dans cette séquence [on compte un énoncé $A(a)$, où a est une constante d'individu, comme sous-énoncé des énoncés $\forall x A(x)$ et $\exists x A(x)$]; pour les calculs des *propositions* classique et intuitionniste, pour l'un et l'autre desquels, par contre, la procédure se termine en un nombre fini d'étapes, elle fournit une méthode de décision; de plus, GENTZEN [$Gtz2$] montre que toute démonstration formelle peut être transformée, par application d'une transformation uniforme, algorithmique, en une démonstration sans coupure, et de plus, pour le calcul classique, organisée, le long de chaque branche, en un segment initial se terminant par une séquence sans quantificateur, suivi d'un segment d'introduction de quantificateurs et d'applications de règles structurelles.

Où gît donc, dans les axiomatiques, la différence entre calcul classique et calcul intuitionniste? *Uniquement* en ce que les seules séquences admises par le calcul intuitionniste sont celles dont le succédent ne comporte *pas plus d'un* énoncé — ce qui impose des restrictions sur deux des règles d'inférence.

Par la suite, d'autres calculs de séquences ont été introduits par Oiva KETONEN [Ktn] en 1944; peu avant 1950, les calculs de séquences sont devenus des objets d'étude pour CURRY (cf. [$Cur14$], p. 245), qui s'y est intéressé dès 1937 ([$Cur6$]) en vue de les appliquer à des recherches de logique combinatoire.

Des origines des conceptions de GENTZEN dans les travaux de théorie de la démonstration de Paul HERTZ

Depuis le début de la décennie précédente, à Göttingen, les preuves formelles linéaires étaient «résolues» en *Beweisfäden* remontant, de la formule finale, aux axiomes réellement utilisés pour la démontrer, en faisant apparaître, à chaque étape à franchir, comment on remonte des conclusions à leurs prémisses; tout cela, ne serait-ce, parfois, que pour ramener à ces axiomes les substitutions effectuées en chemin sur des variables. HILBERT et BERNAYS présentent, dans le premier volume des *Grundlagen der Mathematik* [HbB], les schémas arborescents qui en résultent. Et à Göttingen, dès les alentours de 1920, Paul HERTZ [$Htz1, 2, 3$] avait entrepris l'étude d'axiomatiques abstraites d'un «beliebige Satzsystem». Les énoncés composant un tel système étaient donnés de la forme $(a_1, \ldots, a_n) \to b$, pouvant se lire «wenn $(a_1, \ldots, a_n)$ zusammen ist, so ist b»; ils étaient abstraits en cela que les $a_1, \ldots, a_n, b$ susceptibles d'y figurer étaient des «éléments» dont la nature n'était pas précisée; et dans le notation ci-dessus, le «complexe» $(a_1, \ldots, a_n)$ placé devant la flèche s'appelait déja «l'antécédent», et l'élément b placé derrière, le «succédent» ([*Htz*1]). Les «éléments» intervenaient dans la donnée supposée de règles, selon lesquelles certains énoncés étaient dits résulter d'autres; en fait, dans ce que HERTZ a écrit entre 1922 et 1929, les «preuves» (abstraites) étaient données pour des enchaînements, arborescents sauf cas particuliers, de *Schlüsse* du type: des prémisses $(u_1, \ldots, u_n) \rightarrow a_1$, $(u_1, \ldots, u_n) \rightarrow a_2$, ..., $(u_1, \ldots, u_n) \rightarrow a_m$, ..., $(a_1, \ldots, a_m) \rightarrow b$, conclure à $(u_1, \ldots, u_n) \rightarrow b$ ([*Htz*2]), auxquels venaient s'ajouter, en 1929 ([*Htz*3]), des «inférences immédiates», autorisant de conclure, d'un énoncé, à un énoncé d'antécédent incluant celui de la prémisse. Il a traité, dès 1922 [*Htz*1], de propriétés générales des déductions formelles (bien qu'il les étudiât sous le nom de Beweisen), telles par exemple celles qui disent qu'un segment initial, ou une enchaînée, de déductions formelles, sont des déductions formelles. Il étudiait les systèmes d'énoncés stables par les inférences ci-dessus, un système $\mathfrak{T}$ étant dit «axiomatisé» par une de ses parties $\mathfrak{A}$ telle que tous les éléments de $\mathfrak{T}$ soient finaux dans des preuves de feuilles prises dans $\mathfrak{A}$; il cherchait à déterminer dans quelles conditions un système d'énoncés admet une axiomatisation indépendante, et pour ce faire, il effectuait sur les preuves des transformations les amenant à des formes normales ([*Htz*3]).

GENTZEN n'omet d'en citer aucun des travaux, lorsqu'en 1932 [*Gtz*1] il établit que dans de telles situations abstraites, il y a des systèmes infinis qui n'admettent pas d'axiomatisation indépendante, et caractérise un cas dans lequel cela n'arrive pas. A cette fin, il recourt, pour les preuves, à des formes normales simplifiées, grâce à une décomposition des «Schlüsse» de HERTZ en inférences immédiates et en «coupures»; il emprunte, à HERTZ [*Htz*3], une notion de «satisfaction»: un «complexe» d'éléments satisfait un énoncé donné, s'il ne contient pas l'antécédent de cet énoncé, ou s'il en contient le succédent et l'antécédent; et il introduit ([*Gtz*1]) une notion de *conséquence*, qu'il définit ainsi: un énoncé $\mathfrak{q}$ est une conséquence des énoncés $\mathfrak{p}_1, \ldots, \mathfrak{p}_\nu$, si tout complexe d'éléments satisfaisant $\mathfrak{p}_1, \ldots, \mathfrak{p}_\nu$ satisfait $\mathfrak{q}$. Cette dernière définition anticiperait la notion de conséquence logique introduite par TARSKI trois ans plus tard, si elle comportait des analyses de contructions d'énoncés moins frustres que celles de HERTZ, qui se réduisent à un emploi de l'implication et à un emploi implicite d'une conjonction à un nombre fini de termes; comparée au traitement effectué par TARSKI de la notion de conséquence en 1928 [*Trk*5], qui comporte de telles analyses touchant tous les connecteurs, elle n'apporte pas d'innovation. Mais il se pourrait que le rapprochement de l'idée de telles analyses avec cette conception de la notion de conséquence ait très vite conduit GENTZEN au calcul des séquences, qui en découle tout naturellement, quoique par des raisonnements non-finitistes, mais auxquels une bonne connaissance des démonstrations formelles permet de substituer des preuves finitistes.

Des preuves intuitionnistes de non-contradiction de l'arithmétique de Peano du premier ordre classique

Le plus grand succès de Gentzen va être de produire, en 1936 [*Gtz*3], une preuve intuitionniste de non-contradiction de l'arithmétique classique, qui ne repose nullement, à l'inverse de celle de Gödel en 1933 [*Gdl*8], sur une traduction explicite des formules, mais sur une extension, à un calcul de séquences pour l'arithmétique, du théorème de mise sous forme normale des démonstrations formelles; il en redonnera une version détaillée dans le menu en 1938 [*Gtz*6].

Il faut maintenant maîtriser l'intrusion, comme axiomes, de formules mathématiques sans constante logique et «vérifiables» (en un nombre fini de démarches, pour tout système de valeurs assignées aux variables libres), et celle des applications d'une règle d'inférence supplémentaire, qui est un schéma d'induction complète — des transformations préparatoires adéquates des démonstrations, cependant, ramènent l'usage de chacune de ces applications à l'obtention d'une propriété d'une suite de bâtons, donc à une suite finie de coupures. Analysant les imbrications complexes des transformations finitistes à effectuer sur une hypothétique démonstration formelle de la séquence, *contradictoire*, dont antécédent et succédent sont *vides*, jusqu'à ramener cette démonstration à une forme normale où il n'intervient plus que des inférences de formules vérifiables à formules vérifiables, Gentzen attache, à chaque démonstration hypothétique de cette séquence, un ordinal dénombrable, de telle façon que chaque étape de la réduction fasse nécessairement strictement diminuer l'ordinal associé, et que ce dernier tombe à zéro lorsque la démonstration à laquelle il serait associé serait normale, et donc ne puisse conduire à une séquence intenable, contrairement à l'hypothèse. Gentzen est en mesure de prouver qu'à celà suffisent les ordinaux qui n'atteignent pas un ordinal de la seconde classe des ordinaux de Cantor, l'ordinal ε_0, savoir, le premier ordinal ε tel que $\omega^\varepsilon = \varepsilon$, qui est un ordinal dont Brouwer a parfaitement admis le caractère constructible, et donc l'existence, et la légitimité de l'emploi, implicite dans le raisonnement précédent, du principe de récurrence transfinie appliqué à ses prédécesseurs.

En 1940, ce résultat de Gentzen a été redémontré par Ackermann [*Ack*5], selon une méthode essayée quinze années plus tôt sur la suggestion de Hilbert, et qui, en n'appliquant que de strictes méthodes finitistes, n'avait pas débouché. En 1943 [*Gtz*7], Gentzen a montré que les récurrences transfinies jusqu'à un ordinal inférieur à ε_0 peuvent se représenter dans l'arithmétique; ε_0 est donc le premier ordinal α possèdant la propriété que le principe de récurrence transfinie jusqu'à α permet de démontrer la non-contradiction de l'arithmétique.

De la théorie des ensembles, de 1930 à la veille de la fin de la première moitié du vingtième siècle

Des «grands cardinaux» durant la première moitié du vingtième siècle

Dès le début de leurs carrières, TARSKI et LINDENBAUM ont pris part aux travaux de théorie des ensembles, un des thèmes de recherche mathématique favori en Pologne à cette époque; je ne pourrai, de leurs contributions dans ce domaine, que citer les plus marquantes.

Or, l'école polonaise de mathématiques s'est occupé très tôt du problème de l'existence de «grands cardinaux». HAUSDORFF, qui avait introduit les cardinaux faiblement inaccessibles en 1908 [*Hdf*4], la mettait en doute en 1914 dans ses *Grundzüge der Mengenlehre* [*Hdf*6], en raison de leur taille «exorbitante». Kazimierz KURATOWSKI avait expliqué à la Société Polonaise de Mathématiques, en 1924 [*Kur*2], les raisons qui empêchent de prouver, dans la théorie des ensembles, l'existence d'un tel cardinal: à supposer que la théorie soit non contradictoire, et permette de prouver qu'il en existe un, les ensembles des rangs bornés par le plus petit d'entre eux constituent un domaine sur lequel la restriction de l'appartenance vérifie les axiomes de la théorie, et dans lequel, néanmoins, il ne peut y en avoir.

Vers 1930, TARSKI eut l'idée d'introduire la notion de cardinal *(fortement) inaccessible*: tel que l'ensemble des cardinaux strictement plus petits soit stable par produits indexés par un ensemble de cardinal strictement plus petit. En 1930, tandis que ZERMELO, dans un article intitulé *Grenzzahlen und Mengenbereiche* [*Zml*5], introduisait la notion équivalente de Grenzzahl (et le sigle ZF), SIERPINSKI et TARSKI [*SpT*] montrent que l'inaccessiblité forte implique la faible, et qu'elles sont équivalentes, sous l'hypothèse du continu généralisée.

Ce moment d'intense créativité, et son questionnement, suivent de peu la preuve de ce que, si l'hypothèse du continu est satisfaite, il n'existe sur le continu aucune mesure, même non invariante par translation, fournie en 1929 par BANACH et KURATOWSKI [*BaK*], qui s'étaient demandé si l'existence d'une partie de l'ensemble des nombres réels non mesurable pour la mesure de LEBESGUE, démontrée en 1905 par VITALI [*Vit*], n'était qu'une conséquence fâcheuse de l'invariance par translation. L'année suivante, BANACH [*Bnc*] posait la question de l'existence de cardinaux *mesurables*, c'est-à-dire dont l'ensemble des parties soit porteur d'une mesure, dénombrablement additive, à valeurs dans l'intervalle [0,1] ; Stanisław ULAM [*Ulm*] montrait aussitôt que si un tel cardinal n'est pas au plus de puissance du continu, les mesures qu'ils porte ne prennent que les valeurs 0 et 1; au même moment, ULAM et TARSKI (cf. [*Ulm*]) montraient qu'un cardinal porteur d'une telle mesure est inaccessible [on sait maintenant que la «taille» d'un tel cardinal est beaucoup, beaucoup plus grande que ces résultats ne l'indiquent].

Interrompus par la seconde guerre mondiale, les questionnements et les travaux concernant les «grands cardinaux» (inaccessibles, mesurables, ou d'autres encore, imaginés depuis) ne devaient être repris que lors du second demi-siècle, où l'usage des méthodes inventées par Paul COHEN [*Coh*1, 2] devait leur donner une impulsion considérable.

K. Kuratowski

Des progrès relatifs aux axiomatiques de la théorie des ensembles reposant sur la «limitation de taille», entre 1930 et 1950

L'année 1930 n'était pas terminée, quand GÖDEL se mit à réfléchir (cf. [*Wng*3]) au plan d'attaque du problème de l'hypothèse du continu, esquissé par HILBERT dans sa conférence *Über das Unendliche* [*Hbt*11].

Comme les notions d'ordinal et de fonction définis par applications répétées de spécialisations de variables et de schémas de récurrence n'apparaissaient pas à cette époque comme clairement délimitées, il essaya d'y substituer l'itération de la définissabilité *prédicative* au premier ordre en termes des notions primitives de la théorie des ensembles — ce qui impliquait de traiter de tous les ensembles ainsi définissables, et non uniquement des (graphes de) fonctions, et constituait un retour au procédé d'engendrement de la théorie des types de RUSSELL. La hiérarchie d'ensembles ainsi engendrée part de l'ensemble vide, et procède, d'une étape à la suivante, en adjoignant, aux ensembles préalablement obtenus, les ensembles définissables au premier ordre en termes d'appartenance et des ensembles *préalablement obtenus:* à ne considérer que le passage d'une étape à la suivante en lui-même, les ensembles introduits ont, en termes des précédents, des définitions prédicatives. Aux étapes de passage à la limite, la prédicativité est maintenue en formant la réunion des ensembles d'ensembles obtenus aux étapes antérieures, à la condition que l'ordinal limite, définissant l'étendue de la réunion, soit définissable au premier ordre en termes des ensembles préalablement obtenus.

Moyennant l'axiome de l'infini, c'est bien le cas du premier ordinal infini; cependant, il n'était même pas clair qu'il en allât de même de tous les ordinaux limites dénombrables; la délimitation de ceux qui répondaient à cette condition n'apparaissait pas plus sûre. Ceci conduisit GÖDEL à consentir d'abord à l'imprédicativité résidant dans l'admission de tous les ordinaux dénombrables, et pour en disposer comme *donnés*, à rechercher une preuve de compatibilité *relative*. Mais il fallait alors que les ensembles engendrés selon le processus indiqué satisfassent les axiomes de la théorie des ensembles, et comme ce n'est pas le cas si l'on se restreint aux ordinaux dénombrables, GÖDEL en vint à admettre tous les ordinaux.

Il a qualifié de *constructibles* les ensembles alors obtenus selon le processus indiqué, qui s'exprime au premier ordre, produisant une formule $L(x)$ qui les définit. GÖDEL [*Gdl*13, 14] étudiait alors ce qu'il appelait le *modèle* des ensembles constructibles en *relativisant* les formules à cette définition, c'est-à-dire, en traitant en *interprétation*, dans ce modèle, d'une formule $\varphi(x)$, la formule qui s'en déduit en y préfixant «$L(z)$ET» à tout champ d'un quantificateur «$\exists z$» y apparaissant, et «$L(z) \longrightarrow$» à tout champ d'un quantificateur «$\forall z$» y apparaissant; la méthode de vérification de la *satisfaction*, par le modèle, d'une formule, consistait à produire (le thème d') une démonstration formelle, dans la théorie des ensembles formalisée, de la relativisée de cette formule — dans sa thèse [*Mtw*2], en 1938, MOSTOWSKI attribue à TARSKI, sans doute à l'époque de son séminaire à Varsovie, vers 1926, la paternité de cette méthode de «relativisation des quantificateurs», que TARSKI développera en 1948 en une notion plus large d'«interprétabilité» d'une théorie dans une autre. En 1935, disent encore MOSTOWSKI ([*Mtw*4], p. 203) et GÖDEL lui-même (cf. [*Wng*3]), GÖDEL avait vérifié que les constructibles constituaient bien un modèle de la théorie des ensembles, et de l'axiome de choix — car, à chaque étape de l'engendrement des ensembles constructibles, un bon ordre préalablement obtenu sur les ensembles préalablement obtenus induit, comme l'avait suggéré HILBERT [*Hbt*11], un bon ordre sur les définition des ensembles nouvellement obtenus, et donc sur les ensembles dont ces ensembles sont des éléments. La preuve de la satisfaction de l'hypothèse du continu généralisée, annoncée en 1938 [*Gdl*13], devait encore prendre trois années, et nécessiter l'usage du théorème de

LÖWENHEIM-SKOLEM et l'établissement d'un «lemme de contraction» pour montrer que les ensembles constructibles d'ordinaux finis sont obtenus lors d'une étape de rang dénombrable (éventuellement transfini), et l'analogue pour les minorants stricts des autres cardinaux infinis.

Ainsi, les preuves formelles de la théorie des ensembles avec axiome de choix et hypothèse du continu généralisée sont des tronçons de preuves formelles de la théorie des ensembles sans ces hypothèses supplémentaires, et si la seconde est compatible, ainsi en va-t-il de la première: tel est le théorème de *compatibilité relative* de l'axiome du choix et de l'hypothèse du continu généralisée de GÖDEL [*Gdl*13, 14].

De nos jours, on utilise à sa preuve le «lemme de contraction de MOSTOWSKI», qui est à la fois une simplification et une généralisation, introduites par MOSTOWSKI en 1949 [*Mtw*9], du lemme utilisé par GÖDEL; il dit que si un modèle est constitué d'un ensemble, muni d'une relation binaire s'exprimant par une formule, éventuellement compliquée, de la théorie des ensembles, et si ce modèle satisfait juste (les interprétations de) l'axiome de fondement [la relation est alors qualifiée de «bien fondée», depuis un article [*Zml*6] de ZERMELO, de 1935] et (de) l'axiome d'extension [deux ensembles qui ont les mêmes éléments sont égaux], ce modèle est isomorphe à un modèle constitué d'un ensemble, muni de l'appartenance *primitive* de la théorie des ensembles; si, de plus, ces modèles sont des modèles de la théorie des ensembles, le second en est un modèle «naturel», c'est à dire, un type de la théorie cumulative des types, ensemble des ensembles des rangs aux plus égaux à un rang donné.

Un exemple d'un tel isomorphisme s'obtient à partir du *modèle d'*ACKERMANN [*Ack*4], savoir, du modèle de la théorie des ensembles *sans axiome de l'infini* ayant les nombres entiers naturels pour individus, et dans lequel «u appartient à v» est interprété par «dans la décomposition de v en somme de puissances de 2, le coefficient du terme d'exposant u est égal à 1». La satisfaction, par ce modèle, d'une formule de la théorie des ensembles, se vérifie encore par la production d'une (description d') une preuve formelle, dans la théorie des nombres, de la formule de théorie des nombres en laquelle la formule de théorie des ensembles considérée se *traduit* selon cette interprétation: cette interprétation a servi, en 1937, à prouver la compatibilité de la théorie des ensembles sans axiome de l'infini, relativement à celle de l'arithmétique. Dans la théorie des ensembles (avec axiome de l'infini), se définissent la notion de nombre entier naturel, leurs opérations, et grâce à elles, le «modèle d'ACKERMANN» de cette théorie; celui-ci est isomorphe, dans cette théorie, au modèle des ensembles de rangs finis.

En 1938 aussi, TARSKI a établi [*Trk*19] que postuler l'existence de cardinaux inaccessibles au-delà de tout nombre ordinal donné, comme ZERMELO l'avait proposé en 1930 [*Zml*5], entraîne l'axiome de choix: la proposition équivalente, de postuler que tout ensemble est élément d'un domaine de cardinal inaccessible, revient à postuler que tout ensemble est élément d'un domaine admettant pour éléments toutes celles de ses parties auxquelles il n'est pas équipotent, et stable par passage d'un élément à une partie de cet élément, et par passage d'un élément à un ensemble de parties de cet élément; et cet axiome, lu sans assimiler «puissance» et «cardinal» (d'un ensemble bien ordonnable), entraîne les axiomes de la paire, de l'infini, des parties, et du choix (ce dernier, selon un raisonnement analogue à celui que tenait VON NEUMANN dès 1925 [*vNm*1], à partir de l'axiome désignant comme ensembles toutes les classes non équipotentes à la classe universelle). A la fin de son mémoire, TARSKI remarque que toutes les questions posées sur les nombres inaccessibles resurgissent aussitôt, transférées à des nombres inaccessibles d'ordre supérieur (dont l'hyperinaccessibilité consiste à majorer strictement, en même temps qu'un quelconque de ses minorants stricts, le premier nombre inaccessible majorant ce dernier).

De nouvelles axiomatiques de théorie des ensembles, ne reposant pas sur la «limitation de taille», proposées par QUINE

En 1937, Willard Van Orman QUINE avait proposé, dans un article intitulé *New Foundations for Mathematical Logic* [*Qui*1], une théorie des ensembles, que le sigle NF désigne depuis. Dans ce système, toute formule, pour être regardée comme dotée d'un sens, doit répondre à une condition de *stratification*, qui réside dans la possibilité d'attribuer des niveaux aux variables figurant dans cette formule, de telle façon que pour toute sous-formule de la forme «$u \in v$», le niveau de v soit d'une unité supérieur à celui de u ; il est assuré, par un schéma d'axiomes, d'application restreinte aux fonctions propositionnelles qui s'expriment par une formule stratifiée, que chaque fonction propositionnelle de l'espèce détermine une extension, qui est un ensemble.

Dans sa formulation originelle, ce système souffre d'un défaut ([*Rsr*5]): il a la capacité d'une infinité de preuves d'existence de classes particulières, sans avoir, semble-t-il, celle de prouver l'existence d'une classe infinie. Pour obvier à ce défaut, trois ans plus tard, en 1940, dans un livre intitulé *Mathematical Logic* [*Qui*2], QUINE a proposé une autre axiomatique, désignée depuis sous le sigle ML. Elle dérive de la précédente par l'adjonction de *classes* analogues à celles de la théorie des ensembles de VON NEUMANN-BERNAYS-GÖDEL, et d'un schéma d'axiomes postulant que toute fonction propositionnelle détermine une classe, qui en est l'extension restreinte aux ensembles qui sont éléments d'un autre ensemble.

Le destin ultérieur de ces systèmes ressemble beaucoup à celui des logiques multivalentes: ils ont suscité beaucoup d'intérêt au départ, et bien des contributions intéressantes au progrès général des connaissances, mais en définitive ils n'ont pas été adoptés, en raison de grosses difficultés venues s'ajouter au caractère gênant de la non stabilité du critère de stratification lors de la formation des formules. Les premières de ces difficultés ont été signalées dès 1942: John Barkley ROSSER [*Rsr*7], et, indépendamment ([*Wng*1]), Roger LYNDON, ont montré que le système ML originel permettait de déduire le paradoxe de BURALI-FORTI; certes, d'autres versions exemptes de ce défaut [*Qui*3] ont tout aussitôt proposées par QUINE, mais d'autres difficultés ont été signalées en 1949 [*RrW*]; j'y reviendrai.

De la notion de modèle, jusqu'aux temps de la sémantique

Des origines lointaines, dans l'«Algèbre symbolique», de la notion de modèle en logique

On peut entendre de deux façons l'histoire du concept de *modèle*. On peut s'attacher à retracer les usages de ce concept, tel que nous le connaissons aujourd'hui, ou de concepts qu'il a subsumé ou précisé, alors que le terme de «modèle» n'était pas encore utilisé. Il n'est pas moins intéressant de retracer aussi ce qu'a été, une fois le terme de «modèle» introduit, l'évolution de son emploi, jusqu'à ce qu'un consensus s'établisse sur le concept qu'il désigne de nos jours.

Au dix-neuvième siècle, notre concept actuel de modèle a son origine lointaine dans la lente et douloureuse séparation, achevée en 1833 [*Pck*2] par George PEACOCK, entre une «algèbre symbolique», «science de symboles et de leurs combinaisons» appartenant à un «langage symbolique», d'une part, et d'autre part, de *multiples* «sciences d'application», traitant chacune d'une «interprétation» d'un tel rudiment de formalisme: les notations de l'algèbre seront, désormais, dotées de sens multiples.

Trois courants d'idées, qui n'interfèrent guère que vers la fin du dix-neuvième siècle, transfèrent, jusqu'au vingtième, des concepts issus de cette conception rudimentaire.

Des origines de la notion de modèle en logique: la tradition géométrique et axiomatique

Le premier de ces courants remonte à l'extension, à la géométrie, de la polysémie des représentations, évoquée ci-dessus. Ce pas est effectué par Arthur CAYLEY, en 1858 [*Cay*]. Et ce sera en usant d'«interprétations» des notions primitives de la géométrie par des notions définies en géométrie euclidienne, vérifiant les axiomes de telle ou telle géométrie non-euclidienne, qu'il sera montré que toute contradiction issue de ces axiomes serait issue des axiomes de la géométrie euclidienne — on connaît les «dictionnaires» dont usait POINCARÉ pour expliquer de telles interprétations. Ce courant débouche au vingtième siècle, à travers des conceptions du genre de celles que PADOA exposait en 1900 [*Pad*] au Congrès de Philosophie de Paris, sur la «théorie des postulats» et sur la «méthodologie des sciences déductives». Mais il exerce aussi son influence sur HILBERT: c'est à l'aide d'«interprétations» convenables des notions primitives que HILBERT conduit, dans les *Grundlagen der Geometrie* [*Hbt*1], ses démonstrations d'indépendance d'axiomes et de compatibilité relative à la théorie des nombres réels.

Les origines de la notion de modèle en logique: la tradition de l'algèbre de la logique

Le second des courants d'idées à travers lequel les vues de PEACOCK évoluent vers des concepts modernes est celui de l'algèbre de la logique. Au départ [*Boo*], les classes distinguées dans un «univers du discours» en constituent une interprétation; il en ira de même des relations (binaires), quand on y étendra le traitement algébrique. Mais BOOLE [*Boo*] ramène la validation des figures de syllogismes à la résolution d'équations: la résolution d'équations devient un chapitre de l'algèbre de la logique, qui se traite dans le langage *intuitif* commun aux mathématiciens lorsqu'il parlent d'équations. Ainsi SCHRÖDER [*Sdr*1] dit-il, des «racines» d'une équation, qu'elles la «satisfont» («erfüllen»), et d'une équation qui n'a que des racines, qu'elle est «universellement valide» («allgemeingültig»). Mais SCHRÖDER se rend très bien compte de ce que, de toute une échelle de types concevables, il n'est en mesure d'explorer que le domaine restreint des relations binaires entre les «Dinge», ou «Individuen», qui constituent le «Denkbereich» dont les «Gebieten» représentent une des interprétations de son calcul abstrait, et de ce que l'existence d'une racine peut dépendre du nombre de ces individus. Au début du vingtième siècle, l'accent se déplace, d'une part, sur le nombre des éléments du domaine dans lequel on cherche des solutions, et d'autre part, sur les quantifications portant sur les individus. Or, *discuter*, en fonction d'une famille de «valeurs» attribuées à ses «coefficients», *l'existence de racines* d'une équation booléeienne comportant des «coefficients» individus et des «coefficients» relations (entre individus), et des *inconnues des mêmes natures,* tout en faisant reposer l'attribution de ces valeurs sur le nombre des individus, n'est rien d'autre que rechercher ce que nous entendons aujourd'hui par les «modèles» de l'énoncé (d'ordre pouvant aller jusqu'au second) exprimant cette existence dans le «langage» de ces coefficients, si le premier membre de cette équation était égalé au Vrai (à la classe universelle **1**, dans l'interprétation par les classes), et de la négation de la même formule, si le même premier membre était égalé au Faux (à la classe vide **0**, dans l'interprétation par les classes). De cette manière d'aborder la logique, apprise dans leur jeunesse, deux des plus importants contributeurs du vingtième siècle ne cesseront, leur vie durant, de garder l'empreinte, SKOLEM, et TARSKI bien plus encore, dont nous verrons ce que l'un et l'autre apportent à la théorie des modèles; ils n'arrêteront pas de transposer, dans la façon de voir, et d'écrire, des *Principia Mathematica*, la façon de voir, et d'écrire, des *Vorlesungen über die Algebra der Logik*.

Lorsque, vers la fin de la seconde décennie, on passe des premiers membres des équations aux «expressions», la terminologie se transfère, mutatis mutandis: on dit qu'une «expression» du «calcul fonctionnel» est «satisfaisable dans un domaine d'individus» si ce domaine procure une interprétation des constantes figurant dans cette expression, et une assignation de valeurs convenables (selon leurs types) aux variables y figurant libres, telle que la proposition exprimée par cette expression selon cette interprétation et pour les valeurs assignées soit vraie; avec des variations dans la terminologie (selon les auteurs, et selon les moments, la terminologie de GÖDEL [*Gdl*1] étant la plus cohérente), on dit qu'une expression est «valide dans un domaine d'individus», ou «universellement valide dans un domaine d'individus», si ce domaine procure une interprétation des constantes figurant dans cette expression, telle que pour toute assignation de valeurs à ses variables libres, la proposition exprimée par cette expression selon cette interprétation et pour les valeurs assignées est vraie (en fait, le terme de «domaine» est la plupart du temps utilisé, tacitement, pour désigner le *couple* composé du domaine à proprement parler *et de l'interprétation des constantes* qu'il procure); et l'on dit d'une expression qu'elle est «valide», ou «universellement valide», si elle est valide, ou universellement valide, dans *tout* domaine d'individus (et pour toute interprétation des constantes qu'il procure). Ainsi s'expriment GÖDEL en 1930, dans le traitement [$Gdl1$] du théorème de complétude, et SKOLEM ([$Skm3, 7$] p. ex.), HILBERT ([$Hbt12$] p. ex.), HILBERT et ACKERMANN [HbA], HILBERT et BERNAYS [HbB], jusqu'à la seconde guerre mondiale, quand ils renvoient, en termes de *contenu signifiant* des symboles, à ce que nous appelons des modèles.

Des transpositions formalistes de la notion d'interprétation, avec des termes sans variable libre pour individus, aux alentours de 1930

Les mêmes, quand ils veulent ne référer qu'aux mécanismes formels qui accompagnent les mêmes contenus, s'expriment autrement: ils introduisent, par des règles de construction, des expressions sans variable libre pour servir de noms pour des individus, des règles pour déterminer quand, d'une relation primitive entre ces individus, il doit être dit qu'elle est «vraie», et traitent des «vérifications», requises par la question traitée, à l'aide de purs raisonnements syntaxiques, démarrant par des substitutions de noms aux variables libres, et se continuant par des transformations de formules, remplaçant des formules par des (formules) «réduites», jusqu'à parvenir à des formules exprimant une relation primitive entre des individus représentés par leurs noms: le modèle n'apparaît même pas explicitement. Ainsi procèdent HILBERT et ACKERMANN [HbA] en 1928, GENTZEN [$Gtz4$] en 1936, pour établir la compatibilité, qui du calcul des prédicats du premier ordre, qui de la théorie des types simples; ainsi procèdent encore SKOLEM [$Skm6$] en 1928, et HERBRAND [$Hbd1$] en 1929, le premier, pour établir que la reconnaissance du caractère contradictoire d'une expression peut se faire par étapes procédant sur des domaines finis emboîtés, et qu'en cas d'une suite infinie d'échecs, l'expression testée est satisfaite dans un domaine dénombrable, et le second, pour transformer une éventuelle démonstration formelle d'une contradiction à partir d'une telle expression en démonstration formelle d'une formule du calcul des propositions, algorithmiquement constructible à partir de l'expression testée, et exprimant la réussite de la recherche ainsi conduite. En 1937, quand ACKERMANN [$Ack4$] ramène, à celle de l'arithmétique, la compatibilité de la théorie des ensembles sans axiome de l'infini, il n'emploie pas le terme de «modèle», mais celui d'«interprétation»; l'interprétation par traduction des formules d'un langage dans l'autre, décrite plus haut, dont il se sert, lui permet, elle aussi, des vérifications par de purs raisonnements syntaxiques: le modèle n'est là qu'à titre heuristique, pour guider la recherche des démonstrations formelles.

Cette pratique évolue, par la suite, vers une notion «constructive» de modèle, dont il est fait état, en 1939, dans le second volume des *Grundlagen der Mathematik* [HbB], au moment de redonner des preuves des contributions de HERBRAND, impliquant de tels modèles; enrichie de notions d'effectivité ramenées à des notions de récursivité, elle constitue l'une des sources des travaux consacrés par KLEENE, à partir de 1940, à des interprétations de la logique intuitioniste.

Des origines de la notion de modèle en logique: du rôle de l'émergence de l'algèbre universelle

Le troisième cheminement, des réflexions de Peacock à la notion de modèle, est celui que suit l'algèbre, et qui conduit celle-ci, à la fin du dix-neuvième siècle, à l'introduction d'axiomatiques délibérément non catégoriques, décrivant chacune une espèce de structures, telle la structure d'espace vectoriel (réel), axiomatisée par Peano en 1888 ([*Pno*1]). Au vingtième siècle, les *postulate theorists* poursuivent dans cette voie, qu'ont suivie avant eux des géomètres de la fin du dix-neuvième siècle. Hilbert lui-même, dans les *Grundlagen der Geometrie* [*Hbt*1], traite la géométrie en structure définie sur des domaines d'objets par des relations vérifiant les axiomes; Zermelo [*Zml*3], en 1908, traite de même la théorie des ensembles; Steinitz [*Stz*], en 1910, la notion de corps; Artin et Schreier [*ArS*], en 1926, la notion de corps réel clos. Ainsi se multiplient des axiomatiques dont les structures que nous reconnaissons aujourd'hui pour en être les modèles portent à chaque fois un nom spécifique.

La notion abstraite de *matrice*, introduite par Tarski dans le mémoire [*ŁkT*] qu'il publie conjointement avec Łukasíewícz en 1930, est un pas vers une notion générique, encore restreinte au domaine limité des calculs propositionnels, d'*abstract algebra*, puis de *structure algébrique* (la différence, expliquée par Abraham Robinson [*ARb*2] en 1950, réside en ce que la première est définie par la satisfaction d'équations, tandis que la seconde est définie par la satisfaction d'énoncés du premier ordre quelconques, comme il en apparaît dans le cas de la définition de la notion de corps). Ces notions se constituent durant la décennie qui précède la seconde guerre mondiale: la première est présente dans l'article *On the combination of subalgebras* [*Bkf*1], publié par Garrett Birkhoff en 1933, et dans le titre même de l'article *On the structure of abstract algebras* [*Bkf*2], de 1935, où il introduit la notion d'*algèbre libre*, et prouve un théorème aussi important pour l'algèbre universelle que pour ce qui sera la théorie des modèles, et sur lequel je reviendrai; les structures algébriques sont des cas particuliers des «structures fondamentales de l'analyse» du titre de la première partie des *Eléments de mathématique* de Nicolas Bourbaki; celui-ci trouve, dès 1939 ([*Bbk*], §8), le moyen de définir, sans référence explicite à quelque langage que ce soit, une notion de structure extrêmement générale, susceptible de recouvrir les notions de structures d'interprétation d'une très vaste variété de langages, sur laquelle il serait intéressant de savoir si Tarski a jamais eu l'occasion d'émettre une opinion. Désormais, l'algèbre universelle va se développer et se constituer en branche autonome, offrant, aux travaux des théoriciens des modèles, des paradigmes dont ils s'efforceront de déterminer les ressorts et de délimiter la portée.

De quelques emplois du terme de «modèle» entre 1925 et 1935

Est-ce parce qu'il s'intéresse, autant qu'à la logique, à la physique théorique, et, en particulier, à la mécanique quantique, que l'on voit von Neumann [*vNm*1] employer en 1925 le terme de «modèle de la théorie des ensembles», pour désigner un «sous-système» du «système des domaines d'objets» faisant partie intégrante de la description de son axiomatique de la théorie des ensembles, lorsque cette axiomatique est satisfaite, dans les domaines d'objets de ce sous-système, par la restriction, à ce sous-système, de la relation d'argument à fonction? Est-ce parce qu'il est, à Princeton, le voisin et l'interlocuteur quotidien d'Albert EINSTEIN, que nous voyons GÖDEL — à qui l'on doit aussi, incidemment, deux modèles de l'espace-temps [*Gdl*16, 17] — reprendre ce terme en 1938 [*Gdl*13]? Et est-ce de ce que c'est en *physique* qu'il est habilité, que provient l'usage abondant fait par ZERMELO dans ses articles de 1929 [*Zml*4], et surtout de 1930 [*Zml*5], du terme de Modell — entre guillemets, parfois, mais sans cela, le plus souvent? Il y désigne ainsi une structure caractérisée (à partir d'une autre) par *deux* données, celle qui sera qualifiée, dans la seconde moitié du vingtième siècle, de modèle *naturel* de la théorie des ensembles: inclus dans un «Bereich» donné, lui-même composé d'Urelemente (éléments sans éléments, non ensembles) et d'ensembles, il s'agit d'un «Unterbereich», comprenant une partie donnée des Urelemente, et comprenant les ensembles de rangs inférieurs à un Grenzzahl donné, et tels que toutes les suites descendantes pour l'appartenance qui en sont issues aboutissent, en un nombre fini d'étapes, à un des Urelemente donnés [ou à l'ensemble vide]. Un tel sous-domaine est stable pour l'appartenance, et on le munit de la restriction de celle-ci aux Urelemente et aux ensembles qui en font partie. Ainsi, alors même qu'il ne parvient pas à se faire aux théorèmes de LÖWENHEIM-SKOLEM et de GÖDEL, ZERMELO traite de ce sujet presque comme on le fait encore aujourd'hui. Il répond, en passant, à une objection de SKOLEM [*Skm*4], en disant que les divers Bereichen sont pensés dans une théorie des ensembles d'ordre supérieur, à laquelle s'appliquent tous les axiomes et raisonnements de la théorie des ensembles; et il propose, pour une «Meta-Mengenlehre», l'axiome postulant une suite illimitée de Grenzzahlen, dont il a été question plus haut.

Dès 1930, CARNAP s'essaye à importer en logique la notion de modèle. Dans l'édition anglaise de son livre *The logical syntax of language* [*Cnp*2], parue en 1937, le terme de modèle intervient à peine dans un paragraphe, consacré à la méthode axiomatique, et rédigé en 1933, mais non inclus, «faute de place», dans l'édition originale allemande de 1934. La phrase par laquelle la notion de modèle est introduite: «If it is proved that the axioms are fulfilled for a certain interpretation, *or at least that their fulfilment is not excluded*, we say that by this interpretation a model for the axiom-system is constructed», montre déjà le flou de la notion et le travail d'élaboration qui reste à faire. Plus loin, trois espèces de «modèles» sont distinguées, selon que telle formule est valide, non contravalide, ou non-contradictoire: pendant longtemps, les logiciens ont appelé «modèles» d'un système formel les structures dotées d'opérations et de relations permettant d'interpréter les termes primitifs du langage de ce système, et «modèles d'un système d'axiomes» ceux des «modèles» de ce système où ces axiomes sont satisfaits — mais au même moment, on ne distinguait pas non plus, par une terminologie appropriée, ce qui, dans l'étude d'un système d'axiomes, relevait de son *langage*, savoir, du jeu des symboles signifiants utilisés, classés selon les règles de leur emploi, de ce qui relevait en outre de ce que ces axiomes disent de ses notions primitives; c'est ainsi que HILBERT et BERNAYS [*HbB*] parlent souvent des «axiomes d'un formalisme», alors que le même système de notions primitives permet d'écrire de multiples systèmes d'axiomes, ce qui oblige à distinguer autant de «formalismes» que d'axiomatiques présentes, bien que les formules des uns soient les formules des autres (cela s'illustre sur les axiomes des anneaux et ceux des corps, par exemple).

Encore plus loin dans ce paragraphe, CARNAP recense encore trois avatars de la notion de modèle. Deux sont des versions de la notion d'interprétation par traduction des formules d'un langage en formules d'un autre langage, et la troisième est la notion d'interprétation par substitution de termes et évaluation de «réduites», décrite plus haut.

De l'évolution vers le terme de modèle dans les écrits de TARSKI, GENTZEN et MOSTOWSKI

Les travaux de TARSKI durant cette période portent la marque d'une toute autre influence: ils se situent dans la mouvance de la méthodologie des sciences déductives. Cette discipline a pour objet l'étude des propriétés génériques des ensembles de dires, formalisés en formules d'un système formel adéquat, et relatifs à une science déductive formalisée abstraite; elle a pour champ d'application l'étude des diverses disciplines, ou sciences déductives, «concrètes», formalisées (les dires en langues parlées sont récusés comme affectés de trop d'ambiguïtés et d'imprécisions pour une étude scientifique).

Une science déductive est une théorie d'un domaine de la connaissance; TARSKI [*Trk*9] cite, à titre d'exemples, l'algèbre de la logique, l'arithmétique des nombres réels, la géométrie de la droite, la théorie de l'ordre, la théorie des groupes... A priori, une science déductive se subdivise en de multiples sous-domaines, délimités par des conditions supplémentaires, donnant lieu à des sous-théories, elles aussi déductives. Formalisée, assimilée à l'ensemble des formules qui expriment ses dires, elle se subdivise parallèlement en «systèmes déductifs», dont chacun se compose, *ou bien* des formules exprimant les conséquences déduites, selon des règles d'inférence données, d'hypothèses spécifiques — à l'instar, *en* géométrie, *des* diverses géométries distinguées, disons, dans les *Grundlagen der Geometrie* — ou bien, des formules reconnues valides selon une matrice donnée, assurant à leur ensemble la stabilité selon les mêmes règles d'inférence, propres à la science considérée.

En 1935, lorsqu'il substitue le terme de «théorie déductive» à celui de «science déductive», TARSKI [*Trk*12] prie son lecteur de bien faire la différence; ainsi faut-il le lire; aujourd'hui, on ne fait plus cette différence, et on appelle «théories déductives» d'un système formel, aussi bien ce qu'étaient les «théories déductives» des mémoires de TARSKI de cette époque, que ce qui en était les «systèmes déductifs». En revenant à son langage d'entre 1928 et 1935 [*Trk*5, 6, 9, 10, 12], les systèmes déductifs d'une science déductive sont donc des parties de l'ensemble dénombrable des formules, ordonnées par inclusion, et stables selon une opération de passage aux conséquences. La plus petite de ces parties, précise TARSKI [*Trk*12], «peut être interprétée comme étant l'ensemble de tous les énoncés universellement valides (ou, plus généralement, comme l'ensemble des énoncés que nous reconnaissons d'entrée de jeu comme vrais, lorsque nous entreprenons la construction de la théorie déductive objet de notre recherche métamathématique)». Parmi les autres, sont dits «complets» ceux qui sont maximalement propres ([*Trk*5]); selon un théorème, fondamental, établi par LINDENBAUM en 1927, tout système compatible admet une extension complète.

Dans ses écrits métamathématiques de cette époque, l'attention de TARSKI semble plus tournée vers l'étude des extensions complètes d'une théorie, au nombre desquelles sont les ensembles d'énoncés que satisfait un de ses modèles, que vers l'étude proprement dite de ces modèles; en d'autres termes, il semble s'intéresser plus à ce qu'une théorie complète dit de ses modèles, qu'à ces modèles eux-mêmes. On verra cependant plus loin que cela n'était pas le cas. En 1931, c'est seulement dans son mémoire *Sur les ensembles définissables de nombres réels* [*Trk*8], écrit pour les mathématiciens, qu'apparaissent explicitement, sous la forme de suites de graphes interprétant une suite de notions primitives, des structures, *étrangères* au formalisme d'une axiomatique et à ses formules, comme celles qui constituent, pour nous, les *modèles* de cette axiomatique; dans son mémoire sur le concept de vérité [*Trk*9], même lorsqu'il prend la précaution oratoire de n'introduire le concept de vérité dans un domaine d'individus, par opposition au concept de vérité «absolu» dont il est censé traiter, que juste par souci de complétude, afin de traiter de pratiques d'écoles étrangères, telle celle de Göttingen, il s'abstient soigneusement de mentionner l'ingrédient de ces interprétations des notions primitives, en s'aidant de n'effectuer en détail son traitement que sur un exemple simple de ce point de vue, où l'usage de cet ingrédient passe inaperçu. Et l'on cherche en vain, jusqu'en 1935, le terme de «modèle» dans ses écrits métamathématiques. A posteriori, ce fut là une circonstance heureuse que cette habitude de penser en termes de ce qui est dit, qui lui fit découvrir la notion fondamentale d'équivalence élémentaire, dont je parlerai plus loin, et qui avait, jusque là, échappé aux mathématiciens.

En 1935, au début de son mémoire sur les *Grundzüge des Systemenkalkül* [*Trk*12], il vient à adopter le terme de «modèle» — l'utilisant entre guillemets, comme ZERMELO [*Zml*4] cinq ans auparavant; mais c'est aux *théories déductives* qu'il est là exclusivement réservé: «par théories déductives, j'entends ici les «modèles» (réalisations) du système d'axiomes donné au §1» — il s'agit de l'axiomatique qu'il y donne de la notion de «passage aux conséquences» (la troisième depuis 1928); et cela, alors que dans l'Appendice, il considère les systèmes complets d'énoncés de la théorie élémentaire de l'ordre, vrais des ensembles ordonnés d'un type d'ordre donné, sans qualifier de modèles d'un tel système les ordres du type en question. Enfin, lors du congrès international de philosophie scientifique qui se tient à Paris, la même année, il donne trois conférences dont les rédactions font des textes de vulgarisation remarquables; dans l'une [*Trk*15], il introduit la notion de «sémantique scientifique», en l'opposant au concept de «sémantique» des langues courantes, et en en esquissant la philosophie mathématique sous-jacente et les rudiments de son traitement rigoureux; une autre [*Trk*16] est consacrée à la notion de *conséquence logique*, dont, après avoir critiqué la conception qu'en avait Bernhard BOLZANO en 1837 dans sa *Wissenschaftlehre* [*Blz*], et qui avait été reprise par CARNAP [*Cnp*2, 3], il donne la définition que nous avons retenue: «l'énoncé X résulte logiquement des énoncés de la classe K, si et seulement si tout modèle de la classe K est aussi un modèle de l'énoncé X ». Là, mais surtout dans la troisième conférence *Sur la méthode déductive* [*Trk*17], où il le relie au concept d'*interprétation*, il présente, du concept de *modèle*, ou *réalisation*, d'un ensemble d'*énoncés*, la conception d'une famille d'objets [adéquatement choisis, laisse-t-il implicite], qui satisfont les énoncés issus des précédents par renommage des symboles primitifs de ceux-ci en les noms des objets de cette famille qui leur correspondent, héritée de PADOA [*Pad*] et de KEYSER [*Ksr*2], dont un mot a été dit plus haut. Adaptée au langage de la théorie des types simples, reprise et précisée par CHURCH dans son *Introduction to Mathematical Logic* [*Chu*10, 13], dont la première version imprimée date de 1944, cette définition *sommaire* devient équivalente à celle d'aujourd'hui, si l'on précise, d'abord, comment la notion de satisfaction s'entend pour les formules figurant dans les champs des quantifications (qui peuvent porter sur des variables d'ordres *inférieurs* à ceux des notions primitives), et donc aussi, quels sont les individus, et de quoi se composent les autres types — ce dont l'histoire postérieure m'amènera à parler plus loin.

Chez GENTZEN, qui en 1936 écrit encore dans le style des *Grundzüge* de HILBERT et ACKERMANN [*HbA*] lorsqu'il prouve la non-contradiction de la théorie des types simple [*Gtz*4], on voit apparaître en 1938 [*Gtz*5], quand il ne s'agit pas de théorie de la démonstration, un maniement tout à fait dégagé de la notion de modèle, en un sens qu'il ne précise pas, mais qui convient fort bien à la façon dont nous la comprenons aujourd'hui. Il en va de même chez MOSTOWSKI, dans l'article de 1939 [*Mtw*4] déjà cité plus haut.

Des difficultés d'appréhension des notions de système formel et de modèle, jusqu'aux alentours de 1945

On le voit: en une quinzaine d'années, après 1920, il s'introduit une variété absolument extraordinaire de systèmes formels (et je n'ai pas encore parlé des systèmes de logique combinatoire, de λ-conversion, et de logique modale ...) — CURRY pourra, en 1951, dans son *Outlines of a formalist philosophy of mathematics* [*Cur*13], rédigée en 1939, proposer de définir les mathématiques comme «la science des systèmes formels». Pis: les points de vue sous lesquels sont abordées des notions parentes, guides heuristiques précieux dans la recherche des comportements formels, et heureusement intuitives, sont multiples; tout aussi multiples, les terminologies pour en traiter souffrent de fluctuations, d'un certain flou et de quelque ambiguïté: une mise en ordre s'impose.

En fait, elle est déjà en germe dans des travaux existants, mais peu remarqués. Pour ce qui est des systèmes formels, POST en avait donné, dès son mémoire de 1921 [*Pst*3], une définition générique qui embrasse, en tout cas, toutes les variantes qui en furent utilisées au moins durant la première moitié du vingtième siècle.

Mais ce traitement était passé inaperçu à l'époque, et il reste ignoré jusqu'à ce que paraisse en 1941 un mémoire [*Pst*5] où Post le rappelle. Entre temps, Tarski, dans son mémoire *Sur les ensembles définissables de nombres réels I* [*Trk*8], paru en 1931 et sur lequel je reviendrai, avait déjà montré, sur des exemples, comment manier la notion de satisfaction d'un énoncé par une structure, mais la diffusion du traitement plus théorique qu'il en avait annoncé tardait encore.

De l'introduction, par Tarski et Carnap, de la définition mathématique rigoureuse de la notion de satisfaction

En effet, Tarski, dans un mémoire [*Trk*9] rédigé en 1931, publié en polonais en 1933, puis en allemand, sous le titre *Der Wahrheitsbegriff in den formalisierten Sprachen*, en 1936, esquisse un traitement formalisé de la métamathématique d'une discipline générique, et donne en substance (il ne traite en détail que d'un exemple) la définition — tout à fait conforme aux usages de tout le monde, et que je ne rappellerai donc pas — encore utilisée aujourd'hui, de la notion de *satisfaction* d'une formule par une suite de valeurs convenables assignées aux constantes et aux variables libres que comporte cette formule. La caractéristique essentielle de cette définition est qu'elle procède par récurrence sur la construction de la formule. Une fois fixées les valeurs des constantes, que Tarski escamote allègrement dans ce mémoire, mais qu'il explicite dans le mémoire *Sur les ensembles définissables de nombres réels*, rédigé la même année, un énoncé, qui n'a pas de variables libres, est donc satisfait par toutes les suites de valeurs assignées aux variables, ou par aucune: dans le premier cas, il est dit vrai (pour les valeurs attribuées aux constantes), et dans le second, faux. Ainsi, une définition du concept de vérité est fournie *dans le métalangage*, qui dispose, par rapport au langage propre de la théorie, de *moyens d'expression supplémentaires*; mais, si, comme c'est le cas de l'arithmétique, les instruments du raisonnement métamathématique se formalisent, ou se codent, dans la théorie étudiée, aucune formule du langage *de celle-ci* ne peut y définir le concept de vérité, car l'on pourrait reproduire, formalisé, le raisonnement du paradoxe du Menteur. Il est moins connu que Carnap, indépendamment de Tarski, a lui aussi proposé en 1935 [*Cnp*3] des définitions équivalentes de la satisfaction et de la vérité; mais il n'a pas su en tirer cette conséquence. C'est l'étape capitale, puisque, en 1931, la définissabilité dans une structure a été introduite en termes de la notion de satisfaction, et qu'en 1935, c'est encore en termes de cette notion que s'esquisse la définition de la notion de modèle, et se met en place la notion de conséquence logique: à partir de là, toutes les versions «sémantiques» des autres concepts de la métamathématique se définissent.

Des axiomàtisation, formalisation, et arithmétisation des métadisciplines selon Tarski

Tarski associe en outre, dans le premier des mémoires [*Trk*9] que je viens de rappeler, ainsi que dans un autre [*Trk*10] qui paraît aussi en 1933, à chaque discipline formalisée, sa métadiscipline, qui répond à des spécifications communes à toutes les métadisciplines homologues, mais aussi à des spécifications propres à la discipline objet. Les premières viennent de ce que l'étude des termes formels, formules, démonstrations formelles, ... d'une discipline objet quelle qu'elle soit suppose assez de théorie des ensembles pour traiter d'algèbre des ensembles et des relations, y compris de l'ensemble des parties d'un ensemble; des cardinaux, notamment finis; et des suites, finies et infinies, de symboles par exemple. Les secondes résident dans la liste de ces symboles (qui comporte des symboles utilisés comme connecteurs et comme quantificateurs pour désigner les notions correspondantes de logique, et des symboles spécifiques constituant ce que nous appelons le *langage* de la discipline objet). Ebauchant une axiomatisation d'une métadiscipline générique, et allant jusqu'à en suggérer la formalisation, et l'arithmétisation, grâce à un codage du genre de ceux qu'emploie GÖDEL [*Gdl*3], TARSKI donne, pour la théorie de ce que nous appellerions le semi-groupe libre (des mots) engendré par les symboles constituant le langage, des axiomes, incluant un principe de récurrence (du second ordre), et dont il vante indépendance et catégoricité.

Ces axiomes où expressions et enchaînement des expressions sont traités en notions primitives, avec les nombres entiers naturels et les ensembles d'expressions, sont les premiers du genre; ils ont des interprétations arithmétiques, et des axiomes du même genre permettent aussi, comme le fait pour la première fois peut-être Hans HERMES en 1938 dans sa *Semiotik* [*Hms*], des interprétations de l'arithmétique. Ce courant d'idées a donné naissance à des versions de l'arithématique en termes d'enchaînement de représentations de nombres (dans un système de base donnée), très attractives pour les informaticiens, et dont la première est sans doute due à QUINE en 1946 ([*Qui*4]).

De la notion de modèle dans l'*Introduction to Mathematical Logic* de CHURCH
En 1944 paraît la première version de l'*Introduction to mathematical logic* [*Chu*10] de CHURCH, issue de cours dispensés à Princeton depuis plusieurs années. Elle vise à dégager l'essentiel des formalisations jugées les plus utiles, et comporte un volet sémantique, dans lequel l'accent est mis sur les notions d'interprétation (une par formalisation, mais elles s'emboîtent par prolongement à la façon dont les formalisations s'emboîtent par inclusion) et de *valeur* d'une expression signifiante selon une interprétation. Cette notion-là hérite de la notion de satisfaction son type de définition par récurrence sur la construction; elle permet, en traitant de la satisfaction par une interprétation, pour des langages riches en *termes* représentatifs d'objets, individus ou fonctionnelles de types divers, qui prennent des «valeurs» correspondant à leurs types, de définir la satisfaction par l'attribution, selon cette interprétation, de la valeur «Vrai», indépendamment des valeurs assignées aux variables libres. La notion de modèle, définie selon les indications de TARSKI en 1936 [*Trk*16], et précisée moyennant la description d'un algorithme d'évaluation compliqué, renvoyant à la notion de satisfaction ci-dessus, y occupe une place très modeste, et devait paraître assez obscure aux yeux des contemporains. En tout cas, c'est à Princeton, et à peine à l'issue d'une période de trois années, que vont être soutenues les deux thèses d'où provient la définition de la notion de modèle

à laquelle nous nous référons aujourdhui, celle de John Kemeny en 1946, et celle de Leon Henkin en 1947.

Du mémoire de Kemeny, *Models of logical systems*, de 1948

Pour ce qui est de la notion de modèle, comparativement à ce qu'elle est dans l'ouvrage de Church [*Chu*10], c'est un très grand progrès que constitue l'article de John Kemeny [*Kmn*1], paru en 1948 au *Journal of Symbolic Logic*, sous le titre *Models of logical systems* (et où figure ce qui n'est sans doute pas encore la quinzième occurrence du terme de «modèle» dans les colonnes de cette revue depuis sa création, en y comptant les reviews des travaux de Carnap et de Gödel !). Kemeny y relate les points essentiels de la thèse qu'il a soutenue à Princeton deux ans plus tôt, et où ont été données des définitions impeccables, qu'il rapporte: en premier lieu, d'une notion très large de «système logique», issue, par précision de quelques points, de la notion générique de Post [*Pst*5], et susceptible de recouvrir, par particularisation, une grande majorité des systèmes formels jusque là utilisés; en second lieu, pour le système de théorie des types simple élaboré par Church [*Chu*7] en 1940, des notions fondamentales correspondantes de *semi-modèle*, de *valeur* d'une expression signifiante dans un semi-modèle, de *satisfaction* d'une formule par un semi-modèle, et de semi-modèle, *modèle* d'un système logique, unifiées grâce à la subsomption, sous un concept unique, des diverses formalisations pour les interprétations desquelles la notion de valeur avait été définie par Church [*Chu*10]. Dans sa restriction au premier ordre, la notion de semi-modèle selon la terminologie de Kemeny coïncide avec la notion, qui est encore la nôtre, de structure interprétative du langage d'un système formel, qui se dégage enfin; et la notion de modèle selon cette terminologie, devant s'appliquer à des systèmes formels dont ceux que nous manipulons couramment aujourd'hui sont des particularisations simplifiées, recouvre, dans sa restriction au premier ordre encore, la notion de modèle dont nous nous servons, et dont la définition jouit des mêmes simplifications.

De la formation de la province de la récursivité, jusqu'en 1950

Des travaux de Rôsza Péter sur les récurrences des niveaux supérieurs

Le programme visant à prouver l'hypothèse du continu, esquissé par Hilbert en 1925 dans sa conférence *Über das Unendliche* [*Hbt*11], a été le point de départ de recherches sur les fonctions définies à l'aide d'utilisations itérées de récurrences de niveaux de plus en plus élevés, en particulier, de récurrences multiples, que Rôsza Péter, dont le traité *Rekursive Funktionen* [*Ptr*5] est sorti en 1951, a poursuivies jusqu'au delà de cette époque — notamment, elle a trouvé en 1935 [*Ptr*1] une fonction d'une variable et d'un paramètre, définie par une récurrence double, et qui *énumère* les fonctions récursives primitives d'une variable, en ce sens que celles-ci sont toutes les fonctions d'une variable qui lui sont associées en fixant de toutes les façons possibles le paramètre, et elle a, par la suite ([*Ptr*2, 4]), généralisé ce résultat aux niveaux supérieurs. Pendant ce laps de temps, la conception que l'on se faisait des fonctions récursives et des fonctions calculables a évolué très vite, jusqu'à infirmer les hypothèses sur lesquelles reposait le programme, et à vider, de leur contenu initial, mais non d'un intérêt mathématique intrinsèque, les problèmes concernant ces formes de récurrence, comme cela devait apparaître après 1937.

Des calculs de λ-conversion en 1931

Vers 1925, Veblen, désireux de susciter le développement d'une recherche en logique à Princeton, avait demandé à un étudiant avancé, Alonzo Church, de lui expliquer les idées de Hilbert à ce propos (cf. [*Asp*]) — il n'y avait, à l'époque, que quelques textes de conférences. Très appuyé par Veblen, Church avait soutenu, en 1927, une thèse [*Chu*1] sur les retentissements, sur la théorie des ordinaux de la seconde classe de Cantor, du remplacement de l'axiome de choix par diverses autres hypothèses. Après deux années à Göttingen, il était revenu à Princeton enseigner, et y développait une première version de ce qu'il appelait les *calculs de λ-conversion*, ancêtres de nos «λ-calculs».

Le problème fondamental ainsi abordé est celui de l'exploration du champ des fonctions *génériquement définissables* dans tout domaine d'interprétation, à partir du moment où l'on se donne pour primitives des instructions calculatoires prenant sens dans tout domaine d'individus, comme les permutations et identifications d'arguments, etc ... et l'opération qui, à l'action de deux fonctions, associe la fonction dont l'action résulte de l'introduction de la seconde de ces actions à la place du premier argument de la première de ces actions [cela pourrait s'exprimer aussi, mais anachroniquement, dans la terminologie de la théorie des catégories]. L'idée fondamentale sur laquelle reposent ces «calculs» est donc celle de produire, en termes d'un petit nombre de telles primitives, des représentations de *définitions* de fonctions, composables les unes avec les autres, donc implicitement conçues comme dotées d'un domaine de définition commun, où elles prendraient aussi leurs valeurs, mais *générique*, et pour cela, jamais explicité, de telle sorte que ces fonctions et leurs définitions se présentent comme abstraites (comme c'est le cas, par exemple, de «l'identité», parfois prise pour l'une des primitives). L'objet premier de l'étude de ces «calculs» est la connaissance de l'action, sur ces représentations, des transformations algorithmiques de leurs composantes qui les transforment en représentations de définitions de la même fonction; et enfin, de chercher, dans chacune des classes d'équivalence ainsi déterminées, des «formes normales», s'il en existe.

Les fonctions ainsi déterminées sont dites *λ-définissables*. Dans toutes ces expressions, le préfixe λ- renvoie à une notation introduite par Church [*Chu*2], à la place de notations plus anciennes remontant à Russell et à Frege, et qui consiste, M étant reçu comme une représentation «abstraite» d'un algorithme de calcul, à noter $\lambda x[M]$, ou $\lambda x.M$, l'application qui se note depuis deux ou trois décennies $x \mapsto M$; la variable x, dans ces notations, est liée. On note MN, comme en logique combinatoire, ce qui, selon l'interprétation initialement visée, est la fonction (des arguments de M de rangs différents de 1, et des arguments de N) «valeur en N» de la fonction M. Un renommage de variable liée se représente par la transformation de $\lambda x.M$ en $\lambda y.[y/x]M$, où $[y/x]$ représente la substitution de y à x en toutes ses occurrences dans l'expression qui suit; une «règle de conversion» la permet, s'il n'y a pas d'occurrences libres de y dans M; une autre permet de passer de $(\lambda x.M)N$, valeur en N de la fonction $\lambda x.M$, à $[N/x]M$, et vice-versa: ces exemples sont donnés là à titre d'illustrations de l'esprit dans lequel les quelques «règles de conversion» retenues sont établies. L'introduction de *combinateurs* primitifs va aussi de pair avec celle de règles de conversion qui en expriment le sens; ainsi le sens de «l'identificateur» I s'exprimerait-il à travers la règle permettant de passer de Ix à x, et vice-versa.

L'arithmétique s'introduit dans un tel système, à partir du moment où un composé de combinateurs primitifs joue le rôle d'un *itérateur*; on obtient alors, avec le comportement de représentations des nombres entiers naturels, des expressions qui représentent des définitions de fonctions, et dont on dit qu'elles représentent les nombres entiers naturels dans le système. Tel était, selon le témoignage de Stephen Cole Kleene (cf. [*Crl*2]), le point où Church en était arrivé avec le premier calcul de λ-conversion, lorsqu'il en suivait les cours, durant l'automne 1931. Comme, à l'issue de ces cours, Kleene demandait à Church de s'essayer à quelque chose, Church lui suggéra de développer l'arithmétique dans ce système. Kleene mit d'abord quelques jours pour trouver comment introduire une représentation du prédécesseur, mais, de proche en proche, parvint à faire tant et si bien que, aux abords de 1933, Church commença à se demander si les fonctions arithmétiques λ-définissables ne recouvraient pas la totalité des fonctions arithmétiques calculables.

Des fonctions récursives générales de Herbrand-Gödel

Sur ces entrefaites, Gödel vint à Princeton durant l'année académique 1933-1934, pour y donner une série de conférences [*Gdl*11]. Il accueillit avec scepticisme l'opinion de Church, et fit état d'une notion de fonction récursive qu'il avait élaborée en adaptant une suggestion que Herbrand lui avait faite dans une lettre qu'il lui avait écrite en 1931, peu de temps avant sa mort accidentelle: la notion de fonction récursive dite «générale», par opposition à la notion de fonction récursive, désormais dite «primitive», introduite par GÖDEL auparavant pour la preuve des théorèmes d'incomplétude.

La définition de cette notion use d'une notion de déduction pour des équations fonctionnelles; les symboles acceptés pour écrire celles-ci peuvent soit représenter des fonctions, connues ou non, soit être des indéterminées pour des nombres entiers naturels, soit être le bâton dont les suites représentent les nombres entiers naturels déterminés. Les déductions admises ne reposent que sur deux règles d'inférences: la première permet de conclure, d'une équation, à l'équation qui en résulte par la substitution, à toutes les occurrences d'une indéterminée, d'une suite de bâtons donnée; la seconde permet, partant d'une équation principale, et d'une équation secondaire égalant, à une suite de bâtons, un terme où ne figure aucune indéterminée, de conclure à l'équation qui résulte de l'équation principale en y remplaçant une occurrence du terme, premier membre de l'équation secondaire, par la suite de bâtons, second membre de cette équation. On dit alors d'un système d'équations fonctionnelles, dépendant de fonctions données, supposées connues, qu'il définit récursivement une fonction inconnue f, à partir de ces fonctions données, si toutes les équations de la forme $f(k_1, \ldots, k_n) = k$ entre les suites de bâtons k_1, $\ldots, k_n$ et k, constituant le graphe de f, se déduisent de ce système d'équations, en n'appliquant que les deux règles de déduction admises. Les fonctions récursives générales de HERBRAND-GÖDEL sont alors celles qui sont définies récursivement par les divers systèmes d'équations fonctionnelles dénuées d'occurrences de fonctions données au préalable (en fait, les fonctions définies récursivement à l'aide de fonctions récursives générales sont elles-mêmes récursives générales).

De l'établissement de la coextensivité de trois concepts:
λ-définissabilité, récursivité générale, et μ-récursivité générale

KLEENE et CHURCH montrent bientôt que les mêmes fonctions arithmétiques tombent sous les concepts de λ-définissabilité (en plusieurs versions) et de récursivité générale, et publient ces résultats de leurs travaux en 1936 [*Chu*4], [*Kln*2]. Dans son mémoire *General recursive functions of natural numbers* [*Kln*3], KLEENE développe la théorie des fonctions récursives générales, d'une manière qu'il affinera par la suite ([*Kln*4, 8]). Un des points centraux de ce développement, transposé de la théorie des calculs de λ-conversion, est commun à CHURCH et à KLEENE: en y incorporant un codage analogue à celui que GÖDEL utilise pour la preuve des théorèmes d'incomplétude, et en profitant des caractères récursifs des concepts de (code de) déduction et de (code d'une) suite de bâtons, second membre d'une équation, ils réduisent toute fonction récursive générale à une forme *normale*, où seuls interviennent ces concepts (sous forme codée), le schéma de composition, et le schéma de passage au minimum: savoir, l'opération qui consiste à passer d'une fonction dépendant d'un paramètre, à la fonction, ne dépendant plus de ce paramètre, qui associe, à toute suite de valeurs attribuées aux autres arguments, la première valeur du paramètre qui complète cette suite en une suite de valeurs des arguments de la fonction donnée, pour laquelle cette fonction s'annule. Du coup, le concept de fonction récursive générale s'avère également coextensif à celui de fonction aujourd'hui qualifiée de μ-récursive, qui jouit d'une définition directement arithmétique, et qui reste le préféré dans les exposés d'aujourd'hui: les fonctions μ-récursives sont celles qu'on obtient à partir des constantes, des projections et du successeur, par applications répétées de compositions, de récurrences primitives, et de passages au minimum. Dans la version de 1936 de cette terminologie, une fonction obtenue par passage au minimum n'est introduite comme récursive que sous réserve d'être partout définie (ce qui n'est pas le cas de toutes les fonctions résultant d'un passage au minimum).

De la «Thèse de Church»

Church soutient alors [*Chu*4] le postulat, qu'il avançait depuis au moins deux ans ([*Crl*2]), et que l'on connaît depuis sous le nom de *Thèse de* Church, selon lequel le concept mathématique de «fonction récursive générale» rend convenablement compte du concept intuitif de fonction effectivement calculable (quels que puissent en être les avatars dans le futur). Une remarque que fait Rosser, en relatant, dans une review [*Rsr*3] qui paraît peu après, une note de Gödel *Über die Länge von Beweisen* [*Gdl*12], à propos d'une remarque qu'y fait Gödel, vient apporter de l'eau à son moulin: Gödel considère les arithmétiques des divers ordres entiers naturels dans une théorie des types simple, prétend qu'on trouve toujours une infinité de propositions arithmétiques qui gagnent au passage à l'ordre immédiatement supérieur des preuves formelles aussi dramatiquement raccourcies que l'on peut le souhaiter, mais que l'ensemble des fonctions arithmétiques «effectivement calculables», par contre, semble posséder un caractère «absolu» en ce sens qu'il reste le même pour chacune de ces arithmétiques. Par fonction «effectivement calculable» dans une arithmétique (supposée compatible), Gödel entend là toute fonction $f(x_1, \ldots, x_n)$ des variables $x_1, \ldots, x_n$, pour laquelle une formule $\mathfrak{F}(x_1, \ldots, x_n, y)$ de cette arithmétique est telle que, pour tout $n+1$-uplet de suites de bâtons $(k_1, \ldots, k_n, k)$, on a $f(k_1, \ldots, k_n) = k$ si et seulement si cette arithmétique admet une preuve formelle de l'énoncé $\mathfrak{F}(k_1, \ldots, k_n, k)$. Rosser [*Rsr*3] note qu'en recourant à une argumentation de Church [*Chu*4], qui vient d'établir la même chose pour un système de λ-conversion, on établit aisément que cette classe coïncide avec celle des fonctions récursives générales. En 1939, Hilbert et Bernays [*HbB*] montreront que cette coïncidence est largement indépendante du système dans lequel l'arithmétique est formalisée, pourvu qu'il remplisse un minimum de conditions.

Des notions de fonction calculable par une machine de Turing et de fonction calculable au sens de Post

Et de fait, toutes les définitions de la notion de fonction calculable qui surgiront par la suite — et pour la plupart, d'ailleurs, après 1950 — s'avèreront, en effet, coextensives à la notion de fonction récursive. La première apparue est la notion de fonction calculable par une *machine de* Turing, élaborée par Alan Mathison Turing [*Trg*1] indépendamment des travaux de Princeton, en vue de mathématiser le concept de calculabilité effective. Une machine de Turing est une machine conceptuelle, censée écrire dans un cahier illimité, dont les feuilles juxtaposées forment un ruban. A chaque étape d'un calcul, elle lit une case, puis, selon ce qu'elle lit, et selon celui des états, en nombre fini, qui lui sont permis, dans lequel elle se trouve, elle y efface quelque chose — représentée par un caractère d'un alphabet donné — y écrit une autre, et se déplace d'une case, soit vers la gauche, si elle n'est pas au début du ruban, soit vers la droite, et passe dans un autre état, à moins que l'état précédent ne comporte un ordre d'arrêt. Une machine se représente donc par une suite finie d'instructions, c'est-à-dire par un programme, et la fonction qu'une machine calcule est celle qui associe, à une représentation d'une suite d'arguments (selon un certain code), placée sur le ruban avant de lancer la machine, la représentation déchiffrée sur le ruban au moment de l'arrêt. Or, peu après l'article de présentation de ces conceptions, rédigé en 1936 et paru quelque temps après ceux dont il vient d'être question — et peu après une note [*Pst*4], rédigée postérieurement, et où POST introduisait une notion de calculabilité très voisine de la sienne — TURING [*Trg*2] établissait aussi, au début de 1937, l'équivalence des notions de fonction définissable et de fonction calculable au sens qu'il venait de définir.

De la nouvelle allure prise par les problèmes de décision en raison de la thèse de CHURCH

Ce fut par ce dernier travail que GÖDEL, de son propre aveu ([*vHj*], p. 616), fut convaincu de l'adéquation de la Thèse de CHURCH à la réalité. L'acquiescement à l'identification du concept de fonction récursive comme mise en forme mathématique du concept de fonction effectivement calculable, qu'énonce la Thèse de CHURCH, constitue une étape très importante dans l'évolution de la logique mathématique; il entraîne une nouvelle façon d'en appréhender le rôle et la portée, d'en comprendre l'architecture des différentes branches, et d'en aborder les concepts et les problèmes. Ainsi, les conditions de reconnaissabilité effective imposées aux formules, axiomatiques formalisées, et démonstrations formelles, des systèmes formels, se traduisent par des conditions de récursivité concernant leurs codes respectifs, selon un codage analogue à celui dont GÖDEL avait fait usage dans la preuve des théorèmes d'incomplétude: tout espoir d'échapper, par quelque nouvelle formalisation susceptible de correspondre à une connaissance effective, à ces théorèmes et à leurs conséquences, disparaît.

D'un autre côté, des voies à explorer pour tenter de débloquer des recherches stagnantes s'ouvrent: ainsi, en 1940, CHURCH [*Chu*8] propose de définir une suite de tirages «effectués au hasard» comme une suite dont on ne peut faire varier la fréquence des événements en la remplaçant par une suite partielle déterminant de façon décidable, c'est-à-dire, récursive, son $(n+1)$-ième terme à partir (d'un code, selon un codage comme celui de GÖDEL) de la suite de ses n premiers termes. Il s'avèrera que cette proposition ne résout pas toutes les difficultés de ce genre de questions, et cette suggestion ne sera pas reprise, du moins, durant la première moitié du vingtième siècle.

La thèse de CHURCH conduit aussi à une nouvelle aperception des problèmes de décision: la décidabilité d'une théorie formelle va de pair avec la récursivité de l'ensemble (des codes) de ses théorèmes. Ainsi CHURCH démontre-t-il, sous réserve de leur compatibilité, l'indécidabilité de l'arithmétique de PEANO et de certains de ses sous-systèmes [*Chu*4], y compris le calcul des prédicats du premier ordre [*Chu*5], par la non-calculabilité de l'ensemble de leurs théorèmes. La même approche conduit ROSSER, dès 1936 [*Rsr*2], puis, en 1943 [*Kln*8] et 1950 [*Kln*10], KLEENE, à des extensions raffinées des théorèmes d'incomplétude; par la même méthode, ROSSER [*Rsr*2] montre ce que TARSKI baptisera en 1948 [*Trk*28] *l'indécidabilité essentielle* — savoir, l'indécidabilité de toute extension compatible — de l'arithmétique de PEANO.

La notion de *théorie axiomatisable* va, elle aussi, changer de statut: dès 1936 [*Rsr*2], ROSSER admet comme systèmes d'axiomes des ensembles de formules (dont les codes de GÖDEL sont) récursivement énumérables. On peut être surpris de voir publiée, en 1950, une Note où Antoni JANICZAK [*Jnz*] prouve que toute théorie complète et axiomatisable, même en ce sens, est décidable: TURING avait montré, dans son mémoire *On computable numbers* de 1936 [*Trg*1], que si le(s) système(s) considéré(s) par GÖDEL en 1930 [*Gdl*3] étai(en)t complet(s), il(s) serai(en)t décidable(s), selon un argument facile à adapter à toute théorie complète. Il semble que ce que JANICZAK ait vu, qui passait inaperçu jusque-là, est que cet argument n'était appliqué qu'à des théories axiomatisées: il part de la remarque que les extensions complètes des théories essentiellement indécidables ne sont pas décidables, et précise, au-delà même des conditions de formalisation usuelles, sous quelles hypothèses la complétude entraîne la décidabilité.

Des résultats d'indécidabilité obtenus en combinant interprétabilité et indécidabilité essentielle

En 1950 [*RRb*], Raphael ROBINSON a imaginé une arithmétique, disposant, pour chaque fonction récursive primitive, d'un terme formel exprimant un procédé de calcul des valeurs de cette fonction à partir des valeurs données à ses arguments, et dont l'indécidabilité essentielle s'établit selon l'argumentaire ci-dessus, pour très rudimentaire qu'elle soit — tout juste capable de disposer des preuves formelles établissant la différence entre deux différentes des représentations de suites de bâtons dont elle est supposée disposer, l'énumération des représentations des suites de bâtons qui précèdent une suite de bâtons donnée, et pour chaque fonction récursive primitive, l'égalité ou la différence avec 0 de sa valeur, pour tout système de valeurs de ses arguments. Auparavant, il en avait introduit une extension, encore passablement rudimentaire, mais finiment (et simplement) axiomatisable, en élaguant, d'une arithmétique fragmentaire, finiment axiomatisable et essentiellement indécidable, inventée par TARSKI et MOSTOWSKI en 1939 ([*MwT*3]), les hypothèses non indispensables à la mise en œuvre d'une méthode de preuves d'indécidabilité imaginée par TARSKI: si une théorie compatible T et une théorie finiment axiomatisable et essentiellement indécidable de même langage ont, dans ce langage, une extension compatible commune, la théorie T est indécidable.

C'est essentiellement selon ce schéma que procèdent les démonstrations de toute une rafale de résultats d'indécidabilité relatifs à des théories du premier ordre, annoncés en 1948: TARSKI et MOSTOWSKI prouvent l'indécidabilité des sous-théories de l'arithmétique de PEANO, *de même langage que celle-ci* [cette dernière condition ne peut être omise], et de diverses théories d'anneaux, résultats qu'ils ont obtenus en 1939; TARSKI [Trk29, 30] annonce l'indécidabilité de la théorie des groupes, de diverses théories de réseaux, et de géométries projectives; Julia ROBINSON [JRb1, 2] annonce l'indécidabilité de la théorie du corps des rationnels, de la théorie des corps, et les résultats analogues à ceux de MOSTOWSKI et TARSKI pour les sous-théories de l'arithmétique des nombres entiers positifs dont les langages sont ceux où elle sait définir la somme et le produit; pour les sous-théories de l'arithmétique des nombres entiers dont les langages sont les précédents, augmentés de la notion primitive de nombre positif; et pour les sous-théories de l'arithmétique des nombres réels, dont le langage est celui de la somme, du produit, et de la notion de nombre rationnel. Là, Julia ROBINSON tire simplement les conséquences de définitions qu'elle vient de produire: de la somme en termes du produit et du successeur, dans le cas des nombres entiers positifs et dans celui de tous les nombres entiers; de la somme et du produit en termes de la divisibilité et du successeur, dans le cas des nombres entiers positifs; ainsi que de la notion de nombre entier positif en termes de la notion de nombre rationnel.

Des fonctions partiellement récursives

La forme donnée par TURING [Trg1] à la notion de calculabilité effective introduit tout naturellement des fonctions non nécessairement partout définies — non «totales», dira-t-on — dont il se peut que les valeurs ne soient calculées que pour une partie des suites de valeurs que l'on peut assigner à leurs arguments, et rend clair que la théorie des fonctions récursives — puisqu'on considère désormais que ce sont les mêmes que les fonctions calculables — devrait se développer bien mieux, avec plus de cohérence, en y incluant ces fonctions-là; ce qui revient, dans la définition des fonctions μ-récursives, par exemple, à lever la réserve faite auparavant sur ce point. C'est bien ce qui s'avère être le cas dans le mémoire [Kln8] que KLEENE prépare, après que la notion de fonction récursive ait achevé de se constituer sous sa forme actuelle, lorqu'en 1938 [Kln4] il introduit ainsi la notion de fonction partiellement récursive. Retardée par la seconde guerre mondiale, la publication de ce mémoire n'interviendra qu'en 1943; il comporte d'autres contributions importantes, dont je reparlerai.

Des fonctions récursives élémentaires de KALMÁR

En 1943 [Kmr2], László KALMÁR a défini les fonctions *récursives élémentaires* comme celles qu'on obtient à partir des projections, de l'addition, de la multiplication, de la valeur absolue de la différence, et de la partie entière du quotient euclidien, par applications répétées de compositions et de formations de la fonction dont l'action est la somme, ou le produit, des actions d'une fonction donnée pour un nombre fini de valeurs d'un paramètre. Ces fonctions sont (trivialement) toutes récursives primitives, et il y a des fonctions récursives primitives qui ne sont pas récursives élémentaires, selon un résultat d'Ilona BERECKI présenté en

1950 [*Brc*]; toutes les fonctions qui interviennent pour coder les relations métamathématiques, telles que celle de preuve, dans des questions comme celle des preuves des théorèmes d'incomplétude, n'en sont pas moins récursives élémentaires, ou remplaçables par des fonctions récursives élémentaires, et il existe des formes normales pour les fonctions partiellement récursives qui s'expriment en termes de passage au minimum, et de fonctions récursives élémentaires.

Du développement de la logique combinatoire et du calcul de λ-conversion jusqu'aux alentours de 1950

Les mémoires fondateurs des premières versions élaborées de la logique combinatoire, *Grundlagen der kombinatorischen Logik* [*Cur*2], de Curry, et du calcul de λ-conversion, *A set of postulates for the foundation of logic* [*Chu*2], de Church, sont parus, respectivement, en 1930 et 1932. Tandis que Kleene développait l'étude du contenu arithmétique de la λ-conversion ([*Kln*1]), Curry proposait, de la logique combinatoire, des versions [*Cur*3, 4, 5] prenant en compte des domaines plus larges d'opérateurs attendus, parues entre 1931 et 1934. Puisqu'ils recourent l'un et l'autre à une abstraction de l'opération d'application, qui, à un opérateur et à un opérande, associe la valeur associée par cet opérateur à cet opérande, la parenté des deux systèmes est apparente. Un autre étudiant de Church, John Barkley Rosser, fut donc chargé d'élucider leurs rapports. En 1935 [*Rsr*1], il établit que le système initial de Church était équivalent à un système de logique combinatoire un peu plus faible que la plus étendue des versions qui en avaient été introduites l'année précédente, et proposa des améliorations notables des preuves, moyennant un meilleur choix des combinateurs. Mais alors que la thèse de Rosser était sous presse, Kleene et Rosser [*KlR*] découvrirent une contradiction dans le système de λ-conversion, affectant la version étendue de la logique combinatoire, mais non sa version initiale.

Church et Rosser réagirent en publiant, l'année suivante, une seconde version du calcul de λ-conversion, qui échappe à la contradiction [*ChR*]. En 1941, Rosser [*Rsr*6] et Curry [*Cur*7] proposaient de nouvelles versions, et de nouvelles améliorations, de la logique combinatoire, et Church publiait à Princeton un manuel intitulé *The calculi of λ-conversion* [*Chu*9], où il introduisait une autre version encore du calcul de λ-conversion, changeant de notations pour en adopter de plus claires, qui sont à peu de choses près celles de notre λ-calcul. Entre temps, il avait produit une version du calcul comportant des types: c'est la version [*Chu*7], toute transparente, elle, de la théorie des types simple, de Church, publiée en 1940, déjà évoquée plus haut, et dont Leon Henkin va proposer des modèles en 1950 [*Hnk*3]. Pendant tout ce temps, Curry, cherchant les racines de l'incompatibilité ([*Cur*9]), l'avait tirée d'un autre paradoxe ([*Cur*11]), simplifié à l'extrême, parfois appelé le «paradoxe de Curry» — une variante du paradoxe de Russell. Il en avait conclu ([*CyF*], p. 10) qu'un système de logique combinatoire doit, d'une certaine façon, définir la notion de proposition, et, sur le point de partir à l'armée, avait fait le bilan [*Cur*10, 12] des avancées, des projets à développer, et des conjectures auxquelles il était parvenu. Je ne suivrai pas ici ce qui se passe après la seconde guerre mondiale en ce qui concerne l'une et l'autre des deux disciplines; à la fin de la première moitié du vingtième siècle, elles étaient encore pleines de questionnements, et à la recherche de leur forme définitive.

Du théorème de CHURCH-ROSSER

Pour CHURCH, un terme du calcul de λ-conversion ne «signifiait» vraiment quelque chose que s'il était «convertible» en une «forme normale» ([*CyF*], p. 106). Un terme est dit «sous forme normale» s'il ne comporte aucune composante de la forme $(\lambda x.M)N$ (en particulier, il ne doit pas lui-même être de cette forme). Par une «λ-conversion», il est entendu une succession de transformations d'un terme en un autre, dont chacune est autorisée par une règle de conversion. Et un terme est dit «convertible» en un autre, s'il existe une λ-conversion qui, partant du premier, s'achève sur le dernier.

La preuve de non contradiction du système de λ-conversion proposé en 1936 [*ChR*] par CHURCH et ROSSER recourt au théorème maintenant connu sous le nom de «théorème de CHURCH-ROSSER»; il dit que les λ-conversions possèdent la propriété maintenant connue sous le nom de «propriété de CHURCH-ROSSER», savoir: si un même terme est convertible, selon deux λ-conversions différentes, en deux termes différents, ceux-ci sont à leur tour convertibles en un *même* terme. Ainsi, *si* un terme admet une forme normale, on ne peut manquer de l'atteindre par une recherche systématique, quels que soient l'ordre d'application des règles, et les termes auxquels on les applique.

De l'indécidabilité récursive du calcul de λ-conversion et du calcul des prédicats du premier ordre

Tous les termes sont à l'origine de λ-conversions: celles qui procèdent d'un terme qui n'est pas convertible en une forme normale, ou bien s'arrêtent sur un terme qui n'est pas sous forme normale, ou bien «bouclent», ou bien se développent indéfiniment, de façon désordonnée; et c'est là que l'on découvre le phénomène: les premiers prototypes d'ordinateurs ne sont même pas encore conçus. Un problème se posait donc: comment reconnaître les termes convertibles en une forme normale? (C'est l'équivalent, pour le calcul de λ-conversion, de ce qu'on appelle, depuis l'article de TURING, qui arrive quelques temps après, le «problème de la halte»). Plus généralement, comment reconnaître si deux termes donnés sont convertibles l'un dans l'autre?

Avec celui d'identifier les théorèmes d'une axiomatique formalisée capable d'exprimer un minimum d'arithmétique, tels sont justement les problèmes dont CHURCH découvre, en 1936 [*Chu*4], qu'ils sont «récursivement indécidables», en usant d'un codage des termes par les nombres entiers naturels, analogue à celui que GÖDEL utilise à la preuve des théorèmes d'incomplétude, et en procédant par «diagonalisation», les propriétés testées s'exprimant, par le truchement du codage, par des formules de l'arithmétique élémentaire.

C'est dans l'article [*Chu*4] où il relate la preuve de l'indécidabilité d'un ensemble de formules de cette espèce, que CHURCH énonce la Thèse dont il a déjà été question. L'indécidabilité récursive, qu'il introduit là, consiste en ce qu'il n'y a pas de fonction récursive qui trie les solutions (nécessairement en nombre infini dans ce cas) en attribuant à leurs codes une valeur, des non-solutions (elles aussi en nombre infini), en attribuant à leurs codes une autre valeur. Ce n'est pas du tout incompatible avec l'existence, pour chaque solution, d'une preuve effective de ce que son code est celui d'une solution; cela veut seulement dire que *la preuve elle-même varie avec la solution,* de telle façon qu'il n'y a *pas d'argumentation uniforme,* dépendant d'un paramètre, et qui fournirait, pour chaque égalisation de ce paramètre à un code de solution, une preuve de ce que ce code est celui d'une solution.

Peu après, CHURCH s'aperçoit qu'en appliquant le théorème de la décision, une procédure de décision pour une théorie donnée se transforme en une procédure de décision pour une extension finie de cette théorie, si cette extension se fait sans introduction de nouvelles notions primitives, à langage constant. En sens inverse, le retrait d'un nombre fini d'axiomes, sans retrait de notions primitives, conduit donc d'une théorie dénuée de procédure de décision à une théorie tout aussi indécidable. Comme l'arithmétique qu'il a été amené à considérer comporte justement un nombre fini d'axiomes, l'indécidabilité reconnue s'étend aux fragments qu'on en obtient en enlevant des axiomes — jusqu'au fragment qui, gardant les mêmes notions primitives, n'a plus d'axiome spécifique: le calcul des prédicats du premier ordre. Dans *A note on the Entscheidungsproblem* [*Chu*5], d'un peu plus d'une page, publiée au *Journal of Symbolic Logic*, il fait part de la nouvelle: il n'existe pas de procédé qui permette de trier, selon un algorithme composé d'un nombre fini de règles, et en un nombre fini d'étapes, les formules du calcul des prédicats du premier ordre logiquement vraies.

Ces résultats font naître le soupçon que le dixième problème de HILBERT pourrait aussi être récursivement indécidable. C'est en tout cas dès 1940 [*Ptr*3] qu'il s'exprime par la bouche de Rôsza PÉTER, en une occasion où elle reprend des idées de KALMÁR. Cependant, aucun résultat en la matière ne sera obtenu dans la période qui nous occupe.

Des preuves d'indécidabilité par réduction

Par opposition aux preuve d'indécidabilité, dites *directes*, mises en oeuvre par GÖDEL en 1931 [*Gdl*3] et par CHURCH en 1936 [*Chu*4, 5], et qui reposent sur un codage arithmétique des expressions d'un système formel pour établir, dans l'arithmétique élémentaire, que la présence de certaines démonstrations formelles y introduirait des contradictions, la méthode qui, en présence d'un problème d'indécidabilité, a par la suite été le plus largement appliquée, est celle qui consiste à le *ramener* à un problème dont l'indécidabilité est déjà connue. C'est à cette méthode que ressortit, par exemple, l'usage du théorème de TARSKI sur les théories essentiellement indécidables.

C'est aussi en employant de telles réductions qu'en 1946 POST prouve l'indécidabilité du «problème des mots» pour les semi-groupes, posé par Axel THUE [*Thu*] en 1914: on donne un nombre fini d'égalités entre des suites de signes déterminées (d'un «alphabet» fini donné); ces égalités engendrent une équivalence entre suites finies de signes; donner un algorithme, applicable à toute équivalence ainsi engendrée, et qui trie les couples de suites équivalentes. Ce résultat négatif est obtenu indépendamment par A.A. MARKOV; l'une et l'autre des publications [*Pst*8] de POST et [*Mrk*] de MARKOV interviennent en 1947. En 1950 [*Trg*4], TURING a obtenu le résultat analogue pour les semi-groupes admettant la «loi de simplification» qui impose l'équivalence des suites résultant de la suppression d'un éventuel même premier signe, ou d'un éventuel même second signe, de suites équivalentes.

Ce problème n'est lui-même qu'une version affaiblie du «problème des mots» pour les *groupes définis par générateurs et relations:* founir un algorithme décidant en un nombre fini d'étapes si deux «mots» en les générateurs — savoir, deux composés de suites finies de générateurs — sont égaux, posé par Max DEHN [*Dhn*], en 1911, et qui ne sera résolu, par la négative (en ce qui concerne la *classe* de tous les systèmes de générateurs et de relations), qu'en 1955.

Des ordinaux constructibles de CHURCH **et** KLEENE

Le calcul de λ-conversion a permis à CHURCH et à KLEENE de délimiter, de 1936 [*ChK*] à 1938 [*Kln*4], [*Chu*6], les ordinaux *prédicatifs*, c'est-à-dire, constructibles, ou, comme ils disent d'abord, définissables, *sans référence à la totalité des ordinaux,* à partir de leurs prédécesseurs stricts, à l'aide des opérations de passage au successeur et de passage à la limite d'une suite infinie croissante indexée par les nombres entiers naturels. Plus précisément, ils imposent aux ordinaux constructibles la capacité d'admettre une notation appartenant à un système de notations qui, à partir d'une notation de ce système, permette: de déterminer effectivement si l'ordinal dénoté est l'ordinal zéro, un ordinal successeur, ou un ordinal limite; dans le second cas, de déterminer effectivement une notation du système pour son prédécesseur; et dans le troisième cas, de déterminer effectivement une suite de notations du système, dénotant les termes d'une suite infinie strictement croissante d'ordinaux, dont il soit la limite.

CHURCH et KLEENE démontrent que les ordinaux ainsi délimités constituent un segment initial dénombrable de la seconde classe d'ordinaux de CANTOR. Ils annoncent que le premier ordinal non constructible est limite et dénombrable, qu'il ne possède aucune définition qui ne fasse référence à la totalité des ordinaux, et qu'il dépasse, non seulement ε_0, mais encore le premier ordinal solution de l'équation $f(x) = x$, pour toute application strictement croissante f de la seconde classe d'ordinaux de CANTOR vers elle-même, introduite par une définition explicite dans l'article de VEBLEN de 1908 à ce sujet [*Vbl*6].

En 1939, poursuivant une idée émise en 1935 par CHURCH [*Chu*3], qui espérait, me semble-t-il, y trouver une échappatoire à des théorèmes d'incomplétude à la GÖDEL, TURING [*Trg*3] cherche à réduire au maximum l'incomplétude inhérente à un système formel, en itérant transfiniment, sur les ordinaux constructibles, à partir d'un système de base, le passage d'un système à celui qui en résulte par l'adjonction d'un énoncé vrai non encore formellement démontrable, et en formant la «logique ordinale» constituée de la réunion (malheureusement non axiomatisable) des systèmes ainsi formés.

Des notions de classe récursive et de classe arithmétique, chez GÖDEL en 1931
GÖDEL, dans son mémoire [*Gdl*3] sur les théorèmes d'incomplétude, a introduit le qualificatif de *récursif* [*récursif primitif*, aujourd'hui] pour désigner, sous le nom de *classes* (ou *relations*, ou quelquefois, par abus de langage, *prédicats*), les images réciproques de 0 par une fonction récursive [récursive primitive, aujourd'hui, mais, bien sûr, quand, par la suite, l'acception du concept de fonction récursive s'étend, cette définition du qualificatif de «récursif» dans son application aux classes est conservée, son champ d'application héritant de la même extension], qu'il s'agisse de parties de l'ensemble des nombres entiers naturels, ou d'une de ses puissances entières naturelles supérieures ou égales à 2. Un peu plus loin dans ce mémoire, il introduit le qualificatif d'*arithmétique*, pour désigner les parties des mêmes ensembles qui se définissent au premier ordre exclusivement en termes d'addition, de multiplication, et des constantes logiques. Il est un passage de la preuve de ses théorèmes d'incomplétude où il lui faut montrer que les classes récursives [primitives] sont arithmétiques; il se sert à cette fin d'une fonction qu'il a inventée, et qui code par des nombres entiers naturels, de façon récursive primitive à longueur donnée, les suites finies de nombres entiers naturels.

Des ensembles récursivement énumérables considérés dans leurs rapports avec les ensembles récursifs
Dans son mémoire *General recursive functions of natural numbers* [*Kln*3] de 1936, KLEENE introduit l'appellation d'ensemble *récursivement énumérable* pour toute image de l'ensemble des nombres entiers naturels par une fonction récursive. Ce terme, qui vient tout naturellement, ne fait que reprendre le terme analogue d'ensemble *effectivement énumérable*, lancé par Emile BOREL [*Brl*2] en 1908 et alors très usité, dans une terminologie cohérente avec la Thèse de CHURCH, qui voit l'effectivité dans la récursivité: l'usage en a très vite été universel.

Dans un article [*Pst*6] paru en 1943, mais dont il est peut-être intéressant de noter qu'il l'avait soumis pour publication en 1941, POST réintroduit des systèmes formels génériques (d'une espèce aujourd'hui rangée dans le genre des «grammaires formelles»), déjà présents dans son mémoire [*Pst*3] de 1921, mais qui n'avaient alors pas retenu l'attention. Ils s'appréhendent par les données successives: d'une «suite finie de lettres»; d'un nombre fini d'«assertions primitives»; et d'un nombre fini de «productions». Par rapport à ce qui se fait, par exemple, dans un système formel de calcul des prédicats, les «énonciations», qui s'écrivent avec les lettres de l'alphabet, c'est-à-dire, qui sont des suites finies de ces lettres, jouent le rôle des «formules»; par «assertions primitives», on comprend certaines énonciations; chacune des «productions» particulières à l'espèce de systèmes en cause autorise de «conclure», d'énonciations «prémisses» en nombre fini (caractéristique de la production) et présentant, dans un ordre donné (caractéristique de la production), des segments donnés, éventuellement vides (et caractéristiques de la production), à une énonciation «conclusion» non vide et devant présenter des caractéristiques (attachées à la production) analogues. Ces données permettent de définir une «dérivation» comme une suite finie d'énonciations dont tous les termes sont des assertions primitives ou sont conclusion d'une application d'une production à des prémisses entrées auparavant dans la suite, et d'entendre, par «assertions», les énonciations finales des dérivations. Si l'on reçoit les «lettres» comme des *chiffres*, et les suites finies de lettres comme des représentations de nombres dans le système de numération dont la base est le nombre des lettres, les «assertions» forment un ensemble de nombres entiers naturels dont on montre, selon la même argumentation que celle de GÖDEL dans son mémoire sur les théorèmes d'incomplétude, qu'il est récursivement énumérable. Réciproquement, toute donnée d'une fonction récursive générale, dont un ensemble récursivement énumérable est l'image, permet de construire un système de l'espèce décrite ci-dessus, dont cet ensemble est l'ensemble des «assertions»: les «systèmes de Post» permettent donc de définir la notion d'ensemble récursivement énumérable en termes de descriptions de ces systèmes, dans lesquelles la notion de fonction récursive n'est pas explicitement présente. POST montre qu'on peut, à cette fin, se restreindre à des systèmes dont les descriptions sont plus simples: ceux qu'il qualifie de «canoniques», ou ceux, plus particuliers encore, qu'il qualifie de «normaux» — ces derniers n'ont qu'une seule «assertion primitive», chacune de leurs productions a une seule prémisse, de la forme AX (où A est un suite de lettres donnée, segment initial de l'énonciation AX, et caractéristique de cette production), et une conclusion de la forme XB (où B est un suite de lettres donnée, non vide, caractéristique de cette production), et peut être appliquée à n'importe quelle suite X.

En 1944, dans son mémoire *Recursively enumerable sets of positive integers and their decision problem* [*Pst*7], il développe à partir de là une théorie autonome des ensembles récursivement énumérables. Il prouve qu'elle contient la théorie des ensembles récursifs, en établissant le théorème, en fait déjà obtenu par KLEENE [*Kln*8] l'année précédente sous une autre formulation, et selon lequel un ensemble de nombres entiers naturels est récursif, si et seulement si cet ensemble et son complémentaire sont récursivement énumérables. Il inaugure, enfin, toute une classification des ensembles récursivement énumérables non récursifs, dont font partie les ensembles (de codes) de théorèmes récursivement indécidables, ainsi que ROSSER avait été le premier, dans sa note de 1936 [*Rsr*2] sur les théorèmes d'incomplétude, à le remarquer.

Des travaux de TARSKI concernant la géométrie, les nombres réels, et les algèbres de Boole de parties

En 1927, TARSKI a proposé une axiomatisation très originale de la géométrie, dont un résumé [*Trk*4] est paru en 1929. Cette axiomatisation est fondée sur la méréologie de LEŚNIEWSKI (cf. [*Lnw*2], §6), qui axiomatise la relation du tout à la partie, rangée parmi les notions primitives de logique — c'est sur les propriétés de cette relation que reposent les justifications intuitives des syllogismes aristotéliciens — et sur la seule notion primitive spécifique de sphère. Grâce à une axiomatisation convenable de cette notion, TARSKI parvenait à définir les notions de point et de paire de points équidistants d'un même point, à l'aide de familles de sphères ouvertes concentriques, de telle façon que ces deux notions satisfassent une axiomatique de la géométrie établie en 1908 [*Pie*] par Mario PIERI (HUNTINGTON avait donné en 1916 [*Hun*4] une autre version d'une axiomatique de la géométrie reposant sur les notions primitives d'inclusion et de sphère).

Ses réflexions sur l'axiomatique de la géométrie, dont il enseignait, en 1926, une autre version (cf. [*Szb*]), sont très liées à l'élaboration de sa méthode de décision pour les théories élémentaires des nombres réels et de la géométrie élémentaire, qu'il évoque dans une conférence [*Trk*7] prononcée en 1930 à Łwów; j'ai encore à revenir sur son mémoire *Sur les ensembles définissables de nombres réels* [*Trk*8], de 1931, et où, cette méthode exceptée, il détaille les conceptions qu'il y a exposées.

De ces travaux sont issues, aussi, certaines de ses contributions ultérieures à la théorie des algèbres de BOOLE — celles-ci furent nombreuses, et je ne les passerai pas systématiquement en revue. Dans son mémoire *Zur Grundlegung der Booleschen Algebra* [*Trk*13], paru en 1935, il explique en effet que la méréologie est, au rejet de la classe vide près, la théorie d'une algèbre de BOOLE complète générique, introduit les dénominations d'atome, d'algèbre de BOOLE atomique, d'algèbre de BOOLE complète, et débrouille quelques relations de distributivité faisant intervenir les opérations booléiennes infinies, et les rapports entre certaines de ces relations. Selon les indications qu'il fournit en 1938 dans son mémoire *Der Aussagenkalkül und die Topologie* [*Trk*20], ce sont ces travaux de géométrie qui l'ont amené, aussi, à établir que les ouverts réguliers d'un espace topologique forment une algèbre de BOOLE, si l'on prend comme borne supérieure de deux ouverts réguliers le plus petit ouvert régulier qui contient leur réunion.

De la définissabilité et de la définissabilité arithmétique dans la structure de la droite réelle chez TARSKI, en 1930

Le traitement de la sémantique, dans le mémoire *Sur les ensembles définissables de nombres réels* [*Trk*8], est très en avance par rapport à ce qu'il est dans d'autres contextes: après avoir décrit sommairement la notion de satisfaction, dont il traite au même moment dans un autre écrit, TARSKI définit en fait la notion d'ensemble définissable à un ordre quelconque dans une structure d'interprétation d'un langage. La notion de structure d'interprétation d'un langage s'entend, pour nous, comme celle d'un jeu d'interprétations de ses notions primitives, toutes relatives à un même domaine d'individus donné. Dans le texte de TARSKI, ce jeu est détaillé dans les cas des deux exemples qui seront mentionnés ci-après: une telle structure se présente là comme une suite de graphes composés de suites d'individus de même longueur, et dont chacun interprète une notion primitive de nombre d'argument égal; des individus, il est seulement dit que ce sont des objets, qui «ne devront point [être] des nombres réels» — sans doute faut-il entendre: pas nécessairement. Le traitement *des* cas des nombres réels — car il y en a *deux*: celui du langage du nombre 1 (élément distingué), de l'ordre et de l'addition, traité en détail, et celui du langage de l'addition et de la multiplication, dont le traitement n'est qu'allusif — apparaît, relativement à cette ouverture sur le cas générique, comme paradigmatique. C'est par référence à ce cas que TARSKI — est-ce indépendamment de GÖDEL? — appelle ensembles et relations *arithmétiquement définissables*, de ce que nous appelons maintenant une structure, les parties et graphes de relations de cette structure, définissables *au premier ordre* — quelque vingt ans plus tard, il réintroduira cette terminologie en théorie des modèles.

En ce qui concerne les nombres réels, et plus généralement les corps réels clos, ce mémoire de TARSKI est le mémoire fondateur de ce qu'on appelle aujourd'hui la *géométrie algébrique réelle*. TARSKI établit que les ensembles de nombres réels arithmétiquement définissables en termes du nombre 1, de l'ordre et de l'addition, sont les réunions finies d'intervalles d'extrémités rationnelles; et que les ensembles de nombres réels arithmétiquement définissables en termes de l'addition et de la multiplication, sont les réunions finies d'intervalles d'extrémités algébriques.

A la fin de son mémoire, il donne une formule transcrivant, en opérations ensemblistes pratiquées à partir des graphes interprétant un jeu de notions primitives, une forme normale prénexe d'une formule du langage du premier ordre composé de ces notions primitives. Les quantificateurs existentiels s'y transcrivent en projections, et dualement, les quantificateurs universels en projections de «cylindres inscrits»: KURATOWSKI y a aussitôt reconnu les procédés de construction des ensembles projectifs, tout juste introduits l'année précédente par LUSIN [*Lsn*2]; une note conjointe [*KrT*] de KURATOWSKI et de TARSKI (qui rend elle aussi compte de conférences prononcées l'année précédente à Łwów) vient donc juste à la suite du mémoire de TARSKI; il y est montré, d'une part, que si les graphes de notions primitives sont projectifs, les extensions des formules de leur langage sont des ensembles projectifs, et d'autre part, que c'est le cas des ensembles arithmétiquement définissables de nombres réels. A la suite de cette note, il vient encore un article [*Kur*3] dans lequel KURATOWSKI met à profit la théorie qui précède pour présenter de nombreux exemples d'ensembles projectifs dont il calcule, à partir de l'analyse de leurs définitions, à quelle classe ils appartiennent.

De la théorie de la récursivité: de la hiérarchie arithmétique

L'analogie, qui a frappé KURATOWSKI, entre les rapports que les quantifications et les connections établissent entre les formules, et ceux que les opérations booléiennes et géométriques correspondantes établissent entre les ensembles projectifs, ne cesse de hanter les esprits en Pologne. Sitôt la notion d'ordinal définissable introduite, KURATOWSKI s'empresse, en 1937, de publier une Note sur *Les types d'ordre définissables et les ensembles boréliens* [*Kur*4]. Pendant la seconde guerre mondiale (cf. [*Crl*2]), MOSTOWSKI eut l'idée d'explorer le début d'analogie constaté, en faisant correspondre, selon le passage des nombres réels vers les nombres entiers naturels, aux ensembles boréliens, les ensembles «décidables» — en fait, les ensembles récursifs, pour le déploiement complet de son traitement — puis en retraçant ce que l'on peut savoir, dans l'un et l'autre cas, de la hiérarchie des ensembles engendrés, à partir des précédents, en saturant, alternativement, par projections, puis par passage aux complémentaires, puis par réunions et intersections. Il naît de là un classement des graphes de nombres entiers naturels en classes analogues aux classes de la hiérarchie projective, tout simplement parce que dans un cas comme dans l'autre, à partir d'une classe génératrice (celle des boréliens dans un cas, celle des ensembles «décidables» dans l'autre), on procède de la même manière.

Or, dans un cas comme dans l'autre, les graphes des notions primitives (addition, multiplication, ordre, ...) appartiennent à la classe génératrice; donc, les graphes de nombres entiers naturels *arithmétiquement définissables* se répartissent en classes entretenant entre elles les mêmes rapports que les classes de la hiérarchie projective: c'est pourquoi, de nos jours, on qualifie d'*arithmétique* la hiérarchie des ensembles arithmétiques de nombres entiers naturels. De plus, ceux-ci entretiennent, avec leurs formules de définition, supposées sous forme prénexe, exactement le même type de rappports que les ensembles arithmétiques de nombres réels avec les leurs.

La classe génératrice, dans le cas des nombres entiers naturels, étant celle des ensembles récursifs, qui sont tous arithmétiquement définissables, la hiérarchie ne laisse échapper aucun des graphes engendrés à partir de cette classe. Alors, de plus, on voit que, dans le passage des nombres réels aux nombres entiers naturels, aux ensembles analytiques correspondent les ensembles récursivement énumérables (en 1928 [*Mng*], Karl MENGER avait cru reconnaître, dans les ensembles analytiques, les ensembles de nombres réels admis par BROUWER, mais HEYTING [*Htg*3] n'avait pas trouvé son argumentation recevable, du point de vue intuitionniste); et on peut encore se référer au premier développement [*Kln*3] de la théorie des fonctions récursives de KLEENE, de 1936, qui suggère l'obtention, pour chaque classe de la hiérarchie, de fonctions universelles. Par une «fonction universelle» pour la classe restreinte des ensembles de n-uplets (n fixé) appartenant à une classe donnée de la hiérarchie, MOSTOWSKI [*Mtw*6] entend une surjection de l'ensemble des nombres entiers naturels sur cette classe restreinte, et dont le graphe est arithmétiquement définissable, donc appartient à la hiérarchie — dans le cas des fonctions récursives, considéré par KLEENE, une fonction universelle f pour la classe des fonctions récursives à n arguments (n fixé) associe à chaque nombre entier naturel p une fonction $f(p)$ à n arguments, et se présente souvent, en faisant apparaître le paramètre p comme un argument supplémentaire, sous la forme $f(p, x_1, \ldots, x_n)$. Dans le cas des ensembles, c'est la *formule de définition* $f(p, x_1, \ldots, x_n)$ d'un ensemble de n-uplets générique de la classe, qui entre en jeu; elle permet de tenir un raisonnement «diagonal» à la CANTOR en identifiant le paramètre p à d'autres arguments, puis en altérant uniformément la valeur obtenue, c'est-à-dire, ici, en passant à la négation: on obtient la formule de définition d'un ensemble qui appartient à la hiérarchie, sans pouvoir appartenir à la classe en jeu; on montre ainsi que les inclusions de la hiérarchie sont strictes. Il a pu suffire de vouloir poursuivre l'analogie avec la hiérarchie projective, où l'existence de telles fonctions universelles était connue, en s'appuyant sur l'existence de fonctions universelles pour la classe des ensembles récursifs, pour sentir que des fonctions universelles devaient exister à tous les niveaux.

MOSTOWSKI ayant abandonné son manuscrit lors d'une évacuation forcée de son domicile, doit après la guerre le reconstituer, avant qu'il ne paraisse, en 1947 [*Mtw*6], dans les *Fundamenta Mathematicae*, sous le titre *On definable sets of positive integers*. A ce moment-là, il y a déjà quatre années que KLEENE a publié un mémoire intitulé *Recursive predicates and quantifiers* [*Kln*8], rédigé en 1940 (date à laquelle il a d'ailleurs pris la précaution [*Kln*5] de l'annoncer), et dans lequel, adoptant le point de vue du classement des *prédicats* arithmétiques, c'est-à-dire des *formules* du premier ordre de l'arithmétique, il aboutit à la hiérarchie des formules correspondante, engendrée par la classe des formules qui définissent des graphes récursifs de nombres entiers naturels, en saturant, alternativement, par quantifications existentielles, puis par passage aux négations, puis par disjonctions et conjonctions.

A ceci près qu'il établit qu'un prédicat est récursif si et seulement si lui-même et sa négation sont récursivement énumérable, il développe une théorie analogue à celle de MOSTOWSKI, avec des prédicats universels, qui permettent d'établir que les inclusions de la hiérarchie sont strictes. Il y apparaît que la classe à laquelle appartient une formule est déterminée par *les* couples de caractéristiques des préfixes de *ses* formes normales prénexes (qui sont susceptibles de se présenter sous des formes différentes sous ce rapport), les caractéristiques d'un préfixe en cette matière étant la nature du quantificateur par lequel il débute, et le nombre d'alternances de quantificateurs existentiels et de quantificateurs universels qu'il comporte, *les quantificateurs bornés n'entrant pas en ligne de compte;* chaque architecture de formule impose un nombre minimum d'alternances. Et c'est juste dans une note de bas de page que KLEENE note la ressemblance de cette hiérarchie avec la hiérarchie des ensembles projectifs. Il tire aussi du théorème sur la classification un résultat d'incomplétude pour les «logiques ordinales» proposées quatre ans auparavant par TURING [*Trg*3].

Vers 1950, indépendamment (cf. [*Kln*13]), Martin Davis [*Dvs*1] et MOSTOWSKI [*Mtw*10] sont les premiers à lancer l'idée d'une extension au transfini de la hiérarchie arithmétique. MOSTOWSKI, dont la version couvre les ordinaux constructifs, introduit ainsi la hiérarchie aujourd'hui qualifiée d'*hyperarithmétique*.

Selon la version de la théorie des fonctions partiellement récursives qui fait intervenir les machines de TURING (et qui n'est pas celle que KLEENE utilise dans ses travaux de l'époque, où l'on retrouve, cependant, les équivalents exacts des notions qui suivent), du fait que le mécanisme de fonctionnement des machines de TURING à partir de la donnée de leurs caractéristiques propres leur est commun à toutes, il y a parmi ces machines des machines universelles pour la classe des fonctions partiellement récursives, auxquelles on n'a qu'à indiquer, sous forme de code, quelles sont les caractéristiques propres au fonctionnement d'une machine particulière, et un n-uplet de nombres, pour qu'elles décodent le code de cette machine particulière, selon une procédure indépendante de celle-ci, puis en émulent à partir de là le fonctionnement, et calculent, si elle existe, la valeur en ce n-uplet de la fonction calculée par cette machine particulière. Dans la théorie des fonctions récursives dont KLEENE élabore, durant les années qui précèdent la seconde guerre mondiale, les théorèmes fondamentaux, et dont la mieux venue des versions est celle qui s'applique aux fonctions partiellement récursives, les codes des machines qui calculent une fonction récursive sont appelés les *index* de cette fonction; outre l'existence de formes normales, ces théorèmes rendent compte de relations entre certaines fonctions particulières et les fonctions représentées par les formes normales; le plus important peut-être est une espèce de théorème de point fixe, qui dit que, pour tout argument d'une fonction récursive à $n + 1$ arguments, il y a une valeur telle que si on fixe cet argument à cette valeur, on tombe juste sur la fonction à n arguments dont cette valeur est un index.

L'affinement de cette classification, et l'étude des propriétés des ensembles appartenant à ses diverses classes, ont par la suite été activement poursuivis. Ils se sont avérés d'un niveau de difficulté particulièrement élevé — le problème de l'existence de plusieurs degrés récursivement énumérables, par exemple, n'a été résolu [*Fdb*1, 2], [*Mnk*] qu'en 1956 — et ont donné lieu à de nombreux travaux; toute une branche de la théorie de la récursivité en est issue.

En 1950, KLEENE [*Kln*10] établit l'existence de deux ensembles récursivement énumérables disjoints non vides *récursivement inséparables*, c'est-à-dire tels qu'il n'existe pas d'ensemble récursif qui contienne l'un et soit disjoint de l'autre. Sur ce point, l'analogie avec la hiérarchie projective cesse, puisque cette dernière satisfait au contraire au théorème de séparation de LUSIN [*Lsn*1], selon lequel deux ensembles analytiques disjoints non vides sont toujours séparés par un ensemble borélien; par contre, le fait qu'un ensemble soit récursif si et seulement si lui-même et son complémentaire sont récursivement énumérables est l'exact analogue du théorème de Mikhaïl Iakovlevitch SOUSLIN [*Sln*], selon lequel un ensemble est borélien si et seulement si lui-même et son complémentaire sont analytiques; sur ce point, l'analogie se poursuit aux niveaux supérieurs, où valent des théorèmes d'intersection du même genre, ainsi que POST [*Pst*9] l'a établi en 1948 pour la hiérarchie arithmétique.

Des degrés de non-résolubilité récursive

POST introduit, dans son mémoire de 1944 [*Pst*7], diverses notions de *réductibilité* du problème de décision d'un ensemble de nombres entiers naturels (trouver un algorithme récursif qui en trie les éléments) à celui d'un autre ensemble de nombres entiers naturels. Par exemple, la donnée d'une fonction récursive dont le premier de ces ensembles est l'image réciproque du second par cette fonction, permet de ramener la question de l'appartenance d'un nombre au premier, à celle de l'appartenance, au second, de l'image de ce nombre par cette fonction. POST emprunte au mémoire de TURING de 1939 [*Trg*3] la plus générale de ces notions de réductibilité: le problème de décision d'un ensemble de nombres entiers naturels sera ramené à celui d'un autre, si l'on peut décrire une *machine de* TURING *à oracle* qui résout le premier problème, à partir du moment où son oracle fournit une solution du second. Par une machine (de TURING) à oracle, TURING entend une machine (de TURING) modifiée de façon à pouvoir, lors d'un étape de calcul, soumettre un nombre à un «oracle», qui lui répond en lui indiquant quel caractère elle doit imprimer sur la case du ruban en face de laquelle sa tête de lecture se trouve.

Ces relations de réductibilité sont des préordres sur l'ensemble des ensembles de nombres entiers naturels, et POST introduit, sous l'appellation de *degrés* de non-résolubilité — «récursive», ajoutera-t-il en 1948 — relatifs à l'une d'entre elles, les classes d'ensembles de nombres entiers naturels selon l'équivalence associée à cette relation.

En 1944, il esquisse avant tout une théorie des degrés des ensembles récursivement énumérables; il fait tout de même allusion à l'existence, hors ceux-ci, d'une suite infinie croissante de degrés au sens de TURING, qu'il décèle dans l'article cité de TURING; en 1948, il en exhibera une de façon explicite. Des degrés récursivement énumérables, il n'en décèle qu'un, dont il montre qu'il est certainement le plus grand, en tous les sens, introduits là, de la notion de réductibilité: c'est celui de l'ensemble des codes des machines de TURING qui, appliquées à leur propre code, s'arrêtent. Pour chacune des notions de réductibilité plus faible que celle de TURING, POST présente une notion d'ensemble récursivement énumérable possédant une propriété particulière, sous laquelle il suggère de prospecter des candidats à un degré de résolubilité moindre. Il introduit ainsi toute une classification des ensembles récursivement énumérables, et en compare les représentants les uns aux autres, tout en indiquant quelques-unes de leurs propriétés.

L'affinement de cette classification, et l'étude des propriétés des ensembles appartenant à ses diverses classes, ont par la suite été activement poursuivis. Ils se sont avérés d'un niveau de difficulté particulièrement élevé — le problème de l'existence de plusieurs degrés récursivement énumérables, par exemple, n'a été résolu [*Fdb*1, 2], [*Mnk*] qu'en 1956 — et ont donné lieu à de nombreux travaux; toute une branche de la théorie de la récursivité en est issue.

Point sur divers cas spéciaux du problème de la décision de formules du premier ordre, vers 1950

Jusqu'en 1934, on avait pu trouver des solutions au problème de la décision pour les formules du premier ordre présentant des formes prénexes dont les préfixes comportent certaines successions favorables de quantificateurs. Les recherches qui se poursuivaient alors visaient à examiner si l'on ne pouvait pas trouver aussi des solutions, éventuellement plus élaborées, pour d'autres successions de quantificateurs.

En 1936, le théorème d'indécidabilité de CHURCH [*Chu*5] vint rendre évidente l'existence de préfixes défavorables, tels qu'il n'y a pas de solution au problème de la décision pour les formules de formes normales prénexes présentant ces préfixes. Or, GÖDEL, qui avait entre 1929 et 1933 [*Gdl*4, 10] décrit un préfixe favorable, avait aussi établi que le problème de la décision général pour toutes les formules du premier ordre se ramène au problème de la décision pour les formules de forme normale prénexe présentant un préfixe composé de trois quantificateurs existentiels suivis de quantificateurs universels. Y avait-il été conduit par le hasard de la recherche, ou pressentait-il le résultat de CHURCH? En 1936 en tout cas, ce résultat devait prendre un relief nouveau, car il en résultait qu'il ne peut y avoir de méthode de décision récursive pour les formules présentant des préfixes de ce type. Le problème maintenant posé était de délimiter la frontière entre les préfixes de formules prénexes telles que les formules présentant ces préfixes admettent un procédé de décision, et les autres; une méthode pour détecter des préfixes exclus, était de montrer que le problème de la décision général pour les formules du premier ordre se réduit au même problème pour les formules prénexes présentant un préfixe soupçonné défavorable. Diverses avancées sur cette question ont eu lieu entre 1936 et 1950; les plus efficientes ont été celle de János SURÁNYI, qui a prouvé la réductibilité du problème général à ce même problème, pour les formules prénexes de chacun des préfixes:

$$\exists x \exists y \exists z \forall u \text{ et } \exists x \exists y \forall z \exists u, \text{ en 1943 } [Sur1];$$
$$\exists x \forall y \forall z \exists u \exists v \text{ et } \forall x \exists y \forall z \exists u \exists v, \text{ en 1948 } [Sur2].$$

D'un tout autre type est la forme spéciale du problème de la décision à laquelle répond, en 1950, une Note [*Trb*] de B.A. TRAHTÉNBROT: il montre que le problème de la décision pour les formules valides dans tout domaine fini n'est pas récursivement soluble, en établissant, par une méthode qui s'apparente aux méthodes directes de GÖDEL et de CHURCH, que l'ensemble de ces formules n'est pas récursivement énumérable, bien que son complément le soit.

Des débuts de l'«analyse calculable»

La recherche de notions «récursives» ou «calculables» en analyse remonte aux environs de 1936. Le premier article de TURING [*Trg*1], de 1936, s'intitulait *On computable numbers, with an Application to the Entscheidungsproblem.* Mais ce fut en Pologne qu'avant la guerre BANACH et MAZUR poursuivirent cette idée de la façon la plus conséquente. La seconde guerre mondiale ayant entravé la publication de leurs travaux de cette époque, il n'en reste qu'un résumé, *Sur les fonctions calculables* [*BMz*], paru en 1937. La monographie *Computable analysis* [*Mzr*], de MAZUR, éditée en 1963 à Varsovie par Andrzej GRZEGORCZYK et Helena RASIOWA, d'après des notes prises à des conférences prononcées par MAZUR à Varsovie durant l'année académique 1949-1950, ne permet pas de trier les apports des différentes époques. Entre temps, Ernst SPECKER a publié un essai *Nicht konstruktiv beweisbare Sätze der Analysis* [*Spk*], et certaines des définitions qu'il a envisagées sont incorporées à l'ouvrage.

Je me bornerai à citer l'une des définitions d'une notion de nombre réel récursif présentées, d'ailleurs due à SPECKER. Elle requiert la donnée d'une fonction récursive f telle que la valeur absolue de la différence entre le quotient $f(n)/n$ et la valeur absolue du nombre réel a que l'on veut «récursif» soit inférieure à $1/n$. Les nombres récursifs ainsi définis forment un corps, algébriquement clos dans le corps des nombres réels, et qui contient les nombres algébriques.

S. Mazur

L'essai va jusqu'aux fonctions d'une variable réelle; les propriétés en sont étudiées en considérant des suites de nombres réels récursives. Il apparaît là le type de difficulté qui arrête toutes les reconstructions «constructivistes» de l'analyse: une suite récursive n'a pas toujours pour inverse une suite récursive, et cela s'étend aux fonctions; et l'on n'a pas le théorème de BOLZANO-WEIERSTRASS.

Des progrès et controverses, postérieurs à 1930, touchant mathématiques et logique intuitionnistes

Jusqu'au début de la seconde guerre mondiale, les intuitionnistes sont très peu nombreux. La décennie qui précède est une période de lents progrès pour les mathématiques intuitionnistes. BELINFANTE, après avoir dégagé et étudié diverses notions intuitionnistes de convergence, s'attaque à la théorie intuitionniste des fonctions d'une variable complexe; ses travaux se closent sur une série de communications sur ce sujet, parue en 1941, tandis que HEYTING publie un mémoire sur l'algèbre intuitionniste. Dans l'intervalle, la version intuitionniste de la topologie [*Frd*] est due, en 1936, à Hans FREUDENTHAL.

Entre temps, après la formalisation de HEYTING, la logique intuitionniste retient l'attention: en 1932 [*Kmg*2], KOLMOGOROFF en propose une interprétation comme «logique des problèmes»; ainsi, $A \to B$ représente, selon cette interprétation, le problème de «ramener le problème B au problème A» (une solution du problème A permettant d'obtenir une solution du problème B), $\forall x A$ représente le problème «d'indiquer une méthode de résolution du problème A, indépendante du paramètre x», et $\exists x A$, le problème de «désigner une valeur du paramètre x, solution du problème A»; la même année, et l'année d'après, GÖDEL [*Gd*7, 10] et GENTZEN [*Gtz*2] explorent les rapports entre les mathématiques classiques et les mathématiques intuitionnistes, aux niveaux des calculs propositionnels, des calculs des prédicats, et des arithmétiques formalisées; des méthodes de décision pour le calcul propositionnel intuitionniste sont proposées par GENTZEN [*Gtz*2] en 1934, par JAŚKOWSKI [*Jkk*1] en 1935, par WAJSBERG [*Wbg*] en 1938; une autre encore le sera par MCKINSEY et TARSKI [*MKT*3] en 1948; dans l'intervalle, TARSKI [*Trk*20] fait connaître, en 1937, des rapports, relatés ci-après, entre espaces topologiques et calcul propositionnel intuitionniste.

Parallèlement, un débat se poursuit quant aux significations intuitionnistes de l'implication et de la négation: faut-il, comme HEYTING [*Htg*2], retenir le schéma d'axiomes

$$\text{NON}A \to (A \to B),$$

ou faut-il, comme KOLMOGOROFF [*Kmg*1] en 1925, ne pas prendre position là-dessus ? Le système qui résulte de celui de HEYTING en supprimant ce schéma est étudié en 1936 par Ingebrigt JOHANSSON [*Jhs*] sous le nom de *calcul minimal*; c'est en quelque sorte, la partie du calcul des propositions sur laquelle se fait l'accord de la logique classique et des différentes conceptions de la logique intuitionniste. JOHANSSON est à même d'en établir une version dans un système de déduction naturelle, et une autre sous la forme d'un calcul de séquences.

A partir de 1942, BROUWER reprend la plume (p. ex [*Brw*12]) et approfondit les concepts introduits vingt ans plus tôt. En 1944 [*Grs*], G.F.C. GRISS engage une polémique sur le rôle de la négation en mathématiques; il préconise une mathématique sans négation, une discipline qui n'est plus guère pratiquée maintenant. En 1947, dans un travail [*vDz*] rédigé depuis 1942, VAN DANTZIG, cherchant des domaines où mathématiciens classiques et mathématiciens intuitionnistes pourraient

A. Tarski

se rencontrer, propose aussi celui des mathématiques *affirmatives*, composé des propositions dénuées de négations, et celui des mathématiques *stables*, composé des propositions équivalentes, en logique intuitionniste, à leur double négation, qui recouvre tous les énoncés négatifs (l'ensemble de ces domaines constitue néanmoins une partie très limitée des mathématiques). Pour établir l'impossibilité de se passer de négation, BROUWER présente, au dixième congrès international de philosophie, qui se tient à Amsterdam en 1948, un procédé de construction d'un nombre réel non égal à zéro (savoir, pour lui, dont l'égalité à zéro implique une contradiction) et dont on ne peut prouver, au sens intuitionniste, ni qu'il est strictement supérieur à zéro, ni qu'il lui est strictement inférieur ([*Brw*13]); dans les années qui suivent, par la même méthode, il construit des contre-exemples intuitionnistes à divers théorèmes d'analyse classique: cela va, en 1952 [*Brw*14], jusqu'au théorème de BOLZANO-WEIERSTRASS.

Vers 1940, KLEENE soupçonne l'existence d'interprétations de la logique et des mathématiques intuitionnistes dans la théorie des fonctions récursives (cf. [*Kln*14]). Partant d'une remarque de HILBERT et BERNAYS, interprétant en 1934 [*HbB*] un énoncé existentiel intuitionniste comme une «communication incomplète», appelant à la reconstitution d'une information permettant la production d'un être mathématique répondant à la condition demandée, et d'une preuve de cette assertion-là, KLEENE conjecture alors que cette information est exprimable, après arithmétisation, par la donnée d'une fonction récursive; savoir, que si une formule de la forme $\forall x \exists y A(x,y)$ admet une démonstration formelle intuitionniste, on peut produire une fonction récursive φ telle que la formule $A(x, \varphi(x))$ produit, par toute substitution, à x, d'une représentation d'un nombre entier naturel, un énoncé vrai. En recourant à une fonction récursive universelle énumérant les fonctions à un argument, selon les valeurs attribuées à un paramètre supplémentaire, la fonction φ peut à son tour être codée par un de ses index e, savoir, l'une des valeurs entières naturelles à donner au paramètre pour obtenir φ. KLEENE [*Kln*6, 7, 9] est ainsi parvenu, à partir de ces idées, à donner, du concept «le nombre entier naturel e réalise la formule A (de l'arithmétique)», une définition telle que si une formule de la forme $\forall x \exists y A$ admet une preuve formelle dans l'arithmétique de HEYTING (ayant pour axiomes les axiomes de PEANO et les axiomes du calcul des prédicats intuitionniste), on peut calculer un nombre entier naturel e tel qu'il en aille de même de la formule exprimant, selon cette définition, que e réalise $\forall x \exists y A$. KLEENE, et plus tard David NELSON [*Nls*], en 1947, sont ainsi en mesure de montrer qu'un certain nombre de schémas de théorèmes classiques ne produisent pas des théorèmes intuitionnistes, faute d'être nécessairement réalisables; KLEENE souligne particulièrement que tel est le cas du schéma $\forall x[A(x)$ OU NON $A(x)]$. En raison de la seconde guerre mondiale, la publication [*Kln*9] de ces travaux est intervenue en 1945; ses développements ultérieurs se situent après 1950. Il en va de même des idées analogues émises indépendamment par BETH [*Bth*1] en 1947, de mettre en œuvre une sémantique intuitionniste usant d'une notion de modèle adaptée à la logique intuitionniste, à chercher, et (indépendamment d'une suggestion analogue faite par KLEENE en 1941), d'interpréter la notion des «règles» requises par la construction des suites génératrices de nombres réels de BROUWER [*Brw*5] par celle d'algorithmes récursifs.

Des avancées, après 1930, au carrefour entre logique propositionnelle, algèbre et topologie

Des rapports entre algèbres de BOOLE, systèmes déductifs, calcul des propositions classique et calcul des propositions intuitionniste, dégagés par TARSKI en 1935

Dès 1931, TARSKI (cf. [*Trk*34], p. 421) avait remarqué que, dans l'algèbre de BOOLE des parties de l'ensemble des énoncés, les systèmes déductifs satisfont les axiomes du calcul propositionnel intuitionniste [soit, que ces axiomes prennent pour valeur l'ensemble de tous les énoncés, quels que soient les systèmes déductifs assignés pour valeurs à leurs variables, si l'on interprète: la conjonction par l'intersection; la disjonction par le système engendré par la réunion; l'implication par le plus grand système (qui existe) dont la trace sur l'impliquant est incluse dans l'impliqué; et la négation, par le plus grand système (qui existe) dont la trace est le plus petit des systèmes]; en outre, les sytèmes déductifs «axiomatisables» (finiment), c'est-à-dire, engendrés par un énoncé (conjonction d'un nombre fini d'axiomes), satisfont de même les axiomes de l'algèbre de BOOLE: tels sont les principaux résultats (en dehors de ceux de l' Appendice, qui relèvent de la théorie de modèles, et sur lesquels je reviendrai) de son mémoire sur les *Grundzüge des Systemenkalkül* [*Trk*12], de 1935 et 1936.

Des rapports entre topologie, algèbre de BOOLE et algèbre linéaire, dégagés par STONE aux alentours de 1935

Cette publication se situe entre l'annonce [*Stn*1] faite, par M. H. STONE, en 1934, du théorème de représentation des algèbres de BOOLE par les parties ouvertes et fermées d'un espace topologique compact et totalement discontinu, et la publication de ses mémoires détaillés de 1936 [*Stn*3] et 1937 [*Stn*4], établissant les théorèmes de représentation des algèbres de BOOLE bien connus, avec, dans l'intervalle, en 1935, sa Note [*Stn*2] sur l'interdéfinissabilité des algèbres de BOOLE et des anneaux idempotents — les anneaux booléiens — et la transposition qui s'ensuit de la notion d'idéal aux algèbres de BOOLE: le mémoire de 1936 est consacré à la représentation de toute algèbre de BOOLE par des algèbres de parties de l'ensemble de ses idéaux premiers — indépendamment de TARSKI, STONE s'aperçoit aussi que les idéaux d'une algèbre de BOOLE satisfont les axiomes du calcul des propositions intuitionniste; le mémoire de 1937 établit la dualité entre les algèbres de BOOLE et les espaces topologiques compacts et totalement discontinus, les idéaux entrant en dualité avec les ouverts.

STONE n'a eu à prouver que la définissabilité d'une algèbre de BOOLE sur tout anneau de BOOLE possédant une unité, car Orrin FRINK [*Frk*] avait reconnu, dès le début de 1926, la définissabilité, sur toute algèbre de BOOLE, d'algèbres (linéaires) associatives sur le corps à deux éléments, annoncée dans une communication parue au début de 1928. L'idée de rechercher de telles structures, au moins sur les algèbres de BOOLE de parties d'un ensemble, peut avoir germé de la constatation du caractère de loi de groupe abélien de l'«addition modulo 2», dont FRINK trouve qu' «in the analysis situs ... it plays a more important rôle than logical addition». C'est d'ailleurs un retour à l'addition originellement introduite par BOOLE [*Boo*] lui-même (qui cependant la déclarait non définie quand elle portait sur des termes ne s'excluant pas mutuellement), également effectué (avec son extension), en utilisant le corps à deux éléments, par GÉGALKINE [*Ggk*] en 1927.

Ainsi que STONE l'a dit lui-même ([*Stn*6]), le cheminement d'idées qui l'a conduit à ces travaux trouve sa source dans les problèmes de la physique théorique de l'époque: «The need for investigations of this character was suggested to me by the theory of operator-rings in the HILBERT space: there, as in other rings and linear algebras, the «spectral» representation as a «direct sum» of irreducible subrings reposes in essence upon the construction of an abstract Boolean algebra; and this construction, trivial for rings with strong chain conditions, is not trivial in the case of operator-rings». En effet, des physiciens comme Paul DIRAC et Werner Karl HEISENBERG représentaient les états dont un système (physique) est susceptible, par les points d'un espace de HILBERT complexe, et chaque grandeur observable, par un opérateur sur cet espace, auto-adjoint, non nécessairement borné, et admettant pour valeurs propres, réelles, les divers résultats de mesure dont cette grandeur est susceptible. La théorie spectrale de HILBERT, à laquelle VON NEUMANN avait donné sa forme actuelle en 1931 [*vNm*5], avait introduit les projections orthogonales sur les sous-espaces fermés engendrés par les parties du spectre; sur ces opérateurs, les opérations naturelles: composition, différence d'avec l'identité, déterminent une algèbre de BOOLE, isomorphe au quotient de l'algèbre de BOOLE des parties boréliennes du spectre, selon les ensembles de mesure nulle. Une question de HAAR sur des opérateurs auto-adjoints avait amené VON NEUMANN à publier, en 1931, un court travail intitulé *Algebraische Repräsentanten der Funktionen «bis auf eine Menge vom Mass Null»* [*vNm*4]. C'est à cette Note, et à celle de STONE, parue l'année précédente, que VON NEUMANN et STONE [*Stn*1] renvoient, dans un mémoire intitulé *The determination of representative elements in the residual classes of a Boolean algebra* [*vNS*], qu'ils publient en 1935.

Une fois la notion d'idéal étendue aux algèbres de BOOLE, son extension, jusqu'aux ensembles ordonnés, ne tarde pas: c'est le fait, dès 1935, de H.M. MACNEILLE, à la recherche, sur la suggestion de STONE (cf. [*Stn*6]), de «complétions» d'ensembles ordonnés en un réseau complet aussi proche que possible d'une algèbre de BOOLE. Le plus connu des théorèmes obtenus par MACNEILLE [*McN*] utilise une extension, à un ensemble ordonné quelconque, de la notion de coupure utilisée par DEDEKIND [*Ded*1] pour définir les nombres réels à partir des nombres rationnels. Il fournit un plongement, avec conservation des bornes inférieures et supérieures existantes, d'un ensemble ordonné dans un réseau complet, minimal pour ces propriétés, mais qui n'est pas même nécessairement distributif (mais c'est une algèbre de BOOLE, si l'ensemble plongé en était déjà une). STONE ([*Stn*6]) prisait plus un procédé de complétion plus raffiné de MACNEILLE, qui aboutit à plonger un réseau distributif dans une algèbre de BOOLE complète, avec conservation des bornes inférieures finies et des bornes supérieures; John Charles Chenoweth MCKINSEY et TARSKI en usent en 1945 [*MKT*2] pour établir un théorème de représentation pour les algèbres à fermeture.

Des rapports entre métamathématique générale, algèbre de BOOLE et topologie, dégagés par MOSTOWSKI et TARSKI, en 1938

TARSKI, dans une communication de 1937 [Trk18], MOSTOWSKI, dans son mémoire *Abzählbare Boolesche Körper und ihre Anwendung auf die allgemeine Metamathematik* [Mtw1], de 1938, où il atteste avoir obtenu maints résultats sur les suggestions de TARSKI, prennent acte de l'habillage de la métamathématique en interprétation de l'algèbre de BOOLE, issu des travaux de STONE: une théorie déductive stable par formation de négations et d'implications devient, par «passage au quotient» selon la relation «$x \leftrightarrow y$ appartient au moins inclusif des systèmes déductifs», une algèbre de BOOLE; les systèmes déductifs de la théorie deviennent les idéaux de l'algèbre; les systèmes déductifs complets et compatibles, les idéaux premiers; les systèmes déductifs [finiment] axiomatisables, les idéaux principaux; et le «type structurel» de la théorie, au sens de TARSKI [Trk12], le type d'isomorphisme de l'algèbre.

MOSTOWSKI franchit un pas de plus en renvoyant, de l'espace dual de cette algèbre de BOOLE, aux fermés de l'ensemble de CANTOR auxquels il est homéomorphe. Or, un tel fermé est dénombrable ou de puissance du continu; notant que dans la dualité de STONE, les points isolés correspondent aux idéaux premiers principaux, MOSTOWSKI est en mesure de calculer les nombres de systèmes déductifs complets et axiomatisables et de systèmes déductifs complets et non [finiment !] axiomatisables d'une théorie déductive, dont le couple constitue la «paire caractéristique», introduite par TARSKI [Trk12] en 1936 dans les *Grundzüge des Systemenkalkül, Zweiter Teil,* de cette théorie. Si l'espace dual est dénombrable, selon un théorème démontré en 1920 par MAZURKIEWICZ et SIERPINSKI [MzS], son type d'homéomorphisme, donc le type d'isomorphisme de l'algèbre, est déterminé par cette paire caractéristique; et selon un théorème de MAHLO, de 1911 [Mhl2], cet espace est homéomorphe à un ensemble bien ordonné, déterminé par la paire. Si l'espace dual est de puissance du continu, MOSTOWSKI tire, du théorème de représentation établi par VON NEUMANN et STONE dans leur mémoire commun [vNS] cité plus haut, le corollaire qu'une algèbre de BOOLE dénombrable, dont l'ensemble des idéaux premiers est de puissance du continu, contient des sous-algèbres de tous les types d'isomorphisme d'algèbres dénombrables.

Visiblement, c'est à la lumière de l'avant-dernier résultat cité et de la *Contribution à la topologie des ensembles dénombrables* [MzS], de MAZURKIEWICZ et SIERPINSKI, auquel il renvoie, qu'il faut lire le mémoire commun de MOSTOWSKI et TARSKI, *Boolesche Ringe mit geordneter Basis* [MwT1], paru en 1939. Ils y montrent que si une algèbre de BOOLE est dotée d'un système de générateurs ordonné, (par l'ordre induit), chacun de ses éléments se représente, de façon unique, comme somme de générateurs; l'algèbre est isomorphe à l'algèbre des réunions finies d'intervalles semi-ouverts à droite de la base (parmi lesquels il faut compter l'ensemble vide); et son espace dual est homéomorphe à l'espace des sections initiales non vides de sa base. Leurs algèbres de BOOLE *héréditairement atomiques* sont aujourd'hui qualifiées de *superatomiques*; ils en montrent des propriétés de structure; en particulier, que la superatomicité équivaut à l'existence d'une base dispersée, et qu'alors, toutes les bases ordonnées sont dispersées.

Des rapports entre calcul des propositions intuitionniste, topologie, algèbres de BOOLE et calcul des propositions classique, dégagés par TARSKI en 1938

En 1930, sachant que le calcul des propositions trivalent de ŁUKASÍEWÍCZ n'accepte pas le tiers-exclu, Hans HAHN (cf. [*Gdl*5]) avait demandé si le calcul des propositions intuitionniste n'admettait pas une matrice à trois valeurs et caractéristique, c'est-à-dire telle que les théorèmes de ce calcul soient juste les énoncés ne prenant que la valeur désignée, quelles que soient les valeurs assignées aux variables. En 1931 [*Gdl*5], GÖDEL avait produit une famille infinie emboîtée de calculs propositionnels intermédiaires entre le calcul intuitionniste et le calcul classique, montrant que le calcul des propositions intuitionniste ne peut posséder de matrice caractéristique finie. Au congrès international de philosophie scientifique de Paris de 1935, S. JAŚKOWSKI avait décrit [*Jkk*2] une suite infinie de matrices, caractéristique du calcul des propositions intuitionniste, en ce sens que les théorèmes de ce calcul sont les énoncés satisfaits *simultanément* par toutes ces matrices. Il en tirait aussi un procédé de décision pour ce calcul, différent de celui que GENTZEN [*Gtz*2] avait établi l'année précédente.

En 1937 [*Stn*5], STONE étend, aux réseaux, son théorème de représentation: il en résulte que les matrices de JAŚKOWSKI se représentent par les familles d'ouverts d'espaces topologiques emboîtés. Dans une communication [*Trk*20] présentée la même année et publiée l'année suivante (cf. [*Trk*34], p. 421), sous le titre *Der Aussagenkalkül und die Topologie*, TARSKI fait bien plus que de vérifier un passage à la limite inductive; il établit une série de théorèmes sur les rapports qu'entretiennent familles d'ouverts des topologies, calcul des propositions intuitionniste, algèbres de BOOLE, et calcul des propositions classique. Je me bornerai à mentionner le plus remarquable, et le plus nouveau pour l'époque, où les matrices faisaient plutôt l'objet de constructions ad hoc: si l'on arrange les ouverts d'une topologie en matrice (pour les connecteurs usuels), en convenant d'interpréter la disjonction et la conjonction, respectivement, par l'intersection et par la réunion; l'implication, par le plus grand des ouverts dont la trace sur l'impliquant est incluse dans l'impliqué; la négation, par le passage à l'extérieur; et si l'on prend l'espace entier pour valeur distinguée, alors, tous les théorèmes du calcul des propositions intuitionniste sont satisfaits par cette matrice; la réciproque n'est pas vraie en général, mais il y a des espaces topologiques, dont font partie les espaces euclidiens (et l'ensemble de CANTOR, muni de la topologie induite), qui sont tels que la matrice ainsi obtenue est caractéristique: les énoncés qu'elle satisfait sont exactement les théorèmes du calcul des propositions intuitionniste. A la fin de son mémoire, TARSKI introduit brièvement les réseaux que l'on nomme aujourd'hui des *algèbres de* HEYTING, dénomination dont cependant, il ne fait pas usage dans ses travaux ultérieurs avec J.C.C. MCKINSEY (qui, cependant, en est probablement à l'origine).

Des logiques modales de LEWIS, et des premiers rapports notés entre calcul des propositions intuitionniste, calcul des propositions de la logique S4, et topologie

LEWIS, que choquait la conception de l'implication acceptant que n'importe quelle proposition soit impliquée par une proposition fausse, avait déjà cherché, dans son *Survey of Symbolic Logic* [*Lws*] de 1918, à définir une implication «stricte» en utilisant les modalités du Possible et du Nécessaire, envisagées comme *connecteurs*, cette fois, et adjointes à cet effet aux connecteurs classiques de conjonction, disjonction, et négation. Proposant d'entendre «A ET NON B n'est pas possible» par «A implique strictement B», il en avait alors donné un traitement formel dans l'esprit de celui du calcul propositionnel des *Principia Mathematica.* En 1932, LEWIS et LANGFORD publient une *Symbolic Logic* [*LwL*] où le premier propose cinq systèmes formels de calcul des propositions pour une logique «modale» du Possible et du Nécessaire, numérotés de S1 à S5. Il y a eu pas mal d'études sur ces systèmes, leurs théorèmes, leurs variantes, leur métathéorie, et leurs rapports réciproques, dans le *Journal of Symbolic Logic* et dans d'autres revues, jusqu'aux alentours de 1950. Je ne parlerai que des plus marquantes parmi celles qui concernent le systèmes S4, auquel les informaticiens continuent, aujourd'hui, de s'intéresser, mais plus encore, en raison des liens que l'on va voir qu'il entretient avec le calcul des propositions intuitionniste et la topologie.

Ce genre de questions a intéressé GÖDEL, qui, dès 1933 [*Gdl*9], a présenté deux traductions des énoncés du calcul des propositions intuitionniste en énoncés du calcul des propositions modal S4, tels que les théorèmes formels se traduisent en théorèmes formels — en 1948 [*MKT*3], MCKINSEY et TARSKI ont montré que seuls sont des théorèmes les énoncés dont les traductions sont des théorèmes. Dans sa contribution [*Wyl*3], intitulée *The Ghost of Modality*, au volume des *Philosophical essays in memory of Edmund Husserl*, paru en 1940, Hermann WEYL note en outre la parenté du système S4 avec le traitement à la KURATOWSKI [*Kur*1] de la topologie.

En 1945, indépendamment, Ruth BARCAN MARCUS [*BcM*], puis Rudolf CARNAP [*Cnp*4], ont proposé, à deux mois d'intervalle, des axiomatiques pour étendre, qui les systèmes S2 et S4, qui le système S5, en calculs des prédicats du premier ordre modaux. Ils se rencontrent sur le point central d'une telle axiomatisation, en proposant de retenir, pour schéma d'axiomes, le schéma, aussitôt connu sous le nom de *formule de* BARCAN:

$$\forall x\Box A(x) \to \Box\forall x A(x),$$

où le signe $\Box$ représente la modalité du Nécessaire. Comme l'implication en sens inverse résulte des autres axiomes, ce schéma est en fait équivalent au schéma

$$\forall x\Box A(x) \leftrightarrow \Box\forall x A(x),$$

qui autorise, donc, «à équivalence près», la permutation des actions successives d'une quantification universelle et d'un opérateur de Nécessité. Des autres axiomes

H. Weyl

il résulte aussi que ce schéma est équivalent au schéma analogue, autorisant de permuter, «à équivalence près», les actions successives d'une quantification existentielle et d'un opérateur de Possibilité.

Un calcul propositionnel modal fondé sur le calcul propositionnel intuitionniste a été proposé par FITCH en 1948 [*Fit*].

Des rapports entre calcul des propositions des logiques classique, intuitionniste, S4, algèbres de BOOLE**, algèbres de** BROUWER **et de** HEYTING**, et «closure algebras», dégagés par** TARSKI **et** MCKINSEY

MCKINSEY s'est intéressé aux logiques modales dès 1934 [*McK*1]. Il a été, en 1939 [*McK*2], l'auteur d'une démonstration d'indépendance des connecteurs primitifs de la logique intuitionniste — la première remonte à WAJSBERG, en 1938 [*Wbg*]. En 1941 [*McK*4], il prouve en tous détails (sans toutefois faire référence aux réflexions de Hermann WEYL [*Wyl*3]), que si l'on arrange les espaces topologiques en matrices pour le système S4, en interprétant les connecteurs propositionnels classiques par les opérations booléiennes (finies) qui leur correspondent sur l'ensemble des parties de cet espace, la Possibilité par le passage à la fermeture, et la Nécessité par le passage à l'intérieur, et en prenant l'espace entier pour valeur distinguée, les théorèmes du système S4 sont exactement les énoncés qui sont ainsi satisfaits dans tous les espaces topologiques. Il décrit en outre un algorithme associant, à chaque (construction d'un) énoncé, un nombre entier naturel, tel que cet énoncé est un théorème de S4, si et seulement s'il est satisfait par tous les espaces topologiques finis, de nombre de points au plus égal au nombre ainsi calculé: cela lui donne un procédé de décision pour le calcul S4.

Pour préciser les liens entre topologie et calcul des propositions modal S4, McKINSEY et TARSKI, dans un mémoire intitulé *The Algebra of Topology* [*MKT*1], paru en 1944, introduisent la notion de «closure algebra»: il s'agit d'une algèbre de BOOLE, munie d'une opération supplémentaire, appelée une «fermeture», et qui satisfait aux transcriptions, en termes des opérations primitives de l'algèbre de BOOLE, des axiomes de définition d'un espace topologique par la donnée de son opération de fermeture, dus à KURATOWSKI [*Kur*1]. Afin de préparer un article ultérieur où les questions de logique interviennent, ils introduisent (essentiellement) la notion de «closure-algebraic function»: sans que ce soit dit, il s'agit des fonctions dénotées par les énoncés du calcul S4, lorsqu'on les prend comme représentant des algorithmes de calcul sur une «closure algebra»; chacun des énoncés représente ainsi la fonction qui, à une assignation de valeurs prises dans cette algèbre, associe la valeur obtenue, quand on interprète la conjonction et la disjonction, respectivement, par la borne inférieure et par la borne supérieure; la négation, par le passage au complémentaire; et la Possibilité, par la fermeture. Pour l'étude des identités (appelées «équations» dans le texte) satisfaites par les «closure algebras», tout un arsenal de notions et de propositions est déployé, jusqu'à parvenir à montrer que les identités satisfaites dans toutes les «closure algebras» (y compris dans les espaces topologiques, où elles prennent la forme d'«identités topologiques») sont les mêmes que celles qui sont satisfaites dans toutes les «closure algebras» finies, les mêmes que celles qui sont satisfaites dans un espace euclidien (donné), et les mêmes que les «identités topologiques» (sans opération booléienne infinie !) qui sont satisfaites dans tous les espaces topologiques, et les mêmes encore que celles qui se dérivent des postulats (de KURATOWSKI [*Kur*1]). Il y a même un sous-espace de la droite, susceptible d'une construction effective, qui fournit un contre-exemple effectif à chaque identité qui n'est pas satisfaite dans quelque espace topologique.

Dans une fort intéressante annexe, McKINSEY et TARSKI se collètent avec l'objection d'avoir manié des classes de fonctions non permises par la théorie des ensembles de VON NEUMANN-BERNAYS-GÖDEL, lancent l'idée d'introduire de même des algèbres de BOOLE munies d'une opération de dérivation (analogue à celle de CANTOR-BENDIXSON [*Bdx*]), et introduisent les notions de classe d'algèbres *équationnellement définissables*: définissables comme vérifiant un système d'identités; et de classe d'algèbres *universellement définissables*: définissables comme vérifiant des axiomes sous une forme prénexe qui ne comporte qu'une suite de quantificateurs universels. Ce sont là, avant la lettre, des notions de théorie des modèles!

En 1945 [*MKT*2], McKINSEY et TARSKI étudient les «algèbres de BROUWER», que forment les éléments «clos» d'une «closure algebra»; après avoir montré, en utilisant le théorème de MACNEILLE [*McN*] évoqué plus haut, que toute algèbre de BROUWER peut être ainsi représentée, et peut aussi l'être comme un sous-réseau du réseau des parties fermées d'un espace topologique, ils étudient les rapports entre les «closure algebraic functions» et leurs analogues relatives aux «algèbres de BROUWER»; ils développent, pour ces dernières, une théorie analogue à celle qu'ils ont développée l'année précédente pour les «closure algebraic functions» et leurs liens avec diverses espèces d'espaces topologiques. Au passage, il établissent que, parmi ces analogues — qui ne sont autres que les fonctions dénotées par les énoncés du calcul des propositions intuitionniste, reçus comme représentant des algorithmes de calcul sur une «algèbre de BROUWER» — il y en a une infinité dont les arguments sont en nombre fixé donné (non nul), «a result which was kindly communicated to us by Dr Kurt GÖDEL», et qui revient à montrer que les «algèbres de BROUWER» libres à un nombre fini non nul de générateurs sont infinies.

En 1948 [*MKT*3], ils complètent leurs résultats de 1944; ce qu'ils leur ajoutent revient à prouver qu'un énoncé se prouve formellement dans S4, si et seulement si l'identité de la «closure-algebraic function» qu'il représente et de la fonction constante «Vrai», qui ne prend que la valeur unité, est satisfaite par toute «closure algebra». Ils en profitent pour démontrer deux propriétés de S4: premièrement, si un énoncé admet une preuve formelle, il en va de même de la Nécessité de cet énoncé; deuxièmement — une «vieille» conjecture de GÖDEL [*Gdl*9]): si la disjonction des Nécessités de deux énoncés admet une preuve formelle, il en va de même, soit de l'un, soit de l'autre de ces énoncés.

Dans le même travail, ils traitent aussi du calcul des propositions intuitionniste; ils montrent qu'un énoncé admet une preuve formelle si et seulement si l'identité de la fonction représentée par cet énoncé et de la fonction constante «Vrai» est satisfaite par toute algèbre des ouverts d'une «closure algebra» (savoir, pour nous, par toute algèbre de HEYTING), et si et seulement s'il y a un espace euclidien dont l'algèbre des ouverts satisfait cette identité. Ils en profitent pour établir une autre conjecture de GÖDEL [*Gdl*9]: une disjonction n'a de preuve formelle en logique intuitionniste que si et seulement si tel est le cas d'au moins un de ses termes. Et pour finir, ils décrivent trois procédés de traduction des énoncés du calcul des propositions intuitionniste en énoncés du calcul S4, tels que les énoncés du premier admettent une preuve formelle, si et seulement si tel est le cas de leur traduction.

Des avancées de l'algèbre universelle entre 1935 et 1950

Les derniers travaux cités ressortissent à l'algèbre universelle; conçus en vue de leur application à la logique, mise en œuvre dans le tout dernier, ils appartiennent à la préhistoire de la théorie des catégories.

Le premier théorème important d'algèbre universelle, établi en 1935 [*Bkf*2] par Garrett BIRKHOFF, dit qu'une classe de structures algébriques est *caractérisable* par des «équations» (des identités, pour nous) entre des termes (fonctionnels) construits à l'aide des (représentations des) opérations de définition de ces structures — en d'autres termes, que cette classe «est une variété» — si et seulement si elle est stable par passage des images homomorphes, aux sous-structures, et aux produits. La satisfaction supposée des identités de définition n'est autre qu'un cas particulier de la satisfaction de formules, qui en l'occurrence peuvent être regardées comme des axiomes: vu ainsi, ce théorème est aussi un théorème important *de théorie des modèles,* avant la lettre, et il va participer à en susciter le développement. En fait, c'est un cas particulier du théorème de complétude; pour le démontrer, BIRKHOFF établit une «correspondance de GALOIS» (notion théorisée dix ans plus tard par Öystein ORE [*Ore*3]) entre les variétés et les théories équationnelles stables par application de règles d'inférences tout à fait analogues à celles que GÖDEL [*Gdl*11] introduit dans la théorie des fonctions récursives générales de HERBRAND-GÖDEL.

L'algèbre universelle et son germe de théorie des modèles, qui pour le moment s'occupe de formules satisfaites par les structures algébriques, vont se développer de pair durant les quinze dernières années de la première moitié du vingtième siècle. L'algèbre universelle utilise les théories des groupes, des anneaux, des corps, ..., en particulier les mémoires d'algèbre abstraite d'Emmy NOETHER [*Nth*1, 2, 3], comme paradigmes pour son développement, généralisant les notions d'homomorphisme, de congruence, de sous-structure, d'engendrement libre ..., et en étend les théorèmes; on peut citer KUROSH décelant en 1935 [*Krh*] dans les anneaux et réseaux des décompositions en éléments irréductibles; ORE écrivant en 1935 et 1936 un long mémoire *On the foundations of abstract algebra* [*Ore*1], et traitant des relations d'équivalence en 1942 [*Ore*2]; Samuel EILENBERG et Saunders MACLANE s'occupant des congruences en 1945 [*EML*]; DIAMOND et MCKINSEY étudiant les *Algebras and their subalgebras* en 1947 [*DMK*], GOLDIE introduisant et prouvant *The Jordan-Hölder theorem for abstract algebras* en 1950 [*Gld*]; BIRKHOFF montrant en 1944 [*Bkf*3] que toute algèbre est isomorphe à un produit sous-direct d'algèbres sous-directement irréductibles; et d'autres encore.

Dans ce mouvement d'idées, le «théorème du point fixe» de TARSKI, qui dit que toute fonction croissante d'un réseau complet vers lui-même admet des «points fixes», a été annoncé par TARSKI en 1949 [*Trk*31], et donné comme acquis dès 1939, voire dès 1927 [*Knt*], en collaboration avec KNASTER.

Plus près de ce qui sera la théorie des modèles proprement dite, MCKINSEY établit en 1943 [*McK*5] le premier descendant de ce qui sera une longue lignée de théorèmes façonnés sur le modèle du théorème de BIRKHOFF [*Bkf*2]; il dit qu'un énoncé «universel» (savoir, à forme prénexe dotée d'un préfixe où n'apparaissent que des quantificateurs universels) est préservé par produit direct fini, si et seulement s'il est équivalent à un cas de ce que nous appelons les «énoncés de HORN»; mais ce ne sera qu'en 1951 que paraîtra l'article [*Hrn*] où Alfred HORN montre que les énoncés désignés sous son nom sont préservés par produit directs.

Des axiomatiques du «calcul des relations», des alentours de 1940, et des problèmes relatifs aux algèbres de relations
En 1940 [*McK*3], McKinsey publie une axiomatique pour le calcul des relations binaires, admettant pour seules notions primitives l'inclusion et la composition (des relations), et indépendante. Il développe quelques pages de conséquences de ses axiomes, jusqu'à montrer que, sous l'axiome de choix et sous l'hypothèse du continu généralisée, ils sont «semi-catégoriques»: à isomorphisme près, il n'y en a qu'une réalisation par nombre d'individus.

L'année suivante, Tarski publie sa propre version [*Trk*22] de ce calcul. A l'opposé de celle de McKinsey, qui n'est ni équationnelle, ni du premier ordre, son axiomatisation prend la forme d'un calcul à partir d'identités. C'est la théorie informelle développée par Schröder dans le volume III des *Vorlesungen über die Algebra der Logik* [*Sdr*1] qu'il axiomatise ainsi. Ses identités portent sur deux groupes d'opérations: les interprétations attendues des premières sont les opérations booléiennes sur les graphes; celles des secondes sont les graphes de l'identité, du passage à la réciproque, et de la composition des relations, extension aux relations de la composition des fonctions.

De cette théorie équationnelle, les quantificateurs ont disparu. En réalité, ils réapparaissent si l'on retraduit en termes d'opérations ensemblistes les significations attendues des opérations. Toute équation du calcul des relations exprime de la sorte la même proposition qu'une formule du calcul des prédicats. On sait depuis une date indéterminée du début du siècle (celle d'une lettre de Korselt dont Löwenheim [*Lwh*] fait état) que la réciproque n'est pas vraie: on ne peut exprimer, par exemple, à l'aide des notions primitives du calcul des relations, qu'un domaine comporte quatre éléments. Tarski annonce qu'il sait faire beaucoup mieux, et que cette propriété subsiste si l'on adjoint aux opérations primitives du calcul n'importe quel nombre de relations primitives invariantes sous l'action des permutations des individus.

En 1947 [*JsT*3], Bjarni Jónsson et Tarski transposent les axiomes d'une axiomatique plus élaborée de ce calcul, en identités de définition de la notion d'*algèbre de relations*, à savoir, la structure déterminée sur une algèbre de Boole par un opérateur unaire — dont l'interprétation attendue est le passage à la réciproque — et par un opérateur binaire — dont l'interprétation attendue est la composition des relations — vérifiant des identités adéquates.

Ces algèbres posent un problème de représentation: toutes les structures ainsi déterminées sur une algèbre de Boole sont-elles bien représentables comme algèbres de relations sur un support d'individus, définies sur ce support par les opérations données ci-dessus pour attendues ? Pour le résoudre, Jónsson et Tarski annoncent [*JsT*2] l'introduction des notions d'opérateur sur une algèbre de Boole, d'algèbre de Boole à opérateurs, et un théorème de représentation pour ces dernières: les opérateurs se prolongent, «fonctoriellement», dirions-nous aujourdhui, c'est à dire, en préservant les identités qu'ils satisfont, en opérateurs sur l'algèbre des parties de l'ensemble des idéaux premiers du support; mais dans le cas des algèbres de relations, on n'est pas encore au but, puisqu'il faut représenter à son tour cet ensemble d'idéaux premiers comme ensemble de *couples* d'éléments d'un ensemble. Cette difficulté explique, menés de front, les mémoires *Direct decompositions in finite algebraic systems* [*JsT*1], publié en 1947 par JÓNSSON et TARSKI, et *Projective algebra I* [*EvU*], publié par C. J. EVERETT et ULAM en 1946, et l'étude, qu'ils commencent et que continuent Louise CHIN et TARSKI, qui l'annoncent en 1947 [*ChT*], des *algèbres projectives*, qui sont des algèbres de BOOLE munies d'opérateurs dont l'interprétation attendue est celle des opérations de projections d'un produit cartésien d'ensembles sur ses facteurs.

Des théorèmes et méthodes généraux issus des débuts de la théorie des modèles

Des débuts de la théorie des modèles: premiers théorèmes et méthodes

Enoncé à l'aide des moyens d'expression relatés plus haut, qui ont permis de traiter des modèles de façon intuitive avant que le concept de modèle n'ait été mis en forme et intégré aux métamathématiques, le premier théorème clef (dans le temps) de la théorie des modèles moderne est le théorème de LÖWENHEIM-SKOLEM. Les suivants sont les théorèmes de complétude et de compacité du calcul des prédicats du premier ordre classique, établis par GÖDEL [*Gdl*3] en 1930. Le théorème de compacité est étendu, par GÖDEL lui-même, aussitôt [*Gdl*7], et indépendamment, par Anatoli Ivanovitch MALTSEV, en 1936 [*Mlt*1], au calcul des propositions, dans le cas d'un langage non dénombrable, puis à nouveau par MALTSEV, en 1941 [*Mlt*2], au calcul des prédicats du premier ordre, dans le cas d'un langage non dénombrable, cas auquel MALTSEV étend le théorème de complétude en 1936 [*Mlt*1].

En dehors de cela, de la méthode des matrices, et de la production sporadique de certains modèles particuliers, par exemple en vue de montrer qu'un énoncé ne résulte pas de certains autres, les résultats ressortissant à la théorie des modèles obtenus avant 1930 sont principalement ceux que fournissent des utilisations judicieuses de la méthode d'élimination des quantificateurs décrite plus haut.

Des succès de la méthode d'élimination des quantificateurs après 1930

Mettant en oeuvre cette méthode, et fondé sur une extension du théorème de STURM [*Stm*], le procédé de décision de TARSKI pour la théorie du premier ordre du corps ordonné des nombres réels, annoncé dès 1931 [*Trk*8], n'a été publié qu'en 1948 [*Trk*24] — l'apparition des premiers ordinateurs suscitait un espoir de réalisation pratique — mais il subsiste des épreuves réalisées en 1939 en vue d'une publication empêchée par le déclenchement de la guerre et publiées en 1967 [*Trk*35]. Dans une seconde édition de 1951, ce procédé est appliqué à la théorie des corps réels clos, élaborée par Emil ARTIN et Otto SCHREIER en 1926 [*ArS*], et que TARSKI n'avait pas encore intégrée à ses conceptions en 1939; il est juste pris acte là de son identité formelle avec la théorie complète du corps ordonné des nombres réels ([*Snc*]).

En 1939 [*Lfd*3], la méthode d'élimination des quantificateurs est à nouveau utilisée par LANGFORD pour étendre ses résultats de décidabilité et de complétude à la théorie d'un *fragment du second ordre* de l'ordre dense avec bornes supérieures pour les segments «fondamentaux» — s'entend, aujourd'hui, initiaux. Le fragment de logique du second ordre en question est justement celui qui permet les quantifications sur les segments fondamentaux. Ce travail de pionnier précède de plusieurs décennies des investigations sur des fragments analogues de logique du second ordre, appliqués à des théories diverses.

Des procédures de décision par élimination des quantificateurs seront encore produites en 1948, dont certaines obtenues pendant la guerre et restées non publiées, pour diverses théories du premier ordre, assorties de classifications des modèles des extensions axiomatisables et des extensions complètes de ces théories, selon les ensembles d'énoncés qu'ils satisfont: pour la théorie des algèbres de BOOLE, une de JAŚKOWSKI [*Jkk*3], une autre de TARSKI [*Trk*27]; pour la théorie des bons ordres non vides, une de MOSTOWSKI et TARSKI [*MwT*2]; pour la théorie des groupes abéliens, une de Wanda SZMIELEW [*Szm*].

Du théorème de LÖWENHEIM-SKOLEM «ascendant»

Un autre parmi les théorèmes obtenus précocement est le «théorème de LÖWENHEIM-SKOLEM *ascendant*», qui dit que si des énoncés admettent un modèle infini, ils en admettent un de tout cardinal plus grand donné, publié par MALTSEV en 1936 [*Mlt*1]. TARSKI a dit l'avoir présenté à son séminaire en 1928 — et le fait que l'ordre des nombres rationnels et celui des nombres réels répondent à une même «description structurelle» de la théorie élémentaire de l'ordre dense sans premier ni dernier élément, et qu'en raison de la complétude de cette théorie, aucun énoncé élémentaire ne les différencie, peut bien l'y avoir conduit. Le souvenir qu'en avaient gardé, en 1934, les éditeurs des *Fundamenta Mathematicae*, était qu'il avait établi «dass für jede widerspruchsfreie Zählaussagenmenge, die keine Interpretation im Endlichen besitzt, eine Interpretation (nicht nur im Abzählbar-Unendlichen, sondern auch) im Unabzählbar-Unendlichen existiert», ainsi qu'ils l'ont écrit alors dans une *Note* adjointe à un article de SKOLEM [*Skm*10] dont je reparlerai, et à laquelle MALTSEV [*Mlt*1] a renvoyé.

Sur les modèles de la théorie des types simple et de la théorie des ensembles, durant les quinze dernières années avant 1950

De l'invariance des relations définissables en termes logiques sous l'action des permutations d'individus

La part que TARSKI avait prise, ainsi que le rappelle Hourya SINACEUR [*Snc*], à la révision de la troisième édition [*Fkl*5] de l'*Einleitung in die Mengenlehre* de FRAENKEL, où il a ainsi pu prendre connaissance des conceptions de celui-ci sur l'invariance des ensembles définis en termes d'une paire de non ensembles sous l'action de la transposition des éléments de celle-ci; les travaux que lui-même entreprenait au même moment en géométrie, en même temps qu'en méthodologie des sciences déductives, et la façon qu'il avait d'appréhender la signification des formules selon les diverses façons d'en réaliser les notions primitives par des graphes, lui avaient fait percevoir le rôle que jouent, dans la mise en oeuvre de la méthode de PADOA, les permutations du support d'une structure: pour qu'une notion soit définissable en termes de certaines autres, il faut, quel que soit le domaine d'individus, quels que soient les graphes réalisant ces notions, que le groupe des permutations du support, sous l'action desquelles le graphe réalisant la notion à définir reste invariant, contienne le groupe des permutations du support sous l'action desquelles chacun des graphes réalisant les notions définissantes reste invariant. Cette conception est implicite dans une conférence qu'il prononce à Prague en 1934 [$Trk11$], et dont le texte paraît l'année suivante dans la revue du Cercle de Vienne, *Erkenntnis*, sur les notions de définissabilité et de complétude des concepts. Elle est beaucoup plus explicite dans le texte [$LdT2$], cosigné avec LINDENBAUM, et intitulé *Über die Beschränktheit der Ausdrucksmittel deduktiver Theorien*, d'une autre conférence qu'il prononce à Vienne l'année suivante. Il y montre que l'on peut démontrer formellement que toute relation entre des objets de types divers et qui s'exprime par des moyens purement logiques est invariante par toute permutation du domaine des individus. A titre d'exemple des conséquences de cette proposition, il montre que deux partitions en deux classes dont les nombres d'éléments forment un couple prescrit, sont indiscernables par des moyens purement logiques.

C'est dans la *caractérisation exacte* de la définissabilité d'une notion en termes d'autres par la *coïncidence*, dans toute structure apte à réaliser ces notions, du groupe des permutations du support sous l'action desquelles le graphe réalisant la notion à définir est invariant, avec le groupe des permutations du support sous l'action desquelles les graphes réalisant les notions définissantes sont invariants, que Marc KRASNER voyait le vrai ressort de la théorie de GALOIS, masqué, selon lui, par la «linéarisation» de cette théorie. Dans son mémoire *Une généralisation de la notion de corps* [Krn], paru en 1938, ainsi que dans ses enseignements ultérieurs, il apparaît qu'il établit cette coïncidence pour une notion de définissabilité *relative à un langage infinitaire*, qui n'est pas précisé. Dans la conversation, il attribuait à la finitude de la conjonction et de la disjonction les limitations des capacités de la logique classique finitaire.

Digression sur des langages infinitaires introduits en connaissance de cause avant 1950

Un langage avec disjonctions et conjonctions infinies, mais soumises à des conditions de constructibilité «intuitionnistes», intervient, au niveau du calcul des propositions, dans des publications de Piotr Sergeievitch NOVIKOFF de 1939 [$Nkf1$] et 1943 [$Nkf2$]; il est caractéristique de l'époque que CHURCH, lorsqu'il rend compte [$Chu11$] du premier de ces travaux dans le *Journal of Symbolic Logic*, souligne que ces traits excluent son calcul du champ des logiques «in the proper sense of the word», alors que de tels langages — sans même la restriction d'effectivité de la construction des formules — seront activement étudiés, aux Etats-Unis même, durant la seconde moitié du vingtième siècle. En 1939, NOVIKOFF annonce la compatibilité de son calcul; l'année suivante, D.A. BOCHVAR [*Bvr*] en donne des preuves de compatibilité et de complétude: «jede Formel erfüllbar entweder widerlegbar ist», mais sous l'hypothèse de dénombrabilité des conjonctions et disjonctions infinies, reprise en 1943 par NOVIKOFF lui-même, qui précise alors sa propre preuve de compatibilité.

En 1949 [*Nkf3*] ces tentatives de Novikoff débouchent sur une répartition des formules de la logique élémentaire ordinaire en classes déductivement stables, dont il se sert pour démontrer des théorèmes du genre: le principe de récurrence n'est pas conséquence au premier ordre des axiomes de l'ordre naturel, et des définitions de fonctions par récurrence; il annonce par ailleurs, dans une autre Note [*Nkf4*] dont le texte suit celui de la précédente, que le problème de déterminer les théorèmes dont la démonstration ne fait pas appel au principe de récurrence n'est récursivement soluble que pour la classe des formules sans variables liées, cas pour lequel il dit avoir une méthode de décision.

Un langage de même espèce, cette fois sans négation, mais avec des quantifications, et des équations pour formules indécomposables en formules moins longues, intervient dans un étonnant article précurseur [*Jrd*] du physicien réputé Pascual JORDAN, paru en 1949 aux *Abhandlungen aus dem Mathematischen Seminar der Universität Hamburg*.

Il convient de rappeler que dès 1906 [*Hdm1*], HADAMARD n'exclut pas des «énonciations de complexité infinie»; qu'en 1920 [*Skm3*], encore hors le cadre d'un formalisme, SKOLEM établit le théorème de LÖWENHEIM-SKOLEM pour des conjonction infinies dénombrables de formules résultant chacune d'une suite finie de quantifications universelles portant sur une suite infinie dénombrable de quantifications existentielles s'exerçant à leur tour sur des disjonctions infinies de conjonctions infinies dénombrables de formules du premier ordre sans quantificateurs; et qu'en 1935 [*Zml6*], ZERMELO, plutôt que d'admettre les résultats de SKOLEM et de GÖDEL, esquisse un traitement de la théorie des ensembles usant de moyens d'expression infinitaires.

Des résultats d'indépendance obtenus de 1938 à 1948 avec des modèles à permutations, par MOSTOWSKI **surtout**

En 1938 [*Mtw2*], MOSTOWSKI reprend dans sa thèse la preuve de l'indépendance de l'axiome du choix à l'égard des autres axiomes, dans le système résultant de l'adjonction, à la théorie des types de TARSKI ([*Trk7*]), ou à l'une des théories des ensembles alors existantes, d'un «axiome de l'infini» [différent du nôtre dans son contenu] supposant la présence d'individus non ensembles formant un domaine infini, en faisant dans son raisonnement un usage typique des permutations de ce domaine. C'est dans son application à la théorie des types de TARSKI que son argumentation est la plus facile à suivre: cette théorie n'admet comme axiomes, en dehors de ceux du calcul des propositions, que les axiomes d'extension et de compréhension pour les types des ensembles et les types des classes des divers ordres;

parmi ces axiomes, les derniers sont les seuls à poser l'existence de classes (dans le cas de la théorie des ensembles, les axiomes qui posent des existences de classes sont, l'axiome de choix excepté, des cas particuliers d'un axiome de compréhension qui, dans toute sa généralité, permettrait la tenue du raisonnement menant au paradoxe de RUSSELL); or, il résulte de là que ces classes sont définies en termes de logique (savoir, de connecteurs et de quantificateurs) et des paramètres qui figurent dans leur définition; mais ces paramètres eux-mêmes ne sont légitimes que si leur propre existence est préalablement posée. Ainsi, un ensemble infini et co-infini d'individus ne peut se définir en termes d'un ensemble fini d'individus, car il existe toujours une permutation sous l'action de laquelle ce dernier est invariant, alors que le premier et son complémentaire ne le sont pas. Il en résulte que l'on ne peut déduire, de l'axiome de l'infini, l'existence d'un ensemble infini et co-infini d'individus, puisqu'au cas où l'ensemble infini dont l'existence est supposée, et que l'on peut donc admettre comme paramètre, aurait un complémentaire fini, il serait définissable en termes de ce complémentaire, et les seuls ensembles infinis dont on pourrait reconnaître l'existence, devant donc être définissables aussi en termes de ce complémentaire fini, devraient eux aussi être de complémentaire fini. Or, l'existence d'un ensemble infini et co-infini d'individus est une conséquence de l'axiome de choix, et si cet axiome était déductible, sa conséquence le serait aussi.

Avec des arguments du même genre, MOSTOWSKI montre encore, dans sa thèse, la non-équivalence, en l'absence de l'axiome de choix, de toute une famille de définitions de la finitude d'un ensemble, considérées par TARSKI en 1924 [*Trk*3]. Il annonce, là et dans une Note [*Mtw*3] parue la même année, que dans un système d'axiomes récursif, à toute définition de la finitude d'un ensemble on peut toujours en associer une plus forte en ce sens qu'elle implique la première, alors que la première ne l'implique pas. En 1950 [*Trb*], TRAHTÉNBROT retrouve indépendamment un résultat plus fort: il établit aussi, par des voies que MOSTOWSKI lui-même juge plus simples que les siennes, que la notion de finitude d'un ensemble n'admet pas, dans les théories des ensembles usuelles, de plus faible définition (entendue comme non tautologique et usant d'une caractérisation valide dans tous les domaines finis).

La même année, LINDENBAUM et MOSTOWSKI [*LdM*] annoncent avoir établi, sous l'hypothèse d'existence d'un nombre infini de non ensembles, des preuves d'indépendance de diverses conséquences de l'axiome de choix, de façon plus rigoureuse que FRAENKEL, quand il s'agit de résultats obtenus auparavant par celui-ci. Parmi ces conséquences se trouve le «principe d'ordination», selon lequel tout ensemble admet un ordre (non nécessairement bon). En 1939 [*Mtw*4], MOSTOWSKI montre, toujours sous l'hypothèse de l'existence d'un ensemble infini d'individus non ensembles, que l'axiome de choix ne résulte pas du «principe d'ordination». Il construit cette fois un modèle interne (c'est à dire, analogue au modèle des ensembles «constructibles» de GÖDEL [*Gdl*13, 14]) de la théorie des ensembles et du principe d'ordination. Le domaine d'individus du modèle construit se compose d'ensembles du modèle supposé donné, choisis de telle façon que ce domaine est muni d'un ordre isomorphe à l'ordre des nombres entiers rationnels. A chaque ensemble du modèle construit doit correspondre un nombre fini d'individus de ce modèle, tels que cet ensemble reste invariant sous l'action de toute permutation croissante des individus de ce modèle, sous l'action de laquelle ces individus restent invariants; et la propriété caractéristique des ensembles du modèle construit, est d'être un ensemble du modèle suposé donné, tel que lui-même et chacun des ensembles qui se contruise à partir de lui dans le modèle donné, par les constructions admises par les axiomes de la théorie des ensembles, possède la propriété d'invariance ci-dessus décrite.

En 1945 [*Mtw*5], MOSTOWSKI utilise des raffinements de raisonnements analogues pour établir des liens de dépendance ou d'indépendance entre les principes de choix restreints aux familles d'ensembles de cardinalités finies données. Il se sert encore d'un modèle d'un type analogue pour montrer, en 1948 [*Mtw*8], et toujours sous l'hypothèse de l'existence d'individus non ensembles formant un ensemble infini, que l'axiome de choix ne résulte ni du principe des choix dépendants (dont l'énoncé, introduit par BERNAYS en 1942, a ainsi été baptisé par TARSKI en 1948 [*Trk*23]), ni de l'axiome de choix «restreint», c'est-à-dire, de ce que nous appelons l'«axiome de choix dénombrable», lui-même conséquence du principe des choix dépendants.

De la cristallisation, vers 1946, de la perception du rôle des modèles non standards

Les cinq années qui suivent la fin de la seconde guerre mondiale voient donc une notion de modèle propre à la logique, bien délimitée selon les canons des mathématiques, débarrassée du vague du mode d'expression informel d'intuitions encore incomplètement dégrossies des séminaires, éclore, évoluer sitôt apparue, et justifier bientôt cette évolution par d'importantes avancées en théorie des ensembles.

En effet, à peine soutenue, en 1946, la thèse [*Kmn*1] où KEMENY présente la première définition mathématique précise de la notion de modèle pour une classe très large et elle aussi mathématiquement bien délimitée de systèmes formels, Leon HENKIN soutient à son tour, en 1947 (cf. [*Hnk*2, 3]), la sienne, sur la classe des modèles de la théorie des types de CHURCH [*Chu*7]. Le résumé [*Hnk*1] qu'il en présente à la fin de 1947 à l'*Association for Symbolic Logic* en dit: «Although the higher-order calculi are in one sense incomplete, as follows from the work of GÖDEL, it is found that when the definition of «completeness» of these systems is recast to take account of SKOLEM's ideas on the «relativity of the set concept» a completeness proof can be furnished».

Sous cette formulation un peu sibylline se cache une interrogation du genre de la suivante, qui sous-tend aussi ce que fait MOSTOWSKI dans la première partie d'un article *On absolute properties of relations* [*Mtw*7], paru au *Journal of Symbolic Logic* l'année d'après: la théorie des types de CHURCH, bien qu'elle comporte des types de tous les ordres entiers naturels, s'interprète aisément dans la théorie des ensembles: à «lire» les types attribués aux variables comme autant de symboles de prédicats à une place, constants; puis la convenance d'un prédicat à un sujet, comme une relation, constante, d'appartenance de ce sujet à l'extension de ce prédicat, on «lit» la théorie des types comme un fragment de théorie des ensembles du premier ordre. Or, la complétude de la logique du premier ordre, appliquée à la théorie des ensembles, fait que les formules vraies dans tous les modèles de cette théorie en sont des théorèmes. D'où vient donc que toutes les formules vraies dans tous les modèles de la théorie des types n'en soient pas des théorèmes.

Conformément aux remarques inlassablement ressassées par SKOLEM depuis quelque vingt-cinq ans au moment des travaux de HENKIN, la théorie des ensembles admet, selon le *théorème* de LÖWENHEIM [Lwh], des modèles *dénombrables*, dont SKOLEM explique l'existence par le caractère relatif de la notion d'ensemble, et qui concourent à restreindre le champ des formules vraies dans tous ses modèles. Pour comprendre les idées développées par HENKIN dans le second [$Hnk3$] des articles où il publie l'essentiel de sa thèse, en 1950, il suffit d'examiner ce qui se passe pour le second ordre. La conception que l'on se faisait, à l'époque des théorèmes de GÖDEL [$Gdl1,3$], de la validité d'une formule du second ordre dans un domaine d'individus, voulait que cette formule soit satisfaite par toute donnée d'un individu par constante d'individu (représentation d'un élément distingué), d'un graphe par constante relationnelle (représentation d'une relation primitive entre individus), et ceci étant fixé, par toute assignation de valeurs aux variables libres, assignant des individus aux variables d'individus, et des graphes aux variables de relation: *tous les graphes* sont bons à prendre, et le domaine d'interprétation du type des classes d'individus (du second ordre) du modèle, implicite dans toutes ces considérations, se compose de *toutes les parties* du domaine des individus. Or, dans un modèle dénombrable de théorie des ensembles, l'ensemble des parties d'un ensemble infini ne s'interprète pas par l'ensemble de toutes ses parties dans l'absolu, mais comme celui de ses parties que le modèle reconnaît pour siennes; le domaine d'interprétation du type des classes d'individus du modèle de la théorie des types qui s'extrait naturellement d'un tel modèle ne se compose que d'*une partie* des classes d'individus, répondant à la seule exigence de posséder des propriétés de stabilité par réunion, intersection, ... *imitant,* des propriétés correspondantes du domaine de toutes les classes d'individus, tout ce que l'on peut exprimer en un nombre fini de termes. On n'avait pas, jusque là, admis de tels modèles pour la théorie des types. Que l'on admette de tels modèles, dont fort heureusement les travaux de SKOLEM et de TARSKI exhibent des exemples, et à l'instar de ce que fait MOSTOWSKI dans l'article précité [$Mtw7$] où il en traite en détail, les appelant *non absolus* alors qu'à Princeton on les baptise *non standards*, et l'on retrouve pour la théorie des types la complétude perdue — conclusion que MOSTOWSKI semble laisser échapper.

Des modèles non standards produits par SKOLEM, avant 1945

SKOLEM, en 1922 [$Skm4$], VON NEUMANN, en 1925 [$vNm1$], n'avaient donné, des modèles de la théorie des ensembles aujourd'hui qualifiés de *non standards*, que des descriptions intuitives et limitées.

SKOLEM n'avait pas voulu en rester là. En 1933 [$Skm9$], puis en 1934 [$Skm10$] dans les *Fundamenta Mathematicae*, il avait montré que des axiomes du premier ordre en nombre dénombrable, mais aussi des axiomes d'ordres plus élevés en nombre fini, ne peuvent caractériser complètement «die Zahlenreihe». Car, raisonnait-il (sans pouvoir l'exprimer dans une terminologie telle que celle qui suit), les derniers de ces axiomes n'ont de toutes façons pas plus d'une infinité dénombrable de conséquences du premier ordre; or, disait-il, partons d'une infinité dénombrable d'énoncés du premier ordre, dans un langage de l'ordre et d'un nombre dénombrable de fonctions arithmétiques, vrais dans la suite des nombres entiers positifs; complétons une énumération des fonctions du langage de ces énoncés en une suite infinie de fonctions arithmétiques, stable par composition, et comprenant en outre les polynômes à coefficients entiers positifs et les fonctions (de SKOLEM, pour nous) à intervenir dans la mise sous forme normale (de SKOLEM, pour nous) de ces énoncés; à partir de la (sous-) suite des fonctions à un argument ainsi déterminée, établissait-il, on peut définir, de multiples manières, un préordre sur ces fonctions, donc un ordre sur les classes associées, tel que les prolongements naturels des fonctions du langage de ces énoncés, et cet ordre, définissent, sur le support constitué de ces classes, une structure N*, satisfaisant ces énoncés, et dans laquelle ils induisent, sur les classes des fonctions constantes, un segment initial isomorphe à la structure N déterminée, sur les valeurs des fonctions constantes, par les fonctions dont on est parti, et par l'ordre naturel [la définition d'un tel préordre repose sur une utilisation récurrente de la remarque simple suivante: si f et f' sont deux fonctions arithmétiques à un argument (distinctes), des trois ensembles de nombres entiers naturels n tels que $f(n) < f'(n)$, $f(n) = f'(n)$, $f(n) > f'(n)$, l'un au moins est infini; la fonction g étant celle, strictement croissante, qui en énumère les éléments dans l'ordre naturel, on peut, revenant à la paire constituée de f et de f', affirmer qu'il existe une fonction strictement croissante g, telle que l'on a, soit $fg < f'g$, soit $fg = f'g$, soit $fg > f'g$ (le produit dénotant la composition); comme il se superpose à cela, pour achever la récurrence, une diagonalisation, qui fait perdre ce qui se passe sur des segments initiaux, on débouche sur la conclusion que pour toute paire de fonctions f et f' de la suite de fonctions considérée pour construire N*, il existe une fonction strictement croissante g telle que l'on a, *asymptotiquement*, soit $fg < f'g$, soit $fg = f'g$, soit $fg > f'g$, relations que l'on prend, respectivement, pour (classe de)f <(classe de)f', (classe de)f =(classe de)f' et (classe de)f >(classe de)f']. HILBERT avait déjà employé, dans les *Grundlagen der Geometrie* [$Hbt1$], un modèle dont les individus sont des classes de fonctions, comparées selon leur comportement asymptotique, mais il n'avait pas eu à user d'une pondération semblable.

Des modèles non standards issus, avec la notion de ω-incomplétude, des travaux de TARSKI d'avant 1945

Les éditeurs des *Fundamenta Mathematicae* avaient fait remarquer, dans une Note annexée à cet article [$Skm10$], que dès les années 1926-1928, TARSKI avait établi, dans son séminaire, que les types d'ordre ω de l'ensemble des nombres entiers naturels et $\omega + \omega^* + \omega$ de l'ensemble obtenu en mettant bout à bout l'ensemble des entiers naturels et l'ensemble des entiers rationnels satisfaisaient les mêmes Zählaussagen du langage de l'ordre; ils exhibaient ainsi, sans le savoir, un modèle non standard de l'ordre naturel (mais de l'ordre naturel *seulement*, et non, en outre, d'opérations élémentaires de l'arithmétique). TARSKI lui-même en avait fait état dans l'*Annexe* à ses *Grundzüge des Systemenkalküls, Zweiter Teil*, de 1936 [*Trk*12].

La référence à ce modèle est implicite dans les *Einige Betrachtungen über die Begriffe der ω-Widerspruchsfreiheit und der ω-Vollständigkeit* [*Trk*10] de TARSKI, parues en 1933 et où sont relatés des travaux exposés à la Société Philosophique de Varsovie en 1927, et la découverte faite alors de la notion baptisée trois ans plus tard d'ω-compatibilité par GÖDEL [*Gdl*3], et de la notion réintroduite et baptisée là d'ω-complétude (En 1925 [*Flr*] Paul FINSLER avait eu prescience de l'ω-incompatibilité, mais sans avoir su en faire une étude mathématique précise).

Dans ces *Betrachtungen*, ces notions se rapportent à l'arithmétique des classes d'individus dans la théorie des types simples; l'axiome de l'infini, d'ordre *trois*, y dit qu'aucune classe finie n'épuise tous les individus; une classe est dite finie si elle appartient à toute classe de classes d'individus, stable par adjonction d'un individu à une classe, et à laquelle appartient la classe vide. Par ailleurs, qu'une classe admette au plus, ou au moins, 0, ou 1, ou 2, ... individus, s'exprime par des formules du second ordre, dont je dirai qu'elles *attestent* des nombres naturels dont elles expriment une telle propriété. Une théorie est dite *ω-compatible*, si pour toute propriété exprimée par une formule à un argument individu, et pour chaque nombre entier naturel *attesté n*, on *peut déduire*, des axiomes de cette théorie, que cette propriété est commune aux classes d'au plus n individus (ce qu'une formule exprime), *tandis que*, de ces axiomes, on *ne peut déduire* qu'il existe une classe d'individus finie à laquelle cette propriété n'appartient pas — le passage que nous faisons mentalement des premières conclusions à l'impossibilité de la seconde fait intervenir l'hypothèse tacite, *que les formules et les règles de déduction usuelles, finitaires, ne peuvent exprimer*, selon laquelle il n'y a de nombres entiers naturels qu'attestés; et cette théorie est dite ω-complète si, lorsque, pour tout nombre entier naturel *attesté n*, ces axiomes *permettent de déduire* qu'une propriété est commune aux classes d'au plus n individus, ils *permettent aussi de déduire* que cette propriété est commune aux classes finies.

TARSKI considère la théorie T_ω ayant pour axiomes les énoncés (en nombre infini) $\alpha_0, \alpha_1, \alpha_2, \ldots, \alpha_n, \ldots$ correspondant à tous les nombres attestés 0, 1, 2, ..., n, ..., et où α_n est l'énoncé disant qu'aucune classe à au plus n éléments n'épuise tous les individus. Cette théorie est compatible, car toute preuve formelle ne recourt qu'à un nombre fini d'axiomes, et les axiomes $\alpha_0, \alpha_1, \ldots, \alpha_n$ ont, pour chaque nombre attesté n, des modèles évidents, que de plus on peut prendre modèles de NONα_k pour tout nombre $k > n$, et a fortiori modèles de la négation de l'axiome de l'infini. L'axiome de l'infini n'est donc pas un théorème formel de T_ω ; cette théorie est donc ω-incomplète.

Or, l'adjonction d'un axiome ne détruit la compatibilité d'une théorie que si et seulement si sa négation en est un théorème formel; la théorie qui se déduit de la théorie T_ω par l'adjonction de la négation de l'axiome de l'infini fournit donc un exemple de théorie compatible, non ω-compatible. Comme la négation de l'axiome de l'infini dit qu'il existe une classe finie qui épuise tous les individus, tout modèle de T_ω ne peut être qu'un modèle dans lequel il y a des nombres entiers naturels non attestés, donc non standard.

Digression sur la règle d'induction infinie

Tarski rappelle une de ses remarques de 1927: le phénomène d'ω-incompatibilité serait empêché si l'on pouvait formaliser l'hypothèse tacite de tout à l'heure, par l'adjonction d'une règle de déduction, la règle d'induction infinie (alias, par la suite, pour les anglophones, «ω-règle») posant, pour toute formule $P(x)$ à une variable libre x, que des prémisses $P(0), P(1), P(2), \ldots$, on peut inférer la conclusion $\forall x P(x)$. Mais cette règle à une infinité de prémisses n'est par là pas finitiste; Tarski rappelle qu'elle fait l'objet, peu de temps auparavant — à la fin de 1930 — de méditations [*Hbt*13] de Hilbert, et lui reproche en outre de ne pas opposer d'obstacle au raisonnement par lequel Gödel [*Gdl*3] conclut à l'existence de propositions indécidables. Rosser, dès 1937 [*Rsr*4], et d'autres après celui-ci, dont Tarski [*Trk*21] lui-même, donneront de cette difficulté des preuves métamathématiques précises; il n'y aura pas, avant 1950, d'investigations plus approfondies des conséquences de l'adoption de cette règle.

Des modèles non standards dont Mostowski prouve l'existence en 1946

Dans la seconde partie de son article *On absolute properties of relations* [*Mtw*7], Mostowski *code,* par des points de l'ensemble de Cantor, les graphes de relations entre nombres entiers naturels, les suites (indexées par des nombres entiers naturels) d'ensembles de nombres entiers naturels, les suites de telles suites, les suites de suites de telles suites, etc... Grâce à de bonnes propriétés de continuité de ces codages, Mostowski est en mesure de prouver que les codes des graphes de relations à un nombre donné d'arguments forment des fermés; que les codes des *modèles* dont tous les types sont dénombrables et qui satisfont un ensemble dénombrable, fini ou non, de formules, forment des boréliens de classe ω au plus; et finalement, que les codes des graphes de relations qui vérifient un ensemble dénombrable de formules dans le modèle «absolu», savoir, celui dont chaque type d'un niveau supérieur se compose de toutes les parties du type immédiatement inférieur, forment aussi un borélien de classe ω au plus, lorsque ces formules sont élémentaires ou forment un ensemble «absolu», c'est-à-dire tel que ces formules sont satisfaites par tout modèle si et seulement si elles sont satisfaites par le modèle «absolu». Or le calcul, effectué par Kuratowski en 1937 [*Kur*4], de la classe projective de l'ensemble des codes des graphes de relations qui vérifient la formule qui définit un prédicat binaire comme un bon ordre, montre qu'il n'est pas borélien: Mostowski montre ainsi qu'il y a des modèles non standards dans lesquels un prédicat binaire satisfaisant à cette formule a un graphe interprétatif qui n'est pas celui d'un bon ordre sur les entiers naturels, alors que le graphe interprétatif de ce prédicat dans le modèle absolu est bien celui d'un tel bon ordre.

L'audace de ce type de raisonnement est tout à fait caractéristique de ce que les logiciens vont faire avec le théorème de LÖWENHEIM-SKOLEM durant la seconde moitié du vingtième siècle.

De la théorisation de l'extension de la notion de modèle à celle de modèle non standard

Les principaux développements de la thèse de HENKIN font, en 1949 et 1950, l'objet de deux articles au *Journal of Symbolic Logic*. D'abord, dans le premier, intitulé *The completeness of the first-order functional calculus* [$Hnk2$], HENKIN présente une argumentation qui débouche à la fois sur le théorème de complétude du calcul des prédicats du premier ordre et sur le théorème de LÖWENHEIM-SKOLEM: il part d'un ensemble d'énoncés compatibles et en construit une interprétation à l'aide d'une extension compatible maximale, qu'il obtient en modifiant la preuve d'existence d'une telle extension établie par LINDENBAUM (cette preuve consiste à prendre un à un les énoncés n'appartenant pas à l'ensemble donné, à garder ceux qui sont compatibles avec les énoncés considérés avant lui, et à rejeter les autres). Pour que l'extension construite décrive «bien» cette interprétation, il est essentiel, lors de la construction pas à pas de celle-ci, de veiller, chaque fois qu'une formule de la forme $\exists x A(x)$ vient s'y introduire, à y introduire la formule $A(a)$, où a est une constante devant dénoter un individu distingué et non encore utilisée, afin d'éviter un phénomène analogue à l'ω-incompatibilité; le processus obtenu tend vers un «point fixe», auquel la construction se termine, avec l'effet recherché: en gros, les individus d'un modèle solution, à isomorphisme près, du problème posé, sont les individus distingués à l'aide des constantes utilisées (y compris avant le début du processus).

Dans l'article de 1950, intitulé *Completeness in the theory of types* [$Hnk3$], HENKIN établit, pour toute la théorie des types de CHURCH [$Chu7$], un théorème de *complétude*, qui était de nature à surprendre les esprits non avertis, puisque le théorème d'incomplétude de GÖDEL [$Gdl3$] s'applique aux logiques d'ordre au moins deux, en raison de ce que les nombres entiers naturels 0, 1, 2, ..., y sont *attestés*: ils se définissent comme prédicats du second ordre. Or, les deux résultats coexistent, comme déjà dit, sans se contredire: ce ne sont pas les mêmes modèles que l'on prend en compte dans l'un et dans l'autre cas; il y a à la fois incomplétude, pour la sémantique des modèles «standards» auxquels on pensait jusque-là, et complétude, pour la sémantique de tous les modèles, standards et *non standards,* selon la terminologie employée par HENKIN dans le corps de son article.

HENKIN [$Hnk3$] décrit les types d'ordres des modèles non standards de l'arithmétique: on les obtient en mettant bout à bout le type d'ordre ω des nombres entiers naturels attestés 0, 1, 2, ..., puis autant de fois que l'on veut, des copies du type d'ordre $(\omega^* + \omega)\eta$ du produit lexicographique du type d'ordre $\omega^* + \omega$ des nombres entiers rationnels par le type d'ordre η des nombres rationnels, qui s'obtient en mettant bout à bout, à raison d'une par nombre rationnel, et dans l'ordre où ces nombres se présentent, une copie du type d'ordre des nombres entiers rationnels.

Dans la même livraison du *Journal of Symbolic Logic*, ROSSER et Hao WANG, dans un article [*RrW*] intitulé *Non-standard models for formal logic*, définissent un modèle non standard comme un modèle dans lequel, ou bien l'identité n'est pas interprétée par l'identité des individus du modèle, ou bien la classe interprétant l'ensemble des nombres entiers naturels n'est pas bien ordonnée par l'ordre interprétant l'ordre naturel, ou bien la classe interprétant les ordinaux n'est pas bien ordonnée par l'ordre interprétant l'ordre sur les ordinaux. Ils établissent que le système NF de QUINE [*Qui*1] n'a *pas de modèle standard* du tout — sans dire que cela entraîne le même défaut pour ML — et montrent des théorèmes du genre de celui qui affirme qu'un système formel, assez fort pour qu'on puisse y développer la théorie des nombres, ne peut disposer d'une démonstration formelle de ce que l'hypothèse de sa compatibilité entraîne qu'il a un modèle standard.

De quelques premiers résultats sur les forces comparées de versions de la théorie des ensembles et de la théorie des types

En 1949, dans un article intitulé *An undecidable arithmetical statement* [*Mtw*9], MOSTOWSKI avait montré que, dans la théorie des ensembles de VON NEUMANN-BERNAYS-GÖDEL renforcée par adjonction de l'axiome d'existence d'un ordinal fortement inaccessible: premièrement, il y a un ensemble, de cardinal fortement inaccessible, modèle naturel de la théorie de VON NEUMANN-BERNAYS-GÖDEL non ainsi renforcée — ce qui s'établit en formalisant une argumentation présentée dès 1924 [*Kur*2] par KURATOWSKI pour la théorie des ensembles de ZERMELO; et deuxièmement, il y a donc des ensembles modèles naturels de la théorie non renforcée, dont certains présentent un ensemble dénombrable lui-même modèle naturel de cette théorie, mais dont le plus petit ne peut posséder cette particularité — ce qui s'établit en constatant que dans la théorie renforcée, le raisonnement selon lequel on prouve le théorème de LÖWENHEIM-SKOLEM se formalise, et en utilisant les rapports des modèles naturels avec les types. En conséquence, si l'axiome d'existence d'un ordinal fortement inaccessible est compatible avec la théorie de VON NEUMANN-BERNAYS-GÖDEL, l'existence d'un ensemble dénombrable modèle naturel de la théorie non renforcée n'y est pas décidable; l'adjonction éventuelle, aux règles de déduction, de la règle d'induction infinie, n'y changerait rien; et (ce que MOSTOWSKI laisse implicite) la théorie renforcée est strictement plus «forte» que la théorie non renforcée, puisque la première, prouvant que la seconde admet un modèle, en prouve la compatibilité.

MOSTOWSKI donnait aussi, à l'énoncé indécidable précédent, une forme arithmétique équivalente, du second ordre (ou, si l'on préfère, se rapportant à l'arithmétique des nombres réels), après avoir montré que la preuve de ce qu'un modèle naturel dénombrable est isomorphe à un modèle dont le support est un ensemble de nombres entiers naturels se formalise aussi dans la théorie renforcée; l'énoncé arithmétique visé s'obtient alors en écrivant qu'il existe une partie de l'ensemble des nombres entiers naturels codant le graphe d'une relation d'appartenance d'un modèle isomorphe à un modèle naturel, modèle de la théorie non renforcée. Or, l'écriture de la satisfaction de la condition de fondement introduit alors un quantificateur universel sur des parties de l'ensemble des nombres entiers naturels: l'énoncé obtenu dit qu'un certain complémentaire d'ensemble analytique n'est pas vide. Et c'est pour montrer l'équivalence de cet énoncé avec l'énoncé de théorie des ensembles indécidable qu'il faut, pour revenir à celui-ci, démontrer que la structure déterminée sur le modèle arithmétique par la relation qui y interprète l'appartenance est isomorphe à un modèle «naturel»: c'est à cet effet que MOSTOWSKI introduit et démontre là le «lemme de contraction de MOSTOWSKI», dont il a été question plus haut.

En 1949 [*Kmn*2], à l'encontre de l'idée, tacitement reçue depuis longtemps, que la théorie des types simple et la théorie des ensembles de ZERMELO originelle de 1908 [*Zml*2] se vaudraient, comme fondations des mathématiques, KEMENY produit une définition, dans la théorie de ZERMELO, de la vérité des énoncés de la théorie des types, et en conclut que la seconde est strictement plus «forte» que la première, en ce sens que: par des moyens élémentaires, on peut démontrer la compatibilité de la première dans la seconde, sous réserve de la compatibilité de celle-ci; il y a dans la seconde des théorèmes que la première ne peut pas prouver; et il y a des ensembles de nombres entiers naturels, définissables dans la seconde, et non définissables dans la première.

KEMENY [*Kmn*2] suggère toute une hiérarchie de théories des types, avec des types transfinis, et de théories des ensembles, avec des axiomes d'existence d'ordinaux de plus en plus complexes, entre les termes de laquelle de telles relations s'établissent.

ROSSER et WANG, à la fin de l'article [*RrW*] cité plus haut, montrent que la théorie des ensembles de ZERMELO-FRAENKEL est strictement plus «forte» que la théorie des ensembles de ZERMELO, par un raisonnement analogue, mais usant d'un modèle, comme l'avait fait MOSTOWSKI. On sait alors, depuis deux ans, qu'il n'en va pas de même des théories des ensembles de VON NEUMANN-BERNAYS-GÖDEL et de ZERMELO-FRAENKEL, dont la compatibilité de la seconde relativement à la première est presque immédiate. Beaucoup plus ardue est la preuve de compatibilité *de la première* relativement à la seconde, entièrement conforme aux canons finitistes hilbertiens, qu'apporte Ilse NOVAK en 1948 [*Nvk*1, 2], et qu'elle publie en 1950 [*Nvk*3]: un procédé dû à HILBERT, et décrit dans les *Grundlagen der Mathematik* ([*HbB*], Bd. II), permet d'étendre le langage de la théorie de ZERMELO-FRAENKEL avec des termes, dont l'interprétation attendue est qu'ils nomment des individus; il est montré dans les *Grundlagen* que l'extension est inessentielle — dans le langage originel, il ne se glisse pas de nouveaux théorèmes — et qu'elle n'introduit pas de contradiction, s'il n'y en avait pas auparavant. Ces individus forment un modèle dénombrable (comme le sont les termes d'un langage finitaire) de la théorie de ZERMELO-FRAENKEL, mais la preuve procède sans expliciter celui-ci, juste en triant par récurrence, dans l'ordre lexicographique, les énoncés qui le décrivent; cela se fait selon un procédé effectif, une fois donnés les théorèmes de la théorie; la non-contradiction de la théorie entraîne celle de cette description. On fait la même chose, après «arithmétisation» des termes formels et des formules, en rajoutant à ce modèle les «classes» prévues par la théorie de VON NEUMANN-BERNAYS-GÖDEL, qui, par rapport à lui, seront des ensembles de la théorie de ZERMELO-FRAENKEL, en profitant de ce que ces classes sont *définissables* en termes de formules du langage de la théorie de ZERMELO-FRAENKEL et, pour chacune, d'un nombre fini de variables pour désigner des classes — ce que l'on rajoute, en fait, ce sont les énoncés susceptibles de décrire le comportement de ces classes, triés ensuite comme l'impliquent les définitions de celles-ci.

Il en résulte, remarque MOSTOWSKI [*Mtw*11] (cf. aussi [*RrW*]), que la théorie de VON NEUMANN-BERNAYS-GÖDEL est une extension inessentielle de la théorie de ZERMELO-FRAENKEL; à savoir, si un énoncé T du langage de la seconde, c'est-à-dire, s'exprimant en termes d'appartenance et d'ensembles seulement, sans référence à des classes, est un théorème formel de la première, il est déjà un théorème formel de la seconde. En effet, si T n'était pas un théorème de la théorie de ZERMELO-FRAENKEL, la théorie qui résulte de celle-ci par adjonction de NONT comme axiome serait compatible; on pourrait refaire à partir de cette théorie étendue le raisonnement de NOVAK [*Nvk*3], et cela conduirait à la compatibilité de la théorie qui résulte de la théorie de VON NEUMANN-BERNAYS-GÖDEL par adjonction comme axiome de NONT; contradiction, T se démontrant déjà rien qu'avec les axiomes de la théorie de VON NEUMANN-BERNAYS-GÖDEL.

Comment se fait-il, alors, que, dans la théorie de VON NEUMANN-BERNAYS-GÖDEL, l'on puisse montrer que les axiomes de la théorie de ZERMELO-FRAENKEL sont satisfaits, mais non en montrer la compatibilité? Car, au cas contraire, le raisonnement de NOVAK donnerait une preuve, dans la théorie de VON NEUMANN-BERNAYS-GÖDEL, de sa propre compatibilité, contredisant le théorème d'incomplétude de GÖDEL [*Gdl*3]. Dans un article intitulé *Impredicative definitions in the axiomatic set theory* [*Mtw*11], qui paraît en 1951 aux *Fundamenta Mathematicae,* postérieurement à celui de ROSSER et WANG [*RrW*] dont j'ai déjà parlé, et où ces derniers donnent d'autres preuves, moins transparentes, du résultat de NOVAK et de l'impossibilité de démontrer la compatibilité de la théorie de ZERMELO-FRAENKEL dans la théorie de VON NEUMANN-BERNAYS-GÖDEL, MOSTOWSKI donne, de cette situation surprenante, l'explication ci-après, dont les éléments sont également présents dans une Note [*Wng*2] de WANG, de 1950, et qui suppose compatible la théorie de VON NEUMANN-BERNAYS-GÖDEL, où tous les énoncés cités se formalisent (dans ce qui suit, j'appelle «théorèmes de ZERMELO-FRAENKEL» les théorèmes formels de la théorie de ZERMELO-FRAENKEL, et un tel théorème est dit d'ordre (entier naturel) n, s'il a une preuve faisant appel à au plus n axiomes et au plus n applications de règles d'inférence):

- la théorie de VON NEUMANN-BERNAYS-GÖDEL souffre d'une ω-incomplétude, en ce que l'on peut y prouver tout énoncé de la forme:

«Tout théorème de ZERMELO-FRAENKEL *d'ordre n est vrai»,*

où n est une suite de bâtons, mais *non*, pour autant, l'énoncé:

«Tout théorème de ZERMELO-FRAENKEL, *quelqu'en soit l'ordre entier naturel n, est vrai»;*

- la théorie de VON NEUMANN-BERNAYS-GÖDEL ne satisfait pas la règle de récurrence, car on peut aussi y prouver les énoncés:

  «Tout théorème de ZERMELO-FRAENKEL *d'ordre 1 est vrai», et*

  «Quel que soit le nombre entier naturel n, que tout théorème de ZERMELO-FRAENKEL *d'ordre n soit vrai implique que tout théorème de* ZERMELO-FRAENKEL *d'ordre* $n + 1$*» soit vrai;*

- et pour finir, on ne peut, dans la théorie de VON NEUMANN-BERNAYS-GÖDEL, prouver l'existence d'une classe des ordres entiers naturels des théorèmes de ZERMELO-FRAENKEL qui ne sont pas vrais.

Pendant ce temps, quelques mois après son article avec ROSSER déjà cité, WANG [*Wng*1] établit, en utilisant les parties d'un modèle de NF dont les individus sont les nombres entiers naturels, la compatibilité d'une version relativement peu affaiblie de ML, sous réserve de la compatibilité de NF.

Des premières applications mathématiques et des premières notions théoriques de la théorie des modèles

Des notions de théorie des modèles introduites par TARSKI, peu avant 1950
En 1950, au congrès international des mathématiciens qui s'est tenu à Cambridge, dans le Massassuchetts, l'adresse de TARSKI [*Trk*32] traitait de *Some notions and methods on the borderline of algebra and metamathematics*. Pour une grande part, il y reprend des communications déjà présentées [*Trk*25, 26], une vingtaine de mois plus tôt, à la fin de 1948, à une réunion de l'*American Mathematical Society*. Ici et là, il introduit la notion de *classe arithmétique* ou *élémentaire,* pour désigner une classe de structures homologues — appelées «systèmes» en 1948, «algèbres au sens large» en 1950 — satisfaisant une formule du premier ordre: ce n'est qu'en 1954 [*Trk*33] qu'il parlera de «théorie des modèles»; il *ré*introduit ([*Trk*25]), sous les noms d'*équivalence arithmétique*, ou d'*indistingabilité élémentaire*, la notion, remontant au séminaire de Varsovie des années 1927 à 1929, et déjà présentée en 1936 dans une publication, l'Annexe à ses *Grundzüge des Systemenkalkül* [*Trk*12], sous les noms d'indistingabilité élémentaire, ou d'*équivalence élémentaire* — ce dernier lui est resté — pour désigner la relation entretenue par des structures homologues (en 1936, il ne s'agissait encore que de types d'ordres) satisfaisant les *mêmes énoncés élémentaires*. Il introduit encore, en 1948 [*Trk*25], la notion de *type arithmétique,* pour désigner une classe de structures homologues élémentairement équivalentes, et la notion de *théorie élémentaire d'une classe de structures,* pour

désigner l'ensemble des énoncés élémentaires satisfaits par toutes les structures de cette classe.

En 1948 [*Trk*25], il met l'accent sur l'algèbre de BOOLE dénombrable que forment les classes arithmétiques, et sur ce qu'un type arithmétique n'est autre que l'intersection des classes arithmétiques constituant un ultrafiltre de cette algèbre de BOOLE. En 1950 [*Trk*32], il met l'accent sur la topologie engendrée dans une classe de structurs par la famille EC des classes élémentaires, et dont les fermés forment la famille EC_Δ des classes de structures qui satisfont un ensemble dénombrable d'énoncés — c'est là qu'il apparaît que le «théorème de compacité» exprime la compacité de l'espace obtenu; il introduit la notion de sous-modèle, sous le nom de sous-algèbre au sens large; et donne, en exemple de structures élémentairement équivalentes, le corps des nombres réels et le corps des nombres algébriques. Ces corps ne sont pas isomorphes: ils ne sont même pas équipotents; on ne peut cependant les distinguer par aucune propriété du premier ordre. En fait, le concept d'équivalence élémentaire est un concept aussi fondamental pour les mathématiques que celui d'isomorphisme.

Des anticipations de Pascual JORDAN en 1949

En 1949, dans l'article [*Jrd*] cité plus haut, Pascual JORDAN anticipe quelques notions et théorèmes de théorie des modèles, d'un genre qui ne s'épanouira que plus tard: appelant Verknüpfungsbereiche les algèbres, il établit que, pour qu'une classe de domaines soit celle des domaines qui satisfont une formule universelle (comportant une quantification universelle portant sur une formule sans quantificateur), il faut et il suffit qu'elle contienne les domaines à un individu et qu'elle soit stable par passage aux sous-domaines et aux images isomorphes; pour qu'elle soit la classe des domaines satisfaisant une formule universelle fondamentale (comportant une quantification universelle portant sur une conjonction de disjonctions d'identités), il faut et il suffit qu'elle soit stable par passage aux sous-domaines et aux images homomorphes; et, pour qu'elle soit la classe des domaines satisfaisant une formule universelle-existentielle-universelle fondamentale (comportant une succession d'une quantification universelle, d'une quantification existentielle, puis d'une quantification universelle, portant sur une conjonction de disjonctions d'identités), il faut et il suffit qu'elle soit stable par passage aux sous-domaines et aux images homomorphes. Il illustre ces théorèmes en les appliquant aux groupes, anneaux, et réseaux.

L'apport de MALTSEV: un nouvel abord des applications mathématiques des théorèmes fondamentaux de logique

Les idées qui gouvernaient les articles de MALTSEV de 1936 [*Mlt*1] et 1941 [*Mlt*2] n'ont été comprises, et reconnues, peu à peu, qu'avec le temps. MALTSEV n'a pas fait qu'y introduire le théorème de compacité: il a inauguré un nouvel art de l'utiliser pour établir des propositions des mathématiques à l'aide des théorèmes fondamentaux de la logique.

Dès 1936 [*Mlt*1], il recourt, sans l'exprimer, cependant, de façon adéquate, à l'idée de décrire un modèle, grâce à une extension du langage par adjonction d'une constante individuelle par individu, en présentant une à une des suites d'individus que les graphes composant ce modèle doivent englober, et d'autres qu'ils doivent éviter; les énoncés exprimant les premières de ces conditions se réduisent à un prédicat portant sur des arguments qui sont tous de nouvelles constantes individuelles, et ceux qui expriment les secondes, à des négations de tels énoncés; plus tard, Paul COHEN [*Coh*1, 2] nommera «énoncés de base» les énoncés de l'une et de l'autre de ces espèces. MALTSEV appelle «configuration» tout ensemble de tels énoncés, et introduit sous le nom de «configuration complète», ce que Abraham ROBINSON [*ARb*2] rebaptisera quatorze ans plus tard le *diagramme* d'un modèle.

Pour démontrer le théorème de LÖWENHEIM-SKOLEM *ordinaire*, mais étendu au cas d'un langage de cardinal infini quelconque, il reprend une preuve de SKOLEM [*Skm*4], notant qu'elle fournit un modèle du plus petit cardinal infini au moins égal à celui [du langage] des énoncés. Pour démontrer le théorème de LÖWENHEIM-SKOLEM *«ascendant»*, il reprend cette même preuve, agrémentée d'un raisonnement «par compacité» nouveau, dont c'est la première apparition dans l'histoire, et dont la postérité sera considérable: il s'agit maintenant d'*étendre* un modèle; pour cela, il adjoint au langage une nouvelle constante individuelle b^*, et au diagramme du modèle à étendre, les énoncés de la forme $b \neq b^*$ pour toutes les constantes individuelles b dénotant un individu de ce modèle, et montre que le diagramme ainsi complété est satisfaisable, puisque toute partie finie de ce diagramme est réalisée dans le modèle à étendre.

Comme application mathématique de ce théorème, il note juste là que «tout corps algébrique infini a des extensions» [propres, bien sûr].

En 1941 [*Mlt*2], il formule explicitement le théorème de compacité pour les formules du premier ordre, vu comme généralisation de théorèmes du type: «si telle ou telle propriété est vraie de toute sous-algèbre de type fini de telle ou telle algèbre, elle est vraie de cette algèbre», et définit la notion de propriété *formulable au premier ordre*. Cette note est avant tout consacrée à une dizaine d'applications à la théorie des groupes, du genre de quatre qui résultent par particularisation du corollaire suivant du théorème de compacité: pour qu'un groupe G présente une suite de composition dont chaque terme a pour quotient par le suivant un groupe astreint à posséder un ensemble fixé de propriétés formulables au premier ordre et se transmettant d'un groupe à ses sous-groupes, il faut et il suffit qu'il en aille ainsi pour tout sous-groupe de type fini de ce groupe G.

D'un programme analogue d'Abraham ROBINSON: la métamathématique de l'algèbre

Le thème d'investigations auquel appartient ce genre de théorèmes est rangé, par Abraham ROBINSON, dans la «métamathématique de l'algèbre», comme celui-ci appelle le domaine qui recouvre les thèmes du programme de travail qu'il présente à la fin de 1948 à l'*Association for Symbolic Logic* [*ARb*1], et en 1950 [*ARb*2], au congrès international des mathématiciens de Cambridge, aux Etats-Unis.

A ce moment-là, comme ROBINSON le dit en 1950 [*ARb*2], plusieurs logiciens, dont lui-même, TARSKI, mais aussi (cf. [*Vgm*]) HENKIN dans sa thèse en

1947, ont obtenu indépendamment le métathéorème suivant: soit $\varphi(x_1, \ldots, x_n)$ une formule du premier ordre dans le langage des corps, satisfaite par tous les corps de caractéristique nulle. Or, on ne sait écrire, au premier ordre, qu'un corps est de caractéristique nulle, qu'en écrivant, pour toute suite de bâtons n, l'énoncé exprimant que, pout tout individu (du corps) x, la somme de n termes x n'est pas nulle. Or, la preuve de $\varphi(x_1, \ldots, x_n)$ ne fait intervenir qu'un nombre fini de ces axiomes; si p_0 est la plus longue suite de bâtons p pour laquelle la preuve utilise celui de ces axiomes qui porte sur p termes x, cette preuve va subsister dans la théorie qu'on obtient de la théorie des corps de caractéristique nulle en n'en gardant que les axiomes des corps, et ceux qui expriment que la caractéristique est différente de tous les nombres entiers naturels au plus égaux à p_0; donc, pour tout nombre entier naturel $p > p_0$, la formule $\varphi(x_1, \ldots, x_n)$ est aussi satisfaite par tout corps de caractéristique p.

HENKIN, dans son article de 1949 [$Hnk2$], raisonne aussi «par compacité» pour montrer que si un énoncé α a des modèles finis de nombres d'éléments arbitrairement grands, il a un modèle infini: si M est l'ensemble des nombres d'éléments des modèles finis de α, et si n appartient à M, l'énoncé β_n

$$\exists x_1 \exists x_2 \ldots \exists x_n [x_1 \neq x_2 \text{ ET } x_1 \neq x_3 \text{ ET } \ldots$$
$$\ldots \text{ ET } x_1 \neq x_n \text{ ET } x_2 \neq x_3 \text{ ET } \ldots \text{ ET } x_{n-1} \neq x_n]$$

et l'énoncé α ont un modèle. Or, pour n et p entiers naturels, tels que $n > p$, l'énoncé β_n implique l'énoncé β_p; toute partie finie de l'ensemble composé de tous les énoncés β_n, et de α, admet donc un modèle: il suffit de prendre un modèle de α ayant assez d'éléments pour atteindre, ou dépasser, l'indice du plus grand des β_n appartenant à cette partie finie. Par l'argument de compacité, il y a donc un modèle de tous les énoncés β_n, pour n entier naturel, et de α: mais un tel modèle est un modèle infini de α.

ROBINSON range, dans ce qu'il appelle en 1950 [$ARb2$] les «principes de transfert» — ce nom leur est resté — les métathéorèmes dont la satisfaction de propositions mathématiques est l'objet et qui, résultant de théorèmes de logique, comme ci-dessus du théorème de compacité, énoncent que toute proposition mathématique satisfaite par les structures d'une espèce donnée est nécessairement satisfaite par les structures d'une autre espèce donnée. Comme exemple d'application du principe de transfert énoncé plus haut, il donne, en 1948 [$ARb1$], celui de la formule: «le polynôme à coefficients entiers $q(x_1, \ldots, x_n)$ est irréductible», supposée satisfaite dans toutes les extensions du corps des nombres rationnels, et de ce fait, satisfaite dans tous les corps de caractéristique assez grande; en 1950 [$ARb2$], il donne l'exemple, quelque peu plus sophistiqué, d'un système de polynômes à coefficients entiers dont le nombre des racines dans toute extension du corps des nombres rationnels est borné par un nombre m, et qui possède la même propriété dans tout corps de caractéristique assez grande.

Au vrai, cet exemple combine tacitement le principe de transfert précédent avec un autre, déjà connu de TARSKI depuis les alentours de 1930, et relevé par lui en 1948 dans la première édition de sa méthode de décision pour la théorie élémentaire du corps des nombres réels [*Trk*24]. Ce principe repose sur la complétude de cette théorie, complétude qui va de pair avec l'équivalence élémentaire de tous ses modèles: toute formule $\varphi(x_1, \ldots, x_n)$ du langage de cette théorie, du premier ordre, satisfaite par le corps des nombres réels, est satisfaite par tout corps réel clos — donc, en particulier, par le corps des nombres rationnels. L'exemple donné explicitement par Robinson dans ses exposés de 1948 [*ARb*1] et 1950 [*ARb*2] est celui des corps algébriquement clos: dès qu'un énoncé exprimé dans le langage de la théorie (complète) des corps est satisfait par le corps des nombres complexes, il est satisfait par tout corps algébriquement clos. L'idée que de tels principes de transfert s'appliquent entre corps, voire anneaux ordonnés, archimédiens, et leurs équivalents élémentaires non archimédiens, apparaît tant chez Robinson que chez Tarski. A cette époque, les corps non archimédiens font l'objet d'un regain d'intérêt: en 1948 [*Hwt*], E. Hewitt en fabrique comme quotients d'anneaux de fonctions continues à valeurs réelles par des idéaux maximaux appropriés — mais c'est seulement en 1960 [*Ksl*] (cf. aussi [*CgK*], p. 520) que les logiciens se rendront compte que cette construction relève (sous une autre présentation) d'une opération très intéressante pour eux!

Un second volet du programme de Robinson [*ARb*1, 2] consiste à prendre systématiquement les notions de l'algèbre universelle comme paradigmes en vue de développer la théorie des modèles, et à rechercher quelle extension de sens donner dans cette théorie à des notions telles que celles de polynôme, d'idéal, d'élément algébrique, d'extension séparable, inséparable, transcendante, de clôture algébrique, ..., de façon à redonner, dans leur application à la théorie de l'algèbre universelle dont elles proviennent, le sens que ces notions avaient auparavant. Dans cette intention, Robinson [*ARb*2] introduit la notion de *diagramme* d'un modèle d'une théorie: il s'agit, dans le langage de cette théorie, enrichi de constantes pour désigner chacun des individus du modèle, de l'ensemble des énoncés sans connecteur ni quantificateur satisfaits par le modèle, et des négations des énoncés sans connecteur ni quantificateur non satisfaits par le modèle.

Postface, sur l'état de la logique vers 1950

Je me suis abstenu volontairement, sauf pour indiquer en passant combien tardifs ont été certains accomplissements, de mentionner quelque développement postérieur à 1950, afin de donner à voir de quelle façon la logique mathématique a pu être perçue, au milieu du vingtième siècle: une très vivante branche des mathématiques, riche en enseignements et en problèmes passionnants, pour ceux qui s'y intéressent; mais encore à l'écart du reste des mathématiques, auxquelles elle emprunte bien plus qu'elle ne semble avoir à rendre, elle est reçue comme irritante par sa mathématisation continuelle des *représentations* des objets mathématiques, tout autant que de ces objets eux-mêmes; elle inquiète un peu par sa moisson de curieux résultats négatifs, qui semblent pourtant sans effet notable sur le travail des mathematiciens «ordinaires», et suscite quelque scepticisme par son foisonnement même en entreprises à l'avenir incertain. Il n'est donné qu'à quelques esprits, de tournure philosophique, de croire à l'importance qu'elle prendra dans le futur, par les lumières quelle apportera sur les architectures de la connaissance déductive en général, et des mathématiques en particulier.

Bibliographie

[Ack1] ACKERMANN W. Begründung des «tertium non datur» mittels der Hilbertschen Theorie der Widerspruchsfreiheit, *Mathematische Annalen* **93** (1924), 1-36.

[Ack2] ACKERMANN W. Zum Hilbertschen Aufbau der reellen Zahlen, *Mathematische Annalen* **99** (1928), 118–133. Eng. tr. [vHj], 493–507.

[Ack3] ACKERMANN W. Über die Erfüllbarkeit gewisser Zählausdrücke, *Mathematische Annalen* **100** (1928), 638–649.

[Ack4] ACKERMANN W. Die Widerspruchsfreiheit der allgemeinen Mengenlehre, *Mathematische Annalen* **114** (1937), 305–315.

[Ack5] ACKERMANN W. Zur Widerspruchsfreiheit der Zahlentheorie, *Mathematische Annalen* **117** (1940), 162–194.

[Ack6] ACKERMANN W. *Solvable Cases of the Decision Problem.* North Holland, Amsterdam, 1955, VIII+114 pp.

[Aik] AIKEN H.H. Proposed Automatic Calculating Machine, ed. and pref. by A.G. Oettinger and T.C. Bartee, *IEEE Spectrum*, Institute of Electrical and Electronics Engineers, Aug. 1964, 62–69.

[Ajd] AJDUKIEWICZ K. *Z metodologji nauk dedukcyjnych.* Wydawnictwo Polskiego Towarzystwa Filozoficznego we Łwowie **10**. Łwów, 1921.

[Anl] ANELLIS I.H. The First Russell Paradox, [Drk], 33–46.

[Ank] ÁNOSKAIA S.A. Osnovaniá matématiki i matématičéskaia logika, *Matématiki v SSSR za tridcat'lét 1917–1947.* OGIZ, Moscou et Leningrad, 1948, 9–50.

[Art] ARTIN E. *The Collected Papers.* Ed. by Lang S., Tate J. Addison, Wesley, 1965. Rééd. Springer, Berlin, Heidelberg, New-York, 1982.

[ArS] ARTIN E. und SCHREIER O. Algebraische Konstruktion reeller Körper, *Abhandlungen aus dem Mathematischen Seminar der Hamburgischen Universität* **5** (1926), 85–99. Abg. [Art], 258-272.

[Asp] ASPRAY W. Oswald Veblen and the Origins of Mathematical Logic at Princeton, [Drk], 54–70.

[AsK] ASPRAY W. and KITCHER P. (eds) *History and Philosophy of Modern Mathematics.* University of Minnetosa Press, Minneapolis, 1988.

[Bcm] BACHMANN H. *Transfinite Zahlen.* Springer, Berlin-Göttingen-Heidelberg, 1955, VII+204 S.

[Bld] BALDUS R. Über Hilberts Vollständigkeitsaxiom, *Mathematische Annalen* **100** (1928), 321-333.

[Bnc] BANACH S. Über additive Massfunktionen in abstrakten Mengen, *Fundamenta Mathematicae* **15** (1930), 97–101.

[BaK] BANACH S. et KURATOWSKI K. Sur une généralisation du problème de la mesure, *Fundamenta Mathematicae* **14** (1929), 127–131.

[BMz] BANACH S. et MAZUR S. Sur les fonctions calculables, *Rocznik Polskiego Towarsystwa Matematycznego* (=Annales de la Société Polonaise de Mathématique), **16** (1937), 223.

[BnT] BANACH S. et TARSKI A. Sur la décomposition des ensembles de points en parties respectivement congruentes, *Fundamenta Mathematicae* **6** (1924), 244–277. Repr. [Trk37] **I**, 119–154.

[BcM] BARCAN MARCUS R. A functional calculus of first order based on strict implication, *Journal of Symbolic Logic* **11** (1946), 1–11.

[Bmn] BEHMANN H. Beiträge sur Algebra der Logik, insbesondere zum Entscheidungsproblem, *Mathematische Annalen* **86** (1922), 163–229.

[BlN] BELL C.G. and NEWELL A. *Computer Structures: Readings and Examples*. New-York, 1971.

[Blt] BELTRAMI E. Saggio di interpretazione della geometria non-euclidea, *Giornale di Matematiche*, *6* (1868), 284–312.

[Bem] BEMAN W.W. *Essays on the theory of numbers*. Open Court, Chicago, 1901, 115 pp.

[BnP] BENACERRAF P. and PUTNAM H. *Philosophy of mathematics: selected readings*. Prentice-Hall, Englewood Cliffs, New-Jersey, 1964.

[Bdx] BENDIXSON I. Quelques théorèmes de la théorie des ensembles de points, *Acta mathematica* **2** (1883), 415–429.

[Brc] BERECZKI I. Nem-elemi rekurziv függvény létezése, *Comptes Rendus du Premier Congrès des Mathématiciens Hongrois 1950*, 409–417.

[Brn1] BERNAYS P. Axiomatische Untersuchung des Aussagen-Kalküls der «Principia Mathematica», *Mathematische Zeitschrift* **25** (1926), 305–320.

[Brn2] BERNAYS P. A system of axiomatic set theory, *Journal of Symbolic Logic* **2** (1937), 65–77; **6** (1941), 1–17; **7** (1942), 67–89; **8** (1943), 89–106; **13** (1948), 65–79; **19** (1954), 81–96.

[BnS] BERNAYS P. und SCHÖNFINKEL M. Zum Entscheidungsproblem der mathematischen Logik, *Mathematische Annalen* **99** (1928), 401–419.

[Bth1] BETH E.W. Semantical considerations on intuitionistic mathematics, *Koninklijke Nederlandse Akademie van Weteschappen, Proceedings of the section of sciences*, series A, **50** (1947), 1246–1251 = *Indagationes Mathematicae*, **9** (1947), 572–577.

[Bth2] BETH E.W. On Padoa's method in the theory of definition, *Koninklijke Nederlandse Akademie van Weteschappen, Proceedings of the section of sciences*, series A, **56** (1953) = *Indagationes Mathematicae*, **15** (1953), 330–339.

[Bth3] BETH E.W. *The foundations of mathematics*. North-Holland, Amsterdam, 1959, XXVI+741 pp. 2^{d} rev. ed. 1965, XXVIII+741 pp.

[Btz] BETTAZZI R. Sui punti de discontinuità delle funzioni di variabile reale, *Rendiconti del Circolo Matematico di Palermo* **6** (1892), 173–195

[Bkf1] BIRKHOFF G. On the combination of subalgebras, *Proceedings of the Cambridge Philosophical Society* **29** (1933), 441–464.

[Bkf2] BIRKHOFF G. On the structure of abstract algebras, *Proceedings of the Cambridge Philosophical Society* **31** (1935), 433–454.

[Bkf3] BIRKHOFF G. Subdirect unions in universal algebras, *Bulletin of the American Mathematical Society* **50** (1944), 764–768.

[BvN] BIRKHOFF G. and NEUMANN J. VON. On the logic of quantum mechanics, *Annals of Mathematics* ser. 2, **37** (1936), 823–843. Repr. [vNm7] **IV**, 105–125.

[BlP] BLOK W.J. and PIGOZZI D. Alfred Tarski's work on general metamathematics, *Journal of Symbolic Logic* **53** (1988), 36–50.

[Bvr] BOCHVAR D.A. Über einen Aussagenkalkül mit abzählbaren logischen Summen und Produkten, *Mathematiceskij Sbornik (Recueil mathématique)* n. s. **7** (1940), 65–100.

[Blz] BOLZANO B. *Wissenschaftlehre. Versuch einer ausführlichen und grösstentheils neuen Darstellung der Logik mit steter Rücksicht auf deren bisherigen Bearbeiter.* 4 vol. Sulzbach, 1837. Abdruck, Scientia, Aalen, 1970.

[Boo] BOOLE G. *The mathematical analysis of logic, being an essay toward a calculus of deductive reasoning*. London and Cambridge, 82 pp., 1847.

[Brl1] BOREL E. Remarques sur les principes de la théorie des ensembles, *Mathematische Annalen* **60** (1905), S. 194–195. Repr. [Brl3], 1251–1252.

[Brl2] BOREL E. Sur les principes de la théorie des ensembles, *Atti del IV Congresso Internazionale dei matematici, Roma, 6–11 Aprile 1908*. Castelnuovo, Roma, 1909. **2**, 15–17. Réimp. [Brl3], 1267–1269.

[Brl3] BOREL E. *Oeuvres*. **III**. Ed. du C. N. R. S. Paris, 1972.

[BkS] BORKOWSKI L. and SLUPECKI J. The logical works of J. Łukasíewícz, *Studia Logica* **VIII** (1958), pp. 7–56.

[Bbk] BOURBAKI N. *Eléments de Mathematique. Première partie: Les structures fondamentales de l'Analyse. Livre I, Théorie des Ensembles (Fascicule de résultats)*. Hermann, Paris, 1939, VIII+51pp.

[Brw1] BROUWER L.E.J. *Over de grondslagen der wiskunde.* Maas en van Suchtelen, Amsterdam en Leipzig; Noordhoff, Groningen, 1907, 183 pp. Engl. tr. [Brw15], 11–101.

[Brw2] BROUWER L.E.J. De onbetrouwbaarheid der logische principes, *Tijschrift voor wijsbegeerte* **2** (1908), 152–158. Repr. [Brw6], 5–12. Engl. tra. [Brw15], 107–111.

[Brw3] BROUWER L.E.J. *Intuitionisme en formalisme.* Noordhoff, Groningen, 1912. Repr. [Brw6], 5–29. Engl. tr.: [Brw4].

[Brw4] BROUWER L.E.J. Intuitionism and formalism, *Bulletin of the American Mathematical Society* **20** (1913), 81–96. Repr. [BnP], 66–77; [Brw15], 123–138.

[Brw5] BROUWER L.E.J. Begründung der Mengenlehre unabhängig vom logischen Satz vom ausgeschlossenen Dritten. Erster Teil: Allgemeine Mengenlehre, *Verhandelingen der Koninklijke Akademie van Wetenschappen te Amsterdam*, 1st sectie, **12**, n^r5 (1918). Begründung der Mengenlehre unabhängig vom logischen Satz vom ausgeschlossenen Dritten. Zweiter Teil: Theorie der Punktmengen, *Verhandelingen der Koninklijke Akademie van Weteschappen te Amsterdam*, 1st sectie, **12**, n^r7 (1919). Abg. [Brw15], 150–221.

[Brw6] BROUWER L.E.J. *Wiskunde, waarheid, werkelijkheid.* Noordhoff, Groningen, 1919.

[Brw7] BROUWER L.E.J. Intuitionistische Mengenlehre, *Jahresbericht der deutschen Mathematiker-Vereinigung* **28** (1919), 203–208 = *Koninklijke Akademie van Wetenschappen te Amsterdam, Proceedings of the section of sciences* **23** (1922), 949–954. Abg. [Brw15], 230–235.

[Brw8] BROUWER L.E.J. Besitzt jede reelle Zahl eine Dezimalbruchentwickelung? *Koninklijke Akademie van Wetenschappen te Amsterdam, Verslag van de gewone vergaderingen der wis- en natuurkundige afdeeling* **29** (1921), 803–812 = *Koninklijke Akademie van Wetenschappen te Amsterdam, Proceedings of the section of sciences* **23** (1922), 955–964. Abg. *Mathematische Annalen* **83** (1921), 201–210; [Brw15], 236–245.

[Brw9] BROUWER L.E.J. Begründung der Funktionenlehre unhabhängig vom logischen Satz vom ausgeschlossenen Dritten. Erster Teil: Stetigkeit, Messbarkeit, Derivierbarkeit, *Verhandelingen der Koninklijke Akademie van Wetenschappen te Amsterdam*, 1st sectie, **13**, n^r2 (1923), 24p. Abg. [Brw15], 246–267.

[Brw10] BROUWER L.E.J. Über die Bedeutung des Satzes vom ausgeschlossenen Dritten in der Mathematik, insbesondere in der Funktionentheorie, *Journal für die reine und angewandte Mathematik* **154** (1923) 1–7. Engl. tr. [vHj], 334–341. Abg. [Brw15], 268–274.

[Brw11] BROUWER L.E.J. Intuitionistische splitsing van mathematischer grondbegrippen, *Koninklijke Akademie van Wetenschappen te Amsterdam, Vers-*

lag van de gewone vergaderingen der wis- en natuurkundige afdeeling **32** (1923), 877–880. Deutsche Übersetzung unter: Intuitionistische Zerlegung mathematischer Grundbegriffe, *Jahresbericht der deutschen Mathematiker-Vereinigung* **33** (1924), 251–256. Abg. [Brw15], 275–280.

[Brm12] BROUWER L.E.J. Die repräsentierende Menge der stetigen Funktionen des Einheitskontinuums, *Koninklijke Nederlandse Akademie van Wetenschappen, Proceedings of the section of Sciences*, series A, **45** (1942), 443; = *Indagationes Mathematicae* **4** (1942), 154. Abg. [Brw15], 461.

[Brw13] BROUWER L.E.J. Consciousness, Philosophy and Mathematics, *Proceedings of the Tenth International Congress of Philosophy (Amsterdam, Aug. 11–18, 1948)*. North Holland, Amsterdam, 1949, 1235–1249. Repr. [Brw15], 480–494.

[Brw14] BROUWER L.E.J. Over accumulatiekernen van oneindige kernsoorten, *Koninklijke Nederlandse Akademie van Wetenschappen, Proceedings of the section of sciences*, series A, **55** (1952), 439–441 = *Indagationes mathematicae*, **14** (1952), 439–441. Engl. tr. [Brw15], 516–518.

[Brw15] BROUWER L.E.J. *L.E.J. Brouwer collected works*. Ed. by A. Heyting. **1**, Philosophy and Foundations of Mathematics. North Holland, Amsterdam-Oxford, and American Elsevier, New-York, 1975. XV+628 pp.

[BdL] BROUWER L.E.J. und DE LOOR B. Intuitionistischer Beweis des Fundamentalsatzes der Algebra, *Koninklijke Akademie van Wetenschappen te Amsterdam, Proceedings of the section of sciences*, series A, **27** (1924), 186–188. Abg. [Brw15], 283–285.

[BuF1] BURALI-FORTI C. Una questione sui numeri transfiniti, *Rendiconti del Circolo Matematico di Palermo*, **11** (1897), 154–164. Engl. tr. [vHj], 104–111.

[BuF2] BURALI-FORTI C. Sulle classi numeri ben ordinate, *Rendiconti del Circolo Matematico di Palermo*, **11** (1897), 260. Engl. tr. [vHj], 111–112.

[BGN] BURKS A.W., GOLDSTINE H.H. and NEUMANN J. VON. *Preliminary Discussion of the Logical Design of an Electronic Computing Instrument*. Princeton, 28 June 1946. Repr. [vNm7] **V**, 33–79; and as a chapter of [BlN].

[ClM] CALUDE CH. and MARCUS S. The first example of a recursive function which is not primitive recursive, *Historia Mathematica* **6** (1979), 380–384.

[Cnt1] CANTOR G. Beiträge zur Begründung der transfiniten Mengenlehre I, *Mathematische Annalen* **46** (1895), 481–512; II, **49** (1897), 343–347. Trad. fr. *Mémoires de la Société des Sciences Physiques et Naturelles de Bordeaux* (5) **3** (1899), 343–437; éd. séparée, Hermann, 1899; repr. Gabay, Paris, 1992. Abg. [Cnt2], 282–356. Engl. tr. by Ph. E.B. Jourdain, under: *Contributions to the founding of the theory of transfinite numbers*.

Open Court, Chicago and London, 1915, ix+211 pp; repr. Dover, New-York, 1952.

[Cnt2] CANTOR G. *Georg Cantor Gesammelte Abhandlungen mathematischen und philosphichen Inhalts*. Hrsg. E. Zermelo, nebst einem Lebenslauf Cantors von A. Fraenkel. Springer, Berlin, 1932. Abg. Olms, Hildesheim, 1966. 2. Aufl. 1980.

[Ctr1] CARATHÉODORY C.Über die starken Maxima und Minima bei einfachen Integralen, *Mathematische Annalen* **62** (1906), 449–503. Abg. [Ctr3], 80–142.

[Ctr2] CARATHÉODORY C. *Vorlesungen über reelle Funktionen*. Teubner, Leipzig, 1918. Abg. Chelsea, New-York, 1948.

[Ctr3] CARATHÉODORY C. *Gesammelte mathematische Schriften* **I**. Beck, München, 1954.

[Cnp1] CARNAP R. *Abriss der Logistik, mit besonderer Berücksichtigung der Relationstheorie und ihrer Anwendungen*. Schriften zur Wissenschaftlichen Weltauffassung **2**. Springer, Wien, 1929, vi+114 S. Abg. 1937; 1949; 1951; 1954.

[Cnp2] CARNAP R. *Logische Syntax der Sprache*. Wien, 1934. Rev. and enlarged engl. tr. by A. Smeaton, countess von Zeppelin, under: *The logical syntax of language*. Routledge and Kegan Paul, London, 1937. xvi+352 pp. 2^d impr. 1949; 3^d impr. 1951; 4th impr. 1954.

[Cnp3] CARNAP R. Ein Gültigkeitskriterium für die Sätze der klassischen Mathematik, *Monatshefte für Mathematik und Physik* **42** (1935), 163–190.

[Cnp4] CARNAP R. Modalities and quantification, *Journal of Symbolic Logic* **11** (1946), 33–64.

[CsG] CASSINET J. et GUILLEMOT M. *L'axiome du choix dans les mathématiques, de Cauchy (1821) à Gödel (1940)*. Thèse d'Etat, Université Paul Sabatier (Toulouse III), 1983. xx+5O2 pp.

[Cav] CAVAILLÈS J. *Philosophie Mathématique*. Hermann, Paris, 1962, 274 p.

[Cay] CAYLEY A. A Sixth Memoir upon Quantics, *Philosophical Transactions of the Royal Society* **149** (1859), 61–90. Repr. *The Collected Mathematical Papers of A. Cayley* **VII**, Cambridge University Press, 1889, 561–592.

[CgK] CHANG C.C. and KEISLER H.J. *Model Theory*. North Holland, Amsterdam-London, 1973, XII+550 pp.

[Cvr] CHAUVIRÉ CH. *Ludwig Wittgenstein*. Coll. Les contemporains, éd. du Seuil, Paris, 1989, 285 pp.

[ChT] CHIN L.H. and TARSKI A. Remarks on projective algebras, *Bulletin of the American Mathematical Society* **54** (1948), 80–81. Repr. [Trk37] **IV**, p. 577.

[Chu1] CHURCH A. Alternatives to Zermelo's assumption, *Transactions of the American Mathematical Society* **29** (1927), 178–208.

[Chu2] CHURCH A. A set of postulates for the foundations of logic, *Annals of Mathematics*, ser. 2, **33** (1932), 346–366; **34** (1933), 839–864.

[Chu3] CHURCH A. A proof of freedom of contradiction, *Proceedings of the National Academy of Sciences of the U.S.A.* **21** (1935), 275–281.

[Chu4] CHURCH A. An unsolvable problem of elementary number theory, *American Journal of Mathematics* **58** (1936), 345–363. Repr. [Dvs2], 88–107.

[Chu5] CHURCH A. A note on the Entscheidungsproblem, *Journal of Symbolic Logic*, **1** (1936), 40–41. Correction, *Ibid.* 101–102. Repr. [Dvs2], 108–114.

[Chu6] CHURCH A. The constructive second number class, *Bulletin of the American Mathematical Society* **44** (1938), 224–232.

[Chu7] CHURCH A. A formulation of the simple theory of types, *Journal of Symbolic Logic* **5** (1940), 56–68.

[Chu8] CHURCH A. On the concept of a random sequence, *Bulletin of the American Mathematical Society* **46** (1940), 130–135.

[Chu9] CHURCH A. *The calculi of λ-conversion.* Annals of Mathematical Studies, n° 6. Lithoprinted. Princeton University Press, Princeton, N.J., 1941, ii+77 pp. 2^d pr. 1951, ii+82 pp.

[Chu10] CHURCH A. *Introduction to mathematical logic,* Part I. Annals of Mathematical Studies, n^r 13. Lithoprinted. Princeton University Press, Princeton, N.J., 1944. 118 pp.

[Chu11] CHURCH A. Review of [Nkf1], *Journal of Symbolic Logic* **5** (1940), 69–70.

[Chu12] CHURCH A. Review of the 3^d German ed. of [HbA], *Journal of Symbolic Logic* **15** (1950), 59.

[Chu13] CHURCH A. *Introduction to mathematical logic,* **I**. Princeton University Press, Princeton, N.J., 1956. x+376pp. Revised and much enlarged ed. of [Chu10].

[ChK] CHURCH A. and KLEENE S.C. Formal definitions in the theory of ordinal numbers, *Fundamenta Mathematicae* **28** (1936), 11–21.

[ChR] CHURCH A. and ROSSER J.B. Some properties of conversion, *Transactions of the American Mathematical Society* **39** (1936), 472–482.

[Chw1] CHWISTEK L. Zasady czystej teorji typów, *Przegląd Filozoficzny (= Revue Philosophique)*, **25** (1922), pp. 359–391.

[Chw2] CHWISTEK L. The theory of constructive types (Principles of logic and mathematics), *Rocznik Polskiego Towarsystwa Matematycznego (= Annales de la Société Polonaise de Mathématique)* **2** (1924), 9–48; **3** (1925), 92–141.

[Coh1] COHEN P.J. *The independence of the axiom of choice*. Mimeographied, Stanford University, 1963.

[Coh2] COHEN P.J. The independence of the continuum hypothesis, *Proceedings of the National Academy of Sciences of the U.S.A.* I, **50** (1963), 1143–1148; II, **51** (1964), 105–110.

[Cor1] CORCORAN J. Categoricity, *History and Philosophy of Logic* **1** (1980), 187–207.

[Cor2] CORCORAN J. From Categoricity to Completeness, *History and Philosophy of Logic* **2** (1981), 113–119.

[Crl] COURCELLE B. (Ed.) *Logique et Informatique*. Collection Didactique **8**, I.N.R.I.A., 1990.

[Ctr1] COUTURAT L. La logique mathématique de M. Peano, *Revue de Métaphysique et de Morale* **7** (1899), 616–646.

[Ctr2] COUTURAT L. Sur une définition logique du nombre, *Revue de Métaphysique et de Morale* **8** (1900), 23–36.

[Ctr3] COUTURAT L. L'algèbre universelle de M. Whitehead, *Revue de Métaphysique et de Morale* **8** (1900), 323–362.

[Ctr4] COUTURAT L. Kant et la Mathématique moderne, *Bulletin de la Société française de Philosophie* **4** (1904), 125–234.

[Ctr5] COUTURAT L. Sur l'utilité de la logique algorithmique, *Rapports et Comptes-Rendus du 2e Congrès International de Philosophie*. Genève, 1905, 706–711.

[Ctr6] COUTURAT L. Les Principes des Mathématiques, *Revue de Métaphysique et de Morale* **12** (1904), pp. 19–50, 211–240, 664–698, 810–844; **13** (1905), 224–256. Repr. [Ctr8].

[Ctr7] COUTURAT L. La Philosophie des Mathématiques de Kant, *Revue de Métaphysique et de Morale* **12** (1904), 321–383. Quelque peu modifié, en Appendice de [Ctr8].

[Ctr8] COUTURAT L. *Les Principes des Mathématiques, avec un appendice sur la philosophie des mathématiques de Kant*. Paris, 1905. Repr. de [Ctr6] et de [Ctr7], quelque peu modifié; repr. Blanchard, Paris, 1980.

[Ctr9] COUTURAT L. *L'Algèbre de la Logique*. Paris, 1904, 100 pp. Engl. tr. Open Court, La Salle, 1914, xiv+98 pp. Repr. Blanchard, Paris, 1980.

[Crl1] CROSSLEY J.N. (ed.) *Algebra and Logic*. Lecture Notes 450, Springer, Berlin-Heidelberg-New-York, 1975.

[Crl2] CROSSLEY J.N. Reminiscence of logicians, [Crl1], pp. 1–57.

[Cur1] CURRY H.B. An analysis of logical substitution, *American Journal of Mathematics* **51** (1929), 363–384.

[Cur2] CURRY H.B. Grundlagen der kombinatorischen Logik, *American Journal of Mathematics* **52** (1930), 551–558.

[Cur3] CURRY H.B. The universal quantifier in combinatory logic, *Annals of Mathematics* ser. 2, **32** (1931), 154–180.

[Cur4] CURRY H.B. Apparent variables from the standpoint of combinatory logic, *Annals of Mathematics* ser. 2, **34** (1933), 381–404.

[Cur5] CURRY H.B. Some properties of equality and implication in combinatory logic, *Annals of Mathematics* ser. 2, **35** (1934), 849–860.

[Cur6] CURRY H.B. Some properties of formal deducibility, abstract, *Bulletin of the American Mathematical Society* **43** (1937), 615.

[Cur7] CURRY H.B. A revision of the fundamental rules of combinatory logic, *Journal of Symbolic Logic* **6** (1941), 41–53.

[Cur8] CURRY H.B. A formalisation of recursive arithmetic, *American Journal of Mathematics* **63** (1941), 263–282.

[Cur9] CURRY H.B. The paradox of Kleene and Rosser, *Transactions of the American Mathematical Society* **50** (1941), 454–516.

[Cur10] CURRY H.B. The combinatory foundations of mathematical logic, *Journal of Symbolic Logic* **7** (1942), 49–64.

[Cur11] CURRY H.B. The inconsistency of certain formal logics, *Journal of Symbolic Logic* **7** (1942), 115–117.

[Cur12] CURRY H.B. Some advances in the combinatory theory of quantification, *Proceedings of the National Academy of Sciences of the U.S.A.* **28** (1942), 564–569.

[Cur13] CURRY H.B. *Outlines of a formalist philosophy of mathematics.* North Holland, Amsterdam, 1951, 75 pp.

[Cur14] CURRY H.B. *Foundations of mathematical logic.* McGraw Hill, New-York, San Francisco, Toronto, London, 1963, xii+407 pp.

[CyF] CURRY H.B. and FEYS R. *Combinatory Logic* **I**. North Holland, Amsterdam, 1958. XII+417 pp.

[vDz] DANTZIG D. VAN. On the principles of intuitionistic and affirmative mathematics, *Koninklijke Nederlandse Akademie van Weteschappen, Proceedings of the section of sciences*, series A, **50** (1947), 918–929; 1092–1103 = *Indagationes mathematicae*, **9** (1947), 429–440; 506–517.

[Dvs1] DAVIS M. Relatively recursive functions and the extended Kleene hierarchy (abstract), *Proceedings of the International Congress of Mathematicians, Cambridge, Mass., U.S.A., Aug. 30-Sept 6, 1950,* **I**, American Mathematical Society, Providence, Rhode Island, 1952, p. 723.

[Dvs2] DAVIS M. (ed.) *The Undecidable. Basic Papers on Undecidable Propositions, Unsolvable Problems and Computable Fonctions* Raven Press, Hewlitt,New-York, 1965. v+440 pp.

[Dws] DAWSON J.W. JR. The Reception of Gödel's Incompleteness Theorem, [Drk], 84–100.

[Ded1] DEDEKIND R. *Stetigkeit und irrationale Zahlen*. Braunschweig, 1872. 31 S. Engl. tr. [Bem]; [Ded4]. Abg. [Ded5].

[Ded2] DEDEKIND R. *Was sind und was sollen die Zahlen?* Braunschweig, 1888. 6ter Aufl. Braunschweig, 1930. Engl. tr. [Bem], 31–105; [Ded4]. Abg. [Ded3], 335–391; [Ded5].

[Ded3] DEDEKIND R. *Dedekind gesammelte mathematische Werke* **III**. Vieweg, Braunschweig, 1932.

[Ded4] DEDEKIND R. *Essays on the theory of numbers. I. Continuity and irrational numbers. II. The nature and meaning of numbers*. Dover, New-York, 1948.

[Ded5] DEDEKIND R. *Was sind und was sollen die Zahlen? Stetigkeit und irrationale Zahlen*. Vieweg, Braunschweig, 1965.

[Dhn] DEHN M. Über unendliche diskontinuierliche Gruppen, *Mathematische Annalen* **71** (1911), 116–144.

[DMK] DIAMOND A.H. and MCKINSEY J.C.C. Algebras and their subalgebras, *Bulletin of the American Mathematical Society*, **53** (947), 959–962.

[Ddn+] DIEUDONNÉ J., avec la collaboration de P. DUGAC, W.J. et F. ELLISON, J. GUÉRINDON, M. GUILLAUME, G. HIRSCH, C. HOUZEL, P. LIBERMANN, M. LOÈVE, J.-L. VERLEY. *Abrégé d'histoire des mathématiques 1700–1900*. 2t. Hermann, Paris, 1978. 2e éd. modif. et mise à jour, en 1 vol. Hermann, Paris, 1986.

[Dxn] DIXON M. On well ordered aggregates, *Proceedings of the London Mathematical Society*, Series 2, **4** (1906–1907), 18–20.

[DnH] DONER J. and HODGES W. Alfred Tarski and decidable theories, *Journal of Symbolic Logic* **53** (1988), 20–35.

[Drk] DRUCKER T. (ed.) *Perspectives on the History of Mathematical Logic*. Birkhäuser, Boston, Basel, Berlin, 1991. xxiii+195 pp.

[Erf] EHRÉNFÉST M. Review de [Ctr9], *Žurnal Russkago Fizikohimičéskago Obščéstva, Fizičéskij otděl* **42** (1910), Otděl vtoroj, 382–387.

[EML] EILENBERG S. and MACLANE S. The general theory of natural equivalence, *Transactions of the American Mathematical Society* **58** (1945), 231–194.

[Etm] ETCHEMENDY J. Tarski on truth and logical consequence, *Journal of Symbolic Logic* **53** (1988), 51–79.

[EvU] EVERETT C.J. and ULAM S. Projective algebra. I, *American Journal of Mathematics* **68** (1946), 77–88.

[Flr] FINSLER P. Formale Beweise und die Entscheidbarkeit, *Mathematische Zeitschrift* **25** (1926), 676–682. Engl. tr. [vHj], 438–445.

[Fit] FITCH F.B. Intuitionistic modal logic with quantifiers, *Portugaliae Mathematica* **7** (1948), 113–118.

[Fkl1] FRAENKEL A.A. *Einleitung in die Mengenlehre.* Springer, Berlin, 1919. 156 S.

[Fkl2] FRAENKEL A.A. Zu den Grundlagen der Cantor-Zermeloschen Mengenlehre, *Mathematische Annalen* **86** (1922), 230–237.

[Fkl3] FRAENKEL A.A. Der Begriff «definit» und die Unabhängigkeit des Auswahlaxioms, *Sitzungsberichte des Preussischen Akademie der Wissenschaften, Physikalisch-mathematische Klasse, 1922*, 253–257. Engl. tr. [vHj], 284–289.

[Fkl4] FRAENKEL A.A. *Einleitung in die Mengenlehre, eine elementare Einführung in das Reich des Unendlichgrossen.* 2. erweit. Aufl. Die Grundlehren der mathematischen Wissenschaften **9**. Springer, Berlin, 1923. x+252 S.

[Fkl5] FRAENKEL A.A. *Einleitung in die Mengenlehre, eine elementare Einführung in das Reich des Unendlichgrossen.* 3. umgearbeitete und stark erweiterte Aufl. Die Grundlehren der mathematischen Wissenschaften **9**. Springer, Berlin, 1928. x+424 S.

[Fkl6] FRAENKEL A.A. Über die Ordnungfähigkeit beliebiger Mengen. *Sitzungsberichte des Preussischen Akademie der Wissenschaften, Physikalisch-mathematische Klasse, 1928,* 90–91.

[FBH] FRAENKEL A.A. and BAR-HILLEL Y. *Foundations of set theory.* North-Holland, Amsterdam, 1958. x+415 pp.

[Frg1] FREGE G. *Begriffsschrift. Eine der arithmetischen nachgebildete Formelsprache des reinen Denkens.* Halle, 1879. Part. engl. tr. [vHj], 1–82. Abg. [Frg6].

[Frg2] FREGE G. *Grundgesetze der Arithmetik, begriffsschriftlich abgeleitet.* Pohle, Jena, **1**, 1893; **2**, 1903. Abg. Olms, Hildesheim, 1966. Part. engl. tr. [Frg5].

[Frg3] FREGE G. Über die Begriffsschrift des Herrn Peano und meine eigene, *Berichte über die Verhandlungen der Königlichen Sächsischen Gesellschaft der Wissenschaften zu Leipzig, Mathematisch-physikalische Klasse* **48** (1896), 361–378.

[Frg4] FREGE G. Lettera del sig. G. Frege all'editore, *Revue de Mathématiques* **6** (1898). Repr. [Pno7], **2**, 288–294.

[Frg5] FREGE G. *The basic laws of arithmetic. Exposition of the system.* Tr. and ed., with an introduction, by Montgomery Furth. University of California Press, Berkeley and Los Angeles, California, 1964.

[Frg6] FREGE G. *Begriffsschrift und andere Aufsätze.* Hrsg I. Agnelli, Olms, Hildesheim, 1964.

[Frd] FREUDENTHAL H. Zum intuitionistischen Raumbegriff, *Compositio Mathematica* **4** (1936), pp. 82–111.

[Fdb1] FRIEDBERG R.M. The solution of Post's problem by the construction of two recursively enumerable sets of incomparable degrees of unsolvability (abstract), *Bulletin of the American Mathematical Society* **62** (1956), 260.

[Fdb2] FRIEDBERG R.M. Two recursively enumerable sets of incomparable degrees of unsolvability (solution of Post's problem 1944), *Proceedings of the National Academy of Sciences of the U.S.A.* **43** (1957), 236–238.

[Frk] FRINK O. On the existence of linear algebras in Boolean algebras, *Bulletin of the American Mathematical Society* **34** (1928), 329–333.

[Ggk] GÉGALKINE J.J. O téhniké vyčislénij prédloženij v symboličéskoj logiké, *Matématičéskij Sbornik (Recueil mathématique de la Société Mathématique de Moscou)* **34** (1927), 9–28. Rés. fr. sous: Sur le calcul des propositions de la logique symbolique.

[Gtz1] GENTZEN G. Über die Existenz unhabhängiger Axiomensysteme zu unendlichen Satzsystemen, *Mathematische Annalen* **107** (1932), 329–350. Engl. tr. [Gtz8], 29–52.

[Gtz2] GENTZEN G. Untersuchungen über das logische Schliessen, *Mathematische Zeitschrift* **39** (1934), 176–210; 405–431. Trad. fr. sous: *Recherches sur la déduction logique*, par J. Ladrière (avec notes de R. Feys et J. Ladrière). Presses Universitaires de France, Paris, 1955. XII+170 pp. Engl. tr. [Gtz8], 69–131.

[Gtz3] GENTZEN G. Die Widerspruchsfreiheit der reinen Zahlentheorie, *Mathematische Annalen* **112** (1936), 493–565. Engl. tr. with galley proofs of sect. IV and V app. [Gtz8], 132–213.

[Gtz4] GENTZEN G. Die Widerspruchsfreiheit der Stufenlogik. *Mathematische Zeitschrift* **41** (1936), 357–366. Engl. tr. [Gtz8], 214–222.

[Gtz5] GENTZEN G. Die gegenwärtige Lage in der mathematischen Grundlagenforschung, *Forschungen sur Logik und Grundlegung der exakten Wissenschaften*, n. F. n^r4 (1938), 19–44. Abg. *Deutsche Mathematik* **3** (1939), 255–268. Engl. tr. [Gtz8], 234–251.

[Gtz6] GENTZEN G. Neue Fassung des Widerspruchsfreiheitsbeweises für die reine Zahlentheorie, *Forschungen sur Logik und Grundlegung der exakten Wissenschaften* n. F. n^r4 (1938), 19–44. Engl. tr. [Gtz8], 252–286.

[Gtz7] GENTZEN G. Beweisbarkeit und Unbeweisbarkeit von Anfangsfällen der transfiniten Induktion in der reinen Zahlentheorie, *Mathematische Annalen* **119** (1943), 140–161. Engl. tr. [Gtz8], 287–308.

[Gtz8] GENTZEN G. *The collected papers of Gerhardt Gentzen*. Ed. by M.E. Szabo, North-Holland, Amsterdam-London, 338 pp.

[Gvt] GIVANT S. Bibliography of Alfred Tarski, *Journal of Symbolic Logic* **51** (1986), 913–941.

[Gvk] GLIVENKO V. Sur quelques points de la logique de M. Brouwer, *Académie Royale de Belgique, Bulletin de la Classe des Sciences*, série 5, **15** (1929), 183–188.

[Gdl1] GÖDEL K. Die Vollständigkeit der Axiome des logischen Funktionenkalküls, *Monatshefte für Mathematik und Physik* **37** (1930), 349–360. Abg. [Gdl18] **I**, 102–122. Engl. tr. [vHj], 582–591; [Gdl18] **I**, 103–123.

[Gdl2] GÖDEL K. Einige metamathematische Resultate über die Entscheidungsdefinitheit und Widerspruchsfreiheit, *Akademie der Wissenschaften in Wien, Mathematisch-naturwissenschaftliche Klasse, Akademische Anzeiger* **67** (1930), 214–215. Abg. [Gdl18] **I**, 140–142. Engl. tr. [vHj], 595–596; [Gdl18] **I**, 141–143.

[Gdl3] GÖDEL K. Über formal unentscheidbare Sätze der Principia Mathematica und verwandter Systeme I, *Monastshefte für Mathematik und Physik* **38** (1931), S. 173–198. Abg. [Gdl18], **I**, 144–194. Engl. tr. [Dvs2], 4–38; [vHj], 596–616; [Gdl18] **I**, 145–195.

[Gdl4] GÖDEL K. Ein Spezialfall des Entscheidungsproblems der theoretischen Logik, *Ergebnisse eines mathematischen Kolloquiums*, Heft 2 (1929–30, publ. 1932), 27–28. Abg. [Gdl18] **I**, 230–232. Eng. tr. [Gdl18] **I**, 231–233.

[Gdl5] GÖDEL K. Zum intuitionistischen Aussagenkalkül, *Akademie der Wissenschaften in Wien, Mathematisch-naturwissenschaftliche Klasse, Akademische Anzeiger* **69** (1932), 147–151. Abg. [Gdl18] **I**, 222–224. Engl. tr. [Gdl18] **I**, 223–225.

[Gdl6] GÖDEL K. Über Vollständigkeit und Widerspruchsfreiheit, *Ergebnisse eines mathematischen Kolloquiums*, Heft 3 (1930–31, publ. 1932), 12–13. Abg. [Gdl18] **I**, 234–236. Eng. tr. [vHj], 616–617; [Gdl18] **I**, 235–237.

[Gdl7] GÖDEL K. Eine Eigenschaft der Realisierungen des Aussagenkalküls, *Ergebnisse eines mathematischen Kolloquiums*, Heft 3 (1930–31, publ. 1932), S. 20–21. Abg. [Gdl18] **I**, 238–240. Engl. tr. [Gdl18] **I**, 239–241.

[Gdl8] GÖDEL K. Zur intuitionistischen Arithmetik und Zahlentheorie, *Ergebnisse eines mathematischen Kolloquiums*, Heft 4 (1931–32, publ. 1933), 34–38. Abg. [Gdl18] **I**, 286–294. Engl. tr. [Dvs2], 75–81; [Gdl18] **I**, 287–295.

[Gdl9] GÖDEL K. Eine Interpretation des intuitionistischen Aussagenkalküls, *Ergebnisse eines mathematischen Kolloquiums*, Heft 4 (1931–32, publ. 1933), 39–40. Abg. [Gdl18] **I**, 300. Engl. tr. [Gdl18] **I**, 301.

[Gdl10] GÖDEL K. Zum Entscheidungsproblem des logischen Funktionenkalküls, *Monatshefte für Mathematik und Physik* **40** (1933), 433–443. Abg. [Gdl18] **I**, 306–326. Engl. tr. [Gdl18] **I**, 307–327.

[Gdl11] GÖDEL K. *On undecidable propositions of formal mathematical systems.* Mimeographied. Princeton 1934, 30 pp. Rev. enl. ed. [Dvs2], 39–75. Repr. [Gdl18] **I**, 364–369.

[Gdl12] GÖDEL K. Über die Länge von Beweisen, *Ergebnisse eines mathematischen Kolloquiums*, Heft 7 (1934–35, publ. 1936), 23–24. Abg. [Gdl18] **I**, 396–398. Engl. tr. [Dvs2], 82–83; [Gdl18] **I**, 397–399.

[Gdl13] GÖDEL K. The consistency of the axiom of choice and of the generalized continuum hypothesis, *Proceedings of the National Academy of Sciences of the U.S.A.* **24** (1938), 556–557. Repr. [Gdl18] **I**, 26.

[Gdl14] GÖDEL K. *The consistency of the axiom of choice and of the generalized continuum hypothesis with the axioms of set theory.* Princeton University Press, Princeton, N.J., 1940. Rep. [Gdl18] **I**, 33–101.

[Gdl15] GÖDEL K. Russell's mathematical logic, *The philosophy of Bertrand Russell.* Ed. by P.A. Schilpp. Evanston and Chicago, 1944, 123–153. Repr. [BnP], 211–232; [Gdl18] **II**, 119–143.

[Gdl16] GÖDEL K. An example of a new type of cosmological solutions of Einstein's field equations of gravitation, *Review of modern Physics* **21** (1949), 447–450. Repr. [Gdl18] **II**, 190–198.

[Gdl17] GÖDEL K. Rotating universes in general relativity theory, *Proceedings of the International Congress of Mathematicians, Cambridge, Mass., U.S.A., Aug. 30-Sept 6, 1950*, **I**, American Mathematical Society, Providence, Rhode Island, 1952, 175–181. Repr. [Gdl18] **II**, 175–181.

[Gdl18] GÖDEL K. *Kurt Gödel Collected Works.* Ed. by S. Feferman, J.N. Dawson J^r, S.C. Kleene, G.H. Moore, G.H. Moore, R.M. Solovay, and J. van Heijenoort. Oxford University Press, New-York; Clarendon Press, Oxford. **I**, 1986; **II**, 1990.

[Gfb] GOLDFARB W.D. Logic in the twenties: the nature of the quantifier, *Journal of Symbolic Logic* **44** (1979), 351–368.

[Gld] GOLDIE A.W. The Jordan-Hölder theorem for general abstract algebras, *Proceedings of the London Mathematical Society*, Series 2, **52** (1950), 107–131.

[Gls] GOLDSTINE H.H. *The Computer from Pascal to von Neumann.* Princeton University Press, Princeton, N.J. 1972, xi+378 pp. 2^d pr. 1973.

[Ggf] GRIGORIEFF S. Décidabilité et complexité des théories logiques, [Crl], 7–97.

[Grs] GRISS G.F.C. Negatieloze intuitionistische wiskunde, *Verslagen der Koninklijke Akademie van Wetenschappen te Amsterdam* **53** (1944), 261–268.

[Glm1] GUILLAUME M. *Axiomatique et logique*, [Ddn+] **II**, Chap. XIII, 315–430. 2e éd., Chap. XI, 417–483.

[Glm2] GUILLAUME M. Sur l'histoire des modèles non standards et celle de l'analyse non standard, *Cahiers Fundamenta Scientiae*, *n°* 85, Université Louis Pasteur, Strasbourg, 1978, 1–13.

[Hdm1] HADAMARD J. La logistique et l'induction complète. La notion de correspondance, *Revue générale des sciences pures et appliquées* **17** (1906), 161–162. Réimp. [Hdm6] **IV**, 2157–2160.

[Hdm2] HADAMARD J. Les principes de la théorie des ensembles, *Revue générale des sciences pures et appliquées* **17** (1906).

[Hdm3] HADAMARD J. La logistique et la notion de nombre entier, *Revue générale des sciences pures et appliquées* **17** (1906), 906–909. Réimp. [Hdm6] **IV**, 2145–2156.

[Hdm4] HADAMARD J. Les paradoxes de la théorie des ensembles, *Revue générale des sciences pures et appliquées* **19** (1908), 681.

[Hdm5] HADAMARD J. Le développement et le rôle scientifique du calcul fonctionnel, *Atti del Congresso internazionale dei matematici, Bologna 3–10 settembre 1928*. Bologna, 1929, 142–161.

[Hdm6] HADAMARD J. *Oeuvres de Jacques Hadamard.*, Ed. du C.N.R.S. Paris, 1964. 4 t., 2296 pp.

[HBLB] HADAMARD J., BAIRE R., LEBESGUE H. et BOREL E. Cinq lettres sur la théorie des ensembles, *Bulletin de la Société Mathématique de France* **33** (1905), 261–273. Réimp. [Hdm6] **I**, 335–348; [Brl3], 1253–1265; [Lbg4] **I**, 82–94.

[Hml] HAMEL G. Eine Basis aller Zahlen und die unstetigen Lösungen der Funktionalgleichung $f(x+y) = f(x)+f(y)$, *Mathematische Annalen* **60** (1905), 459–462.

[Hdy] HARDY G. The continuum and the second number class, *Proceedings of the London Mathematical Society*, Series 2, **4** (1907), 10–17. Repr. *Collected papers of G.H. Hardy* **VII**. Clarendon Press, Oxford, 1979, 438–445.

[Hrt] HARTOG'S F. Über das Problem der Wohlordnung, *Mathematische Annalen* **76** (1915), 438–433.

[Hdf1] HAUSDORFF F. Der Potenzbegriff in der Mengenlehre, *Jahresbericht der deutschen Mathematiker-Vereinigung* **15** (1904), 569–571.

[Hdf2] HAUSDORFF F. Untersuchungen über Ordnungstypen, *Berichte über die Verhandlungen der Königlichen Sächsischen Gesellschaft des Wissenschaften zu Leipzig, Mathematisch-physikalische Klasse* **59** (1907), 84–159.

[Hdf3] HAUSDORFF F. Über dichte Ordnungstypen, *Jahresbericht der deutschen Mathematiker-Vereinigung* **16** (1907), 541–546.

[Hdf4] HAUSDORFF F. Grundzüge einer Theorie der geordneten Mengen, *Mathematische Annalen* **65** (1908), 435–505.

[Hdf5] HAUSDORFF F. Die Graduierung nach dem Endverlauf, *Abhandlungen der Mathematisch-physikalischen Klasse der Königlichen Sächsischen Gesellschaft der Wissenschaften zu Leipzig* . **31** (1909), 297–334.

[Hdf6] HAUSDORFF F. *Grundzüge der Mengenlehre*. De Gruyter, Leipzig, 1914, 476 S.

[vHj] HEIJENOORT J. VAN. (ed.) *From Frege to Gödel, A Source Book in Mathematical Logic, 1879–1931*. Harvard University Press, Cambridge, Mass., 1967.

[Hzm1] HEINZMANN G. *Entre Intuition et Analyse. Poincaré et le concept de prédicativité.* Blanchard, Paris, 1985. 104 pp.

[Hzm2] HEINZMANN G. (éd.) *Poincaré, Russell, Zermelo et Peano. Textes de la discussion (1906–1912) sur les Fondements des mathématiques: des antinomies à la prédicativité.* Blanchard, Paris, 1986. 332 pp.

[Hnk1] HENKIN L. The completeness of formal systems (abstract) *Journal of Symbolic Logic* **13** (1948), 61.

[Hnk2] HENKIN L. The completeness of the first order functional calculus, *Journal of Symbolic Logic* **14** (1949), 159–166.

[Hnk3] HENKIN L. Completeness of the theory of types, *Journal of Symbolic Logic* **15** (1950), 81–91.

[Hbd1] HERBRAND J. *Recherches sur la théorie de la démonstration.* Thèse, Université de Paris. = *Prace Towarzystwa Naukowego Warzsawskiego, Wydział III Nauk-Matematyczno-fizycznych (= Travaux de la Société des Sciences et des Lettres de Varsovie, Classe III Sciences Mathématiques et Physiques),* n^o33. Part. engl. tr. [vHj], 525–581. Repr. [Hbd2].

[Hbd2] HERBRAND J. *Ecrits logiques.* Ed. par J. van Heijenoort. Presses Universitaires de France, Paris, 1968. Engl. tr. under: *Logical writings by J. Herbrand.* Ed. by W.D. Goldfarb. Reidel, Dordrecht; Harvard University Press, Cambridge, 1971, vii+312 pp.

[Hms] HERMES H. Semiotik, *Forschungen sur Logik und Grundlegung der exakten Wissenschaften* n. F. n^r5 (1938), 5–22.

[Htz1] HERTZ P. Über Axiomensysteme für beliebige Satzsysteme. I Sätze ersten Grades, *Mathematische Annalen* **87** (1922), 246–269.

[Htz2] HERTZ P. Über Axiomensysteme für beliebige Satzsysteme. II Sätze höheren Grades, *Mathematische Annalen* **89** (1923), 76–102.

[Htz3] HERTZ P. Über Axiomensysteme für beliebige Satzsysteme. *Mathematische Annalen* **89** (1929), 457–514.

[Hbg1] HESSENBERG G. Grundbegriffe der Mengenlehre, *Abhandlungen der Fries'schen Schule* N. F. **1** (1906), 479–706. Sonderabdruck, Vandenhoek und Ruprecht, Göttingen, 1906.

[Hbg2] HESSENBERG G. Potenzen transfiniter Ordnungszahlen, *Jahresbericht der deutschen Mathematiker-Vereinigung* **16** (1907), 130.

[Hbg3] HESSENBERG G. Kettentheorie und Wohlordnung, *Journal für die reine und angewandte Mathematik* **135** (1909), 81–133.

[Hwt] HEWITT E. Rings of real-valued continuous functions, *Transactions of the American Mathematical Society* **64** (1948), 54–99.

[Htg1] HEYTING A. *Intuitionistische axiomatiek der projectieve meetkunde.* Thèse, Université d'Amsterdam. Groningen, 1925.

[Htg2] HEYTING A. Die formalen Regeln der intuitionistischen Logik, *Sitzungsberichte der Preussischen Akademie der Wissenschaften, Physikalisch-mathematische Klasse 1930*, 42–56.

[Htg3] HEYTING A. *Mathematische Grundlagenforschung. Intuitionismus. Beweistheorie.* Springer, Berlin, 1934. Trad. fr. par P. Destouches-Février, revue et augm. sous: *Les fondements des mathématiques. Intuitionnisme. Théorie de la démonstration.* Gauthier-Villars, Paris, et Nauwelaerts, Louvain, 1955.

[Htg4] HEYTING A. *Intuitionism: An Introduction.* North Holland, Amsterdam, 1956, viii +133pp.

[Hbt1] HILBERT D. *Grundlagen der Geometrie. Festschrift zur Feier der Enthüllung des Gauss-Weber Denkmals in Göttingen.* Teubner, Leipzig, 1899. Engl. tr. Open Court, La Salle, 1899.

[Hbt2] HILBERT D. Über den Zahlbegriff, *Jahresbericht der deutschen Mathematiker-Vereinigung* **8** (1900) 180–194. Abg. [Hbt7].

[Hbt3] HILBERT D. Mathematische Probleme. Vortrag, gehalten auf dem internationalen Mathematiker-Kongress zu Paris, 1900. *Nachrichten von der Königlichen Gesellschaft der Wissenschaften zu Göttingen, mathematische-physikalische Klasse 1900*, 253–297. Trad.fr. par L. Laugel sous: Sur les problèmes futurs des mathématiques, *Compte-Rendu du 2e Congrès International de Mathématiciens* Paris, 1902, 58–114; repr. éd. sép. Gabbay, Paris, 1992. Abg. [Hbt14].

[Hbt4] HILBERT D. *Les principes fondamentaux de la géométrie.* Trad. fr. de [Hbt1] par L. Laugel. Gauthier-Villars, Paris, 1901.

[Hbt5] HILBERT D. *Grundlagen der Geometrie.* 2e Auf. Teubner, Leipzig, 1903.

[Hbt6] HILBERT D. Über die Grundlagen der Logik und der Arithmetik, *Verhandlungen des Dritten Internationalen Mathematiker-Kongresses in Heidelberg vom 8. bis 13. August 1904.* Teubner, Leipzig, 1905, 174–185. Abg. [Hbt7]. Trad. fr. par P. Boutroux *L'enseignement mathématique* **7** (1905), 89–103; repr. éd. sép. Gabay, Paris, 1992. Engl. tr. under: On the foundations of logic and arithmetic, *The monist* **15** (1905), 325–352; [vHj], 129–138.

[Hbt7] HILBERT D. *Grundlagen der Geometrie.* 3. Aufl. Teubner, Leipzig und Berlin, 1909. 4. Aufl. Teubner, Leipzig und Berlin, 1913. 5. Aufl. Teubner, Leipzig und Berlin, 1922. 6. Aufl. Teubner, Leipzig und Berlin, 1923. 7. Aufl. Teubner, Leipzig und Berlin, 1930.

[Hbt8] HILBERT D. Axiomatisches Denken, *Mathematische Annalen* **78** (1918), 405–415. Trad. fr. *L'enseignement mathématique* **20** (1918), 122–136. Abg. [Hbt14], 146–156.

[Hbt9] HILBERT D. Neubegründung der Mathematik. Erste Mitteilung, *Abhandlungen aus dem Mathematischen Seminar der Hamburgischen Universität* **1** (1922), 157–177. Abg. [Hbt14], 157–177.

[Hbt10] HILBERT D. Die logischen Grundlagen der Mathematik, *Mathematische Annalen* **88** (1923), 151–165. Abg. [Hbt14], 178–191.

[Hbt11] HILBERT D. Über das Unendliche, *Mathematische Annalen* **95** (1926), 161–190. Engl. tr. [BnP], 134–151; [vHj], 367–392. Aufzug, *Jahresbericht der deutschen Mathematiker-Vereinigung* **36** (1927), 201–215; abg. [Hbt7], 7^e Aufl.

[Hbt12] HILBERT D. Probleme der Grundlegung der Mathematik, *Atti del Congresso internazionale dei matematici, Bologna 3–10 settembre 1928*, **I**. Bologna, 1929, 135–141. Abg. *Mathematische Annalen* **102** (1930), 1–9.

[Hbt13] HILBERT D. Die Grundlegung der elementaren Zahlenlehre, *Mathematische Annalen* **104** (1931), 485–494.

[Hbt14] HILBERT D. *Gesammelte Abhandlungen* **III**. Springer, Berlin, 1935.

[HbA] HILBERT D. und ACKERMANN W. *Grundzüge der theoretischen Logik*. Springer, Berlin, 1928, viii+120 S. 2. Aufl. Springer, Berlin 1938, viii+133 S. Repr. by Dover Publications, New-York, 1946. 3. Aufl. Springer, Berlin-Göttingen-Heidelberg, 1949, viii+155 S. Engl.h tr. of the 2^d german ed. by L.M. Hammond, G.G. Leckie and F. Steinhardt, with notes by R.E. Luce, under: *Principles of mathematical logic*. Chelsea, New-York, 1950, xii+172 pp.

[HbB] HILBERT D. und BERNAYS P. *Grundlagen der Mathematik*. **I**, Springer, Berlin, 1934, XII+471 S. Repr. by J.W. Edwards, Ann Arbor, 1944. 2. Aufl. Springer, Berlin-Heidelberg-New-York, 1968, XII+567 S. **II**, Springer, Berlin, 1939, XII+497 S. Repr. by J.W. Edwards, Ann Arbor, 1944. 2. Aufl. Springer, Berlin-Heidelberg-New-York, 1970, XII+561 S.

[Hbs1] HOBSON E.W. On the general theory of transfinite numbers and order types, *Proceedings of the London Mathematical Society*, Series 2, **3** (1905), 170–188.

[Hbs2] HOBSON E.W. On the arithmetic continuum, *Proceedings of the London Mathematical Society*, Series 2, **4** (1906–1907), 21–28.

[Hbs3] HOBSON E.W. *The theory of a function of a real variable and the theory of Fourier's series*. Cambridge University Press, 1907. 2^d ed. 1921; 3^d ed. 1927.

[Hrn] HORN A. On sentences which are true of direct unions of algebras, *Journal of Symbolic Logic* **16** (1951), 14–21.

[Hun1] HUNTINGTON E.V. A complete set of postulates for the theory of the absolute continuous magnitude, *Transactions of the American Mathematical Society* **3** (1902), 264–279.

[Hun2] HUNTINGTON E.V. Sets of independent postulates for the algebra of logic, *Transactions of the American Mathematical Society* **5** (1904), 288–309

[Hun3] HUNTINGTON E.V. A set of postulates for ordinary complex algebra, *Transactions of the American Mathematical Society* **6** (1905), 209–229.

[Hun4] HUNTINGTON E.V. A set of postulates for abstract geometry exposed in termes of the simple relation of inclusion, *Mathematische Annalen* **123** (1916), 522–559.

[Hun5] HUNTINGTON E.V. A new set of independent postulates for the algebra of logic with special reference to Whitehead and Russell's Principia Mathematica, *Proceedings of the National Academy of Sciences of the U.S.A.* **18** (1932), 179–180.

[Hun6] HUNTINGTON E.V. New sets of independent postulates for the algebra of logic, with special reference to Whitehead and Russell's Principia Mathematica, *Transactions of the American Mathematical Society* **35** (1933), 274–304; correction, *Ibid.*, 557–558 and 971.

[Hsl1] HUSSERL E. *Logische Untersuchungen* **2**, I. Teil. 2. Aufl. Halle, 1913.

[Hsl2] HUSSERL E. Husserliana, Edmund Husserl Gesammelte Werke. Auf Grund des Nachlasses veröffentlicht in Gemeinschaft mit dem Husserl-Archiv an der Universität Köln vom Husserl-Archiv (Louvain) unter Leitung von H.L. van Bred. Nijhoff, Den Haag. T. **12**, Philosophie der Arithmetik. Mit ergänzenden Texten 1890–1901. Hrsg. von Lothar Eley, 1970, 585S.

[Jnz] JANICZAK A. A remark concerning decidability of complete theories, *Journal of Symbolic Logic* **15** (1950), 277–279.

[Jkk1] JAŚKOWSKI S. On the rules of suppositions in formal logic, *Studia Logica* **I** (1934), 5–32.

[Jkk2] JAŚKOWSKI S. Recherches sur le système de la logique intuitionniste, *Actes du Congrès International de Philosophie Scientifique* **6**. Actualités Scientifiques et Industrielles, **393**. Hermann, Paris, 1936, 58–61.

[Jkk3] JAŚKOWSKI S. Z badań nad rozstrzygalnością rozszerzonej algebry Boole'a, *Časopis pro pėstováni matematiky a fysiky* **74** (1949), 136–137.

[Jhs] JOHANSSON I. Der Minimalkalkül, ein reduzierter intuitionistischer Formalismus, *Compositio mathematica* **4** (1936), 119–136.

[Jsn] JÓNSSON B. The contributions of Alfred Tarski to general algebra, *Journal of Symbolic Logic* **51** (1986), 883–889.

[JsT1] JÓNSSON B. and TARSKI A. *Direct decompositions of finite algebraic systems*. Notre Dame Mathematical Lectures, **5**. University of Notre Dame Press, Notre Dame, Indiana, 1947, vi+64 pp. 2^d pr. 1964. Repr. [Trk37] **III**, 73–143.

[JsT2] JÓNSSON B. and TARSKI A. Boolean algebras with operators, *Bulletin of the American Mathematical Society* **54** (1948), 79–80. Repr. [Trk37] **IV**, 575.

[JsT3] JÓNSSON B. and TARSKI A. Representation problem for relation algebras, *Bulletin of the American Mathematical Society* **31** (1948), 80 and 1192. Repr. [Trk37] **IV**, 575.

[Jrd] JORDAN P. Zur Axiomatik der Verknüpfungsbereiche, *Abhandlungen aus dem Mathematischen Seminar der Universität Hamburg* **16** (1949), 54–70.

[Kmr1] KALMÁR L.Über die Erfüllbarkeit derjenigen Zählausdrücke, welche in der Normalform zwei benachbarte Abzeichen enthalten, *Mathematische Annalen* **108** (1933), 466–484.

[Kmr2] KALMÁR L. Egyszerü példa eldönthetetlen aritmetikai problémára, *Matematikai és Fizikai Lapok* **50** (1943), 1–23. Deutscher Aufzug unter: Ein einfaches Beispiel für ein unentscheidbares arithmetisches Problem.

[Ksl] KEISLER H.J. Isomorphisms of ultraproducts, *Notices of the American Mathematical Society* **7** (1960), 70–71.

[Kmn1] KEMENY J.G. Models of logical systems, *Journal of Symbolic Logic* **13** (1948), 16–30.

[Kmn2] KEMENY J.G. *Type theory versus set theory.* Thesis, Princeton University, 1949. Abstract in *Journal of Symbolic Logic* **15** (1950), 78.

[Ktn] KETONEN O. Untersuchungen zum Prädikatenkalkül, *Annales Academiae Scientiarum Fennicae*, ser. A I, **23**. Helsinki, 1944, 71 pp.

[Ksr1] KEYSER C. The axiom of infinity: a new presuppositions of thought, *Hibbert journal* **2** (1904), 532–552. Repr. [Ksr2].

[Ksr2] KEYSER C. *The human worth of rigorous thinking. Essays and addresses.* New-York, 1916. 2. ed. New-York, 1925.

[Ksr3] KEYSER C. Doctrinal functions, *Journal of Philosophy* **15** (1918), 262–267.

[Kln1] KLEENE S.C. A theory of positive integers in formal logic, *American Journal of Mathematics* **57** (1935), 153–173 and 219–244.

[Kln2] KLEENE S.C. λ-definability and recursiveness, *Duke mathematical Journal* **2** (1936), 340–353.

[Kln3] KLEENE S.C. General recursive functions of natural numbers, *Mathematische Annalen* **112** (1936), 727–742. For an erratum and a simplification, see *Journal of Symbolic Logic* **2** (1937), 38; **3** (1938), 152; **4** (1939), p. IV. Repr. [Dvs2], 236–253.

[Kln4] KLEENE S.C. On notation for ordinal numbers, *Journal of Symbolic Logic* **3** (1938), 150–155.

[Kln5] KLEENE S.C. Recursive predicates and quantifiers (abstr.), *Bulletin of the American Mathematical Society* **46** (1940), 885. Repr. [Dvs2], 254–287.

[Kln6] KLEENE S.C. On the interpretation of intuitionistic number theory (abstr.), *Bulletin of the American Mathematical Society* **48** (1942), 51.

[Kln7] KLEENE S.C. On the interpretation of intuitionistic number theory (abstr.), *Journal of Symbolic Logic* **7** (1942),47.

[Kln8] KLEENE S.C. Recursive predicates and quantifiers, *Transactions of the American Mathematical Society* **53** (1943), 41–73. Repr. [Dvs], 254–287.

[Kln9] KLEENE S.C. On the interpretation of intuitionistic number theory, *Journal of Symbolic Logic* **10** (1945), 109–124.

[Kln10] KLEENE S.C. A symmetric form of Gödel's theorem, *Koninklijke Nederlandse Akademie van Wetenschappen, Proceedings of the section of sciences* **53** (1950), 800–802. = *Indagationes Mathematicae* **12** (1950), 244–246.

[Kln11] KLEENE S.C. *Introduction to Metamathematics.* North-Holland, Amsterdam; Noordhoff, Groningen; Van Nostrand, New-York, 1952, X+550 pp. 2^d pr. 1957. 3^d pr. 1959. 4th pr. 1962. 5th pr. 1964.

[Kln12] KLEENE S.C. *Mathematical Logic.* Wiley, New-York, 1967. Trad. fr. par J. Largeault, sous: *Logique mathématique.* Armand Colin, 1971, 413 pp.

[Kln13] KLEENE S.C. Arithmetical predicates and function quantifiers, *Transactions of the American Society* **79** (1955), 312–340.

[Kln14] KLEENE S.C. Realizability: A retrospective survey, [MRg], 95–112.

[KlR] KLEENE S.C. and **Rosser J.B.** The inconsistency of certain formal logics, *Annals of Mathematics*, ser. 2, **36** (1935), 630–636.

[Knt] KNASTER B. Un theorème sur les fonctions d'ensembles, *Rocznik Polskiego Towarsystwa Matematycznego (= Annales de la Société Polonaise de Mathématique)* **6** (1927 — publ. 1928), 133–134.

[Kmg1] KOLMOGOROFF A.N. O principe tertium non datur, *Matématičéskij Sbornik (Recueil mathématique de la Société Mathématique de Moscou)* **32** (1924–25), 646–667 (rés. fr.). Engl. tr. [vHj], 414–437.

[Kmg2] KOLMOGOROFF A.N. Zur Deutung der intuitionistischen Logik, *Mathematische Zeitschrift* **35** (1932), 58–65.

[DKg] KÖNIG D. Sur les correspondances multivoques des ensembles, *Fundamenta Mathematicae* **8** (1926), 114–134.

[GKg1] KÖNIG G. Zum Kontinuum-Problem, *Verhandlungen des Dritten Internationalen Mathematiker-Kongresses in Heidelberg vom 8. bis 13. August 1904.* Teubner, Leipzig, 1905, 144–147.

[GKg2] KÖNIG G. Zum Kontinuum-Problem, *Mathematische Annalen* **60** (1905), 177–180; Berichtigung dazu, *Ibid.*,, 462.

[GKg3] KÖNIG G.Über die Grundlagen der Mengenlehre und das Kontinuum Problem, *Mathematische Annalen* **61** (1905), 156–160. Trad. fr. sous: Sur le fondement de la théorie des ensembles et le problème du continu, *Acta Mathematica* **30** (1906), 329–334. Engl. tr. [vHj], 145–149.

[GKg4] KÖNIG G.Über die Grundlagen der Mengenlehre und das Kontinuum Problem, *Mathematische Annalen* **63** (1906), 217–221. Trad. fr. sous: Sur le fondement de la théorie des Ensembles et le problème du continu, *Acta Mathematica* **31** (1908), 89–93.

[GKg5] KÖNIG G. *Neue Grundlagen der Logik, Arithmetik and Mengenlehre.* Veit, Leipzig, 1914.

[Krn] KRASNER M. Une généralisation de la notion de corps, *Journal de Mathématiques Pures et Appliquées*, sér. 9, **17** (1938), 367–385.

[Krl1] KREISEL G. Bertrand Arthur William Russell, Earl Russell, 1872–1970, *Biographical Memoirs of Fellows of The Royal society* **19** (1973), 582–620.

[Krl2] KREISEL G. Kurt Gödel, 1906–1978, *Biographical Memoirs of Fellows of The Royal society* **26** (1980), 148–224.

[Kur1] KURATOWSKI K. Sur l'opération $\overline{A}$ de l'analysis situs, *Fundamenta Mathematicae* **3** (1922), 182–199.

[Kur2] KURATOWSKI K. Sur l'état actuel de l'axiomatique de la théorie des ensembles, *Rocznik Polskiego Towarsystwa Matematycznego (= Annales de la Société Polonaise de Mathématique)* **3** (1925), 146–147.

[Kur3] KURATOWSKI K. Evaluation de la classe borélienne ou projective d'un ensemble de points à l'aide des symboles logiques, *Fundamenta Mathematicae* **17** (1931), 249–272.

[Kur4] KURATOWSKI K. Les types d'ordre définissables et les ensembles boréliens, *Fundamenta Mathematicae* **29** (1937), 97–100.

[KrT] KURATOWSKI K. et TARSKI A. Les opérations logiques et les ensembles projectifs, *Fundamenta Mathematicae* **17** (1931), 240–248. Rev. engl. tr. [Trk34], 143–151. Texte revu, [Trk36], 147–156. Réimp. [Trk37], 551–559.

[Krh] KUROSH A.G. Durchschnittsdarstellungen mit irreduziblen Komponenten in Ringen und sogenannten Dualgruppen, *Matématičéskij Sbornik (Recueil mathématique de la Société Mathématique de Moscou)* **42** (1935), 613–616.

[Lfd1] LANGFORD C.H. Some theorems on deducibility, *Annals of Mathematics*, ser. 2, **28** (1927), 16–40.

[Lfd2] LANGFORD C.H. Theorems on deducibility (second paper), *Annals of Mathematics*, ser. 2, **28** (1927), 459–471.

[Lfd3] LANGFORD C.H. A theorem on deducibility for second-order functions, *Journal of Symbolic Logic* **4** (1939), 77–79.

[Lgt] LARGEAULT J. (éditeur et traducteur) *Logique mathématique — Textes*. Armand Colin, Paris, 1971, 281 pp.

[Lbg1] LEBESGUE H. Sur les transformations ponctuelles transformant les plans en plans qu'on peut définir par des procédés analytiques, *Atti della Reale Accademia delle Scienze di Torino* **42** (1907), 3–10. Repr. [Lbg4] **III**, 219–226.

[Lbg2] LEBESGUE H. Contribution à l'étude des correspondances de M. Zermelo, *Bulletin de la Société Mathématique de France* **35** (1917), 202–212. Repr. [Lbg4], **III**, 227–238.

[Lbg3] Lebesgue H. Sur certaines démonstrations d'existence, *Bulletin de la Société Mathématique de France* **45** (1907), 132–144. Repr. [Lbg4], **III**, 279–290.

[Lbg4] Lebesgue H. *Oeuvres scientifiques*. L'Enseignement Mathématique, Genève, 1972. **I, II, III**, 1972; **IV** et **V**, 1973.

[Lbz1] Leibniz G. W. von. *Philosopische Schriften*, ed. C.I. Gerhardt, Berlin, 1840–1890.

[Lbz2] Leibniz G. W. von. *Mathematische Schriften*, ed. C.I. Gerhardt, Ascher-Schmidt, Berlin-Halle, 7 Bd. 1849–1863.

[Lel] Lelièvre G. *L'axiomatisation de la théorie des ensembles et le traitement de la notion de propriété définie*. Thèse de 3e Cycle, Université de Paris-Sorbonne (Paris IV), 1980.

[Lnw1] Leśniewski S. *Podstawy ogólnej teoryi mnogości I*. Moscou, 1916.

[Lnw2] Leśniewski S. O podstawach matematiki, *Przegląd Filozoficzny (= Revue Philosophique)*. Introduction et §§1–3, **30** (1927), 164–206; §4, **31** (1928), 261–291; §5, **32** (1929), 60–101; §§6–9, **33** (1930), 77–105; §§10–11, **34** (1931), 142–170.

[Lnw3] Leśniewski S. Grundzüge eines neuen Systems der Grundlagen der Mathematik, §§1–11, *Fundamenta Mathematicae* **14** (1929), 1–81.

[Lnw4] Leśniewski S. Einleitende Bemerkungen zur Fortsetzung meiner Mitteilung u.d.T. «Grundzüge eines neuen Systems der Grundlagen der Mathematik", *Collecteana Logica* **1** (1938), 1–60; journal détruit par fait de guerre, dont il ne reste que des tirés à part à la Harvard College Library et à la Biliothèque de Logique de l'Université de Münster.

[Lnw5] Leśniewski S. Grundzüge eines neuen Systems der Grundlagen der Mathematik, §12, *Collecteana Logica* **1** (1938), 61–144; journal détruit par fait de guerre, dont il ne reste que des tirés à part à la Harvard College Library et à la Biliothèque de Logique de l'Université de Münster.

[Lvi] Levi B. Intorno alla teoria degli aggregati, *Reale istituto lombardo di scienze e lettere, Rendiconti,* 2^a ser. **35** (1902), 863–868.

[PLv] Lévy P. Remarques sur un théorème de Paul Cohen, *Revue de Métaphysique et de Morale* **69** (1964), 88–94.

[ALv] Lévy A. Alfred Tarski's work in set theory, *Journal of Symbolic Logic* **53** (1988), 2–6.

[Lws] Lewis C.I. *A Survey of symbolic Logic*. University of California Press, Berkeley, 1918. 406 pp. Repr. Dover, New-York, 1960.

[LwL] Lewis C.I. and Langford C.H. *Symbolic Logic*. The Century Co., New-York and London, 1932. xi+506 pp. 2^d pr. Dover, New-York, 1951, 7pp.+pp. 3–506. 2e ed. Dover, New-York, 1959, x+518 pp.

[LdM] Lindenbaum A. et Mostowski A. Über die Unabhängigkeit des Auswahlaxioms und einiger seiner Folgerungen, *Sprawozdania z Posied-*

zeń Towarzystwa Naukowego Warzsawskiego, Wydział III Nauk-Matematyczno-fizycznych (= Comptes-Rendus des Séances de la Société des Sciences et des Lettres de Varsovie, Classe III) **31** (1938), 27–32. Engl. tr. [Mtw15] **II**, 70–74.

[LdT1] LINDENBAUM A. et TARSKI A. Communication sur les recherches de la Théorie des Ensembles, *Srawozdania z Posiedzeń Towarzystwa Naukowego Warzsawskiego, Wydział III Nauk-Matematyczno-fizycznych (= Comptes-Rendus des Séances de la Société des Sciences et des Lettres de Varsovie, Classe III)* **19** (1926), 299–330. Repr. [Trk37] **I**, 171–204.

[LdT2] LINDENBAUM A. et TARSKI A. Über die Beschränktheit der Ausdrucksmittel deduktiver Theorien, *Ergebnisse eines Mathematischen Kolloquiums* **7** (1936), 15–22. Rev. engl. tr. [Trk34], 384–392. Trad. fr. revue [Trk36], 109–120. Repr. [Trk37] **II**, 205–213.

[Lwh] LÖWENHEIM L. Über Möglichkeiten im Relativekalkül, *Mathematische Annalen* **76** (1915), 447–470. Engl. tr. [vHj], 228–251.

[Łuk1] ŁUKASÍEWÍCZ J. O potrzebie zalozenia institytutu metodologicznego *Ruch Filozoficzny* **2** (1912), 17–19.

[Łuk2] ŁUKASÍEWÍCZ J. Farewell lecture delivered in the Warsaw University Lecture Hall on March 7, 1918. Warszawa, 1918. Repr. [Łuk9], 84–86.

[Łuk3] ŁUKASÍEWÍCZ J. O logice trójwartościowej, *Ruch Filozoficzny* **5** (1920) 170–171. Engl. tr. [Łuk9], 87–88.

[Łuk4] ŁUKASÍEWÍCZ J. *Elementy logiki matematycznej*. Lithog. Warzsaw, 1929.

[Łuk5] ŁUKASÍEWÍCZ J. Philosophische Bemerkungen zu mehrwertigen Systemen des Aussagenkalküls, *Sprawozdania z Posiedzeń Towarzystwa Naukowego Warzsawskiego, Wydział III Nauk-Matematyczno-fizycznych (= Comptes-Rendus des Séances de la Société des Sciences et des Lettres de Varsovie, Classe III)* **23** (1930), 51–77. Engl. tr. [Łuk9], 153–178.

[Łuk6] ŁUKASÍEWÍCZ J. Z historii logiki zdań, *Przegląd Filozoficzny (= Revue Philosophique)* **37** (1934), 417–437.

[Łuk7] ŁUKASÍEWÍCZ J. Zur Geschichte der Aussagenlogik, *Erkenntnis* **5** (1935), 111–131.

[Łuk8] ŁUKASÍEWÍCZ J. *Z zagadnień logiki i filozofii. Pisma wybrane*. Ed. J. Słupecki, Państwowe Wydawnictwo Naukowe, Warszawa, 1967.

[Łuk9] ŁUKASÍEWÍCZ J. *Selected Works*. Ed. by L. Borkowski. North Holland, Amsterdam-London, 1970. xii+405 pp.

[ŁkT] ŁUKASÍEWÍCZ J. und TARSKI A. Untersuchungen über den Aussagenkalkül, *Srawozdania z Posiedzeń Towarzystwa Naukowego Warzsawskiego, WydziałIII Nauk-Matematyczno-fizycznych (= Comptes-Rendus des Séances de la Société des Sciences et des Lettres de Varsovie, Classe III)* **23** (1930), 30–50. Rev. engl. tr. [Trk34], 38–59; repr. [Łuk9], 131–152. Trad. fr. revue, [Trk36], 45–65. Trad. polonaise de E. Vielrose, [Łuk8], 129–143. Repr. [Trk37] **I**, 323–343.

[Lus] LUSCHEI E.C. *The logical systems of Leśniewski*. North Holland, Amsterdam, 1962. VII+361 pp.

[Lsn1] LUSIN N.N. Sur les ensembles analytiques, *Fundamenta Mathematicae* **10** (1927), 1–95.

[Lsn2] LUSIN N.N. *Leçons sur les ensembles analytiques et leurs applications*. Gauthier-Villars, Paris, 1930, xv+328 pp. 2^d pr. Chelsea, New-York, 1972

[McN] MACNEILLE H.M. Partially ordered sets, *Transactions of the American Mathematical Society* **42** (1937), 416–460.

[MCP] MCCULLOCH W.S. and PITTS W. A logical calculus of the ideas immanent in nervous activity, *Bulletin of Mathematics and Biophysics* **5** (1943), 115–133.

[McK1] MCKINSEY J.C.C. A reduction in number of the postulates for C.I. Lewis' system of strict implication, *Bulletin of the American Mathematical Society* **40** (1934), 425–427.

[McK2] MCKINSEY J.C.C. Proof of the independence of the primitive symbols of Heyting's calculus of propositions, *Journal of Symbolic Logic* **4** (1939), 155–158.

[McK3] MCKINSEY J.C.C. Postulates for the calculus of binary relations, *Journal of Symbolic Logic* **5** (1940), 85–97.

[McK4] MCKINSEY J.C.C. A solution of the decision problem for the Lewis systems $S2$ and $S4$, with an application to topology, *Journal of Symbolic Logic* **6** (1941), 117–134.

[McK5] MCKINSEY J.C.C. The decision problem for some classes of sentences without quantifiers, *Journal of Symbolic Logic* **8** (1943), 61–76.

[MKT1] MCKINSEY J.C.C. and TARSKI A. The algebra of topology, *Annals of Mathematics* ser. 2, **45** (1944), 141–191. Repr. [Trk37] **II**, 609–659.

[MKT2] MCKINSEY J.C.C. and TARSKI A. On closed elements in closure algebras, *Annals of Mathematics* ser. 2, **47** (1946), 122–162. Repr. [Trk37] **III**, 21–61.

[MKT3] MCKINSEY J.C.C. and TARSKI A. Some theorems about the sentential calculi of Lewis and Heyting, *Journal of Symbolic Logic* **13** (1948), 1–15. Repr. [Trk37] **III**, 147–162.

[McN] MCNULTY G.F. Alfred Tarski and undecidable theories, *Journal of Symbolic Logic* **51** (1986), 890–898.

[Mhl1] MAHLO P. Über lineare transfinite Mengen, *Berichte über die Verhandlungen der Königlichen Sächsischen Gesellschaft der Wissenschaften zu Leipzig, Mathematisch-physikalische Klasse* **63** (1911), 187–225.

[Mhl2] MAHLO P. Über die Dimensiontypen des Herrn Fréchet im Gebiete der linearen Mengen, *Berichte über die Verhandlungen der Königlichen Säch-*

sischen Gesellschaft der Wissenschaften zu Leipzig, Mathematisch-physikalische Klasse **63** (1911), 319–347.

[Mhl3] MAHLO P. Zur Theorie und Anwendung der ρ_0-Zahlen I, *Berichte über die Verhandlungen der Königlichen Sächsischen Gesellschaft der Wissenschaften zu Leipzig, Mathematisch-physikalische Klasse* **64** (1912), 108–112.

[Mhl4] MAHLO P. Zur Theorie und Anwendung der ρ_0-Zahlen II, *Berichte über die Verhandlungen der Königlichen Sächsischen Gesellschaft der Wissenschaften zu Leipzig, Mathematisch-physikalische Klasse* **65** (1913), 268–282.

[Mlt1] MALTSEV A.I. Untersuchungen aus dem Gebiet der mathematischen Logik, *Matématičéskij Sbornik n.s.* **1** (1936), 323–336. Engl. tr. [Mlt3], 1–14.

[Mlt2] MALTSEV A.I. Ob odnom obščém métodé polučéniá lokal'nyh téorém téorii grupp, *Učényé zapiski, Fiziko-matématičékié nauki, Ivanovskij Gosudarstvénnyj Pédagogičéskij Institut, Fiziko-matématičéskié Fakulteta* **1** (1941), 3–9. Engl. tr. [Mlt3], 15–21.

[Mlt3] MALTSEV A.I. *The metamathematics of algebraic systems: collected papers 1936–1967.* (B.F. Wells, III, editor). North Holland, Amsterdam-London, 1971.

[Mrk] MARKOV A.A. Névozmožnost' nékotoryh algorifmov v téorii associativnyh sistém, *Doklady Akadémii Nauk SSSR* n.s. **55** (1947), pp. 587–590; **58** (1947), 353–356. Engl. tr. of the first part under: On the impossibility of certain algorithms in the theory of associative systems, *Comptes-Rendus (Doklady) de l'Académie des Sciences de l'URSS*, n.s. **55** (1947), 583–586.

[MRg] MATHIAS A.R.D. and ROGERS H. (editors) *Cambridge Summer School in Mathematical Logic.* Lecture Notes in Mathematics **337**, Springer, Berlin-Heidelberg-New-York, 1973, IX+660 pp.

[Mzr] MAZUR S. Computable analysis, *Rozprawy Matematyczne* **33** (1963), 1–111.

[MzS] MAZURKIEWICZ S. et SIERPINSKI W. Contribution à la topologie des ensembles dénombrables, *Fundamenta Mathematicae* **1** (1920), 17–27. Repr. [Srp3] **II**, 294–302.

[Mng] MENGER K. Bemerkungen zu Grundlagenfragen, *Jahresbericht der deutschen Mathematiker-Vereinigung* **37** (1928), 213–226, 298–325.

[Mmf] MIRIMANOFF D. Les antinomies de Russell et de Burali-Forti et le problème fondamental de la théorie des ensembles, *L'enseignement mathématique* **19** (1917), 37–52.

[Moj] MOOIJ J.J.A. *La philosophie des mathématiques de Henri Poincaré.* Gauthier-Villars, Paris, et Nauwelaerts, Louvain, 1966. ii+175 pp.

[Mor1] MOORE G.H. Beyond first-order logic: the historical interplay between mathematical logic and axiomatic set theory, *History and Philosophy of Logic* **1** (1980), 95–137.

[Mor2] MOORE G.H. The Emergence of First-order Logic, [AsK], 95–135.

[Mrg] MORGAN A. DE. On the syllogism, *n°* IV, and on the logic of relations (read 23 April 1860), *Transactions of the Cambridge Philosophical Society* **10** (1864), 331–358.

[Mtw1] MOSTOWSKI A. Abzählbare Boolesche Körper und ihre Anwendung auf die allgemeine Metamathematik, *Fundamenta Mathematicae* **29** (1937), 34–53. Engl. tr. [Mtw15] **II**, 1–17.

[Mtw2] MOSTOWSKI A. O niezalnezności definicji skończoności w systemie logiki, *Dodatek do Rocznika Polskiego Towarzystwa Matematycznego* **11** (1938), 1–54. Engl. tr. under: On the independence of definitions of finiteness in a system of logic, [Mtw15] **II**, 18–67.

[Mtw3] MOSTOWSKI A. Über den Begriff einer endlichen Menge, *Sprawozdania z Posiedzeń Towarzystwa Naukowego Warzsawskiego, Wydział III Nauk-Matematyczno-fizycznych = Comptes-Rendus des Séances de la Société des Sciences et des Lettres de Varsovie, Classe III* **31** (1938), 13–20.

[Mtw4] MOSTOWSKI A. Über die Unabhängigkeit des Wohlordnungssatzes vom Ordnungsprinzip, *Fundamenta Mathematicae* **32** (1939), 201–252. Engl. tr. [Mtw15] **I**, 290–338.

[Mtw5] MOSTOWSKI A. Axiom of choice for finite sets, *Fundamenta Mathematicae* **33** (1945), 137–168. Repr. [Mtw15] **II**, 92–123.

[Mtw6] MOSTOWSKI A. On definable sets of positive integers, *Fundamenta Mathematicae* **34** (1947), 81–112. Repr. [Mtw15] **I**, pp. 339–370.

[Mtw7] MOSTOWSKI A. On absolute properties of relations, *Journal of Symbolic Logic* **12** (1947), 33–42. Repr. [Mtw15] **II**, 124–133.

[Mtw8] MOSTOWSKI A. On the principle of dependent choices, *Fundamenta Mathematicae* **35** (1948), 127–130. Repr. [Mtw15] **II**, 134–137.

[Mtw9] MOSTOWSKI A. An undecidable arithmetical statement, *Fundamenta Mathematicae* **36** (1949), 143–164. Repr. [Mtw15] **I**, 531–552.

[Mtw10] MOSTOWSKI A. A classification of logical systems, *Studia philosophica* **4** (1949–50, publ. 1951), 237–274. Repr. [Mtw15] **II**, 154–191.

[Mtw11] MOSTOWSKI A. Some impredicative definitions in the axiomatic set-theory, *Fundamenta Mathematicae* **37** (1951), 111–124. Correction: **38** (1952), 238. Repr. [Mtw15] **I**, 479–493.

[Mtw12] MOSTOWSKI A., in collaboration with GRZEGORCZYK A., JAŚKOWSKI S., ŁOS J., MAZUR S., RASIOWA H. and SIKORSKI R. The present state of the investigations in the foundations of mathematics, *Rozprawy Matematyczne*, **9** (1955), 48 pp.

[Mtw13] MOSTOWSKI A. L'oeuvre scientifique de Jan Łukasíewícz dans le domaine de la logique mathématique, *Fundamenta Mathematicae* **44** (1957), 1–11.

[Mtw14] MOSTOWSKI A. Thirty years of foundational studies, Lectures on the development of mathematical logic and the study of foundations of mathematics in 1930–1964, *Acta Philosophica Fennica XVII* (1965) pp. 1–180. Repr. Blackwell, Oxford, 1966; [Mtw15] **I**, 1–176.

[Mtw15] MOSTOWSKI A. *Fundational Studies. Selected Works.* Edited by K. Kuratowski, W. Marek, L. Pacholski, H. Rasiowa, C. Ryll-Nardzewski and P. Zbierski. North Holland, Amsterdam-New-York-Oxford; Państwowe Wydawnictwo Naukowe, Warszawa, 1979. **I**, XLIV+633 pp.; **II**, VIII+605 pp.

[MwT1] MOSTOWSKI A. and TARSKI A. Boolesche Ringe mit geordneter Basis, *Fundamenta Mathematicae* **32** (1939), pp. 69–86. Engl. tr. [Mtw15] **II**, 75–91. Repr. [Trk37] **II**, 531–548.

[MwT2] MOSTOWSKI A. and TARSKI A. Arithmetical classes and types of well ordered systems, *Bulletin of the American Mathematical Society* **55** (1949), 65; 1192. Repr. [Mtw15] **II**, 148; [Trk37] **IV**, 583.

[MwT3] MOSTOWSKI A. and TARSKI A. Undecidability in the arithmetic of integers and in the theory of rings, *Journal of Symbolic Logic* **14** (1949), 76. Repr. [Trk37] **IV**, 586–587.

[Mnk] MUCNIK A.A. Nérazrésimost' problémy svodimosti téorii algoritmov, *Doklady Akadémii Nauk SSSR* n.s. **108** (1956), 194–197.

[NkH] NAKASIMA A. et Hanzawa M. The theory of equivalent transformation of simple partial paths in the relay circuits, *Nippon electrical communication engineering* **9** (Fév. 1938), 32–39. Trad. angl. de deux articles en japonais, parus au «Journal de l'école d'ingénieurs en communications électriques du Japon» **165** (Déc. 1936) et **167** (Fév. 1937).

[Nls] NELSON D. Recursive functions and intuitionistic number theory, *Transactions of the American Mathematical Society* **61** (1947), 307–368.

[vNm1] NEUMANN J. VON. Eine Axiomatisierung der Mengenlehre, *Journal für die reine und angewandte Mathematik* **154** (1925), 219–240. Berichtigung, *Ibid.* **155** (1926), 128. Abg. [vNm7] **I**, 35–56. Engl. tr. [vHj], 393–413.

[vNm2] NEUMANN J. VON. Zur Hilbertschen Beweistheorie, *Mathematische Zeitschrift* **26** (1927), 1–46. Abg. [vNm7] **I**, 339–422.

[vNm3] NEUMANN J. VON. Über eine Widerspruchsfreiheitsfrage in der axiomatischen Mengenlehre, *Journal für die reine und angewandte Mathematik* **160** (1929), 227–241. Abg. [vNm7] **I**, 494–508.

[vNm4] NEUMANN J. VON. Algebraische Repräsentanten der Funktionen «bis auf eine Menge vom Mass Null», *Journal für die reine und angewandte Mathematik* **165** (1931), 109–115. Abg. [vNm7] **II**, 213–219.

[vNm5] NEUMANN J. VON. *Mathematische Grundlagen der Quantenmechanik.* Springer, Berlin, 1931.

[vNm6] NEUMANN J. VON. *First Draft of a Report on the EDVAC.* Philadelphia, 1945.

[vNm7] NEUMANN J. VON. *Collected works.* Pergamon Press, New-York. **I, II, III**, 1961; **IV**, 1962; **V, VI**, 1963.

[vNS] NEUMANN J. VON. and STONE M.H. The determination of representative elements in the residual classes of a Boolean algebra, *Fundamenta Mathematicae* **25** (1935), 353–378. Repr. [vNm7] **IV**, 65–90.

[Ncd] NICOD J. A reduction in the number of the primitive propositions of logic, *Proceedings of the Cambridge Philosophical Society* **19** (1917–1920), 32–41 («received 30 October 1916»).

[Nth1] NOETHER E. Idealtherie in Ringbereichen, *Mathematische Annalen* **83** (1921), 24–66. Abg. [Nth4], 354–396.

[Nth2] NOETHER E. Abstrakter Aufbau der Idealtheorie in algebraischen Zahl- und Funktionenkörper, *Mathematische Annalen* **96** (1926–27), 26–61. Abg. [Nth4], 493–528.

[Nth3] NOETHER E. Hyperklompexe Größen und Darstellungstheorie, *Mathematische Zeitschrift* **30** (1929), 641–692. Abg. [Nth4], 563–614.

[Nth4] NOETHER E. *Gesammelte Abhandlungen.* Hrsg. Jabocson N. Springer, Berlin-Heidelberg-New-York, 1983.

[NCv] NOETHER E. und CAVAILLÈS J. (eds) *Briefwechsel Cantor-Dedekind.* Hermann, Paris, 1937. Tr. fr. [Cav], 187–251.

[Nvk1] NOVAK I.L. The relative consistency of von Neumann's and Zermelo's axioms for set theory, *Bulletin of the American Mathematical Society* **55** (1949), 300.

[Nvk2] NOVAK I.L. The relative consistency of von Neumann's and Zermelo's axioms for set theory, *Journal of Symbolic Logic* **14** (1949), 79.

[Nvk3] NOVAK I.L. A construction for models of consistent systems, *Fundamenta Mathematicae* **37** (1950), 87–110.

[Nkf1] NOVIKOFF P.S. Sur quelques théorèmes d'existence, *Comptes-Rendus (Doklady) de l'Académie des Sciences de l'URSS*, n.s. **23** (1939), 438–440.

[Nkf2] NOVIKOFF P.S. On the consistency of certain logical calculi, *Matématičéskij Sbornik (Recueil mathématique)* **12 (54)** (1943), 231–261.

[Nkf3] NOVIKOFF P.S. O klassah régulárnosti, *Doklady Akadémii Nauk SSSR* n.s. **64** (1949), 293–295.

[Nkf4] NOVIKOFF P.S. Ob aksiomé polnoj indukcii, *Doklady Akadémii Nauk SSSR* n.s., **64** (1949), 457–459.

[Ore1] ORE Ö. On the foundations of abstract algebra, *Annals of Mathematics* ser. 2; I, **36** (1935), 406–437; II, **37** (1936), 265–292.

[Ore2] ORE Ö. Theory of equivalence relations, *Duke mathematical Journal* **9** (1942), 573–627.

[Ore3] ORE Ö. Galois connexions, *Transactions of the American Mathematical Society* **55** (1944), 493–513.

[Pad] PADOA A. Essai d'une théorie algébrique des nombres entiers, précédé d'une introduction logique à une théorie déductive quelconque. *Bibliothèque du Congrès International de Philosophie,t. 3*. Armand Colin, Paris, 1901, 309-365. Engl. tr. [vHj], 118–123.

[Pch] PASCH M. *Vorlesungen über neuere Geometrie*. Leipzig, 1882. 2. Aufl. Teubner, Leipzig 1912.

[Pck1] PEACOCK G. *A Treatise on Algebra*. Cambridge University Press, 1830.

[Pck2] PEACOCK G. Report on the recent progress and present state of certain branches of Analysis. *Report of the third meeting of the British Association for the Advancement of Science, held at Cambridge in 1833*, London, 1834

[Pno1] PEANO G. *Calcolo geometrico secondo l'Ausdehnungslehre di H. Grassmann, preceduto dalle Operazioni della logica deduttiva*. Turin, 1888. Repr. part. [Pno7] **2**, 3–19.

[Pno2] PEANO G. *Arithmetices Principia, nova methodo exposita*. Torino, 1889. Reimpr. [Pno7] **2**, 20–55. Part. engl. tr. [vHj], 83–97.

[Pno3] PEANO G. Démonstration de l'intégrabilité des équations différentielles ordinaires. *Mathematische Annalen* **37** (1890), 182–228. Réimpr. [Pno7] **1**, 119–170.

[Pno4] PEANO G. Sul concetto di numero, *rivista di matematica* **1** (1891), 87–102, 256–267. Repr. [Pno7] **3**, 80–109.

[Pno5] PEANO G. *Formulaire de mathématiques*. **1**, Turin, 1895; **2**, §1, Turin, 1897; §2, Turin, 1898; §3, Turin, 1899; **3**, Turin, 1901; **4**, Bocca, Turin, Clausen, Turin, 1903.

[Pno6] PEANO G. Super theorema di Cantor-Bernstein e Additione, *Rivista di matematica* **8** (1906), 136–157. Reimpr. [Pno7] **1**, 337–358.

[Pno7] PEANO G. *Opere Scelte*. Edizione Cremonese, Roma. **1**, 1957; **2**, 1958; **3**, 1959.

[Prc1] PEIRCE C.S. On the algebra of logic. Ch. I: Syllogistic. Ch II: The logic of non-relative terms. Ch. III: The logic of relatives. *American Journal of Mathematics* **3** (1880), 15–57. Repr. [Prc3], **III**, 104–157.

[Prc2] PEIRCE C.S. On the algebra of logic: A contribution to the philosophy of notation, *American Journal of Mathematics* **7** (1885), 180–202. Repr. [Prc3], **III**.

[Prc3] PEIRCE C.S. *Collected Papers*. Harvard University Press. **I**, ed. by C. Hartshorne and P. Weiss, 1931; **II**, ed. by C. Hartshorne and P. Weiss,

1932; **III**, ed. by C. Hartshorne and P. Weiss, 1933; **IV**, ed. by C. Hartshorne and P. Weiss, 1933; **V**, ed. by C. Hartshorne and P. Weiss, 1934; **VI**, ed. by C. Hartshorne and P. Weiss, 1935; **VII**, ed. by A.W. Burks, 1958; **VIII**, ed. by A.W. Burks, 1958.

[Ptr1] PÉTER R. Konstruktion nichtrekursiver Funktionen, *Mathematische Annalen* **111** (1935) 42–60.

[Ptr2] PÉTER R. Über die mehrfache Rekursion, *Mathematische Annalen* **113** (1936) 489–527.

[Ptr3] PÉTER R. Az axiomatikus módszer korlátai, *Matematikai és Fizikai Lapok* **48** (1941), pp. 120–143. Deutscher Auszug, unter: Die Schranken der axiomatischen Methode.

[Ptr4] PÉTER R. Zusammenhang der mehrfachen und transfiniten Rekursionen, *Journal of Symbolic Logic* **15** (1950) 248–272.

[Ptr5] PÉTER R. *Rekursive Funktionen.* Akadémiai Kiado, Budapest, 1951, 206 pp. Engl. ed.: *Recursive functions.* Academic Press, New-York, 1967.

[Pie] PIERI M. La geometria elementare instituita sulle nozione di «punto» e «sfera», *Memorie di Matematica e di Fisica della Societa Italiana delle Scienze, Serie Terza,* **15** (1908), 345–450.

[Pnc1] POINCARÉ H. Des Fondements de la géométrie, à propos d'un livre de M. Russell, *Revue de Métaphysique et de Morale* **7** (1899), 251–279.

[Pnc2] POINCARÉ H. Sur les Principes de la géométrie, *Revue de Métaphysique et de Morale* **8** (1900), 73–86.

[Pnc3] POINCARÉ H. Du rôle de l'intuition et de la logique en mathématiques, *Compte-Rendu du 2e Congrès International de Mathématiciens,* Paris, 1902, 119–130.

[Pnc4] POINCARÉ H. Les Mathématiques et la Logique, *Revue de Métaphysique et de Morale* **13** (1905), pp. 815–835; **14** (1906), 17–34. Repr. avec qqes var. dans [Pnc6]. Repr. avec les var. [Hzm2], 11–53.

[Pnc5] POINCARÉ H. Les Mathématiques et la Logique, *Revue de Métaphysique et de Morale* **14** (1906), 294–317. Repr. avec qqes var. dans [Pnc6]. Repr. avec les var. [Hzm2], 79–104.

[Pnc6] POINCARÉ H. *Science et Méthode.* Flammarion, Paris, 1908, 311pp. Deutsche Übersetzung Teubner, Leipzig, 1914, 311pp. Engl. tr. Nelson, Walton on Thames, 1914, 288pp.

[Ppr] POPPER K. Der wichtigste Beitrag seit Aristoteles, *Wissenschaft aktuell* n^r4, 1980, 50–51.

[Por1] PORTE J. *Collectifs et fonctions récursives.* Multigraphié, Institut Blaise Pascal, 1963, iv+19 pp.

[Por2] PORTE J. Systèmes de Post, algorithmes de Markov, *Cybernetica* **2** (1958), 114–149.

[Pst1] POST E.L. Introduction to a general theory of elementary propositions (abstract), *Bulletin of the American Mathematical Society* **26** (1920), 437.

[Pst2] POST E.L. Determination of all closed systems of truth tables, *Bulletin of the American Mathematical Society* **26** (1920), 437.

[Pst3] POST E.L. Introduction to a general theory of elementary propositions, *American Journal of Mathematics* **43** (1921), 163–185. Repr. [vHj], pp. 264–283.

[Pst4] POST E.L. Finite combinatory processes — formulation I, *Journal of Symbolic Logic* **1** (1936), 103–105. Repr. [Dvs2], 288–291.

[Pst5] POST E.L. *The two-valued iterative systems of mathematical logic*. Annals of mathematics series $n^r 5$, Princeton University Press, Princeton, N.J., 1941.

[Pst6] POST E.L. Formal reductions of the general combinatorial decision problem, *American Journal of Mathematics* **65** (1943), 197–215.

[Pst7] POST E.L. Recursively enumerable sets of positive integers and their decision problem, *Bulletin of the American Mathematical Society* **50** (1944), 284–316. Repr. [Dvs2], 304–337.

[Pst8] POST E.L. Recursive unsolvability of a problem of Thue, *Journal of Symbolic Logic* **12** (1947), 1–11. Repr. [Dvs2], 292–303.

[Pst9] POST E.L. Degrees of recursive unsolvability. Preliminary report (abstract), *Bulletin of the American Mathematical Society* **54** (1948), 641–642.

[Pst10] POST E.L. Absolutely unsolvable problems and relatively undecidable propositions. Account of an anticipation, [Dvs2], 340–433.

[Pbg] PRESBURGER M. Über die Vollständigkeit eines gewissen Systems der Arithmetik ganzer Zahlen, in welchem die Addition als einzige Operation hervortritt, *Sprawozdanie z I Kongresu Matematikow Krajow Slowianskich (= Comptes-rendus du I Congrès des Mathématiciens des Pays Slaves.)* Warszawa, 1930, 92–101.

[Qui1] QUINE W.V.O. New foundations for the mathematical logic, *The American Mathematical Monthly* **44** (1938), 70–80.

[Qui2] QUINE W.V.O. *Mathematical Logic*. Norton, New-York, 1940, xiii+348 pp. 2^d rev. ed. Harvard University Press, 1951, xii+346 pp.

[Qui3] QUINE W.V.O. Element and number, *Journal of Symbolic Logic* **6** (1941), 135–149.

[Qui4] QUINE W.V.O. Concatenation as a Basis for Arithmetic, *Journal of Symbolic Logic* **11** (1946), 105–114.

[Rms1] RAMSEY F.P. The foundations of mathematics, *Proceedings of the London Mathematical Society*, Series 2, **25** (1925), 338–384. Repr. [Rms3], 1–61.

[Rms2] RAMSEY F.P. On a problem of formal logic, *Proceedings of the London Mathematical Society,* Series 2, **30** (1930), 264–286. Repr. [Rms3], 82–111.

[Rms3] RAMSEY F.P. *The foundations of mathematics and other logical essays.* Ed. Richard Bevan Braithwaite. Paul, Trench, Trubner, London; Harcourt, Brace, New York, 1931. Repr. Humanities Press, New-York, 1950.

[RgT] RANG B. and THOMAS W. Zermelo's discovery of the «Russell paradox», *Historia Mathematica* **8** (1981), 15–22.

[Rch] RICHARD J. Les principes des mathématiques et le problème des ensembles, *Revue générale des sciences pures et appliquées* **16** (1905), 541. Engl. tr. [vHj], 142–144.

[ARb1] ROBINSON A. On the metamathematics of algebra, *Journal of Symbolic Logic* **14** (1949), 74.

[ARb2] ROBINSON A. On the application of symbolic logic to algebra, *Proceedings of the International Congress of Mathematicians, Cambridge, Mass., U.S.A., Aug. 30-Sept 6, 1950,* **I**, American Mathematical Society, Providence, Rhode Island, 1952, 686–694. Repr. [ARb3] **I**, 3–11.

[ARb3] ROBINSON A. *Selected Papers.* Ed. Keisler H.J., Körner S., Luxemburg W.A.J., Young A.D. **I, II, III**, 1979, Yale University Press, New Haven and London.

[JRb1] ROBINSON J. Undecidability in the arithmetic of integers and rationals and in the theory of fields, *Journal of Symbolic Logic* **14** (1949), 77.

[JRb2] ROBINSON J. Definability and decision problems in arithmetic, *Journal of Symbolic Logic* **14** (1949), 98–114.

[RRb] ROBINSON R. An essentially undecidable axiom system, *Proceedings of the International Congress of Mathematicians, Cambridge, Mass., U.S.A., Aug. 30-Sept 6, 1950,* **I**, American Mathematical Society, Providence, Rhode Island, 1952, 729–730.

[Rsr1] ROSSER J.B. A mathematical logic without variables I, *Annals of Mathematics,* ser. 2, **36** (1935), 127–150; II, *Duke mathematical Journal* **1** (1935), 328–355.

[Rsr2] ROSSER J.B. Extensions of some theorems of Gödel and Church, *Journal of Symbolic Logic* **1** (1936), 87–91. Repr. [Dvs2], 231–235.

[Rsr3] ROSSER J.B. Review of [Gdl11], *Journal of Symbolic Logic* **1** (1936), 116.

[Rsr4] ROSSER J.B. Gödel theorems for non-constructive logics, *Journal of Symbolic Logic* **2** (1937), 129–137.

[Rsr5] ROSSER J.B. On the consistency of Quine's New Foundations for mathematical logic, *Journal of Symbolic Logic* **4** (1939), 15–24.

[Rsr6] ROSSER J.B. New sets of postulates for combinatory logic, *Journal of Symbolic Logic* **7** (1942), 18–27.

[Rsr7] ROSSER J.B. The Burali-Forti paradox, *Journal of Symbolic Logic* **7** (1942), 1–17.

[RrT] ROSSER J.B. and TURQUETTE A.R. *Many valued logics*. North Holland, Amsterdam, 1952.

[RrW] ROSSER J.B. and WANG H. Non-standard models for formal logics, *Journal of Symbolic Logic* **15** (1950), 113–129. Errata, *ibid.* p. IV.

[Rtg] RUITENBURG W. The Unintended Interpretations of Intuitionistic Logic, [Drk], 134–160.

[Rsl1] RUSSELL B. *An Essay on the Foundations of Geometry*. Cambridge University Press, 1897. Trad. fr. A. Cadenat, revue et annotée par l'Auteur et par L. Couturat. Gauthier-Villars, Paris, 1901, X+274 pp.

[Rsl2] RUSSELL B. Sur les axiomes de la géométrie, *Revue de Métaphysique et de Morale* **7** (1899), 684–707.

[Rsl3] RUSSELL B. *The Principles of Mathematics*, **I**. Cambridge,1903, 534 pp. 2^d ed. George Allen and Unwin, London, 1937, W.W. Norton, New-York, 1938, xxxix+534 pp. Repr. George Allen and Unwin, London, 1942, 1948, 1950, 1951, 1956.

[Rsl4] RUSSELL B. On some difficulties in the theory of transfinite numbers and order types, *Proceedings of the London Mathematical Society*, Series 2, **4** (1907), 29–53. Repr. [Hzm2], 54–78.

[Rsl5] RUSSELL B. Les paradoxes de la logique. *Revue de Métaphysique et de Morale* **14** (1906), 627–650. Repr. [Hzm2], 121–144.

[Rsl6] RUSSELL B. Mathematical logic as based on the theory of types, *American Journal of Mathematics* **30** (1908), 222–262. Repr. [Rsl7], 59–102; [vHj], 150–182. Repr. part. [Hzm2], 200–223.

[Rsl7] RUSSELL B. *Logic and knowledge, Essays 1901–1950*. Ed. Robert Charles Marsh, Allen and Unwin, London, 1956.

[Rsl8] RUSSELL B. My mental development, *The Philosophy of Bertrand Russell*. Ed. Paul Arthur Schilpp. Tudor, New-York, 1944, 3–20.

[Sfs] SCHOENFLIES A. Zur Erinnerung an Georg Cantor, *Jahresbericht der deutschen Mathematiker-Vereinigung* **31** (1922), 97–106.

[Slz] SCHOLZ H. In memoriam Jan Łukasíewícz, *Archiv für Mathematische Logik und Grundlagenforschung* **3** (1957), 3–18.

[Sfk] SCHÖNFINKEL M. Über die Bausteine der mathematischen Logik, *Mathematische Annalen* **92** (1924), 305–316. Engl. tr. [vHj], 355–366.

[Sdr1] SCHRÖDER E. *Vorlesungen über die Algebra der Logik (exakte Logik)*. Leipzig. **1**, 1890; **2**, Erster Teil, 1891; **3**, Erster Teil, 1895; **2**, Zweiter Teil, 1905. Abg. [Sdr3].

[Sdr2] SCHRÖDER E. *Abriß der Algebra der Logik*, ed. E. Müller. Teubner, Leipzig-Berlin, 1909–1910. Abg. [Sdr3].

[Sdr3] SCHRÖDER E. *Vorlesungen über die Algebra der Logik (exakte Logik).* Chelsea, New-York, 1966 (Abdrucke von [Sdr1] und [Sdr2]).

[Sct1] SCHÜTTE K. Untersuchungen zum Entscheidungsproblem der mathematischen Logik, *Mathematische Annalen* **109** (1934), 572–603.

[Sct2] SCHÜTTE K. Über die Erfüllbarkeit einer Klasse von logischen Formeln, *Mathematische Annalen* **110** (1934), 161–194.

[Shn] SHANNON C.E. A symbolic analysis of relay and switching circuits, *Transactions of the American Institute of Electrical Engineers* **57** (1938), 713–723. Repr. under: *A symbolic analysis of relay and switching circuits.* Berkeley, New-York, 1952, i+11 pp.

[Shf] SHEFFER H.M. A set of five independent postulates for Boolean algebras, with application to logical constants, *Transactions of the American Mathematical Society* **14** (1913), 481–488.

[Srp1] SIERPINSKI W. L'axiome de M. Zermelo et son rôle dans la Théorie des Ensembles et l'Analyse, *Bulletin de l'Académie des Sciences de Cracovie, Classe des Sciences Mathématiques, Série A,* 1918, 99–152. Repr. [Srp3] **II**, 208–285.

[Srp2] SIERPINSKI W. L'hypothèse du continu généralisée et l'axiome de choix, *Fundamenta Mathematicae* **34** (1945), 1–5. Repr. [Srp3] **III**, 484–488.

[Srp3] SIERPINSKI W. *Oeuvres choisies.* Publié par les soins de St. Hartmann, K. Kuratowski, E. Marczewski, A. Mostowski, et A. Schinzel. Académie Polonaise des Sciences, Institut de Mathématiques. Państwowe Wydawnictwo Naukowe, Warszawa. **I**, 1974, 300pp.; **II**, 1975, 780 pp.; **III**, 685 pp.

[SpT] SIERPINSKI W. et TARSKI A. Sur une propriété caractéristique des nombres inaccessibles, *Fundamenta Mathematicae* **15** (1930), 292–300. Repr. [Srp3] **III**, 29–35; [Trk37] **I**, 289–297.

[Snc] SINACEUR H. *Corps et Modèles.* Mathesis, Vrin, Paris, 1991, 496 pp.

[Skm1] SKOLEM TH. *Indersækelser innenfor logikkens algebra.* Thèse, Université d'Oslo, 1913.

[Skm2] SKOLEM TH. Untersuchungen über die Axiome des Klassen-kalküls und über die Produktations- und Summationsprobleme, welche gewisse Klassen von Aussagen betreffen, *Skrifter utgit av Videnskapsselskapet i Kristiania, I. Matematisk-naturvidenskabelig Klasse 1919,* n^r3. 37 pp. Abg. [Skm11], 67–101.

[Skm3] SKOLEM TH. Logisch-kombinatorische Untersuchungen über die Erfüllbarkeit oder Beweisbarkeit mathematischer Sätze nebst einem Theorem über dichte Mengen, *Skrifter utgit av Videnskapsselskapet i Kristiania, I. Matematisk-naturvidenskabelig Klasse 1920,* n^r4. 36 pp. Engl. tr. [vHj], 254–263. Abg. [Skm11], 103–136.

[Skm4] SKOLEM TH. Einige Bemerkungen zur axiomatischen Begründung der Mengenlehre, *Matematikerkongressen i Helsingfors des 4–7 Juli 1922,*

Den femte skandinaviska matematiker kongressen, Redegörelse. Akademiska Bokhandeln, Helsinki, 1923, 217–232. Engl. tr. [vHj], 290–301. Abg. [Skm11], 137–152.

[Skm5] SKOLEM TH. Begründung der elementaren Arithmetik durch die rekurrierende Denkweise ohne Anwendung scheinbarer Veränderlichen mit unendlichem Ausdehnungsbereich, *Skrifter utgit av Videnskapsselskapet i Kristiania, I. Matematisk-naturvidenskabelig Klasse 1923, n^r 6.* 38 pp. Engl. tr. [vHj], 302–333. Abg. [Skm11], 153–188.

[Skm6] SKOLEM TH. Über die mathematische Logik, *Norsk matematisk tidsskrift* **10** (1928), 125–142. Engl. tr. [vHj], 508–524. Abg. [Skm11], 189–206. Trad. fr. et présentation, [Lgt], 91–108.

[Skm7] SKOLEM TH. Über einige Grundlagenfragen der Mathematik, *Skrifter utgitt av Det Norske Videnskaps-Akademi i Oslo I. Matematisk-naturvidenskapelig klasse 1929, n^r 4*, 1930, 3–21. Abg. [Skm11], 227–273.

[Skm8] SKOLEM TH. Über einige Satzfunktionen in der Arithmetik, *Skrifter utgitt av Det Norske Videnskaps-Akademi i Oslo I. Matematisk-naturvidenskapelig klasse 1930, n^r 7*, 1931. 28 pp. Abg. [Skm11], 281–306.

[Skm9] SKOLEM TH. Über die Unmöglichkeit einer vollständigen Charakterisierung der Zahlenreihe mittels eines endlichen Axiomensystems, *Norsk matematisk forenings skrifter* ser. 2, *n^o* 10 (1933), 73–82. Abg. [Skm11], pp. 345–354.

[Skm10] SKOLEM TH. Über die Nicht-charakterisierbarkeit der Zahlenreihe mittels endlich oder abzählbar unendlich vieler Aussagen mit ausschliesslich Zahlenvariablen, *Fundamenta Mathematicae* **23** (1934), pp. 150–161. Abg. [Skm11], pp. 355–366.

[Skm11] SKOLEM TH. *Selected Works in Logic by Thoralf Skolem.* Ed. J.E. Fenstad. Universitetforlage, Oslo, 1970, 732 pp.

[Słp1] SŁUPECKI J. Kriterium pelnosci wielowartościowych system'ow logiki zdán, *Comptes-Rendus des Séances de la Société des Sciences et des Lettres de Varsovie* **32** (1939), 102–109.

[Słp2] SŁUPECKI J. Dowod aksjomatyzowalności pelnych systemów wielowartościowych, *Comptes-Rendus des Séances de la Société des Sciences et des Lettres de Varsovie* **32** (1939), 110–128. Deutscher Aufzug.

[Smr] SMORYNSKI C. *Logical Number Theory I. An Introduction.* Universitext, Springer, Berlin-Heidelberg-New-York-London-Paris-Tokio-Hong-Kong-Barcelona-Budapest 1991, x+4O5 pp.

[Sln] SOUSLIN M.I. Sur une définition des ensembles mesurables *B* sans nombres transfinis, *Comptes-Rendus hebdomadaires des Séances de l'Académie des Sciences* Paris **164** (1917), 11–30.

[Spk] SPECKER E. Nicht konstruktiv beweisbare Sätze der Analysis, *Journal of Symbolic Logic* **14** (1949), 145–158.

[Stz] STEINITZ E. Algebraische Theorie der Körper, *Journal für die reine und angewandte Mathematik* **137** (1910), 167–309.

[Stn1] STONE M.H. Boolean algebras and their application to topology, *Proceedings of the National Academy of Sciences of the U.S.A.* **20** (1934), 197–202.

[Stn2] STONE M.H. Subsumption of the theory of Boolean algebras under the theory of rings, *Proceedings of the National Academy of Sciences of the U.S.A.* **21** (1935), 103–105.

[Stn3] STONE M.H. The theory of representations for Boolean algebras, *Transactions of the American Mathematical Society* **40** (1936), 37–111.

[Stn4] STONE M.H. Applications of the theory of Boolean rings to general topology, *Transactions of the American Mathematical Society* **41** (1937), 475–481.

[Stn5] STONE M.H. Topological representations of distributive lattices and Brouwerian logics, *Časopis pro péstovani matematiky a fysiky* **67** (1937–1938), 1–25.

[Stn6] STONE M.H. The representation of Boolean algebras, *Bulletin of the American Mathematical Society* **44** (1938), 807–816.

[Stm] STURM CH. F. Mémoire sur la résolution des équations numériques, *Mémoires présentés par divers Savans étrangers à l'Académie Royale des Sciences, section Sciences mathématiques et physiques* **VI**, 273–318.

[Sud] SUDAN G. Sur le nombre transfini ω^ω, *Bulletin mathématique de la Société Roumaine des Sciences* **30** (1927), 11–30.

[Sur1] SURÁNYI J. Zur Reduktion des Entscheidungsproblems des logischen Funktionenkalküls, *Matematikai és fizikai lapok* **50** (1943), 51–74.

[Sur2] SURÁNYI J. Reduction of the decision problem to formulas containing a bounded number of quantifiers only, *Proceedings of the Tenth International Congress of Philosophy (Amsterdam, Aug. 11–18, 1948),* North-Holland, Amsterdam, 1949, 756–758.

[Szb] SZCZERBA L.W. Tarski and Geometry, *Journal of Symbolic Logic* **51** (1986), 907–912.

[Szm] SZMIELEW W. Decision problem in group theory, *Proccedings of the Tenth International Congress of Philosophy (Amsterdam, Aug. 11–18, 1948),* North-Holland, Amsterdam, 1949, 763–766.

[Trk1] TAJTELBAUM A. O wyrazie pierwotnym logistyki, Thèse de doctorat, *Przegląd Filozoficzny (= Revue Philosophique)* **26** (1923), 68–89. Trad. fr. revue, en deux parties: TAJTELBAUM A. Sur le terme primitif de la logistique, *Fundamenta Mathematicae* **4** (1923), 196–200, repr. [Trk37] **I**, 15–20; TAJTELBAUM-TARSKI A. Sur les *truth-functions* au sens de MM. Russell et Whitehead, *Fundamenta Mathematicae* **5** (1924), 59–74, repr. [Trk37] **I**, 23–38. Rev. engl. tr. of both [Trk34], 1–23. Version fr. revue d'après la préc. [Trk36], 1–25.

[Trk2] Tajtelbaum-Tarski A. Sur quelques théorèmes qui équivalent à l'axiome du choix, *Fundamenta Mathematicae* **5** (1924), 147–154. Repr. [Trk37] **I**, 41–48.

[Trk3] Tarski A. Sur les ensembles finis, *Fundamenta Mathematicae* **6** (1924), 45–95. Repr. [Trk37] **I**, 67–117.

[Trk4] Tarski A. Les fondements de la géométrie des corps, *Księga Pamiatkowa Pierwszego Polskiego Zjazdu Matematycznego,* Supplément aux *Rocznik Polskiego Towarsystwa Matematycznego (= Annales de la Société Polonaise de Mathématique),* Krakow, 1929, pp. 29–33. Rev. engl. tr. [Trk34], 24–29. Trad. fr. de la préc. [Trk36], 27–34. Repr. [Trk37] **I**, 227–231.

[Trk5] Tarski A. Über einige fundamentale Begriffe der Metamathematik, *Sprawozdania z Posiedzeń Towarzystwa Naukowego Warzsawskiego, Wydział III Nauk-Matematyczno-fizycznych (= Comptes-Rendus des Séances de la Société des Sciences et des Lettres de Varsovie, Classe III)* **23** (1930), 22–29. Abg. [Trk37] **I**, 313–320. Rés. sous: Remarques sur les notions fondamentales de la méthodologie des mathématiques, *Rocznik Polskiego Towarsystwa Matematycznego (= Annales de la Société Polonaise de Mathématique)* **7** (1928), 270–272. Rev. engl. tr. [Trk34], 30–37. Trad. fr. revue [Trk36], 35–44. Repr. [Trk37] **IV**, 552–554.

[Trk6] Tarski A. Fundamentale Begriffe der Methologie der deduktiven Wissenschaften I, *Monatshefte für Mathematik und Physik* **37** (1930), 361–404. Rev. engl. tr. [Trk34], 60–109. Trad. fr. revue [Trk36], 67–116. Abg. [Trk37] **I**, 347–390.

[Trk7] Tarski A. Über definierbare Zahlenmengen, *Rocznik Polskiego Towarsystwa Matematycznego (= Annales de la Société Polonaise de Mathématique)* **9** (1930), 206–207. Abg. [Trk37] **IV**, 560–561.

[Trk8] Tarski A. Sur les ensembles définissables de nombres réels I, *Fundamenta Mathematicae* **17** (1931), 210–239. Rés. sous [Trk7]. Rev. engl. tr. [Trk34], 110–142. Version fr. revue d'après la préc. [Trk36], 117–146. Repr. [Trk37] **I**, 519–548.

[Trk9] Tarski A. Pojęcie prawdy językach nauk dedukcyjnych, *Prace Towarzystwa Naukowego Warszawskiego, Wydział III Nauk Matematycznofizycznych (= Travaux de la Société des Sciences et des Lettres de Varsovie, Classe III Sciences Mathématiques et Physiques) n°* 34, Warszawa, 1933, vii+116 pp. Trad. fr. revue [Trk36], 157–269.

Rés. sous: O pojęciu prawdy w odniesieniu do sformalizowanych nauk dedukcyjnych, *Ruch Filozoficzny* **12** (1930–31), 210–211. Repr. [Trk37] **IV**, 555–559.

Deutsche Übersetzung bei L. Blaustein, mit einem Nachwort, unter: Der Wahrheitsbegriff in den formalisierten Sprachen, *Studia Philosophica* **1**

(1935), 261–405. Sonderabdruck, Lwów, 1935, 145 pp. Abg. [Trk37] **II**, 53–198. Rev. engl. tr. [Trk34], 152–278.

Deutsche Aufzug, unter: Der Wahrheitsbegriff in den deduktiven Disziplinen, *Akademie der Wissenschaften in Wien, Mathematisch-naturwissenschaftliche Klasse, Akademische Anzeiger* **69** (1932), 23–25. Abg. [Trk37] **I**, 615–617.

Cf. [Gvt] pour d'autres éd. postérieures à 1950.

[Trk10] TARSKI A. Einige Betrachtungen über die Begriffe der ω-Widerspruchsfreiheit und der ω-Vollständigkeit, *Monatshefte für Mathematik und Physik* **40** (1933), 97–112. Rev. engl. tr. [Trk34], 279–295. Trad. fr. revue [Trk36], 5–22. Abg. [Trk37] **I**, 621–636.

[Trk11] TARSKI A. Z badań metodologicznych nad definjowalnością terminów, *Przegląd Filozoficzny (= Revue Philosophique)* **37** (1934), 438–460. Deutsche Übersetzung unter: Einige methodologische Untersuchungen über die Definierbarkeit der Begriffe, *Erkenntnis* **5** (1935),174–175. Rev. engl. tr. [Trk34], 296–319. Trad. fr. [Trk36], 23–46. Abg. [Trk37] **I**, 639–659.

[Trk12] TARSKI A. Grundzüge des Systemenkalküls. Erster Teil, *Fundamenta Mathematicae* **25** (1935), 503–526; Zweiter Teil, *Fundamenta Mathematicae* **26** (1936), 283–301. Rev. engl. tr. [Trk34], 342–383. Trad. fr. [Trk34], 69–108. Abg. [Trk37] **II**: Erster Teil, 27–50; Zweiter Teil, 225–243.

[Trk13] TARSKI A. Zur Grundlegung der Booleschen Algebra, *Fundamenta Mathematicae* **24** (1935), 177–198. Rev. engl. tr. [Trk34], 320–341. Trad. fr. [Trk36], 47–68. Abg. [Trk37] **II**, 3–24.

[Trk14] TARSKI A. *O logice Matematycznej i metodzie dedukcyjnej.* Bibljotecka Matematyczna, **3–5**, Ksiaznica-Atlas, Lwow-Warszawa, 1936. 167 pp.

Deutsche Übersetzung unter: *Einführung in die mathematische Logik und die Methodologie der Mathematik.* Julius Springer, Wien, 1937. x+166 S.

Enlarged and rev. engl. tr. by O. Helmer under: *Introduction to logic and to the methodology of deductive sciences.* Oxford University Press, Oxford and New-York, 1941. xviii+239 pp. 2^d, revised edition, 1946.

Cf. [Gvt] pour les nombreuses éditions ultérieures.

[Trk15] TARSKI A. O ugruntowaniu naukowej semantyki, *Przegląd Filozoficzny (= Revue Philosophique)* **39** (1936), 50–57. Deutsche Übersetzung unter: Grundlegung der wissenschaftlichen Semantik, *Actes du Congrès International de Philosophie Scientifique* **3**. Actualités Scientifiques et Industrielles, **390**, Hermann, Paris, 1936, 1–8. Rev. engl. tr. [Trk34], 401–408. Trad. fr. revue [Trk36], 131–139. Abg. [Trk37] **II**, 261–268.

Cf. [Gvt] pour d'autres éd. postérieures à 1950.

[Trk16] Tarski A. O pojęciu wynikania logicznego *Przegląd Filozoficzny (= Revue Philosophique)* **39** (1936), 58–68. Deutsche Übersetzung unter: Über den Begriff der logischen Folgerung, *Actes du Congrès International de Philosophie Scientifique* **7**. Actualités Scientifiques et Industrielles, **394**, Hermann, Paris, 1936, 1–11. Rev. engl. tr. [Trk34], 409–420. Trad. fr. revue [Trk36], 141–152. Abg. [Trk37] **II**, 271–281.

Cf. [Gvt] pour d'autres éd. postérieures à 1950.

[Trk17] Tarski A. Sur la méthode déductive, *Travaux du IXe Congrès International de Philosophie* **3**. Actualités Scientifiques et Industrielles, **535**, Hermann, Paris, 1937, 95–103. Repr. [Trk37] **II**, 325–333.

[Trk18] Tarski A. Ideale in der Mengenkörpern, *Rocznik Polskiego Towarsystwa Matematycznego (= Annales de la Société Polonaisede Mathématique)* **15** (1937), 186–189. Abg. [Trk37] **II**, 285-288.

[Trk19] Tarski A. Über unerreichbare Kardinalzahlen, *Fundamenta Mathematicae* **30** (1938), 68–89. Abg. [Trk37] **II**, 359–380.

[Trk20] Tarski A. Der Aussagenkalkül und die Topologie, *Fundamenta Mathematicae* **31** (1938), 103–134. Rev. engl. tr. [Trk34], 421–454. Trad. fr. revue [Trk36], 153–190. Abg. [Trk37] **II**, 475–506.

[Trk21] Tarski A. On undecidable statements in enlarged systems of logic and the concept of truth, *Journal of Symbolic Logic* **4** (1939), 105–112. Trad. fr. [Trk36], 191–202. Repr. [Trk37] **II**, 561-568.

[Trk22] Tarski A. On the calculus of relations, *Journal of Symbolic Logic* **6** (1941), 73–89. Trad. fr. [Trk36], 243–264. Repr. [Trk37] **II**, 571–587.

[Trk23] Tarski A. Axiomatic and algebraic aspects of two theorems on sums of cardinals, *Fundamenta Mathematicae* **35** (1948), 79–104. Repr. [Trk37] **III**, 173–198.

[Trk24] Tarski A. *A decision method for elementary algebra and geometry.* (prep. for publ. by J.C.C. Mc Kinsey), U.S. Air Force Project RAND, R-109, the RAND Corp. Santa Monica, California, 1948, iv+60 pp. 2^d rev. ed. (prep. for publ. with the assistance of J.C.C. McKinsey), University of California Press, Berkeley and Los Angeles, California, 1951, iii+63 pp. Repr. [Trk37] **III**, 299–368.

[Trk25] Tarski A. Arithmetical classes and types of mathematical systems, *Bulletin of the American Mathematical Society* **55** (1949), 63, 1192. Repr. [Trk37] **IV**, 579.

[Trk26] Tarski A. Metamathematical aspects of arithmetical classes and types, *Bulletin of the American Mathematical Society* **55** (1949), 63–64. Repr. [Trk37] **IV**, 580.

[Trk27] Tarski A. Arithmetical classes and types of Boolean algebras, *Bulletin of the American Mathematical Society* **55** (1949), 64. Repr. [Trk37] **IV**, 581.

[Trk28] TARSKI A. On essential undecidability, *Journal of Symbolic Logic* **14** (1949), 75–76. Repr. [Trk37] **IV**, 584–585.

[Trk29] TARSKI A. Undecidability of group theory, *Journal of Symbolic Logic* **14** (1949), 76–77. Repr. [Trk37] **IV**, 588–589.

[Trk30] TARSKI A. Undecidability of the theories of lattices and projective geometries, *Journal of Symbolic Logic* **14** (1949), 77–78. Repr. [Trk37] **IV**, 590–591.

[Trk31] TARSKI A. A fixpoint theorem for lattices and its applications, *Bulletin of the American Mathematical Society* **55** (1949), 1051–1052. Repr. [Trk37] **IV**, 592.

[Trk32] TARSKI A. Some notions and methods on the borderline of algebra and metamathematics, *Proceedings of the International Congress of Mathematicians, Cambridge, Mass., U.S.A., Aug. 30-Sept 6, 1950,* **I**, American Mathematical Society, Providence, Rhode Island, 1952, 705–720. Repr. [Trk37] **III**, 461–468.

[Trk33] TARSKI A. Contributions to the theory of models. *Koninklijke Akademie van Wetenschappen te Amsterdam, Proceedings,* series A, **57** (1954); **58** (1955) = *Indagationes Mathematicae* **16** (1954), 572–588; **17** (1955), 56–64. Repr. [Trk37] **III**, 517–525; 529–535; 539–547.

[Trk34] TARSKI A. *Logic, semantics, metamathematics. Papers from 1923 to 1938.* Clarendon Press, Oxford, 1956. xii+471 pp. 2^d rev. edi. with editor's introduction and an analytic index (J. Corcoran, ed.). Hackett Publishing Company, Indianapolis, 1983, xxx+506 pp.

[Trk35] TARSKI A. *The completeness of elementary algebra and geometry.* Institut Blaise Pascal, Paris, 1967. Repr. [Trk37] **IV**, 291–346.

[Trk36] TARSKI A. *Logique, sémantique, métamathématique 1923–1944.* Philosophies pour l'Age de Science. (Gilles-Gaston Granger, éd.) Armand Colin, Paris. **I**, 1972, viii+276 pp. **II** 1974, 331 pp.

[Trk37] TARSKI A. *Collected Papers.* Ed. by S. R. Givant and R. N. McKenzie. 4 t. Birkhäuser, Basel-Boston-Stuttgart 1986.

[Thu] THUE A. Probleme über Veränderungen von Zeichenreihen nach gegebenen Regeln, *Skrifter utgit av Videnskapsselskapet i Kristiania, I. Matematisk-naturvidenskabelig Klasse 1914,* n^r 10. 34 pp.

[Trb] TRAHTÉNBROT B.A. Névozmožnost' algorifma dlá problémy razréšimosti na konéčny klassah, *Doklady Akademii Nauk SSSR* **70** (1950), 569–572.

[Trg1] TURING A.M. On computable numbers, with an application to the Entscheidungsproblem, *Proceedings of the London Mathematical Society* Series 2, **42** (1936–37), 230–265; A correction, *Ibid.* **43** (1937), 544–546. Repr. [Dvs2] 115–153.

[Trg2] TURING A.M. Computability and λ-definability, *Journal of Symbolic Logic* **2** (1937), 153–163.

[Trg3] TURING A.M. Systems of logic based on ordinals, *Proceedings of the London Mathematical Society,* Series 2, **45** (1939), 161–228. Repr. [Dvs2] 154–222.

[Trg4] TURING A.M. The word problem in semi-groups with cancellation, *Annals of Mathematics* ser. 2, **52** (1950), 491–505.

[Ulm] ULAM S. Zur Masstheorie in der allgemeinen Mengenlehre, *Fundamenta Mathematicae* **16** (1930), 140–150.

[Vgn] VAUGHAN H.E. Review of [ARb2] and [Trk32], *Journal of Symbolic Logic* **18** (1953), 182.

[Vgt1] VAUGHT R.L. Model Theory before 1945, *Proceedings of the Tarski Symposium.* (ed. L. Henkin et alii), American Mathematical Society, Providence, Rhode Island, 1974, 153–172.

[Vgt2] VAUGHT R.L. Alfred Tarski's work in model theory, *Journal of Symbolic Logic* **51** (1986), 869–882.

[Vbl1] VEBLEN O. A system of axioms for geometry, *Transactions of the American Mathematical Society* **5** (1904), 343–384.

[Vbl2] VEBLEN O. The Heine-Borel theorem, *Bulletin of the American Mathematical Society* **10** (1904), 436–439.

[Vbl3] VEBLEN O. Definitions in terms of order alone in the linear continuum and in well-ordred sets, *Transactions of the American Mathematical Society* **6** (1905), 165–171.

[Vbl4] VEBLEN O. Theory of plane curves in nonmetrical analysis situs, *Transactions of the American Mathematical Society* **6** (1905), 83–98.

[Vbl5] VEBLEN O. The square root and relations of order *Transactions of the American Mathematical Society* **7** (1906), 197–199.

[Vbl6] VEBLEN O. Continuous increasing functions of finite and transfinite ordinals, *Transactions of the American Mathematical Society* **93** (1908), 280–292.

[Vbl7] VEBLEN O. *Analysis Situs.* American Mathematical Society Colloquium Publications, **5**, part 2 (1922).

[Vbl8] VEBLEN O. Remarks on the foundations of geometry, *Bulletin of the American Mathematical Society* **31** (1925), 121–141.

[VWh1] VEBLEN O. and WHITEHEAD J.H.C. A set of axioms for differential geometry, *Proceedings of the National Academy of Sciences of the U.S.A.* **17** (1931), 551–561.

[VWh2] VEBLEN O. and WHITEHEAD J.H.C. *Foundations of Differential Geometry.* Cambridge Tracts in Mathematics and Mathematical Physics, n^r 29. Cambridge University Press, 1932.

[VbY1] VEBLEN O. and YOUNG J.W. A set of assumptions for projective geometry, *American Journal of Mathematics* **30** (1908), 347–380.

[VbY2] VEBLEN O. and YOUNG J.W. *Projective Geometry.* Ginn, Boston. **1**, 1910; **2**, 1918.

[Vit] VITALI G. *Sul problema della misura dei gruppi di punti di una retta.* Tip. Gamberini e Parmegiani, Bologna, 1905.

[Wbg] WAJSBERG M. Untersuchungen über den Aussagenkalkül von A. Heyting, *Wiadomości matematyczne* **46** (1938), 45–101.

[Wng1] WANG H. A formal system of logic, *Journal of Symbolic Logic* **15** (1950), 25–31.

[Wng2] WANG H. Remarks on the comparison of axiom systems, *Proceedings of the National Academy of Sciences of the U.S.A.* **36** (1950), 448–453.

[Wng3] WANG H. Gödel's and Some Other Examples of Problem Transmutation, [Drk], 101–109.

[Wyl1] WEYL H. Über die Definitionen der mathematischen Grundbegriffe, *Mathematisch-naturwissenschaftliche Blätter* **7** (1910), 93–95 und 109–113. Abg. [Wyl4] **I**, 298–304.

[Wyl2] WEYL H. *Das Kontinuum. Kritische Untersuchungen über die Grundlagen der Analysis.* Veit, Leipzig, 1918. VI+83 S. Abg. [Wyl+].

[Wyl3] WEYL H. The ghost of modality, *Philosophical essays in memory of Edmund Husserl.* Harvard University Press, Cambridge, Mass., 1940, 278–303. Repr. [Wyl4], **III**, 684–709.

[Wyl4] WEYL H. *Gesammelte Abhandlungen,* ed. K. Chandrasekharan, Springer, Berlin, 1968, 4 Bd.

[Wyl+] WEYL H. et alii, *Das Kontinuum und andere Monographien.* De Gruyter, Berlin und Leipzig, 1932. Chelsea, New-York, sans date.

[Wnr] WIENER N. A simplification of the logic of relations, *Proceedings of the Cambridge Philosophical Society* **17** (1912–1914), 387–390. Repr. [vHj], 224–227.

[Whd] WHITEHEAD A.N. On cardinal numbers, *American Journal of Mathematics* **24** (1902), 367–394.

[WhR] WHITEHEAD A.N. and RUSSELL B. *Principia Mathematica.* Cambridge University Press, Cambridge, England. **I**, 1910; **II**, 1912; **III**, 1913. 2^d ed. **I**, 1925; **II**, 1927; **III**, 1927.

[Wtg] WITTGENSTEIN L. Logisch-philosophische Abhandlungen, *Annalen der Naturphilosophie* **14** (1921), 185–262. Repr. with engl. tr. by C.K. Ogden and F.P. Ramsey under: *Tractatus Logico-philophicus.* With an introduction by B. Russell. Routledge and Kegan Paul, London, 1922, 189 pp. Engl. tr. by D. Pears and B.McGuinness, Routledge and Kegan Paul, London, 1971. Trad. fr. par P. Klossowski, Gallimard, Paris, 1961; collection «Tel», 1986.

[Zml1] ZERMELO E. Beweis, daß jede Menge wohlgeordnet werden kann, *Mathematische Annalen* **59** (1904), 514–516. Engl. tr. [vHj], 139–141.

E. Noether, M.-P. Dubreil, P. Dubreil

[Zml2] Zermelo E. Neuer Beweis für die Möglichkeit einer Wohlordnung *Mathematische Annalen* **65** (1908), 107–128. Abg. [Hzm2], 157–178. Engl. tr. [vHj], 183–198.

[Zml3] Zermelo E. Untersuchungen über die Grundlagen der Mengenlehre I, *Mathematische Annalen* **65** (1908), 261–281. Abg. [Hzm2], 179–199. Eng. tr. [vHj], 199–215.

[Zml4] Zermelo E. Über den Begriff der Definitheit in der Arithmetik, *Fundamenta mathematicae* **14** (1929), 339–344.

[Zml5] Zermelo E. Über Grenzzahlen und Mengenbereiche, *Fundamenta mathematicae* **16** (1930), pp. 29–47.

[Zml6] Zermelo E. Grundlagen einer allgemeinen Theorie der mathematischen Satzsysteme, *Fundamenta mathematicae* **25** (1935), 136–146.

Marcel Guillaume
Département de mathématiques
Université Blaise-Pascal
F-63177 Aubière Cedex

P.S. Novikov, 1967

W. Sierpinski

Function Theory 1900–1950

Walter K. Hayman
University of York

1 Introduction

The theory of functions was born in the 19th century. Gauss had proved that algebraic equations of degree N have N complex roots, but the key names are probably Cauchy with his integral Theorem and the Theorem of residues and Riemann with Riemann surfaces and the conformal mapping Theorem in 1851, although a satisfactory proof had to wait until 1900. The Cauchy-Riemann equations are also fundamental.

The subject sprang to life in the 20th century and there is more material in our period than can be adequately covered here. I would like to report on 3 areas which reflect my personal interests. They are entire functions, meromorphic functions and functions in the unit disk.

2 Entire functions

Weierstrass showed that an entire function, i.e. one regular in the whole plane and so an everywhere convergent power series,

$$f(z) = \sum_{0}^{\infty} a_n z^n,$$

can be expressed as a product in terms of its zeros by means of the Weierstrass factors

$$E(z, q) = (1 - z) \exp\left(z + \frac{z^2}{2} + \ldots + \frac{z^q}{q}\right).$$

Hadamard (1893) used the maximum modulus

$$M(r, f) = \sup_{|z|=r} |f(z)|$$

to define the order

$$\lambda = \limsup_{r \to \infty} \frac{\log \log M(r, f)}{\log r}.$$

If λ is finite and $q = [\lambda]$, then Hadamard showed that

$$f(z) = e^{P(z)} z^a \prod_{\nu=1}^{\infty} E(z/z_\nu, q), \tag{1}$$

where $P(z)$ is a polynomial of degree at most q. In particular, if λ is not an integer, f has infinitely many zeros. If $0 \leq \lambda < 1$, then $q = 0$, and

$$f(z) = Az^a \prod_{\nu=1}^{\infty} (1 - z/z_\nu).$$

Applying these ideas to Riemann's function $\zeta(s)$ Hadamard and de la Vallée Poussin were later able to prove the prime number Theorem.

The functions

$$\frac{\sin\sqrt{z}}{\sqrt{z}} = 1 - \frac{z}{3!} + \frac{z^2}{5!} + \ldots, \cos\sqrt{z} = 1 - \frac{z}{2!} + \frac{z^2}{4!} + \ldots,$$

have order 1/2 and so we obtain the well known product expansions

$$\sin z = \pi z \prod_{\nu=1}^{\infty} (1 - \frac{z^2}{\pi^2 \nu^2}) \text{ and } \cos z = \prod_{\nu=1}^{\infty} (1 - \frac{4z^2}{(2\nu - 1)^2 \pi^2}).$$

In order to prove the uniqueness, Hadamard considered $f(z)/\Pi(z)$, where $\Pi(z)$ is the right hand side of (1) and showed that the order of this is less than $q + 1$. To do this he had to estimate the minimum modulus

$$\mu(r, f) = \inf_{|z|=r} |f(z)|.$$

He showed that for most values of r

$$\mu(r, f) > \exp(-r^{\lambda+\varepsilon}).$$

Thus $f(z)/\Pi(z)$ is entire without zeros and of order less than $q + 1$, and from this it follows without difficulty that $f/\Pi = e^P$, where $P(z)$ is a polynomial of degree at most q.

In the early years of the century there began an intensive study of $\mu(r, f)$ compared with $M(r, f)$. A key role was played by the Swedish Mathematician Anders Wiman. He noted (1903) that, if $0 < \lambda < 1$, then

$$f(z) = \prod_{\nu=1}^{\infty} (1 + \frac{z}{\nu^{1/\lambda}}) \tag{2}$$

has order λ and, except near the negative axis, we have for $z = re^{i\theta}$, $|\theta| < \pi$,

$$\log|f(z)| = \frac{\pi}{\sin(\pi\lambda)} r^{\lambda} \cos(\lambda\theta) + O\{\log r + \log(\frac{1}{\pi - \theta})\},$$

so that

$$\frac{\log \mu(r)}{\log M(r)} \to \cos(\pi\lambda) \text{ as } r \to \infty \text{ outside a small set.}$$

This led Wiman to conjecture and to prove (1905) that, if $\lambda < 1/2$, $\mu(r)$ is unbounded as $r \to \infty$. Littlewood (1908) conjectured that

$$\limsup_{r\to\infty} \frac{\log \mu(r)}{\log M(r)} \geq \cos(\pi\lambda), 0 < \lambda < 1,$$

and proved this result with 2λ instead of λ. He also showed that, if λ is finite,

$$\limsup_{r\to\infty} \frac{\log \mu(r)}{\log M(r)} \geq -C(\lambda),$$

where $C(\lambda)$ is a constant. Valiron (1913) and Wiman (1915) independently proved Littlewood's conjecture, that $-C(\lambda) = \cos(\pi\lambda)$, $0 \leq \lambda < 1$. The value of $C(\lambda)$ is still unknown for $\lambda > 1$. Wiman (1918) conjectured $C(\lambda) = 1$ for $\lambda > 1$, and proved this if $f(z) \neq 0$. Beurling (1949) proved it when $f(z)$ attains its minimum on a ray, as is the case for the extremal functions (2) when $\lambda < 1$. Nevertheless Wiman's conjecture is false and the order of $C(\lambda)$ is $\log\lambda$ as $\lambda \to \infty$. For infinite order we have

$$\mu(r) > M(r)^{-A \log\log\log M(r)} \text{ for some arbitrarily large } r,$$

if $A = 2.2$, but not if $A = .09$ (Hayman 1952). Recently Fryntov (1993) has shown that $C(\lambda) > 1$ whenever $\lambda > 1$.

Besicovitch (1927) proved the first density theorem, namely that, if $0 < \lambda < \lambda' < 1$, the set where

$$\log \mu(r) > \cos(\pi\lambda') \log M(r)$$

has upper density at least $1 - \lambda/\lambda'$. The area has continued to flourish since 1950 with particularly notable contributions by Kjellberg, starting with his thesis (1948).

2.1 Asymptotic values

If Γ is a path going from a finite point to ∞, and

$$f(z) \to a, \text{ as } z \to \infty \text{ along } \Gamma$$

we say that $f(z)$ has the asymptotic value a along the path Γ. If $\lambda < 1/2$, $\mu(r)$ is unbounded by Wiman's Theorem and so $f(z)$ can have no finite asymptotic value in this case. On the other hand, if k is a natural number $k \geq 2$, the function

$$f(z) = \int_0^z \frac{\sin t^{k/2}}{t^{k/2}} dt$$

has order $k/2$ and approaches k distinct asymptotic values

$$w_\nu = w_0 e^{2\pi i\nu/k}, 0 < \nu \leq k$$

along the rays

$$\Gamma_\nu : z = re^{2\pi i\nu/k}.$$

If $k = 1$, the function $(\sin z^{1/2})/z^{1/2}$ has order 1/2 and also has the asymptotic value 0 along the positive axis.

This led Denjoy (1907) to conjecture that an entire function of order λ can have at most 2λ distinct finite asymptotic values. He proved his conjecture when the paths are straight lines. Carleman (1921) proved this with 5λ instead of 2λ. Ahlfors (1930) proved Denjoy's conjecture completely and was awarded one of the first 2 Fields medals for this achievement. He mapped the domains D_ν between Γ_ν and $\Gamma_{\nu+1}$ onto a half plane. His estimates for the mapping function were crucial, though similar techniques had been developed independently by Grötzsch from 1928. The technique led in the hands of Ahlfors and Beurling (1950) to the notion of extremal length. This seminal idea plays a role in quasiconformal mapping and other areas. A key idea used in Ahlfors' work was the Phragmén-Lindelöf principle (1908), which asserts that if $f(z)$ is analytic in the right half plane P, and bounded on the imaginary axis, then either $f(z)$ is bounded in P or grows at least like e^{az}. If $f(z) \to 0$ as $z \to \infty$ on the positive or negative imaginary axis, then in the former case $f(z) \to 0$ as $z \to \infty$ in $| \to z| < \pi/2 - \varepsilon$. Thus, if $f(z)$ tends to different limits as $z \to \infty$ on the positive and negative imaginary axes, $f(z)$ cannot be bounded in P and so must grow at least like e^{az}.

The Phragmén-Lindelöf principle has many applications. It was for instance used very effectively by Littlewood (1914) on the Riemann zeta function to show that $\pi(x) - Li(x)$ changes sign infinitely often.

Ahlfors' Theorem led to a great deal of further research, particularly in the case when $f(z)$ attains its minimal growth with k asymptotic values, namely

$$\liminf_{r\to\infty} \frac{\log M(r)}{r^{k/2}} < \infty.$$

For instance Heins (1948b) proved that in this case

$$\log\log M(r) \sim \frac{k}{2}\log r, \quad \text{as } r \to \infty.$$

This was improved by Kennedy (1955) to

$$\log\log M(r) = \frac{k}{2}\log r + o(\log r)^{1/2};$$

Kennedy (1956) also showed that in this form the result is sharp. If $k = 1$, and more generally if $\mu(r)$ is bounded above Heins (1948a) proved more strongly that

$$\log\log M(r) = \frac{1}{2}\log r + A + o(1).$$

These results were perhaps the earliest examples of regularity theorems. If under certain hypotheses functions have a limited growth, then the extremal functions have regular growth. Such regularity theorems have been proved in a wide variety of circumstances by Heins, Kennedy, Kjellberg and others.

While finite asymptotic values are excluded for functions of order less than 1/2 by Wiman's Theorem, Iversen (1915–1916) proved that every nonconstant entire function has infinity as an asymptotic value. In particular the set $E_K : |f(z)| > K$ has unbounded components for every K. If f has k distinct finite asymptotic values, where $k \geq 2$, then E_K has at least k components for large K. From this fact alone Heins deduced (1959) Ahlfors's conclusion, but this and other generalisations, for instance to subharmonic functions in R^m, come outside our period.

Among many other important conclusions about entire functions, I should like to single out the Wiman-Valiron method (Wiman 1918 & Valiron 1928). Wiman showed that, for most values of r and in a neighbourhood of a point $z_0 = re^{i\theta}$ such that $f(z_0) = M(r, f)$, the function $f(z) = \sum_0^\infty a_n z^n$ is given to a good approximation by

$$\sum_{N-p}^{N+p} a_n z^n$$

where $\mu(r) = |a_N| r^N$ here denotes the largest term (not the minimum modulus) of the series and $p = O\ (N^{1/2+\varepsilon})$. He deduced that if $f = u + iv$, and

$$A(r, u) = \inf_{|z|=r} u(z), \quad B(r, u) = \sup_{|z|=r} u(z),$$

then we have for most r

$$-A(r) \sim B(r) \sim M(r, f).$$

If $f(z) \neq 0$ these conclusions can be applied to $\log f(z)$ and lead to Wiman's Theorem about the minimum modulus of nonzero entire functions. Wiman also used the method to prove that, for most r, the maximum term $\mu(r)$ satisfies

$$M(r) = O\{\mu(r)\big(\log \mu(r)\big)^{1/2+\varepsilon}\}.$$

3 Meromorphic functions and Nevanlinna Theory

It is said that in the latter half of the 19th century all Ph.D. theses in the Ecole normale had to begin with the words

«*Il suit du théorème célèbre de M. Picard...*»

The result (1879) asserts that a transcendental entire function assumes all values with at most one exception infinitely often. The function e^z, which is never zero, shows that the exceptional value can actually occur. The question naturally arises of giving this result a quantitative form and to extend it to meromorphic functions. To some extent this process had been started by Hadamard (1893), Borel (1897) and Blumenthal (1910). But Rolf Nevanlinna achieved a precision and beauty in his (1925) paper which led Hermann Weyl to call it (1943) «*One of the few great mathematical events of our century*».

Nevanlinna's start was the formula of Jensen (1899), which actually goes back to Jacobi (1827). Suppose that

$$f(z) = c_p z^p + c_{p+1} z^{p+1} + \dots$$

is meromorphic in $|z| \leq r$ with zeros a_μ and poles b_ν in $0 < |z| \leq r$. Then

$$\log |c_p| = \frac{1}{2\pi} \int_0^{2\pi} \log |f(re^{i\theta})| d\theta + \Sigma \log \frac{r}{|b_\nu|} - \Sigma \log \frac{r}{|a_\mu|} - p \log r.$$

Let $n(t,a)$ denote the number of roots of $f(z) = a$ in $|z| \leq t$. Following Valiron (1913), Nevanlinna wrote

$$p^+ \log r + \Sigma \log \frac{r}{|a_\mu|} = \int_0^r \frac{n(t,0) - n(0,0)}{t} dt + n(0,0) \log r = N(r,0)$$

$$p^- \log r + \Sigma \log \frac{r}{|b_\nu|} = \int_0^r \frac{n(t,\infty) - n(0,\infty)}{t} dt + n(0,\infty) \log r = N(r,\infty).$$

Here

$$p^+ = \max(p,0), p^- = \max(-p,0).$$

We also write $\log^+ x = \max(\log x, 0)$, so that

$$\log x = \log^+ x - \log^+ \frac{1}{x}, \text{ if } x \geq 0.$$

Hence

$$\frac{1}{2\pi} \int_0^{2\pi} \log |f(re^{i\theta})| d\theta = \frac{1}{2\pi} \int_0^{2\pi} \log^+ |f(re^{i\theta})| d\theta$$
$$- \frac{1}{2\pi} \int_0^{2\pi} \log^+ |\frac{1}{f(re^{i\theta})}| d\theta.$$

This was the apparently simple, but vital new step. Writing

$$m(r,\infty) = \frac{1}{2\pi}\int_0^{2\pi} \log^+ |f(re^{i\theta})| d\theta$$

and, for every finite complex a

$$m(r,a) = \frac{1}{2\pi}\int_0^{2\pi} \log^+ \left|\frac{1}{f(re^{i\theta}) - a}\right| d\theta,$$

Nevanlinna obtained Jensen's formula in the symmetrical form

$$m(r,\infty) + N(r,\infty) = m(r,0) + N(r,0) + \log |c_p|. \tag{3}$$

Writing $f - a$ instead of f in (3) increases the left hand side, and so the right hand side, by at most a bounded term.

We now define the characteristic function $T(r, f)$ by

$$T(r, f) = m(r,\infty) + N(r,\infty).$$

Then we obtain

Nevanlinna's First Fundamental Theorem. *For every complex a and $a = \infty$, we have as $r \to \infty$*

$$T(r, f) = m(r,a) + N(r,a) + O(1).$$

It turns out that $m(r,a)$ is in general small. Thus the expected size of $N(r,a)$ is its upper bound $T(r)$. The second fundamental theorem tells us how often and how much $N(r,a)$ can be below expectation. We define the deficiency

$$\delta(a) = \liminf_{r\to\infty} \frac{m(r,a)}{T(r)} = 1 - \limsup_{r\to\infty} \frac{N(r,a)}{T(r)}$$

and

$$\theta(a) = 1 - \limsup_{r\to\infty} \frac{\bar{N}(r,a)}{T(r)},$$

where in $\bar{N}$ multiple roots are counted only once. Then the second fundamental theorem yields the deficiency relation

$$\Sigma\theta(a) \leq 2. \tag{4}$$

If f leaves out a value a then $\theta(a) = 1$, so that this can happen for at most two distinct values of a. If f is entire, $\theta(\infty) = 1$, so there can be at most one finite omitted value. Thus we get Picard's Theorem in a much more precise form.

Suppose next that f is entire and that for some finite values a all the roots of $f = a$ are multiple. Then $\bar{N} \le \frac{1}{2}N \le \frac{1}{2}T + O(1)$, so that $\theta(a) \ge \frac{1}{2}$. Since also $\theta(\infty) = 1$, there can be at most two such values a. The function $\sin z$ shows that two such multiple values can actually occur, since the equations $\sin z = \pm 1$ have only double roots. A meromorphic function can have at most 4 multiple values by a similar argument and here the elliptic function $\wp(z)$ does indeed have 4 such values ∞, e_1, e_2, e_3, since

$$\wp'(z)^2 = 4(\wp - e_1)(\wp - e_2)(\wp - e_3).$$

As another application of his theory Nevanlinna showed that if the equations $f = a$, $g = a$ have the same roots for 5 distinct values of a, even without counting multiplicity, then $f \equiv g$. On the other hand the functions e^z and e^{-z} assume the four values $0, \infty, \pm 1$ at the same points.

Nevanlinna Theory also deals very successfully with derivatives. The following result is due to Bureau (1932) for entire f and to Hayman (1959) for meromorphic f.

If f is transcendental then either f assumes every finite value infinitely often or every derivative of f assumes every finite value except possibly zero infinitely often.

The functions $e^z - a$ show that zero does indeed play an exceptional role for derivatives.

Nevanlinna Theory allows very simple manipulations. Thus if f_1 to f_q are distinct functions we have

$$T(r, \Pi f_\nu) \le \Sigma T(r, f_\nu),$$
$$T(r, \Sigma f_\nu) \le \Sigma T(r, f_\nu) + \log q$$

and, by (3),

$$T(r, \frac{1}{f}) = T(r, f) + \text{ constant}.$$

The latter result eliminates the need to use the minimum modulus in Hadamard's Representation Theorem. Nevanlinna developed his theory in the two books (1929, 1936).

Ahlfors (1935) extended Nevanlinna Theory to coverings of the Riemann sphere by Riemann surfaces. Suppose that D_ν, $\nu = 1$ to p are simply connected domains on the Riemann sphere, whose closures are disjoint and suppose further that f is transcendental meromorphic in the plane and maps compact regions Δ_ν^j onto the D_ν in a q to 1 manner. Denote by q_ν the minimum of q. Then Ahlfors' islands Theorem states that

$$\sum_{\nu=1}^{p} \left(1 - \frac{1}{q_\nu}\right) \le 2.$$

Thus if $p \geq 5$ at least one of the domains Δ_ν^j for some ν is mapped (1,1) conformally onto D_ν. Ahlfors' Theorem generalises the deficiency relation and so Picard's Theorem. We may say that $f(z)$ has a simple island over at least one of 5 domains D_ν. On the other hand the elliptic function $\wp(z)$ has no simple island over any domain containing one of the 4 multiple values ∞, e_1, e_2 and e_3.

As a final variant of Nevanlinna Theory I should like to mention its extension by Cartan (1933) to meromorphic curves, i.e. parametric curves $(f_1, f_2, \ldots, f_n)$ in $\mathbb{R}^n$, where f_1 to f_n are entire functions not all of which vanish for any complex z. The technique makes it possible to tackle Waring's problem for analytic functions. Nevanlinna Theory also plays an important role in algebraic and differential geometry and in potential theory.

4 Functions in the unit disk

Riemann's mapping theorem (1851) states that, given a simply connected domain D other than the whole plane and a point w_0 of D, there exists a unique function $f(z)$ regular in $|z| < 1$ and mapping $|z| < 1$ (1,1) conformally onto D so that $f'(0) > 0$. However the first rigourous proof was only given by Osgood (1900). The proof that is currently used is due to Radó (1922/23), based on Montel's (1912, 1927) idea of normal families, in which the Vitali (1903) selection principle plays a key role. A family $\tilde{\mathcal{F}}$ of meromorphic functions in a domain D is said to be normal if from every sequence f_n in $\tilde{\mathcal{F}}$ we can select a subsequence f_{n_p} converging locally uniformly in D in the metric of the Riemann sphere.

The corresponding mapping theorem for simply connected Riemann surfaces was achieved independently by Koebe (1907) and Poincaré (1908). A landmark in this area was Weyl (1913). Cecioni (1908) first proved that a planar surface, i.e., one which is cut in 2 pieces by a simple closed curve can be mapped onto a region bounded by parallel line segments.

The success of regular (1,1) mappings directed attention increasingly to univalent functions. Let S be the class of functions

$$f(z) = z + a_2 z^2 + \ldots \tag{5}$$

regular in $\Delta : |z| < 1$ and univalent, so that for each w the equation $f(z) = w$ has at most one root in Δ. Bieberbach (1916) introduced the class S and proved that $|a_2| \leq 2$. Suppose that $f(z) \neq w$. Then

$$\phi(z) = \frac{w f(z)}{w - f(z)} = z + \left(a_2 + \frac{1}{w}\right) z + \ldots \in S.$$

Thus

$$|a_2 + \frac{1}{w}| \leq 2, \quad |\frac{1}{w}| \leq 4, \quad |w| \geq \frac{1}{4}.$$

Thus Bieberbach obtained in sharp form a theorem of Koebe (1907), that the image of Δ by $f(z)$ always covers the disk $|w| < 1/4$. Equality holds for the Koebe functions

$$f_\theta(z) = \frac{z}{(1 - ze^{i\theta})^2} = z + 2e^{i\theta}z^2 + \ldots + ne^{i(n-1)\theta}z^n + \ldots \qquad (6)$$

which map Δ onto the plane cut radially from $-\frac{1}{4}e^{-i\theta}$ to ∞.

Bieberbach conjectured that $|a_n| \leq n$ for all n. This was proved by Löwner (1923) for $n = 3$ and also for various subclasses of S by other authors. A full proof of Bieberbach's conjecture was only given by de Branges (1985). Löwner's theory plays an essential part in the Branges' proof.

Bieberbach's inequality $|a_2| \leq 2$ leads directly to sharp upper and lower bounds for $|f(z)|$, $|f'(z)|$ and $|f'(z)/f(z)|$ for a fixed point z in Δ. In all cases the bounds are attained only for the Koebe functions.

The study of analytic functions is closely related to topology. Jordan's theorem arose from Cauchy's theorem and really started topology. Later Carathéodory (1913) developed the theory of prime ends or generalised boundary elements of a simply connected domain D and showed that in a conformal map of Δ onto D the prime ends of D correspond to the boundary points of Δ. If D is a Jordan domain, i.e. the interior of a simple closed curve Γ, the map from to D extends to a continuous (1,1) map of $|z| \leq 1$ to $\bar{D} = \Gamma \cup D$. Thus in this case $\bar{D}$ is topologically equivalent to a closed disk.

Koebe (1910) also proved that a finitely connected domain may be mapped conformally onto a plane domain whose complement consists of disks or points. This is still a hard theorem (see e.g. Grunsky (1978, pp. 114, 194)). It is an open problem whether the result extends to arbitrary domains, though there is a rumour that it has recently been proved for domains whose complement has countably many components. There is a strong link between conformal and quasi-conformal mappings.

4.1 p-valent functions

At the beginning of the century it came to be realised that if

$$f(z) = \sum_{0}^{\infty} a_n z^n$$

is regular or meromorphic in Δ and something is known about the values assumed by $f(z)$ then we can obtain information about the growth of the maximum modulus $M(r, f)$ and the coefficients a_n of $f(z)$. In particular some growth properties of univalent functions generalise to p-valent functions. If p is a positive integer then $f(z)$ is called p-valent in Δ if the equation $f(z) = w$ has at most p roots in Δ for varying w. Cartwright (1935) proved that in this case

$$M(r, f) < A(p) \left(\sum_{\nu=0}^{p} |a_\nu| \right) (1 - r)^{-2p}, \quad 0 \leq r < 1.$$

From this Biernacki (1936) deduced the corresponding bounds

$$|a_n| < A(p) \left(\sum_{\nu=1}^{p} |a_\nu| \right) n^{2p-1}. \tag{7}$$

Here $A(p)$ denotes a constant depending only on p. These results were generalised by Spencer (1940) to mean p-valent functions, i.e. those for which the equation $f(z) = w$ has at most p roots on the average, when we average over disks $|w| < R$. Now p can be any positive number, though (7) fails when $p < 1/4$, but holds for $p > 1/4$. The case $p = 1/4$ is still open. The proofs depend on the conformal mapping technique of Ahlfors (1930). For univalent functions (7) is due to Littlewood (1925).

4.2 Covering Theorems

Schottky (1904) and Landau (1904) obtained local versions of Picard's Theorem. Suppose that $f(z)$ is regular and $f(z) \neq 0, 1$ in Δ. Then Landau proved

$$|a_1| \leq \Omega(a_0) \tag{8}$$

and a few months later Schottky proved

$$M(r, f) \leq \Omega(a_0, r), \quad 0 \leq r < 1 \tag{9}$$

where the constants Ω depend only on the quantities indicated.

The actual form of the bounds was gradually improved. Landau (1906) proved that

$$\Omega(a_0) \leq 2|a_0|\{\left|\log |a_0|\right| + A_0\}$$

in (8) but the best value of the constant

$$A_0 = \frac{\Gamma(1/4)^4}{4\pi^2} = 4.37\ldots$$

was only obtained independently by Hempel and Lai (1979). In (9) Hempel (1980) proved

$$\Omega(a_0, r) \leq \frac{1}{16}\left(e^\pi \max(1, |a_0|)\right)^{(1+r)/(1-r)}, \quad 0 \leq r < 1$$

where the constants $1/16, e^\pi$ cannot be replaced by smaller quantities. An important consequence of the Schottky-Landau Theorem is that a family of meromorphic functions omitting 3 values is normal.

Another important covering Theorem is that of Bloch (1926). If

$$f(z) = z + \ldots \tag{10}$$

is regular in Δ then $f(z)$ has a simple island over some disk of radius B, i.e. $f(z)$ maps some subdomain of Δ (1,1) conformally onto a disk of radius B. Ahlfors

(1938) proved that the constant B satisfies $B \geq \frac{1}{4}\sqrt{3}$, which was the best result known until Bonk (1990) proved $B > \frac{1}{4}\sqrt{3} + 10^{-14}$. In the opposite direction Ahlfors and Grunsky (1937) proved by an example that $B \leq .472\ldots$ and this example is conjectured to give the best constant.

Another example of a covering Theorem is the following due to Landau (1922). If $f(z)$ given by (10) is regular in Δ, then $f(z)$ assumes there all values on some circle $|w| = r$, where $r > L - \varepsilon$. If $f(z)$ is univalent then the image of Δ contains the disk $|w| < L$, and so Landau's Theorem contains that of Koebe. I showed (1951) that $L = 1/4$ with equality only for the Koebe functions (6).

In conclusion I would like to record my debt for a large number of written and spoken helpful comments. For section 1 I have made particular use of Kjellberg (1948) and in section 3 the survey by Gaier (1990) has been invaluable. Professor Gaier also supplied me with a variety of other historical references. I owe to Professor von Renteln the knowledge that Jacobi anticipated Jensen in Jensen's formula and to Professor Bureau information regarding Picard values of derivatives.

Bibliography

Books and papers listed here are referred to in the text by the author and year of publication. When two papers are quoted by the same author in the same year, they are distinguished by a, b. The figures in brackets at the end of a reference refer to the pages in the text where the work in question is referred to.

AHLFORS, L.V., Untersuchungen zur Theorie der konformen Abbildungen und der ganzen Funktionen. *Acta Soc. Sci. Fenn. Nova Ser.* **I** n^0 9 (1930). [4, 11]

AHLFORS, L.V., Zur Theorie der Überlagerungsflächen. *Acta Math.* **65** (1935), 157–194. [8]

AHLFORS, L.V., An extension of Schwarz's Lemma. *Trans. Amer. Math. Soc.* **43** (1938), 359–364. [12]

AHLFORS, L.V. and BEURLING, A., Conformal invariants and functiontheoretic Null-sets. *Acta Math.* **83** (1950), 101–129. [4]

AHLFORS, L.V. and GRUNSKY, H., Über die Blochsche Konstante. *Math. Zeit.* **42** (1937), 671–673. [12]

BESICOVITCH, A.S., On integral functions of order < 1. *Math. Ann.* **97** (1927), 677–695. [3]

BEURLING, A., Some theorems on boundedness of analytic functions. *Duke Math. J.* **16** (1949), 355–359. [3]

BIEBERBACH, L., Über die Koeffizienten derjenigen Potenzreihen, welche eine schlichte Abbildung des Einheitskreises vermitteln. *S. B. Preuss. Akad. Wiss.* **138** (1916), 940–955. [9]

BIERNACKI, M., Sur les fonctions multivalentes d'ordre *p*. *C. R. Acad. Sci. Paris* **203** (1936), 449–451. [11]

BLOCH, A., Les théorèmes de M. Valiron sur les fonctions entières et la théorie de l'uniformisation. *Ann. Fac. sci. Univ. Toulouse* (3) **17** (1926), 1–22. [12]

BLUMENTHAL, O., Principes de la théorie des fonctions entières d'ordre infini. (Gauthier-Villars, Paris 1910). [6]

BONK, M., On Bloch's constant. *Proc. Amer. Math. Soc.* **110** (1990), n^0 4, 889–894. [12]

BOREL, E., Sur les zéros des fonctions entières. *Acta Math.* **20** (1897), 357–396. [6]

BUREAU, F., Mémoire sur les fonctions uniformes à point singulier isolé. *Mém. Soc. royale sci. Liège* (3) **17** (1932), 1–52. [8]

CARATHÉODORY, C., Über die gegenseitige Beziehung der Ränder bei der Abbildung des Innern einer Jordanschen Kurve auf einen Kreis. *Math. Ann.* **73** (1913), 305–320. [10]

CARATHÉODORY, C., Über die Begrenzung einfachzusammenhängender Gebiete. *ibid.* 323–370. [10]

CARLEMAN, T., Sur les fonctions inverses des fonctions entières. *Arkiv. för mat. och fys.* **15**, n^0 10 (1921). [4]

CARTAN, H., Sur les zéros des combinaisons linéaires de *p* fonctions holomorphes données. *Mathematica (Cluj)* **7** (1933), 5–29. [9]

CARTWRIGHT, M.L., Some inequalities in the theory of functions. *Math. Ann.* **111** (1935), 98–118. [11]

CECIONI, F., Sulla rappresentazione conforme delle aree piane pluriconnesse su un piano in cui siano eseguiti dei tagli paralleli. *Rend. Circ. mat. Palermo* **25** (1908), 1–19. [9]

DE BRANGES, L., A proof of the Bieberbach conjecture. *Acta Math.* **154** (1985), 137–152. [10]

DENJOY, A., Sur les fonctions entières de genre fini. *C. R. Acad. Sci. Paris* **145** (1907), 106–108. [4]

FRYNTOV, A.E., An example of a subharmonic function that relates to a question of W. Hayman. *Proc. Amer. Math. Soc.* (1993). [3]

GAIER, D., Über die Entwicklung der Funktionentheorie in Deutschland. 1890–1950. Ein Jahrhundert Mathematik 1890–1990. Festschrift zum Jubiläum der DMV, 361–420. Deutsche Mathematiker Vereinigung Friedr. Vieweg & Sohn, Braunschweig/Wiesbaden, Editors Gerd Fischer, Friedrich Hirzebruch, Winfried Scharlau and Willi Törnig. [12]

GRÖTZSCH, H., Über einige Extremalprobleme der konformen Abbildung. Ber. sächs. Akad. Wiss. Leipzig, *Math. Phys. Kl.* **80** (1928), 367–376. [4]

GRUNSKY, H., Lectures on Theory of Functions in Multiply connected domains. Vandenhoek and Ruprecht, Göttingen 1978. [10]

HADAMARD, J., Etude sur les propriétés des fonctions entières et en particulier d'une fonction considérée par Riemann. *J. Math. pures appl.* **9** (1893), 171–215. [1, 6]

HAYMAN, W.K., Some applications of the transfinite diameter to the theory of functions. *J. Analyse Math.* **1** (1951), 135–154. [12]

HAYMAN, W.K., The minimum modulus of large integral functions. *Proc. London Math. Soc.* (3) **2** (1952), 469–512. [3]

HAYMAN, W.K., Picard values of meromorphic functions and their derivatives. *Ann. Math.* **70** (1959), 9–42. [8]

HEINS, M.H., Entire functions with bounded minimum modulus; subharmonic function analogues. *Ann. Math.* **49** (1948a), 200–213. [4]

HEINS, M.H., On the Denjoy-Carleman-Ahlfors theorem. *ibid.* (1948b), 533–537. [4]

HEINS, M.H., On a notion of convexity connected with a method of Carleman. *J. Analyse Math.* **7** (1959), 53–77. [5]

HEMPEL, J., The Poincaré metric on the twice punctured plane and the theorems of Landau and Schottky. *J. London Math. Soc.* (2) **20** (1979), 435–455. [11]

HEMPEL, J., Precise bounds in the theorems of Schottky and Picard. *J. London Math. Soc.* (2) **21** (1980), 279–286. [11]

IVERSEN, F., Sur quelques propriétés des fonctions monogènes au voisinage d'un point singulier. *åfv. af Finska Vet. Soc. Forh.* **58A**, n^0 25 (1915–1916). [5]

JACOBI, C.G.J., Über den Ausdruck der verschiedenen Wurzeln einer Gleichung durch bestimmte Integrale. *J. reine angew. Math.* **2** (1827), 1–8. [6]

JENSEN, J.L.W.V., Sur un nouvel et important théorème de la théorie des fonctions. *Acta Math.* **22** (1899), 359–364. [6]

KENNEDY, P.B., On a conjecture of Heins. *Proc. London Math. Soc.* (3) **5** (1955), 22–47. [4]

KENNEDY, P.B., On a class of integral functions bounded on certain curves. *Proc. London Math. Soc.* **(3)** 6 (1956), 518–547. [4]

KJELLBERG, B., On certain integral and subharmonic functions: a study in minimum modulus. Thesis, Uppsala (1948). [3, 12]

KOEBE, P., Über die Uniformisierung beliebiger analytischer Kurven. Göttinger Nachr. 1907, 197–210. [10]

KOEBE, P., Über die konforme Abbildung mehrfach zusammenhängender Bereiche. *J.-Ber. Deutsche Math.-Vereinig.* **19** (1910), 339–348. [10]

LAI, W.T., The exact value of Hayman's constant in Landau's theorem. *Scientia Sinica* **22** (1979), 129–134. [11]

LANDAU, E., Über eine Verallgemeinerung des Picardschen Satzes. *S.B. Preuss. Akad. Wiss.* (1904), 1118–1133. [11]

LANDAU, E., Über den Picardschen Satz. *Vierteljahrsschr. der Naturf. Ges. Zürich.* **51** (1906), 252–318. [11]

LANDAU, E., Zum Koebeschen Verzerrungssatz. *Palermo Rend.* **46** (1922), 347–348. [12]

LITTLEWOOD, J.E., A general theorem on integral functions of finite order. *Proc. London Math. Soc.* **6** (1908), 189–204. [2]

LITTLEWOOD, J.E., Sur la distribution des nombres premiers. *C. R. Acad. Sci. Paris* **158** (1914), 1869–1872. [4]

LITTLEWOOD, J.E., On inequalities in the theory of functions. *Proc. London Math. Soc.* (2) **23** (1925), 481–519. [11]

LÖWNER, K., Untersuchungen über schlichte konforme Abbildungen des Einheitskreises I. *Math. Ann.* **89** (1923), 103–121. [10]

MONTEL, P., Sur les familles de fonctions analytiques qui admettent des valeurs exceptionnelles dans un domaine. *Ann. de l'Ecole normale* (3) **23** (1912), 487–535. [9]

MONTEL, P., Leçons sur les familles normales de fonctions analytiques et leurs applications. Paris 1927. [9]

NEVANLINNA, R., Zur Theorie der meromorphen Funktionen. *Acta Math.* **46** (1925), 1–99. [6]

NEVANLINNA, R., Le théorème de Picard-Borel et la théorie des fonctions méromorphes. (Gauthier-Villars, Paris 1929). [8]

NEVANLINNA, R., Eindeutige analytische Funktionen. (Springer, Berlin 1936). [8]

OSGOOD, W.F., On the existence of the Green's function. *Trans. Amer. Math. Soc.* **1** (1900), 310–314. [9]

PHRAGMÉN, E. and LINDELÖF, E., Sur une extension d'un principe classique de l'analyse. *Acta Math.* **31** (1908), 381–406. [4]

PICARD, E., Sur une propriété des fonctions entières. *C. R. Acad. Sci. Paris* **88** (1879), 1024–1027. [6]

POINCARÉ, H., Sur l'uniformisation des fonctions analytiques. *Acta Math.* **31** (1908), 1–64. [9]

RADÓ, T., Sur la représentation conforme de domaines variables. *Acta Sci. Szeged* **1** (1922/23), 180–186. [9]

RIEMANN, B., Grundlagen für eine allgemeine Theorie der Funktionen einer veränderlichen komplexen Grösse. Inaugural Dissertation, Göttingen 1851, 45 pp, especially page 40. Werke Dover, New York 1953. [1, 9]

SCHOTTKY, F., Über den Picardschen Satz und die Borelschen Ungleichungen. *S. B. Preuss. Akad. Wiss.* (1904), 1244–1263. [11]

SPENCER, D.C., On finitely mean valent functions II. *Trans. Amer. Math. Soc.* **48** (1940), 418–435. [11]

VALIRON, G., Sur les fonctions entières d'ordre fini et d'ordre nul et en particulier les fonctions à correspondance régulière. *Ann. Fac. sci. Toulouse* (3) **5** (1913), 117–257. [3, 6]

VALIRON, G., Lectures on the general Theory of Integral functions. Ed. Privat Toulouse 1928, reprinted Chelsea, New York 1949. [5]

VITALI, G., Sopra le serie de funzioni analitiche. *Rend. Ist. Lombardo* **2.** Ser. 36 (1903), 772–774. [9]

WEYL, H., Die Idee der Riemannschen Fläche. Teubner, Leipzig and Berlin 1913. [9]

WEYL, H. and WEYL, F.J., Meromorphic functions and analytic curves. Princeton University Press 1943. [6]

WIMAN, A., Über die angenäherte Darstellung von ganzen Funktionen. *Ark. Mat. Astr. o Fysik* **1** (1903). [2]

WIMAN, A., Sur une extension d'un théorème de M. Hadamard. *Ark. Mat. Astr. o Fysik* **2** (1905), n^0 14. [2]

WIMAN, A., Über eine Eigenschaft der ganzen Funktionen von der Höhe Null. *Math. Ann.* **76** (1915), 197–211 [3]

WIMAN, A., Über den Zusammenhang zwischen dem Maximalbetrage einer analytischen Funktion und dem grössten Betrage bei gegebenem Argumente der Funktion. *Acta Math.* **41** (1918), 1–28. [3, 5]

WALTER K. HAYMAN
Departement of Mathematics
University of York
Heslington, York, YO15DD, UK

La préhistoire des conjectures de Weil

Christian Houzel
Université Paris VII

Les conjectures de Weil, formulées en 1949 et résolues par Deligne en 1973, ont été entre ces deux dates la principale incitation à développer la géométrie algébrique abstraite. Le but de cet article est de retracer l'histoire des théories qui ont précédé l'énoncé des conjectures de Weil et de tenter d'expliquer comment elles ont conduit à cet énoncé; ces théories ont toutes un certain caractère métaphorique; elles s'inspirent d'abord de celle des nombres algébriques (Artin) pour passer à celle des fonctions algébriques d'une variable (Schmidt) puis, avec A. Weil, à la géométrie des courbes algébriques. Les conjectures elles-mêmes postulent l'existence, pour les variétés algébriques sur un corps fini, d'une théorie cohomologique ayant des propriétés analogues à celles de la cohomologie des variétés algébriques sur le corps des complexes.

Dès 1857 Dedekind consacrait un article à l'étude arithmétique des polynômes à coefficients dans le corps $\mathbf{F}_p$ des entiers *modulo* un nombre premier p (Dedekind 1857), en soulignant l'analogie de cette théorie avec l'arithmétique usuelle des nombres entiers. Dans son langage, il parle plutôt des propriétés *modulo* p des polynômes à coefficients entiers; la congruence $mod.p$ de deux polynômes à coefficients entiers signifie que les coefficients de leur différence sont tous divisibles par p et cette relation est compatible avec les opérations usuelles sur les polynômes. Après avoir défini la divisibilité $mod.p$ et développé la division euclidienne et l'algorithme d'Euclide $mod.p$ qui permet d'obtenir un p.g.c.d., Dedekind définit les polynômes irréductibles $mod.p$ (appelés par lui *Primfunktionen*) et il établit la décomposition unique $mod.p$ d'un polynôme en produit de tels polynômes irréductibles. Il passe ensuite à l'étude des congruences *doubles, modulo* p et un polynôme M, notées $A \equiv B\ (modd.p, M)$ (pour $A - B$ est divisible par $Mmod.p$) qui sont analogues aux congruences introduites par Gauss dans l'arithmétique des entiers; en particulier, il y a p^μ classes $(modd.p, M)$ si M est de degré μ $mod.p$. Dedekind établit, par exemple, l'existence d'une *racine primitive* $(modd.p, M)$, c'est-à-dire d'un polynôme dont la classe engendre le groupe multiplicatif des classes $(modd.p, M)$ étrangères à M. De telles congruences, nommées *höheren Kongruenzen* par Dedekind, avaient déjà été considérées par E. Galois (Galois 1830).

H. Kornblum

Le travail de Dedekind a été poursuivi à la veille de la première guerre mondiale par un étudiant de Göttingen, Heinrich Kornblum (1890–1914) malheureusement tué à la guerre. Kornblum démontre, pour les polynômes $mod.p$, l'analogue du théorème de Dirichlet sur les nombres premiers dans une progression arithmétique:

si A et M sont des polynômes non congrus à 0 *mod.p* et *étrangers mod.p*, il existe une infinité de classes de polynômes P irréductibles *mod.p* et $\equiv A$ (*modd.p*, M). Ce travail devait servir de thèse à Kornblum et il a été édité par Landau après la guerre (Kornblum 1919). Kornblum suit pas à pas la méthode de Dirichlet en introduisant les *caractères* χ (*modd.p*, M), c'est-à-dire les caractères du groupe multiplicatif des classes (*modd.p*, M) étrangères à M (prolongés par 0 pour les polynômes non étrangers à M), ainsi qu'une fonction $L(s, \chi)$ pour chacun de ces caractères; il pose

$$L(s,\chi) = \sum_F \frac{\chi(F)}{p^{[F]s}}$$

où F parcourt l'ensemble des classes *mod.p* à coefficient dominant égal à 1 (appelées *primaires* par Dedekind et jouant un rôle analogue à celui des entiers positifs) et où $[F]$ note le *degré* de F. La variable s est réelle et la série converge pour $s > 1$ car il y a p^n classes F telles que $[F] = n$.

Lorsque χ n'est pas le caractère «principal» χ_0, $L(s, \chi)$ est en fait un polynôme en p^{-s} de degré $< m = \deg M$; en effet, dans chaque classe (*mod.p*) il y a le même nombre p^{n-m} de polynômes $\in \mathbf{F}_p[x]$ «primaires» et de degré $n \geq m$ et la somme des valeurs de χ sur toutes ces classes est nulle. Pour étudier $L(s, \chi_0)$, Kornblum établit d'abord que

$$L(s,\chi) = \prod_P \frac{1}{1 - \frac{\chi(P)}{p^{[P]s}}}$$

où P parcourt l'ensemble des classes primaires irréductibles *mod.p* (χ quelconque, $s > 1$); si $\chi = \chi_0$, $\chi_0(P) = 0$ ou 1 selon que P divise M ou non et on trouve ainsi

$$L(s,\chi_0) = \frac{1}{1 - \frac{1}{p^{s-1}}} \prod_{P|M} \left(1 - \frac{1}{p^{[P]s}}\right),$$

où le premier facteur, mettant en évidence un pôle d'ordre 1 en $s = 1$, est l'analogue de la fonction *zêta* de Riemann.

En notant B la classe inverse (*modd.p*, M) de A, on a

$$\sum_\chi \chi(BP^n) = 1$$

ou 0 selon que P^n est ou non congru à A (*modd.p*, M) et ceci permet d'établir la formule

$$\sum_\chi \chi(B) \frac{L'}{L}(s,\chi) = h \log p \sum_{P^n \equiv A} \frac{[P]}{p^{[P]ns}}$$

où h est le nombre des caractères, c'est-à-dire le nombre des classes (*modd.p*, M) étrangères à M. Le pôle en $s = 1$ de $\frac{L'}{L}(s, \chi_0)$ ne disparaît pas du premier membre car $L(1,\chi) \neq 0$ pour $\chi \neq \chi_0$ et, comme la partie du second membre correspondant à des exposants $n > 1$ reste finie en $s = 1$,

$$\sum_{P \equiv A} \frac{[P]}{p^{[P]s}}$$

tend vers l'infini lorsque s tend vers 1 d'où le résultat cherché. Le théorème se ramène ainsi à établir que $L(1,\chi) \neq 0$ pour $\chi \neq \chi_0$, ce que Kornblum fait en suivant Dirichlet. Dans la suite de son travail Kornblum démontre encore l'infinitude des classes de P irréductibles *mod.p* avec un degré $[P]$ de parité donnée et $\equiv A$ $(modd.p, M)$; Landau remarque d'ailleurs que la démonstration s'étend au cas où on impose que $[P] \equiv l$ $(mod.k)$ où k et l sont des entiers donnés, $k > 1$.

E. Artin

La thèse d'E. Artin, terminée en 1921 (Artin 1924), se propose d'étendre l'étude arithmétique de Dedekind aux extensions quadratiques du corps K des fractions rationnelles à coefficients dans $\mathbf{F}_p$ (p nombre premier impair); le modèle est ici l'arithmétique des extensions quadratiques de $\mathbf{Q}$ développée elle aussi par Dedekind (Dedekind 1893).

A tout polynôme $F \in \mathbf{F}_p[t]$, Artin associe le nombre $|F| = p^n$ où n est le degré de F; c'est le nombre de classes $(modd.p, F)$ au sens de Dedekind. Il est clair que $F \mapsto |F|$ se prolonge en une valeur absolue ultramétrique de K et que le complété de K pour cette valeur absolue est le corps des séries formelles en t^{-1} à coefficients dans $\mathbf{F}_p$. Ce corps joue, dans la théorie d'Artin, un rôle analogue à celui du corps des nombres réels dans l'arithmétique ordinaire: si F est l'un de ses éléments, Artin note sgn F son coefficient dominant (celui de la plus haute puissance de t dans F).

Artin s'intéresse à un corps $\Omega = K(\sqrt{D})$ où $D \in \mathbf{F}_p[t]$ est sans facteur carré. Ses éléments s'écrivent (d'une manière unique) $\alpha = A + B\sqrt{D}$ avec $A, B \in K$; le *conjugué* d'un tel élément α est $\alpha' = A - B\sqrt{D}$ et sa *norme* est $N(\alpha) = \alpha\alpha' = A^2 - B^2D$. Les éléments α de Ω entiers sur $\mathbf{F}_p[t]$ sont ceux pour lesquels A et B sont des polynômes. Si $\mathbf{a}$ est un idéal de l'anneau des entiers de Ω, il est engendré comme $\mathbf{F}_p[t]$-module par deux éléments ω_1 et ω_2 que l'on peut choisir de la forme $\omega_1 = T$ p.g.c.d. des polynômes appartenant à $\mathbf{a}$ et $\omega_2 = R + S\sqrt{D}$ où S est le p.g.c.d. des coefficients de $\sqrt{D}$ dans les éléments de $\mathbf{a}$ et $|R| < |T|$; dans ce cas Artin dit que (ω_1, ω_2) est une *base adaptée* de $\mathbf{a}$. Il existe alors des polynômes A, B, C tels que $T = 2CS$, $R = BS$ et $D = B^2 - 4AC$; inversement, si T, R et S sont ainsi choisis, le module engendré par T et $R + S\sqrt{D}$ sur $\mathbf{F}_p[t]$ est un idéal dans l'anneau des entiers de Ω.

Lorsque des éléments ω_1 et ω_2 engendrent sur $\mathbf{F}_p[t]$ un idéal $\mathbf{a}$, l'expression $\left(\det(\omega_1; \omega_2; \omega_1'; \omega_2')\right)^2$ est définie par $\mathbf{a}$ à un facteur près qui est un carré de $\mathbf{F}_p$;

dans le cas d'une base adaptée, on trouve $(4CS^2)^2D$. Artin note $N\,\mathbf{a}$ (*norme* de $\mathbf{a}$) le polynôme «primaire» proportionnel à CS^2; l'idéal engendré par $N\,\mathbf{a}$ n'est autre que $\mathbf{a}\,\mathbf{a}'$ où $\mathbf{a}'$ est l'idéal conjugué de $\mathbf{a}$. La norme est multiplicative et $|N\,\mathbf{a}|$ est le nombre de classes d'entiers de Ω modulo $\mathbf{a}$.

On établit sans peine les propriétés de divisibilité des idéaux; en particulier, tout idéal se décompose d'une manière essentiellement unique en produit d'idéaux *premiers*. Chaque idéal premier $\mathbf{p}$ divise un unique polynôme irréductible $P \in \mathbf{F}_p[t]$ et on a $N\,\mathbf{p} = P^2$ ou P; le premier cas se présente lorsque la congruence $X^2 \equiv D$ $(mod.P)$ n'a pas de solution dans $\mathbf{F}_p[t]$ et le second dans l'éventualité contraire. On obtient donc tous les idéaux premiers en décomposant dans Ω les polynômes irréductibles P. Si P divise D, c'est le carré de l'idéal $(P\sqrt{D})$. Sinon, on pose $[\frac{D}{P}] = 1$ ou -1 selon que la congruence $X^2 \equiv D$ $(mod.P)$ a ou non une solution; dans le premier cas (P) se décompose en le produit de deux idéaux premiers conjugués $(P, B \pm \sqrt{D})$ (où $B^2 \equiv D$ $(mod.P)$) et dans le second (P) reste premier dans Ω.

Deux idéaux $\mathbf{a}$ et $\mathbf{b}$ de Ω sont *équivalents* s'il existe des entiers α et β (non nuls) de ce corps tels que $(\beta)\mathbf{a} = (\alpha)\mathbf{b}$; les classes d'équivalence d'idéaux forment un groupe multiplicatif dont l'élément neutre ou *classe principale* est la classe des idéaux principaux. Artin s'intéresse surtout à établir la finitude de ce groupe et à évaluer son ordre h. Si $\mathbf{a}$ est un idéal de Ω engendré par des entiers ω_1 et ω_2 comme $\mathbf{F}_p[t]$-module, Artin lui associe l'élément $\omega = \frac{\omega_2}{\omega_1}$ de Ω; lorsqu'on change la base (ω_1, ω_2), ω subit une transformation homographique $\omega \mapsto \frac{\alpha\omega+\beta}{\gamma\omega+\delta}$ à coefficients dans $\mathbf{F}_p[t]$ et deux idéaux équivalents donnent la même classe de Ω *modulo* ces transformations homographiques. Inversement, on voit que chaque classe de Ω *modulo* homographie provient d'une unique classe d'idéaux; la classification des idéaux se ramène donc à celle des éléments ω qu'on leur associe. Dans le cas *imaginaire*, c'est-à-dire où $\sqrt{D}$ n'appartient pas au corps des séries formelles en t^{-1} (ceci se produit lorsque le degré de D est impair ou bien lorsqu'il est pair et que le coefficient dominant n'est pas un carré de $\mathbf{F}_p$), chaque classe de fonctions irrationnelles $\omega = X + Y\sqrt{D}$ *modulo* homographie (X, Y dans le corps des séries formelles) contient une fonction *réduite* pour laquelle $|X| < 1$, $|\omega\omega'| \geq 1$ et sgn $2Y = 1$; si $X = \frac{B}{2C}$ et $Y = \frac{1}{2C}$ avec $B^2 - 4AC = D$, cela s'écrit $|B| < |C| \leq |\sqrt{D}|$ et sgn $C = 1$ et on dit que l'idéal $(2C, B + \sqrt{D})$ est *réduit*. Il n'y a qu'un nombre fini de ω réduites pour D donné, d'où la finitude du nombre de classes d'idéaux dans le cas imaginaire; à titre d'exemples, Artin trouve qu'il n'y a qu'une classe si D est de degré 0 ou 1 tandis qu'il y en a deux si D est de degré 2. Les classes d'idéaux sont regroupées en *genres* comme dans la théorie des nombres; le genre *principal* est le sous-groupe du groupe des classes formé des carrés et les genres sont les classes *modulo* ce sous-groupe. Les genres correspondent bijectivement aux classes *ambiguës* ou égales à leurs conjugées (qui est leur inverse dans le groupe des classes d'idéaux); si s est le nombre de facteurs irréductibles de D, le nombre des classes ambiguës ou des genres est 2^{s-1} ou 2^s selon que D admet ou non un facteur irréductible de degré impair. Le nombre de classes est donc impair seulement dans le cas où D est irréductible de degré impair.

Dans le cas *réel* où D est de degré pair avec un coefficient dominant carré, les fonctions irrationnelles $\omega = X + Y\sqrt{D}$ s'écrivent comme des séries formelles en t^{-1}; Artin dit que ω est *réduite* si $|\omega'| < 1 < |\omega|$, soit $|B-\sqrt{D}| < |C| < |B+\sqrt{D}|$ si $\omega = \frac{B+\sqrt{D}}{2C}$ et il démontre que chaque classe *modulo* homographie contient une fonction réduite en se servant de la théorie des fractions continues. Il en déduit encore la finitude du nombre de classes; par exemple $K(\sqrt{t^2 + a_1 t + a_2})$ (avec $a_2 - \frac{1}{4}a_2^1 \neq 0$) a une seule classe d'idéaux. Si D est irréductible («primaire» de degré pair), le nombre de classes de $K(\sqrt{D})$ est impair.

Les unités de Ω, c'est-à-dire les éléments ε tels que $N(\varepsilon) = c$ soit un élément inversible de $\mathbf{F}_p$, sont en nombre *fini* w dans le cas imaginaire; plus précisément $w = p^2 - 1$ si D est de degré 0 (élément non carré de $\mathbf{F}_p$) et $w = p - 1$ dans les autres cas. Dans le cas réel les unités sont en nombre infini et de la forme $\varepsilon = a\varepsilon_0^k$ où a est inversible dans $\mathbf{F}_p$, ε_0 est une unité fondamentale et k parcourt $\mathbf{Z}$; si D est irréductible («primaire» de degré pair) $N(\varepsilon_0)$ n'est pas un carré dans $\mathbf{F}_p$ et on peut choisir ε_0 de manière que ce soit une racine primitive $g\, mod.p$.

Artin établit une loi de réciprocité quadratique énoncée sans démonstration par Dedekind: si P et Q sont des polynômes irréductibles «primaires» de degré respectifs μ et ν, $[\frac{P}{Q}][\frac{Q}{P}] = (\frac{-1}{p})^{\mu.\nu}$ où le second membre contient le symbole de Legendre. Il faut y ajouter la loi complémentaire $[\frac{a}{P}] = (\frac{a}{p})^{\mu}$ pour $a \in \mathbf{F}_p^*$. Artin donne une autre forme à la loi de réciprocité, à l'aide d'un symbole

$$[\frac{M}{N}] = \prod_i [\frac{M}{Q_i}]$$

pour $N = Q_1 Q_2 \ldots$ (Q_i irréductibles) analogue à celui de Jacobi; on a

$$[\frac{M}{N}][\frac{N}{M}] = (-1)^{\frac{|M|-1}{2}\frac{|N|-1}{2}} \text{ et } [\frac{a}{M}] = a^{\frac{|M|-1}{2}}$$

pour M et N «primaires» étrangers et $a \in \mathbf{F}_p^*$.

L'évaluation du nombre h de classes d'idéaux se fait au moyen de la fonction

$$\mathbf{Z}(s) = \sum_{a} \frac{1}{|N\mathbf{a}|^s}$$

où la somme est étendue à l'ensemble des idéaux (non nuls) de $K(\sqrt{D})$ et converge pour Re$s > 1$; cette fonction, analogue à la fonction *zêta* de Dedekind pour un corps de nombres, est aussi notée $\mathbf{Z}_D(s)$. On a encore

$$\mathbf{Z}(s) = \prod_{\mathbf{p}} \frac{1}{1 - \frac{1}{|N\mathbf{p}|^s}}$$

avec un produit étendu à tous les idéaux premiers; en regroupant les facteurs qui correspondent aux idéaux premiers **p** divisant un polynôme irréductible P fixé on trouve $\frac{1}{1-\frac{1}{|P|^s}}\frac{1}{1-\frac{[\frac{D}{P}]}{|P|^s}}$ si P ne divise pas D et $\frac{1}{1-\frac{1}{|P|^s}}$ si P divise D. Ceci donne finalement

$$\mathbf{Z}(s) = \frac{1}{1-p^{-(s-1)}}\sum_F [\frac{D}{F}]\frac{1}{|F|^s}$$

où la somme est étendue à tous les polynômes «primaires»; le premier facteur joue le rôle de la fonction *zêta* de K. Artin regroupe les termes pour lesquels F a un degré donné ν et trouve

$$\mathbf{Z}(s) = \frac{1}{1-p^{-(s-1)}}\sum_{\nu=0}^{\infty}\frac{\sigma_\nu}{p^{\nu s}}$$

avec

$$\sigma_\nu = \sum_{|F|=p^\nu}[\frac{D}{F}]$$

(sgn $F=1$, $(D,F)=1$); à l'aide de la loi de réciprocité quadratique il établit que $\sigma_\nu = 0$ pour $\nu \geq n$ lorsque D est de degré $n \geq 1$. Pour $D = g$ (racine primitive $mod.p$), $\sigma_\nu = p^\nu(-1)^\nu$; on a donc

$$\mathbf{Z}_D(s) = \frac{1}{1-p^{-(s-1)}}\sum_{\nu=0}^{n-1}\frac{\sigma_\nu}{p^{\nu s}}$$

pour $D \neq g$ et

$$\mathbf{Z}_g(s) = \frac{1}{1-p^{-(s-1)}}\frac{1}{1+p^{-(s-1)}}.$$

Dans le cas réel (D de degré pair, sgn $D=1$), la loi de réciprocité donne encore $\sigma_0+\sigma_1+\ldots+\sigma_{n-1}=0$ et on voit donc que $\mathbf{Z}(0)=0$; comme le changement de D en gD multiplie σ_ν par $(-1)^\nu$, on a

$$(*) \qquad \mathbf{Z}_D\left(s+\frac{\pi i}{\log p}\right) = \frac{1-p^{-(s-1)}}{1+p^{-(s-1)}}\mathbf{Z}_{gD}(s)$$

et on voit que $\mathbf{Z}_D(\frac{\pi i}{\log p}) = 0$ lorsque D est de degré pair et que sgn $D = g$. Les zéros «triviaux» ainsi obtenus pour $\mathbf{Z}(s)$ en impliquent d'autres car cette fonction admet la période $\frac{2\pi i}{\log p}$.

Le lien de la fonction $\mathbf{Z}(s)$ avec le nombre de classes s'obtient en considérant, pour chaque classe **K** d'idéaux, une fonction partielle

$$\mathbf{Z}(s,\mathbf{K}) = \sum_{a\in K}\frac{1}{|\mathbf{Na}|^s} = |N\,\mathbf{a}|^s\Sigma\frac{1}{|N\,\alpha|^s}$$

où $\mathbf{a} \in \mathbf{K}$ et la dernière somme est étendue aux idéaux principaux (α) divisibles par $\mathbf{a}$. Dans le cas imaginaire, cela donne

$$\mathbf{Z}(s,\mathbf{K}) = \frac{|N\,\mathbf{a}|^s}{w} \sum_{\alpha\equiv 0(mod.\mathbf{a})} \frac{1}{|N\,\alpha|^s} = (\frac{p^r}{|A|^s} + \frac{1}{|C|^s})\frac{p-1}{w(1-p^{-(2s-1)})}$$

$$+(\frac{p^{-(r+1)(2s-1)}}{|C|^s}) + \frac{p^r}{|A|^s}\frac{(p-1)^2}{w(1-p^{-(2s-1)})(1-p^{-(2s-1)})}$$

en prenant $\mathbf{a} = (2C, B+\sqrt{D})$ réduit et en posant $r = [\frac{a-b}{2}]$ où a et b sont les degrés de A et de C; on trouve ainsi $\mathbf{Z}(0,\mathbf{K}) = -\frac{1}{w}$ et, pour le nombre de classes, $h = -w\,\mathbf{Z}(0)$ soit $h = 1$ si $D = g$ et $h = \sigma_0 + \sigma_1 + \ldots + \sigma_{n-1}$ dans les autres cas imaginaires. En considérant la limite de $(s-1)\mathbf{Z}(s,\mathbf{K})$ pour $s = 1$ on obtient une autre expression de h; cette limite est en effet $\frac{1}{2\log p}$ si $D = g$, $\frac{\sqrt{p}}{\sqrt{|D|}\log p}$ si D est de degré impair et $\frac{p+1}{2\sqrt{|D|}\log p}$ si D est de degré pair ≥ 2 (sgn $D = g$) et on a donc, dans le cas imaginaire:

$$h = \sqrt{\frac{|D|}{p}}\sum_{\nu=0}^{n-1}\frac{\sigma_\nu}{p^\nu}$$

pour deg D impair,

$$h = \frac{2\sqrt{|D|}}{p+1}\sum_{\nu=0}^{n-1}\frac{\sigma_\nu}{p^\nu}$$

pour deg D pair. On voit ainsi que la fonction $\mathbf{Z}(s)$ a un pôle d'ordre 1 en chacun des points $s = 1 + \frac{2k\pi i}{\log p}$.

Dans le cas réel, chaque idéal principal peut être engendré par une fonction α telle que $|\varepsilon_0| \leq |\alpha| < |\varepsilon_0|^2$ où ε_0 est une unité fondamentale et on peut choisir α de $p-1$ manières; ainsi

$$\mathbf{Z}(s,\mathbf{K}) = \frac{|N\,\mathbf{a}|^s}{p-1}\Sigma\frac{1}{|N\,\alpha|^s}$$

où $\mathbf{a} \in \mathbf{K}$ et α parcourt $\mathbf{a}$ en respectant les inégalités $|\varepsilon_0| \leq |\alpha| < |\varepsilon_0|^2$. On choisit encore $\mathbf{a} = (2C, B+\sqrt{D})$ réduit et on a $\alpha = 2CX + (B+\sqrt{D})Y$; on voit que, sauf pour un nombre *fini* de termes, $|2CX| = |(B+\sqrt{D})Y|$, ce qui revient à dire que $Y = -E(\frac{2CX}{B+\sqrt{D}}) + F$ où E prend la partie entière de la série formelle en t^{-1} et F est un polynôme tel que $\frac{|\varepsilon_0|}{|\sqrt{D}|} \leq |F| < \frac{|\varepsilon_0|^2}{|\sqrt{D}|}$. A un nombre fini de termes près $\mathbf{Z}(s,\mathbf{K})$ est ainsi

$$\frac{1}{(p-1)|\sqrt{D}|^s}(\Sigma\frac{1}{|X|^s})(\Sigma\frac{1}{|F|^s})$$

$$= \frac{p-1}{|\sqrt{D}|^s}(1-p^{-(s-1)})\frac{1-p^{-R(s-1)}}{1-p^{-(s-1)}}p^{-(R-\frac{n}{2})(s-1)}$$

en posant $R = \frac{\log|\varepsilon_0|}{\log p}$ (analogue au *régulateur* de la théorie des nombres) et $n = \frac{\log|D|}{\log p} = \deg D$. Finalement $(s-1)\mathbf{Z}(s,\mathbf{K})$ tend vers $\frac{(p-1)R}{|\sqrt{D}|\log p}$ lorsque s tend vers 1 et on a

$$h = \frac{|\sqrt{D}|}{(p-1)R}\sum_{\nu=0}^{n-1}\frac{\sigma_\nu}{p^\nu}.$$

Les calculs précédents donnent le résidu de $\mathbf{Z}(s)$ en $s = 1$ sous la forme $\frac{h}{\kappa \log p} \neq 0$, avec $\kappa = \frac{2\sqrt{|D|}}{p+1}$ pour D de degré pair ≥ 2, sgn $D = g$, $\kappa = \sqrt{\frac{|D|}{p}}$ pour D de degré impair et $\kappa = \frac{|\sqrt{D}|}{(p-1)R}$ pour D de degré pair, sgn $D = 1$; la relation $(*)$ montre alors que $\mathbf{Z}_D(s)$ ne s'annule pas en $1 + \frac{\pi i}{\log p}$ ni en aucun $1 + \frac{(2k+1)\pi i}{\log p}$. Artin en déduit alors que $\mathbf{Z}(s)$ ne s'annule pas sur la droite Re$s = 1$ et que la borne supérieure des parties réelles de zéros de cette fonction (périodique) est donc < 1.

L'expression de $\mathbf{Z}(s,\mathbf{K})$ dans le cas imaginaire conduit à une *équation fonctionnelle*

$$\mathbf{Z}(1-s,\mathbf{K}) = \frac{1-p^{-(s-1)}}{1-p^s}\left(\sqrt{\frac{|D|}{p}}\right)^{2s-1}\mathbf{Z}(s,\mathbf{K})$$

si D est de degré impair et

$$\mathbf{Z}(1-s,\mathbf{K}) = \frac{1-p^{-(s-1)}}{1-p^{2s}}(\sqrt{|D|})^{2s-1}\mathbf{Z}(s,\mathbf{K})$$

si D est de degré pair (sgn $D = g$); pour $\mathbf{Z}(s)$ cela donne une équation fonctionnelle analogue. On passe au cas réel au moyen de la relation $(*)$:

$$\mathbf{Z}(1-s) = \left(\frac{1-p^{-(s-1)}}{1-p^s}\right)^2(\sqrt{|D|})^{2s-1}\mathbf{Z}(s).$$

Cette équation fonctionnelle permet de ramener le calcul de $\mathbf{Z}'(0)$ à celui de la limite de $s\,\mathbf{Z}(1-s)$ en $s = 0$ et on obtient ainsi

$$h = -\frac{p-1}{R\log p}\mathbf{Z}'(0) = -\frac{1}{R}(\sigma_1 + 2\sigma_2 + \ldots + (n-1)\sigma_{n-1}).$$

L'équation fonctionnelle peut s'exprimer par des relations entre les coefficients σ_ν de $\mathbf{Z}(s)$; pour $\deg D = 2m+1$ impair on a $\sigma_{2m-\nu} = p^{m-\nu}\sigma_\nu$ et pour $\deg D = 2m$ pair on a $\sigma_{2m-\nu} \pm p\sigma_{2m-\nu-1} = p^{m-\nu}(\sigma_\nu \pm p\sigma_{\nu-1})$ avec le signe $+$ dans le cas imaginaire et le signe $-$ dans le cas réel. Ces relations permettent de simplifier le calcul du nombre de classes h à partir des σ_ν; les cas où $\deg D \leq 2$ étant déjà connus, Artin établit des tables pour $\deg D = 3$, $p = 3, 5$ ou 7 et pour $\deg D = 4$, $p = 3$. Ces tables permettent aussi d'établir l'*hypothèse de Riemann* dans chacun des cas considérés (une quarantaine); cette hypothèse dit que les zéros non triviaux de $\mathbf{Z}(s)$ sont tous sur la droite $\operatorname{Re} s = \frac{1}{2}$ et on peut encore l'exprimer en disant que les racines $\beta_\nu \neq \pm 1$ de l'équation $z^{n-1} + \sigma_1 z^{n-2} + \ldots + \sigma_{n-1} = 0$ sont toutes de valeur absolue $p^{1/2}$. Lorsque $n = 3$, l'équation précédente se réduit à $z^2 + \sigma_1 z + p = 0$ et l'hypothèse de Riemann signifie que les racines de cette équation sont imaginaires conjuguées, c'est-à-dire que $|\sigma_1| < 2\sqrt{p}$; les tables donnent $|\sigma_1| \leq 3$ pour $p = 3$, $|\sigma_1| \leq 4$ pour $p = 5$ et $|\sigma_1| \leq 5$ pour $p = 7$, donc l'hypothèse de Riemann est vérifiée dans ces cas. Lorsque $n = 4$, l'équation s'écrit

$$z^3 + \sigma_1 z^2 + \sigma_2 z + p = 0$$

avec $\sigma_2 = p - 1 + \sigma_1$ et elle a une racine triviale $z = -1$; les autres racines sont données par $z^2 + (\sigma_1 - 1)z + p = 0$ et l'hypothèse de Riemann revient à $|\sigma_1 - 1| < 2\sqrt{p}$, inégalités qu'Artin vérifie pour $p = 3$ ou 5.

Le développement de $\log \mathbf{Z}(s)$ selon les puissances de p^{-s} permet d'établir la formule

$$\sum_{d|\nu} d\pi(p^d) = p^\nu - \beta_1^\nu - \beta_2^\nu - \ldots - \beta_{n-1}^\nu$$

où $\pi(x)$ note le nombre des idéaux premiers $\mathbf{p}$ tels que $|N\,\mathbf{p}| = x$; si $\theta \in [\frac{1}{2}, 1[$ désigne la borne supérieure des parties réelles des zéros de $\mathbf{Z}(s)$ (en excluant les cas où $\deg D \leq 2$), on en déduit une relation analogue au théorème des nombres premiers: $\pi(p^\nu) = \frac{p^\nu}{\nu} + O(\frac{p^{\theta\nu}}{\nu})$, avec $\theta = \frac{1}{2}$ si l'hypothèse de Riemann est vraie (le nombre total d'idéaux $\mathbf{a}$ tels que $|N\,\mathbf{a}| = p^\nu$ est $\frac{h}{\kappa} p^\nu$).

Les formules pour le nombre de classes à partir de $\mathbf{Z}(0)$ (cas imaginaire) ou de $\mathbf{Z}'(0)$ (cas réel) s'écrivent encore

$$h = \prod_{\nu=1}^{n-1} (\beta_\nu - 1)$$

pour $n = \deg D$ impair,

$$h = 2\prod_{\nu=1}^{n-1} (\beta_\nu - 1)$$

pour n pair et $\operatorname{sgn} D = g$,

$$h = \frac{1}{R}\prod_{\nu=1}^{n-2} (\beta_\nu - 1)$$

pour n pair et sgn $D = 1$. Si h' est le nombre de classes de $K(\sqrt{gD})$, la relation $(*)$ permet alors d'établir que

$$hh' = \prod_{\nu=1}^{n-1}(\beta_\nu^2 - 1) \qquad \text{pour } n \text{ impair}$$

$$\text{et } hh' = \frac{2}{R}\prod_{\nu=1}^{n-1}(\beta_\nu^2 - 1) \quad \text{pour } n \text{ pair,}$$

d'où Artin déduit que, si l'hypothèse de Riemann est vraie, h est compris entre $(\sqrt{p}-1)^{n-1}$ et $(\sqrt{p}+1)^{n-1}$ (resp. $2(\sqrt{p}-1)^{n-1}$ et $2(\sqrt{p}+1)^{n-1}$) pour n impair (resp. pair; on suppose toujours que $n \geq 3$). Dans le cas où $n = 3$, la minoration précédente n'apprend rien et Artin la remplace par une autre, qu'il déduit d'une majoration élémentaire pour h' : $h \geq \frac{1}{n}(\frac{p-1}{\sqrt{p}})^{n-1}$ pour n impair, $h \geq \frac{1}{(n-1)^2}(\frac{p-1}{\sqrt{p}})^{n-2}$ pour n pair. Artin se sert de ces inégalités pour établir (en admettant l'hypothèse de Riemann) qu'il n'y a qu'un nombre *fini* de corps $K(\sqrt{D})$ ayant un nombre de classes h donné (p et $n \geq 3$ variant); par exemple il n'y en a pas avec $h = 1$ lorsque $p > 5$ et $n > 3$. Pour $p = 3$, Artin conjecture que $K(\sqrt{t^3 - t - 1})$ est le seul corps ayant un nombre de classes égal à 1. Il complète ces résultats par des minorations du nombre de classes dans un genre donné; il établit ainsi que, si l'hypothèse de Riemann est vraie, il n'existe qu'un nombre fini de corps (avec $n \geq 3$) pour lesquels ce nombre a une valeur donnée. Par exemple, ce nombre ne peut être 1 que lorsque $p = 3$, 5 ou 7.

Artin termine sa thèse en précisant le théorème de Kornblum. Une étude soigneuse du groupe multiplicatif (*mod.M*) et la loi de réciprocité quadratique lui permettent de démontrer que tout caractère réel χ du groupe multiplicatif *mod.M* s'écrit $\chi(A) = [\frac{D}{A}]$ où $D = (\frac{-1}{p})^n QR^2$, Q désignant le produit d'une partie des polynômes irréductibles distincts divisant M, R le produit des autres et n le degré de Q. Il trouve ainsi, pour χ réel $\neq \chi_0$, que

$$L(s,\chi) = \prod_{\nu=1}^{\kappa}(1 - [\frac{D_1}{R_\nu}]|R_\nu|^{-s}).(1 - p^{-(s-1)}.\mathbf{Z}_{D_1}(s)$$

où $D_1 = (\frac{-1}{p})^n Q$ et les R_ν $(1 \leq \nu \leq \kappa)$ sont les facteurs irréductibles de R, d'où il déduit que $L(s,\chi)$ ne s'annule pas sur la droite Re$s = 1$; cette propriété est facile dans le cas du caractère principal χ_0 et Artin l'établit encore dans le cas d'un caractère imaginaire en suivant la méthode employée par Landau pour le cas de la théorie des nombres. Ceci lui permet de démontrer que le nombre $\pi_A(p^\nu)$ des polynômes irréductibles $P \equiv A$ (*mod.M*) et tels que $|P| = p^\nu$ (c'est-à-dire de degré ν) est

$$\frac{p^\nu}{\nu\Phi(M)} + O\left(\frac{p^{\nu\theta}}{\nu}\right)$$

où $\theta \in [\frac{1}{2}, 1[$ désigne la borne supérieure des parties réelles des zéros des $L(s, \chi)$ et $\Phi(M)$ désigne l'ordre du groupe multiplicatif *mod.M*; en particulier, les polynômes irréductibles sont asymptotiquement répartis d'une manière uniforme entre les diverses classes *mod.M*.

F.K. Schmidt

Dans sa thèse Friedrich Karl Schmidt s'intéressait à une théorie arithmétique plus générale que celle d'Artin puisqu'il étudiait les idéaux et les unités d'une extension finie quelconque K du corps $k(z)$ des fonctions rationnelles à coefficients dans un corps fini k (Schmidt 1925); cette théorie avait été aussi abordée par P. Sengenhorst (Sengenhorst 1925). Considérant un corps K de caractéristique $p_0 > 0$ et de degré de transcendance 1 sur la fermeture algébrique k de $\mathbf{F}_{p_0}$ dans K, Schmidt démontre par exemple que, si $z \in K$ n'appartient pas à k, la fermeture intégrale $\Im$ de $k[z]$ dans K est finie sur $k[z]$ et que c'est un anneau de Dedekind (*Multiplikationsring* dans sa terminologie).

Mais la généralisation correspondante de la fonction *zêta* présente des difficultés et Schmidt a été conduit à changer de point de vue pour aller plus loin; il a adopté celui introduit par Dedekind et Weber dans l'étude des corps de fonctions algébriques d'une variable *complexe* (Dedekind-Weber 1882). En vue de définir précisément les points de la surface de Riemann associée à une fonction algébrique u de z définie par une relation $F(z, u) = 0$ (F polynôme en z et u à coefficients complexes), ces auteurs faisaient une construction algébrique à partir du corps K des fonctions rationnelles en z et u liés par la relation précédente; le corps K est une extension finie de $\mathbf{C}(z)$ et une *place* $\mathbf{P}$ de K est, par définition, un anneau de valuation de K qui s'interprète comme l'anneau des fonctions régulières en un point de la surface de Riemann. L'idéal maximal $\mathbf{p}$ de $\mathbf{P}$ correspond à l'ensemble des fonctions nulles en $\mathbf{P}$ et l'homomorphisme $\mathbf{P} \to \mathbf{C}$ de $\mathbf{P}$ dans son corps résiduel correspond à l'évaluation d'une fonction en $\mathbf{P}$; cette théorie est complètement développée dans le livre classique de Hensel et Landsberg (Hensel-Landsberg 1902).

Schmidt (Schmidt 1931) considère donc, au lieu des idéaux premiers d'un anneau d'entiers $\Im \subset K$ construit comme ci-dessus, les places $\mathbf{P}$ de K qu'il définit comme les anneaux intégralement clos admettant K comme corps des fractions; cette notion ne dépend que de K et non pas du choix de la variable z. Lorsque z est choisi dans une place $\mathbf{P}$, l'intersection de la fermeture entière $\Im$ de $k[z]$ avec l'idéal maximal $\mathbf{p}$ de $\mathbf{P}$ est un idéal premier $\tilde{\mathbf{p}}$ de $\Im$ et le localisé $\Im_{\tilde{p}}$ coïncide avec $\mathbf{P}$; on obtient ainsi toutes les places comme localisés d'anneaux $\Im$ en des idéaux premiers pour les divers choix de z et il suffit d'ailleurs de prendre un z et son inverse $\frac{1}{z}$. Les places qui ne proviennent pas d'idéaux premiers de $\Im$ s'interprètent comme des points à l'infini relativement à la variable z; l'utilisation des places au lieu des idéaux premiers revient ainsi à faire jouer aux points à l'infini le même rôle qu'aux autres points.

Au lieu des idéaux d'un anneau d'entiers $\mathfrak{J}$, Schmidt considère les *diviseurs*, c'est-à-dire les expressions formelles $\mathbf{c} = \mathbf{p}_1^{e_1}\mathbf{p}_2^{e_2}\dots\mathbf{p}_s^{e_s}$ où les $\mathbf{p}_i$ sont des idéaux maximaux de places $\mathbf{P}_i$ et les $e_i \in \mathbf{Z}$; les diviseurs forment un groupe multiplicatif (commutatif) D, engendré librement par les diviseurs *premiers* de la forme $\mathbf{c} = \mathbf{p}$. On dit qu'un diviseur est *entier* si tous ses exposants e_i sont ≥ 0 et qu'un diviseur $\mathbf{c}'$ est multiple d'un autre $\mathbf{c}''$ si le quotient $\frac{\mathbf{c}'}{\mathbf{c}''}$ est entier; il est clair que tout diviseur est quotient de deux diviseurs entiers, bien déterminés si on leur impose d'être étrangers. A tout élément $z \neq 0$ de K est associé un diviseur formé des $\mathbf{p}_i$ tels que $\mathbf{P}_i z \neq \mathbf{P}_i$ avec les exposants e_i définis par $\mathbf{P}_i z = \mathbf{p}_i^{e_i}$ (ces $\mathbf{p}_i$ sont en nombre fini); les éléments de k sont caractérisés par le fait d'avoir un diviseur nul. A tout idéal (fractionnaire) $\tilde{\mathbf{c}}$ d'un sous-anneau de Dedekind $\mathfrak{R} \subset K$ est associé un diviseur $\mathbf{c} = \mathbf{p}_1^{e_1}\mathbf{p}_2^{e_2}\dots\mathbf{p}_s^{e_s}$ où $\tilde{\mathbf{c}} = \tilde{\mathbf{p}}_1^{e_1}\tilde{\mathbf{p}}_2^{e_2}\dots\tilde{\mathbf{p}}_s^{e_s}$ est la décomposition de $\tilde{\mathbf{c}}$ en produit d'idéaux premiers dans $\mathfrak{R}$ et $\mathbf{p}_i$ est l'idéal maximal de $\mathfrak{R}_{\tilde{\mathbf{p}}_i}$ pour $i = 1, 2, \dots, s$; inversement, tout diviseur ne faisant intervenir que des places $\mathbf{P}_i$ contenant $\mathfrak{R}$ est de cette forme.

L'*ordre* f d'un diviseur premier $\mathbf{p}$ est, par définition, le degré de l'idéal $\mathbf{p}$ c'est-à-dire celui de l'extension résiduelle $(\mathbf{P}/\mathbf{p} : (k[z]\cap\mathbf{P})/(k[z]\cap\mathbf{p}))$ (où z appartient à $\mathbf{P}$ mais pas à k); on définit ensuite l'ordre d'un diviseur quelconque de manière qu'il en dépende additivement (on dit plutôt maintenant le *degré*). Le diviseur d'un élément z de K est toujours d'ordre 0; l'ordre est donc le même pour tous les diviseurs d'une même *classe* de $\mathbf{D}$ *modulo* le sous-groupe $\mathbf{H}$ formé des diviseurs d'éléments de K. Le *genre* g de K est défini par la formule $g = \frac{w_z}{2} - m_z + 1$ où z est un élément de K non dans k et tel que K soit séparable sur $k(z)$, $m_z = (K : k(z))$ et w_z est l'indice de *ramification* de K par rapport à $k(z)$, c'est-à-dire l'ordre d'un diviseur *différente* ∂_z de K par rapport à $k(z)$ convenablement défini; Schmidt établit que le genre ne dépend pas du choix de z grâce à la notion, qu'il introduit, de dérivée d'un élément de K par rapport à un autre.

Le point central de la théorie de Schmidt est la démonstration d'un théorème analogue à celui de *Riemann-Roch* dans la théorie classique des fonctions algébriques; il lui permet, dans le cas où k est *fini*, d'évaluer le nombre des diviseurs entiers appartenant à une classe donnée de diviseurs $\mathbf{C} \in \mathbf{D}/\mathbf{H}$. Les diviseurs équivalents à un diviseur donné $\mathbf{c}$ sont en effet de la forme $(\alpha)\mathbf{c}$ où $\alpha \in K$ et on doit écrire qu'un tel diviseur est multiple du diviseur neutre $\mathbf{e}$, c'est-à-dire que α est multiple de $\frac{\mathbf{e}}{\mathbf{c}}$; Schmidt commence par établir que l'espace vectoriel (*Modul*) des éléments de K multiples d'un diviseur donné est de rang fini sur k (c'est assez difficile). Le rang r (*Rang*) de l'espace des éléments multiples de $\frac{\mathbf{e}}{\mathbf{c}}$ ne dépend que de la classe $\mathbf{C}$ du diviseur $\mathbf{c}$ et Schmidt l'appelle la *dimension* $\{\mathbf{C}\}$ (*Dimension*) de $\mathbf{C}$; il est alors clair que, si k a p éléments, la classe $\mathbf{C}$ contient $\frac{p^r-1}{p-1}$ diviseurs entiers. Le théorème de Riemann-Roch donne une information sur la dimension $\{\mathbf{C}\}$; pour l'énoncer, Schmidt introduit la *classe différentielle* $\mathbf{W}$ définie par les diviseurs $\frac{\partial_z}{n_z^2}$ où z est un élément de K non dans k, n_z est le dénominateur de son diviseur et ∂_z est la différente de K par rapport à $k(z)$. Pour toute classe de diviseurs $\mathbf{C}$, la classe $\mathbf{C}' = \frac{\mathbf{W}}{\mathbf{C}}$ est appelée la classe *complémentaire* de $\mathbf{C}$; le théorème de Riemann-Roch, pour la démonstration duquel Schmidt renvoie à Hensel et Landsberg s'énonce en disant que $\{\mathbf{C}\} = \{\frac{\mathbf{W}}{\mathbf{C}}\} + q - g + 1$ où q est l'ordre de $\mathbf{C}$ et g est le genre de K. Utilisant le fait que l'ordre de $\mathbf{W}$ est $2g-2$, Schmidt donne une forme plus symétrique $\{\mathbf{C}\} - \frac{q}{2} = \{\mathbf{C}'\} - \frac{q'}{2}$ (où q' est l'ordre de $\mathbf{C}'$) pour le théorème de Riemann-Roch. Lorsque $q \geq 2g-2$, $q' \leq 0$ et par suite $\{\frac{\mathbf{W}}{\mathbf{C}}\} = 0$ ce qui donne $\{\mathbf{C}\} = q - g + 1$ (> 0 si de plus $q > 0$).

Dans la suite de son mémoire, Schmidt suppose toujours que le corps des constantes k a un nombre fini p d'éléments (puissance de p_0); alors il n'y a qu'un nombre fini de polynômes de degré donné et on en déduit qu'il n'y a qu'un nombre fini de diviseurs entiers d'ordre donné. Grâce au théorème de Riemann-Roch, Schmidt démontre alors que, pour $n \geq 2g-2$ et divisible par $m = (K : k(z))$, tout diviseur d'ordre 0 est quotient de deux diviseurs entiers d'ordre n; il en déduit que le groupe $\mathbf{D}_0/\mathbf{H}$ des classes de diviseurs d'ordre 0 est *fini* et que tout sous-groupe $\mathbf{D}'$ de $\mathbf{D}$ non contenu dans $\mathbf{D}_0$ est d'indice fini.

A tout diviseur $\mathbf{c}$ d'ordre f, Schmidt associe le nombre $|\mathbf{c}| = p^f$; il pose

$$\mathbf{Z}(s) = \prod_{\mathbf{p}} \frac{1}{1 - |\mathbf{p}|^{-s}}$$

où le produit est étendu à tous les diviseurs premiers de K. Ce produit converge pour $\mathrm{Re}\, s > 1$ et il est aussi égal à la somme de la série

$$\sum_{\mathbf{c}} \frac{1}{|\mathbf{c}|^s}$$

étendue à tous les diviseurs entiers. Le théorème de Riemann-Roch donne le nombre de termes de cette série pour lesquels $\mathbf{c}$ est dans une classe donnée d'ordre $\geq 2g-2$ et il permet donc de sommer cette série à un nombre fini de termes près; Schmidt démontre ainsi que la fonction $\mathbf{Z}(s)$ se prolonge à tout le plan avec comme seules singularités des pôles simples sur les droites $\mathrm{Re}\, s = 1$ et $\mathrm{Re}\, s = 0$ et il établit en même temps que le p.g.c.d. des ordres des diviseurs premiers de K est 1. Explicitement on a

$$\mathbf{Z}(s) = \frac{1}{p-1} \sum_{q=1}^{q_0-1} \sum_{i=1}^{h} \frac{p^{\{\mathbf{C}_q^{(i)}\}}}{p^{qs}} + \frac{hp^{-(g-1)}}{p-1} \frac{p^{(2g-2)(1-s)}}{1-p^{1-s}} + \frac{h}{p-1} \frac{1}{1-p^s}$$

où g est le genre de K, $q_0 = 2g-2$ et h est le nombre de classes de diviseurs d'ordre 0. En 0 il y a un pôle simple de résidu $-\frac{h}{(p-1)\log p}$ et en 1 un pôle simple de résidu $\frac{hp^{1-g}}{(p-1)\log p}$; naturellement d'autres pôles s'en déduisent par la périodicité de $\mathbf{Z}(s)$ (période $\frac{2\pi i}{\log p}$). La formule explicite précédente donne l'*équation fonctionnelle* $\mathbf{Z}(1-s) = p^{(g-1)(2s-1)}\mathbf{Z}(s)$ en utilisant le théorème de Riemann-Roch sous la forme $\{\mathbf{C}\} - \frac{q}{2} = \{\mathbf{C}'\} - \frac{q'}{2}$.

Si z est un élément de K non dans k et que $\mathfrak{J}$ désigne la fermeture entière de $k(z)$ dans K, Schmidt note

$$\mathbf{Z}_{\mathfrak{J}}(s) = \prod_{\tilde{p}} \frac{1}{1 - |\tilde{\mathbf{p}}|^{-s}}$$

la fonction *zêta* d'Artin correspondante; on a

$$\mathbf{Z}(s) = \prod_{\mathbf{u}} \frac{1}{1 - |\mathbf{u}|^{-s}} \mathbf{Z}_{\mathfrak{J}}(s)$$

où $\mathbf{u}$ parcourt l'ensemble (fini) des diviseurs premiers distincts intervenant dans $\frac{1}{z}$. On trouve ainsi que $\mathbf{Z}_{\mathfrak{J}}(s)$ n'a comme singularités que des pôles d'ordre 1 en 1 et aux points qui s'en déduisent par périodicité et qu'elle s'annule en 0 sauf dans le cas où le dénominateur du diviseur de z est une puissance d'un unique diviseur premier $\mathbf{u}$; dans ce dernier cas $\mathbf{Z}_{\mathfrak{J}}(0) = -\frac{u_z h}{p-1}$ où u_z est l'ordre de $\mathbf{u}$. Le groupe des classes d'idéaux de $\mathfrak{J}$ est isomorphe à $\mathbf{D}/\mathbf{HU}$ où $\mathbf{U}$ désigne le sous-groupe de $\mathbf{D}$ engendré par les diviseurs premiers intervenant dans $\frac{1}{z}$; on obtient ainsi la finitude de ce groupe et la formule $h = \frac{h_z r_z}{u_z}$ où h_z est le nombre de classes d'idéaux de $\mathfrak{J}$, u_z est le p.g.c.d. des ordres des diviseurs premiers qui interviennent dans $\frac{1}{z}$ et r_z est le *régulateur* de $\mathfrak{J}$, c'est-à-dire l'indice du groupe $\mathbf{H} \cap \mathbf{U}_0$ des diviseurs d'unités de $\mathfrak{J}$ dans le groupe $\mathbf{U}_0 = \mathbf{D}_0 \cap \mathbf{U}$. A partir du résidu de $\mathbf{Z}(s)$ en 1, Schmidt calcule celui de $\mathbf{Z}_{\mathfrak{J}}(s)$ sous la forme

$$\frac{h_z r_z}{p^{\frac{w_z}{2}}} u_z \frac{(p^{f_1} - 1)^{e_1} \dots (p^{f_s} - 1)^{e_s}}{(p-1)\log p}$$

où les exposants f_i, e_i proviennent du dénominateur $\mathbf{u}_1^{e_1}\mathbf{u}_2^{e_2} \dots \mathbf{u}_s^{e_s}$ du diviseur de z, f_i désignant l'ordre du diviseur premier $\mathbf{u}_i$; dans le cas où le degré

$$m_z = (K : k(z)) = e_1 f_1 + \dots + e_s f_s$$

vaut 2 on retrouve les formules d'Artin.

H. Hasse

Au congrès international d'Oslo en 1936, Hasse a présenté un travail dans lequel il établit l'hypothèse de Riemann pour la fonction *zêta* de Schmidt dans le cas du genre 1 (Hasse 1936). Il démontre que cette hypothèse est équivalente à la majoration $|N_1 - (q+1)| \leq 2\sqrt{q}$ où N_1 est le nombre de «points», c'est-à-dire de diviseurs premiers de degré 1 (ordre au sens de Schmidt) et q est le nombre d'éléments du corps des constantes; ce type d'inégalité renforce des encadrements obtenus par Davenport et par Mordell pour le nombre N des solutions d'une équation diophantienne $y^2 = f(x)$ sur $\mathbf{F}_p$ où f est un polynôme de degré 3 à coefficients dans $\mathbf{F}_p$ (Davenport $|N-p| \leq cp^{3/4}$; Mordell $|N-p| \leq cp^{2/3}$ et ces auteurs conjecturaient que l'exposant $\frac{1}{2}$ était le meilleur possible).

Hasse part d'un polynôme irréductible $f(x, y)$ à coefficients dans $\mathbf{F}_p$ (p premier) et il considère un facteur irréductible f_0 de f sur la clôture algébrique k de $\mathbf{F}_p$; il désigne par k_0 le corps engendré par les coefficients de f_0, par K_0 le corps $k_0(x, y)$ où x et y sont liés par l'équation $f_0(x, y) = 0$ et par g le genre de K_0. Le nombre d'éléments de k_0 est $q = p^f$ et le nombre de *points* (diviseurs premiers d'ordre 1) de K_0 est N_1; dans le cas de Davenport et Mordell on a $N_1 = N + 1$ à cause du point à l'infini. La fonction *zêta* est

$$\mathbf{Z}(s) = \prod_{\mathbf{p}} \frac{1}{1 - \frac{1}{(\mathbf{Np})^s}} = \prod_{n \geq 1} \left(\frac{1}{1 - \frac{1}{q^{ns}}}\right)^{N_n} = 1 + \frac{N_1}{q^s} + \dots$$

où le premier produit s'étend à tous les diviseurs premiers, $\mathbf{Np}$ désigne la «norme» du diviseur $\mathbf{p}$ et N_n est le nombre de diviseurs premiers de degré n; Hasse la décompose en deux facteurs, $\mathbf{Z}(s) = \mathbf{Z}_0(s)L(s)$ où

$$\mathbf{Z}_0(s) = \frac{1}{1 - \frac{1}{q^s}} \frac{1}{1 - \frac{q}{q^s}} = 1 + \frac{q+1}{q^s} + \dots$$

est la fonction *zêta* de $k_0(x)$ tandis que

$$L(s) = 1 + \frac{N_1 - (q+1)}{q^s} + \dots + \frac{q^g}{q^{2gs}} = \prod_{i=1}^{2g} \left(1 - \frac{\omega_i}{q^s}\right)$$

est un polynôme de degré $2g$ en $\frac{1}{q^s}$. On a ainsi

$$N_1 - (q+1) = -\sum_{i=1}^{2g} \omega_i;$$

l'*hypothèse de Riemann*, soit $|\omega_i| = q^{1/2}$ pour tout i, implique donc la majoration $|N_1 - (q+1)| \leq 2g\sqrt{q}$.

Dans le cas où $g = 1$ on a simplement $L(s) = 1 + \frac{N_1-(q+1)}{q^s} + \frac{q}{q^{2s}}$ et l'hypothèse de Riemann signifie que les racines ω_1 et ω_2 sont imaginaires conjuguées. Hasse étudie la structure de l'ensemble $\mathbf{A}$ des points du corps $K = k(x, y)$; une fois choisie une «origine» $\mathbf{o}$ dans $\mathbf{A}$, cet ensemble se trouve muni d'une structure de groupe d'élément neutre $\mathbf{o}$, car chaque diviseur de degré 0 est équivalent à un diviseur $\frac{\mathbf{p}}{\mathbf{o}}$ où $\mathbf{p}$ est premier. Pour n non multiple de p, les solutions de l'équation $n\mathbf{p} = \mathbf{o}$ dans $\mathbf{A}$ forment un sous-groupe d'ordre n^2; mais pour $n = p^k$, ce sous-groupe est d'ordre 1 ou n selon que l'*invariant de Hasse* A de K est nul ou non. Hasse conjecture des résultats analogues dans le cas d'un genre g quelconque pour le groupe des classes de diviseurs de degré 0: si n n'est pas divisible par p, la division par n de l'élément neutre conduit à un sous-groupe d'ordre n^{2g} et si $n = p^k$, elle donne un sous-groupe d'ordre n^r où r est le rang d'une matrice A carrée d'ordre g associée à K.

Les éléments de k_0 sont caractérisés, parmi ceux de k, par l'équation $a^q = a$. Les points de K_0 peuvent donc être identifiés à ceux de K qui sont invariants relativement à l'opération de *Frobenius* π ainsi définie: on considère l'isomorphisme $\pi : \phi(x, y) \mapsto \phi(x^q, y^q)$ de K sur son sous-corps $K\pi = k(x^q, y^q)$ et on pose $\pi\mathbf{p} = (N_{K/K\pi}\mathbf{p})^{\pi^{-1}}$ pour tout diviseur premier $\mathbf{p}$ de K. Hasse explicite l'analogie de π avec la multiplication complexe des fonctions elliptiques; le rapport τ des périodes d'une fonction elliptique est quadratique imaginaire dans le cas où l'invariant $J = \frac{g_2^3}{\Delta}$ vérifie une certaine équation algébrique qui se réduit *mod.p* à $J^q \equiv J$ (*mod.p*) avec $q = p^f$. Dans ce cas la fonction $h(u) = \frac{g_2 g_3}{\Delta}\wp(u)$ vérifie $h(\pi u_0) \equiv h(u_0)^q$ (*mod.*$\mathbf{P}$) où $\pi \in \mathbf{Q}(\tau)$ est tel que $q = \pi\bar{\pi}$ et $\mathbf{P}$ est un idéal de $\mathbf{Q}(\tau, J, \wp(u_0))$ au-dessus de π.

Ceci amène Hasse à construire la théorie des *méromorphismes* de K; un tel méromorphisme est un isomorphisme μ de K sur l'un de ses sous-corps $K\mu$. Pour tout diviseur premier $\mathbf{p}$ on pose $\mu\mathbf{p} = (N_{K/K\mu}\mathbf{p})^{\mu^{-1}}$ de manière à faire opérer μ sur le groupe $\mathbf{A}$; la loi de $\mathbf{A}$ définit une loi (notée additivement) sur l'ensemble M des méromorphismes μ tels que $\mu\mathbf{o} = \mathbf{o}$ et l'élément neutre de cette loi est le méromorphisme O tel que $O\mathbf{p} = \mathbf{o}$ quel que soit $\mathbf{p}$. Cette loi et la composition des méromorphismes font de M un *anneau* (non nécessairement commutatif!) contenant $\mathbf{Z}$, dont les éléments correspondent aussi aux points du corps $\mathbf{K} = K(\xi, \eta)$ (avec $f_0(\xi, \eta) = 0$). Hasse définit la *norme* $N(\mu)$ de $\mu \in M$ comme le degré $[K : K\mu]$; on a $N(\mu\mu') = N\mu.N\mu'$ et $N(O) = 0$ et il en résulte que M n'a pas de diviseur de 0 non nul. La propriété fondamentale $N(\mu + \nu) + N(\mu - \nu) = 2N(\mu) + 2N(\nu)$ implique les identités

$$N(m\mu + n\nu) = m^2 N(\mu) + mn(N(\mu + \nu) - N(\mu) - N(\nu)) + n^2 N(\nu) \text{ et}$$
$$N(n) = n^2$$

d'où l'inégalité $(N(\mu \pm \nu) - N(\mu) - N(\nu))^2 \leq 4N(\mu)N(\nu)$ (cf. l'inégalité de Cauchy-Schwarz) qui donne en particulier $(N(\mu \pm 1) - N(\mu) - 1)^2 \leq 4N(\mu)$; il en résulte qu'un méromorphisme $\mu \in M$ vérifie une équation du second degré

$$\mu^2 + l\mu + m = 0 \text{ avec } l = N(\mu - 1) - N(\mu) - 1 = \mu + \bar{\mu} \text{ et } m = N(\mu) = \mu\bar{\mu}$$

($\bar{\mu}$ «conjugué» de μ) et $l^2 \leq 4m$. L'anneau M des méromorphismes de K est de l'un des trois types suivants:

I. *l'ordre principal* $\mathbf{Z}$ (anneau des entiers);

II. un ordre d'un corps quadratique imaginaire $\Omega = \mathbf{Q}(\delta)$ ($\delta^2 = d < 0$);

III. un ordre d'un corps gauche quadratique imaginaire $A = \mathbf{Q}(\delta, \delta')$ où $\delta^2 = d < 0$, $\delta'^2 = d' < 0$ et $\delta'\delta = -\delta\delta'$.

Cette théorie donne immédiatement l'hypothèse de Riemann car les points de K_0 s'identifient aux points $\mathbf{p}_1$ de K tels que $\pi\mathbf{p}_1 = \mathbf{p}_1$; ce sont donc les points annulés par $\pi - 1$ et leur nombre est $N_1 = N(\pi - 1)$. L'équation vérifiée par π s'écrit $\pi^2 + l\pi + q = 0$ avec $l = N(\pi - 1) - N(\pi) - 1 = N_1 - (q+1)$ et $q = N(\pi)$; comme on sait que $l^2 < 4q$, les racines de cette équation sont imaginaires conjuguées π et $\bar{\pi}$ (cas limites $N_1 - (q+1) = 0$ ou $\pm 2\sqrt{q}$ dans le cas dégénéré). Or $\pi^2 + l\pi + q$ est le numérateur de la fonction *zêta*; le résultat précédent signifie donc que les racines de $\mathbf{Z}$ sont imaginaires conjuguées et, comme leur produit est q, leur valeur absolue est $q^{1/2}$.

Hasse termine son article en signalant une suggestion de Deuring pour étendre la théorie au cas d'un genre quelconque $g \geq 1$; il s'agit de construire, pour les corps K de genre g et de caractéristique $p > 0$, l'analogue de la théorie des *correspondances* développée par Hurwitz pour les surfaces de Riemann (Hurwitz 1886). Si $\mu : K \xrightarrow{\sim} K_0^*$ est un isomorphisme de corps, on considère le corps composé $K^* = K_0^* K$ et on pose, pour tout diviseur premier $\mathbf{p}$, $\mu\mathbf{p} = (N_{K^*/K_0^*}\mathbf{p})^{\mu^{-1}}$; en posant $m = [K^* : K_0^*]$ et $n = [K^* : K]$, la correspondance associe un n-uple de $\mu\mathbf{p}$ à chaque m-uple de $\mathbf{p}$. La norme de μ est $N(\mu) = m^g$; les correspondances μ sont associées aux diviseurs premiers $\mathbf{P}$ de l'extension $\mathbf{K}$ construite comme plus haut (le degré de $\mathbf{P}$ est n). Le passage des méromorphismes aux correspondances permet de retrouver une loi additive et d'avoir un anneau de correspondances; en effet l'ensemble des points de K n'a pas de structure de groupe si $g \geq 2$.

A. Weil

C'est précisément à l'aide de la théorie des correspondances qu'André Weil est finalement parvenu à démontrer l'hypothèse de Riemann pour un genre g quelconque; il a annoncé son résultat dès 1940 dans une note aux *Comptes rendus* de l'Académie des Sciences, mais la démonstration complète n'a été publiée qu'en 1948 (Weil 1948a). Entre-temps il a élaboré la géométrie algébrique abstraite dont il avait besoin (Weil 1946); car Weil se place dans un cadre entièrement géométrique (avec un corps de base fini) au lieu du cadre inspiré de la théorie des fonctions algébriques qui servait à Schmidt et à Hasse.

Weil considère donc une *courbe algébrique complète* Γ sans point multiple définie sur un corps de base k; une courbe est, par définition, une variété abstraite de dimension 1 au sens des *Foundations*. Lorsque k est parfait, un corps K de fonctions algébriques d'une variable sur k tels que ceux considérés par Schmidt peut toujours s'identifier au corps abstrait Ω_k des fonctions sur une courbe complète Γ sans point multiple qui ont k comme corps de définition. De plus, les notions de diviseur sur Γ rationnels par rapport à k (au sens de la géométrie algébrique: ce sont des combinaisons linéaires formelles à coefficients entiers de points) et de diviseur de Ω_k se correspondent de manière à respecter le degré d'un diviseur et le diviseur associé à une fonction; contrairement à l'usage de Schmidt et Hasse, Weil adopte une notation *additive* pour le groupe des diviseurs. Lorsque k est fini à q éléments, on définit la fonction *zêta* de Γ par

$$Z(u) = \sum_{\mathbf{a}} u^{\deg(\mathbf{a})} = \prod_{\mathbf{p}} (1 - u^{\deg(\mathbf{p})-1})$$

avec une somme étendue à tous les diviseurs positifs rationnels par rapport à k et un produit étendu à tous les diviseurs premiers rationnels par rapport à k; la convergence a lieu pour $|u| < \frac{1}{q}$ et on a encore

$$Z(u) = \sum_{m=0}^{\infty} D_m u^m$$

où D_m est le nombre de diviseurs positifs de degré m rationnels par rapport à k. Comme Schmidt, Weil s'appuie sur le théorème de Riemann-Roch pour évaluer le nombre des diviseurs positifs rationnels par rapport à k équivalents à un diviseur $\mathbf{a}$ donné; ce nombre est $\frac{q^{l(\mathbf{a})}-1}{q-1}$ où $l(\mathbf{a})$ désigne le rang de l'espace vectoriel $L(\mathbf{a})$ sur le corps des constantes formé de 0 et des fonctions φ dont le diviseur (φ) majore $-\mathbf{a}$. Le théorème de Riemann-Roch dit que $l(\mathbf{a}) = \deg(\mathbf{a}) - g + 1 + r(\mathbf{a})$ avec $r(\mathbf{a}) = l(\mathbf{k} - \mathbf{a})$ où g est le *genre* de Γ (défini comme le maximum de $\deg(\mathbf{a}) - l(\mathbf{a}) + 1$ lorsque $\mathbf{a}$ varie) et $\mathbf{k}$ est un diviseur *canonique*, projection sur Γ d'un cycle intersection $X = \Delta.[(\theta) - \Delta]$ dans lequel Δ désigne la diagonale du produit $\Gamma \times \Gamma$ et θ désigne une fonction sur $\Gamma \times \Gamma$ ayant k pour corps de définition et telle que $\nu_\Delta(\theta) = 1$ (θ s'annule à l'ordre 1 sur Δ). La classe $\omega = \{\theta\}$ d'une telle fonction ω *modulo* les fonctions θ' pour lesquelles $\nu_\Delta(\theta') > 1$ s'appelle une *différentielle* et $\mathbf{k}$ ne dépend que de la différentielle ω; on dit que c'est le diviseur (ω) de ω. Tous les diviseurs canoniques sont équivalents; ils sont de degré $2g - 2$ et on a $l(\mathbf{k}) = g$, $r(\mathbf{k}) = 1$. La démonstration du théorème de Riemann-Roch occupe la première partie du travail de Weil; comme chez Schmidt et Hasse, ce théorème permet de démontrer que $Z(u)$ est une fonction rationnelle de la forme $\frac{P(u)}{(1-u)(1-qu)}$ où $P(u)$ est un polynôme de degré $2g$ et que l'on a une équation fonctionnelle $Z(\frac{1}{qu}) = q^{1-g}u^{2-2g}Z(u)$.

Weil considère la différentielle logarithmique

$$(**) \qquad d[\log Z(u)] = \sum_{\mathbf{p}} \sum_{\rho=1}^{\infty} \deg(\mathbf{p}) u^{\rho \deg(\mathbf{p})} . \frac{du}{u} = \sum_{n=1}^{\infty} \nu_n u^n \frac{du}{u}$$

où, pour chaque n,

$$\nu_n = \sum_{\deg(\mathbf{p})|n} \deg(\mathbf{p})$$

s'interprète encore comme le nombre des points P de Γ tels que $k(P)$ soit contenu dans l'extension k_n de degré n de k (on rappelle que cette extension est unique dans une clôture algébrique donnée et qu'elle a q^n éléments). Pour calculer un coefficient ν_n, Weil utilise le fait que les points P qu'il compte sont les points fixes d'une *correspondance* I_n sur la courbe Γ; la deuxième partie de son travail développe la théorie des correspondances sur une courbe.

Par définition, une correspondance X sur Γ est un diviseur sur le produit $\Gamma \times \Gamma$ (combinaison linéaire formelle à coefficients entiers de courbes). On écrit $X \sim Y$ lorsque $X - Y$ est le diviseur d'une fonction $\varphi \neq 0$ sur $\Gamma \times \Gamma$ et $X \equiv Y$ lorsqu'il existe des diviseurs $\mathbf{a}$ et $\mathbf{b}$ sur Γ tels que $X \sim Y + \Gamma \times \mathbf{a} + \mathbf{b} \times \Gamma$; dans ce dernier cas on dit que X et Y sont équivalents et l'objet principal de la théorie est l'étude de l'ensemble des *classes de correspondances* relativement à cette notion d'équivalence. La symétrie $(P \times Q)' = Q \times P$ de $\Gamma \times \Gamma$ induit une involution $X \mapsto X'$ sur l'ensemble des correspondances et cette involution est compatible avec les relations $\sim$ et $\equiv$. Weil définit une forme bilinéaire (à valeurs entières) $(X, Y) \mapsto I(X.Y)$ sur le groupe additif des correspondances de manière que $I(X.Y) = \deg(X.Y)$ lorsque le produit d'intersection $X.Y$ est défini; cette forme passe au quotient par la relation $\sim$. A une correspondance X sont associés deux entiers $d(X)$ et $d'(X)$ tels que les projections de X sur les deux facteurs de $\Gamma \times \Gamma$ soient respectivement $d(X)\Gamma$ et $d'(X)\Gamma$. L'image d'un point P par la correspondance X est définie comme le diviseur $X(P)$ sur Γ tel que $X_0.(P \times \Gamma) = P \times X(P)$ où X_0 n'a pas de composante de la forme $A \times \Gamma$ et $X = X_0 + \mathbf{a} \times \Gamma$ ($\mathbf{a}$ diviseur sur Γ). Par linéarité on étend l'opération de X aux diviseurs $\mathbf{a}$ sur Γ de manière que $X(\mathbf{a})$ soit la deuxième projection de $X_0.(\mathbf{a} \times \Gamma)$; le degré de $X(\mathbf{a})$ est $d(X) \deg(\mathbf{a})$. L'application $\mathbf{a} \mapsto X(\mathbf{a})$ est compatible avec la relation $\sim$ et $X \equiv 0$ équivaut à $X(\mathbf{m}) \sim 0$ pour tout diviseur $\mathbf{m}$ de degré 0 sur Γ.

Weil définit alors la composition $X \circ Y$ de deux correspondances; c'est une correspondance pour laquelle $(X \circ Y)(\mathbf{a}) = X(Y(\mathbf{a}))$ quel que soit le diviseur $\mathbf{a}$ sur Γ. On a $(X \circ Y)' = Y' \circ X'$, $\Delta \circ X = X \circ \Delta = X$ en notant Δ la correspondance diagonale pour laquelle $\Delta(P) = P$, $d(X \circ Y) = d(X)d(Y)$ et $d'(X \circ Y) = d'(X)d'(Y)$. La composition des correspondances est compatible avec la relation $\equiv$; c'est la multiplication pour une structure d'anneau sur l'ensemble A des classes de correspondances. Cet anneau (non commutatif) a pour élément unité la classe δ de Δ et il est muni d'une «anti-involution» $\xi \mapsto \xi'$ provenant de la symétrie de $\Gamma \times \Gamma$. Si X et Y sont des correspondances quelconques $I(X.Y') = I[\Delta.(X \circ Y)]$. Comme dans la théorie de Hurwitz, on associe à toute correspondance X un entier $S(X) = d(X) + d'(X) - I(X.\Delta)$ qui dépend linéairement de X. Il ne dépend que de la classe de X dans A, ce qui définit une forme linéaire $\sigma : A \mapsto \mathbf{Z}$ ayant des propriétés d'une *trace*: $\sigma(\xi') = \sigma(\xi)$ et $\sigma(\xi.\eta) = \sigma(\eta.\xi)$; de plus $\sigma(\delta) = 2g$. La propriété essentielle de σ est la positivité de la forme quadratique $\sigma(\xi.\xi')$: si $\xi \neq 0$ on a $\sigma(\xi.\xi') > 0$; c'est l'analogue, en géométrie algébrique abstraite, d'un résultat de Castelnuovo et Severi (Castelnuovo-Severi 1926). Dans le cas du genre 1, Weil le démontre grâce au fait que l'on a alors $\xi.\xi' = N\delta$ où N est un entier > 0; pour $g \geq 2$ il est obligé de travailler sur la variété $\Omega = \Gamma \times \Gamma \times \ldots \times \Gamma$

produit de $d = d(X)$ facteurs identiques à Γ (où X est une correspondance de la classe ξ).

Pour démontrer l'hypothèse de Riemann, Weil applique la théorie des correspondances à la correspondance I définie par l'automorphisme $\omega : x \mapsto x^q$ du domaine universel et à ses itérées I_n; si $(x_1, \ldots, x_m)$ sont les coordonnées d'un point P sur un représentant de Γ, celles de $I_n(P)$ sont $(x_1^{q^n}, \ldots, x_m^{q^n})$. La correspondance I_n est une courbe sur $\Gamma \times \Gamma$; on a $d(I_n) = 1$, $d'(I_n) = q^n$ et $\deg(I_n.\Delta) = \nu_n$; en notant ι la classe de I, celle de I_n est ι^n et on a $\sigma(\iota^n) = 1+q^n-\nu_n$. La positivité de $\sigma(\xi.\xi')$ appliquée à $\xi = x\delta + y\iota^n$ implique alors celle de $2gx^2 + 2\sigma(\iota^n)xy + 2gq^n y^2$ quels que soient x et y et on a donc $|\sigma(\iota^n)| = |1 + q^n - \nu_n| \leq 2gq^{n/2}$. Or le numérateur

$$P(u) = (1-u)(1-qu)Z(u)$$

de la fonction *zêta* a pour dérivée logarithmique

$$d[\log P(u)] = -\sum_{n=1}^{\infty} \sigma(\iota^n)u^n.\frac{du}{u}$$

en vertu de la formule $(**)$ et l'inégalité précédente prouve que la série converge pour $|u| < q^{-1/2}$; ainsi $P(u)$ n'a ni zéro ni pôle dans ce disque, donc $Z(u)$ n'a pas de zéro et a un seul pôle $u = \frac{1}{q}$. Par l'équation fonctionnelle on en déduit que $Z(u)$ n'a pas non plus de zéro à l'extérieur de ce disque et que son seul pôle y est $u = 1$; on a bien établi que tous les zéros de $Z(u)$ sont sur le cercle $|u| = q^{-1/2}$. En écrivant

$$P(u) = \prod_{i=1}^{2g}(1-\alpha_i u),$$

les α_i sont des entiers algébriques de valeur absolue $q^{1/2}$ et on a

$$(***) \qquad \sigma(\iota^n) = \sum_{i=1}^{2g} \alpha_i^n.$$

Weil termine son mémoire en définissant des fonctions L associées à des caractères non nécessairement commutatifs de groupes de Galois; dans la théorie des nombres algébriques, de telles fonctions L avaient été introduites par Artin (Artin 1923) qui avait formulé à leur sujet de célèbres conjectures. Weil établit, pour ses fonctions L, l'analogue de l'hypothèse de Riemann, une équation fonctionnelle et il annonce un résultat permettant de démontrer les conjectures d'Artin dans son cas. Ce résultat se trouve dans un deuxième travail de Weil consacré à la théorie des variétés abéliennes, c'est-à-dire aux variétés de groupe qui sont complètes (Weil 1948b); la loi de groupe est alors nécessairement commutative. Pour toute courbe Γ complète sans point multiple de genre $g > 0$, Weil construit une variété abélienne J de dimension g, la *jacobienne* de Γ; par définition elle est birationnellement équivalente au «produit symétrique» de g facteurs identiques à Γ sur lequel on définit une loi de composition en se servant de l'addition des diviseurs de degré g (on choisit un diviseur $\mathbf{a}$ de degré g rationnel sur le corps de base k pour jouer le rôle d'élément neutre). Il existe une fonction φ définie (à une constante additive près) sur Γ à valeurs dans J telle que, si K est un corps de définition pour Γ, J et φ et si $M_1, \ldots, M_g$ sont des points génériques indépendants de Γ par rapport à K,

$$z = \sum_{i=1}^{g} \varphi(M_i)$$

soit un point générique de J par rapport à K. Pour qu'un diviseur $\mathbf{m}$ de degré 0 sur Γ soit équivalent à 0, il faut et il suffit que $S(\varphi(\mathbf{m})) = 0$ en notant, pour tout cycle

$$X = \sum_i m_i(x_i)$$

sur J, par $S(X)$ le point

$$\sum_i m_i x_i$$

de J; l'application $\mathbf{m} \mapsto S(\varphi(\mathbf{m}))$ induit un isomorphisme du groupe des classes de diviseurs de degré 0 (sur Γ) sur le groupe des points de J. Si f est une fonction définie sur Γ à valeurs dans une variété abélienne A, il existe un unique homomorphisme λ de J dans A tel que $f = \lambda\varphi + a$ où a est une constante. Soit C une correspondance sur Γ; en appliquant la dernière propriété à $f(M) = S(\varphi(C(M))$, on trouve un endomorphisme ξ de J tel que $S(\varphi(C(\mathbf{m}))) = \xi.S(\varphi(\mathbf{m})) + \deg(\mathbf{m})a$ pour tout diviseur $\mathbf{m}$ sur Γ (a constante). L'application $C \mapsto \xi$ ainsi obtenue définit un isomorphisme de l'anneau A des classes de correspondances sur l'anneau des endomorphismes de J et la trace $\sigma(\xi)$ d'une classe de correspondance ξ reçoit alors une interprétation remarquable.

A tout homomorphisme λ d'une variété abélienne A dans une autre B de même dimension n est attaché un entier $\nu(\lambda)$ tel que $\lambda(A) = \nu(\lambda)B$ où le premier membre est, par définition, le cycle projection sur B du graphe de λ; si δ désigne l'homomorphisme identique de A, $\nu(a\delta) = a^{2n}$ quel que soit l'entier a. Soient J la jacobienne d'une courbe Γ, λ un endomorphisme de J et a, b des entiers; alors $\nu(a\delta + b\lambda)$ est un polynôme homogène en a et b de degré $2g$ et de la forme $a^{2g} + \sigma(\lambda)a^{2g-1}b + \ldots + \nu(\lambda)b^{2g}$. Si l est un nombre premier distinct de la caractéristique p, le groupe $\mathbf{g}_l(J)$ des points de J dont l'ordre est une puissance de l est isomorphe à $\mathbf{Q}_l^{2g}/\mathbf{Z}_l^{2g}$ et on peut faire opérer λ sur ce groupe; une fois choisi un isomorphisme, il existe une matrice bien déterminée $M_l(\lambda)$ carrée d'ordre $2g$ et à coefficients entiers l-adiques relevant dans $\mathbf{Q}_l^{2g}$ les opérations de λ. Le polynôme P tel que $P(a) = \nu(a\delta - \lambda)$ pour a entier coïncide avec le polynôme caractéristique de cette matrice; en particulier $\sigma(\lambda) = \operatorname{Tr} M_l(\lambda)$ et $\nu(\lambda) = \det M_l(\lambda)$. Dans le cas où on prend pour λ l'endomorphisme de J associé à la correspondance ι sur Γ, on a, pour tout $n \geq 0$,

$$\sigma(\iota^n) = \sum_{i=1}^{2g} \omega_i^n$$

où les ω_i sont les racines du polynôme caractéristique et on voit donc que les ω_i coïncident avec les α_i en comparant avec $(***)$; ainsi le numérateur P de la fonction *zêta* est le polynôme caractéristique de ι opérant sur la jacobienne.

Vers les conjectures de Weil

Weil s'est ensuite attaqué à l'étude de la fonction *zêta* attachée à une variété algébrique X_0 de dimension ≥ 2 sans point multiple définie sur un corps fini k à q éléments. Pour tout entier $m \geq 1$, il note ν_m le nombre de points P de X_0 tels que $k(P) \subset k_m$ et il définit la fonction $Z(t)$ par $Z(0) = 1$ et

$$t \frac{d}{dt} \log Z(t) = \sum_{m=1}^{\infty} \nu_m t^m.$$

Par exemple, si X_0 est l'espace affine de dimension r, on a $\nu_m = q^{mr}$ donc $t \frac{d}{dt} \log Z(t) = \frac{q^r t}{1-q^r t}$ et $Z(t) = \frac{1}{1-q^r t}$; si X_0 est l'espace projectif de dimension r, $\nu_m = 1 + q^m + \ldots + q^{mr} = \frac{q^{m(r+1)}-1}{q-1}$ et $t \frac{d}{dt} \log Z(t) = \frac{t}{1-t} + \frac{qt}{1-qt} + \ldots + \frac{q^r t}{1-q^r t}$ d'où $Z(t) = \frac{1}{(1-t)(1-qt)\ldots(1-q^r t)}$. Weil calcule aussi la fonction *zêta* de la grassmannienne des sous-espaces de dimension r dans l'espace de dimension n; on a alors

$$\nu_m = \frac{q^{m(n+1)} - 1}{q - 1} \cdots \frac{q^{m(n+1)} - q^{mr}}{q^{m(r+1)} - q^{mr}} = 1 + b_1 q^m + \ldots + b_d q^{dm}$$

avec $d = (n - r)(r + 1)$ (dimension de la grassmannienne) et

$$b_i = \frac{i^{n+1} - 1}{i^{r+1} - 1} \cdots \frac{i^{n+1} - i^r}{i^{r+1} - i^r} \text{ pour } 1 \leq i \leq r$$

(c'est le nombre de Betti en dimension $2i$ de la grassmannienne complexe) et on tire $t \frac{Z'(t)}{Z(t)} = \frac{t}{1-t} + b_1 \frac{qt}{1-qt} + \ldots + b_d \frac{q^d t}{1-q^d t}$ d'où

$$Z(t) = \frac{1}{(1-t)(1-qt)^{b_1} \ldots (1-q^d t)^{b_d}}.$$

Dans un article de 1949 (Weil 1949a), Weil étudie le cas d'une *hypersurface monomiale*, d'équation $a_0 x_0^{n_0} + \ldots + a_r x_r^{n_r} = b$ dans l'espace affine de dimension

$r+1$; il utilise, pour compter les points de l'hypersurface, une méthode déjà mise en oeuvre par Hardy et Littlewood dans leur étude du problème de Waring (Hardy-Littlewood 1922) et reprise par Hasse et Davenport à propos de l'équation $ax^m+by^n+cz^r=0$ (Davenport-Hasse 1935). Comme le remarque A. Weil, Gauss avait déjà utilisé une méthode analogue en traitant la congruence $ax^3-by^3\equiv 1\ (mod.p)$ avec p premier $\equiv 1\ (mod.3)$ (Gauss 1801, n^{0}358) et la congruence $ax^4-by^4\equiv 1\ (mod.p)$ avec p premier $\equiv 1\ (mod.4)$ (Gauss 1832). En prenant d'abord $b=0$, Weil transforme l'équation de son hypersurface en le système $x_i^{n_i}=u_i\ (0\leq i\leq r)$ et

$$L(u)=\sum_{i=1}^{r}a_iu_i=0;$$

le nombre de points s'écrit donc

$$N=\sum_{L(u)=0}N_0(u_0)\dots N_r(u_r)$$

où, pour $0\leq i\leq r$ et pour $u\in k$, on désigne par $N_i(u)$ le nombre des $x\in k$ tels que $x^{n_i}=u$. Pour $u=0$ on a $N_i(0)=1$ et pour $u\neq 0$, $N_i(u)$ est nul ou bien égal au p.g.c.d. d_i de n_i et de $q-1$. On considère un générateur w du groupe multiplicatif k^* et un nombre rationnel α dont le dénominateur divise $q-1$; en posant $\chi_\alpha(w)=e^{2\pi i\alpha}$ on définit un caractère multiplicatif d'ordre $q-1$ que l'on prolonge à k entier en posant $\chi_\alpha(0)=0$ si α n'est pas entier et $\chi_\alpha(0)=1$ si $\alpha\in\mathbf{Z}$. Alors

$$N_i(u)=\sum_\alpha\chi_\alpha(u)$$

où la somme est étendue aux α rationnels $\in[0,1[$ dont le dénominateur divise d_i; on obtient donc

$$N=\sum_{u,\alpha}\chi_{\alpha_0}(u_0)\dots\chi_{\alpha_r}(u_r)$$

avec une somme étendue aux $u=(u_i)$ tels que $L(u)=0$ et aux $\alpha_i\in[0,1[$ tels que $d_i\alpha_i\in\mathbf{Z}$. Les termes où tous les α_i sont nuls donnent une somme égale à q^r; ceux où le nombre s des α_i nuls est $<r$ donnent une somme nulle et il reste les termes pour lesquels $0<\alpha_i<1$. Ainsi

$$N=q^r+\sum_\alpha\chi_{\alpha_0}(a_0^{-1})\dots\chi_{\alpha_r}(a_r^{-1})S(\alpha)\quad(0<\alpha_i<1,\quad d_i\alpha_i\in\mathbf{Z})$$

avec

$$S(\alpha)=\sum_{\Sigma u_i=0}\chi_{\alpha_0}(u_0)\dots\chi_{\alpha_r}(u_r);$$

ceci se transforme en

$$N=q^r+(q-1)\sum_\alpha\chi_{\alpha_0}(a_0^{-1})\dots\chi_{\alpha_r}(a_r^{-1})j(\alpha)$$

où

$$j(\alpha) = \sum_{1+v_1+\ldots+v_r=0} \chi_{\alpha_1}(v_1)\ldots\chi_{\alpha_r}(v_r)$$

est une *somme de Jacobi* et où les α_i sont assujettis aux conditions $0 < \alpha_i < 1$, $d_i\alpha_i \in \mathbf{Z}$ et

$$\sum_i \alpha_i \in \mathbf{Z}.$$

Si ψ est un caractère additif non trivial de k, on développe chaque χ en «série de Fourier»

$$\chi(x) = \frac{g(\chi)}{q}\sum_t \bar{\chi}(t)\bar{\psi}(tx)$$

avec

$$g(\chi) = \sum_x \chi(x)\psi(x)$$

somme de Gauss; ceci donne $j(\alpha) = \frac{1}{q}g(\chi_{\alpha_0})\ldots g(\chi_{\alpha_r})$ de valeur absolue $q^{\frac{r-1}{2}}$. Tous ces calculs conduisent enfin à l'inégalité $|N - q^r| \leq M(q-1)q^{\frac{r-1}{2}}$ où M note le nombre de solutions du système $n_i\alpha_i \equiv 0$ $(mod.1)$,

$$\sum_i \alpha_i \equiv 0$$

$(mod.1)$ et $0 < \alpha_i < 1$ pour $0 \leq i \leq r$.

Pour le nombre N_1 de solutions de l'équation

$$\sum_i a_i x_i^{n_i} + 1 = 0$$

on peut employer la relation $N' = (q-1)N_1 + N$ où N' est le nombre de solutions de

$$\sum_{i=0}^{r} a_i x_i^{n_i} + x_{r+1}^{q-1} = 0;$$

on obtient

$$N_1 = q^r + \sum_\alpha \chi_{\alpha_0}(a_0^{-1})\ldots\chi_{\alpha_r}(a_r^{-1})j\left(\alpha_0,\ldots,\alpha_r,-\sum_{i=0}^{r}\alpha_i\right)$$

d'où l'inégalité $|N_1 - q^r| \leq M_1 q^{r/2}$ avec $M_1 = (d_0-1)\ldots(d_r-1) < n_0 n_1\ldots n_r$.

Dans le cas où tous les exposants n_i ont la même valeur n, on peut considérer la variété projective d'équation homogène $a_0x_0^n + \ldots + a_rx_r^n = 0$; si $\bar{N}$ est son nombre de points on a $N = 1 + (q-1)\bar{N}$ d'où

$$\bar{N} = 1 + q + \ldots + q^{r-1} + \sum_\alpha \chi_{\alpha_0}(a_0^{-1})\ldots\chi_{\alpha_r}(a_r^{-1})j(\alpha).$$

Weil utilise un théorème de Hasse et Davenport pour évaluer de même le nombre $\bar{N}_\nu$ de points rationnels sur k_ν; ce théorème exprime pour μ multiple de ν, les sommes de Gauss de k_μ au moyen de celles de k_ν. Il trouve ainsi la dérivée logarithmique de la fonction *zêta* sous la forme

$$\sum_{\nu=1}^{\infty} \bar{N}_\nu U^{\nu-1} = -\sum_{h=1}^{r-1} \frac{d}{dU} \log(1 - q^h U)$$
$$+ (-1)^r \sum_{\alpha} \frac{1}{\mu(\alpha)} \frac{d}{dU} \log(1 - C(\alpha) U^{\mu(\alpha)})$$

où la dernière somme est étendue aux $\alpha_i \in]0,1[$ tels que

$$n\alpha_i \equiv \sum_i \alpha_i \equiv 0$$

(*mod*.1) et $\mu(\alpha)$ désigne le plus petit μ tel que $(q^\mu - 1)\alpha_i \in \mathbf{Z}$ tandis que $C(\alpha) = (-1)^{r-1} \bar{\chi}_{\alpha_0}(a_0) \dots \bar{\chi}_{\alpha_r}(a_r) j(\alpha)$. Ainsi

$$Z(U) = \frac{P_{r-1}(U)^{(-1)^r}}{(1-U)(1-qU)\dots(1-q^{r-1}U)}$$

où P_{r-1} est un polynôme de degré M dont tous les zéros ont pour valeur absolue $q^{-\frac{r-1}{2}}$. Or Weil avait demandé à Dolbeault de calculer les nombres de Betti de la variété projective définie sur $\mathbf{C}$ par une équation du type considéré; le résultat est $B_h = 0$ ou 1 selon que $h < r - 1$ est impair ou pair et $B_{r-1} = M$. Ceci a conduit Weil à formuler les conjectures suivantes:

1. La fonction *zêta* d'une variété algébrique X_0 de dimension n sur k est *rationnelle*.
2. Elle satisfait une *équation fonctionnelle* $Z(\frac{1}{q^n t}) = \pm q^{\frac{n\chi}{2}} t^\chi \mathbf{Z}(t)$ où $\chi = (\Delta \cdot \Delta)$ joue le rôle de la *caractéristique d'Euler-Poincaré* de X_0.
3. On a $\mathbf{Z}(t) = \frac{P_1(t)P_3(t)\dots P_{2n-1}(t)}{P_0(t)P_2(t)\dots P_{2n}(t)}$ où $P_0(t) = 1 - t$, $P_{2n}(t) = 1 - q^n t$ et, pour $1 \leq h \leq 2n-1$, $P_h(t)$ est un polynôme dont les racines sont des entiers algébriques de valeur absolue $q^{-h/2}$.
4. Si B_h est le degré de P_h, ou *nombre de Betti* de X_0 en dimension h, on a

$$\chi = \sum_{h=0}^{2n} (-1)^h B_h.$$

De plus, si X est une variété algébrique sans point multiple sur un corps de nombres algébriques K, pour presque tout idéal premier $\mathbf{p}$ de K les nombres de Betti ainsi définis pour la réduction $X_{\mathbf{p}}$ de $X mod.\mathbf{p}$ coïncident avec ceux de X (au sens classique).

Weil indique enfin comment on pourrait déduire ces conjectures de l'existence d'une théorie cohomologique convenable pour les variétés algébriques sur un corps fini k; une telle théorie associe, d'une manière fonctorielle, à toute variété X sans point multiple sur k une suite d'espaces vectoriels $H^i(X)$ sur un corps K de *caractéristique* 0 et elle doit vérifier les conditions suivantes:

1) Dualité de Poincaré: si X est de dimension n, $H^i(X)$ est nul sauf pour $0 \leq i \leq 2n$ et $H^{2n}(X) \approx K$; il y a de plus un accouplement bilinéaire

$$H^i(X) \times H^{2n-i}(X) \to H^{2n}(X) \approx K$$

qui est une dualité parfaite.

2) Formule de Künneth: $H^*(X) \otimes_K H^*(Y) \approx H^*(X \times Y)$ pour deux variétés X et Y sans point multiple.

3) Classe d'un cycle: il y a un homomorphisme $\gamma_X : C^i(X) \to H^{2i}(X)$ où $C^i(X)$ désigne le groupe des classes de cycles pour l'équivalence numérique; il est compatible avec la multiplication et «fonctoriel».

4) Théorèmes de Lefschetz sur les sections hyperplanes.

Dans une telle théorie on dispose en effet de la formule de Lefschetz pour calculer le nombre de points fixes d'une correspondance: si $f \in H^0(X \times X)$ est une correspondance,

$$\langle f \cdot \Delta \rangle = \sum_{i=0}^{2n} (-1)^i \text{ Tr } (f_i)$$

où $f_i : H^i(X) \to H^i(X)$ est l'homomorphisme induit par f; cette formule généralise, en dimension quelconque, la formule de Hurwitz pour les correspondances sur une courbe (la jacobienne d'une courbe joue d'ailleurs le rôle d'un premier groupe de cohomologie). On peut donc calculer le nombre ν_m des points de X rationnels sur k_m comme

$$\sum_{i=0}^{2n} (-1)^i \text{ Tr } F_i^m$$

où F est la *correspondance de Frobenius* (notée plus haut ι dans le cas des courbes) et on a

$$t\frac{d}{dt} \log Z(t) = \sum_{i=0}^{2n} (-1)^i \sum_m \text{ Tr } (F_i^m) t^m = \sum_{i=0}^{2n} (-1)^{i+1} t \frac{d}{dt} \log \det(1 - tF_i)$$

d'où

$$Z(t) = \prod_{i=0}^{2n} \det(1 - tF_i)^{(-1)^{i+1}}$$

ce qui prouve la première conjecture (rationalité); la dualité de Poincaré donne alors l'équation fonctionnelle (conjecture 2) et on obtient l'hypothèse de Riemann (conjecture 3) à l'aide des théorèmes de Lefschetz sur les sections hyperplanes qui permettent de faire une récurrence sur la dimension.

Weil lui-même a démontré ses conjectures par voie directe pour certaines variétés autres que celles déjà considérées, par exemple les intersections de deux quadriques $F = 0$ et $G = 0$ (Weil 1954a; selon une idée de Carlitz il ramène le décompte des points à celui des points d'une courbe hyperelliptique $y^2 = \varphi(x, 1)$ où $\varphi(u, v)$ est le déterminant de $uF + vG$) ou les surfaces cubiques (Weil 1954b).

La première démonstration de la conjecture 1 dans le cas général n'utilise pas non plus la cohomologie. Elle est due à Dwork qui compte les points selon une méthode analogue à celle de (Weil 1949a) en remplaçant toutefois le théorème de Hasse-Davenport par des relèvements multiplicatifs *p- adiques* (Dwork 1959); il obtient ainsi la fonction *zêta* sous une forme de quotients de produits de déterminants de Fredholm d'opérateurs nucléaires p-adiques et il peut conclure grâce à un critère de rationalité de séries formelles à coefficients entiers qui généralise un théorème d'E. Borel.

La construction d'une théorie cohomologique satisfaisant aux conditions de Weil était un des principaux buts de la remise en chantier de la géométrie algébrique entreprise par Grothendieck à partir de 1957. Il y est parvenu en développant la cohomologie *étale l-adique* où l est un nombre premier distinct de la caractéristique p du corps de base; son travail est la base de la démonstration des conjectures de Weil par Deligne. La démonstration de la rationalité par Dwork a d'ailleurs été interprétée d'une manière cohomologique par divers auteurs (Washnitzer et Monsky); ce genre d'interprétation est englobé dans une autre théorie cohomologique inventée par Grothendieck, la cohomologie *cristalline* qui joue le rôle d'une cohomologie p-adique. On notera qu'il n'y a pas de cohomologie à coefficients *entiers* ou rationnels; un exemple dû à Serre prouve en effet que la cohomologie à coefficients entiers d'une variété algébrique complexe n'est pas un invariant algébrique. Les diverses cohomologies introduites n'ont donc pas de source commune et le problème de leur comparaison se pose. La théorie conjecturale des *motifs* de Grothendieck devrait donner une source commune à toutes les cohomologies; elle a pu être mise en oeuvre dans des cas particuliers, par exemple pour une démonstration des conjectures de Weil pour les variétés unirationnelles (Manin 1968).

Bibliographie

[1] ARTIN, E., 1923, *Über eine neue Art von L-Reihen.*

[2] ARTIN, E., 1924, *Quadratische Körper in Gebiete der höheren Kongruenzen,* Math. Zeits. 19 (1924) p. 153–246.

[3] CASTELNUOVO, G. et SEVERI, F., 1926, *Trattato di geometria algebrica*, 1926.

[4] Davenport, H. et Hasse, H., 1935, *Die Nullstellen der Kongruenzzetafunktionen in gewissen zyklischen Fällen*, Journ. für die reine und angew. Math. 172 (1935) p. 151–182.

[5] Dedekind, R., 1857, *Abriss einer Theorie der höheren Kongruenzen in Bezug auf einen reellen Primzahl-Modulus*, Journ. für die reine und angew. Math. 54 (1857) p. 1–26 = Ges. Math. Werke I p. 40–67.

[6] Dedekind, R., 1893, *Supplément XI aux Vorlesungen über Zahlentheorie* de Dirichlet, 4ème éd., Braunschweig 1893 = Ges. Math. Werke III.

[7] Dedekind, R. et Weber, H., 1882, *Theorie der algebraischen Funktionen einer Veränderlichen*, Journ. für die reine und angew. Math. 92 (1882) p. 181–290 = Dedekind, Ges. Math. Werke II, p. 238–349.

[8] Deligne, P., 1973, *La conjecture de Weil I*, Publ. math. de l'IHES 43 (1973) p. 273–307.

[9] Dwork, B., 1960, *On the rationality of the zeta function*, Amer. J. of math. 82 (1960) p. 631–648.

[10] Galois, E., 1830, *Sur la théorie des nombres*, Bull. Sc. Math. XIII (1830) p. 428 = Oeuvres, Paris 1976, p. 113–127.

[11] Gauss, C.F., 1801, *Disquisitiones arithmeticae*, Leipzig 1801 = Werke 1.

[12] Gauss, C.F., 1832, *Theorie residuorum biquadraticorum, Comm. sec.*, Comm. Soc. Reg. Sc. Gott., Rec. 7 (1832) = Werke 2, p. 67–148.

[13] Grothendieck, A., 1964, *Formule de Lefschetz et rationalité des fonctions L*, Séminaire Bourbaki n^{0}279.

[14] Hardy, G.H. et Littlewood, J.E., 1922, *Some problems of partitio numerorum.*

[15] Hasse, H., 1936, *Über die Riemannsche Vermutung in Funktionenkörpern* (Congrès intern. Oslo).

[16] Hensel, K. et G. Landsberg, 1902, *Theorie der algebraischen Funktionen einer Variabeln*, Leipzig 1902, réimpr. New York 1965.

[17] Hurwitz, A., 1886, *Über algebraische Korrespondenzen und das verallgemeinerte Korrespondenzprinzip*, Berichte k. sächs. Gesellsch. der Wissenschaften, math. phys. Klasse (1886) p. 10–38 = Math. Ann. 28 (1887) p. 561–585 = Math. Werke I, p. 163–188.

[18] Kornblum, H., 1919, *Über die Primfunktionen in einer arithmetischen Progression*, Math. Zeits. 5 (1919) p. 100–111.

[19] Lejeune Dirichlet, G., 1837, *Beweis des Satzes, dass jede unbegrenzte arithmetische Progression, deren erstes Glied und Differenz ganze Zahlen ohne gemeinschaftlichen Factor sind, unendlich viele Primzahlen enthält*, Abh. k. Preuss. Ak. der Wiss. 1837, p. 45–81 = Math. Werke p. 313–342.

[20] Manin, Y., 1968, *Correspondances, motifs et transformations monoïdales*, Mat. Sbornik 77 (1968).

[21] RIEMANN, B., 1859, *Über die Anzahl der Primzahlen unter einer gegebenen Grösse*, Monatsb. der Berliner Ak. 1859 = Ges. Math. Werke, p. 145–153.

[22] SCHMIDT, F.K., 1925, *Allgemeine Körper im Gebiet der höheren Kongruenzen* (Dissert. Freiburg).

[23] SCHMIDT, F.K., 1931, *Analytische Zahlentheorie in Körpern der Charakteristik p*, Math. Zeits. 33 (1931) p. 1–32.

[24] SENGENHORST, P., 1925, *Körper von der Charakteristik p*, Math. Zeits. 24 (1925) p. 1–39.

[25] WEIL, A., 1940, *Sur les fonctions algébriques à corps de constantes fini*, CRAS 1940.

[26] WEIL, A., 1946, *Foundations of Algebraic Geometry*, Amer. Math. Soc. Coll. 29, New York 1946.

[27] WEIL, A., 1948a, *Sur les courbes algébriques et les variétés qui s'en déduisent*, Actual. Sc. et Indus. 1041, Paris 1948.

[28] WEIL, A., 1948b, *Variétés abéliennes et courbes algébriques*, Actual. Sc. et Indus., Paris 1948.

[29] WEIL, A., 1949a, *Number of solutions of equations in finite fields*, Bull. Amer. Math. Soc. 55 (1949) p. 497–508.

[30] WEIL, A., 1954a, *Footnote to recent paper*, Amer. J. of math. 76 (1954) p. 347–350.

[31] WEIL, A., 1954b, *Abstract versus classical algebraic geometry*, Proc. Intern. math. Congress Amsterdam III, p. 550–558.

CHRISTIAN HOUZEL
CNRS-REHSEIS
156, Avenue Parmentier
F-75010 Paris
Université Paris VII
2, Place Jussieu
F-75251 Paris Cedex 05

A.O. Gel'fond, 1967

Des séries de Taylor au mouvement brownien, avec un aperçu sur le retour

Jean-Pierre Kahane
Université de Paris-Sud, Orsay

Cet article propose une promenade. Nous partirons de Borel et de la provocation qu'il lance en 1896: *«une série de Taylor admet, en général, son cercle de convergence comme coupure»*. Nous parcourerons quelques-uns des travaux suscités par cette affirmation — son effet sur Borel lui-même, les interprétations au moyen de la théorie de Baire, les interprétations probabilistes, la démarche de Steinhaus et sa rencontre avec la théorie de Wiener du mouvement brownien. Arrivés là — nous sommes en 1933 — le paysage change, avec les Grundbegriffe de Kolmogorov. Mais la théorie du mouvement brownien s'affirme et s'affine avec Paul Lévy. L'invariance des trajectoires du mouvement brownien plan par représentation conforme explique la dernière partie du titre: un aperçu sur le retour.

Les travaux de Borel de 1896–1897 sur les séries de Taylor, tels que Borel les analyse en 1912

La phrase citée ci-dessus n'appartient pas à la note de 1896 [Bo3], ni à l'article de 1897 qui la suit [Bo4]. Elle est empruntée à la Notice de Borel de 1912, où il fait la première somme de ses travaux [Bo6].

Pourquoi est-ce une provocation? C'est que, jusque là, les séries de Taylor non prolongeables font figure d'exceptions. Poincaré en a donné un exemple, à savoir la série $\sum_1^\infty 2^{-n}x^{3^n}$ [Po], et Hadamard, la même année 1892, donne un énoncé un peu plus général: «la série $\sum b_\mu x^{c_\mu}$ admet son cercle de convergence comme ligne singulière si le rapport $(c_{\mu+1} - c_\mu)/c_\mu$ est constamment supérieur à un nombre fixe s» [Ha]. La condition est loin d'être optimale (un théorème de Fabry dit que la conclusion demeure sous l'hypothèse $\lim(c_\mu/\mu) = \infty$ [Fa1] p. 382), mais elle présente un intérêt dans d'autres questions, et elle est restée dans la littérature comme la «condition de lacunarité d'Hadamard» [Zy].

L'article de Poincaré se réfère explicitement à la théorie de Weierstrass du prolongement analytique [We], et, pour mettre en évidence des fonctions non prolongeables sur un ensemble donné, il a recours à des séries de fractions rationnelles $\sum A_n(x - a_n)^{-1}$. La thèse d'Hadamard [Ha] porte en grande partie sur la détection des pôles et sur le comportement comparé de la suite des coefficients de

At S. Mandelbrojt's father's house, Warsaw, 1930, S. Mandelbrojt, . , A. Denjoy, J. Hadamard, S. Mandelbrojt's father, P. Montel

Taylor et de la fonction au voisinage du cercle de convergence; le cas «naturel» y est donc celui du prolongement analytique. La thèse de Borel elle-même, inspirée de Poincaré, traite du prolongement (dans un sens tout à fait nouveau, qui devait à son tour inspirer Denjoy dans ses travaux de 1921 sur la quasi-analyticité, voir [De] pp. 76–84, notes pp. 34–37) des séries de fractions rationnelles [Bo1]. Fin 1896, il publie une note «sur la région de sommabilité d'une série de Taylor» [Bo2], et il introduit à cette occasion la transformation et le procédé de sommation qui portent son nom. Quelques semaines ensuite, en décembre, c'est sous le titre cryptique «sur les séries de Taylor» qu'il affirme que, lorsque les coefficients sont quelconques, le cercle de convergence est une coupure [Bo3]. Il s'en explique dans une lettre à Mittag-Leffler, qui constitue l'article [Bo4].

Dans sa notice de 1912, Borel indique nettement, dans le chapitre *fonctions d'une variable complexe*, que c'est le prolongement analytique des fonctions représentées par les séries de Taylor qui a motivé l'introduction de ses procédés de sommation, de la théorie des séries divergentes, et de la «méthode nouvelle» qui est «la considération de la fonction entière associée à une série de Taylor donnée» — c'est-à-dire la transformation de Borel. Vu d'aujourd'hui, il s'agit là de progrès majeurs. Cependant, pour Borel, le plus important n'est pas là. Voici son analyse [Bo6] I, p. 154).

«Un résultat qui me paraît plus important est le suivant: une série de Taylor admet, en général, son cercle de convergence comme coupure...

La difficulté principale était d'en préciser le sens avant d'en donner la démonstration... On peut partager la série en une infinité de groupes successifs de termes et, à chaque groupe, faire correspondre un point du cercle de convergence qui ne dépende que des coefficients de ce groupe; ces points forment un ensemble E; tout point de l'ensemble dérivé E' est un point singulier. Il est clair, dès lors, que si les coefficients successifs sont choisis au hasard, c'est-à-dire d'une manière indépendante des coefficients précédents, la probabilité pour que l'ensemble E soit dense sur tout le cercle est égale à l'unité.»

Dans tout ce passage, les mots clés sont *en général, au hasard, indépendant, probabilité égale à l'unité.* Borel en avait une intuition claire, mais il a fallu des dizaines d'années, à lui et à d'autres, pour que ces mots clés deviennent des concepts mathématiques formalisés. Pour Borel, ç'allait être le point de départ de sa théorie des *probabilités dénombrables*, c'est-à-dire de la conception moderne des probabilités comme fonctions σ-dénombrables d'événements [Bo5].

Ajoutons une remarque sur la démonstration de Borel, sous la forme où il l'écrit à Mittag-Leffler [Bo4]. A la fin de sa lettre se trouve la phrase suivante: «On a donc sur le cercle une infinité d'arcs indépendants, dont la somme dépasse tout nombre donné, donc, en général, tout point du cercle appartiendra à une infinité d'arcs.»

Cette phrase constitue, en fait, la partie difficile du "lemme de Borel-Cantelli»: si des événements sont indépendants et si la somme de leurs probabilités diverge, il est presque sûr qu'une infinité d'entre eux se produise. Ainsi le lemme de Borel-Cantelli se trouve déjà utilisé par Borel en 1896–97; ordinairement c'est à l'article de Borel de 1909 et un article de Cantelli de 1916 [Bo5] [Ca] que renvoient les traités de probabilités [Re].

Avant de continuer la promenade, dégageons une première leçon sur la façon dont se forgent les théories et les concepts. A chaque époque, il y a les grands problèmes du moment. Puis les théorèmes, prévus ou imprévus, qui répondent à ces problèmes, et que la postérité, quelquefois, oublie. Puis les outils de ces théorèmes, les lemmes, dont l'intérêt s'affirme progressivement hors du champ qui leur a donné naissance; la postérité les retient comme lemmes (Borel-Cantelli, Fatou) ou les appelle théorèmes (Lebesgue). Puis les concepts qui donnent le bon cadre et l'usage commode de ces outils (la σ-additivité, la probabilité, l'indépendance pour ce qui concerne Borel; la mesure, l'intégrale, pour Lebesgue et Fatou). Ces concepts, formalisés, donnent les définitions dont l'enseignement fait le point de départ des théories. Mais, historiquement, les définitions sont des points d'aboutissement d'efforts séculaires, et il n'est possible de comprendre leur richesse qu'en retrouvant, dans les théories et les applications, au moins une partie des résultats et des réflexions qui leur ont donné naissance.

Ayant ainsi respiré, abordons pour un moment un chemin de traverse.

Que veut dire «en général»?

Sans aucun doute, pour Borel, «en général» signifie *presque sûrement* au sens actuel; c'est un concept *probabiliste*. Il en est de même pour Fabry, qui, en 1897 (en fait, quelques jours après la note de Borel de décembre 1896), donne une nouvelle démonstration (disons: amorce de démonstration) de l'assertion de Borel [Fa2] (sur l'approche de Fabry, on peut consulter [HM] p. 61). Cependant, en se restreignant aux séries de Taylor à coefficients entiers, «en général» peut s'entendre au sens *ensembliste* de la cardinalité. En 1918, Hausdorff observe qu'en se restreignant de la sorte, les séries de Taylor prolongeables forment un ensemble dénombrable, et les séries de Taylor non prolongeables un ensemble non dénombrable, donc beaucoup plus riche [Hau2].

Enfin, «en général» peut s'entendre au sens *topologique*. C'est ce que fait Pólya, également en 1918 [Pol1]: pour une topologie convenable, l'ensemble des séries de Taylor non prolongeables forment un ouvert dense (voir aussi [HM], p. 68).

Dans le prolongement de Pólya, le travail le plus achevé me paraît être celui de St. Kierst et E. Szpilrajn de 1933 [KS]. L'ensemble des fonctions holomorphes dans le disque unité ouvert est muni de la topologie de la convergence uniforme sur tout compact, et c'est alors un espace de Fréchet. Suivant une inspiration constante des mathématiciens polonais de l'époque, particulièrement illustrée par Mazurkiewicz, les auteurs se placent au point de vue de la catégorie de Baire, c'est-à-dire cherchent des propriétés qui ont lieu sur une intersection dénombrable d'ouverts denses — on dit aussi aujourd'hui propriétés *génériques*, ou *quasi-sûres*. Mazurkiewicz avait un résultat inédit: le non prolongement est une propriété générique. Les auteurs indiquent une série d'autres propriétés génériques, pour les fonctions holomorphes dans le disque unité, pour les fonctions entières, pour les fonctions méromorphes.

A partir de là s'offrirait à nous une autre promenade. Comment interviennent ces trois points de vue — probabiliste, ensembliste, topologique — dans différentes théories mathématiques ? Ne fût-ce qu'en analyse harmonique, ou dans la théorie des séries de Dirichlet, ou dans les théorèmes de superposition, il y a là matière à de jolies excursions.

Mais revenons à notre itinéraire principal, de Borel à Steinhaus, Wiener et Paul Lévy.

Le point du vue de Steinhaus

Malgré l'analogie aujourd'hui évidente entre la mesure de Lebesgue (1904) et les probabilités dénombrables de Borel (1909), la traduction de l'une aux autres ne s'est pas faite d'un seul coup. La première formalisation qui unit la probabilité sur l'ensemble des parties illimitées de pile ou face et la mesure de Lebesgue sur l'intervalle $[0, 1]$ est due à H. Steinhaus (1923) et il vaut la peine de citer son introduction [St1]:

«M. E. Borel a montré le premier l'intérêt qui s'attache à l'étude de probabilités dénombrables et il a donné des applications arithmétiques trouvées sur cette voie, parmi lesquelles le théorème suivant connu comme «le paradoxe de Borel» a attiré l'attention des analystes:

La probabilité *pour que la fréquence du chiffre 0 dans le développement dyadique d'un nombre pris au hasard soit égale à* $1/2$ a pour valeur l'unité; on

H. Steinhaus

appelle *fréquence* du chiffre 0 la valeur limite du quotient par n du nombre de fois que ce chiffre figure parmi les n premiers chiffres du développement.

On trouve chez divers auteurs l'énoncé suivant du même théorème:

Presque tous les nombres α ont la propriété que la fréquence des zéros dans leurs développements dyadiques est égale à $1/2$; *presque tous* signifie que la *mesure* de Lebesgue de l'ensemble des α privés de la propriété en question est *nulle*. Pour obtenir cet énoncé il suffit de modifier verbalement la démonstration primitive de M. Borel sans toucher à son idée.

Le but de cette Note est d'établir un système des postulats pour les probabilités dénombrables qui permettra une fois pour toutes de passer d'une interprétation à l'autre dans les recherches de ce genre.»

Comme référence à Borel, Steinhaus cite [Bo5] pp. 259–260, et comme référence à «d'autres auteurs» [Hau1] pp. 419–422.

C'est en 1929, dans un article publié en 1930 [St2], que Steinhaus revient à ce sujet, pour traiter de la probabilité pour qu'une série de Taylor admette le cercle de convergence comme coupure. La référence principale est Borel 1896, et c'est la première fois que l'énoncé de Borel prend un sens mathématique précis. Steinhaus considère une série

$$\sum_0^\infty c_n z^n$$

dont le rayon de convergence est 1, et les séries obtenues par des déphasages aléatoires

$$(S) \qquad \sum_0^\infty c_n e^{2\pi i \varphi_n} z^n$$

où les φ_n sont des variables aléatoires indépendantes uniformément distribuées sur $[0,1]$. Du moins, c'est ainsi que nous pouvons dire les choses aujourd'hui. Le travail de Steinhaus consiste a) à construire les φ_n comme fonctions d'une variable réelle $\varphi \in [0,1]$; b) à montrer que le non-prolongement de (S) correspond à un *ensemble borélien* de valeurs de φ; c) à montrer que cet ensemble est de *mesure pleine*.

L'étape a) est simple et ingénieuse — et inspirée du jeu de pile ou face. On écrit en développement binaire

$$\begin{aligned} \varphi_1 &= 0, \varphi_{11}\varphi_{12}\varphi_{13}\cdots \\ \varphi_2 &= 0, \varphi_{21}\varphi_{22}\varphi_{23}\cdots \\ \varphi_3 &= 0, \varphi_{31}\varphi_{32}\varphi_{33}\cdots \\ &\cdots \qquad \text{et} \\ \varphi &= 0, \varphi_{11}\varphi_{21}\varphi_{12}\varphi_{31}\varphi_{22}\varphi_{13}\varphi_{41}\varphi_{32}\cdots \end{aligned}$$

Ainsi, φ permet de coder toute la suite (φ_n). C'est le moyen de transférer la mesure de Lebesgue sur $[0,1]$ en une probabilité sur $[0,1]^{\mathbb{N}}$.

L'étape b) ne présente pas de difficulté conceptuelle. Il s'agit de traduire en formules le fait qu'un point donné du cercle de convergence soit régulier pour (S).

L'étape c), par contre, est très originale. Elle consiste à montrer que si la probabilité du fait précédent est > 0, c'est nécessairement 1. En d'autres termes, dans ce cas particulier (mais il est substantiellement équivalent au cas général) c'est la fameuse loi du zéro-un de Kolmogoroff [Ko].

La conclusion est bien que (S) a presque sûrement son cercle de convergence pour coupure.

Mais l'intérêt de la démarche de Steinhaus va bien au delà. La clé en est le codage de (φ_n) à l'aide de φ. La suite (φ_n) est un prototype de variables aléatoires indépendantes. Toute suite de variables aléatoires indépendantes peut s'interpréter comme une suite $X_n(\varphi_n)$, toute propriété probabilisable comme un ensemble mesurable en φ, et sa probabilité comme la mesure de Lebesgue de cet ensemble de φ. A ce stade, toutes les notions utilisées par Borel en 1896 avec un contenu intuitif se trouvent parfaitement formalisées, et la démonstration est impeccable. De plus, essentiellement, tous les problèmes de la théorie des probabilités peuvent être réduits, concrètement, à la mesure de Lebesgue.

Au cours des années 1920 le prototype des séries de variables aléatoires indépendantes était fourni par les séries de Rademacher, introduites en 1922 [Ra] (le cas de convergence p. p. avait été traité par Rademacher, le cas de divergence p. p. par Khintchine et Kolmogoroff en 1925 [KK]). A partir de 1930 ce sont les séries de Steinhaus qui attirent l'attention, et les variables φ_n de Steinhaus permettent de traiter des fonctions indépendantes en toute généralité [SKR].

A la rencontre du mouvement brownien

La théorie physique du mouvement brownien est due à Einstein et date de 1905 [Ei]. Son application à la détermination des dimensions atomiques, c'est-à-dire du nombre d'Avogadro, est l'oeuvre de Jean Perrin, et elle se trouve exposée de façon magistrale dans un article de 1909 [Pe1] et dans son livre de 1912 [Pe2]. Sur les conseils de Russell, Norbert Wiener lit Einstein, et il retient particulièrement de Jean Perrin sa description des trajectoires browniennes. Citons Perrin [Pe1]:

«Les enchevêtrements de la trajectoire sont si nombreux et si rapides, qu'il est impossible de les suivre et que la trajectoire notée est toujours infiniment plus simple et plus courte que la trajectoire réelle. De même la vitesse moyenne d'un grain pendant un temps donné (quotient du déplacement par le temps) varie *follement* en grandeur et en direction sans tendre aucunement vers une limite quand le temps de l'observation décroît... C'est un des cas où l'on ne peut s'empêcher de penser à ces fonctions continues qui n'admettent pas de dérivées, qu'on regarderait à tort comme de simples curiosités mathématiques, puisque la nature peut les suggérer aussi bien que les fonctions à dérivées.»

En termes actuels, la physique impose un modèle. Le mouvement brownien doit être décrit comme un processus gaussien (X_t), à accroissements stationnaires, les accroissements du futur étant indépendants du passé, et cela impose une normalisation, à savoir que la variance de X_t pour $t > 0$ est t (à un facteur près).

La loi de (X_t) décrit l'évolution globale du phénomène. Les trajectoires individuelles sont celles des $X(t, \omega)$ quand ω est fixé. Le programme de Wiener, inspiré par Einstein et Jean Perrin, consiste à mettre en évidence une version $X(t, \omega)$ du processus (X_t) qui soit presque sûrement continue, et presque sûrement nulle part dérivable. Dans de nombreux articles sur le sujet (en particulier [Wi1], [PW] chap. 9), Wiener cite des passages de Jean Perrin évoquant les fonctions sans dérivées à propos des trajectoires browniennes.

Le début du programme est réalisé en 1923, dans l'article fondamental *Differential space* [Wi1]. A vrai dire, Wiener ne construit pas à proprement parler une fonction $X(t,\omega)$, mais il construit une mesure sur l'ensemble des fonctions continues, qui en fait un espace probabilisé; c'est ainsi chaque fonction de l'espace qui joue le rôle de ω. Tout l'arsenal de la mesure et de l'intégration dans les espaces fonctionnels se trouve mobilisé. Les propriétés de la série de Fourier du processus, qu'on appelle aujourd'hui série de Fourier-Wiener, sont établies: il s'avère que les coefficients sont des variables aléatoires gaussiennes indépendantes, dont la variance est l'inverse de la fréquence.

A l'occasion d'une autre grande étude, *Generalized harmonic analysis*, en 1930, Wiener revient au mouvement brownien, en simplifiant un peu l'exposé, et en montrant que les fonctions continues sur lesquelles la probabilité est concentrée satisfont p. s. une condition de Hölder d'ordre $\frac{1}{4}$ [Wi2].

La fin du programme n'est achevée qu'en 1933, par une collaboration avec Paley et Zygmund. Dans [PWZ], il est en effet montré que p.s. le module de continuité des $X(t,\omega)$ est $0(h^\alpha)$ pour tout $\alpha < \frac{1}{2}$, et en sens opposé que p.s. pour tout t on a

$$\overline{\lim_{h\to 0}} \frac{|X(t+h,\omega)-X(t,\omega)|}{|h|^\alpha} = \infty,$$

pour tout $\alpha > \frac{1}{2}$, c'est-à-dire la non-dérivabilité dans un sens très fort.

L'introduction à [PWZ] a été sans nul doute rédigée par Wiener, et elle met bien en évidence le rôle séminal de Borel et l'importance du point de vue de Steinhaus. En voici le début:

«The introduction of the notion of the random into analysis is in the first instance the work of Borel. His theory of *probabilités dénombrables* concerns itself with quantities depending upon an infinite sequence of choices and with their average values. In the simplest case, the choices in question are between two alternatives, which may be taken as the signs + and −. Thus questions as to the probability of convergence of the series

$$\sum_{n=0}^{\infty} \pm C_n$$

and of the distribution of its sums belong to this order of ideas. Such questions have been much discussed in the pages of *Fundamenta Mathematicae* and *Studia Mathematica*, and are associated with the names of Steinhaus and Rademacher, among others. To Steinhaus in particular is due the reduction of such questions to questions concerning the Lebesgue integral.»

A. Zygmund

Puis sont indiqués à grands traits les travaux de Paley et Zygmund sur les séries de fonctions aléatoires du type

$$\sum_{1}^{\infty} \pm f_n(t)$$

(séries de Fourier-Rademacher et séries de Taylor-Rademacher en particulier), ou du type

$$\sum_{1}^{\infty} e^{2\pi i \varphi_n} f_n(t)$$

(séries de Fourier-Steinhaus et séries de Taylor-Steinhaus en particulier), qui datent de 1930 et 1932 [PZ]. A juste titre, leur intérêt pour la production de contre-exemples est souligné.

Enfin le mouvement brownien se présente.

«There is however a way in which functions with an element of randomness present themselves to the physicist. The Brownian movement, the motion of a suspended particle in a fluid in response to the irregularities of the molecular collisions to which it is subjected, is of so strikingly random a character that it does not even seem to possess a velocity. The physicist Perrin has commented on the striking way in which this motion suggests the non-differentiable continuous functions of Weierstrass, and Borel has commented on this in connection with his theory of denumerable probabilities. The components of the displacement of such a particle, regarded as functions of the time, are random functions in a sense differing considerably from that of Zygmund and Paley. Wiener has reduced the theory of such random functions to a mathematically definite form, also depending upon a Lebesgue integration.

. . . . . .

The purpose of this paper is to bridge the gap between the PZ and the W theories.»

A cette occasion, Wiener va modifier son point de départ. Dans [PWZ], il part encore du mouvement brownien pour étudier la série de Fourier-Wiener. Mais, comme résultat de sa collaboration avec Paley, en 1934, c'est maintenant la série de Fourier-Wiener qui définit $X(t, \omega)$. Elle est de la forme

$$\sum_{1}^{\infty} \xi(\omega_n) f_n(t),$$

où les $\xi(\omega_n)$ sont des variables gaussiennes normalisées indépendantes, et les $f_n(t)$ sont des multiples de sinus ou cosinus d'arcs multiples. Formellement, elle ressemble aux séries considérées par Paley et Zygmund, et se prête à un traitement voisin.

Conceptuellement, c'est un saut considérable. Au lieu de construire une mesure sur un espace de fonctions (alors identifié à l'espace probabilisé), Wiener part du modèle de Steinhaus, qui lui fournit tout de suite des ω_n indépendantes uniformément réparties sur $[0, 1]$ (ce sont les φ_n de Steinhaus), et il prend comme espace probabilisé l'intervalle $[0, 1]$ de la droite réelle, muni de la mesure de Lebesgue, à la manière de Steinhaus. C'est ainsi qu'il présente la construction de $X(t, \omega)$, appelée *the fundamental random function*, et désignée alors par $\psi(t, \alpha)$, dans les derniers chapitres de son livre avec Paley de 1934 [PW].

En apparence, toute la théorie des probabilités se ramène ainsi à l'analyse des fonctions d'une variable réelle et à la mesure de Lebesgue sur $[0, 1]$. Wiener s'est converti à Steinhaus, Lebesgue justifie Borel et Einstein.

Le nouveau rôle moteur des probabilités

L'apparence est trompeuse. D'abord, 1933 est l'année des *Grundbegriffe* de Kolmogorov [Ko], où les probabilités s'affirment comme théorie indépendante. Ensuite et surtout, la richesse des objets de cette théorie, qu'ils viennent du monde physique, de la biologie, de l'économie, des assurances, tend à jouer un rôle moteur dans l'analyse classique elle-même. C'est ce que nous voulons dire en tentant de donner un *aperçu sur le retour*. Pour ne pas nous égarer, nous nous bornons au mouvement brownien et aux fonctions analytiques.

La dernière grande oeuvre de Wiener sur le mouvement brownien est *Homogeneous chaos* [Wi3], la théorie du chaos probabiliste, qu'aujourd'hui on connaît moins que le chaos déterministe, mais dont le rôle fut considérable.

C'est alors que Paul Lévy entre en scène. Dès sa première note sur le sujet, c'est le mouvement brownien dans le plan qui le passionne et dont il entreprend l'étude géométrique [Le2]. Son premier grand mémoire sur ce sujet date de 1939 [Le3]. A vrai dire, Paul Lévy a déjà fait un exposé du mouvement brownien linéaire comme cas particulier de processus à accroissements indépendants dans son livre de 1937 [Le1], et il a établi le rôle de la fonction $\sqrt{2h \log \frac{1}{h}}$ dans l'estimation presque sûre du module de continuité. Il se réfère à Bachelier [Ba1,2] et à Wiener [Wi1] comme fondateurs de la théorie, et cite Khintchine [Kh] pour la loi du logarithme itéré. Il ignore Smoluchowski. Il ne semble pas connaître les travaux de Paley et Wiener; en tous cas il ne les prend pas pour point de départ. Paul Lévy est porté par une intuition géométrique et probabiliste, et pour lui la réduction de la probabilité à la mesure de Lebesgue, ou tout autre fondement des probabilités, doit présenter à peu près le même intérêt que l'axiomatique de Hilbert pour la géométrie euclidienne.

L'article de 1939 introduit des novations fondamentales, comme les intégrales stochastiques $\int_0^1 (dX(t))^2$ et $\int_0^1 Y(t)dX(t)$, où $X(t)$ et $Y(t)$ sont deux versions indépendantes du mouvement brownien linéaire (donc $(X,Y)(t)$ une version du mouvement brownien plan), il montre que la courbe décrite est d'aire nulle, et il en précise la géométrie.

Cependant ce n'est que dans un court article de 1947, dans une revue peu accessible aux mathématiciens [Le4], qu'il signale l'invariance des trajectoires par représentation conforme, qu'apparemment il connaissait fort bien depuis longtemps. Comme corollaire, il donne l'interprétation de la mesure harmonique comme probabilité d'atteinte par le mouvement brownien — qu'indépendamment Kakutani avait déjà mise en évidence [Ka], et qui d'ailleurs faisait partie du folklore depuis les travaux de Pólya [Pol2]. L'influence des conversations avec son gendre Laurent Schwartz est sensible, comme l'indique une note à la suite du corollaire:

«D'après une conversation que j'ai eue en 1941 avec M. L. Schwartz, ce corollaire était déjà connu. Mais M. Schwartz n'a pas pu me donner de référence, et, comme ce corollaire peut s'établir directement, je ne sais pas si le théorème sur la représentation conforme est nouveau ou non. C'est pour cela que j'ai attendu jusqu'à ce jour pour le publier.»

En dehors des deux livres de Paul Lévy [Le1], [Le5], l'exposé le plus attrayant est celui du mémorial des sciences mathématiques de 1954, qui se lit comme un roman, pourvu qu'on veuille bien se laisser porter par l'intuition de l'auteur [Le6].

L'intérêt propre du mouvement brownien plan n'a fait que croître dans la seconde moitié du siècle. Pour apprécier combien le sujet est vivant, je renvoie aux exposés et travaux de Le Gall [LG]. Le lien à la mesure harmonique a pris un aspect nouveau, dynamique, avec les simulations de croissance d'objets plans bombardés par des particules browniennes qui viennent s'y agréger [WS].

D'autre part l'intérêt du mouvement brownien en analyse classique s'est constamment affirmé [Kah], [Bu], [Pe], [Da]. Par exemple, le théorème de Picard sur les fonctions entières correspond à une propriété d'enroulement des trajectoires browniennes autour de deux points. Les classes de Hardy du disque, H^p, peuvent se définir et s'étudier à l'aide du comportement de leurs éléments sur des trajectoires browniennes. Grosso modo, tous les invariants conformes peuvent être considérés comme des invariants des trajectoires browniennes considérées sans loi du temps de parcours. C'est pourquoi, dans la géométrie de la variable complexe, il arrive que le mouvement brownien soit plus pratique que les géodésiques (par exemple, si, au moyen de la fonction modulaire, on représente le disque unité sur le revêtement universel du plan percé de deux points, les géodésiques de la géométrie de Poincaré se transforment en chemins compliqués; mais les trajectoires browniennes se transforment simplement en trajectoires browniennes). Le mouvement brownien est un moyen de "voir» à la fois la mesure harmonique, les ensembles de capacité logarithmique nulle, la frontière de Martin, et même les géodésiques. A titre d'exemple, voici un exercice que je dois à Adrien Douady. On considère, en géométrie de Poincaré dans une bande, une géodésique joignant deux points A et B du même bord de la bande; montrer qu'elle est contenue dans le disque (euclidien) de diamètre AB. Preuve: une fourmi partant d'un point M de la géodésique et qui s'arrête quand elle sort de la bande a autant de chance de s'arrêter sur le segment AB ou sur le reste de la frontière. Si on l'autorise à se promener dans tout le demi plan limité par la droite AB, elle a maintenant plus de chance de rencontrer le segment AB que le reste de la frontière, donc M se trouve entre le segment AB et la géodésique du demi-plan de Poincaré joignant A et B, qui est le demi-cercle euclidien de diamètre AB.

Il est sans doute inutile de poursuivre. Il y a bien aujourd'hui réhabilitation de l'intuition probabiliste jusque dans l'analyse classique. Au terme de cette promenade à travers une tranche de l'histoire des mathématiques de la première moitié du siècle, nous avons vu un problème d'analyse stimuler le développement des probabilités, et nous voyons maintenant le mouvement brownien stimuler l'analyse. J'ajouterai que la promenade au hasard a du bon, dans beaucoup de domaines de la vie, de la science, et, sans doute, de l'histoire des mathématiques.

Références

[Ba1] L. BACHELIER: Théorie de la spéculation. *Ann. Sci. Ec. Norm. Sup.* 17 (1900), 21–86.

[Ba2] L. BACHELIER: Théorie mathématique du jeu. *Ann. Sci. Ec. Norm. Sup.*18 (1901), 143–210.

[Bo1] E. BOREL: Sur quelques points de la théorie des fonctions. *Ann. Sci. Ec. Norm. Sup.* 12 (1885), 9–55.

[Bo2] E. BOREL: Sur la région de sommabilité d'une série de Taylor. *C. R. Acad. Sci. Paris* 123 (oct. 1896), 348.

[Bo3] E. BOREL: Sur les séries de Taylor. *C. R. Acad. Sci. Paris* 123 (déc. 1896), 1051–1052.

[Bo4] E. BOREL: Sur les séries de Taylor. *Acta Math.* 20(1897), 243–247.

[Bo5] E. BOREL: Sur les probabilités dénombrables et leurs applications arithmétiques. *Rend. Circ. Mat. Palermo* 26 (1909), 247–271.

[Bo6] E. BOREL: *Oeuvres*, tome I. Paris, CNRS, 1972.

[Bu] D.L. BURKHOLDER: Brownian motion and classical analysis. *Probability. Proceedings of symposia in pure mathematics* 31 (1977), 5–14.

[Ca] F.P. CANTELLI: La tendenza ad un limite nel senzo del calculo delle probabilita. *Rend. Circ. Mat. Palermo* 16 (1916), 191–201.

[Da] B. DAVIS: Brownian motion and analytic functions (special invited paper). *Ann. Prob.* 7 (1979), 913–932.

[De] A. DENJOY: *Un demi-siècle (1907–1956) de notes communiquées aux Académies de Paris, d'Amsterdam, des Lincei*, volume I: *la variable complexe*, 221+58 p. Paris, Gauthier-Villars, 1957.

[Ei] A. EINSTEIN: Ueber die von der molekularkinetischen Theorie der Wärme geforderten Bewegung von in ruhenden Flüssigkeiten suspendierter Teilchen. *Annalen der Physik* 19 (1905), 371–381.

[Fa1] E. FABRY: Sur les points singuliers d'une fonction donnée par son développement en série et sur l'impossibilité du prolongement analytique dans des cas très généraux. *Ann. Sci. Ec. Norm. Sup.* (3), 13 (1896), 367–399.

[Fa2] E. FABRY: Sur les séries de Taylor. *C. R. Acad. Sci. Paris* 124 (janv. 1897), 142–143.

[Ha] J. HADAMARD: Essai sur l'étude des fonctions données par leur développement de Taylor. *J. Math. Pures & Appl.* (4) 8 (1892), 101–186.

[HM] J. HADAMARD & S. MANDELBROJT: La série de Taylor et son prolongement analytique. *Scientia*, Gauthier-Villars, Paris, 1926.

[Hau1] F. HAUSDORFF: *Grundzüge der Mengenlehre*. Leipzig, 1914.

[Hau2] F. HAUSDORFF: Zur Verteilung der fortsetzbaren Potenzreihen. *Math. Zeitschrift* 4 (1919), 98–103.

[Kah] J.-P. KAHANE: Brownian motion and classical analysis. *Bull. London Math. Soc.* 8 (1976), 145–155.

[Ka] S. KAKUTANI: Two dimensional Brownian motion and harmonic functions. *Proc. Math. Inst. Osaka* 21 (1945), 138–140.

[Kh] A. KHINTCHINE: *Asymptotische Gesetze der Wahrscheinlichkeitsrechnung*. Ergebnisse der Math. und ihrer Grenzgebiete, Springer, t. 2, 4 (1933), 1–77.

[KK] A. KHINTCHINE & A.N. KOLMOGOROV: Ueber Konvergenz von Reihen deren Glieder durch den Zufall bestimmt werden. *Mat. Sbornik* 32 (1925), 668–77.

[KS] S. KIERST & E. SZPILRAJN: Sur certaines singularités des fonctions analytiques uniformes. *Fund. Math.* 21 (1933), 276–294.

[Ko] A.N. KOLMOGOROV: *Grundbegriffe der Wahrscheinlichkeitsrechnung*. Berlin, Springer, 1933.

[Le1] P. LÉVY: *Théorie de l'addition des variables aléatoires*. Paris, Gauthier-Villars, 1937.

[Le2] P. LÉVY: Mouvement brownien et schémas géométriques. *C. R. Acad. Sci. Paris* 207 (1938), 1152–1154.

[Le3] P. LÉVY: Le mouvement brownien plan. *Amer. J. Math.* 62, 3 (1940), 487–550.

[Le4] P. LÉVY: Trois théorèmes sur le mouvement brownien. *Ass. franc. avanc. sciences*, Supplt. à l'Intermédiaire des recherches mathématiques, fasc. 9 (1947), 124–126.

[Le5] P. LÉVY: *Processus stochastiques et mouvement brownien*. Paris, Gauthier-Villars, 1948.

[Le6] P. LÉVY: *Le mouvement brownien*. Mémorial Sc. Math. 126 (1954), Paris, Gauthier-Villars.

[LG] J.F. LE GALL: *Some properties of planar Brownian motion*. Cours à la XXe école d'été de probabilités de Saint-Flour (1990).

[PW] R.E.A.C. PALEY & N. WIENER: *Fourier transforms in the complex domain*. A.M.S. Coll. Publ. 19, Providence, 1934.

[PWZ] R.E.A.C. PALEY, N. WIENER & A. ZYGMUND: Notes on random functions. *Math. Z.* 37 (1933), 647–668.

[PZ] R.E.A.C. PALEY & A. ZYGMUND: On some series of functions.
(1) *Proc. Cambridge Philos. Soc.* 26 (1930), 337–357.
(2) *Ibid*, 458–474.
(3) *Ibid*, 28 (1932), 190–205.

[Pe] K.C. PETERSEN: Brownian motion, Hardy spaces and bounded mean oscillation. *London Math. Soc.* Lecture Note Series 28 (1977), Cambridge Univ. Press.

[Pe1] J. PERRIN: Mouvement brownien et réalité moléculaire. *Annales de Chimie et de Physique* (2) 18 (1909), 5–114.

[Pe2] J. PERRIN: *Les atomes*. Paris, 1912.

[Po] H. POINCARÉ: Sur les fonctions à espaces lacunaires. *Amer. J. Math.* 14 (1892), 201–221.

[Pol1] G. PÓLYA: Ueber die Potenzreihen, deren Konvergenzkreis natürliche Grenze ist. *Acta Math.* 41 (1916–1918), 99–118.

[Pol2] G. PÓLYA: Ueber eine Aufgabe der Wahrscheinlichkeitsrechnung betreffend die Irrfahrt im Strassennetz. *Math. Annalen* 84 (1921), 149–160.

[Ra] H. RADEMACHER: Einige Sätze über Reihen von allgemeinen Orthogonalfunktionen. *Math. Ann.* 87 (1922), 112–138.

[Re] A. RENYI: *Calcul des probabilités*. Paris, Dunod, 1966.

[SKR] H. STEINHAUS, M. KAC & C. RYLL-NARDZEWSKI: Sur les fonctions indépendantes. *Studia Math.* 6 (1936), 46–58, 59–66, 89–97, 7 (1938), 1–15, 96–100, 9 (1940), 121–132, 10 (1948), 1–20, 11 (1949), 133–144, 12 (1951), 102–107, 13 (1953), 1–17.

[St1] H. STEINHAUS: Les probabilités dénombrables et leur rapport à la théorie de la mesure. *Fund. Math.* 4 (1923), 286–310.

[St2] H. STEINHAUS: Ueber die Wahrscheinlichkeit dafür, dass der Konvergenzkreis einer Potenzreihe ihre natürlich Grenze ist. *Math. Zeitschrift* 31 (1930), 408–416.

[We] K. WEIERSTRASS: *Zur Funktionenlehre*. Monatsberichte der Akademie der Wissenschaften zu Berlin, août 1880, 719–743.

[Wi1] N. WIENER: Differential space. *J. Math. and Phys.* 7 (1923), 131–174.

[Wi2] N. WIENER: Generalized harmonic analysis. *Acta Math.* 55 (1930), 117–258.

[Wi3] N. WIENER: The homogeneous chaos. *Amer. J. Math.* 60 (1938), 897–936.

[WS1] T.A. WITTEN & L.M. SANDER: Diffusion limited aggregation, a kinetic critical approach. Phys. Rev. Lett. 47 (1981), 1400–1463.

[WS2] T.A. WITTEN & L.M. SANDER: Diffusion limited aggregation. Phys. Rev. B. 27 (1983), 5686–5697.

[Zy] A. ZYGMUND: On a theorem of Hadamard. *Ann. Soc. Polon. Math.* 21 (1948), 52–69.

JEAN-PIERRE KAHANE
Mathématique, Bât. 425
Université Paris-Sud
F-91405 Orsay Cedex

Géométrie et relativité

André Lichnerowicz
Collège de France, Paris

C'est à un survol de l'histoire des interactions entre géométrie différentielle et physique théorique, particulièrement Relativité, pendant la période de référence que j'aimerais consacrer cet exposé.

L'élément primitif de la géométrie différentielle est la notion de *variété différentiable* et celle-ci provient historiquement de la Mécanique. Le premier exemple rencontré par l'homme de variétés différentiables naturelles, à un nombre arbitraire (fini) de dimension nous a été donné par la Mécanique Analytique de Lagrange. Dans ses études reposant sur les espaces (ou espaces-temps) de configuration des systèmes dynamiques, étaient déjà présents non seulement le concept de variété différentiable, mais, plus ou moins implicitement beaucoup d'autres concepts géométriques traités en termes de coordonnées locales q^i. On doit noter qu'il n'y a aucune référence à un plongement dans un espace affine, que le traitement est local, mais que l'interprétation mécanique conduit naturellement à une volonté globale, presque explicitée, mais dont il faudra attendre bien longtemps la réalisation. Les p_i mêmes sont, ici ou là, présents et, avec eux, quelque chose comme le cotangent à l'espace de configuration qui est l'espace de phase.

Mais nous devons attendre Riemann et sa célèbre Dissertation inaugurale *«Sur les hypothèses qui servent de fondement à la géométrie»* pour voir partiellement formulé ce qui était précédemment largement implicite. La géométrie des espaces de Riemann avait commencé à naître. L'apparition due au cristallographe Voigt du concept de *tenseur*, conçu comme le système de ses composantes avec les règles de transformations par changement de bases, devait donner à la première et la plus importante des grandes géométries différentielles, la géométrie riemannienne, l'un de ses instruments essentiels. L'algorithme développé par Christoffel dans le domaine des mathématiques pures et dans la recherche d'invariant semblait quelque peu formel et un peu gratuit. Mais avec Ricci et Levi-Civita (Math. Ann. 1900) cette approche d'une part se géométrise, d'autre part donne naissance à de multiples applications à la dynamique classique, à l'hydrodynamique et á l'élasticité. Grossmann, collaborateur et ami d'Einstein, connaissait bien ce mémoire. C'est là que grâce à Grossmann et à partir de 1911, Einstein devait trouver un cadre tout prêt à accueillir le développement de la *théorie de la Relativité Générale.*

En préparant cet exposé, j'ai relu l'extraordinaire mémoire de 75 pages, en français, dû à Ricci et Levi-Civita et intitulé *«Méthodes de calcul différentiel absolu et leurs applications»*. Comme nous sommes au début de la période de référence, j'aimerais analyser brièvement quelques caractères de ce Mémoire. Dans sa partie purement mathématique, il s'agit essentiellement d'une synthèse originale de tout ce qui est connu en ce temps, de la géométrie riemannienne, y compris des articles fort peu antérieurs de Ricci qui est le géomètre de l'équipe et y compris des résultats nouveaux. La terminologie réserve des surprises: le mot *tenseur* est absent et il lui est substitué le terme de *système* (systèmes covariants ou contravariants de fonctions); au lieu de forme quadratique fondamentale (définie positive) ou métrique, on parle de quadrique fondamentale. A un système covariant est associé par la métrique un système contravariant et inversement; pour cette correspondance on parle de *systèmes réciproques*. Les concepts d'espace vectoriel ou de module sont à la fois absents et présents puisqu'algébriquement, les notions d'addition, de multiplication par un scalaire, de composition c'est-à-dire de produit tensoriel, de contraction, de produit intérieur sont explicitées.

La dérivation covariante d'un tenseur est définie à partir des symboles de Christoffel, ses propriétés reconnues et le tenseur de courbure est présent sans qu'il soit nommé courbure. Les invariants différentiels gradient, divergence, rotationnel sont décrits dans ce contexte ainsi que le laplacien dit de Beltrami.

Sous le titre de *géométrie intrinsèque et instruments de calcul*, on rencontre d'une part une méthode mettant en oeuvre des n-uples orthogonaux mobiles et l'apparition des *coefficients de rotation de Ricci*, d'autre part des liens avec la géométrie différentielle classique à l'époque, à la Darboux. Il s'y ajoute, sous le nom d'application analytique du calcul différentiel absolu, l'usage des dérivées covariantes pour la recherche systématique d'invariants différentiels scalaires. Les *applications mécaniques* comportent la traduction de la dynamique analytique du système mécanique le plus banal (système holonome fermé à contraintes indépendantes du temps) en termes de géométrie riemannienne. L'étude du mouvement d'un tel système dynamique devient l'étude du mouvement d'un point dans un espace de Riemann dont la variété de base est l'espace de configuration et dont la métrique est donnée par l'énergie cinétique. Les applications de ce point de vue qui est lagrangien dans son essence sont remarquables, par exemple en ce qui concerne des problèmes de stabilité qu'il est possible d'étudier en termes du tenseur de courbure. Les travaux de Stächel, le célèbre problème des transformations des équations de la dynamique sont traités dans ce cadre. Les *applications physiques* comportent en particulier la traduction, en coordonnées arbitraires de l'espace euclidien de dimension 3, des équations de l'électrodynamique, de la théorie de la chaleur et de l'élasticité.

On peut dire en somme qu'Einstein a eu la possibilité d'élaborer relativement rapidement son admirable et courageuse théorie de 1915, grâce à l'édifice construit préalablement par Ricci et Levi-Civita, un édifice qui était une sorte de palais mathématique offert à la volonté et aux intuitions d'un génie. Einstein lui-même a toujours reconnu tout ce qu'il devait à ces géomètres et mécaniciens.

En fait, quand parurent les travaux de Ricci, ils ne furent accueillis par les mathématiciens de l'époque que par une estime polie. Bien qu'ils aient excité l'intérêt de quelques spécialistes à travers le monde, leur champ était malheureusement trop loin des centres d'intérêt de la communauté mathématique principalement préoccupée d'analyse, de théorie de la mesure ou de théorie des fonctions

A. Lichnerowicz, Luxembourg 1981

de variables réelles ou complexes. Quant aux géomètres différentiels, ils polissaient studieusement la théorie des surfaces plongées dans l'espace euclidien ou celle des congruences de courbes, ou se préoccupaient de géométrie projective ou conforme, ou plus généralement des espaces de Klein venus du programme d'Erlangen et ils trouvèrent l'oeuvre de Ricci «quelque peu formelle et insuffisamment géométrique» selon un mot de Darboux. Seul Poincaré y voyait une possibilité de «géométrisation globale des problèmes dynamiques».

La situation change complètement avec l'apparition de la théorie de la Relativité Générale. A partir de la fin de la première guerre mondiale, cette théorie a fourni un champ d'intérêt commun aux mathématiciens et aux physiciens théoriciens. Autour d'elle et stimulés par elle et par sa vogue prodigieuse, les savants se mettent à créer de nouvelles structures géométriques et l'on peut dire, je crois, qu'elle fut l'une des sources authentiques du développement de la géométrie différentielle jusqu'aux années 1935–1940.

D'abord vint la découverte presque simultanée par Levi-Civita (1917) et Schouten (1918) de la notion de «parallélisme» dans une variété riemannienne considérée comme plongée dans un espace euclidien. Il s'agissait en fait d'une interprétation géométrique de la condition de nullité de la dérivée covariante d'un vecteur. Mais on voyait là se révéler un instrument, un langage, des intuitions indiscutablement géométriques.

D'autre part, la physique elle-même voulait briser le cadre trop restrictif de la géométrie riemannienne. Les physiciens du temps ne connaissaient que deux champs physiques, le champ électromagnétique et le champ gravitationnel qui sont, pour nous, les deux champs à longue portée. C'était essentiellement l'étude du champ électromagnétique et du groupe de Lorentz qui avait conduit les physiciens à construire la théorie de la Relativité restreinte (Einstein 1905) et à s'intéresser à cet étrange espace-temps plat de Minkowski (1907), avec sa métrique hyperbolique. Dans le cadre de la théorie de la Relativité générale, qui est essentiellement, une théorie relativiste de la gravitation, le champ électromagnétique n'intervenait pas naturellement, mais semblait être surimposé artificiellement de l'extérieur, la liaison s'effectuant d'une part par l'écriture des équations de Maxwell en termes de dérivation covariante, d'autre part par la contribution du champ électromagnétique aux sources du champ gravitationnel en termes de tenseur d'énergie du champ électromagnétique. Il y avait là scandale, aux yeux d'Einstein lui-même, un scandale qui ne nous affecte plus maintenant de la même façon, à cause de la multiplicité et de la richesse des champs physiques que nous avons été obligés de prendre au sérieux. Mais nous y reviendrons.

Ainsi la physique théorique s'est sentie dans l'obligation de découvrir une «théorie du champ unitaire». Pendant bien des années, elle va chercher à développer une théorie qui puisse apporter l'unification des champs gravitationnels et électromagnétiques au sein d'un unique hyperchamp dont la donnée soit équivalente à celle d'une structure géométrique pour l'univers.

La recherche va s'engager dans deux voies. En premier lieu, l'addition à l'espace-temps einsteinien de dimension 4 de nouvelles dimensions conduit à un premier succès: en 1921 Kaluza introduit un espace hyperbolique (W_5, g) de dimension 5 et l'étude est raffinée en 1926 par Oscar Klein. Pour parler un langage anachronique, (W_5, g) est un espace fibré en cercles sur l'espace-temps W_4 comme base et il existe sur W_5 un champ de vecteurs tangent aux fibres cercles et unitaire qui est une isométrie infinitésimale de (W_5, g). En écrivant les équations d'Einstein sur (W_5, g) et en les décomposant par quotient, on obtient à la fois les équations de Maxwell et la liaison du tenseur d'Einstein de l'espace-temps de base avec le tenseur électromagnétique de Maxwell. Ainsi l'électromagnétisme peut être parfaitement intégré aux propriétés géométriques, la dimension supplémentaire étant inobservable. La découverte peu après d'autres champs physiques fit perdre son intérêt à la théorie de Kaluza-Klein; de plus il y avait en ce temps une grande méfiance des physiciens pour cette dimension supplémentaire. Cette démarche a retrouvé son intérêt dans les années 1970 avec les théories de jauge et en particulier de grande unification. On parle alors de la théorie de Kaluza-Klein généralisée non abélienne.

L'autre voie, plus fructueuse géométriquement, fut ouverte dès 1919 par Hermann Weyl qui élargit la structure géométrique en introduisant une géométrie dans laquelle le groupe des similitudes se substitue au groupe de Lorentz. En même temps apparaît l'espace d'Eddington, premier exemple de ce que nous nommons un espace à connexion affine. Ces premières ébauches, maladroites du point de vue physique, ont été importantes pour la géométrie. Mais il s'agissait de constructions isolées et il en était de même pour les premiers essais d'Einstein lui-même. Elles furent englobées à partir de 1922 dans la synthèse réalisée par le grand théoricien des groupes, armé du calcul différentiel extérieur, qu'était Elie Cartan. Pour celui-ci ses «espaces généralisés», comme il les nommait devaient avoir une importance considérable en *Physique Mathématique*. Devant le développement présent des théories de jauge, il ne nous appartient pas de le contredire.

Qu'on me permette de citer, sur ce sujet, un passage remarquable de la Notice de travaux scientifiques (1931) d'Elie Cartan qui met clairement en évidence cette interaction entre géométrie et physique relativiste: «Les espaces que j'ai imaginés», dit Elie Cartan, «sont une généralisation à laquelle il était impossible de parvenir en suivant les idées directrices de Riemann, Weyl et Eddington. C'est ma conception de la structure des groupes qui m'a guidé et montré la voie... Je me contenterai de parler des espaces qui touchent directement la théorie de la relativité».

On sait que, par la théorie de la Relativité générale, Einstein a bouleversé la conception qu'on se faisait jusqu'à lui d'un espace-temps homogène, préexistant aux phénomènes et dans lequel ces phénomènes viennent s'insérer sans l'altérer. La théorie de la relativité restreinte respectait cette conception et avait tout au plus pour résultat de changer la nature du groupe fondamental de l'univers. Il n'en est plus de même avec la relativité générale pour laquelle les propriétés géométriques et cinématiques de l'espace-temps sont en quelque sorte contingentes et dépendent de la distribution de la matière... Est-ce dire que la notion de groupe soit exclue de la Physique? Il n'en est rien, car l'hypothèse fondamentale de la théorie d'Einstein est non pas, comme beaucoup de personnes l'ont cru, qu'il est possible de formuler les lois de la Physique dans tout système arbitraire de coordonnées, *ce qui serait une simple tautologie*, mais que, dans toute région suffisamment petite de l'espace-temps, les lois de la relativité restreinte sont vraies en première approximation.

«D'après l'hypothèse d'Einstein, les phénomènes physiques doivent permettre de révéler au physicien le caractère euclidien d'une très petite portion de l'Univers. Or... le caractère euclidien infinitésimal de l'Univers se révèlera dans le déplacement euclidien infinitésimal (qui amène le repère orthonormé d'un observateur en coïncidence avec le repère orthonormé d'un observateur infiniment voisin). Par suite non seulement *la distance* des deux observateurs infiniment voisins, mais encore *la comparaison des directions* issues de ces deux observateurs sont l'une et l'autre du domaine de l'expérience».

E. Cartan, at his countryhouse, 1935 (courtesy H. Cartan)

De ces considérations, Elie Cartan déduit que l'espace-temps le plus général compatible avec l'hypothèse d'Einstein est ce qu'il nomme un espace à connexion euclidienne (nous dirions lorentzienne) pouvant comporter une torsion et il montre que, dans ses recherches concernant le champ unitaire, les idées d'Einstein ont évolué précisément dans cette voie qu'il avait indiquée dès 1922. La lecture de la correspondance Cartan-Einstein ne montre pas une pleine intelligence par Einstein des idées de Cartan. «Ce que je viens de dire», ajoute Cartan, «s'applique à tout autre groupe que celui des déplacements euclidiens et conduit aussi à la création des classes extrêmement étendues d'espaces qui interviennent en Analyse et Mécanique... J'ai en particulier recherché en 1923 si les phénomènes mécaniques seuls permettaient la comparaison des directions dans l'espace-temps et j'ai montré qu'il en serait ainsi si chaque élément de matière avait un moment cinétique non infiniment petit par rapport à la quantité de mouvement» (présence d'un spin classique dans la terminologie actuelle).

On voit clairement sur ce passage combien historiquement géométrie et physique relativiste se sont appuyées l'une sur l'autre en se communiquant aussi bien leurs exigences propres que leurs intuitions respectives. Physiquement nous nous trouvons ici aux sources de la théorie dite d'Einstein-Cartan (1924) qui fut renouvelée et développée dans les années 1965 par Trautman, des astrophysiciens et des cosmologistes. Mathématiquement, à partir de 1922 Elie Cartan a réussi, en rapprochant les différentes géométries différentielles existantes du programme d'Erlangen, à montrer comment à tout groupe de Lie, il est possible d'associer «une classe d'espaces à connexion de ce groupe»? Le programme offert par Elie Cartan dépassait de beaucoup toutes les réalisations effectivement construites à cette époque; il y oeuvra lui-même en créant des espaces à connexion projective ou à connexion conforme en un nouveau sens (1934). Ce programme est encore pour une part notre programme.

Dans les années 1925–1935 beaucoup de géomètres différentiels travaillent et publient à travers le monde. Parmi les plus notables, citons Bompiani en Italie, Schouten en Hollande, Eisenhart, Veblen aux Etats-Unis. Mais, sauf en ce qui concerne les groupes de Lie globaux qui émergent les premiers (H. Weyl, E. Cartan), cette géométrie différentielle, qui contient beaucoup de faits intéressants et utiles, est essentiellement *locale*. Deux voies différentes ouvrent un chemin vers la géométrie globale: d'une part les rares variétés globales W_n rencontrées sont soit définies comme espaces homogènes, soit plongées globalement dans $\mathbb{R}^N$. D'autre part, topologie algébrique et topologie différentielle se développent et finissent par aboutir (voir Alexandroff et Hopf (1936)) à une notion de variété différentiable globale décrite, à la démarche et à la terminologie près, en termes équivalents à nos cartes et atlas. Il revenait à Whitney de réaliser en 1935–36 la fusion des deux voies par son théorème de plongement global dans un $\mathbb{R}^N$ d'une variété différentiable au sens des topologues. La géométrie différentielle globale qui est la nôtre était vraiment née.

Dès lors les choses vont très vite. Avec principalement Stiefel et Ehresmann, la notion d'espace fibré s'axiomatise et se développe dans les années 1937–39 et le programme d'Elie Cartan peut s'énoncer en termes d'espaces fibrés, une structure qu'Elie Cartan semblait appréhender presque intuitivement et il n'est pas étonnant que son élève Ehresmann ait apporté la définition et les propriétés essentielles. Le langage des fibrés n'a jamais été explicitement utilisé par Cartan, mais je l'ai souvent entendu parler de «la variété des repères», ces repères pouvant correspondre à différents groupes.

Les espaces fibrés naturels les plus simples sont ceux définis par l'espace des vecteurs ou celui des covecteurs tangents aux différents points d'une variété différentiable, ou bien entendu l'espace des repères linéaires aux différents points de la variété. Dans ce dernier cas, la fibre en un point est *difféomorphe au groupe structural* qui est ici le groupe linéaire. Un tel fibré est appelé un fibré principal et aux espaces à connexion du groupe G de la terminologie de Cartan, correspondent dans la terminologie contemporaine *les connexions sur un fibré principal de fibre* G. Dans le cas du fibré des repères linéaires, on obtient des connexions linéaires (improprement appelées affines parce qu'à chacune d'elles est canoniquement associée une connexion affine). Si la variété est munie d'une métrique, riemannienne ou lorentzienne, au fibré principal des repères orthonormés correspondent les connexions métriques dont une et une seule, celle dite de Levi-Civita, est sans torsion. Selon la signature de la métrique, le groupe structural est soit le groupe orthogonal, soit le groupe de Lorentz.

Quand de 1937 à 1939, sur les conseils d'Elie Cartan, j'ai commencé à travailler sur la théorie de la Relativité générale, je me suis vite rendu compte qu'il était nécessaire de la considérer comme une théorie globale et cela a abouti à ma thèse de 1939 intitulée «problèmes globaux en Mécanique relativiste». Le système d'Einstein lui-même

$$S^i_j \equiv R^i_j - \frac{1}{2}g^i_j R = \chi T^i_j \qquad (i, j = 0, 1, 2, 3)$$

apparaissait complètement insuffisant en lui-même. Il nous fallait décrire la structure différentielle de la variété lorentzienne W_4, analyser la structure mathématique du système d'Einstein de manière à proposer des conditions initiales et des conditions aux limites. Notons que localement sur une hypersurface $\Sigma(x^0 = 0)$, S^0_j ne dépend que des métriques initiales et de leurs dérivées premières le long de l'hypersurface. On voit ainsi que la théorie d'Einstein est de ce point de vue une théorie de la propagation par ondes.

L'analyse de la structure du système d'Einstein imposait finalement à la variété différentiable d'être (C^2, C^4 par morceaux), c'est-à-dire dont les cartes sont à dérivées 3e et 4e à discontinuités finies à la traversée locale de morceaux d'hypersurfaces, la métrique g étant de classe (C^1, C^3 par morceaux) au même sens. Les atlas comportent donc un large système de cartes. En précisant que (W_4, g) est C^∞ ce qui est parfois utile mathématiquement, on perd une partie du contenu physique qu'il faut apprendre à récupérer en particulier en ce qui concerne les ondes gravitationnelles. On établit facilement que les caractéristiques du système d'Einstein sont des hypersurfaces partout tangentes au champ fondamental de cônes défini par la métrique et donc identiques aux fronts d'onde électromagnétique donnés par les équations de Maxwell; de même les rayons gravitationnels (ou bicaractéristiques) coïncident mathématiquement avec les rayons électromagnétiques.

Le phénomène fondamental de la gravitation est l'interdépendance des mouvements des masses matérielles. Pour décrire cette interdépendance, Newton s'adressa au jeu des forces attractives et obtint effectivement une description d'une admirable précision. Pour décrire cette interdépendance dans un contexte relativiste, Einstein lui abolit le jeu des forces et écrit ses équations de champ qui signifient que ce sont les masses elles-mêmes (ou plus généralement les distributions d'énergie) qui courbent l'espace-temps, qui sont sources de courbure, chaque masse engrenant son propre champ sur un champ extérieur ($T_{ij} = 0$) commun. C'est cet engrenage, ce raccordement qui crée l'interdépendance des mouvements des masses et *la condition de ce raccordement est que la métrique totale sur l'espace-temps soit* C^1. Les conditions aux limites sont données par le fait que topologiquement $W_4 = W_3 \times \mathbb{R}$ où les feuilles W_3 sont orientées dans l'espace et que W_3 est soit compacte, soit que W_4 est à compartiment asymptotique euclidien à l'infini dans l'espace, en un sens convenable. Physiquement, il n'y a pas ainsi d'énergie provenant de l'infini. On doit noter que des équations d'Einstein et de la condition de raccordement on peut déduire par passage à la limite *comme un théorème* ce qu'Einstein avait appelé *le principe des géodésiques*: la trajectoire spatio-temporelle d'une particule d'épreuve ponctuelle dans un champ de gravitation extérieur donné est une géodésique orientée dans le temps de la métrique extérieure.

Le modèle d'univers le plus simple, le modèle de Schwarzschild, satisfait aux conditions précédentes. Dans le cas particulier d'un champ extérieur stationnaire, il doit être impossible d'introduire une distribution énergétique compatible avec ce champ dans un domaine où celui-ci est régulier. On doit donc étudier la validité de la proposition suivante:

Proposition A. *L'introduction de distributions énergétiques dans un champ extérieur stationnaire donné ne peut s'effectuer que dans des domaines où ce champ n'est pas régulier.*

Si l'on admet la nécessité de cette proposition, on voit que dans un modèle d'univers constitué par une métrique extérieure partout régulière dans W_4 (avec ses conditions aux limites) stationnaire, il est impossible d'introduire une distribution énergétique. Un tel modèle d'univers doit être le modèle d'un univers *vide* et par suite sans gravitation, donc plat. Nous sommes amenés à la proposition suivante:

Proposition B. *Toute métrique extérieure stationnaire satisfaisant aux axiomes de la relativité générale et partout régulière doit être plate (localement euclidienne).*

J'ai démontré ces deux propositions en 1939 et 1946, fortement approuvé par Einstein et Pauli pour lesquels elles renforçaient la cohérence de toute la théorie. On doit noter que la seconde proposition a un lien étroit avec le théorème de la masse gravitationnelle positive tel qu'il a été établi récemment par E. Witten, Yau et Schoen.

Quittons la relativité. En 1952 nous avons établi avec Armand Borel les premiers théorèmes concernant les groupes d'holonomie dans le cas riemannien, notion qui avait été introduite fort intuitivement par Elie Cartan dès 1927. Cela a permis peu de temps après, à partir de la théorie des représentations d'Elie Cartan, à Marcel Berger de donner son fameux catalogue des groupes d'holonomie des variétés riemanniennes irréductibles et non symétriques. Il en résultait qu'il s'agissait exactement des groupes orthogonaux opérant transitivement sur la sphère. Le cas de $U(n)$ caractérisait les variétés kähleriennes ainsi dénommées par une erreur historique d'André Weil puisque Schouten (1925) les avait introduites et étudiées de manière plus profonde que Kähler.

Quittons la géométrie riemannienne provisoirement pour explorer d'autres géométries devenues importantes. Même en Mécanique Analytique classique, le concept d'espace fibré s'est révélé l'instrument fondamental pour une géométrisation globale des problèmes dynamiques, pour reprendre l'expression de Poincaré. Il s'agit en fait *d'assurer à la représentation mathématique une objectivité qui corresponde à l'objectivité des phénomènes physiques étudiés* et c'est sans doute cette exigence qui rend nécessaire la liaison entre géométrie et physique pour assurer une pleine intelligence de celle-ci.

Le formalisme lagrangien n'apporte pas grand chose à la géométrie. Il n'en est pas de même du *formalisme hamiltonien*. Pour un système à liaisons indépendantes du temps, l'espace de base est le fibré cotangent à l'espace de configuration: un covecteur élément de ce fibré, définit un état dynamique du système. Le fibré n'est autre que l'espace de phase des physiciens ou mécaniciens. Il est bien connu désormais qu'un tel fibré admet une structure symplectique naturelle exacte définie par la 2-forme de Poincaré fermée et non dégénérée qui s'écrit localement $\Sigma dp_i \wedge dq^j$. L'étude des équations de la Mécanique a montré, depuis Hamilton et Jacobi, qu'il est essentiel de pouvoir effectuer des changements des coordonnées (p_i, q^i) ne respectant pas la structure cotangente. On est ainsi conduit à prendre pour espace de phase une variété symplectique (W, F) qui définit la géométrie du système dynamique. *Un mouvement est alors décrit par une courbe intégrale $C(t)$ d'un champ hamiltonien X_H automorphisme infinitésimal de la structure symplectique défini par l'hamiltonien H*, t désignant le temps classique. Telle est la signification géométrique des classiques équations de Hamilton. L'étude de l'algèbre de Lie des automorphismes infinitésimaux symplectiques conduit naturellement au crochet de Poisson et à une forme intrinsèque de l'équation de la dynamique. L'analyse des propriétés des variétés symplectiques a essentiellement commencé avec Ehresmann et P. Libermann (1949) et elle s'est puissamment développée. Il existe sur une variété symplectique des connexions symplectiques mais aucune n'est canonique. L'analyse des déformations de l'algèbre de Lie de Poisson sur une variété symplectique a conduit à une meilleure intelligence de la quantification sur l'espace de phase et à l'élaboration d'une quantification intrinsèque (1974–1978). Les espaces homogènes symplectiques ont fait l'objet d'études approfondies (à partir de 1953).

L'approche hamiltonienne d'un système dynamique dont les liaisons dépendent du temps est moins connue. Dans ce cas, l'espace de base naturel est le fibré des directions cotangentes à l'espace-temps de configuration. Cette variété W, de dimension $(2n + 1)$, admet une structure de contact naturelle π et ceci a justifié l'intérêt porté aux variétés de contact qui peuvent être déduites par quotient de variétés symplectiques convenables de dimension $(2n + 2)$. *Un mouvement est alors décrit sur la variété de contact (W, π) par une courbe intégrale $C(s)$ d'un automorphisme infinitésimal de la structure de contact*; le paramètre s n'est autre que *l'action* à un facteur constant près. On notera que cet énoncé est indépendant de tout repérage de l'espace-temps de configuration et englobe, à lui seul, tous les principes de la mécanique classique. On peut rendre cet énoncé plus maniable en passant à la symplectisation naturelle de la variété de contact, qui peut être identifiée ici au fibré cotangent de l'espace-temps de configuration.

Un objet géométrique remarquable, de nature métrique et distinct du tenseur, était né en 1913 des recherches d'Elie Cartan sur les représentations irréductibles du groupe orthogonal $SO(n)$. Mais cet objet était resté un peu dans l'ombre, soigneusement caché derrière la transparence des mathématiques pures. Ce n'est que lorsque le génie de Dirac cherchant à établir une théorie relativiste de l'électron, le redécouvrit quelque quinze ans plus tard, qu'il acquit un nom et une reconnaissance dans le domaine de la Science. Ainsi le concept de spineur est né deux fois, d'abord entre les mains d'un mathématicien, puis dans celles d'un physicien. Elie Cartan dans ses *leçons sur la théorie des spineurs* (1938) reprit la question en dimension quelconque, d'abord algébriquement, puis en consacrant quelques pages aux spineurs en géométrie riemannienne et à l'opérateur de Dirac sur une variété riemannienne ou lorentzienne. Ces quelques pages ont paru énigmatiques tant qu'elles n'ont pas été traduites en termes d'espaces fibrés (existence d'un fibré principal de groupe $\mathrm{Spin}(n)$, revêtement à deux feuillets du fibré des repères orthonormaux directs, c'est-à-dire existence d'une structure spinorielle). Ceci n'a été fait que vers 1958. S'ils ont été depuis longtemps un instrument fondamental de la théorie quantique des champs, les spineurs sont devenus au cours des toutes dernières années un outil nouveau et puissant de la géométrie différentielle globale, principalement en liaison avec les travaux remarquables d'Atiyah-Singer (1960–1965). Mais nous sortons de la période de référence.

J'aimerais terminer par une observation: depuis quinze ans beaucoup des cours d'été de physique théorique commencent par six ou sept leçons de géométrie différentielle concernant différentielle extérieure, fibrés et connexions, leçons qui semblent banales aux mathématiciens depuis quarante ans au moins, mais qui apparaissent désormais comme devenues partie intégrante de la physique. Permettez-moi une boutade: il arrive que des physiciens, trop récents convertis, prennent la découverte par eux-mêmes d'une situation géométrique connue pour un résultat physique nouveau. Peut-être conviendrait-il de les mettre en garde?

ANDRÉ LICHNEROWICZ
Collège de France
11, Place M. Berthelot
F-75231 Paris Cedex 05

Boundary value problems for nonlinear ordinary differential equations: from successive approximations to topology

Jean Mawhin
Université de Louvain

1 Introduction

The history of the Cauchy problem for ordinary differential equations can be found in the surveys of Paul PAINLEVÉ [58] and of Max MÜLLER [49], which cover the period going from Augustin CAUCHY's pioneering contributions to the late twenties. Much less material is available concerning another important aspect of the basic theory of ordinary differential equations: *boundary value problems*. The origin of the theory of boundary value problems for *linear* ordinary differential equations can be traced to the work of Brook TAYLOR and Leonard EULER on vibrating strings in the eighteenth century and the development of a general theory is associated with the important contributions of Charles STURM and Joseph LIOUVILLE in the first part of the nineteenth century. One can refer to the excellent monograph [46] for the corresponding historical details, and to [5] and [20] for more recent developments of the Sturm-Liouville theory and its importance in the development of functional analysis.

Nonlinear boundary value problems have not been less important in the outgrowth of modern mathematics and the period 1900–1950 has been particularly interesting. Many techniques and methods which underly the present development of this area have been indeed created during this period and have important links not only with close areas like the calculus of variations and nonlinear integral or partial differential equations, but also with algebraic topology and functional analysis. The aim of this paper is to give a brief description of the main trends in this development, with the hope that it will stimulate the much more detailed study that this important area deserves. For the sake of brevity, we shall concentrate on results about the *existence* of solutions, leaving aside the important questions of *uniqueness, multiplicity, bifurcation* and *approximation*.

2 Picard's pioneering work

Although it belongs to the last decade of the nineteenth century, it is impossible to describe the development of the theory of boundary value problems for nonlinear ordinary differential equations without starting by the fundamental work of Emile PICARD in this area.

It is well known that one of the main contributions of PICARD to mathematics is his systematic use of the *method of successive approximations* for the study of nonlinear ordinary and partial differential equations. Introduced already in 1830 by LIOUVILLE in the study of linear boundary value problems occuring in one-dimensional heat conduction problems, the method of successive approximation consists in trying to solve an equation of the form

$$x = F(x)$$

with F continuous, by starting from some x_0 and defining the sequence (x_n) of successive approximations through

$$x_{n+1} = F(x_n).$$

It is then immediate that if (x_n) converges to some x^*, x^* will be a solution of the equation above.

The first memoir of PICARD devoted to the method of successive approximations, and which is anounced by a note in the *Comptes Rendus de l'Académie des Sciences de Paris* of January 13 1890, is a substantial work of sixty-five pages published the same year [59]. This paper is mostly devoted to semilinear elliptic differential equations but ends with a section entitled *A few remarks on ordinary differential equations*. After a proof, by the method of successive approximations, of the unique solvability of the Cauchy problem for a first order equation, this section is devoted to the study of the two-point boundary value problem (sometimes called nowadays the *Picard problem*)

$$y''(x) + f(x, y(x), y'(x)) = 0, \ (x \in [a, b]), \ x(a) = A, x(b) = B, \tag{1}$$

or the corresponding problem for systems of such equations. For simplicity of exposition in this paper, we will restrict ourself to the case of a single equation, as systems do not really introduce any new idea or method in the works we shall analyze, and to the case of second order equations, which are the most important ones for applications. It is assumed that, for $t \in [a, b]$, $|u| < L$, $|v| < L'$, f is continuous and satisfies the *Lipschitz condition*

$$|f(x, u, v) - f(x, w, z)| < \alpha|u - w| + \beta|v - z|. \tag{2}$$

Starting from an arbitrary function y_0, and integrating recursively the sequence of (linear non homogeneous) equations

$$-y_n''(x) = f(x, y_{n-1}(x), y_{n-1}'(x)), \ (n = 1, 2, \ldots),$$

P. Painlevé

with the boundary conditions,

$$y_n(a) = A, \ y_n(b) = B,$$

PICARD shows that the sequence (y_n) is well defined and converges to a limit y when $b - a$ is sufficiently small, and that y is a solution of the given boundary value problem. In the particular case of the problem

$$y''(x) + f(x, y(x)) = 0, \ y(a) = A, y(b) = B. \tag{3}$$

with f continuous and $f(x, .)$ non increasing for each fixed $x \in [a, b]$, PICARD also proves that there exists *at most* one solution, a result which is then extended to problem (1) when f has continuous partial derivatives with respect to u and v

which satisfy the inequality

$$-4\frac{\partial f}{\partial u}(x,u,v) > \left[\frac{\partial f}{\partial v}(x,u,v)\right]^2.$$

In the special case of the linear problem

$$y''(x) = 2A(x)y'(x) + B(x)y(x),\ y(a) = 0 = y(b),$$

the corresponding condition $B(x) > A^2(x)$ can be improved into the existence of a continuous function λ on $[a,b]$ such that

$$\lambda'(x) - (\lambda(x) + A(x))^2 + B(x) > 0.$$

This remark is developed in [61] in the special case of the problem

$$y''(x) = B(x)y(x),\ y(a) = 0 = y(b),$$

with $0 > B(x) > -h$ for all $x \in [a,b]$. By solving the nonlinear inequality above with $A = 0$ and B replaced by $-h$, it is shown that the linear boundary value problem above has a unique solution if

$$b - a < \frac{\pi}{\sqrt{|h|}},$$

i.e. if

$$|h| < \frac{\pi^2}{(b-a)^2}.$$

The special case of a differential equation with constant coefficients easily shows that this result is optimal. Some extensions will be given later by Alfred ROSENBLATT [68], using a similar approach, and by Masuo FUKUHARA [24] using a comparaison with a nonlinear autonomous conservative second order equation and estimates of the time duration along of a trajectory of such an equation.

This last section of [59] will be developed in the substantial memoir [60] anounced by a note in the *Comptes Rendus* of October 17 1892. As PICARD writes it in the introduction, *les méthodes d'approximation dont nous faisons usage sont théoriquement susceptibles de s'appliquer à toute équation, mais elles ne deviennent vraiment intéressantes pour l'étude des propriétés des fonctions définies par les équations différentielles que si l'on ne reste pas dans les généralités et si l'on envisage certaines classes d'équations. Le champ où l'approximation converge est alors susceptible d'être agrandi; on verra même des exemples où les approximations seront convergentes pour toute valeur réelle de la variable indépendante ... La plus grande partie de ce travail est consacrée à une méthode d'approximations permettant d'obtenir les intégrales d'un système de n équations du second ordre, ces intégrales ayant des valeurs données pour deux valeurs de la variable*. In [60], PICARD assumes, without loss of generality, that, in equation (1), $a = A = 0$ and $b > 0$, and he makes more precise his statement of [59] by showing that the method of successive approximations of this paper converges when the following conditions

$$|B| < L, \quad \frac{Mb^2}{2} + |B| < L, \quad Mb + \left|\frac{B}{b}\right| < L', \tag{4}$$

and

$$\frac{\alpha b^2}{2} + \beta b < 1, \tag{5}$$

hold, where M is the maximum of $|f(x,u,v)|$ on $[a,b] \times [-L,L] \times [-L',L']$. If, in particular, the Lipschitz condition (2) holds on $[a,b] \times \mathbb{R} \times \mathbb{R}$, then one can always find L and L' so that (4) holds and the convergence of successive approximations is insured under the sole condition (5). Notice that this remark, which will inspire a lot of further work as we shall see, is not explicitely made by PICARD.

In the volume III of his famous *Traité d'analyse* [61], PICARD improves his conditions by a more careful analysis of the estimates and is able to replace (5) by

$$\frac{\alpha b^2}{8} + \frac{\beta b}{2} < 1. \tag{6}$$

This is the condition which will still be proposed in [62] which, as far as boundary value problems for ordinary differential equations are concerned, does not differ essentially from [61]. Notice that in the special case of a linear equation independent of y', the above condition is more restrictive than the optimal one

$$\alpha < \frac{\pi^2}{b^2}$$

found earlier.

It may be interesting to remark that in all the above existence proofs using successive approximations, PICARD never uses *explicitely* the equivalent form of (1) as an *integral equation*

$$y(x) = A\frac{b-x}{b-a} + B\frac{x-a}{b-a} + \int_a^b G(x,s) f(s, y(s), y'(s))\, ds$$

where G is the *Green function* defined by

$$G(x,s) = \frac{(b-x)(s-a)}{b-a} \quad \text{if } a \le s \le x \le b,$$

$$G(x,s) = \frac{(b-s)(x-a)}{b-a} \quad \text{if } a \le x \le s \le b.$$

although he writes the solution of the linear problem

$$-y''(x) = \varphi(x),\ y(a) = A,\ y(b) = B$$

in the form of the simple integral

$$y(x) = A\frac{b-x}{b-a} + B\frac{x-a}{b-a} + \frac{b-x}{b-a}\int_a^x (s-a)\varphi(s)\,ds + \frac{x-a}{b-a}\int_x^b (b-s)\varphi(s)\,ds,$$

which is nothing but

$$y(x) = A\frac{b-x}{b-a} + B\frac{x-a}{b-a} + \int_a^b G(x,s)\varphi(s)\,ds.$$

He uses this expression to obtain the required estimates of y and y'. In [61] and in a subsequent note [63] devoted to some singular examples of successive approximations, and in the corresponding detailed memoir [64], which is from 1900 and is his last research work on boundary value problems for nonlinear differential equations, PICARD will explicitely name and use the Green function and refer to a memoir of Heinrich BURKHARDT [7] of 1894 for its introduction.

The Picard method will be extended by Onorato NICOLETTI [55] and E. GAU [25] to differential equations of order n with various types of multipoint or two-point linear boundary conditions, and by ROSENBLATT [67] to second order differential equations which may be singular at the boundary of the corresponding interval. By combining Picard's method of successive approximations with the modification of the right-hand member of the equation outside a suitable interval, Thomas H. GRONWALL [28] will prove the existence of a solution for (3) when $-f$ is locally Lipschitizian and increasing in y, has the same sign as y and is such that the integral

$$\int_a^\infty f(x,y)\,dx$$

diverges.

3 Sharp existence and uniqueness conditions using successive approximations

As mentioned earlier, the existence conditions obtained by PICARD in the above mentioned paper are not sharp in the sense that, when applied to the special case of the linear problem

$$y''(x) + \lambda y(x) = h(x), \quad y(a) = 0 = y(b), \tag{7}$$

for which the Lipschitz constants are respectively $\alpha = |\lambda|$ and $\beta = 0$, they give the sufficient condition

$$\frac{|\lambda|(b-a)^2}{8} < 1, \tag{8}$$

although a direct resolution of (7) shows that a unique solution exists for all right-hand members h if and only if

$$\lambda \neq \frac{k^2\pi^2}{(b-a)^2}, \quad (k = 1, 2, \dots),$$

which provides the sharp sufficient condition

$$\frac{|\lambda|(b-a)^2}{\pi^2} < 1. \tag{9}$$

A way of obtaining a sufficient condition of existence and uniqueness which would reduce to (9) in the linear situation was already introduced in 1922 by Georg HAMEL [30] in the special case of the two-point boundary value problem for the *forced pendulum equation*

$$y''(x) + \alpha \sin y(x) = b \sin x, \quad y(0) = 0 = y(\pi), \tag{10}$$

where $\alpha \geq 0$ and $b \in \mathbb{R}$. The existence of periodic solutions for this equation had been first considered heuristically in 1918 by Georg DUFFING [23], and, because of the symmetry of the nonlinearity and of the forcing term, the solutions of (10) can be extended by symmetry arguments to some 2π-periodic solutions. Using the Green function of the associated linear problem to reduce (10) to the nonlinear integral equation

$$y(x) = \alpha \int_0^\pi G(x,\tau) \sin y(\tau)\, d\tau - b \sin x,$$

HAMEL uses the method of successive approximations

$$y_n(x) = \alpha \int_0^\pi G(x,\tau) \sin y_{n-1}(\tau)\, d\tau - b \sin x,$$

with the first approximation $y_1(x) = -b \sin x$. From the inequalities

$$|y_{n+1}(x) - y_n(x)| \leq \alpha \int_0^\pi G(x,\tau)|y_n(\tau) - y_{n-1}(\tau)|\, d\tau,$$

he deduces that, for $n \geq 2$, one has

$$|y_{n+1}(x) - y_n(x)| \leq \alpha^{n-1} \int_0^\pi G_{n-1}(x,\tau)|y_2(\tau) - y_1(\tau)|\, d\tau,$$

where

$$G_n(x,\tau) = \int_0^\pi \dots \int_0^\pi G(x,\tau_1)G(\tau_1,\tau_2)\dots G(\tau_{n-1},\tau)\,d\tau_1\dots d\tau_{n-1}.$$

A direct computation then shows that

$$\lim_{n\to\infty} \int_0^\pi G_{n-1}(x,\tau)\,d\tau = 1,$$

so that the uniform convergence of the sequence (y_n) over $[0,\pi]$ follows easily from the condition

$$\alpha < 1. \tag{11}$$

As observed by HAMEL, this uniqueness condition is sharp in that, for $\alpha > 1$, the equation with $b = 0$ may have more than one solution. HAMEL's estimates only depend upon the fact that the function $\alpha \sin$ is Lipschitzian with Lipschitz constant α, so that his approach provides also the existence and uniqueness of the solution of the boundary value problem

$$y''(x) + f(x,y(x)) = 0, \quad y(a) = 0 = y(b), \tag{12}$$

when

$$|f(x,y) - f(x,z)| \leq \alpha|y-z|,$$

when the condition (11), which corresponds to the sharp one (9) when $a = 0$ and $b = \pi$, holds.

The same year 1922 had seen the publication of the Ph.D. thesis of Stefan BANACH [1] containing his abstract version of the method of successive approximations, the so-called *contraction mapping principle* or *Banach fixed point theorem*, stating that if E is a complete normed vector space and $\mathcal{U} : E \to E$ a mapping such that, for some $0 \leq M < 1$ and all $y \in E$ and $z \in E$, one has

$$\|\mathcal{U}(y) - \mathcal{U}(z)\| \leq M\|y - z\|,$$

then there exists a unique element $v \in E$ such that $v = \mathcal{U}(v)$, and v is the limit of the sequence (v_n) defined by $v_{n+1} = \mathcal{U}(v_n)$, with v_0 arbitrary in E. BANACH only applied his result to linear problems. As a preliminary lemma to prove the implicit function theorem by the method of successive approximations, Edouard GOURSAT [27] had already stated and proved a similar result twenty years before for a contraction $\mathcal{U}$ from a closed ball of $\mathbb{R}^n$ into itself. In 1920, K.W. LAMSON [39] had extended Goursat's result to the case where $\mathcal{U}$ is defined on a set M of real functions defined over a set P. A *norm* $\|.\|$ was defined on M, satisfying all the axioms (that BANACH and others will later make classic), which make M a complete normed vector space. The Goursat's case is recovered by the

S. Banach

choice $P = \{1, 2, \ldots, n\}$ and other special cases correspond to various spaces of continuously differentiable functions.

Interpreted in an abstract way, HAMEL's approach provides the extension of the Banach fixed point theorem (which will be stated and proved abstractly in the sixtees only!) assuming only that *some iterate* $\mathcal{U}^m$ of the operator $\mathcal{U}$ is a contraction on E. Here E is the space of continuous functions over $[a, b]$ with the supremum norm and $\mathcal{U}$ is the operator in E associated to the equivalent nonlinear integral equation.

Hamel's explicit conclusions are restricted to problem (10) and the conclusion for the general case (12) will be given explicitely in 1944 by Fritz LETTENMEYER [43]. In contrast to HAMEL's treatment, the approach adopted in [43], written in the language of functional analysis, consists essentially in showing that, under condition (11), the operator $\mathcal{U}$ associated to the equivalent nonlinear integral equation is a contraction in the space E of continuous functions y over $[a, b]$ such that $\frac{y(x)}{\sin\frac{\pi(x-a)}{b-a}}$ is bounded over $[a, b]$, which is a complete normed space for the norm

$$\|y\| = \sup_{x \in [a,b]} \left| \frac{y(x)}{\sin\frac{\pi(x-a)}{b-a}} \right| .$$

LETTENMEYER will also prove in a similar way the existence and uniqueness of a solution for (1) under the condition

$$\frac{\alpha(b-a)^2}{\pi^2} + \frac{4\beta(b-a)}{\pi^2} < 1 \tag{13}$$

which clearly improves PICARD's condition (6) and reduces to the sharp condition (11) when f does not depend upon y'. The condition (13) is not sharp for right-hand members depending upon y' and its refinement will be the object of further studies after 1950.

Applied to the linear problem (7), the Hamel-Lettenmeyer condition provides the unique solvability for each right-hand member whenever

$$\lambda \in] - \frac{\pi^2}{(b-a)^2}, \frac{\pi^2}{(b-a)^2} [,$$

and leaves open the question of the obtention of results in the nonlinear case which would reduce to the general *non-resonance conditions*

$$\lambda \in] - \infty, \frac{\pi^2}{(b-a)^2} [,$$

or

$$\lambda \in] \frac{k^2\pi^2}{(b-a)^2}, \frac{(k+1)^2\pi^2}{(b-a)^2} [,$$

for some integer $k \geq 1$, insuring the unique solvability of the linear problem for an arbitrary forcing term. After some pioneering work of Jean LERAY [41] which will be mentioned later, the nonlinear extension of the last situation was completely treated for the first time by Charles L. DOLPH in his Ph.D. thesis at Princeton University, the results of which were announced in 1945 [21] and the complete proofs published in 1949 [22]. DOLPH considers more generally *nonlinear integral equations* of the form

$$y(x) = \int_a^b K(x,s) f(s, y(s))\, ds, \tag{14}$$

where K is such that the associated operator

$$\mathcal{K} : y \to \int_a^b K(.,s)y(s)\,ds$$

is a completely continuous Hermitian operator in the Hilbert space $L^2(a,b)$ with (real) characteristic values

$$\ldots \leq \lambda_{-2} \leq \lambda_{-1} \leq \lambda_0 \,(< 0 <) \lambda_1 \leq \lambda_2 \ldots.$$

The function $f : [a,b] \times \mathbb{R} \to \mathbb{R}$ is continuous and there exist an integer k and real numbers μ_k and μ_{k+1} such that

$$\lambda_k < \mu_k \leq \mu_{k+1} < \lambda_{k+1}, \tag{15}$$

and

$$\mu_k \leq \frac{f(x,y) - f(x,z)}{y - z} \leq \mu_{k+1}, \tag{16}$$

for all $x \in [a,b]$ and all $y \neq z$ in $\mathbb{R}$. The idea of DOLPH consists in considering the equation (14) as a fixed point problem in the Hilbert space $L^2(a,b)$ (in contrast to the previous treatments which deal with the Banach space $C([a,b])$). Letting

$$\mu = \frac{\mu_{k+1} + \mu_k}{2}, \quad \gamma = \frac{\mu_{k+1} - \mu_k}{2},$$

DOLPH writes (14) in the equivalent form

$$y(x) = \mu \int_a^b K(x,s)y(s)\,ds + \int_a^b K(x,s)g(s,y(s))\,ds, \tag{17}$$

or abstractly

$$y = \mu\mathcal{K}y + \mathcal{K}\mathcal{G}(y), \tag{18}$$

where $g(x,y) = f(x,y) - \mu y$ now satisfies, by (16), the condition

$$-\gamma \leq \frac{g(x,y) - g(x,z)}{y - z} \leq \gamma, \tag{19}$$

for all $x \in [a,b]$ and $y \neq z$ in $\mathbb{R}$, and where $\mathcal{G}$ is the nonlinear operator defined on $L^2(a,b)$ by

$$\mathcal{G}(y)(x) = g(x, y(x)).$$

Now g is Lipschitzian in y with Lipschitz constant γ and, as μ is obviously not a characteristic value of $\mathcal{K}$, equation (18) can be rewritten as

$$y = (I - \mu\mathcal{K})^{-1}\mathcal{K}\mathcal{G}(y), \tag{20}$$

and spectral analysis in $L^2(a,b)$ implies that, for the corresponding operator norm,

$$\|(I-\mu\mathcal{K})^{-1}\mathcal{K}\| = \sup_i |\frac{1}{\lambda_i - \mu}| < \frac{1}{\gamma}.$$

Consequently, the right hand member of (20) is a contraction in the space $L^2(a,b)$, and the existence and uniqueness of the solution follows immediately from the contraction mapping theorem. Although DOLPH does not mention it explicitely (he restricts the applications to ordinary differential equations to the forced pendulum equation), his theorem applied to problem (12) immediately provides existence and uniqueness of the solution under the conditions of Hamel-Lettenmeyer and also when there exists an integer $k \geq 1$ and real numbers μ_k, μ_{k+1} such that

$$\frac{k^2\pi^2}{(b-a)^2} < \mu_k \leq \mu_{k+1} < \frac{(k+1)^2\pi^2}{(b-a)^2} \tag{21}$$

and

$$\mu_k \leq \frac{f(x,y)-f(x,z)}{y-z} \leq \mu_{k+1} \tag{22}$$

for all $x \in [a,b]$ and $y \neq z \in \mathbb{R}$. This paper of DOLPH will motivate a lot a further work in the second half of the century.

4 Variational methods

In his famous paper of 1900 [34], David HILBERT had rehabilitated the famous *Dirichlet principle*, by showing that, under reasonable conditions upon the data, a solution of the two-dimensional *Dirichlet problem*

$$\Delta u(x) = 0, \ (x \in \Omega), \ u(x) = \varphi(x), \ (x \in \partial\Omega),$$

could be associated to an achieved minimum of the corresponding *Dirichlet integral*

$$\int_\Omega |\nabla u(x)|^2 \, dx,$$

over all function u in $C^1(\Omega) \cap C(\overline{\Omega})$ equal to φ on $\partial\Omega$. The application of this *direct method of the calculus of variations* to boundary value problems for second order nonlinear ordinary differential equations was made in 1915 by Leon LICHTENSTEIN [45], in a paper which contains many important new features. If we denote by F the indefinite integral of f with respect to y,

$$F(x,y) = \int_0^y f(x,s)\, ds,$$

the direct method of the calculus of variations for solving problem (12), where, without loss of generality, we take $a = 0$ and $b = \pi$, consists in observing that (12) is just the Euler-Lagrange equation associated to the functional

$$\Phi(y) = \int_0^\pi \left[\frac{(y'(x))^2}{2} - F(x, y(x))\right] dx,$$

over the space $C_0^1([0,\pi])$ of continuously differentiable functions y over $[0,\pi]$ such that $y(0) = 0 = y(\pi)$. Thus, if one can show that Φ reaches a minimum over such a space at some y, then y will be a solution of (12). The basic assumption made in [45] is that F is bounded from above, i.e. that

$$F(x, y) \leq A \tag{23}$$

for some real number A and all $x \in I = [0,\pi]$ and $y \in \mathbb{R}$. This implies immediately that the functional Φ is bounded from below on $C_0^1(I)$ and its infimum is denoted by d. If (y_n) denotes a *minimizing sequence* (i.e. $\Phi(y_n) \to d$ if $n \to \infty$), then, writing

$$y_n(x) = \sum_{j=1}^{\infty} \frac{y_{n,j}}{j} \sin jx,$$

so that $\sum_{j=1}^{\infty} y_{n,j} \sin jx$ is the Fourier series of y_n', LICHTENSTEIN proves the existence of a subsequence such that, for each positive integer j, $(y_{n_k,y})$ converges to some y_j. He then shows that the corresponding function y defined by

$$y(x) = \sum_{j=1}^{\infty} \frac{y_j}{j} \sin jx$$

belongs to $C^2(I)$, vanishes at 0 and π, minimizes Φ over $C_0^1(I)$ and hence is a solution to (12). To realize this program, LICHTENSTEIN observes that $(\|y_n'\|_{L^2})$ is bounded, so that the same is true for

$$\left(\sum_{j=1}^{\infty} y_{n,j}^2\right),$$

and a diagonal process gives the existence of convergent subsequences $(y_{n_k,j})$ and (y_{n_k}). He shows then that (y_{n_k}) converges uniformly on I to the function y given above, and hence that

$$\frac{1}{2}\int_0^\pi (y_{n_k}'(x))^2\, dx \to d + \int_0^\pi F(x, y(x))\, dx = d + \Delta.$$

This easily implies that

$$\frac{\pi}{4}\sum_{j=1}^{\infty} y_j^2 \leq d + \Delta,$$

and an argument by contradiction based upon the definition of d implies that the equality indeed holds. Thus the infimum d of Φ is reached at y, and this implies the relations

$$\frac{\pi}{2}y_l - l^{-1}\int_0^{\pi} f(x, y(x)) \sin lx \, dx = 0,$$

and

$$\frac{\pi}{2}\sum_{l=1}^{\infty} y_l v_l - \int_0^{\pi} f(x, y(x)) v(x) \, dx = 0,$$

for all positive integers l and all $v(x) = \sum_{j=1}^{\infty} \frac{v_j}{j} \sin jx$ such that $\sum_{j=1}^{\infty} v_j^2$ converges. Choosing for $v(x)$ the Green function $G(s, x)$ of the associated linear problem, LICHTENSTEIN deduces from those relations that

$$y(s) = \int_0^{\pi} G(s, x) f(x, y(x)) \, dx,$$

for all $s \in I$, so that $y \in C_0^1(I) \cap C^2(I)$ is a solution of (12) such that $\Phi(y) = d$.

Thus LICHTENSTEIN's treatment contains all the basic ingredients of the present version of the direct method of the calculus of variations for the study of (12), namely:

1. Considering a minimizing sequence (y_k).
2. Showing the boundedness of (y_k) in the Sobolev space $H_0^1(I)$ and using it to obtain a subsequence converging weakly in $H_0^1(I)$ and uniformly on I to some y.
3. Using the weak lower semicontinuity of the square of the H_0^1 norm to show that y minimizes Φ.
4. Using the Euler-Lagrange equation to show that the (weak) solution y is indeed a classical one.

LICHTENSTEIN also gives interesting special cases of his existence result like

$$y'' = y^{2n+1} + \sum_{k=0}^{2n} a_{2n+1-k}(x) y^k,$$

where n is a nonnegative integer and the a_j are continuous functions over I, and

$$y'' = k(x) \exp y,$$

where k is a positive continuous function over I. In the special case of the linear problem (7), one has

$$F(x, y) = \frac{\lambda y^2}{2} - h(x)y,$$

and then F is bounded from above for all h if and only if $\lambda < 0$. Combined with the results of HAMEL-LETTENMEYER and DOLPH, Lichtenstein's theorem completes therefore the nonlinear analogue of the non-resonance conditions of the linear case, but, under his assumptions, the uniqueness and the possibility of getting the solutions by successive approximations are of course lost.

Notice that LICHTENSTEIN uses also the same approach in [45] to study the corresponding *periodic boundary value problem*

$$y''(x) + f(x, y(x)) = 0, \quad y(0) = y(2\pi), \quad y'(0) = y'(2\pi),$$

when F is not only bounded from above but moreover sign $yf(x, y) < 0$ for sufficiently large $|y|$. This seems to be the first occurence of a variational existence proof for periodic solutions of non-autonomous differential equations.

The direct method of the calculus of variations will be applied in 1922 by Georg HAMEL, in the above mentioned paper [30], to the existence of 2π-periodic solutions of the forced pendulum equation

$$y''(x) + \alpha \sin y(x) = b \sin x \tag{24}$$

for all values of α and b. HAMEL, in a rather sketchy way, and without reference to LICHTENSTEIN's paper, applies HILBERT's method to the corresponding functional

$$\Phi(y) = \int_0^{2\pi} [\frac{(y'(x))^2}{2} + \alpha \cos y(x) + by(x) \sin x]\, dx,$$

on the space of 2π-periodic continuously differentiable functions. To show that Φ is bounded below, HAMEL writes it in the equivalent form

$$\begin{aligned}\Phi(y) &= \int_0^{2\pi} \left[\frac{(y'(x))^2}{2} + \alpha \cos y(x) + by'(x) \cos x\right] dx \\ &= \int_0^{2\pi} \left[\frac{1}{2}(y'(x) + b \cos x)^2 + \alpha \cos y(x) - \frac{1}{2} b^2 \cos^2 x\right] dx,\end{aligned}$$

through an integration by part of the last term. In the more general situation of the equation

$$y''(x) + \alpha \sin y(x) = h(x),$$

the above argument remains valid if and only if the primitive $H(x)$ of h is 2π-periodic, i.e. if and only if

$$\int_0^{2\pi} h(x)\, dx = 0.$$

Such a result will be independently rediscovered in the early eighties where the study of the multiplicity of the periodic solutions of pendulum type equations will be a challenging problem for the most recent techniques of *critical point theory*.

The direct method of the calculus of variations will be used in 1930 by Adolf HAMMERSTEIN [32] (after a preliminary announcement [31] in 1927), to prove the existence of solutions of the so-called *Hammerstein integral equations* which, in the one-dimensional case, are of the form considered above

$$y(x) = \int_a^b K(x,s) f(s, y(x))\, ds, \tag{25}$$

where K is symmetrical, positive definite, piecewise continuous, bounded and f is continuous. If

$$(0 <)\, \lambda_1 < \lambda_2 < \dots$$

denote the characteristic values of the linear problem

$$\lambda \int_a^b K(x,s) y(s)\, ds = y(x),$$

and

$$\varphi_1(x),\ \varphi_2(x), \dots,$$

an orthogonal system of corresponding eigenfunctions, HAMMERSTEIN assumes that the indefinite integral F of f satisfies the condition

$$F(x,y) \leq \frac{\alpha y^2}{2} + C \tag{26}$$

for some

$$\alpha < \lambda_1, \quad C > 0 \tag{27}$$

and all $x \in [a,b]$ and $y \in \mathbb{R}$, which clearly generalizes LICHTENSTEIN's one. To prove the existence of a solution for (25) under those assumptions, HAMMERSTEIN uses an approach initiated by Walter RITZ [66] for linear problems by considering the sequence of finite-dimensional minimization problems for the functionals H_m defined over $\mathbb{R}^m$ by

$$H_m(\alpha_1, \alpha_2, \dots, \alpha_m) = \frac{1}{2} \sum_{j=1}^{m} \lambda_j \alpha_j^2 - \int_a^b F(x, \sum_{j=1}^{m} \alpha_j \varphi_j(x))\, dx.$$

Using (26) and (27), it is easy to show that

$$H_m(\alpha_1, \alpha_2, \dots, \alpha_m) \geq \frac{1}{2} \sum_{j=1}^{m} (\lambda_j - \alpha)\alpha_j^2 - C(b-a),$$

so that each H_m is bounded below. It has therefore a minimal value d_m reached for some

$$c^{(m)} = (c_1^{(m)}, \ldots, c_m^{(m)}) \in \mathbb{R}^m,$$

which provides the sequence of functions (y_m) defined by

$$y_m(x) = \sum_{j=1}^{m} c_j^{(m)} \varphi_j(x).$$

HAMMERSTEIN proves then the existence of a subsequence which converges uniformly over $[a, b]$ to a solution y of the equation (25). The uniqueness is proved if condition (26) is replaced by

$$\frac{\partial f}{\partial y}(x, y) \leq \alpha,$$

for all $x \in [a, b]$ and $y \in \mathbb{R}$.

HAMMERSTEIN observes also that by setting

$$z = -y'',$$

or, equivalently, taking in account the boundary conditions $y(a) = 0 = y(b)$,

$$y(x) = \int_a^b G(x, s) z(s)\, ds,$$

the functional Φ associated to (12) can be written

$$\begin{aligned}\Phi(y) &= -\int_a^b [\frac{y(x)y''(x)}{2} + F(x, y(x))]\, dx \\ &= \frac{1}{2}\int_a^b \int_a^b G(x, s) z(s) z(x)\, ds\, dx \\ &\quad - \int_a^b F(x, \int_a^b G(x, s) z(s)\, ds)\, dx = \Psi(z),\end{aligned}$$

so that every solution y of (7) will be such that z minimizes Ψ.

A version of Hammerstein theorem will be proved in 1935 by Michael GOLOMB [26] for the abstract equation

$$y - \mathcal{H}^2 \mathcal{G}(y) = 0,$$

where $\mathcal{H} : E \to E$ is a bounded operator in the Hilbert space E, so that, for some $h > 0$ and all $y \in E$, one has

$$(\mathcal{H}y, \mathcal{H}y) \leq \frac{1}{h}(y, y).$$

Moreover, $\mathcal{G}$ is defined and continuous on the closed hull E' of the range of $\mathcal{H}$ and is on E' the gradient of a real function $\mathcal{F}$ such that

$$\mathcal{F}(y) \leq \frac{\alpha}{2}(y,y) + C,$$

for all $y \in E'$ and some $\alpha < h$. The solution is obtained by minimizing the functional

$$\Psi(y) = (y,y) - 2\mathcal{F}(\mathcal{H}y)$$

over the space E'.

A variant of Hammerstein's approach and result will be given in 1936 by H.O. HIRSCHFELD [35] by replacing the integral in the equation by an Euler type approximation. This complicated approach provides error bounds between the minimum points of the approximate functionals and the solution of the problem. In 1938, Silvio CINQUINI [12] will return to the original LICHTENSTEIN's simple approach and show that the existence of a solution for (12) which minimizes the corresponding functional Φ still holds under HAMMERSTEIN's condition

$$F(x,y) \leq \frac{\alpha y^2}{2} + C,$$

where $\alpha < \frac{\pi^2}{(b-a)^2}$. The new ingredient with respect to Lichtenstein's treatment consists in the use of the POINCARÉ's inequality

$$\frac{\pi^2}{(b-a)^2}\int_a^b (y(x))^2\,dx \leq \int_a^b (y'(x))^2\,dx,$$

which is valid for all function y such that $y(a) = 0 = y(b)$ and $y' \in L_2(a,b)$.

HAMMERSTEIN's result is of course a generalization of the so-called HAMEL-LETTENMEYER condition where the Lipschitz condition on f is replaced by a growth condition on its indefinite integral F. In the already quoted papers [21], [22], DOLPH has considered the much more difficult problem of extending similarly his conditions (16). A candidate for the generalization would be that there exist an integer k and real numbers μ_k and μ_{k+1} such that condition (15) and condition

$$\frac{\mu_k y^2}{2} - C \leq F(x,y) \leq \frac{\mu_{k+1} y^2}{2} + C' \tag{28}$$

hold for all $x \in [a,b]$ and all y and z in $\mathbb{R}$. Assuming that K is symmetric and positive definite, so that it can be written as

$$K(x,s) = \int_a^b H(x,u)H(u,s)\,du,$$

for some positive and symmetric function $H(x,s)$, DOLPH, by setting

$$y(x) = \int_a^b H(x,s)z(s)\,ds,$$

rewrites, like in [26], equation (14) equivalently as

$$z(x) = \int_a^b H(x,s)f(s, \int_a^b H(s,u)z(u)\,du)\,ds.$$

which is the Euler-Lagrange equation of the functional

$$\Psi(z) = \frac{1}{2}\int_a^b (z(x))^2\,dx - \int_a^b \int_0^{\int_a^b H(x,s)z(s)\,ds} f(x,v)\,dv\,dx.$$

Assuming in addition of (15) and (28) that

$$|f(x,y)| \leq C|y| + D,$$

for all $x \in [a,b]$, all $y \in \mathbb{R}$ and some $C, D \geq 0$, and that the functional Ψ possesses only one maximum on any n-dimensional linear manifold in $L^2(a,b)$ parallel to the one containing the origin and the first k eigenfunctions of $K(x,y)$, DOLPH proves the existence of at least one solution y for (14) such that the corresponding z satisfies the *minimax property*

$$\Psi(z) = \inf_{M_k \in \Gamma} \sup_{u \in M_k} \Psi(z),$$

where Γ denotes the class of k-dimensional linear manifolds M_k defined above. Despite a large number of papers written on the subject since the late seventies and using modern tools of *critical point theory*, it is still unknown if conditions (15) and (28) are sufficient to insure the solvability of (14).

5 Topological methods

A natural approach to prove the existence of a solution to the problem (1) consists in first proving the existence of a solution $y(x;c)$ defined over $[a,b]$ of the initial value problem

$$y''(x) + f(x, y(x), y'(x)) = 0, \quad y(a) = A, \quad y'(a) = c,$$

and then showing that the scalar equation

$$y(b;c) = B$$

has at least one solution $\hat{c}$; the corresponding function $y(x;\hat{c})$ will then be a solution of (1). This approach is called the *shooting method* and the existence of a solution for the scalar equation above will follow in general from a very simple topological existence principle: the *intermediate value theorem.* Such an approach was introduced in 1905 by Carlo SEVERINI [85] [86] to prove that (1), with f continuous, had at least one solution when $b-a$ and $\frac{B-A}{b-a}$ are sufficiently small. SEVERINI approached f by a sequence of polynomials (P_k) so that the corresponding initial value problems

$$y''(x) + P_k(x, y(x), y'(x)) = 0, \quad y(a) = A, \quad y'(a) = c,$$

have a unique solution. He then used the shooting method to obtain a solution y_k of the corresponding boundary value problem

$$y''(x) + P_k(x, y(x), y'(x)) = 0, \quad y(a) = A, \quad y(b) = B,$$

and then the Ascoli-Arzelà theorem to deduce from (y_k) a solution y of (1).

In a paper published in 1912 and proving some results already anounced in 1904 in a note in the *Comptes Rendus de l'Académie des Sciences de Paris*, Sergej N. BERNSTEIN [3] used the shooting method to prove that if

$$\frac{\partial f}{\partial y}(x, y, z) \leq \gamma < 0, \tag{29}$$

for all $x \in [a,b]$, $y \in \mathbb{R}$ and $z \in \mathbb{R}$, and if, for each $R > 0$, there exist $C \geq 0$ and $D \geq 0$ such that

$$|f(x, y, z)| \leq Cz^2 + D, \tag{30}$$

for all $x \in [a,b]$, $|y| \leq R$ and $z \in \mathbb{R}$, then the problem (1) with f real analytic has a unique solution. This paper contains in particular the famous *Bernstein's lemma* showing that if y is a bounded solution on $[a,b]$ of the equation

$$y''(x) + f(x, y(x), y'(x)) = 0,$$

and

$$|f(x, y(x), y'(x))| \leq C|(y'(x))^2| + D$$

for some $C \geq 0, D \geq 0$ and all $x \in [a,b]$, then $|y'(x)|$ is uniformly bounded over $[a,b]$. The existence part of BERNSTEIN's result will be partially extended in 1932 by Renato CACCIOPPOLI [10], where (29) is replaced by the non increasing character of $f(x,.,z)$ and (30) is replaced by a uniform Lipschitz condition with respect to z.

In a celebrated paper published in 1922, George D. BIRKHOFF and Oliver D. KELLOGG [4] have extended to some function spaces the famous *Brouwer fixed point theorem* by proving the existence of at least one fixed point for continuous mappings from a compact subset of $C([a,b])$, $C^k([a,b])$ or of $L^2(a,b)$ into itself. They apply their fixed point theorem to *nonlinear multipoint boundary value problems* of the form

$$y^{(n)}(x) + f(x, y(x), y'(x), \ldots, y^{(n-1)}(x)) = 0,$$

$$\int_a^b \sum_{j=0}^{n-1} p_{ij}(x) y^{(j)}(x)\,dx + \sum_{j=0}^{n-1}\sum_{k=1}^{m} q_{ijk} y^{(j)}(x_k) = c_i, \tag{31}$$

$$(i = 1, 2, \ldots, n; \quad a \leq x_1, x_2, \ldots, x_m \leq b),$$

where f and the p_{ij} are continuous and the boundary conditions are such as to determine uniquely a polynomial y of degree $n-1$. Problem (31) is equivalent to proving the existence of a fixed point of the transformation S defined by

$$S(y)(x) = -\int_0^x \int_0^{x_{n-1}} \cdots \int_0^{x_1} f(s, y(x), \ldots, y^{(n-1)}(s))\,dx_0\,dx_1 \ldots dx_{n-1}$$
$$+ a_0 + a_1 x + \ldots + a_{n-1} x^{n-1},$$

where the coefficients a_j are explicitely defined in term of y by the boundary conditions. The application of the Birkhoff-Kellogg fixed point theorem in $C^{n-1}([a,b])$ to S implies that if there exists a set of n constants $B_0, B_1, \ldots, B_{n-1}$ such that when

$$\|y\| \leq B_0, \|y'\| \leq B_1, \ldots, \|y^{(n-1)}\| \leq B_{n-1},$$

the same inequalities hold for $S(y)$, then S has a fixed point, and thus (31) has at least one solution. The conditions will be in particular satisfied if $b-a$ is sufficiently small. A similar result, with some explicit estimates for $b-a$, will be independently proved in 1928 by FUKUHARA for the problem (1). Existence is insured if

$$b - a \leq \min\left\{\sqrt{\frac{8L}{M}}, \frac{2L'}{M}\right\},$$

when $|f(x,y,z)| \leq M$ for $|y| \leq L$, $|z| \leq L'$. The method of proof consists in using an Euler-Cauchy-Lipschitz approximation method like in the Cauchy problem, insuring the boundary condition at each approximation by a shooting type argument and the intermediate value theorem, and then going to the limit using the Ascoli-Arzelà theorem.

The assumptions of the Birkhoff-Kellogg's paper are satisfied for $b-a$ arbitrary if there exists $B > 0$ such that

$$|f(x, y_1, \ldots, y_n)| \leq B,$$

for all $x \in [a,b]$, $y_1 \in \mathbb{R}, \ldots, y_n \in \mathbb{R}$. This result will be explicitely stated and proved, without any reference to [4], by CACCIOPPOLI in [8] where he reproves

J. Schauder

the Birkhoff-Kellogg fixed point theorem in the space $C^{(n-1)}([a,b])$. Notice that Julius SCHAUDER [72] [73] had already extended the Birkhoff-Kellogg fixed point theorem to the case of a continuous operator S mapping into itself a convex compact subset of a Banach space with a Schauder basis, had eliminated this last condition in 1930 [74] and had given applications of his fixed point theorem to the Cauchy problem for ordinary differential equations and to some boundary value problems for partial differential equations. Applications of the Birkhoff-Kellogg-Schauder fixed point theorem to nonlinear Hammerstein integral equations will also be given by Viktor V. NIEMYTZKI in [56] [57].

In particular, although it is not explicitely mentioned in [4], the problem (1) will have at least one solution for all $A, B \in \mathbb{R}$ if there exist $B > 0$ such that

$$|f(x, y, z)| \leq B,$$

for all $x \in [a, b], y \in \mathbb{R}, z \in \mathbb{R}$.

This last result will be reproved in 1931, using the shooting method, by Giuseppe SCORZA DRAGONI [75] under the supplementary assumption that the functions $\frac{\partial f}{\partial y}$ and $\frac{\partial f}{\partial z}$ exist and are bounded on bounded sets. In another paper published the same year [76], SCORZA DRAGONI will observe that *this result can be considered as known but the contribution is in the simplicity of the approach* and he refers to [4] and to [8]. SCORZA DRAGONI adds that *it is not difficult to free this result from the assumptions upon the partial derivatives of f, for example by using the methods of Caccioppoli and Birkhoff.* The case of f satisfying the *Carathéorody conditions* (i.e. $f(., y, z)$ measurable for each y, z, $f(x, ., .)$ continuous for almost every x and $|f(x, y, z)| \leq g(x)$ for some Lebesgue integrable function g) will be first treated by SCORZA DRAGONI [79] and then, in a simpler way, by Giuseppe ZWIRNER [89].

In his paper [9], CACCIOPPOLI acknowledges the priority of BIRKHOFF-KELLOGG [4] and of SCHAUDER [72], [73] in the extensions of the Brouwer fixed point theorem to Banach spaces and applies it to the proof of the existence of a solution of (31) when the boundedness of f is replaced by the more general condition

$$\lim_{y \to \infty} \frac{f(x, y_1, y_2, \ldots, y_n)}{(\sum_{j=1}^{n} y_j^2)^{1/2}} = 0.$$

Topological methods will also be used to prove results related to HAMMERSTEIN's one [32]. In [10], CACCIOPPOLI proves that if $\mathcal{S}$ is a completely continuous mapping from a Banach space E into itself such that $\mathcal{T} = \mathcal{I} - \mathcal{S}$ is locally invertible at each point and $\mathcal{T}(u) \to \infty$ if $\|u\| \to \infty$, then $\mathcal{S}$ has a unique fixed point. This result generalizes earlier ones of Jacques HADAMARD [29] in $\mathbb{R}^n$ and of Paul LÉVY [44] in some function spaces. CACCIOPPOLI then applies this result to the nonlinear integral equation (25) with K symmetrical and f such that $\frac{\partial f}{\partial y}(x, y)$ exists, is continuous and such that

$$\left| \frac{\partial f}{\partial y}(x, y) \right| \leq \alpha < \lambda_1,$$

where λ_1 is the smallest characteristic value associated to the nucleus K. The same theorem is used to prove that the problem (1) has a solution when $f(x, ., z)$ is non increasing for each $x \in [a, b]$ and $z \in \mathbb{R}$ and $f(x, y, .)$ is globally Lipschitzian uniformly in $x \in [a, b]$ and $y \in \mathbb{R}$. The two results of CACCIOPPOLI will be reproved, using the approximation and shooting method, by SCORZA DRAGONI [78] and unified and extended to the case where

$$f(x, y, z) = \varphi(x, y) + \psi(x, y, z),$$

where $\varphi(x,.)$ is non increasing and

$$\frac{\psi(x,y,z)}{|y|+|z|} \to 0 \text{ if } |y|+|z| \to \infty.$$

Problem (1) is also considered by HAMMERSTEIN in [33] who proves the existence of at least one solution when $a=0$, $b=\pi$ and

$$|f(x,y,z)| \leq \alpha|y|,$$

for all $x \in [a,b]$, all $z \in \mathbb{R}$, all $y \in \mathbb{R}$ with $|y|$ sufficiently large and some

$$\alpha < \frac{\pi^{3/2}\sqrt{3}}{(b-a)^2},$$

which is not a sharp condition but is better than PICARD's one. HAMMERSTEIN uses a *Galerkin's method* by considering first the approximate equations

$$F_j(\alpha_1,\dots,\alpha_n) := j\alpha_j - \frac{2}{\pi}\int_0^\pi f\left(x, \sum_{k=1}^n \frac{\alpha_j}{j}\sin jx, \sum_{j=1}^n \alpha_j \cos jx\right)\sin jx\,dx = 0,$$
$$(j = 1,2,\dots,n),$$

and showing that the assumptions imply the existence of some $d>0$ such that F_j is positive for $\alpha_j = d$, $|\alpha_k| \leq d$ if $k \neq j$, and F_j is negative for $\alpha_j = -d$, $|\alpha_k| \leq d$ if $k \neq j$. He then refers to the *Brouwnschen Verschiebungssatz* (sic) in [6] to conclude to the existence of a zero $\alpha^{(n)}$ of $(F_1,\dots,F_n)$ in $[-d,d]^n$. This n-dimensional version of the intermediate value theorem is not explicitely stated in [6], which however contains a close statement and introduces some techniques (namely *topological degree* theory), which allow an easy proof of this theorem. Indeed, as early as in 1883, the result was already stated and claimed to be an easy consequence of the *Kronecker index* (a forerunner of degree theory), by Henri POINCARÉ [65], and is presently known as *Miranda's theorem*, after Carlo MIRANDA proved in 1940 its equivalence with the *Brouwer fixed point theorem* [47]. HAMMERSTEIN then uses a limiting argument to construct, from the $\alpha^{(n)}$, a solution to the original problem. By modifying the vector field f in a suitable region, HAMMERSTEIN shows that a similar result holds when f is nonnegative for nonnegative u and satisfies for those u a similar growth condition.

Other results based upon the existence of some growth condition upon f will be proved in 1939 by Silvio CINQUINI [13] using the SEVERINI's approach and versions of the Gronwall and Bernstein lemmas (attributed to Leonida TONELLI!) to prove the existence of one solution to (1) when f satisfies on $[a,b]\times\mathbb{R}^2$ conditions of the type

$$|f(x,y,z)| \leq \varphi(x) + \psi(x)|z|,$$

with φ and ψ integrable over (a, b), or of the type

$$|f(x, y, z)| \le \alpha(x)|y| + \beta(x)|z| + \gamma(x),$$

where α, β, γ are non-negative integrable functions over (a, b) such that

$$\int_a^b [[2(b-a) + |B - A|]\alpha(x) + \beta(x)]\, dx < 1,$$

or of the type

$$|f(x, y, z)| \le \psi_0(y)z^2 + \psi_1(x),$$

where ψ_0 and ψ_1 are nonnegative functions respectively integrable over $\mathbb{R}$ and (a, b). Those results will be immediately generalized to the case of one-sided growth conditions by TONELLI [87] who will in particular prove the existence of a solution when, for each $Y \ge 0$, one can find three functions α, β and ψ like above such that

$$|f(x, y, z)| \le \beta(x)|z| + \psi(x),$$

when $|y| \le Y$ and

$$\text{sign } yf(x, y, z) < \alpha(x)|y| + \beta(x)|z| + \psi(x).$$

when $|y| > Y$. Related results will be given the same year by CINQUINI [14] [15], SCORZA DRAGONI [81] and ZWIRNER [91]. Those papers and their *Aggiunte* reveal the existence of a polemics about errors, methods and priority questions between CINQUINI and SCORZA DRAGONI and this polemics will last for more than ten years, as shown by the sequence of papers [16] [17] [18] and [82] [83] [84].

6 Continuation and Leray-Schauder methods

Motivated by the study of the solvability of (12) in the case of $f(x, y)$ real analytic in y, Rudolf IGLISCH [36] has used in 1933 a *continuation method* which can be traced to BERNSTEIN's work [2] and Torsten CARLEMAN's paper [11] on semilinear elliptic equations, to prove the existence of solutions for the nonlinear integral equation (25) when $f(x, y)$ is real analytic with respect to y. Under the assumption that

$$\frac{f(x, y) - f(x, 0)}{y} \le \alpha < \lambda_1,$$

for all $|y|$ sufficiently large, with λ_1 the smallest characteristic value associated to the symmetric kernel K, IGLISCH introduces the one-parameter family of equations

$$y(x) - \int_a^b K(x, s) \left[\tau f(s, y(s)) + (1 - \tau)\frac{\lambda_1}{2} y(s)\right] ds, \quad \tau \in [0, 1], \tag{32}$$

which reduces for $\tau = 1$ to the given equation and for $\tau = 0$ to a linear equation having only the trivial solution. He then shows that there exists $R > 0$ such that, for all $\tau \in [0, 1]$, each possible solution u of (32) satisfies the a priori bound

$$\max_{x \in [a,b]} |u(x)| \leq R.$$

He also shows, by an implicit function theorem argument that, for each $\tau_0 \in [0, 1]$, there exists $\rho(\tau_0) > 0$ such that (32) is uniquely solvable for

$$\tau \in [0, 1] \cap]\tau_0 - \rho(\tau_0), \tau_0 + \rho(\tau_0)[.$$

As mentioned by IGLISCH in a note added in proof, similar results had been obtained by LERAY [41] [40], who had suggested several improvements in IGLISCH paper. In [41], using a continuation approach that he calls the *Arzelà-Schmidt method*, LERAY proves the existence of at least one solution for nonlinear integral equations of the form

$$y(x) = h \int_0^1 K(x, s) y(s)\, ds + k \int_0^1 H(x, s, y(s))\, ds, \tag{33}$$

when h is not a characteristic value λ_i of the corresponding linear problem

$$y(x) = \lambda \int_0^1 K(x, s) y(s)\, ds,$$

H is bounded on $[0, 1] \times [0, 1] \times \mathbb{R}$ or such that

$$\frac{H(x, s, y)}{y} \to 0 \text{ if } |y| \to \infty,$$

uniformly on $[0, 1] \times [0, 1]$, and k is a real number. In particular, this result implies the solvability of the boundary-value problem (12) when

$$\frac{f(x, y)}{y} \to h \text{ if } |y| \to \infty,$$

uniformly in $[a, b]$, under the condition that h is not an eigenvalue of the corresponding linear problem. As another example, LERAY also proves the existence of a nonnegative solution of (33) for all $h \geq 0$ and $k \geq 0$ when $K \geq 0$ and $H > 0$ for all $u \geq 0$. Finally, he applied his method to the nonlinear integral equation

$$y(x) + h \int_0^1 K(x, s) g(y(s))\, ds = 0,$$

when $h > 0$, $K(x, s) + K(s, x)$ has all its characteristic values positive and there exist some $C \geq 0$ such that $|g(y)| \leq C$ whenever $yg(y) < 0$.

One year later, LERAY and SCHAUDER will give to this approach its natural topological frame in their fundamental paper [42] which generalizes *Brouwer degree theory* to completely continuous perturbations of identity in Banach spaces and encompasses the Schauder fixed point and the continuation method. This theory will be applied by LERAY and SCHAUDER to various problems for partial differential equations. In 1937 Erich ROTHE [69] will use Leray-Schauder theory to extend the Schauder fixed point theorem to the case of a completely continuous mapping sending the boundary of a ball into the ball and will give applications to nonlinear integral equations, and in particular to (25) when K is not necessarily symmetric and

$$|f(x,y)| \leq G(|y|),$$

with G nondecreasing and such that the equation

$$MG(r) - r = 0$$

has at least one positive root, where

$$M = \max_{x \in [a,b]} \int_a^b |K(x,s)|\, ds.$$

DOLPH [21], [22] will use the Leray-Schauder method to prove the existence of at least one solution to the nonlinear integral equation (25) when K is such that the associated operator

$$\mathcal{K} : y \to \int_a^b K(.,s)y(s)\, ds$$

is a completely continuous Hermitian operator in the Hilbert space $L^2(a,b)$ with (real) characteristic values

$$\ldots \leq \lambda_{-2} \leq \lambda_{-1} \leq \lambda_0 (< 0 <) \leq \lambda_1 \leq \lambda_2 \ldots.$$

The function $f : [a,b] \times \mathbb{R} \to \mathbb{R}$ is assumed to be continuous and such that there exist an integer k and real numbers μ_k and μ_{k+1} such that condition (15) and

$$\begin{aligned} \mu_k y - A \leq f(x,y) \leq \mu_{k+1} y + A \quad & \text{for } y > y_0, \\ \mu_{k+1} y - A \leq f(x,y) \leq \mu_k y + A \quad & \text{for } y < -y_0, \end{aligned}$$

hold for all $x \in [a,b]$. Mark A. KRASNOSELS'KII [37] [38] will also use Leray-Schauder theory to extend some of the results of LERAY on nonlinear integral equations mentioned above. The second half of the century will see a tremendous development of the applications of the Leray-Schauder method to nonlinear boundary value problems.

7 Lower and upper solutions and related results

In the already mentioned paper [75], SCORZA DRAGONI considers the solvability of problem (1) under the assumption of the existence of two functions y_0 and y_1 of class C^2 with $y_0(x) \leq y_1(x)$ over $[a,b]$ such that the following conditions are satisfied:

a) f is defined, continuous and bounded over the cylinder

$$C = \{(x,y,z) : x \in [a,b], y_0(x) \leq y \leq y_1(x), z \in \mathbb{R}\};$$

b) $f(x, y_0(x), .)$ and $f(x, y_1(x), .)$ are monotone for each fixed $x \in [a,b]$;

c) y_0 and y_1 are solutions of the differential equation

$$y''(x) + f(x, y(x), y'(x)) = 0;$$

d) $y_0(a) \leq A \leq y_1(a)$, $y_0(b) \leq B \leq y_1(b)$.

The proof introduces the *modification of the equation* to an equation defined over $[a,b] \times \mathbb{R}^2$

$$y''(x) + g(x, y(x), y'(x)) = 0, \tag{34}$$

where g is bounded, equal to f for $y_0(x) \leq y \leq y_1(x)$ and decreasing for $y < y_0(x)$ and $y > y_1(x)$. By a theorem mentioned in Section 5, equation (34) will have at least one solution y satisfying the boundary conditions

$$y(a) = A, \; y(b) = B. \tag{35}$$

The assumptions and constructions above then imply, using a maximum principle type argument, that this solution satisfies the inequality

$$y_0(x) \leq y(x) \leq y_1(x), \; x \in [a,b],$$

and therefore is a solution of the original problem (1).

In his paper [76] published the same year, SCORZA DRAGONI will generalize this result by replacing condition c above by the more general condition:

$$-y_0''(x) \leq f(x, y_0(x), z) \;\text{ if }\; x \in [a,b],\; z \leq y_0'(x),\; (\text{resp. } z \geq y_0'(x)),$$
$$-y_1''(x) \geq f(x, y_1(x), z) \;\text{ if }\; x \in [a,b],\; z \geq y_1'(x),\; (\text{resp. } z \leq y_1'(x)).$$

Thus this paper came close to the concept of *lower* and *upper solution* for problem (1) that Mitio NAGUMO [50] introduced in 1937 in a more general and almost definite way by proving, using also the shooting method, the existence of at least one solution for (1) when there exist two functions α and β of class C^1 such $\alpha(x) < \beta(x)$ over $[a,b]$, α' and β' have left and right derivatives $D_\pm\alpha'$, $D_\pm\beta'$, and the following conditions are satisfied:

a′) f is continuous as well as $\frac{\partial f}{\partial y}$ and $\frac{\partial f}{\partial z}$ over the cylinder

$$C = \{(x,y,z) : x \in [a,b], \alpha(x) \leq y \leq \beta(x), z \in \mathbb{R}\};$$

b′) $|f(x,y,z)| \leq \varphi(|z|)$ for all $(x,y,z) \in C$ and some positive continuous function φ such that

$$\int_0^\infty \frac{u\,du}{\varphi(u)} = +\infty; \tag{36}$$

c′) $-D_\pm\alpha'(x) < f(x,\alpha(x),\alpha'(x))$, $-D_\pm\beta'(x) > f(x,\beta(x),\beta'(x))$ for all $x \in [a,b]$;

d′) $\alpha(a) \leq A \leq \beta(a)$, $\alpha(b) \leq B \leq \alpha(b)$.

This paper contains as first theorem the famous *Nagumo's lemma* which generalizes the *Bernstein's lemma* mentioned above and which replaces condition (30) by the more general one

$$|f(x,y,z)| \leq \varphi(|z|),$$

with φ satisfying (36). Notice that neither [3] nor [76] are mentioned in [50]. This Nagumo's lemma will be improved by SATO [71] who calls *Nagumo's condition* the condition

$$|f(x,y,z)| \leq \psi(y)\varphi(|z|),$$

where φ satisfies condition (36) and ψ is integrable over $[\alpha_m, \beta_M]$, with $\alpha_m = \min_{[a,b]} \alpha$ and $\beta_M = \max_{[a,b]} \beta$, and some geometric variants and extensions of the results of [50] will be given by NAGUMO in [51] [52] [53] [54]. This type of results will be the object of many studies after 1950.

Nagumo's paper will suggest to SCORZA DRAGONI [78] the following generalization proving the existence of at least one solution for (1) is insured when there exist two functions α and β of class C^2 with $\alpha(x) \leq \beta(x)$ over $[a,b]$ such that the following conditions are satisfied:

a″) f is locally Lipschitzian with respect to y and z and continuous over the cylinder

$$C = \{(x,y,z) : x \in [a,b], \alpha(x) \leq y \leq \beta(x), z \in \mathbb{R}\};$$

b″) $|f(x,y,z)| \leq \varphi(z)$ for all $(x,y,z) \in C$ and some positive continuous function φ such that

$$\int_0^\infty \frac{u\,du}{\varphi(u)} = +\infty, \quad \int_{-\infty}^0 \frac{u\,du}{\varphi(u)} = -\infty; \tag{37}$$

c″) $-\alpha''(x) \leq f(x,\alpha(x),\alpha'(x))$ and $-\beta''(x) \geq f(x,\beta(x),\beta'(x))$ for all $x \in [a,b]$;

d″) $\alpha(a) \leq A \leq \beta(a)$, $\alpha(b) \leq B \leq \alpha(b)$.

A function α (resp. β) satisfying the conditions c'' and d'' will be called a *lower solution* (resp. *upper solution*) for the problem (1).

SCORZA DRAGONI will extend his result in [79] and [80] to the case of problem (1) when f satisfies only the Carathéodory conditions over C. In this case, the concept of lower and upper solution is generalized by assuming that α and β are of class C^1 and such that the functions

$$-\alpha'(x) - \int_a^x f(s,\alpha(s),\alpha'(s))\,ds \text{ and } \int_a^x f(s,\beta(s),\beta'(s))\,ds + \beta'(x),$$

are non increasing on $[a,b]$. He also replaces the Nagumo's condition by Sato's one. Slight extensions will be given one year later by ZWIRNER [88], [90], CINQUINI [14] and SCORZA DRAGONI himself [81] in the context of the polemics mentioned above. Good surveys of those contributions of the Italian school to nonlinear boundary value problems for ordinary differential equations can be found in [70], [19] and [47].

References

[1] S. BANACH, Sur les opérations dans les ensembles abstraits et leur application aux équations intégrales, *Fundamenta Math.* 3 (1922), 133–181

[2] S.N. BERNSTEIN, Sur la généralisation du problème de Dirichlet, I, II, *Math. Ann.* 62 (1906), 253–271; 69 (1910), 82–136

[3] S.N. BERNSTEIN, Sur les équations du calcul des variations, *Ann. Scient. Ecole Normale Sup.* 29 (1912), 431–465

[4] G.D. BIRKHOFF and O.D. KELLOGG, Invariant points in function space, *Trans. Amer. Math. Soc.* 23 (1922), 96–115

[5] M. BÔCHER, Boundary problems in one dimension, in *Proc. International Congress of Mathematicians*, Cambridge, Mass., 1912, 163–195

[6] L.E.J. BROUWER, Beweis der Invarianz der Dimensionenzahl, *Math. Ann.* 70 (1911), 161–165

[7] H. BURKHARDT, Sur les fonctions de Green relatives à un domaine à une dimension, *Bull. Soc. Math. France* 22 (1894), 71–75

[8] R. CACCIOPPOLI, Un teorema generale sull'esistenza di elementi uniti in una trasformazione funzionale, *Rend. Accad. Naz. Lincei* (6) 11 (1930), 794–799

[9] R. CACCIOPPOLI, Sugli elementi uniti delle trasformazioni funzionali: un'osservazione sui problemi di valori ai limiti, *Rend. Accad. Naz. Lincei* (6) 13 (1931), 498–502

[10] R. CACCIOPPOLI, Sugli elementi uniti delle trasformazioni funzionali: un teoremi di esistenza e di unicità ed alcune sue applicazioni, *Rend. Sem. Mat. Padova* 3 (1932), 1–15

[11] T. CARLEMAN, Ueber eine nichtlineare Randwertaufgabe bei der Gleichung $\Delta u = 0$, *Math. Z.* 9 (1921), 35–43

[12] S. CINQUINI, Sopra i problemi di valori al contorno per equazioni differenziali non lineari, *Boll. Un. Mat. Ital.* (1) 17 (1938), 99–105

[13] S. CINQUINI, Problemi di valori al contorno per equazioni differenziali (non lineari) del secondo ordine, *Ann. R. Scuola Norm. Sup. Pisa* (2) 8 (1939) 1–22

[14] S. CINQUINI, Sopra i problemi di valori al contorno per equazioni differenziali del secondo ordine, *Ann. R. Scuola Norm. Sup. Pisa* 8 (1939), 271–283

[15] S. CINQUINI, Problemi di valori al contorno per equazioni differenziali ordinarie, *Rend. Sem. Mat. Fis. Milano* 14 (1940), 157–170

[16] S. CINQUINI, Un' osservazione sopra i problemi di valori al contorno per l'equazione $y'' = f(x, y, y')$, *Boll. Un. Mat. Ital.* (2) 2 (1940), 322–325

[17] S. CINQUINI, Sopra una recensione del Signor Scorza Dragoni. *Ann. R. Scuola Norm. Sup. Pisa* (2) 11 (1942), 105–106

[18] S. CINQUINI, Sopra un'osservazione del Signor Scorza Dragoni su un problema per le equazioni differenziale ordinarie, *Ann. R. Scuola Norm. Sup. Pisa* (2) 11 (1942), 217–221

[19] S. CINQUINI, L. AMERIO and A. GHIZZETTI, Analisi matematica in Italia nel campo reale (dal 1939 al 1945), *Pontif. Acad. Scienze, Relationes Auctis, Sci. Temp. Belli* 22 (1948), 3–85

[20] J. DIEUDONNÉ, *History of Functional Analysis*, Math. Studies 49, North-Holland, Amsterdam, 1981

[21] C.L. DOLPH, Non-linear integral equations of the Hammerstein type, *Proc. Nat. Acad. Sci. USA* 31 (1945), 60–65

[22] C.L. DOLPH, Nonlinear integral equations of the Hammerstein type, *Trans. Amer. Math. Soc.* 66 (1949), 289–307

[23] G. DUFFING, *Erzwungene Schwingungen bei veränderlicher Eigenfrequenz*, Vieweg, Braunschweig, 1918

[24] M. FUKUHARA, Quelques recherches sur les équations différentielles du second ordre, *Japanese J. Math.* 5 (1928), 351–367

[25] E. GAU, Sur un théorème de M. E. Picard, *Bull. Soc. Math. France* 43 (1915), 62–69

[26] M. GOLOMB, Zur Theorie der nichtlinearen Integralgleichungen, Integralgleichungssysteme und allgemeinen Funktionalgleichungen, *Math. Z.* 39 (1935), 45–75

[27] E. GOURSAT, Sur la théorie des fonctions implicites, *Bull. Soc. Math. France* 31 (1903) 184–192

[28] T.H. GRONWALL, On the existence and properties of the solutions of a certain differential equation of the second order, *Annals of Math.* 28 (1927), 355–364

[29] J. HADAMARD, Sur les transformations ponctuelles, *Bull. Soc. Math. France* 34 (1906), 71–84

[30] G. HAMEL, Über erzwungene Schwingungen bei endlichen Amplituden, *Math. Ann.* 86 (1922), 1–13

[31] A. HAMMERSTEIN, Nichtlineare Integralgleichungen und direkte Methoden der Variationsrechnung, *Sitzungsber. Berliner Math. Ges.* 26 (1927), 66–77

[32] A. HAMMERSTEIN, Nichtlineare Integralgleichungen nebst Anwendungen, *Acta Math.* 54 (1930), 117–176

[33] A. HAMMERSTEIN, Die erste Randwertaufgabe für nichtlineare Differentialgleichungen 2ter Ordnung, *Sitzungsber. Berliner Math. Gesellschaft* 30 (1932), 3–10

[34] D. HILBERT, Ueber des Dirichletsche Prinzip, *Jahresber. Deutsche Math. Ver.* 8 (1900), 184–188

[35] H.O. HIRSCHFELD, A generalization of Picard's method of successive approximation, *Proc. Cambridge Phil. Soc.* 32 (1936), 86–95

[36] R. IGLISCH, Existenz- und Eindeutigkeitssätze bei nichtlinearen Integralgleichungen, *Math. Ann.* 108 (1933), 161–189

[37] M.A. KRASNOSEL'SKII, On some fixed point principle for completely continuous operators in functions spaces (Russian), *Dokl. Akad. Nauk SSSR* (NS) 73 (1950), 13–15

[38] M.A. KRASNOSEL'SKII, Eigenfunctions of nonlinear operators asymptotically close to linear operators (Russian), *Dokl. Akad. Nauk SSSR* (NS) 74 (1950), 177–179

[39] K.W. LAMSON, A general implicit function theorem with an application to problems of relative minima, *Amer. J. Math.* 42 (1920), 243–256

[40] J. LERAY, Sur certaines classes d'équations intégrales non linéaires, *Comptes Rendus Acad. Sci. Paris* 194 (1932), 1627–1629

[41] J. LERAY, Etude de diverses équations intégrales non linéaires et de quelques problèmes que pose l'hydrodynamique, *J. Math. Pures Appl.* 12 (1933), 1–82

[42] J. LERAY and J. SCHAUDER, Topologie et équations fonctionnelles, *Ann. Sci. Ecole Norm. Sup.* 51 (1934) 45–78

[43] F. LETTENMEYER, Über die von einem Punkt ausgehenden Integralkurven einer Differentialgleichung zweiter Ordnung, *Deutsche Math.* 7 (1944), 56–74

[44] P. LÉVY, Sur les fonctions de lignes implicites, *Bull. Soc. Math. France* 48 (1920), 13–27

[45] L. LICHTENSTEIN, Über einige Existenzprobleme der Variationsrechnung. Methode der unendlichvielen Variabeln, *J. Reine Angew. Math.* 145 (1915), 24–85

[46] J. LÜTZEN, *Joseph Liouville 1809–1882. Master of Pure and Applied Mathematics,* Studies in the History of Mathematics and Physical Sciences 15, Springer, New York, 1990

[47] C. MIRANDA, Un'osservazione su un teorema di Brouwer, *Boll. Un. Mat. Ital.* (2) 3 (1940), 527

[48] C. MIRANDA, *Problemi di esistenza in analisi funzionale,* Litografia Tacchi, Pisa, 1948–49

[49] M. MÜLLER, Neuere Untersuchungen über den Fundamentalsatz in der Theorie der gewöhnlichen Differentialgleichungen, *Jahresber. Deutsche Math.Ver.* 37 (1927), 33–48

[50] M. NAGUMO, Ueber die Differentialgleichung $y'' = f(x, y, y')$, *Proc. Phys. Math. Soc. Japan* (3) 19 (1937), 861–866

[51] M. NAGUMO, Dai-2-kai Jôbibunhôteishiki no Kyôkaichi Mondai I, II, *Kansû Hôteishiki* 5 (1939), 27–34; 6 (1939), 37–44

[52] M. NAGUMO, $y'' = f(x, y, y')$ no Kyôkaichi Mondai ni tsuite I, II, *Kansû Hôteishiki* 13 (1941), 36–46; 17 (1942), 50–52

[53] M. NAGUMO, Ueber das Randwertproblem der nichtlinearen gewöhnlichen Differentialgleichungen zweiter Ordnung, *Proc. Phys. Math. Soc. Japan* (3) 24 (1942), 845–851

[54] M. NAGUMO, *Shazôdo to Sonzaiteiri (Degree of Mapping and Existence Theorems),* Kawade, Tokyo, 1948

[55] O. NICOLETTI, Sulle condizioni iniziali che determinano gli integrali delle equazioni differenziali ordinarie, *Atti R. Acc. Sci. Torino* 33 (1897–98) 746–759

[56] V. NIEMYTZKI, Théorèmes d'existence et d'unicité des solutions de quelques équations intégrales non-linéaires, *Mat. Sbornik* 41 (1934) 421–438

[57] V. NIEMYTZKI, Sur les équations intégrales non linéaires comparables aux équations linéaires, *Dokl. Akad. Nauk SSSR* 15 (1937), 17–19

[58] P. PAINLEVÉ, *Existence de l'intégrale générale. Détermination d'une intégrale particulière par ses valeurs initiales,* Encyclopédie des sciences mathématiques, tome II, vol. 3, fasc. 1, Teubner, Leipzig, 1910

[59] E. PICARD, Mémoire sur la théorie des équations aux dérivés partielles et la méthode des approximations successives, *J. de Math. Pures Appl.* 6 (1890), 145–210, 231

[60] E. PICARD, Sur l'application des méthodes d'approximations successives à l'étude de certaines équations différentielles ordinaires, *J. Math. Pures Appl.* 9 (1893), 217–271

[61] E. PICARD, *Traité d'analyse,* Gauthier-Villars, Paris, 1896

[62] E. PICARD, *Leçons sur quelques problèmes aux limites de la théorie des équations différentielles,* Gauthier-Villars, Paris, 1930

[63] E. PICARD, Sur certains exemples singuliers d'approximations successives, *Comptes Rendus Acad. Sci. Paris* 126 (1898), 497–500

[64] E. PICARD, Sur un exemple d'approximations successives divergentes, *Bull. Soc. Math. France* 28 (1900), 137–143

[65] H. POINCARÉ, Sur certaines solutions particulières du problème des trois corps, *C.R. Acad. Sci. Paris* 97 (1883), 251–252; *Bull. Astronomique* 1 (1884), 65–74

[66] W. RITZ, Ueber eine neue Methode zur Lösung gewisser Variationsprobleme der mathematischen Physik, *J. Reine Angew. Math.* 135 (1909), 1–61

[67] A. ROSENBLATT, Sur les théorèmes de M. Picard dans la théorie des problèmes aux limites des équations différentielles ordinaires non linéaires, *Bull. Sci. Math.* (2) 57 (1933), 100–106

[68] A. ROSENBLATT, Sur les problèmes aux limites des équations différentielles ordinaires du second ordre, *Bull. Soc. Roy. Sci. Liège* 2 (1933), 87–90

[69] E. ROTHE, Zur Theorie der topologischen Ordnung und der Vektorfelder in Banachschen Räumen, *Compositio Math.* 5 (1937), 177–197

[70] G. SANSONE, *Equazioni differenziali nel campo reale,* Parte seconda, Zanichelli, Bologna, 1941

[71] T. SATÔ, Sur l'équation différentielle $y'' = f(x, y, y')$, *Proc. Imper. Acad. Japan* 13 (1937), 348–350

[72] J. SCHAUDER, Zur Theorie stetiger Abbildungen in Funktionalräumen, *Math. Z.* 26 (1927), 47–65

[73] J. SCHAUDER, Bemerkungen zu meiner Arbeit 'Zur Theorie stetiger Abbildungen in Funktionalräumen', *Math. Z.* 26 (1917), 417–431

[74] J. SCHAUDER, Der Fixpunktsatz in Funktionalräumen, *Studia Math.* 2 (1930), 171–180

[75] G. SCORZA DRAGONI, Il problema dei valori ai limiti studiato in grande per gli integrali di una equazione differenziale del secondo ordine, *Giornale di Mat. (Battaglini)* (3) 69 (1931), 77–112

[76] G. SCORZA DRAGONI, Il problema dei valori ai limiti studiato in grande per le equazioni differenziali del secondo ordine, *Math. Ann.* 105 (1931), 133–143

[77] G. SCORZA DRAGONI, A proposito di alcuni teoremi relativi ad un problema ai limite per un' equazione differenziale del secondo ordine, *Rend. R. Accad. Naz. Lincei* (6) 22 (1935), 44–48

[78] G. SCORZA DRAGONI, Su un problema di valori ai limite per le equazioni differenziali ordinarie del secondo ordine, *Rend. Semin. Mat. R. Univ. Roma* (4) 2 (1938), 177–215. Aggiunta, *ibid.*, 253–254

[79] G. SCORZA DRAGONI, Elementi uniti di transformazioni funzionali e problemi di valori ai limiti, *Rend. Sem. Mat. R. Univ. Roma* (4) 2 (1938), 255–275

[80] G. SCORZA DRAGONI, Intorno a un criterio di esistenza per un problema di valori ai limiti, *Rend. R. Accad. Naz. Lincei* (6) 28 (1938), 317–325

[81] G. SCORZA DRAGONI, A proposito di un teorema sulle equazioni differenziali ordinarie, *Rend. Sem. Mat. Univ. Padova* 10 (1939), 90–100

[82] G. SCORZA DRAGONI, Un'osservazione su un problema per le equazione differenziali ordinarie, *Ist. Veneto Sci. Let. Arti, Parte II, Cl. Sci. Mat. Nat.* 101 (1942), 203–212

[83] G. SCORZA DRAGONI, A proposito di alcuni teoremi sulle equazioni differenziali ordinarie, *Rend. Sem. Mat. Univ. Padova* 15 (1946), 60–131. Rettifica, *ibid.*, 16 (1947), 1–2

[84] G. SCORZA DRAGONI, Ancora a proposito di alcuni teoremi sulle equazioni differenziali ordinarie, *Rend. Sem. Mat. Univ. Padova* 18 (1949), 115–139

[85] C. SEVERINI, Sopra gli integrali delle equazioni differenziali ordinarie di ordine superiore al primo con valori prestabiliti in punti dati, *Atti R. Acc. Torino* 40 (1904–1905), 853–869

[86] C. SEVERINI, Sopra gli integrali delle equazioni differenziali del secondo ordine con valori prestabiliti in due punti dati, *Atti R. Acc. Torino* 40 (1904–1905) 1035–1040

[87] L. TONELLI, Sull'equazione differenziale $y'' = f(x, y, y')$, *Ann. Scuola Norm. Sup. Pisa, Sc. Fis. Mat.* (2) 8 (1939), 75–88

[88] G. ZWIRNER, Su un problema di valori al contorno per equazioni differenziali del secondo ordine, *Rend. Sem. Mat. R. Univ. Roma* (4) 3 (1939), 57–70

[89] G. ZWIRNER, Un'osservazione su un problema ai limiti per le equazioni differenziali, *Boll. Un. Mat. Ital.* (2) 1 (1939), 334–336

[90] G. ZWIRNER, Un criterio di esistenza per un problema di valori al contorno per equazioni differenziali del secondo ordine, *Rend. Sem. Mat. Univ. Padova* 10 (1939), 55–64

[91] G. ZWIRNER, Sopra un teorema sulle equazioni differenziali del secondo ordine, *Rend. Sem. Math. Univ. Padova* 10 (1939), 65–68

JEAN MAWHIN, Institut de mathématique, Université catholique de Louvain, 2, Chemin du Cyclotron, B-1348 Louvain-La-Neuve.

Partial Differential Equations in the First Half of the Century

Louis Nirenberg
Courant Institute, New York

The field of partial differential equations (PDE) is large and also rather technical. I shall give only some highlights of the period, with a few indications of later developments. The choice of topics reflects my personal taste and limitations of knowledge. No doubt there are important omissions; others in the field would undoubtedly make different choices.

Professor Gelfand in his talk here referred to two faces of mathematics: one related to problems in physics and other sciences, the other coming from questions within mathematics itself. PDE is connected mainly with the first face. Much work was initiated by problems in physics; in fact many problems in science as well as engineering, may be expressed mathematically in terms of differential equations. In addition, physicists have contributed new mathematical ideas. General relativity theory led to a particular system of nonlinear PDE — in which much work is still going on. In addition, quantum theory led to many problems, for example the Schrödinger equation (1927). But I will not discuss those topics here.

In addition to the sciences, many problems in PDE arose from geometry.

For simplicity I will confine the discussion mainly to equations of order at most two. A general differential equation for a function u defined in a domain Ω (open connected set) in $\mathbb{R}^n$, or an n-dimensional manifold, with local coordinates $x = (x_1, \ldots, x_n)$ takes the form

$$F(x, u(x), \partial u(x), \partial^2 u(x)) = 0\,, \quad \forall x \in \Omega\,. \tag{1}$$

Here $\partial u = (\partial_1 u, \ldots, \partial_n u) = \left(\frac{\partial u}{\partial x_1}, \ldots, \frac{\partial u}{\partial x_n}\right)$ denote the first derivatives of u;

$$\partial^2 u = \left\{\frac{\partial^2 u}{\partial x_i \partial x_j}\right\} = \{u_{ij}\}$$

denote the second derivative; $\partial^j u$ will denote derivatives of order j. F is a (smooth) function of the various variables, and the relationship (1) is to hold at every point $x \in \Omega$. A *linear* partial differential operator takes the form (we use summation convention)

$$Lu = a_{ij}(x)u_{ij} + b_i(x)u_i + c(x)u\,. \tag{2}$$

In case u is vector valued, the coefficients may be matrix valued; in (1) F may be vector-valued. I recall some classical notions (only for the case of one function u, and one equation or operator). A *characteristic direction* is a vector $\xi \neq 0$ (in the cotangent space) at x for which

$$a_{ij}(x)\xi_i\xi_j = 0 .$$

A hypersurface Σ in Ω, defined locally by $\phi(x) = 0$, with $\nabla\phi \neq 0$ on Σ, is called noncharacteristic at a point x if

$$a_{ij}\phi_i\phi_j(x) \neq 0 .$$

For a nonlinear operator F as in (1), a hypersurface Σ is called noncharacteristic at x — relative to a particular function u satisfying (1) at x — if for the linearization of F about u, i.e. for

$$L = \frac{\partial F}{\partial u_{ij}}(x, u, \partial u, \partial^2 u)\partial_{ij} + \frac{\partial F}{\partial u_j}(\ )\partial_j + \frac{\partial F}{\partial u}(\ ) ,$$

Σ is noncharacteristic:

$$\frac{\partial F}{\partial u_{ij}}(x, u(x), \partial u(x), \partial^2 u(x))\phi_i\phi_j \neq 0 .$$

The operator L in (2) is called *elliptic* at x if it has no real characteristic directions. In that case (assuming the a_{ij} are real) the matrix $\{a_{ij}(x)\}$ is definite, say positive definite. The best known example is the Laplace operator

$$\Delta u = \Sigma u_{jj} .$$

In two dimensions, in the (x, y) plane, this is associated with the beloved Cauchy-Riemann equation (for complex f):

$$f_{\bar{z}} = \frac{1}{2}(\partial_x + i\partial_y)f = 0 ,$$

$z = x + iy$.

The operator L is called *hyperbolic* if $\{a_{ij}\}$ has signature $(-, -, -, +)$ — the well known example being the wave equation

$$u_{tt} - \sum u_{x_j x_j} = 0 .$$

Here, one of the independent variables denoted by t is distinguished and called time. Another classic example of a so-called *parabolic* operator is the heat equation

$$u_t - Lu = 0 \ , \ \text{with } L \text{ elliptic, e.g. } u_t - \Delta u = 0 .$$

These equations arose in simple models describing certain physical situations.

Main Topics

I will mainly discuss problems related to existence of solutions under various boundary conditions (BC), or initial conditions (IC), i.e., conditions at some initial time t_0; also uniqueness of solutions; estimates, and regularity of solutions. The calculus of variations is intimately tied to differential equations and I will say a bit about that field. See also the expository article [185] by Petrowsky. In addition I will discuss some work in hydrodynamics, and also some problems in geometry, though others will speak on geometry.

I. Gelfand, Luxembourg, 1992

First I would like to call attention to an excellent recent historical paper on which I have greatly relied, and which I warmly recommend: [Bemelmans – Hildebrandt – von Wahl]. In addition, my thanks to Peter Lax and S. R. S. Varadhan for a number of interesting conversations.

In studying PDE the heart of the subject lies in *inequalities* — estimates of all kinds. (In that subject, the (1934) book Inequalities, Hardy-Littlewood-Polya [76], continues to be extremely useful.) Progress in the field has usually been in jumps — because some new inequality has been established. In the first half century suitable function spaces were discovered for each class of problem above (elliptic,

F. Riesz

etc.) so that the methods of functional analysis, developed at the same time, could be used in a systematic way. For example, the implicit function theorem in infinite dimensional space [84] given by Hildebrandt and Graves in 1927, could be used to solve perturbation problems for nonlinear equations. The F. Riesz [199] theory of compact operators could be applied to elliptic boundary value problems. The von Neumann [175] theory of 1929 for unbounded self adjoint operators in Hilbert space. This theory was developed independently at the same time by C. Loewner but never published by him. Concurrently, there were of course developments in other fields, topology, etc., which interacted with PDE.

In his talk, Professor Hayman spoke about various inequalities involved in the theory of entire functions. I should say that the inequalities used in PDE, up to about 25 years ago, are usually not as delicate. They are rougher, though, in recent years the use of sharp constants in various inequalities has played an increasing role, such as the sharp constants in Sobolev inequalities. (Friedrichs, under whom I studied, often stressed the applicability of rough inequalities to various problems!)

I. General Equations

In the last century there was much work in PDE in connection with particular problems from physics and geometry. In particular, methods were introduced for finding explicit solutions. There was, however, one completely general theorem, the Cauchy-Kowalewski theorem. It is a local result for the initial value problem for (1), in which the initial data — called Cauchy data — of u are prescribed on a noncharacteristic hypersurface Σ, i.e., on Σ, u and its normal derivative $\phi_i(x)u_{x_i}$ are prescribed. Let us describe it for an equation of the form (in which one variable is called t)

$$u_{tt} = f(x, t, u(x, t), \partial u, \partial_x^2 u, \partial_{xt}^2 u) \ . \tag{3}$$

The hyperplane $t =$ const. is then automatically noncharacteristic. In the initial value problem, or Cauchy problem, one prescribes

$$u = u_0(x) \ , \quad u_t = u_1(x) \quad \text{at } t = t_0 \ , \tag{4}$$

for x near x_0 and seeks a solution on a neighbourhood of (x_0, t_0).

Cauchy-Kowalewski Theorem. *If all the data are analytic, i.e. u_0, u_1 are analytic, and f is analytic in all its arguments, then there exists an analytic solution of (3) in some neighbourhood of (x_0, t_0); there is only one analytic solution.*

Observe that from (4) we may compute all the x-derivatives of u and u_t on $t = t_0$ (near x_0). In addition, if u is a solution of (3), we may compute u_{tt} and all its x-derivatives on $t = t_0$. By differentiating (3) further, we may in fact compute *all* the derivatives of u on $t = t_0$ near x_0. In particular they are all determined at (x_0, t_0). The classical proof then consists in showing that the Taylor series for u, using all these computed derivatives at (x_0, t_0), converges in some neighbourhood, and yields a solution of (3), (4). Since every derivative of u is known at (x_0, t_0) there is only one analytic solution.

Open problem

Can there exist a nonanalytic (say C^∞) solution?

This is still unsettled after all these years. However, very recently, Métivier constructed a nonlinear analytic system having two different C^∞ solutions with the same Cauchy data on an initial noncharacteristic hypersurface.

In the linear case

$$u_{tt} = a_{jk}u_{x_jx_k} + a_i u_{x_it} + b_i u_{x_i} + bu_t + cu \ , \tag{5}$$

in which all the coefficients are analytic functions of (x, t) the answer is no. This was proved by Holmgren [88] in 1901 (for first order systems in two dimensions, but the method works in general). John [101] in 1949 generalized the method and the result in a very significant way. He uses a globalized version of Holmgren's result: if the Cauchy data are given on a noncharacteristic region U of a hypersurface then the solution is uniquely determined at all points which can be reached by deforming U, keeping its boundary in U, by a one parameter family of analytic noncharacteristic hypersurfaces. [101] contains a generalization of «noncharacteristic» for lower dimensional surfaces; it has many interesting results.

A natural attempt to settle the open problem (in the negative) is the following: Suppose in addition to the analytic solution u, of (3), (4), there is a C^∞ one v. Using the integral theorem of the mean one sees that $w = u - v$ satisfies a linear equation of the form (5) — but with C^∞ coefficients, not analytic. In addition, w has zero Cauchy data on $t = t_0$, near x_0. Thus if one could prove uniqueness for the Cauchy problem for a general linear equation $Lw = 0$, with the initial surface Σ noncharacteristic for the operator L, it would follow that the answer to the open problem is no.

Many people tried to do this. While I was a graduate student, in 1947 or 8, E. Hopf visited New York University for some time. He was working on the problem of uniqueness in the (noncharacteristic) Cauchy problem for a general linear equation. He felt he had almost proved it — there was still something not quite proved — and he gave a lecture on his ideas. The lecture was extremely interesting even though (as he acknowledged) the proof was not complete. In 1955 De Giorgi [41] presented a counterexample.

For various classes of linear equations, uniqueness for the Cauchy problem has been proved. In particular, in 1939, Carleman [25] proved uniqueness for first order elliptic systems in the plane. He introduced a technique of establishing L^2 estimates, for derivatives of functions u with compact support, in terms of Lu. The L^2 estimates involved powers of weight functions. Estimates of «Carleman type» have, since, become the traditional technique in many cases for proving uniqueness in the Cauchy problem. For very general results using this technique, see Hörmander [96] section 17.2 and Chapter 28.

Once we leave the analytic category the Cauchy problem is not well posed in general, even if one has uniqueness.

Well posed problem — introduced in 1932 by Hadamard [75]:
A problem is considered well posed if (i) a solution exists, (ii) it is unique or, if not, the set of solutions forms a finite dimensional manifold, (iii) the solution (if unique) depends continuously on the data — in some topology involving only

control on a finite number of derivatives. That is, under small perturbation of the equation, or of the boundary (or initial) data, the solution changes slightly.

For example, the initial value problem for a holomorphic function in a domain in $\mathbb{C}$ is not well posed. Consider a half disc D^+ in the complex plane $|z| < 1$, $\text{Im } z = y > 0$. We cannot prescribe C^∞ values of a complex function u on $y = 0$ and expect to find a holomorphic function in D^+ with this initial data. Indeed if we prescribe $\text{Im } u = 0$ on $y = 0$, then any holomorphic function in D^+ satisfying this can be reflected across, as holomorphic to the full disk $D = \{|z| < 1\}$, and therefore its real part is necessarily real analytic in $y = 0$. In fact, if u is holomorphic in D^+ it is impossible to estimate $|u|$ in, say, $|z| < \epsilon < 1$, $y \geq 0$, in terms of any norm involving a finite number of derivatives, of the values of u on $y = 0$. So the problem is ill posed.

Nevertheless, it is possible to obtain such an estimate if we know some global bound on u, say $|u| < 1$ in D^+. This is an exercise in analytic function theory in the (1925) collection of problems of Polya-Szegö [189]. These beautiful books helped shaping many of us when we were students.

This exercise has been extended by many people in the study of **Ill Posed Problems**: One wishes to estimate solutions in terms of the data, assuming some a priori bound for the solution. This is useful to engineers who may measure certain data for which the problem is ill posed. See for example John [101], and the book Lavrent'ev [118] for other references.

Returning to the Cauchy-Kowalewski theorem, Cartan [27], [28], [29], [30], and Kähler [104] used it to treat classes of systems of PDE — even in case the number of equations differs from the number of unknowns, as often occurs in problems of geometry. In case there are more equations than unknowns, so called overdetermined systems, one usually needs compatibility conditions. For example consider the inhomogeneous Cauchy-Riemann equations in n complex variables $z_j = x_j + iy_j$, $j = 1, \ldots, n$:

$$u_{\bar{z}_j} = \frac{1}{2}(u_{x_j} + iu_{y_j}) = f_j \tag{6}$$

Obvious compatibility conditions are

$$f_{j_{\bar{z}_k}} = f_{k_{\bar{z}_j}} . \tag{7}$$

Cartan-Kähler theory requires analyticity of the equations, and the theory is used to obtain local solutions. Attempts have been made to generalize the theory to C^∞, but very few general results are known. A *major open problem* for linear PDE is to find local solutions of various classes of overdetermined — even elliptic — systems. Yang [230] has obtained results for a class of overdetermined systems. See the recent book by Bryant, Chern, Gardner, Goldschmidt, Griffiths [19].

For global solvability, say for equations (6), assuming (7), one needs conditions on the domain such as pseudo-convexity, strong pseudo-convexity etc. — introduced by Levi [129] in 1910. Today this is a very active subject.

A last word about Cauchy-Kowalewski. In 1941 Nagumo [170] observed that one did not need F (or f in (3)) to be analytic in t, continuity sufficed. Assuming analyticity in the other variables, one obtains a solution which is analytic in these variables and merely of class C^2 in t. This result has been extended to abstract equations in «scales of Banach spaces» by various people; see for example Nishida [179] for general nonlinear equations.

One more general topic before turning to special classes of equations. It is connected with IFT, the implicit function theorem. In 1956 Nash [171] developed a deep extension of the classical IFT, in his proof of the result that any compact Riemannian manifold may be isometrically embedded in some Euclidean space (of much higher dimension). His ideas have been extended by various people to a technique now called the Nash-Moser technique. It is involved in KAM (Kolmogorov, Arnold, Moser) theory in dynamical systems, see for example Moser [168], [169]. In its simplest form it involves solving an equation of the form

$$F(u,\lambda) = 0\,, \qquad F(0,0) = 0$$

for a function u; here λ is in a parameter space, λ small. One assumes $F_u(u,\lambda)$ is invertible for u and λ «close» to their respective origins — but with some loss of derivatives when computing the inverse. The usual Picard iteration scheme doesn't work, because of this loss. Instead one uses an iterative scheme which combines Newton's method with smoothing processes.

An influential book in the period is Courant-Hilbert [39] on PDE and Mathematical Physics, Vol. 1 published in 1924, Vol. 2 in 1937. (Much expanded English editions appeared in 1953 and 1962). These books have been and still are much used.

II. Elliptic Equations

(a) Linear equations with regular coefficients

The best known boundary value problem for elliptic equations is the Dirichlet problem

$$\Delta u = f \quad \text{in} \quad \Omega \tag{8}$$

$$u = \phi \quad \text{on} \quad \partial\Omega \tag{9}$$

For $f = 0$, the solution is obtained by minimizing the Dirichlet integral

$$\min_{u=\phi \text{ on } \partial\Omega} \int_\Omega |\nabla u|^2\,. \tag{10}$$

Riemann and Weierstrass had worked on this approach but it was Hilbert [82] in 1904 who first gave a rigorous proof of the existence of a minimum, for a weak solution in H_0^1, i.e. functions in L^2, vanishing on $\partial\Omega$, with derivatives in L^2. After the work of Fredholm [60] in 1903, using integral equations, Hilbert [82] extended the theory and used it to treat the Laplace equation in 1905. His basic book [83] on the subject appeared in 1912.

In that same year Weyl [225] presented his analysis on the asymptotic behaviour of the eigenvalues for the Dirichlet problem for the Laplace operator. This was extended by Carleman [26] in 1934, Minakshisundaram and Pleijel [158] in 1949. Much work has been done since on estimating remainder terms.

The idea of generalised solution was extended (after Hilbert) to many different problems. Sobolev [210], in 1938, proved the «Sobolev inequalities»: If u has derivatives of order m in L^p then derivatives $\partial^j u$ of order $j < m$ belong to L^q for suitable $q(p, n)$. The «Sobolev spaces», $W^{m,p}$, completion of C^∞ under the norm

$$\|u\|_{m,p} = \left(\int_{j\leq m} |\partial^j u|^p\right)^{1/p}.$$

were introduced. Here ∂^j represents all derivatives of order j. When using, for example, the theory of self adjoint operators, people went to great lengths to describe the domain of some particular differential operator. Friedrichs [61], [62], [63] did away with much of the intricate real variable by considering generalized solutions — obtained always by completion in some norm. So did Schauder [205] in his work on hyperbolic equations. The generalized functions and their derivatives were always functions. Bochner [12] on the other hand, in work on Fourier series, considered derivatives of continuous functions; see also [13]. These generalized notions of functions led finally to the introduction of more general objects, namely distributions, by L. Schwartz, in 1945. Distributions included delta functions which had been introduced by Dirac in 1926. The theory of distributions is summarized in the book [208] of 1950 (it also discusses in more detail the earlier work on generalized functions). A related notion, «currents», a generalized form of a manifold, was introduced by de Rham in [198]. In 1944 Friedrichs [63] had proved the identity of weak and strong extensions of partial differential operators, and he introduced into PDE the (now standard) smoothing technique of mollifiers.

Related to these generalized functions are various inequalities and theorems on continuous embedding of one function space in another. For example Rellich [196] (1930) proved that the injection $H^1(\Omega) \hookrightarrow L^2(\Omega)$, is compact for bounded domains Ω having smooth boundaries.

Let us now turn to elliptic equations (2) with variable coefficients. In 1907 Levi [128] had constructed a parametrix, i.e. it behaved like a fundamental solution. To obtain general existence theorems one had to find suitable function spaces in which to operate, i.e. spaces X, Y so that L maps X continuously into Y and L^{-1} (if it exists) maps Y continuously in X. One might think that if $\Delta u = f$ and $f \in C(\Omega)$ then $u \in C^2(\Omega)$, but (fact of life) that is not the case. Hölder spaces were found to be suitable: $C^{j+\alpha}$, $j \geq 0$ integer, $0 < \alpha < 1$, is the space of functions with continuous derivatives up to order j satisfying a Hölder condition with exponent α, i.e.

$$|\partial^j u(x) - \partial^j u(y)| \le C|x - y|^\alpha$$

This is a particular modulus of continuity of the $\partial^j u$. This was first done by Hölder [87] in the last century, then Korn [110] 1909, followed by Lichtenstein [140] in 1912. They obtained local solutions of elliptic equations in the Hölder spaces. Kellogg [105] established the Hölder estimates for solutions of (8), (9) up to the boundary. His 1929 book [106] is still a very useful reference. The global theory for equations with variable coefficients, in the Hölder spaces, now called Schauder theory, and used all the time, was developed by Schauder [204], [205] and was also announced around the same time by Caccioppoli [20]. It proved the existence of solutions in $C^{2+j+\alpha}(\overline{\Omega})$ in case $f \in C^{j+\alpha}(\overline{\Omega})$. Also Hopf [90] treated Hölder spaces.

Only much later did people extend the global theory to Sobolev spaces $W^{m,p}$, $1 < p < \infty$ — with the aid of singular integral (pseudo-differential) operators; see for example Lions-Magenes [148].

In 1924, Wiener [229] presented his criterion on boundary points ensuring that the solution of the Dirichlet problem, in a domain with nonsmooth boundary, takes on the given boundary value there. The Laplace equation in two dimensions is closely tied to analytic function theory. In 1950–51 Bers [8] introduced pseudo-analytic functions. These were complex valued solutions of an elliptic linear first order equation (with complex coefficients) similar to the Cauchy-Riemann equations. The solutions have many properties of analytic functions. These became an essential tool in later developments concerning Riemann surfaces started by Teichmüller.

An aside here. About general equations, L. Schwartz raised the question whether every equation

$$P(\partial)u = f \in C^\infty$$

with constant coefficients can be solved in $\mathbb{R}^n$. This was proved independently by Malgrange [152] and Ehrenpreis [50] (1954–56). People believed that locally, near a point, any general equation which does not «degenerate» there should be solvable. But in 1957 Lewy [138] produced his remarkable, and simple, example of a first order equation with affine (complex) coefficients $Lu = f \in C^\infty$ which has no solution in any open set. This led to considerable study, see Hörmander [98] Chapter 26.

(b) A priori estimates

A method that has been much used in solving nonlinear problems, primarly elliptic ones, is the so called *continuity method.* One wishes to solve some nonlinear equation for a function u; it may be a differential, or integral, equation, expressed schematically in the form

$$F[u] = 0$$

The method is to connect it by a one-parameter family of equations

$$F_t[u] = 0\,, \qquad 0 \le t \le 1\,, \qquad F_1[u] = F[u]\,,$$

to a simpler equation $F_0[u] = 0$ for which one has a solution u_t. Then for each t one tries to prove the existence of a solution u_t by proving two things:

(i) The set of t for which a solution exists is (relatively) open.

(ii) The set of t for which a solution exists is closed.

Then this set is the whole interval [0,1].

Step (i) is usually proved with the aid of the implicit function theorem (IFT) in a suitably chosen space. To prove (ii) one derives a priori estimates for the solutions u_t and their derivatives — assuming the solutions exist.

Bernstein [5] used this method in 1906 and 1910 to treat boundary value problems for nonlinear elliptic equations in two dimensions. He estimated derivatives of various orders of the solutions u_t. In fact in 1904, by estimating derivatives of all orders, he had proved, [4], the analyticity of solutions of (analytic) elliptic equations in the plane. He derived the estimates by applying the maximum principle to suitable combinations of derivatives. One formulation of that principle is the following:

The **maximum principle** is said to hold for the operator L (of (2)) in a domain Ω, if the conditions on a function w:

$$\left.\begin{array}{rl} Lw \ge 0 & \text{in} \quad \Omega \\ w \le 0 & \text{on} \quad \partial\Omega \end{array}\right\} \quad \Rightarrow \quad w \le 0 \quad \text{in} \quad \Omega\,.$$

General forms of the maximum principle were presented by Hopf [89] in 1927. (The Hopf lemma, which asserts that in the situation above, if $w < 0$ in Ω and $w = 0$ at some smooth boundary point, then the exterior normal derivative w_ν is positive there, was proved by him 25 years later [95].)

Related to the maximum principle are the estimates of Phragmén and Lindelöf [186] in 1908, giving bounds on holomorphic or harmonic functions in unbounded domains (or domains with corners). These have been extended to elliptic operators in unbounded domains in $\mathbb{R}^n$.

Over the years, many variations and new ways of applying the maximum principle have been discovered, in order to prove estimates, and to derive properties of solutions of elliptic (and also parabolic) equations. It has become an indispensable tool. See the books Protter-Weinberger [190], Ladyzhenskaya-Uraltseva [114], Ladyzhenskaya-Solonnikov-Uraltseva[115], Gilbarg-Trudinger [70], Krylov [112].

In 1938, Leray [124] recognized that the geometry of the domain may affect the estimates one obtains. In 1968, Jenkins and Serrin [99] showed that if one seeks a function u in a domain Ω, with $u = g$ on $\partial\Omega$, such that the graph of u is a minimal surface, then this is solvable for *all* g in $C^2 \Leftrightarrow$ the mean curvature h of $\partial\Omega$ satisfies $h \geq 0$.

Returning to a priori estimates, even if one has them, and so can prove step (ii) of the continuity method, the method may still fail because, in step (i), one may not be able to apply IFT for some t. (Nowadays one may try to apply Nash-Moser.) What replaces the continuity method is a topological method: degree theory. It is related to homotopy theory which had, in fact, been originally developed as a tool to attack nonlinear problems.

Degree theory

In 1912 Brouwer [18] proved his fixed point theorem and introduced the degree of a map from an oriented n dimensional manifold to another. In 1922 Birkhoff and Kellogg [11] extended the fixed point theorem to some function spaces. Schauder [203] extended it to very general infinite dimensional spaces.

The seminal 1934 paper Leray-Schauder [127] extended Brouwer's degree theory to maps in a Banach space, which are of the form: Identity minus a compact operator. They applied this to second order quasilinear elliptic equations. Since then, the ideas developed in this paper have played an essential role in the study of nonlinear problems in infinite dimensional spaces. They are used all the time.

A useful paper in studying degree theory, and important for the theory of differential topology, is Sard [202].

In recent years attempts have been made to extend degree theory to operators which are more general than Id-compact — in particular to Fredholm maps. (A nonlinear map is called Fredholm, if at any point u, its derivative A, is a linear Fredholm map, i.e. kernel A has finite dimension d, and range A is closed and has finite codimension d^*; $d - d^*$ is the index of A. See Elworthy-Tromba [53], Fitzpatrick-Pejsachowicz [59] and Li [139].

III. Elliptic Equations and Calculus of Variations

(i) As described above, Hilbert solved (8) — with $f = 0$ — and (9) — by minimizing the Dirichlet integral in (10). This led to a generalized solution belonging to the space $H^1(\Omega) = \{u;\ u \text{ and } \nabla u \in L^2(\Omega)\}$. The regularity of the generalized solution was then proved.

Hilbert did much work in the *direct methods* of the Calculus of Variations, i.e. to seek a minimum of, say, a functional of the form

$$I(u) = \int_\Omega f(x, u, \nabla u)\, dx\ , \tag{11}$$

by considering minimizing sequence u_j and showing that a subsequence converges to a generalized solution. Ellipticity of the Euler-Lagrange equation associated with I follows if $f(x, u, p)$, $p \in \mathbb{R}^n$, is convex in p and, say,

$$c_0|\xi|^2 \leq f_{p_i p_j}\ \xi_i \xi_j\ , \qquad \forall\ \xi \in \mathbb{R}^n\ ,\ c_0 > 0\ . \tag{12}$$

The direct method was then developed by many others, for example Lebesgue [120] in 1907, Lichtenstein [142] in 1914, Tonelli [219] 1915, who studied functions of bounded variation — see his book [220] of 1921. A basic, and still useful book of the period, on Calculus of Variations, though on different subjects, is Carathéodory [24].

(ii) A problem that attracted much interest was the Plateau problem: to find a surface of minimal area spanning a given simple closed curve in $\mathbb{R}^3$. This was first solved by Douglas [48] in 1931 — for surfaces of the type of the disc. For this he was awarded the first Field Medal. Douglas used the direct methods of Calculation of Variations. Courant [35] in 1936–40 developed a simpler approach using Dirichlet integral and isothermal (conformal) coordinates; this method could extend to other problems. In 1939, Douglas [49] considered surfaces of more complicated topological type, which are bounded by several curves. Rado [194] obtained a solution via approximation by polyhedral surfaces. For more recent work on Plateau problems and minimal surfaces, see Courant [36], Nitsche [180], Struwe [214] and Dierkes-Hildebrandt-Küster-Wohlrab [45].

One more remark about minimal surfaces in R^3, of the form $z = u(x, y)$. Such a surface has the property that its graph locally minimizes area among all surfaces spanning a curve on it. The corresponding equation for $z = u(x)$, $x \in \mathbb{R}^n$, is

$$\Delta u - \frac{u_i u_j u_{ij}}{1 + |\nabla u|^2} = 0 . \tag{13}$$

In 1915 Bernstein [6] proved that if z is a solution of this equation on all of $\mathbb{R}^2$, then z is an affine function; no condition at infinity is required. This inspired much further work. In particular, after many years it was proved that the result holds for $n \leq 7$. But then in 1969 Bombieri, De Giorgi and Guisti [14] presented examples showing the result is false for $n \geq 8$.

Bernstein's original proof was based on a geometric theorem which says that if $u(x, y)$ is defined on all of $\mathbb{R}^2$, if the Gauss curvature of its graph is nonpositive, i.e.,

$$u_{xx} u_{yy} - u_{xy}^2 \leq 0 ,$$

and if $u = o(\sqrt{x^2 + y^2})$ at infinity, then

$$u_{xx} u_{yy} - u_{xy}^2 \equiv 0 .$$

This result caught the attention of a number of people; in particular, T. Rado liked it so much he arranged that a German translation be published in 1927 in Math. Zeitschrift, see [6]. When I was a graduate student, during the period of E. Hopf's visit, mentioned earlier, I gave a seminar talk on this paper. At a certain point in the talk Hopf said «I don't see why *that's* true.» In fact the proof was incorrect. Up to then no one before had observed the error, and I did not see how to fix it. After some time Hopf [94] gave a correct proof, which involved considerably more work.[1]) I have tried many times to find a simple analytic proof — without success. Bers [9] proved a local version of Bernstein's result for solutions of (13) in a disc in $\mathbb{R}^2$ punctured at the origin; the origin is a removable singularity.

(iii) Regularity and estimates Returning to functions u which minimize (11), assuming (12), the nineteenth of the 23 open problems of Hilbert [81] posed in 1901, was to prove the regularity of generalized solutions (stationary points). Bernstein [4] proved analyticity in Ω of the solution — in case f is analytic — provided one knows that $u \in C^3$. Lichtenstein [141] in 1913 required u to be in C^2. Lewy [132], using the solvability of the initial value problem for hyperbolic equations, proved analyticity for fully nonlinear elliptic equations in the plane in (1927) and in [134] (1934). Hopf [90] proved analyticity for $n > 2$.

Finally, in 1940, for $n = 2$, Morrey [161] proved analyticity of generalized stationary points $u \in H^1_{loc}$ of I in (11), i.e. u and ∂u are in L^2_{loc}. He did this with the aid of estimates of the growth, in balls, of the Dirichlet integral of solutions of suitable linear problems of the form (for $n = 2$)

$$Lu = -\frac{\partial}{\partial x_i}(a_{ij}\frac{\partial u}{\partial x_j}) + b_i u_i + cu = 0 \text{ in } \Omega \,. \tag{14}$$

Here L has bounded measurable coefficients, and

$$c_0|\xi|^2 \le a_{ij}\xi_i\xi_j \le C_0|\xi|^2 \;\forall\, \xi \in \mathbb{R}^n \;,\; c_0, C_0 > 0 \tag{15}$$

These estimates yielded Hölder continuity of u. In 1957 De Giorgi [42] extended this result to $n > 2$. This was done at the same time by Nash [172] — also for parabolic equations — using different arguments. Their work has led to much further development, beginning with Moser [166], who gave a new proof of the results of De Giorgi. In [167] he also proved an important Harnack inequality for positive solutions u of (14); namely, for any compact subset K of Ω,

$$\max_K u \le C \min_K u \,, \tag{16}$$

where C depends only on K, Ω and the constants c_0, C_0, $|b_i|_{L^\infty}$, $|c_i|_{L^\infty}$; see Stampachia [211]. Some years later Krylov and Safonov [113] proved the Hölder continuity and Harnack inequality for elliptic operators L of the form (2), having bounded measurable coefficients, and which are uniformly elliptic, i.e. (16) holds (they also proved the Harnack inequality for parabolic operators). These new inequalities have led to much further development, especially for fully nonlinear elliptic equations. See for example the book by Krylov [112].

Returning to Morrey's solution of Hilbert's nineteenth problem, in 1968 De Giorgi [43] constructed a simple example showing that the result need not hold

1) At this point in the lecture here, Prof. Gelfand rose and said that at about the same time, in the 40's, in his seminar, a student had talked on Bernstein's paper. Gelfand observed the error, and told Bernstein about it and he then published a correction in [7].

for elliptic systems; solutions may have singularities. Much work has since been done on «partial regularity», on estimating the Hausdorff dimension (introduced by Hausdorff [77] in 1919) of the set of possible singular points of solutions. A good reference is Giaquinta [69]. In this connection, the book by Federer [55] on geometric measure theory developed many new ideas and techniques. It will no doubt be cited in a survey of mathematics in this second half century. In addition to the books on elliptic equations already mentioned, we should add the 1966 book by Morrey [162]; it contains many deep results and develops many techniques.

(iv) Further remarks about the Calculus of Variations. Courant [34] and Weyl [225] proved the min-max characterizations of all eigenvalues of $-\Delta$ under Dirichlet boundary conditions in a bounded domain $\Omega \subset \mathbb{R}^n$. These are the analogues of the corresponding min-max characterizations of Fischer [58] for the eigenvalues of a Hermitian matrix. In 1923–24 Faber [54] and Krahn [111] proved Rayleigh's conjecture that among all membranes with fixed area the disc has the smallest first frequency, i.e. the disc has the smallest principal eigenvalue for the Laplace operator for functions vanishing on the boundary.

Of greatest impact in the Calculus of Variations, in connection with differential equations, is, of course, the work of Morse and of Lyusternik and Schnirelman [151]. The latter generalized the min-max theory of Birkhoff [10]. The papers of Morse are fundamental. In [163] he considered C^2 functions on a domain in $\mathbb{R}^n$ and on a finite dimensional manifold. The functions are required to have nondegenerate critical points, i.e., at any critical point, the Hessian matrix of second derivatives of the function is nonsingular. The Morse index of a nondegenerate critical point is the number of negative eigenvalues of the Hessian. Already [163] contains the «Morse Lemma», which states that, locally, one may introduce new coordinates near a nondegenerate critical point of a smooth function, in terms of which the function becomes a pure quadratic. In [163] he introduced what is now called «Morse deformations». These are homeomorphisms of the manifold obtained by flow along the negative gradient of the function. [163] also contains the important Morse inequalities connecting the indices of critical points and the Betti numbers of the manifold.[2])

Excellent introductions to the theory, and further developments, may be found in Milnor [157], Bott [16], [17] and Mawhin-Willem [154] Chapter 8.

Morse had many papers extending the theory to infinite dimensions in the context of functionals defined on spaces of curves — leading to ordinary differential equations. In particular, he studied geodesics on manifolds; I mention here only [164] and [165]. The theory of [163] has been extended to general infinite

---

2) At this point in the talk, Professor P. Dombrowski pointed out that already in the 1860's Möbius [159], [160] had studied height functions (e.g. the x_3 coordinate) on closed surfaces in $\mathbb{R}^3$, with nondegenerate critical points. He related the number and nature of critical points with the topology of the surface. A description of this is contained in [46], a general article on geometry in the period 1890–1990, together with suggestive drawings from [159]. Mobius' proofs were not quite complete; in particular, he did not use «negative gradient flow».

dimensional spaces: Hilbert manifolds etc. See for example Chang [31] and [154] Chapter 8, where other references may be found, in particular, to work of Palais and Smale. In addition, the notion of «Morse index» has been generalized in a very significant way in Conley [33] to general vector fields which are not gradients of a function. It is an element of a homotopy group, attached to a connected set where the vector field vanishes. This work has found important applications.

In 1930 Lyusternik [150] proved a far reaching generalization of the result of Fisher, mentioned above for matrices, namely, that a C^1 even function on the sphere S^{n-1} has at least n pairs of critical points. This is related to the antipodal theorem of Borsuk [15]. Lyusternik and Schnirelman [151] in 1934 introduced important min-max methods for obtaining multiple critical points of functions on manifolds. The arguments involved the notion of «category» of a set Σ: the smallest number of contractible sets whose union covers Σ. In contrast to Morse theory which assumes that the function is in C^2, only C^1 is required. The methods lead readily to infinite dimensional generalizations, which have been much developed in the last 15 years. Excellent references to much current work on variational methods in differential equations are Rabinowitz [191], [192], [193], Mawhin-Willem [154], Struwe [214] and Ekeland [52]. Other useful books on nonlinear functional analysis are Deimling [44] and Zeidler [232].

To conclude this section we mention three more papers: (i) The von Neumann [175] min-max theorem in game theory of 1928; (ii) the extension in 1949 by Fenchel [56] of the Legendre transform to nonsmooth convex functions; (iii) a recent, simple looking, but very useful principle of Ekeland [51]; Let (X,d) be a complete metric space and $F\colon X \to \mathbb{R} \cup \{+\infty\}$ a lower semi-continuous function which is bounded below. If $z \in X$ and $\epsilon > 0$ are such that

$$F(z) \le \inf F + \epsilon$$

then $\exists\, y \in X$ such that

$$F(y) + \epsilon\, d(y,z) \le F(z)$$

and

$$F(z) \le F(y) - \epsilon\, d(x,y)\,, \quad \forall\, x \in X\,.$$

(v) Miscellaneous remarks. In connection with second order elliptic equations, and the maximum principle, there is an extensive probabilistic theory. I will just mention a few papers, since I am ill informed on the subject. Random walks (which led to Brownian motion) are related to difference schemes for the Laplace equation. In Courant-Friedrichs-Lewy [38] it was proved that solutions of the standard difference scheme approximating the Laplace equation, converge, in fact, to a solution of the Laplace equation as the mesh size vanishes. In 1923 Wiener [228] gave a rigorous mathematical definition of Brownian motion as a measure. For some years, people in probability theory were not aware of this; they relied on nonrigorous notions of Brownian motion and investigated its connection with the Laplace equation; see Khintchine [107], Petrowsky [183], Lévy

[131]. Finally in 1944, Kakutani [103] proved that the «hitting measure» (to hit the boundary) of Brownian motion in two dimensions, is the Poisson kernel for the Laplace equation. Lévy [131] had discovered some of Kakutani's results. For probabilistic and potential theory see the book Doob [47].

A paper that has had much influence is Feynman [57], on path integrals and the Feynman integral.

Brownian motion is tied up with the heat, as well as with Laplace, equation. Due to lack of time I've said nothing about the heat equation. The study of the initial value problem (with possible boundary conditions) for the heat equation — more generally, for parabolic equations — is closely tied to the abstract theory of semigroups. I would like to mention the 1948 papers Hille [85] and Yosida [231] for the so-called Hille-Yosida theorem. It gives a complete characterization of the infinitesimal generators of continuous semigroups. I wish to mention Kolmogorov-Petrowsky-Piskinov [109], a paper on semilinear parabolic equations in one space variable which has led to much further work.

IV. Hyperbolic Equations

This subject, as well as the remaining ones will be condensed due to lack of time.

In the early part of the century Hadamard did much work on second order hyperbolic equations, see for example [73]. Much of the work at that time on linear equations is presented in his subsequent books [74], [75]. In the latter he worked with fundamental solutions and developed his theory of «finite parts» of integrals. This work was developed a great deal in M. Riesz [201]. In 1928, Perron [181] worked on first order semilinear systems. As mentioned earlier, in 1927 Lewy [132] used holomorphic extension to the complex, for a solution of the initial value problem in two dimensions for nonlinear second order equations, to prove the analyticity of solutions of nonlinear (analytic) elliptic equations in two dimensions. See also [134]. In [133] he solved the initial value problem for nonlinear hyperbolic equations which are not analytic, in two dimensions; he used characteristics.

Schauder [206], in 1935, proved the solvability of the Cauchy problem for a quasilinear hyperbolic second order equation in any number of variables. He did this by approximating by an analytic problem, and then using energy estimates. Under certain conditions he obtained solutions for large time, not just locally.

There is much activity today on finding global solutions for all time, or on determining lower bounds for the time when solutions may blow up.

In 1937 Petrowsky [184] introduced a general notion of hyperbolicity for higher order equations and systems, and proved existence and uniqueness for the Cauchy problem. He developed energy estimates. Leray [125] observed that Petrowsky's existence argument works only if the space dimension is ≤ 8, and in [126] he treated the general linear case of strictly hyperbolic equations and

M. Riesz

systems. See also Gårding [68]. The proceedings of a 1953 conference in Arden House, where [125] appeared, may be of historical interest.

In 1954–58 Friedrichs [63], [64] developed a general theory of symmetric systems, hyperbolic and otherwise, and used energy estimates to treat initial and also boundary value problems. In 1962 Calderón [22] showed how one could derive energy estimates for general strictly hyperbolic equations, and hence solve the Cauchy problem — rather easily — by reducing the equations to a symmetric system of singular integral (pseudo-differential) operators. This was a striking use of such operators in PDE.

The study of linear hyperbolic operators also led to the theory of **Fourier Integral Operators** developed in Hörmander [97], see [98], vol. 4. It grew out of the work on initial value problems for hyperbolic equations in Lax [119]; he developed ideas of geometrical optics.

We should not forget the seminal paper Courant-Friedrichs-Lewy [38], which is concerned with *difference schemes* for partial differential equations and energy estimates. In particular it investigated the stability of difference schemes for hyperbolic equations. Here an important discovery was made: that in the scheme, the time step cannot be larger than some constant times the steps in space.

V. Other Topics

V.1. Fluid Dynamics

Since Newton, there has been much interest in the study of rotating fluids of fixed shape. In a series of papers Lyapounoff [149] studied perturbation problems: the existence of solutions near known ones, depending on some parameter ϵ. At about the same time, 1908, Schmidt [207] developed another approach. One of the important general features of their work, which involved branching of solutions, was the reduction of a perturbation problem in infinite dimensions to one in finite dimensions — to the so-called bifurcation equation. This is the «Lyapounoff-Schmidt» reduction. Much further work was done by Lichtenstein [143], [144], [145] and his students (see section 5 in Bemelmans-Hildebrandt-von Wahl [3]). For global versions of bifurcation solutions see Rabinowitz [191], and the books connected with nonlinear functional analysis mentioned earlier.

In connection with bifurcation theory, one should also mention the important «Hopf bifurcation» for periodic solutions of systems of ordinary differential equations, of 1942, Hopf [91]. The method has been developed for parabolic equations, etc.; see Marsden-McCracken [153].

One of the principal mathematical achievements in fluid dynamics of the first half century was the work of Leray [121], [122], [123] in 1933–34 on the Navier-Stokes equations. In n space dimensions, these are n equations for the flow velocity of the fluid $u = (u^1, u^2, \dots, u^n)$ in a domain in $\mathbb{R}^n$, and the pressure p,

$$u_t^i - \nu\,\Delta u^i + u^j \partial_j u^i + \partial_i p = f_i \text{ forcing term}, \quad i = 1.2, 3$$

$$\operatorname{div} u = \sum u_i^i = 0 \;;$$

ν is viscosity. For this problem one wishes to prescribe, say, regular, initial values of u, and possibly boundary conditions $u = 0$ on the spatial boundary. No initial or boundary conditions are imposed on p. If one projects the first equations for u onto a part orthogonal to gradients of functions — thereby making a nonlocal operation — one may get rid of the pressure, but then the problem becomes nonlocal. This special feature makes the problem a difficult one, which is still far from settled. In his papers, Leray used Hilbert space methods equations, introducing the relevant energy. For $n = 2$ he obtained unique global (all $t > 0$) solutions, while in $\mathbb{R}^3$ he obtained a generalized solution which is not known to be smooth or unique. In 1950–51 [93] obtained weak solutions by approximation by finite dimensional

M. Krasner, J. Leray, Luxembourg, 1981

problems — the procedure of Galerkin. The results of Leray and Hopf yield strong (and regular) solutions for small time in case $n \geq 3$. But uniqueness of these (weak) solutions is not known. (The uniqueness of more regular solutions — if they exist — is known.) Various functional analytic techniques have been brought to bear on the problem; see for example Temam [217]. Leray's work contained some study of the possible singular set of the solution, for $n = 3$. Following work of V. Scheffer, Caffarelli, Kohn and Nirenberg [21] showed in 1982 that the set of singular points (for *suitable* weak solutions) in a domain in $\mathbb{R}^3$ i.e. points (x, t), $x \in \mathbb{R}^3$ time $t \in R_+$ has 1-dimensional Hausdorff measure equal to zero. They made use of ideas introduced in nonlinear elliptic theory to obtain partial regularity.

The 1932 book Lamb [116] is a classic, and continues to be used. considerable part of it was concerned with the study of surface, or water, waves. Here the surface of the water is not known a priori, it is a so-called free boundary. The associated mathematical problems attracted a considerable number of workers. In 1921, Nekrassov [173] proved the existence of progressing periodic waves of small amplitude in a two dimensional channel of infinite depth; see also [174]. This was proved independently by Levi-Civita [130] in 1925. In 1926, the corresponding result on periodic solutions in a channel of finite depth was proved by Struik [213]. In 1978 Toland [218] proved the existence of a periodic progressing solution of maximal amplitude in a canal of infinite depth, solving a problem of Stokes. Amick, Fraenkel and Toland [2] showed that at the crest, there is a sharp angle of 120° — as conjectured by Stokes.

In 1948 Lavrent'ev [117] proved the existence in such a channel of a progressing wave with a single crest — a solitary wave. This was proved independently by Friedrichs and Hyers [66]. An excellent reference for work in water waves up to 1957 is the book by Stoker [212].

The question of whether there is a progressing wave with two or more, but a finite number, of crests, was settled in 1988, by Craig and Sternberg [40]; they showed there were none.

Two more topics, briefly. In the 40's there was considerable activity on the theory of shock waves i.e., flow in which discontinuities across certain fronts, the shock fronts, occur. This is summarized in the basic 1948 book Courant-Friedrichs [37]. I will just add two papers, at that time, which led to considerable further work: Hopf [92], concerned with so-called viscosity solutions, and von Neumann-Richtmeyer [177]. In 1965 there appeared the deep paper of Glimm [72]. For more recent developments in this subject, which is a very active one at present, and in which there are many open problems, see the book by Smoller [209].

Fluid flow in which there are subsonic and supersonic regions involve equations which are of **mixed type**, respectively, elliptic, and hyperbolic in these regions. The first work on such equations, linear ones, is that of Tricomi [222] in 1922. It has led to much further work.

V.2. Singular integral (pseudo-differential) operators and Fourier transform
Fourier transform (and series) had been used in the previous century to solve particular equations with constant coefficients. Surprisingly, it was not used very extensively in PDE in the first half of this century (see however Herglotz [78] — not till the questions on solvability of general equations with constant coefficients of no special type, posed by L. Schwartz towards its end (see the end of section II(b)). Since then, Fourier transform has played a key role. In 1953 Gårding [67] made essential use of Fourier transform in studying higher order elliptic equations, in the so called «Gårding Inequality». See also Hörmander [96]. Fourier transform enters in the important work Calderon-Zygmund [23], and their subsequent papers on singular integral operators. In some sense the first example of such an operator is the Hilbert transform on a curve of 1904, Hilbert [82]. That it is bounded in L_p, for $1 < p < \infty$ was proved by M. Riesz [200]. «Riesz transforms» are higher dimensional analogues.

Pseudo-differential operators grew out of singular integral operators in an attempt to deal with the operators as an algebra, see Kohn-Nirenberg [108]. (Singular integral operators are pseudo-differential operators with homogeneous symbol of order zero.) They give meaning to operators of the form $a(x, \partial_x)$ where $a(x, \xi)$

(called the symbol) belongs to some class of functions, but are not just a polynomial in ξ with coefficients depending on x — this would correspond to partial differential operators. Various classes of functions a are considered. In fact, in 1927, Weyl [225] had already defined such operators, in a formal way, in his work on quantum mechanics. There were other tentative precursors of the theory in the first half century. In 1927–28 Tricomi [223], [224] studied singular integral operators in $\mathbb{R}^2$, with a independent of x and homogeneous in ξ of degree -2, in connection with PDE. In a series of papers 1932–37, G. Giraud studied singular integral operators on manifolds in connection with his work on the oblique derivative problem for elliptic equations, and other problems. He studied the compositon of such operators; see for example Giraud [71]. Mikhlin [156] also had many papers on the subject. He introduced the term «symbol» of an operator.

In addition to Hörmander [98], excellent expositions of pseudo-differential, and Fourier integral operators may be found in Taylor [216] and Treves [221].

V.3. Geometry

A few remarks about PDE and geometry. I will just discuss isometric embedding of a Riemannian manifold in some Euclidean space. But first, one should mention the de Rham [197] theory (1931) of cohomology and differential forms on compact manifolds; it led to the Hodge [86] theory which yields a harmonic form to represent each cohomology class. I would also like to mention a geometric result (of 1948) even though it does not involve PDE. It occurs in John [100] — a fundamental paper in the field of nonlinear programming. The result states that the boundary of a bounded open convex set Ω in $\mathbb{R}^n$ lies between homothetic ellipsoids of ratio $1/n$. The outer ellipsoid is the one of least volume containing Ω.

Turning to the problem of isometric embedding of a Riemannian manifold in some Euclidean space. In 1916, Weyl [226] studied the problem of embedding S^2 with a given Riemannian metric having positive Gauss curvature, as a closed convex surface in E^3. He tried to use the continuity method, starting with the canonical metric on S^2. The problem involves a nonlinear elliptic equation of Monge-Ampère type. With the aid of the maximum principle he obtained estimates for the C^2 norm of the embedding, but these were not sufficient to complete the argument. Lewy [136] solved the problem in 1938 in case the given metric is analytic. (In addition [137] solved the Minkowski problem: that of finding a closed convex surface in $\mathbb{R}^3$ with the property that the Gauss curvature at any point is a given positive function K (on S^2) of the normal to the surface, i.e. its spherical image, at the point; he did this in case K is analytic.) See also [135]. In 1942 Alexandroff [1] obtained a generalized solution of Weyl's problem as a limit of convex polyhedra. In case the given Riemannian metric g is smooth but not analytic, Pogorelov [187], [188] proved that Alexandroff's solution is regular. Independently, in the smooth case, Nirenberg [178] obtained a priori estimates extending those of Weyl, enabling one to complete the continuity method. These made use of ideas of Morrey mentioned earlier, involving the growth of Dirichlet integrals of certain functions, in discs.

Uniqueness in the Weyl problem (modulo a rigid translation) was proved in 1927 by Cohn and Vossen [32] in case the given metric is analytic. In 1943 Herglotz [79] proved uniqueness in case g is merely smooth. A related long standing open problem is to determine whether *every* closed surface in $\mathbb{R}^3$ is rigid, i.e. cannot be deformed continuously and isometrically to some other — or is it possible that there exists nature's accordion?

The local question of embedding a neighbourhood of the origin in $\mathbb{R}^2$, with a given smooth Riemannian metric, isometrically into $\mathbb{R}^3$ is still not settled in the general case. If the metric is analytic, then isometric embeddability follows with the aid of the Cauchy-Kowalewski theorem. In the smooth case, if the Gauss curvature K of the metric is positive, or negative, at the origin, then again one knows isometric embeddability by using local solvability of a nonlinear elliptic, respectively hyperbolic equation. In case $K(0) = 0$ and $\nabla K(0) \neq 0$ local solvability was proved by Lin [147] in 1986. In his proof he described, and solved, a new well posed problem for mixed equations of Tricomi-type, as mentioned in section V.1. In [146] he also solved the problem in case $K \geq 0$ near the origin; the proof uses the Nash-Moser procedure.

In recent years, nonlinear PDE have played a central role in many problems in geometry.

Acknowledgment
This was supported under NSF grant DMS-8806731.

Bibliography

[1] Alexandroff, A.D., Existence of a polyhedron and of a convex surface with given metric, Mat. Sbornik (NS) 11(53) (1942) 15–65.

[2] Amick-Fraenkel-Toland, C.J.-L.E.-J.F., On the Stokes conjecture for the wave of extreme form, Acta Math. (148) (1982) 193–214.

[3] Bemelmans-Hildebrandt-von Wahl, J.-S.-W., Partielle Differentialgleichungen und Variationsrechnung in Ein Jahrhundert Mathematik 1890–1990, Festschrift zum Jubiläum der DMV, Vieweg. Braunschweig (1990) 149–230.

[4] Bernstein, S.N., Sur la nature analytique des solutions de certaines équations aux dérivées partielles du second ordre, Math. Ann. 59 (1904) 20–76.

[5] Bernstein, S.N., Sur la généralisation du problème de Dirichlet, I, II, Math. Ann. 62 (1906) 253–271; 69 (1910) 82–136.

[6] Bernstein, S., Sur un théorème de géométrie et son application aux équations aux derivées partielles du type elliptique, Comm. Soc. Math. Kharkov 15 (1915) 38–45. German trans. Math. Zeitschr. 26 (1927) 551–558.

[7] Bernstein, S.N., An addition to my paper: strengthening of a theorem on surfaces with negative curvature, Izvestia Akad. Nauk SSSR 7 (1943) 297–298.

[8] Bers, L., Partial differential equations and generalized analytic functions, Proc. Nat. Acad. Sci. USA 36 (1950) 130–136 and 37 (1951) 42–47.

[9] Bers, L., Isolated singularities of minimal surfaces, Annals of Math. 2, Ser. 53 (1951) 364–386.

[10] Birkhoff, G.D., Dynamical systems with two degrees of freedom, Trans. Amer. Math. Soc. 18 (1917) 199–300.

[11] Birkhoff-Kellogg, G.D.-O., Invariant points in function spaces, Trans. Amer. Math. Soc. 23 (1922) 96–115.

[12] Bochner, S., Vorlesungen über Fouriersche Integrale, Leipzig, 1932.

[13] Bochner, S., Linear partial differential equations with constant coefficients, Annals of Math. 47 (1946) 202–212.

[14] Bombieri-De Giorgi-Guisti, E.-E.-E., Minimal cones and the Bernstein theorem, Invent. Math. 7 (1969) 243–269.

[15] Borsuk, K., Drei Sätze über die n-dimensionale Euklidische Sphäre, Fund. Math. 21 (1933) 177–190.

[16] Bott, J., Marston Morse and his mathematical works, Bull. Amer. Math. Soc. (NS) 3 (1980) 907–950.

[17] Bott, R., Lectures on Morse Theory, old and new, Bull. Amer. Math. Soc. 7 (1982) 331–358.

[18] Brouwer, L.E.J., Über Abbildung von Mannigfaltigkeiten, Math. Ann. 71 (1912) 97–115.

[19] Bryant-Chern-Gardner-Goldschmidt-Griffiths, R.L.-S.S.-R.B.-H.L.-P.A., Exterior Differential Systems, Math. Sc. Res. Inst. Publ. 18, Springer, 1991.

[20] Caccioppoli, R., Sulle equazioni ellittiche a derivate parziali con n variabili independenti, Rend. Accad. Lincei 19 (1934) 83–89.

[21] Caffarelli-Kohn-Nirenberg, L.-R.-L., Partial regularity of suitable weak solutions of the Navier-Stokes equations, Comm. Pure Appl. Math. 35 (1982) 771–831.

[22] Calderón, A.P., Existence and uniqueness theorems for systems of partial differential equations, Fluid Dynamics and Appl. Math. (Proc. Symp. Univ. of Maryland, 1961), New York (1962) 147–195.

[23] Calderón-Zygmund, A.P.-A., On the existence of certain singular integrals, Acta Math. 88 (1952) 85–139.

[24] Carathéodory, C., Variationsrechnung partieller Differentialgleichungen erster Ordnung (Leipzig) 1935.

[25] Carleman, T., Sur un problème d'unicité pour les systèmes d'équations aux dérivées partielles à deux variables indépendantes, Arkiv. für Math., Astr. och Fysik, 26 B (1939) 1–9.

[26] Carleman, T., Propriétés asymptotiques des fonctions fondamentales des membranes vibrantes, C. P. Cong. des Math. Scand. Stockholm (1934) 34–44 (Lund 1935).

[27] Cartan, E., Sur certaines expressions différentielles et le problème de Pfaff., Ann. Sc. Ec. Norm. 3 (16) (1899) 239–332.

[28] Cartan, E., Sur l'intégration des systèmes d'équations aux différentielles totales, Ann. Sci. Ec. Norm. (3) 18 (1901) 241–311.

[29] Cartan, E., Leçons sur les invariants intégraux, Hermann, Paris, 1922.

[30] Cartan, E., Les systèmes différentiels extérieurs et leurs applications géometriques, Hermann, Paris, 1945.

[31] Chang, K.C., Infinite dimensional Morse theory and multiple solution problems, Progress in Nonlinear Differential Equations and Their Applications, Birkhäuser, Boston, 1992.

[32] Cohn-Vossen, E., Zwei Sätze über die Starrheit der Eiflächen, Nachr. von der Gesell. Wiss. zu Göttingen, Math. Phys. Klasse (1927) 125–134.

[33] Conley, C.C., Isolated invariant sets and the Morse index, CBMS Regional Conf. Ser. 38, Amer. Math. Soc., Providence, 1978.

[34] Courant, R., Zur Theorie der kleinen Schwingungen, Zeit. f. angew. Math. und Mech. 2 (1922) 278–285.

[35] Courant, R., Plateau's problem and Dirichlet's principle, Ann. of Math. 38(2) (1937) 679–724.

[36] Courant, R., Dirichlet's principle, conformal mapping and minimal surfaces, Interscience-Wiley, New York (1970).

[37] Courant-Friedrichs, R.-K.O., Supersonic flow and shock waves, Interscience-Wiley, New York, 1948.

[38] Courant-Friedrichs-Lewy, R.-K.O.-H., Über partielle Differenzengleichungen der mathematischen Physik, Math. Ann. 100 (1928) 32–74.

[39] Courant-Hilbert, R.-D., Methoden der mathematischen Physik, Berlin, Vol. 1, 1924; Vol. 2 (1937).

[40] Craig-Sternberg, W.-P., Symmetry of solitary waves, Comm. Partial Diff. Eqs. 13 (1988) 603–633.

[41] De Giorgi, E., Un esempio di non-unicita della soluzione del problema di Cauchy relativo ad una equazione differenziale lineare a derivate parziale di tipo parabolico, Rend. Mat. 14 (1955) 382–387.

[42] De Giorgi, E., Sulla differenziabilità e l'analiticità delle estremali degli integrali multipli regolari, Mem. Accad. Sci. Torino Cl. Fis. Mat. Natur. 3(3) (1957) 25–43.

[43] De Giorgi, E., Un esempio di estremali discontinue per un problema variazionale di tipo ellittico, Boll. UMI 4 (1968) 135–137.

[44] Deimling, K., Nonlinear Functional Analysis, Springer, Berlin, 1985.

[45] Dierkes-Hildebrandt-Küster-Wohlrab, U.-S.-A.-O., Minimal Surfaces, 2 vols., Grundlehr. der Math. Wiss. 295, Springer, Berlin, 1992.

[46] Dombrowski, P., Differentialgeometrie, Jahrhundert Mathematik 1890–1990, Festschrift zum Jubiläum der DMV, Vieweg, Braunschweig (1990) 323–360. See pages 330–332.

[47] Doob, J.L., Classical Potential Theory and its Probabilistic Counterparts, Springer, New York, 1984.

[48] Douglas, J., Solution of the problem of Plateau, Trans. Amer. Math. Soc. 33 (1931) 263–321.

[49] Douglas, J., Minimal surfaces of higher topological structure, Ann. of Math. 2(40) (1939) 205–298.

[50] Ehrenpreis, L., Solutions of some problems of division, Amer. J. Math. I, 76 (1954) 883–903; II, 77 (1955) 286–292; III, 78 (1956) 685–715; IV, 82 (1960) 522–588.

[51] Ekeland, I., On the variational principle, J. Math. Anal. Appl. 47 (1974) 324–353.

[52] Ekeland, I., Convexity methods in Hamiltonian mechanics, Ergebn. der math. und ihrer Grenzgeb. 3, Vol. 19, Springer-Verlag, Berlin, 1990.

[53] Elworthy-Tromba, K.D.-A.J., Differential structures and Fredholm maps, Proc. Symp. Global Anal. 15, Amer. Math. Soc., Providence (1970) 45–94.

[54] Faber, G., Beweis dass unter allen homogenen Membranen von gleicher Fläche und gleicher Spannung die Kreisförmige den tiefsten Grundton gibt, Sitzungsberichte der Bayrischen Akad. der Wiss.(1923) 169–172.

[55] Federer, H., Geometric measure theory, Springer-Verlag, Heidelberg and New York (1969).

[56] Fenchel, W., On conjugate convex functions, Canad. J. Math. 1 (1949) 73–77.

[57] Feynman, R.P., Space-time approach to nonrelativistic quantum mechanics, Rev. Mod. Phys. 20 (1948) 367–387.

[58] Fischer, E., Über quadratische Formen mit reellen Koeffizienten, Monatsh. f. Math. v. Phys. 16 (1905) 234–239.

[59] Fitzpatrick-Pejsachowicz, P.M.-J., An extension of the Leray-Schauder degree for fully nonlinear elliptic problems, Symp. Pure Math. AMS Proceedings Vol. 45 Part 1 (1968) 423–439.

[60] Fredholm, J., Sur une classe d'équations fonctionelles, Acta Math. 27 (1903) 365–390.

[61] Friedrichs, K.O., Spektraltheorie halbbeschränkter Operatoren und Anwendung auf die Spektralzerlegung von Differentialoperatoren Math. Annalen I. 109 (1934) 465–487; II 685–713, Berichtigung 110 (1935) 777–779.

[62] Friedrichs, K.O., On differential operators in Hilbert spaces, Amer. J. Math. 61 (1939) 523–544.

[63] Friedrichs, K.O., The identity of weak and strong extensions of differential operators, Trans. Amer. Math. Soc. 55 (1944) 132–151.

[64] Friedrichs, K.O., Symmetric hyperbolic linear differential equations, Comm. Pure Appl. Math. 7 (1954) 517–550.

[65] Friedrichs, K.O., Symmetric positive linear differential equations, Ibid. 11 (1958) 333–418.

[66] Friedrichs-Hyers, K.O.-D.H., The existence of solitary waves, Comm. Pure Appl. Math. 7 (1954) 517–550.

[67] Gårding, L., Dirichlet's problem for linear elliptic partial differential equations, Math. Scand. 1 (1953) 55–72.

[68] Gårding, L., Solution directe du problème de Cauchy pour les équations hyperboliques, La Théorie des Équations aux Dérivées partielles, Colloq. Int. CNRS Paris 1956.

[69] Giaquinta, M., Multiple integrals in the calculus of variations and nonlinear elliptic systems, Annals of Math. Studies 105, Princeton Univ. Press, 1983.

[70] Gilbarg-Trudinger, D.-N.S., Elliptic Partial Differential Equations of Second Order, Springer, Berlin, 1nd Ed., 1983.

[71] Giraud, G., Équations à intégrales principales, Ann. Sci. Ec. Norm. Sup. Paris 51 (1934) 251–372.

[72] Glimm, J., Solutions in the large for nonlinear hyperbolic systems of equations, Comm. Pure Appl. Math. 18 (1965) 95–105.

[73] Hadamard, J., Leçons sur la propagation des ondes, Hermann, Paris, 1903.

[74] Hadamard, J., Lectures on Cauchy's problem in linear partial differential equations, Yale Univ. Press (1923).

[75] Hadamard, J., Le problème de Cauchy et les équations aux dérivées partielles linéaires hyperboliques, Hermann et Cie., Paris, 1932.

[76] Hardy-Littlewood-Polya, G.H.-J.E.-G., Inequalities, Cambridge Univ. Press, 1934.

[77] Hausdorff, F., Dimension und äusseres Mass, Math. Annalen 79 (1919) 157–179.

[78] Herglotz, G., Über die Integration linearer partieller Differentialgleichungen mit konstanten Koeffizienten II, Anwendung Fouriersche Integrale, Berl. Math. Phys. Kl. Sächs. Akad. Wiss. Leipzig 78 (1926) 287–318; III, Anwendungen Ibid. 80 (1928) 69–114.

[79] Herglotz, G., Über die Starrheit der Eiflächen, Abh. Math. Sem. der Hansischen Univ. 15 (1943) 127–129.

[80] Hilbert, D., Über das Dirichletsche Prinzip. Jber. Deutsch. Math. Verein. 8 (1900) 184–188.

[81] Hilbert, D., Mathematische Probleme Archiv f. Math. u. Phys. 3(1) (1901) 44–63, 213–237.

[82] Hilbert, D., Über eine Anwendung Verh. III. Intern. Math. Kongresses, Heidelberg, 1904.

[83] Hilbert, D., Grundzüge einer allgemeinen Theorie der linearen Integralgleichungen, Leipzig, 1912.

[84] Hildebrandt-Graves, T.H.-L.M., Implicit functions and their differentials in general analysis, Trans. Amer. Math. Soc. 29 (1927) 127–153.

[85] Hille, E., Functional Analysis and Semigroups, Colloq. Publ. Amer. Math. Soc., 1948.

[86] Hodge, W.V.D., The Theory and Applications of Harmonic Integrals, Cambrdige Univ. Press, 1941.

[87] Hölder, O., Beiträge zur Potentialtheorie, Stuttgart, 1882.

[88] Holmgren, E., Über Systeme von linearen partiellen Differentialgleichungen, Ofversigt af kongl. Vetenkaps, Akad. Forh. 58 (1901) 91–103.

[89] Hopf, E., Elementare Bemerkungen über die Lösungen partieller Differentialgleichungen zweiter Ordnung vom elliptischen Typus, Sitz. Preuss. Akad. Wiss. 19 (1927) 197–152.

[90] Hopf, E., Über den funktionalen, insbesondere den analytischen Charakter der Lösungen elliptischer Differentialgleichungen zweiter Ordnung, Math. Zeitschr. 34, II (1931) 194–233.

[91] Hopf, E., Abzweigung einer periodischen Lösung von einer stationären Lösung eines Differentialgleichungssystems, Ber. Sächs. Akad. Wiss. Leipzig. Math.-Phys. Kl. 94 (1942) 1–22.

[92] Hopf, E., The partial differential equation $u_t + uu_x = \mu u_{xx}$, Comm. Pure Appl. Math. 3 (1950) 201–230.

[93] Hopf, E., Über die Anfangswertsaufgabe für die hydrodynamischen Grundgleichungen, Math. Nachr. 4 (1950/51) 213–231.

[94] Hopf, E., On S. Bernstein's theorem on surfaces $z(x, y)$ of nonpositive curvature, Proc. Amer. Math. Soc. 1 (1950) 80–85.

[95] Hopf, E., A remark on linear elliptic differential equations of second order, Proc. Amer. Math. Soc. 3 (1952) 791–793.

[96] Hörmander, L., On the theory of general partial differential equations, Acta Math. 94 (1955) 161–248.

[97] Hörmander, L., Fourier integral operators I, Acta Math. 127 (1971) 79–183.

[98] Hörmander, L., The Analysis of Linear Partial Differential Operators, Vols. 1–4, Springer-Verlag, 1983–85.

[99] Jenkins-Serrin, H.-S., The Dirichlet problem for the minimal surface equation in higher dimensions, J. Reine Angew. Math. 229 (1968) 170–187.

[100] John, F., Extremum problems with inequalities as subsidiary conditions, Courant Anniversary volume (1948) 187–204.

[101] John, F., On linear partial differential equations with analytic coefficients: Unique continuation of data, Comm. Pure Appl. Math. 2 (1949) 209–253.

[102] John, F., Continuous dependence on data for solutions of PDE with a prescribed bound. Comm. Pure Appl. Math. 13 (1960) 551–585.

[103] Kakutani, S., On Brownian motions in n space, Proc. Imp. Acad. Tokyo 20 (1944) 648–652, Ibid., p. 706–714.

[104] Kähler, E., Einführung in die Theorie der Systeme von Differentialgleichungen, B.G. Teubner, Leipzig, 1934.

[105] Kellogg, O.D., Foundations of Potential Theory, Springer, Berlin, 1929.

[106] Kellogg, O.D., On the derivatives of harmonic functions at the boundary, Trans. Amer. Math. Soc. 33 (1931) 486–510.

[107] Khintchine, A., Asymptotische Gesetze der Wahrscheinlichkeitsrechnung. Erg. Math. Grenzgebiete 2/4 (1933), Springer, Berlin.

[108] Kohn-Nirenberg, J.J.-L., On the algebra of pseudo-differential operators, Comm. Pure Appl. Math. 18 (1965) 219–305.

[109] Kolmogorov-Petrowsky-Piskinov, A.N.-I.-N.S., Étude de l'équation de la diffusion avec croissance de la quantité de matière et son application à un problème biologique, Bgul. Mosk. Gas. Univ. Sev. Intern. A1 (1937) 1–26. English translation in Dynamics of Curved Fronts, R. Pelcé, ed., Perspectives in Physics Series, Academic Press, New York (1988) 105–130.

[110] Korn, A., Über Minimalflächen deren Randkurven wenig von ebenen Kurven abweichen, Abh. Kgl. Preuss. Akad. Wiss. Berlin, 1909.

[111] Krahn, E., Über eine Rayleigh formulierte Minimaleigenschaft des Kreises, Math. Ann. 94 (1924) 97–100.

[112] Krylov, N.V., Nonlinear elliptic and parabolic equations of the second order, Math. and its Applications, D. Reidel Pub. Co. (1987).

[113] Krylov-Safonov, N.V.-M.V., Certain properties of solutions of parabolic equations with measurable coefficients, Izvestia Akad. Nauk SSSR 40 (1980) 161–175.

[114] Ladyzhenskaya-Uraltseva, O.A.-N.N., Linear and Quasilinear Elliptic Equations, Izdat Nauka, Moscow, 1964, Engl. Trans., Academic Press, New York, 2nd Russian ed., 19731.

[115] Ladyzhenskaya-Solonnikov-Uraltseva, O.A.-V.A.-N.N., Linear and Quasilinear Equations of Parabolic Type, Amer. Math. Soc., Providence, 1968.

[116] Lamb, H., Hydrodynamics, Cambridge Univ. Pres, 1932.

[117] Lavrent'ev, M.A., On the theory of long waves, Akad. Nauk Ukraine R.S.R. Zb Proc. Inst. Math. 1 (1948).

[118] Lavrent'ev, M.A., Some improperly posed problems of Math. and Physics, Springer tracts in National Philosophy, 11, 1967.

[119] Lax, P.D., Asymptotic solutions of oscillatory initial value problems, Duke Math. J. 24 (1957) 627–646.

[120] Lebesgue, H., Sur le problème de Dirichlet, Rend. Circ. Mat. Palermo 24 (1907) 371–402.

[121] Leray, J., Etude des diverses équations integrales non linéaires et de quelques problèmes que pose l'hydrodynamique, J. de Math. 126 (1933) 1–82.

[122] Leray, J., Essai sur les mouvements plans d'un liquide visqueux que limitent des parois, J. Math. Pures Appl. Ser. IX 13 (1934) 331–418.

[123] Leray, J., Sur le mouvement d'un liquide visqueux emplissant l'espace, Acta Math. 63 (1934) 193–248.

[124] Leray, J., Discussion d'un problème de Dirichlet, J. Math. Pures Appl. 18 (1939) 249–284.

[125] Leray, J., On linear hyperbolic differential equations with variable coefficients on a vector space, Contributions to the theory of Partial Differential Equations, ed. L. Bers et al., Annals of Math. Studies 33, Princeton Univ. Press, 1954.

[126] Leray, J., Hyperbolic equations with variable coefficients, Lecture Notes, Inst. Adv. Study, Princeton, 1952.

[127] Leray-Schauder, J.-J., Topologie et équations fonctionelles, Ann. Sci. Ec. Norm. Sup. 51 (1934) 45–78.

[128] Levi, E.E., Sulle equazioni lineare totalmente ellittiche, Rend. Circ. Mat. Palermo 24 (1907) 275–317.

[129] Levi, E.E., Studi sui punti singolari essenziali delle funzioni analitiche di due o pìu variabili complesse, Annali di Mat. pura ed appl. 17(3) (1910) 61–87.

[130] Levi-Civita, T., Détermination rigoureuse des ondes permanentes d'ampleur finie, Math. Ann. 93 (1925) 264–314.

[131] Lévy, P., Le mouvement brownien plan, Amer. J. Math. 62 (1940) 487–550.

[132] Lewy, H., Über den analytischen Charakter der Lösungen elliptischer Differentialgleichungen, Nachr. Ges. Wiss. Gottingen (1927) 178–186.

[133] Lewy, H., Über das Anfangswertsproblem einer hyperbolischen, nichtlinearen partiellen Differentialgleichung, zweiter Ordnung mit zwei unabhängigen Veränderlichen, Math. Ann. 98 (1928) 179–191.

[134] Lewy, H., Neuer Beweis des analytischen Charakters der Lösungen elliptischer Differentialgleichungen, Math. Ann. 101 (1929) 609–619.

[135] Lewy, H., A priori limitations for solutions of elliptic Monge-Ampère equations, I. II, Trans. A. M. S. 37 (1935) 417–434; 41 (1937) 365–374.

[136] Lewy, H., On the existence of a closed convex surface realizing a given Riemannian metric, Proc. NAS, USA 24 (1938) 104–106.

[137] Lewy, H., On differential geometry in the large, I. (Minkowski's problem) Trans. AMS 43 (1938) 258–270.

[138] Lewy, H., An example of a smooth linear partial differential equation without solution, Ann. of Math. 66 (1957) 155–158.

[139] Li, Y.Y., Degree theory for second order nonlinear elliptic operators and its applications, Comm. PDE 14 (1989) 1541–1578.

[140] Lichtenstein, L., Über das Poissonsche Integral und über die partiellen Ableitungen zweiter Ordnung des logarithmischen Potentials, J. reine angew. Math. 141 (1912) 12–42.

[141] Lichtenstein, L., Über den analytischen Charakter der Lösungen regulärer zweidimensionaler Variationsprobleme, Bull. Acad. sc. Cracovie (1913) 915–941.

[142] Lichtenstein, L., Über einige Existenzprobleme der Variationsrechnung. Methode der unendlichen Variablen, J. reine angew. Math. 145 (1914) 24–85.

[143] Lichtenstein, L., Über einige Eigenschaften der Gleichgewichtsfiguren rotierender homogenen Flüssigkeiten deren Teilchen einander nach dem Newtonschen Gesetz anziehen, Sitz.-ber. Preuss. Akad. Wiss. (1918) 1120–1135.

[144] Lichtenstein, L., Vorlesungen über einige Klassen nichtlinearer Integralgleichungen und Integro-Differentialgleichungen, Berlin, 1931.

[145] Lichtenstein, L., Gleichgewichtsfiguren rotierender Flüssigkeiten, Berlin, 1933.

[146] Lin, C.S., The local isometric embedding in $\mathbb{R}^3$ of 2-dimensional Riemannian manifolds with nonnegative curvature, J. Diff. Geom. 21 (1985) 213–230.

[147] Lin, C.S., The local isometric embedding in $\mathbb{R}^3$ of two dimensional Riemannian manifolds with Gaussian curvature changing sign cleanly, Comm. Pure Appl. Math. 39 (1986) 867–887.

[148] Lions-Magenes, J.L., – E., Problèmes aux Limites non Homogènes et Applications, Dunod Paris, Vol. I and II, 1968; Vol. III, 1970.

[149] Lyapounoff, A., Sur les figures d'équilibre peu différentes des ellipsoides d'une masse liquide homogène donnée d'un mouvement de rotation. I. Etude générale du problème, Mem. Acad. Imp. Sc. St. Petersburg (1906) 1–225; II, Figures d'équilibre dérivées des ellipsoides de Mclaurin, Ibid. (1909), 1–202; III. Figures d'équilibre dérivées des ellipsoides de Jacobi, Ibid. (1912) 1–227; IV. Nouvelles formules pour la recherche des figures d'équilibre, Ibid. (1914) 1–12.

[150] Lyusternik, L.A., Topologische Grundlagen der allgemeinen Eigenwerttheorie, Monatsh. Math. Phys. 37 (1930) 125–130.

[151] Lyusternik-Schnirelman, L.A.-L., Méthodes topologiques dans les problèmes variationnels, Hermann et Cie, Paris, 1934.

[152] Malgrange, B., Existence et approximation des solutions des équations aux dérivées partielles et des équations de convolution, Ann. Inst. Fourier (Grenoble) 6 (1955–56) 271–355.

[153] Marsden-McCracken, J.E.-M., The Hopf bifurcation and its Applications, New York, 1976.

[154] Mawhin-Willem, J.-M., Critical point theory and Hamiltonian systems, Applied Math. Sci. 74, Springer-Verlag, New York, 1989.

[155] Métivier, G., Counterexamples to Holmgren's uniqueness for analytic non linear problems. Preprint.

[156] Mikhlin, S.G., Equivalent problems in the theory of singular integral equations, Mat. Sbornik (NS) 3(45) (1938) 121–140.

[157] Milnor, J., Morse Theory, Princeton Univ. Press, Princeton, 1963.

[158] Minakshisundaram-Pleijel, S.-A., Some properties of the eigenfunctions of the Laplace operator on Riemannian manifolds, Canad. J. Math. 1 (1949) 242–256.

[159] Möbius, A.F., Theorie der elementaren Verwandtschaft, Berichte über die Verhandl. Kgl. Sächs Ges. Wiss. math.–phys. Klasse 15 (1863) 18–57.

[160] Möbius, A.F., Über die Bestimmung des Inhalts eines Polyeders, Ibid. 17 (1865) 31–68.

[161] Morrey, C.B., Existence and differentiability theorems for the solutions of variational problems for multiple integrals, Bull. Amer. Math. Soc. 46 (1940) 439–458.

[162] Morrey, C.B., Multiple Integrals in the Calculus of Variations, Grundlehren 130, Springer, Berlin, 1966.

[163] Morse, M., Relations between the critical points of a real function of n independent variables, Trans. Amer. Math. Soc. 27 (1925) 345–396.

[164] Morse, M., The foundations of a theory of the calculus of variations in the large, Trans. Amer. Math. Soc. 30 (1928) 213–274.

[165] Morse, M., The Calculus of Variations in the Large, Amer. Math. Soc. Colloq. Publ. 18 (1934).

[166] Moser, J., A new proof of De Giorgi's theorem concerning the regularity problem for elliptic differential equations, Comm. Pure Appl. Math. 13 (1960) 457–468.

[167] Moser, J., On Harnack's theorem for ellptic differential equations, Ibid., 14 (1961) 577–591.

[168] Moser, J., A new technique for the construction of solutions of nonlinear differential equations, Proc. Nat. Acad. Sci. 47 (1961) 1824–1831.

[169] Moser, J., A rapidly convergent iterative method and nonlinear differential equations, Ann. Sc. Norm. Sup. Pisa 20 (1966) I. 265–315, II. 499–535.

[170] Nagumo, M., Über das Anfangswertproblem Partieller Differentialgleichungen, Japan J. Math. (18) 1941, 41–47.

[171] Nash, J., The imbedding problem for Riemannian manifolds, Ann. of Math. 63 (1956) 20–63.

[172] Nash, J., Continuity of solutions of parabolic and elliptic equations, Amer. J. Math. 80 (1958) 931–954.

[173] Nekrassov, A.I., On steady waves, Izv. Ivanova-Voznesensk. Politekh. In-ta 3 (1921).

[174] Nekrassov, A.I., The exact theory of steady waves on the surface of a heavy fluid, Izdat. Akad. Nauk, SSSR, Moscow, 1951.

[175] von Neumann, J., Zur Theorie der Gesellschaftsspiele, Math. Ann. 100 (1928) 295–320.

[176] von Neumann, J., Allgemeine Eigenwerttheorie Hermitescher Funktionaloperatoren, Math. Ann. 102 (1929) 49–131.

[177] von Neumann-Richtmeyer, J.-R.D., A method for the numerical calculation of hydrodynamical shocks, J. Appl. Phys. 21 (1950) 232.

[178] Nirenberg, L., The Weyl and Minkowski problems in Differential Geometry in the Large, Comm. Pure Appl. Math. 6 (1953) 337–394.

[179] Nishida, T., A note on a theorem of Nirenberg, J. Diff. Geom. 12 (1977) 629–633.

[180] Nitsche, J.C.C., Vorlesungen über Minimalflächen., Berlin (1975).

[181] Perron, O., Über Existenz und Nichtexistenz von Integralen partieller Differentialgleichungssysteme im reellen Gebiete, Math. Zeitsch. 27 (1928) 549–564.

[182] Perron, O., Über den analytischen Charakter der Lösungen elliptischer Differentialgleichungen, Nachr. Ges. Wiss. Göttingen (1927) 178–186.

[183] Petrowsky, I., Über das Irrfahrtproblem, Math. Ann. 109 (1933–34) 425–444.

[184] Petrowsky, I., Über das Cauchysche Problem für ein System von partieller Differentialgleichungen, Rec. math. (Mat. Sbornik) 2(44) (1937)815–868.

[185] Petrowsky, I., On some problems of the theory of partial differential equations, Uspekhi Mat. Nauk N.S. 1, no. 3–4 (1946) 44–70, Amer. Math. Soc. Translation no. 12, 1949.

[186] Phragmén-Lindelöf, E.-E., Sur une extension d'un principe classique de l'analyse et sur quelques propriétés des fonctions monogènes dans le voisinage d'un point singulier, Acta Math. 31 (1908) 381–406.

[187] Pogorelov, A.V., Intrinsic estimates for the derivatives of the radius vector of a point on a regular convex surface, Dokl. Akad. Nauk SSSR (N.S.) 66 (1949) 805–808.

[188] Pogorelov, A.V., On the regularity of convex surfaces with regular metric, Ibid., p. 1051–1053.

[189] Polya-Szegö, G.-G., Aufgaben und Lehrsätze aus der Analysis, Vol. I, II, Springer, Berlin, 1925. American edition, Dover, 1945.

[190] Protter-Weinberger, M.H.-H.F., Maximum principles in differential equations, Prentice Hall, PDE Series, Englewood Cliffs, 1967.

[191] Rabinowitz, P.H., Some aspects of nonlinear eigenvalue problems, Rocky Mountain J. Math. 3 (1973) 161–202.

[192] Rabinowitz, P.H., Variational methods for nonlinear eigenvalue problems, Eigenvalues of Nonlinear Problems (ed. G. Prodi) C.I.M.E. Ediz. Cremonese, Roma (1975) 141–195.

[193] Rabinowitz, P.H., Minimax Methods in Critical Point Theory with Applications to Differential Equations, CBMS 65, Amer. Math. Soc., Providence, 1986.

[194] Rado, T., The problem of least area and the problem of Plateau, Math. Zeitschr. 32 (1930) 763–796.

[195] Rado, On Plateau's problem, Ann. of Math. (2)31 (1930) 437–469.

[196] Rellich, R., Ein Satz über mittlere Konvergenz, Nachr. Akad. Wiss. Göttingen Math. Phys. Kl. (1930) 30–35.

[197] de Rham, G., Relations entre la topologie et la théorie des intégrales multiples, Enseign. Math. 3–4 (1936) 213–228.

[198] de Rham, G., Über mehrfache Integrale, Abh. Math. Sem. Hansischen Univ. 12 (1938) 313–339.

[199] Riesz, F., Über lineare Funktionalgleichungen, Acta Math. 41 (1918) 71–98.

[200] Riesz, M., Sur les fonctions conjugées, Math. Z. 27 (1927) 218–244.

[201] Riesz, M., L'intégrale de Riemann-Liouville et le problème de Cauchy, Acta Math. 81 (1949) 1–223.

[202] Sard, A., The measure of the critical points of differentiable maps, Bull. Amer. Math. Soc. 48 (1942) 883–890.

[203] Schauder, J.P., Der Fixpunktsatz in Funktionalräumen, Studia Math. 2 (1930) 171–180. See Oeuvres, p. 168–176.

[204] Schauder, J.P., Über lineare elliptische Differentialgleichungen zweiter Ordnung, Math. Z. 38 (1934) 237–282.

[205] Schauder, J.P., Numerische Abschätzungen in elliptischen linearen Differentialgleichungen, Studia Math. 5 (1935) 34–42.

[206] Schauder, J.P., Das Anfangswertproblem einer quasilinearen hyperbolischen Differentialgleichung zweiter Ordnung in beliebiger Anzahl von unabhängigen Veränderlichen, Fundamenta Math. 24 (1935) 213–246. See Oeuves, PWN Editions Sci. de Pologne Warsaw (1978) 380–409.

[207] Schmidt, E., Zur Lösung der linearen und nichtlinearen Integralgleichungen III, Über die Auflösung der nichtlinearen Integralgleichungen und die Verzweigung ihrer Lösungen, Math. Ann. 65 (1908) 370–399.

[208] Schwartz, L., Théorie des Distributions I, II, Hermann, Paris 1950, 1951.

[209] Smoller, J., Shock waves and reaction-diffusion equations, Springer, New York, 1983.

[210] Sobolev, S.L., Sur un théorème d'analyse fonctionelle, Math. Sb. 4(46) (1938) 471–497. (Russian, French summary) Amer. Math. Soc. Transl. (2)34 (1963) 39–68.

[211] Stampachia, G., Équations Elliptiques du Second Ordre à Coefficients Discontinus, Les Presses de L'Université de Montreal, 1966.

[212] Stoker, J.J., Water Waves, Interscience-Wiley, New York, 1957.

[213] Struik, D.J., Determination rigoureuse des ondes irrotationnelles périodiques dans un canal à profondeur finie, Math. Ann. 95 (1926) 595–634.

[214] Struwe, M., Plateau's problem and the Calculus of Variations, Math. Notes 35, Princeton Univ. Press, Princeton, 1988.

[215] Struwe, M., Variational Methods. Applications to Nonlinear Partial Differential Equations and Hamiltonian Systems, Springer-Verlag, Berlin, 1990.

[216] Taylor, M., Pseudodifferential Operators, Princeton Univ. Press, Princeton, 1981.

[217] Temam, R., Navier Stokes Equations, Theory and Numerical Analysis, North-Holland, Amsterdam, Revised Ed., 1979.

[218] Toland, J.F., On the existence of a wave of greatest height and Stokes' conjecture, Proc. Roy. Soc. London, A363 (1978) 469–485.

[219] Tonelli, L., Sur un méthode directe du calcul des variations, Rend. Palermo 39 (1915) 233–264.

[220] Tonelli, L., Fondamenti di calcolo delle variazioni, I. II, N. Zanichelli Bologna 1921 and 1923.

[221] Treves, F., Introduction to Pseudodifferential and Fourier Integral Operators, Vol. 1 and 2, Plenum Press, New York, 1980.

[222] Tricomi, F.G., Formula d'inversione dell'ordine di due integrazione doppie con asteristico, Rend. Accad. Naz. Lincei 3 (1926) 535–539.

[223] Tricomi, F.G., Equazioni integrali contenti il valor principale di un integrale doppio, Math. Zeit. 27 (1928) 87–133.

[224] Tricomi, F.G., Sulle equazioni lineari alle derivate parziali di 2° ordine di tipo misto, Paris I–VII, Mem. Reale. Accad. Naz. dei Lincei, cl. sc. fisiche, [5], 14 (1922) 133–247.

[225] Weyl, H., Das asymptotische Verteilungsgesetz der Eigenwerte linearer partieller Differentialgleichungen (mit einer Anwendung auf die Theorie der Hohlraumstrahlung) Math. Ann. 71 (1912) 441–479.

[226] Weyl, H., Über die Bestimmheit einer geschlossenen konvexen Fläche durch ihr Linienelement. Vierteljahresschrift der nat.-Forsch. Ges. Zürich 61 (1916) 40–72.

[227] Weyl, H., Quantenmechanik und Gruppentheorie, Z. f. Physik 46 (1927) 1–16.

[228] Wiener, N., Differential space, J. Math. Phys. Mass. Inst. Tech. 2 (1923) 131–174.

[229] Wiener, N., The Dirichlet problem, J. Math. Phys. 3 (1924) 127–146.

[230] Yang, D., Local solvability of overdetermined systems defined by commuting first-order differential operators, Comm. Pure Appl. Math. 39 (1986) 401–421.

[231] Yosida, K., On the differentiablity and the representation of one-parameter semigroups of linear operators, J. Math. Soc. Japan 1 (1948) 15–21.

[232] Zeidler, E., Nonlinear Functional Analysis and its Applications, Vol. I–IV, Springer, New York, 1985–88.

[233] Zygmund, A., Trigonometrical Series, Warsaw, 1935. Trigonometric Series (2nd ed.), 2 vols., Cambridge Univ. Press, 1959.

Louis Nirenberg
Courant Institute
251, Mercer Street
New York NY 10012 USA

Intégration et mesure 1900–1950

Jean-Paul Pier
Centre universitaire de Luxembourg

Les concepts de mesure et de mesurabilité sont entrevus par Cantor, Peano, et forgés par Borel.

Cantor [14] considère $\mathbb{R}^n$ ($n \in \mathbb{N}^*$). Pour l'ensemble borné E de $\mathbb{R}^n$, soit V_ρ la réunion des boules de rayon ρ centrées aux points de $\bar{E}$. Cantor appelle volume de E le nombre

$$\lim_{\rho \to 0^+} \int \dots \int_{V_\rho} dx_1 \dots dx_n.$$

Dans cette terminologie, la *mesure* d'une partie coïncide avec celle de sa fermeture. La mesure de la réunion de deux parties disjointes peut être strictement inférieure à la somme des mesures de ces parties.

Dans l'édition des oeuvres de Cantor par Zermelo en 1932 on lit ce commentaire à propos de la définition due à l'auteur de la théorie des ensembles:

«Der 'Inhalt' einer Punktmenge [wird] in einer Weise definiert, die, so naheliegend sie erscheint, sich doch bisher als wenig fruchtbar erwiesen hat. Der erste wirklich brauchbare Inhaltsbegriff ist m.E. der des Lebesgueschen 'Maßes' als der unteren Grenze einer die Punktmenge einschließenden abzählbaren Intervallmenge» ([15] p. 244).

Peano [58] envisage, pour une partie E de $\mathbb{R}$ ($\mathbb{R}^2, \mathbb{R}^3$) toute réunion finie d'intervalles (polygones, prismes) contenus dans [resp. contenant] E. Il appelle mesure intérieure [resp. extérieure] de E la limite supérieure [resp. inférieure] de la somme des longueurs (aires, volumes) correspondante. Si ces valeurs coïncident, la partie E est dite mesurable; la valeur commune est la mesure de E. Alors, la mesure de la réunion de deux parties mesurables disjointes est la somme des mesures de ces parties. Toutefois, un ouvert borné, un ensemble borné de nombres rationnels ne sont pas nécessairement mesurables.

Dès 1898, Borel introduit son concept de mesure. Sur l'ensemble ouvert borné (dans la pratique par $]0, 1[$) qui est une réunion disjointe finie ou dénombrable d'intervalles, Borel adopte comme mesure la somme des longueurs de ces intervalles. Puis, il étudie les ensembles qui s'obtiennent à partir d'ouverts bornés au moyen des opérations de réunion dénombrable et de différence d'ensembles; ces parties seront appelées boréliennes. La mesure d'une partie borélienne est définie grâce à l'*additivité complète*: Pour une réunion dénombrable de boréliens X_n, deux à deux disjoints, la mesure est la somme des mesures des X_n. Borel se rend compte qu'un ensemble admettant pour mesure zéro peut être non dénombrable, mais que tout ensemble dénombrable a pour mesure zéro. Il écrit:

«Les ensembles dont on peut définir la mesure en vertu des définitions précédentes sont dits par nous mesurables, sans que nous entendions impliquer par là qu'il n'est pas possible de donner une définition de la mesure d'autres ensembles; mais une telle définition nous serait inutile; elle pourrait même nous gêner, si elle ne laissait pas à la mesure les propriétés fondamentales que nous lui avons attribuées dans les définitions que nous avons données» ([5] p. 48).

Les progrès du calcul intégral à la fin du XIXe siècle sont dus aux contributions de Riemann, Darboux, Stieltjes; ils représentent une avancée notable par rapport à la théorie de Cauchy. Le fondateur du calcul intégral moderne, Lebesgue, écrira plus tard:

«Cauchy n'appliquait son procédé de définition qu'à des fonctions considérées a priori comme intéressantes: les fonctions continues; maintenant au contraire, toute fonction sera intéressante à laquelle s'appliquera le procédé de définition» ([45] p. 24).

A l'orée de ce siècle, Hilbert [38] examine, parmi ses fameux problèmes, une question de calcul des variations. Il veut déterminer une fonction f de la variable x telle que l'intégrale définie

$$J = \int_a^b F(f'(x), f(x); x)dx$$

soit minimale, les valeurs prises par la fonction aux extrémités de l'intervalle étant imposées. Il introduit l'*intégrale invariante*

$$J^* = \int_a^b \big(F(y, f(x); x) + (f'(x) - y)F'_y(y, f(x); x)\big)dx;$$

il étudie la forme que y doit prendre afin que cette dernière intégrale soit indépendante du choix de f.

Le 29 avril 1901 est présentée à l'Académie des sciences de Paris une note de Lebesgue qui — chronologiquement parlant — peut être considérée comme un tout premier résultat important en mathématiques au cours du vingtième siècle. L'auteur débute ainsi:

«Dans le cas des fonctions continues, il y a identité entre les notions d'intégrale et de fonction primitive. Riemann a défini l'intégrale de certaines fonctions discontinues, mais toutes les fonctions dérivées ne sont pas intégrables, au sens de Riemann. Le problème de la recherche des fonctions primitives n'est donc pas résolu par l'intégration, et l'on peut désirer une définition de l'intégrale comprenant comme cas particulier celle de Riemann et permettant de résoudre le problème des fonctions primitives.

Pour définir l'intégrale d'une fonction continue croissante

$$\gamma(x) \quad (a \leq x \leq b),$$

on divise l'intervalle (a, b) en intervalles partiels et l'on fait la somme des quantités obtenues en multipliant la longueur de chaque intervalle partiel par l'une des

H. Lebesgue (courtesy L. Félix)

valeurs de y quand x est dans cet intervalle. Si x est dans l'intervalle (a_i, a_{i+1}), y varie entre certaines limites m_i, m_{i+1}, et réciproquement si y est entre m_i et m_{i+1}, x est entre a_i et a_{i+1}. De sorte qu'au lieu de se donner la division de la variation de x, c'est-à-dire de se donner les nombres a_i, on aurait pu se donner la division de la variation de y, c'est-à-dire les nombres m_i. De là deux manières de généraliser la notion d'intégrale. On sait que la première (se donner les a_i) conduit à la définition donnée par Riemann et aux définitions des intégrales par excès et par défaut données par M. Darboux. Voyons la seconde» ([43] p. 1025–1026).

Si la fonction

$$f: \quad x \longmapsto f(x) = y$$
$$[a, b] \longrightarrow \mathbb{R}$$

prend des valeurs comprises entre m et M, pour $m = m_0 < m_1 < m_2 < \ldots < m_p = M$, Lebesgue considère les ensembles

$$E_0 = \{x \in [a, b] : f(x) = m\},$$
$$E_i = \{x \in [a, b] : m_{i-1} < f(x) \leq m_i\},$$

$i = 1, \ldots, p$. Un peu plus loin, il précisera la définition des mesures λ_0, λ_i de ces ensembles. Il examine les sommes

$$m_0\lambda_0 + \sum_{i=1}^{p} m_i\lambda_i,$$

$$m_0\lambda_0 + \sum_{i=1}^{p} m_{i-1}\lambda_i;$$

il écrit:

«Si, quand l'écart maximum entre deux m_i consécutifs tend vers zéro, ces sommes tendent vers une même limite indépendante des m_i choisis, cette limite sera par définition l'intégrale des y qui sera dite intégrable» ([43] p. 1026).

Toute partie E de $[a, b]$ peut être enfermée dans une réunion au plus dénombrable d'intervalles; la limite inférieure de la somme des longueurs de ces intervalles est la mesure de E. La partie E est dite mesurable si

$$\text{mes}\, E + \text{mes}([a, b]\backslash E) = \text{mes}[a, b].$$

Si on y ajoute

«des ensembles de mesures nulles convenablement choisis, on a des ensembles mesurables au sens de M. Borel» ([43] p. 1026).

Lebesgue note que, dans son langage, toute réunion dénombrable d'ensembles mesurables est mesurable; s'ils sont deux à deux disjoints, la mesure de la réunion est la somme des mesures des ensembles. Toute intersection dénombrable de parties mesurables est mesurable.

Une fonction f, bornée (en valeur absolue), intégrable sur toute partie mesurable $\{x \in [a, b] : \alpha < f(x) \leq \beta\}$ est dite sommable. Toute fonction intégrable au sens de Riemann est sommable car l'ensemble des points de discontinuité est de mesure nulle. Soient deux fonctions continues f et φ, la dernière n'étant pas nulle. Une fonction qui ne diffère de f qu'aux points d'un ensemble de mesure nulle, partout dense, et y vaut $f(x) + \varphi(x)$, est sommable, mais non intégrable au sens de Riemann. Lebesgue cite le cas de la fonction $1_{\mathbb{Q}}$.

L'ensemble des fonctions sommables est stable pour l'addition, la multiplication. Si une suite de fonctions sommables admet une limite, cette dernière est sommable. Ainsi, l'ensemble contient toutes les limites de polynômes, donc toutes les fonctions continues, puis toutes les limites de fonctions continues, c'est-à-dire les fonctions de première classe de Baire, puis toutes celles de deuxième classe, etc.

Lebesgue remarque enfin:

«En particulier, toute fonction dérivée, limitée supérieurement en valeur absolue, étant de première classe, est sommable, et l'on peut démontrer que son intégrale, considérée comme fonction de sa limite supérieure, est une de ses fonctions primitives» ([43] p. 1027).

Borel n'a pas rédigé explicitement tous ses travaux sur la mesure réalisés à partir de 1894. Dans sa thèse [44] Lebesgue fournit les détails.

Lebesgue introduit aussi de nouveaux concepts. Il nomme mesurable (B) tout borélien. Ces ensembles, définis par une infinité dénombrable de conditions, ont la puissance du continu. En hommage à Jordan, il appelle (J) tout ensemble sur lequel les mesures intérieure et extérieure coïncident.

L'ensemble fermé E des points

$$x = \frac{a_1}{3} + \frac{a_2}{3^2} + \frac{a_3}{3^3} + \dots$$

où $a_i \in \{0, 2\}$ pour $i = 1, 2, \dots$, est mesurable (B). Le complémentaire par rapport à $[0, 1]$ est formé des intervalles $]\frac{1}{3}, \frac{2}{3}[$, $]\frac{1}{9}, \frac{2}{9}[$, $]\frac{7}{9}, \frac{8}{9}[$, etc.; il est de mesure $\frac{1}{3} + 2 \cdot \frac{1}{3^2} + 4 \cdot \frac{1}{3^3} + \dots = 1$. Ainsi, E doit être de mesure nulle. Comme E a la puissance du continu, on peut à l'aide des points de E constituer une infinité d'ensembles de mesure extérieure nulle qui sont mesurables. La puissance de ces ensembles est celle de l'ensemble des sous-ensembles de $[0, 1]$. Lebesgue en conclut qu'il existe des ensembles mesurables (J) qui ne sont pas mesurables (B).

Dans un autre ordre d'idées, en 1903, Hadamard [34] considère l'espace vectoriel $\mathscr{C}([a, b])$ des fonctions numériques continues sur $[a, b]$. Une forme linéaire F sur $\mathscr{C}([a, b])$ est dite continue si quand (f_n) converge vers f uniformément dans $\mathscr{C}([a, b])$, on a $\lim_{n\to\infty} F(f_n) = F(f)$. On peut choisir une fonction φ telle que

$$f(x) = \lim_{n\to\pm\infty} n \int_a^b f(t)\varphi(n(t - x))dt$$

uniformément pour $x \in [a, b]$. Alors,

$$F(f) = \lim_{n\to\pm\infty} \int_a^b f(t)\Phi_n(t)dt,$$

où $\Phi_n(t) = F(g_n)$ avec $g_n : x \mapsto n\varphi(n(t - x))$.

L'ouvrage *Leçons sur l'intégration* [45] dont la première version date de 1903, réunit l'enseignement fait par Lebesgue au Cours Peccot du Collège de France en 1902 et 1903. Après avoir passé en revue les apports de Cauchy, Riemann, Darboux, Jordan, au chapitre VII il énonce son problème comme celui d'attacher à toute fonction f bornée définie sur $[a, b]$ un nombre fini $\int_a^b f(x)dx$. Cette intégrale doit être invariante par les translations, vérifier l'additivité, l'homogénéité, la positivité, et aussi être normalisée, i.e., $\int_0^1 dx = 1$. Enfin, Lebesgue ajoute une sixième condition dont il dit ne pas savoir si elle est indépendante des cinq précédentes, qui sont manifestement indépendantes entre elles: Si, quand l'indice n croît, $f_n(x)$ tend en croissant vers $f(x)$, l'intégrale de $f_n(x)$ tend vers celle de $f(x)$.

Soit f une fonction réelle, bornée, définie sur $[a, b]$. Lebesgue considère une subdivision $\inf f = \ell_0 < \ell_1 < \ldots < \ell_n = \sup f$, avec $\ell_{i+1} - \ell_i \leq \varepsilon, i = 0, 1, \ldots, n-1$. Pour $i = 0, 1, \ldots, n-1$, on pose

$$\begin{aligned} \varphi_i(x) &= 1 \quad \text{si } \ell_i \leq f(x) < \ell_{i+1}, \\ \varphi_i(x) &= 0 \quad \text{sinon,} \\ \psi_i(x) &= 1 \quad \text{si } \ell_{i-1} < f(x) \leq \ell_i, \\ \psi_i(x) &= 0 \quad \text{sinon.} \end{aligned}$$

Alors,

$$\varphi = \sum_{i=0}^{n} \ell_i \varphi_i \leq f \leq \sum_{i=0}^{n} \ell_i \psi_i = \psi.$$

Les fonctions φ et ψ tendent uniformément vers f quand ε tend vers 0. Lebesgue se rend compte qu'afin de pouvoir définir l'intégrale de f comme limite des intégrales de φ ou de ψ il faut que, si petit que soit ε, on puisse trouver les nombres ℓ_i tels que les fonctions φ_i, ψ_i soient associées à des ensembles mesurables.

Il suffit de savoir intégrer des fonctions caractéristiques, ne prenant que les valeurs 0 ou 1. Traduisant en *langage géométrique* le problème de l'intégration de ces fonctions particulières, Lebesgue arrive à un *problème de la mesure des ensembles*. Il se propose d'attacher à chaque partie bornée E de la droite la mesure de E, à savoir $m(E) \in \mathbb{R}_+$. Les conditions suivantes doivent être satisfaites: 1°) Deux ensembles qui peuvent être mis en coïncidence par le déplacement de l'un d'eux ont même mesure. 2°) La mesure d'une réunion finie ou dénombrable de parties deux à deux disjointes est la somme des mesures de ces parties. 3°) La mesure de l'intervalle unité vaut 1.

La véritable nouveauté que Lebesgue apporte par rapport à l'intégration de Riemann au tournant du siècle consiste dans le fait qu'au lieu de partionner le domaine de définition il partitionne celui des valeurs de la fonction.

Lebesgue dit que la fonction f, bornée ou non, est mesurable si, quels que soient $\alpha, \beta, \alpha < \beta$, l'ensemble $\{x \in [a, b] : \alpha < f(x) < \beta\}$ est mesurable. Alors, l'ensemble

$$\{x \in [a, b] : f(x) = \alpha\} = \lim_{h \to 0^+} \{x \in [a, b] : \alpha - h < f(x) < \alpha + h\}$$

est mesurable. La fonction f est mesurable si et seulement si pour tout α, $\{x \in [a, b] : \alpha < f(x)\}$ est mesurable ou encore $\{x \in [a, b] : \alpha \leq f(x)\}$ est mesurable. La classe des fonctions mesurables est stable pour les opérations algébriques ordinaires.

Là-dessus, Lebesgue introduit, pour une fonction mesurable bornée f, définie sur $[a, b]$ $(a < b)$,

$$\sum_{i=0}^{n-1} \ell_i m\{x \in [a, b] : \ell_i \leq f(x) < \ell_{i+1}\},$$

$$\sum_{i=1}^{n-1} \ell_i m\{x \in [a, b] : \ell_{i-1} < f(x) \leq \ell_i\},$$

m désignant la mesure. La limite commune de ces expressions quand ε tend vers 0 est $\int_a^b f(x)dx$. On pose encore

$$\int_b^a f(x)dx = -\int_a^b f(x)dx.$$

Lebesgue vérifie directement que ces définitions permettent de satisfaire aux six conditions posées au départ.

On ne peut s'empêcher de citer Borel disant en 1909 à propos de l'intégrale de Lebesgue:

«Elle est en quelque sorte faite sur mesure et s'adapte exactement aux propriétés de chaque fonction particulière» ([6] p. 1301–1302).

Lebesgue étend sa définition aux fonctions mesurables non bornées; la fonction mesurable f est dite sommable si $|\int_a^b f(x)dx| < +\infty$. De plus, l'intégrale peut se concevoir pour une fonction définie sur une partie mesurable arbitraire.

Un point central de [44] est le théorème sur le passage à la limite dans l'intégrale, le *théorème de Lebesgue*. Une variante de la dernière condition imposée à son intégrale par Lebesgue s'énonce ainsi: Si, quand l'indice n croît, la fonction $f_n(x)$, sommable dans un ensemble mesurable E, tend en croissant vers la fonction $f(x)$, sommable dans E, l'intégrale de $f_n(x)$ dans E tend en croissant vers celle de $f(x)$. Une conséquence en est le fait suivant: Une série convergente de fonctions positives, dont tous les termes et la somme sont sommables dans un ensemble E, est intégrable terme à terme dans E.

Lebesgue étudie l'*intégrale indéfinie* de la fonction mesurable f;

$$F(x) = \int_a^x f(t)dt + C,$$

C étant une constante. Si la fonction f est bornée, la fonction F est continue grâce au théorème des accroissements finis. Lebesgue obtient la même conclusion pour une fonction f sommable non bornée, par des approximations successives. On a

$$F(b) - F(a) = \int_a^b f(t)dt.$$

Lebesgue considère une fonction d'ensemble: Si E est une partie mesurable,

$$\Psi(E) = \int_E f(x)dx.$$

La fonction Ψ est *complètement additive*, i.e.,

$$\Psi\left(\bigcup_{n=1}^{\infty} E_n\right) = \sum_{n=1}^{\infty} \Psi(E_n)$$

pour une suite (E_n) de parties mesurables deux à deux disjointes. Dans la seconde édition, Lebesgue ajoutera ce commentaire:

«On peut distinguer, avec M. de la Vallée Poussin, le cas de l'additivité restreinte, dans lequel l'égalité précédente n'est assurée que si les E_i sont en nombre fini, et le cas de l'additivité complète, où l'égalité a lieu même s'il y a une infinité dénombrable de E_i. L'additivité complète sera seule importante pour nous. Une intégrale indéfinie est complètement additive» ([45] p. 143–144).

Lebesgue utilise l'importante définition suivante: La fonction d'ensemble Ψ est dite absolument continue si à tout $\varepsilon \in \mathbb{R}_+^*$ on peut faire correspondre $\eta \in \mathbb{R}_+^*$ tel que la condition $m(E) \leq \eta$ entraîne $|\Psi(E)| \leq \varepsilon$. Toute intégrale indéfinie possède cette propriété. Lebesgue peut alors prouver que deux fonctions f_1 et f_2 de même intégrale diffèrent tout au plus sur un ensemble de mesure nulle.

La dérivée f d'une fonction $\mathcal{F}$ est de première classe au sens de Baire; elle est donc mesurable. Si elle est aussi bornée, Lebesgue peut écrire la formule

$$\int_a^x f(t)dt = \lim_{h \to 0} \frac{\int_a^x \mathcal{F}(t+h) - \mathcal{F}(t)}{h} dt = \mathcal{F}(x) - \mathcal{F}(a),$$

la fonction $\mathcal{F}$ étant continue; il résout ainsi le *problème fondamental du calcul intégral*.

Lebesgue démontre aussi que pour qu'une fonction F soit l'intégrale indéfinie d'une fonction inconnue f il faut et il suffit que la fonction F soit absolument continue. Faisant allusion à [79], il écrit en 1926:

«Dans la première édition de ce livre, j'avais signalé cet énoncé, en note [...], de façon tout à fait incidente et sans démonstration. M. Vitali a retrouvé ce théorème et en a publié la première démonstration. C'est à l'occasion de ce théorème que M. Vitali a introduit, pour les fonctions d'une variable, la dénomination de fonction absolument continue et qu'il a montré la simplicité et la clarté que prend toute la théorie quand on met cette notion à sa base» ([45] p. 188).

Lebesgue forge encore les formulations suivantes: Pour qu'une fonction d'ensemble ou d'intervalle soit l'intégrale indéfinie d'une fonction inconnue, il faut et il suffit qu'elle soit complètement additive et absolument continue. Une intégrale indéfinie admet presque partout pour dérivée la fonction intégrée.

En 1905 Lebesgue rappelle que si sa

«définition de l'intégrale comprend celle de Riemann comme cas particulier lorsqu'il s'agit de fonctions bornées, il n'en est pas de même lorsqu'il s'agit de fonctions non bornées» ([46] p. 251).

Aussi, dans la théorie de Lebesgue, les fonctions f et $|f|$ sont-elles simultanément sommables ou non. Riemann a montré qu'une fonction admet une série de Fourier à coefficients convergeant vers 0 si elle est bornée et intégrable; pour Lebesgue, la fonction ne doit pas nécessairement être bornée.

Guidé par le travail de Fejér [28] sur la sommation d'une série de Fourier d'après le procédé de la moyenne arithmétique, Lebesgue prouve que de cette manière il récupère la fonction considérée, sauf peut-être aux points d'un ensemble de mesure nulle.

Borel a développé une théorie *constructive* de l'intégration. On peut relativiser le caractère polémique [24] des questions de parenté en citant Borel lui-même:

«Il n'est cependant pas inutile de faire observer que, si la théorie de l'intégrale de M. Lebesgue était loin d'être une conséquence évidente de ma théorie de la mesure, du moins entre ces deux théories, il n'y a pas de discontinuité logique, tandis qu'il y a un véritable fossé entre ma définition de la mesure et les définitions de la mesure et de l'intégrale qui étaient classiques et universellement admises, avec l'autorité des plus grands noms» ([8] p. 87).

A son tour, Lebesgue reconnaît souvent à quel point ses propres travaux sont redevables de ceux de Borel [29].

Dans la seconde édition de son traité rédigée en 1926, Lebesgue ajoute cette note concernant les ensembles mesurables:

«C'est seulement pour ces ensembles que nous étudierons le problème de la mesure. Je ne sais si l'on peut définir, ni même s'il existe d'autres ensembles que les ensembles mesurables; s'il en existe, ce qui est dit dans le texte ne suffit pas pour affirmer ni que le problème de la mesure est possible, ni qu'il est impossible pour ces ensembles [...]. Quant à la question de l'existence d'ensembles non mesurables, elle n'a guère fait de progrès depuis la première édition de ce livre. Toutefois cette existence est certaine pour ceux qui admettent un certain mode de raisonnement basé sur ce que l'on a appelé l'axiome de Zermelo. Par ce raisonnement, on arrive en effet à cette conclusion: il existe des ensembles non mesurables; mais cette affirmation ne devrait pas être considérée comme contredite si l'on arrivait à montrer que jamais aucun homme ne pourra nommer un ensemble non mesurable!» ([45] p. 114).

Lebesgue observe que l'*ensemble des ensembles mesurables contient l'ensemble des ensembles mesurables* J, mais que la classe est bien plus vaste puisqu'elle est stable pour les réunions et intersections dénombrables. D'autre part, la mesure de la réunion d'une famille croissante de parties mesurables est la limite des mesures de ces parties.

Vitali [78] prouve, via l'axiome du choix, l'existence d'un sous-ensemble, non mesurable au sens de Lebesgue. Plus généralement, il établit la non-existence d'une mesure positive μ, complètement additive, invariante par translations, sur toutes les parties bornées de la droite réelle, telle que $\mu([0,1]) = 1$. A cet effet, il suppose qu'il existe une telle mesure. Pour $x \in \mathbb{R}$, soit $A_x = \{x + b : b \in \mathbb{Q}\}$. Dans chaque $A_x \cap [0, \frac{1}{2}]$ distinct, on choisit p_x. Soit G_0 l'ensemble de ces points. Les parties $G_q = \{q_0 + q : q_0 \in G_0\}$ $(q \in \mathbb{Q})$, deux à deux disjointes, admettent toutes la même mesure et $\sum_{n=2}^{\infty} \mu(G_{1/n}) \leq \mu([0,1]) = 1$. Donc $\mu(G_q) = 0$ pour tout $q \in \mathbb{Q}$ et $\mu(\mathbb{R}) = 0$; on aboutit à une contradiction.

En 1906, Levi [50] illustre l'importance du concept d'intégrale de Lebesgue en l'adaptant au problème de Dirichlet, c'est-à-dire à la détermination d'une fonction u, définie sur un domaine Γ et son contour C, coïncidant sur C avec une fonction donnée, continue sur $\Gamma \cup C$, dérivable sur Γ, et pour laquelle l'expression

$$\iint_{\Gamma} \Big(\big(\frac{\partial u}{\partial x}\big)^2 + \big(\frac{\partial u}{\partial y}\big)^2 \Big) dx\, dy$$

est minimale.

En 1906 aussi, Fatou [27] observe que Lebesgue a démontré que si une série trigonométrique converge et est représentée par une fonction bornée, les coefficients en sont donnés par la *formule d'Euler-Fourier*; d'autre part, il existe des fonctions bornées, non intégrables au sens de Riemann, qui peuvent être représentées par une série trigonométrique convergente. Fatou prouve alors que si une fonction harmonique régulière est bornée à l'intérieur d'un cercle, elle s'exprime par une intégrale de Poisson, prise au sens de Lebesgue, i.e., de la forme

$$F(r,\theta) = \frac{1}{2\pi} \int_{-\pi}^{\pi} \frac{1 - r^2}{1 - 2r\cos(\theta - u) + r^2} f(u) du.$$

Dans le cours de son travail, Fatou établit le résultat qui sera connu sous le nom de *lemme de Fatou*. Soit (f_n) une suite de fonctions positives, bornées, sommables, tendant vers une fonction f, bornée ou non, telle que

$$\int_a^b f_n(x) dx \leq M$$

pour tout n, alors la fonction f est intégrable et

$$\int_a^b f(x) dx \leq \liminf_{n \to \infty} \int_a^b f_n(x) dx.$$

En effet, la fonction f étant mesurable, on considère l'ensemble E des $x \in [a,b]$ pour lesquels $f(x) \leq L$; on définit

$$\varphi_n(x) = f_n(x) \quad \text{si} \quad f(x) \leq L,$$
$$\varphi_n(x) = f(x) \quad \text{si} \quad f(x) > L.$$

P. Fatou

La suite (φ_n) est bornée sur E et tend vers f; donc

$$\lim_{n\to\infty} \int_E \varphi_n(x)dx = \int_E f(x)dx.$$

Comme $\varphi_n \leq f_n$ sur E pour tout n, on en déduit que

$$\int_E f(x)dx \leq \liminf_{n\to\infty} \int_E f_n(x)dx \leq \liminf_{n\to\infty} \int_a^b f_n(x)dx.$$

Si on fait tendre L vers $+\infty$, on obtient bien

$$\int_a^b f(x)dx \leq \liminf_{n\to\infty} \int_a^b f_n(x)dx.$$

Ce résultat demeure valable si sur un ensemble de mesure nulle, la suite (f_n) n'admet pas de limite ou a une limite infinie.

Fatou ajoute que Lebesgue lui a fait remarquer que l'inégalité établie peut être stricte. Riesz écrira plus tard à propos de son travail sur les relations orthogonales dans l'espace hilbertien:

«L'idée et le courage d'essayer d'appliquer [la] notion [de l'intégrale de Lebesgue] à des problèmes dont je m'occupais, me sont venus en lisant, en 1906, l'excellent Mémoire de Fatou [...]» ([66] p. 29), en particulier le lemme de Fatou.

Le premier énoncé du théorème de Fubini, et ce dans le cadre de la théorie de l'intégration de Lebesgue, est formulé ainsi:

«Se $f(x,y)$ è una funzione di due variabili x, y, limitata o illimitata, integrabile in un'area Γ del piano (x,y), allora si ha sempre:

$$\int_\Gamma f(x,y)d\sigma = \int dy \int f(x,y)dx = \int dx \int f(x,y)dy,$$

quando con $d\sigma$ si intenda l'elemento d'area di Γ» ([31] p. 608).

La partie Γ du plan est supposée être *linéairement mesurable*, i.e., toute intersection de Γ avec une parallèle à un axe de coordonnées est mesurable. Une fonction f est dite *linéairement mesurable* si toutes les fonctions $x \mapsto f(x,y)$ et $y \mapsto f(x,y)$ sont mesurables.

Si f est mesurable, il existe une fonction φ mesurable et linéairement mesurable telle que l'ensemble des points en lesquels f et φ diffèrent soit contenu dans un ensemble E de mesure nulle, linéairement mesurable, et les coordonnées $x = c$, $y = d$ correspondant à des droites coupant E en un ensemble de mesure non nulle forment un ensemble de mesure nulle.

Au cas où la fonction f est bornée on obtient ainsi

$$\int_\Gamma f d\sigma = \int_\Gamma \varphi d\sigma = \int dy\Big(\int \varphi dx\Big) = \int dx\Big(\int \varphi dy\Big),$$

d'où on déduit le résultat. Par des approximations successives, Fubini étend sa formule aux fonctions non bornées. Il l'explicite aussi en coordonnées polaires.

En 1909, au tome 1 de la troisième édition de son Cours d'analyse, Jordan définit encore les mesures extérieure et intérieure d'une manière analogue à celle de Peano; il fait usage de carrés à côtés parallèles aux axes. Si E est un ensemble fermé borné de $\mathbb{R}^2$, Jordan dit que l'ensemble des carrés, à côtés parallèles aux axes, intérieurs à E, forment un domaine S; l'ensemble de tels carrés intérieurs à E ou rencontrant la frontière de E forment un nouveau domaine $S \cup S'$ auquel E est intérieur. Jordan explique que les aires de ces domaines tendent vers des *limites fixes* si les carrés sont pris de plus en plus petits; ce sont l'*aire intérieure* et l'*aire extérieure* de E. Si elles coïncident, l'aire E est dite quarrable; la valeur commune est l'aire de E.

Jordan ajoute:

«Les considérations qui précèdent sont évidemment applicables aux ensembles d'un nombre quelconque de dimensions. On pourra déterminer, pour chacun d'eux, une étendue intérieure et une étendue extérieure. Si elles coïncident, l'ensemble sera mesurable» ([39] p. 31).

Evoquant la vie et les travaux de Jordan, Lebesgue écrira:

«Si le Traité de Jordan est riche d'innovations, on y trouve aussi les trésors du passé et même ses souvenirs; il conserve la division surannée en calcul différentiel et calcul intégral; mais, comme ses réflexions lui ont fait reconnaître en l'intégrale la plus simple, la plus intuitive, la plus primitive de toutes les notions de l'analyse, il commence l'exposé du calcul différentiel par la définition de l'intégrale» ([48] p. *ℓxi*).

Le livre de Jordan se veut en grande partie un exposé rigoureux de la théorie de l'intégration et de la mesure dans $\mathbb{R}^2$. Cependant, Bourbaki peut dire qu'on atteint ainsi

«l'étape finale du calcul infinitésimal classique, celle qui est représentée par les grands Traités d'Analyse de la fin du XIXe siècle; [...] celui de Jordan occupe parmi eux une place éminente, pour des raisons esthétiques d'une part, mais aussi parce que, s'il constitue une admirable mise au point des résultats de l'analyse classique, il annonce à bien des égards l'analyse moderne et lui prépare la voie. Après Jordan vient Lebesgue [...] ([11] p. III.66).

Dans le cadre de son étude, motivée par la théorie des équations intégrales de Hilbert et de Schmidt, Hellinger [37] rappelle que si f constitue une fonction à variation bornée au sens de Jordan, l'ensemble des nombres

$$\sum_{i=1}^{n} |\Delta_i f| = \sum_{i=1}^{n} |f(x_i) - f(x_{i-1})|$$

est borné. Si, de plus, la fonction f est continue, d'après les travaux de Lebesgue, quand la longueur maximale des longueurs des subdivisions sur $[a, b]$ converge vers 0, on obtient une limite

$$\int_a^b |df(x)|,$$

la *variation totale* de f sur $[a, b]$. Plus généralement, Hellinger considère une fonction monotone, continue, bornée g. Si pour la fonction f, l'ensemble des nombres

$$\sum_{i=1}^{n} \frac{(\Delta_i f)^2}{\Delta_i g}$$

est borné, on peut déterminer de même la valeur limite

$$\int_a^b \frac{df(x)^2}{dg(x)},$$

l'*intégrale de Hellinger*.

En 1909 aussi, Riesz [62], poursuivant les investigations entreprises par Hadamard, nomme fonctionnelle linéaire sur $\mathscr{C}([0,1])$ toute forme linéaire A sur cet espace vectoriel telle que si (f_i) converge uniformément vers f dans $\mathscr{C}([0,1])$, alors $(A(f_i))$ converge aussi uniformément vers $A(f)$. Comme exemple il peut signaler l'application $f \mapsto \int_0^1 f(x)d\alpha(x)$, où α est une fonction à variation bornée. L'intégrale s'interprète comme une limite de sommes $\sum\limits_i f(\xi_i)(\alpha(x_{i+1}) - \alpha(x_i))$ correspondant à des subdivisions de [0,1] en un nombre fini d'intervalles partiels $[x_i, x_{i+1}], \xi_i \in [x_i, x_{i+1}]$; la longueur des intervalles tend uniformément vers 0.

En vue de la démonstration de la réciproque, Riesz se donne une telle forme linéaire A; on pose

$$F^{(\xi)}(x) = x \quad \text{pour} \quad 0 \leq x \leq \xi,$$
$$F^{(\xi)}(x) = \xi \quad \text{pour} \quad \xi \leq x \leq 1.$$

Les *nombres dérivés* de la fonction $\alpha : \xi \mapsto A(F^{(\xi)})$ constituent des fonctions à variation bornée. La forme initiale du théorème de Riesz est la suivante:

«Etant donnée l'opération $A[f(x)]$, on peut déterminer la fonction à variation bornée $\alpha(x)$, telle que, quelle que soit la fonction continue $f(x)$, on ait

$$A[f(x)] = \int_0^1 f(x)d\alpha(x)»$$

([62] p. 976).

Gray peut faire cette observation pertinente:

«Only rarely does the mathematical community pay a theorem the accolade of transforming it into a tautology. The Riesz representation theorem has received this accolade» ([32] p. 127).

En 1910, Riesz [63] étudie le problème de la résolution de l'équation

$$\int_a^b f(x)\xi(x)dx = c,$$

ξ étant la fonction inconnue; l'intégrale s'entend au sens de Lebesgue. Riesz fournit les énoncés généraux de l'*inégalité de Hölder* et de l'*inégalité de Minkowski*. Il note $[L^p]$ la classe des fonctions f pour lesquelles $|f|^p$ est intégrable.

Dès 1910 Lebesgue [47] se met à généraliser ses résultats au cas $\mathbb{R}^n$.

En 1911, Young [83] décrit une nouvelle approche de la théorie de l'intégration. Sa méthode se différencie de celle de Lebesgue au départ en ce qu'il interprète un ensemble ouvert comme la *limite supérieure* des parties fermées. Young dit alors:

«It follows that the generalised lower integral is the upper bound of the integrals of upper-semi-continuous functions not greater than the function, in the case in which the function is positive (i.e., ≥ 0) and bounded. It at once follows, by subtracting the function from a constant function, that the upper generalised integral is the lower bound of the integrals of lower semi-continuous functions not less than the function» ([83] p. 16).

Toute limite d'une suite croissante [resp. décroissante] de fonctions semi-continues inférieurement [resp. supérieurement] est encore semi-continue inférieurement [resp. supérieurement]; toute fonction semi-continue inférieurement [resp. supérieurement] minorée [resp. majorée] est la limite d'une suite monotone de fonctions continues.

Par définition, l'intégrale généralisée supérieure d'une fonction bornée f est la limite inférieure des intégrales inférieures d'une suite décroissante de fonctions semi-continues inférieurement, minorée par f. L'intégrale généralisée inférieure s'introduit d'une manière corrélative. La fonction f est dite sommable si les deux intégrales généralisées coïncident. Ces définitions sont étendues aux fonctions non bornées.

Young explicite de nombreux résultats supplémentaires; par exemple, si f, g sont des fonctions sommables, bornées ou non,

$$\left(\int_0^c fg\,dx\right)^2 \leq \int_0^c f^2 dx \int_0^c g^2 dx$$

$(c \geq 0)$.

Soit (f_n) une suite de fonctions mesurables sur $[a, b]$, convergeant vers f en mesure m, au sens de Borel: Pour tout $\varepsilon \in \mathbb{R}_+^*$, $\lim_{n\to\infty} m(\{x \in [a, b] : |f(x) - f_n(x)| \geq \varepsilon\}) = 0$. Adoptant une idée apparue chez Weyl, en 1911 Egorov [26] considère, pour $x \in [a, b]$ et $n \in \mathbb{N}^*$,

$$r_n(x) = \limsup_{p\to\infty}\big(|f(x) - f_{n+p}(x)|\big);$$

soit aussi

$$\mu(n, \varepsilon) = m(\{x \in [a, b] : r_n(x) \geq \varepsilon\}).$$

Egorov se donne maintenant une suite (ε_q) décroissant vers 0 et une série à termes strictement positifs $\Sigma\eta_q$ convergente. Quel que soit $q \in \mathbb{N}^*$, il existe $n_q \in \mathbb{N}^*$ tel que $\mu(n_q, \varepsilon_q) \leq \eta_q$. Ainsi, Egorov peut établir qu'en dehors d'un sous-ensemble de $[a, b]$, de mesure arbitrairement petite, la suite (f_n) converge uniformément vers f.

Le problème décrivant la correspondance entre les intégrales et les primitives reçoit une solution dans la théorie compliquée de la *totalisation* mise au point par Denjoy à partir de 1912 [21]. L'auteur fait remarquer d'abord que l'intégrale de Riemann a un sens pour toute fonction continue et toute fonction dont les points de discontinuité constituent un ensemble de mesure nulle, alors que l'intégrale de Lebesgue concerne toute fonction mesurable bornée. L'une et l'autre, définies entre le point fixe a et la variable x, pour une fonction f, sont des fonctions continues de x, et leur dérivée est f, sauf sur un ensemble de mesure nulle au plus. Par contre, une fonction dérivée n'est pas nécessairement intégrable selon l'un des deux modes. Denjoy développe une théorie s'appliquant à toute fonction dérivée f et conduisant à une fonction continue dont f est la dérivée.

Une fonction est dite sommable en un point si elle l'est sur un intervalle admettant ce point en son intérieur. La théorie s'échaffaude de la manière suivante: a) Si f est une fonction sommable sur un intervalle I, la totalisation $V(I)$ de f est l'intégrale de Lebesgue sur cet intervalle. b) La totalisation étant définie pour une suite finie d'intervalles juxtaposés, la somme en est la totalisation sur la réunion. c) Soit f une fonction sommable sur une partie parfaite P de $[a,b]$. Pour tout intervalle J *contigu* à P, on considère l'ensemble des intervalles I contenus dans J n'ayant aucun point de P en leur intérieur et $W(J) = \sup_I |V(I)|$. Si la série de terme général $W(J)$ converge, on pose

$$V(a,b) = \int_P f + \sum_J V(J),$$

$V(J)$ étant la totalisation de f entre les bornes de J.

Maintenant, la fonction f est dite totalisable dans l'intervalle $[a,b]$ si elle satisfait aux conditions suivantes:

1°) Quel que soit le sous-ensemble parfait P de $[a,b]$, l'ensemble des points de P où f n'est pas sommable, n'est pas dense dans P.

2°) Si $V(c',d')$ a été calculé pour tout intervalle $[c',d']$ intérieur à $[c,d]$, $V(c',d')$ tend vers une limite quand c' tend vers c et d' tend vers d. On pose, par définition,
$$V(c,d) = \lim V(c',d').$$

3°) Quel que soit l'ensemble parfait P, l'ensemble des points où la série considérée dans $c)$ n'est pas convergente, constitue une partie non dense dans P.

Ainsi, $V(a,b)$ se calcule toujours à l'aide d'une infinité dénombrable d'intégrations de Lebesgue. Plus tard, on parlera de l'intégrale *spéciale* de Denjoy.

Lusin [55], guidé par le fait que l'intégrale de Lebesgue est liée à la notion de fonction à variation bornée, associe à l'intégrale de Denjoy une fonction *à variation bornée généralisée*. Soit F une fonction continue sur $[a,b]$ et soit P un sous-ensemble parfait de $[a,b]$. On considère une famille Σ d'intervalles $\Delta_1, \dots, \Delta_N$, deux à deux disjoints, telle que $P \subset \bigcup_{i=1}^{N} \Delta_i$ et $P \cap \Delta_i^0 \neq \phi$ pour tout $i = 1, \dots, N$. Si M_i [resp. m_i] est le maximum [resp. minimum] de F sur Δ_i ($i = 1, \dots, N$), on pose

$$\nu_\Sigma = \sum_{i=1}^{N} (M_i - m_i).$$

La fonction F est dite *à variation bornée* pour P s'il existe $k \in \mathbb{R}_+$ tel que $\nu_\Sigma < k$ quel que soit le choix de Σ. On obtient une limite μ_P pour ν_Σ quand la mesure de $\bigcup_{i=1}^{N} \Delta_i$ se rapproche de celle de P et la longueur maximale des Δ_i tend vers 0; μ_P constitue la *variation totale* de F pour P.

Une fonction F, continue sur $[a, b]$, est dite à variation bornée généralisée si quel que soit le sous-ensemble parfait P de $[a, b]$, il existe un intervalle Δ tel que $P_\Delta = P \cap \Delta$ soit parfait et la fonction F soit à variation bornée pour P_Δ. Lusin prouve que la fonction continue F, à variation bornée généralisée, est l'intégrale indéfinie de sa dérivée au sens de Denjoy si et seulement si $\mu_P = 0$ pour tout ensemble parfait P de mesure nulle.

Au même moment Lusin [54] résume des résultats publiés déjà un an auparavant à Moscou. Partant du théorème d'Egorov, il dit que la fonction φ, définie sur [0,1], possède la (C)-propriété si pour tout $\delta \in \mathbb{R}_+^*$ il existe un sous-ensemble parfait non dense P dans [0,1] tel que mes $P > 1 - \delta$ et la fonction $\varphi|_P$ soit continue. Soit maintenant (f_n) une suite de fonctions possédant la propriété (C), convergeant vers la fonction f, sauf éventuellement sur une partie de mesure nulle. Soit $\varepsilon \in \mathbb{R}_+^*$. D'après le théorème d'Egorov, on peut déterminer un ensemble P_0 parfait, non dense tel que mes $P_0 > 1 - \frac{\varepsilon}{2^2}$ et la convergence soit uniforme sur P_0. Pour tout $n \in \mathbb{N}^*$, l'hypothèse permet de considérer P_n avec mes $P_n > 1 - \frac{\varepsilon}{2^{n+2}}$. Alors, la fonction f admet la (C)-propriété pour $P = \bigcup_{n=0}^{\infty} P_n$; mes $P > 1 - \sum_{n=0}^{\infty} \frac{\varepsilon}{2^{n+2}} = 1 - \varepsilon$. Lusin vérifie ainsi que la (C)-propriété est satisfaite par toutes les fonctions de la classification de Baire. Plus généralement, la (C)-propriété équivaut à la mesurabilité.

Lusin peut énoncer encore que si f est une fonction mesurable sur [0,1], prenant des valeurs finies, sauf éventuellement sur un ensemble de mesure nulle, il existe une fonction F telle que $F' = f$ sur un ensemble de mesure 1. Il conclut:

«On peut ainsi dire que le problème de la recherche des fonctions primitives est résolu quand on néglige les ensembles de mesure nulle» ([54] p. 1690).

Riesz estime que le raisonnement de Borel repose sur la notion de convergence en mesure. Il préfère remplacer les polynômes par des fonctions simples, i.e., des fonctions f n'admettant qu'un nombre fini de points de discontinuité et en un tel point x il choisit *pour fixer les idées*

$$f(x) = \frac{1}{2}(f(x^+) + f(x^-)).$$

Il écrit:

«Pour ces fonctions, le problème d'intégration devient trivial. De plus, leur emploi est conforme soit aux idées de Riemann, soit à celles de M. Lebesgue» ([64] p. 641).

Riesz ne veut pas faire intervenir des ensembles mesurables généraux; il considère uniquement un ensemble de mesure nulle, i.e., un ensemble qui peut être enfermé dans une réunion finie ou dénombrable d'intervalles de longueur totale arbitrairement petite. Il envisage des fonctions bornées. Une suite (f_n) de fonctions définies sur $[a, b]$ est dite tendre vers la fonction f d'une façon simple si 1) les fonctions (f_n) sont uniformément bornées, 2) (f_n) converge vers f, sauf éventuellement sur une partie de mesure nulle. Si une suite (f_n) de fonctions simples converge simplement vers la fonction f sur $[a, b]$, cette dernière est dite sommable et on pose

$$\int_a^b f(x)dx = \lim_{n\to\infty} \int_a^b f_n(x)dx.$$

La définition est justifiée grâce aux démonstrations déjà connues dans des situations plus générales. Ces fonctions sommables sont précisément les fonctions sommables au sens de Lebesgue. D'un autre côté, toute fonction intégrable au sens de Riemann est la limite simple de fonctions simples.

Borel revient sur les procédés d'intégration. Il considère l'*intégrale élémentaire* d'un polynôme P, à savoir l'expression

$$\int_a^x \int_b^y \int_c^z P(x, y, z)dx\,dy\,dz.$$

Si une suite (P_n) de polynômes converge *asymptotiquement* vers la fonction bornée f dans un domaine délimité par un parallélipipède rectangle, et est uniformément bornée, la limite des intégrales correspondantes est, par définition, l'intégrale de f. Entre la présentation et la publication de son travail paraissent les résultats de Riesz [64] et de Lusin [54]. Borel ajoute en note:

«Ce qui me paraît distinguer le point de vue de ces auteurs de celui que j'ai adopté, c'est que la définition de ce que j'appelle la convergence asymptotique des suites de polynomes ne repose que sur la mesure des domaines algébriques, dans le cas d'une variable, domaines formés d'un nombre fini d'intervalles» ([7] p. 197).

Faisant allusion aux travaux de Denjoy [21], il écrit:

«Je signale en corrigeant ces épreuves, deux Notes fort intéressantes de M. Denjoy, où ce jeune savant indique comment on peut résoudre d'une manière complète le problème inverse de la dérivation» ([7] p. 198).

En 1913, Radon se propose de généraliser la théorie des équations intégrales linéaires et celle des formes linéaires et bilinéaires au cas de plusieurs variables, prolongeant les travaux de Hilbert, Riesz, Hellinger. Il écrit:

«Will man [...] die angedeutete Theorie in möglichster Allgemeinheit aufbauen, so erscheint die Entwicklung einer Theorie der absolut additiven Mengenfunktionen als unabweisliche Vorarbeit. Das in dieser Richtung bisher Geleistete besteht hauptsächtlich in einer Arbeit von Lebesgue, der sich aber auf 'absolut stetige' Mengenfunktionen einerseits und 'fonctions d'intervalle' andrerseits beschränkt. Das erstere ist für den vorliegenden Zweck zu speziell, das zweite zu wenig schmiegsam» ([61] p. 1295–1296).

Dans $\mathbb{R}^n$, Radon considère les pavés, dits intervalles, qui sont contenus dans $J = [-M, M] \times \ldots \times [-M, M]$ $(M > 0)$. Il s'intéresse à une classe $\mathcal{J}$ de parties de $\mathbb{R}^n$ contenant tous ces pavés telle que $E_1 \cap E_2 \in \mathcal{J}$, $E_1 \backslash E_2 \in \mathcal{J}$ si $E_1, E_2 \in \mathcal{J}$, et $\bigcup_{m=1}^{\infty} E_m \in J$ pour toute suite (E_m) de $\mathcal{J}$; cette classe comprend les boréliens. Une fonction $f : \mathcal{J} \to \mathbb{R}$ est dite vérifier l'*additivité absolue* si, pour toute suite (E_m) de $\mathcal{J}$ constituée de parties deux à deux disjointes, on a

$$f(\bigcup_{m=1}^{\infty} E_m) = \sum_{m=1}^{\infty} f(E_m).$$

Une telle fonction est à variation bornée: Pour tout $E \in \mathcal{J}$, il existe $N \in \mathbb{R}_+^*$ tel que pour toute décomposition finie $(E_p)_{p=1,\ldots,k}$ de E formée d'éléments de $\mathcal{J}$ on ait

$$\sum_{p=1}^{k} |f(E_p)| < N.$$

Pour simplifier, Radon décrit sa procédure dans $\mathbb{R}^2$. Se donner f sur les pavés contenus dans J, c'est se donner une fonction $f : \mathbb{R}^2 \to \mathbb{R}$ vérifiant les propriétés suivantes:

$$F(a', b') - F(a', b) - F(a, b') + F(a, b) \geq 0$$

si $-M \leq a \leq a' \leq M$, $-M \leq b \leq b' \leq M$.

$$\lim_{\substack{\delta \to 0^+ \\ \varepsilon \to 0^+}} F(a - \delta, b - \varepsilon) = F(a, b)$$

si $-M < a \leq M$, $-M < b \leq M$.

Soit E une partie quelconque de J. On note $\bar{f}(E)$ la limite inférieure de l'ensemble des nombres $\sum_{m=1}^{\infty} f(I_m)$, où (I_m) est une suite d'intervalles dont la réunion contient E. On pose encore

$$\underline{f}(E) = f(J) - \bar{f}(J \backslash E).$$

Alors, $\underline{f}(E) \leq \bar{f}(E)$. Aussi

$$\bar{f}(A \cup B) \leq \bar{f}(A) + \bar{f}(B)$$

quelles que soient les parties A et B de J.

Pour $\mathcal{J}$ Radon choisit maintenant l'ensemble des parties E de J pour lesquelles $\underline{f}(E) = \bar{f}(E)$. Quels que soient la partie E de J et $\varepsilon \in \mathbb{R}_+^*$, il existe une partie fermée E_1 telle que $E_1 \subset E$ et $\underline{f}(E) \leq f(E_1) + \varepsilon$.

Radon annonce que toute la théorie de l'intégrale de Lebesgue et toute celle de l'intégrale de Lebesgue-Stieltjes peuvent se déployer dans ce cadre général. Il fait l'impasse des démonstrations qui se transposent ne varietur du travail de Lebesgue; il renvoie ses lecteurs au Cours d'analyse de de la Vallée Poussin.

Dans la seconde édition de son traité, Lebesgue rend cet hommage:

«On n'a pénétré vraiment au fond de cette notion [d'intégrale de Stieltjès] que grâce à la définition qu'en a donnée M. Radon et aux travaux de M. de la Vallée Poussin sur l'extension de la notion de mesure [...].

Si, en ce qui concerne les fonctions d'une seule variable, le Mémoire de M. Radon ne nous a apporté qu'une définition nouvelle, particulièrement heureuse en vérité, de l'intégrale de Stieltjès, pour le cas de plusieurs variables, ce Mémoire fournit une véritable extension de la notion d'intégrale» ([45] p. 263).

En 1914, Perron présente une nouvelle théorie de l'intégration qu'il introduit en ces termes:

«Im folgenden stelle ich eine Definition des bestimmten Integrals zur Diskussion, die, wie ich zeigen werde, mindestens so weittragend ist, wie die Lebesguesche. Sie ist aber viel elementarer, und auch der Nachweis der fundamentalen Gesetze gestaltet sich viel einfacher» ([59] p. 3).

Soit θ une fonction réelle continue sur $[a, b]$; on pose

$$\bar{D}\theta(x) = \limsup_{h \to 0} \frac{\theta(x+h) - \theta(x)}{h} \in \overline{\mathbb{R}},$$

$$\underline{D}\theta(x) = \liminf_{h \to 0} \frac{\theta(x+h) - \theta(x)}{h} \in \overline{\mathbb{R}}.$$

La fonction réelle bornée f est intégrable au sens de Perron sur $[a, b]$ si pour tout $\varepsilon \in \mathbb{R}_+^*$, il existe des fonctions $\varphi_\varepsilon, \psi_\varepsilon$ continues sur $[a, b]$ telles que $\varphi_\varepsilon(a) = \psi_\varepsilon(a) = 0$, $\psi_\varepsilon(b) - \varphi_\varepsilon(b) < \varepsilon$ et, pour tout $x \in [a, b]$, $\bar{D}\varphi_\varepsilon(x) \leq f(x) \leq \underline{D}\psi_\varepsilon(x)$. Alors, l'intégrale de Perron est

$$(\mathcal{P}) \int_a^b f(t)dt = \inf_\varepsilon \psi_\varepsilon(b) = \sup_\varepsilon \varphi_\varepsilon(b).$$

Perron établit les propriétés classiques pour son intégrale, notamment l'additivité et la positivité. Si f est intégrable au sens de Perron, on pose

$$F(y) = (\mathcal{P}) \int_a^y f(x)dx$$

pour $y \in [a, b]$; F est une fonction continue. Si l'application f est continue en y_0, alors $F'(y_0) = f(y_0)$. Si la suite (f_n) de fonctions intégrables au sens de Perron converge uniformément vers f, alors

$$(\mathcal{P})\int_a^b f(x)dx = \lim_{n\to\infty}(\mathcal{P})\int_a^b f_n(x)dx.$$

Perron démontre que toute fonction f, intégrable au sens de Lebesgue sur $[a,b]$, l'est au sens de la nouvelle définition. Il signale que la preuve a en fait déjà été donnée par de la Vallée Poussin dans la troisième édition de [74]. Si $m < f(x) < M$, on considère $m = m_0 < m_1 < \ldots < m_{n-1} < m_n = M$. Les parties $E_i = \{x \in [a,b] : m_i \leq f(x) < m_{i+1}\}$ $(i = 0,1,\ldots,n-1)$ sont mesurables puisque la fonction f est intégrable au sens de Lebesgue; soit L_i la mesure de E_i. Alors, $\sum_{i=0}^{n-1} L_i = b-a$ et si $\varepsilon \in \mathbb{R}_+^*$, d'après la définition de l'intégrale de Lebesgue $(\mathcal{L})$,

$$\sum_{i=0}^{n-1} m_{i+1}L_i < (\mathcal{L})\int_a^b f(x)dx + \varepsilon$$

pour une subdivision suffisamment fine de $[m,M]$. La partie E_i se laisse recouvrir par une réunion dénombrable R d'intervalles deux à deux disjoints dont la longueur totale est majorée strictement par $L_i + \frac{\varepsilon}{n}$. Soit $\mu_i(x)$ la mesure de $R \cap [a,x]$. Alors, $\mu_i(a) = 0$, $\mu_i(b) < L_i + \frac{\varepsilon}{n}$. Si $x \in E_i$, $\underline{D}\mu_i(x) = \bar{D}\mu_i(x) = 1$. Posons maintenant

$$\psi(x) = m_0(x-a) + \sum_{i=0}^{n-1}(m_{i+1} - m_0)\mu_i(x);$$

alors $\psi(a) = 0$ et si $x \in E_i$,

$$\underline{D}\psi(x) \geq m_0 + (m_{i+1} - m_0) = m_{i+1} > f(x).$$

Par ailleurs,

$$\begin{aligned}\psi(b) &= m_0(b-a) + \sum_{i=0}^{n-1}(m_{i+1} - m_0)\mu_i(b)\\ &< m_0(b-a) + \sum_{i=0}^{n-1}(m_{i+1} - m_0)(L_i + \frac{\varepsilon}{n})\\ &= \sum_{i=0}^{n-1} m_{i+1}L_i + \sum_{i=0}^{n-1}(m_{i+1} - m_0)\frac{\varepsilon}{n}\\ &< (\mathcal{L})\int_a^b f(x)dx + (M-m)\varepsilon + \varepsilon.\end{aligned}$$

C. Carathéodory

On procède de même à une minoration de f. On détermine une fonction φ telle que

$$(\mathcal{L}) \int_a^b f(x)dx - \varepsilon - (M - m)\varepsilon < \varphi(b).$$

La première version des résultats de Carathéodory constitue l'objet d'une longue communication adressée à l'Académie des sciences de Göttingen en 1914. Voici l'exposé de son plan:

«Ich habe es [...] für zweckmässig gehalten, meine Darstellung mit einer rein formalen Theorie der Messbarkeit zu beginnen. Dabei wird eine Definition der

Messbarkeit zu Grunde gelegt, die einerseits allgemeiner ist, als die gewöhnliche, weil sie sich auch auf Punktmengen von unendlichem äusserem Maße erstreckt, andrerseits aber scheinbar viel enger. Diese Definition ist aber viel bequemer als die ältere: sie erlaubt sämtliche in Betracht kommenden Sätze ohne tiefliegende Kunstgriffe zu beweisen; und sie ist der gewöhnlichen Definition vollständig äquivalent» ([16] p. 404–405).

Carathéodory introduit sur $\mathbb{R}^q$ la mesure extérieure qui est définie au moyen de cinq axiomes. (I) La fonction μ^* associe à toute partie de $\mathbb{R}^q$ une valeur dans $\bar{\mathbb{R}}_+$. (II) Si $B \subset A \subset \mathbb{R}^q$, alors $\mu^*(B) \leq \mu^*(A)$. (III) Si (A_n) est une suite finie ou dénombrable de parties de $\mathbb{R}^q$, on a $\mu^*(\bigcup_n A_n) \leq \sum_n \mu^*(A_n)$. La partie A est dite mesurable si, pour toute partie W,

$$\mu^*(W) = \mu^*(A \cap W) + \mu^*(W \backslash (A \cap W)).$$

Cette classe possède les propriétés de stabilité usuelles. S'ajoutent maintenant deux conditions supplémentaires: (IV) Si $A_1, A_2 \subset \mathbb{R}^q$ et $\inf\{d(x,y) : x \in A_1, y \in A_2\} > 0$, d désignant la distance dans $\mathbb{R}^q$, alors

$$\mu^*(A_1 \cup A_2) = \mu^*(A_1) + \mu^*(A_2).$$

(V) La mesure extérieure d'une partie A est la limite inférieure des nombres $\mu^*(B)$, B décrivant l'ensemble des parties mesurables contenant A. La mesure intérieure A est définie par

$$\mu_*(A) = \mu^*(A) - \mu^*(B \backslash A)$$

où B est une partie mesurable contenant A telle que $\mu^*(B) = \mu^*(A)$. La partie A est mesurable si et seulement si $\mu_*(A) = \mu^*(A)$, valeur notée $\mu(A)$.

En 1915, de la Vallée Poussin [76] rédige un exposé complet de la théorie de l'intégrale de Lebesgue; c'est une part de ses cours professés à Harvard. Il insiste, en particulier, sur l'importance de la notion de continuité absolue. Si f est une fonction continue sur $[a,b]$ et $\omega = [\alpha, \beta] \subset [a,b]$, en posant $F(\omega) = f(\beta) - f(\alpha)$ on définit une *fonction d'intervalle*. La fonction f est dite absolument continue si $F(\omega)$ tend vers 0 avec mes ω.

Hahn [35] apporte des compléments à la méthode d'intégration selon Riemann, mise au point par Borel [7]; il la considère comme une généralisation heureuse d'une procédure due à Harnack. Sa propre définition est un peu plus générale que celle de Borel; alors que Borel écarte une suite de points singuliers, Hahn se place à l'extérieur d'un ensemble de mesure nulle. Il vérifie notamment que si f et $|f|$ sont intégrables $(\mathcal{B})$, au sens de Borel, alors la fonction f est intégrable $(\mathcal{L})$, au sens de Lebesgue, et

$$(\mathcal{B}) \int_a^b f(x)dx = (\mathcal{L}) \int_a^b f(x)dx.$$

Bauer [2] étend la théorie de Perron aux fonctions de plusieurs variables.

M. Fréchet (courtesy Académie des sciences, Paris)

En 1915, Fréchet réalise sa théorie de l'intégration pour une mesure abstraite, non nécessairement définie sur des ensembles mesurables au sens de Lebesgue. Il observe que l'intégrale de Radon $\int F(P)dh(P)$, où F est une fonction définie dans $\mathbb{R}^n$ et h est une fonction à variation bornée définie dans $\mathbb{R}^n$, résulte d'une fusion de l'intégrale de Lebesgue et de l'intégrale de Stieltjes. L'intégrale de Radon se réduit à celle de Lebesgue au cas où la fonction h est *linéaire*, à celle de Stieltjes au cas où la fonction F est continue. Fréchet écrit:

«D'ailleurs, l'intégrale de Radon peut aussi s'écrire

$$\int_E f(P)df(e)$$

où $f(e)$ est une fonction additive du sous-ensemble variable e de E.

Or, c'est sous cette forme qu'apparaît ce qui me semble être le grand avantage de la définition de M. J. Radon, avantage que celui-ci ne paraît pas avoir remarqué. M. J. Radon avait pour but de réaliser un progrès dans la Théorie des fonctions en unifiant les définitions de Stieltjes et de M. Lebesgue. Mais, en fait, on remarque que, moyennant quelques légères modifications, la définition et les propriétés de l'intégrale de M. J. Radon s'étendent bien au-delà du Calcul intégral classique, elles sont presque immédiatement applicables au domaine infiniment plus vaste du calcul fonctionnel.

En d'autres termes, on peut conserver la majeure partie des définitions et des raisonnements de M. J. Radon en négligeant l'hypothèse faite sur la nature de l'argument P à savoir que P est un point de l'espace à n dimensions» ([30] p. 248–249).

La mesure de Fréchet est définie sur une famille d'ensembles qui est stable pour les réunions dénombrables et les différences d'ensembles. Elle doit vérifier la propriété de l'additivité complète. Ces travaux, comme ceux de Carathéodory, préparent le terrain approprié au cadre qui sera créé en calcul des probabilités par Kolmogorov.

En 1916, Denjoy [22] développe une étude fine et technique sur la *dérivation et son calcul inverse*.

Sierpiński et Lusin [70] décomposent l'intervalle $[0, 1]$ en une infinité, de la puissance du continu, d'ensembles deux à deux disjoints, tels que la mesure extérieure de chacun soit 1.

En 1917, Daniell [20] prolonge la théorie de Young aux fonctions définies sur un ensemble arbitraire.

L'ouvrage de Carathéodory, publié en 1918, constitue un jalon important dans l'histoire de la théorie de l'intégration [60]. L'auteur s'explique sur ses motivations:

«Die Umwälzung, welche die Theorie der reellen Funktionen durch die Untersuchungen von H. Lebesgue erfahren hat, ist ein Prozess der heute in seinen Hauptzügen als abgeschlossen gelten kann. Ein Versuch diese Theorie von Grund aus und systematisch aufzubauen scheint mir daher notwendig geworden zu sein [...].

In einigen [...] Lehrbüchern [...] erscheint [die Lebesguesche Theorie] meistens neben den älteren Integrationstheorien und ist dadurch ihres grössten Vorzugs beraubt, der darin besteht, dass sie den kürzeren und bequemeren Weg darstellt, da wo die alte Fahrstraße oft unnötige Umwege macht» ([17] p. v).

Carathéodory explicite son travail présenté dans [16]. Comme exemple d'une mesure à laquelle sa théorie s'applique il cite la mesure ponctuelle δ_a ($a \in \mathbb{R}^q$), qui sera dite mesure de Dirac; $\delta_a(A) = 1$ si $a \in A$, $\delta_a(A) = 0$ si $a \notin A$. Carathéodory définit encore la mesurabilité d'une fonction au moyen du graphe. Toute fonction n'est pas mesurable. Carathéodory fournit une interprétation de l'intégrale définie. Soient une partie mesurable E de $\mathbb{R}^q$ et $f : E \to \mathbb{R}_+$. Si l'ensemble

$$\{(x_1, \ldots, x_q, y) \in \mathbb{R}^{q+1} : P = (x_1, \ldots, x_q) \in E, 0 \leq y \leq f(P)\}$$

est mesurable dans $\mathbb{R}^{q+1}$ et de mesure finie, cette dernière est notée

$$\int_E f(P)d\omega;$$

la fonction est dite sommable sur E.

Bourbaki décrit ainsi la situation au moment de l'apparition de l'ouvrage de Carathéodory:

«C'est avec ce livre [...] que la notion d'intégrale [...] cède le pas pour la première fois à celle de mesure, qui avait été chez Lebesgue (comme avant lui chez Jordan) un moyen technique auxiliaire [...]. Depuis lors, les auteurs qui ont traité d'intégration se sont partagés entre ces deux points de vue [...]. Les uns ont suivi Carathéodory; dans leurs exposés sans cesse plus abstraits et plus axiomatisés, la mesure, avec tous les raffinements techniques auxquels elle se prête, non seulement joue le rôle dominant, mais encore elle tend à perdre contact avec les structures topologiques auxquelles en fait elle est liée dans la plupart des problèmes où elle intervient. D'autres exposés suivent de plus ou moins près une méthode déjà indiquée en 1911 par W.H. Young, dans un mémoire malheureusement peu remarqué, et développé ensuite par Daniell» ([10] p. 257–258).

En 1919, Denjoy décrit trois procédés destinés à

«montrer comment la méthode de Riemann, secondée par la notion de mesure, conduit à des définitions de l'intégrale équivalentes à celle de M. Lebesgue ou même plus générales» ([23] p. 219).

Il soupçonne sa méthode (A) d'être équivalente à l'intégration de Lebesgue, la méthode (C) d'être plus générale. La procédure (B) est effectivement plus générale; elle s'applique à la fonction f, non intégrable au sens de Lebesgue, définie sur $]-1, 1[$ par

$$f(x) = \frac{1}{x\,\mathrm{Log}|x|} \text{ pour } x \neq 0,$$
$$f(0) = 0.$$

Si f est une fonction intégrable au sens de Lebesgue, mais non au sens de Riemann, peut-on tirer parti des sommes de Riemann pour déterminer la valeur numérique de l'intégrale de Lebesgue ? Les méthodes de Denjoy conduisent à l'intégrale de Lebesgue par des procédés d'intégration riemannienne. Dans son travail [4], effectué sous la direction de Denjoy, Boks fournit les démonstrations pour la méthode (B) de Denjoy.

Soit $f : [a, b] \to \mathbb{R}$. On considère la fonction F de période $b - a$ coïncidant avec f sur $[a, b[$. Soient maintenant

$$x_0 = a \leq \zeta_1 \leq x_1 \leq \zeta_2 \leq \ldots \leq x_{i-1} \leq \zeta_i \leq x_i \leq \ldots \leq \zeta_n \leq x_n = b$$

avec $0 < x_i - x_{i-1}$ ($i = 1, \ldots, n$). Le pas ω de cette subdivision est le plus grand des nombres $x_i - x_{i-1}$. On pose $y_i = x_i + t, \eta_i = \zeta_i + t$ et on forme

$$s(t) = \sum_{i=1}^{n} (y_i - y_{i-1}) F(\eta_i).$$

On dit que la fonction f est intégrable (B) et que l'intégrale de f au sens de (B) a pour valeur I si la mesure de l'ensemble des $t \in \mathbb{R}$ vérifiant $|I - s(t)| > \alpha, 0 \leq t < b - a$, tend vers 0 avec ω quel que soit $\alpha \in \mathbb{R}_+^*$. Cette méthode ne permet guère de conclure à l'intégrabilité de f sur $[a, b]$ à partir de l'intégrabilité sur $[a, c]$ et $[c, b]$ $(a < c < b)$, du moins si $c - a$ et $b - c$ sont incommensurables entre eux. Boks dit que la fonction f est intégrable au sens (B_1) si les sommes de Riemann intervenant sont de la forme

$$\Sigma(y_i - y_{i-1}) f(\eta_i)$$

où la sommation est étendue à toutes les valeurs de i telles que η_i appartienne à $[a, b]$. Il peut vérifier que toute fonction sommable est intégrable au sens (B_1), alors que la réciproque n'a pas lieu. Soit, en effet, f une fonction définie sur $]-a, a[$, impaire, décroissante sur $]-a, 0[$ et $]0, a[$ telle que $x \mapsto xf(x)$ soit une fonction positive, paire, bornée sur $]-a, a[$, nulle et continue en 0. On suppose aussi que ni $\int_{-a}^{-\varepsilon} f(x)dx$ ni $\int_{\varepsilon}^{a} f(x)dx$ ne convergent quand ε tend vers 0. Cette fonction f n'est pas sommable; elle est intégrable au sens de Riemann dans tout segment ne contenant pas l'origine, et intégrable (B_1). Des propriétés de l'intégrale (B_1) Boks déduit une approximation de l'intégrale de Lebesgue au moyen de sommes de Riemann en dehors d'un ensemble de mesure nulle.

Hake [36] rappelle que si f est une fonction continue sur $[a, b]$ et $F(x) = \int_a^x f(t)dt$, alors $F'(x) = f(x)$; tel n'est plus le cas si la fonction f est uniquement supposée être intégrable au sens de Riemann, et encore moins, si la fonction n'est qu'intégrable au sens de Lebesgue. Hake introduit une nouvelle définition en vue de récupérer le théorème fondamental du calcul intégral. Une fonction $f : [a, b] \to \mathbb{R}$ étant donnée, il appelle sur-fonction de f toute fonction g sur $[a, b]$ qui est continue et telle que $g(a) = 0$, $g'_+(x) \neq -\infty$ pour tout $x \in [a, b[$, $g'_+ \geq f$ sur un ensemble de mesure $b - a$. Il définit de même une sous-fonction h. La fonction f est dite (S)-intégrable si pour tout $\varepsilon \in \mathbb{R}_+^*$ il existe une sur-fonction g et une sous-fonction h telles que $0 \leq g(b) - h(b) < \varepsilon$. Alors, $0 \leq g(x) - h(x) < \varepsilon$ pour tout $x \in [a, b]$. Faisant tendre ε vers 0, on arrive ainsi à définir une *fonction intégrale*, notée $S_a^x(f)$. Hake prouve que l'intégrale de Denjoy constitue un cas particulier de la (S)-intégrale. Il se demande si son intégrale est équivalente à celle de Denjoy.

Le travail de Tonelli de 1923 se veut être un exposé de la théorie de l'intégrale de Lebesgue qui ne présuppose pas la connaissance de la théorie de la mesure:

«Il presente lavoro è un tentativo per rendere più simplice, dirò così più elementare, e quindi più facilmente accetabile da tutti, la teoria dell'integrale del

Lebesgue, liberandola completamente della teoria della misura dei gruppi di punti» ([73] p. 105).

Sa méthode est une extension de l'intégrale de Cauchy définie pour la fonction continue; la difficulté principale réside dans le passage à la limite. Une fonction f réelle définie sur $[a,b]$ est dite quasi-continue s'il existe une loi faisant correspondre à tout $n \in \mathbb{N}^*$ une réunion Δ au plus dénombrable d'intervalles ouverts de longueur inférieure à $\frac{1}{n}$ telle que la fonction f soit continue sur $[a,b]\backslash\Delta$. On note ${}_\Delta f$ la fonction continue, coïncidant avec f en dehors de Δ et prolongée linéairement sur Δ. Au cas où la fonction quasi-continue f est bornée, $\int_a^b f(x)dx$ est la limite — si elle existe — des $\int_a^b {}_\Delta f(x)dx$ quand la longueur totale de Δ tend vers 0. Au cas où elle n'est pas bornée, pour $p, q \in \mathbb{R}_+$ on pose

$$\begin{aligned} f_{p,q}(x) &= f(x) && \text{si} \quad -q \le f(x) \le p, \\ &= -q && \text{si} \quad f(x) < -q \\ &= p && \text{si} \quad f(x) > p. \end{aligned}$$

Alors, $\int_a^b f(x)dx$ est la limite — si elle existe — des $\int_a^b f_{p,q}(x)dx$ quand p, q tendent vers $+\infty$.

Si (f_m) est une suite de fonctions quasi-continues convergeant presque partout vers f et si la suite $(|f_m|)$ est majorée par M, alors

$$\lim_{m\to\infty} \int_a^b f_m(x)dx = \int_a^b f(x)dx$$

au cas où ces expressions ont un sens. Il s'agit de remarquer que

$$\lim_{m\to\infty} \int_a^b {}_\Delta f_m(x)dx = \int_a^b {}_\Delta f(x)dx;$$

d'autre part, si la longueur totale de Δ est inférieure à $\frac{1}{n}$,

$$\left| \int_a^b f_m(x)dx - \int_a^b {}_\Delta f_m(x)dx \right| \le \frac{2M}{n}$$

et

$$\left| \int_a^b f(x)dx - \int_a^b {}_\Delta f(x)dx \right| \le \frac{2M}{n}.$$

Tonelli explicite les propriétés habituelles de l'intégrale. Il traite de l'intégration par parties, de l'intégration par substitution d'une variable; il étudie l'intégrale de la dérivée d'une fonction absolument continue.

En 1924, Alexandrov [1] publie une démonstration de l'équivalence de l'intégrale de Perron avec celle de Denjoy. La preuve que cette dernière l'est au sens précédent est obtenue indépendamment de Hake. Alexandrov écrit, d'autre part, qu'au moment de la publication il apprend que la réciproque a aussi été obtenue par Looman, mais que cette démonstration n'est pas encore publiée.

Pour établir la première implication, compte tenu de la définition de l'intégrale de Denjoy et du fait que toute fonction intégrable au sens de Lebesgue l'est au sens de Perron, il suffit à Alexandrov de prouver directement que si la fonction f est intégrable au sens de Perron sur tout sous-intervalle $[\alpha, \beta]$ de $[a, b]$ et

$$J_{a,b} = \lim_{\substack{\alpha \to a \\ \beta \to b}} \int_\alpha^\beta f(t)dt$$

existe, alors f est intégrable au sens de Perron sur $[a, b]$;

$$F(x) = \lim_{\alpha \to a} \int_\alpha^x f(t)dt$$

constitue l'intégrale indéfinie de Perron alors que $F(b) = J_{a,b}$ est l'intégrale définie de Perron.

Quant à la réciproque, Alexandrov commence par dire qu'une fonction continue F est à *variation bien définie* sur une partie parfaite P d'intérieur vide si sur la suite des *intervalles complémentaires* $]a_n, b_n[$ la somme des variations de F converge; avec $\alpha = \inf P$, $\beta = \sup P$, il pose

$$V_P F = F(\beta) - F(\alpha) - \sum_{n=1}^{\infty} (F(b_n) - F(a_n)).$$

Les travaux de Denjoy et de Lusin [55] lui permettent d'énoncer le fait suivant: Supposons que pour toute partie parfaite P d'intérieur non vide de $[a, b]$ il existe $\gamma, \delta \notin P$ tels que la fonction $F\Big|_{P \cap [\gamma, \delta]}$ soit à variation bien définie et que si $P \cap [\gamma, \delta]$ est de mesure nulle, $V_{P \cap [\gamma, \delta]} F = 0$. Alors, F admet presque partout une dérivée f et F est l'intégrale indéfinie de Denjoy de f. Là-dessus, Alexandrov vérifie que pour $F(x)$ il peut choisir l'intégrale indéfinie de Perron $\int_a^x f(t)dt$.

Looman [53] démontre que toute fonction f, *intégrable au sens de Perron-Hake*, l'est au sens de Denjoy et que les deux intégrales coïncident. A cet effet, il commence par vérifier que pour toute partie parfaite P de $[a, b]$ il existe une partie $P \cap [\alpha, \beta]$ sur laquelle f est sommable. Il peut alors procéder par induction transfinie. Les résultats de Looman sont indépendants de ceux d'Alexandrov.

Carson [18] rappelle que le calcul opérationnel est relatif à l'étude d'un système d'équations

$$\begin{aligned} a_{11}x_1 + \ldots + a_{1n}x_n &= f_1(t) \\ &\vdots \\ a_{n1}x_1 + \ldots + a_{nn}x_n &= f_n(t) \end{aligned} \tag{1}$$

où

$$a_{jk} = \alpha_{jk} + \beta_{jk}\frac{d}{dt} + \gamma_{jk}\frac{d^2}{dt^2} + \ldots, \tag{2}$$

$\alpha_{jk}, \beta_{jk}, \gamma_{jk}, \ldots$ étant des constantes. On exige aussi des conditions au bord: Les fonctions f_i et x_i s'annulent pour $t < 0$. Par linéarité on peut se borner à considérer le système (3) qu'on déduit de (1) en posant $f_1 = f, f_2 = \ldots = f_n = 0$. On introduit le système auxiliaire

$$\begin{aligned} a_{11}h_1 + \ldots + a_{1n}h_n &= 1 \\ a_{21}h_2 + \ldots + a_{2n}h_n &= 0 \\ &\vdots \qquad\qquad (t \geq 0) \\ a_{n1}h_1 + \ldots + a_{nn}h_n &= 0. \end{aligned} \tag{4}$$

De la solution de (4) on peut passer à celle de (3) grâce aux relations

$$x_i(t) = \frac{d}{dt}\int_0^t f(t-\tau)h_i(\tau)d\tau \quad (i = 1, \ldots, n). \tag{5}$$

Heaviside a remplacé $\frac{d^n}{dt^n}$ par p^n et abouti à l'*équation opérationnelle*

$$h_i = \frac{1}{H_i(p)} \quad (i = 1, \ldots, n). \tag{6}$$

Carson écrit:

«I have called (6) an equation; as a matter of fact it is not an equation except in a purely symbolic sense, since the left hand side is a function of t and the right hand side is a function of p. Heaviside's point of view, however, was that (6) is the full equivalent of (4), and that the functional form of $H_i(p)$ contains all the information necessary for the explicit solution» ([18] p. 45).

Pour Carson cependant, il s'agit uniquement de l'équivalent symbolique de l'équation intégrale

$$\frac{1}{pH_i(p)} = \int_0^\infty h_i(t)e^{-pt}dt;$$

il ajoute:

«This uniquely determines a continuous function $h_i(t)$ in terms of $H_i(p)$. From this integral equation the rules and formulas of the operational calculus are directly and immediately deducible» ([18] p. 45).

L'idée de Carson consiste à poser $f(t) = e^{pt}$ avec $p \in \mathbb{C}$, $\Re e\, p > 0$, à écrire

$$x_i(t) = \frac{e^{pt}}{H_i(p)} + y_i(t),$$

où le choix de y_i est dicté par la condition au bord. Grâce à (5), on a

$$x_i(t) = \frac{d}{dt}\int_0^t e^{p(t-\tau)} h_i(\tau) d\tau.$$

Donc

$$\frac{1}{H_i(p)} + y_i(t)e^{-pt} = h_i(t)e^{-pt} + p\int_0^t h_i(\tau)e^{-p\tau} d\tau.$$

Il suffit alors de faire tendre t vers $+\infty$.

Le travail [82] de Wiener, datant de 1926, est prémonitoire de la définition ultérieure de la dérivation d'une distribution. Wiener, développant son calcul opérationnel, se rend compte que le calcul symbolique de Heaviside, tout en se déployant sur un *corpus vil*, en particulier la fonction $1_{\mathbb{R}^+}$, est pourvu d'une grande *évidence expérimentale*.

Abandonnant le point de vue des quadratures pour celui des fonctions primitives, dans la seconde édition de son ouvrage, Lebesgue se dit obligé d'exposer le procédé de totalisation de Denjoy; il le fait *à la manière dont Stieltjès avait généralisé l'intégration ordinaire*. Au sujet du choix de sa terminologie, Lebesgue commente dans la préface:

«J'ai dit l'intégrale de Stieltjès; n'aurais-je pas dû dire l'intégrale de Cauchy ? Cauchy a, en effet, très nettement exposé l'importance et la signification physique de la nouvelle intégration prise dans toute sa généralité, alors que Stieltjès a surtout défini logiquement la nouvelle opération, dans le cas seulement d'une variable. Je n'ai pas cru devoir changer la dénomination adoptée. Si je l'avais fait, aurait-il fallu prendre le nom du premier inventeur actuellement connu ou le nom de celui qui a donné à l'intégrale la définition la plus large, actuellement connue ? De toute façon l'attribution eût été inexacte et injuste; autant s'en tenir aux inexactitudes consacrées» ([45] p. vi).

D'ailleurs, dans une lettre adressée à Borel en 1909 ([24] p. 181), Lebesgue écrit que l'intégrale de Stieltjes peut être remplacée par la sienne propre, mais que la réciproque est inexacte.

Dans la seconde édition de sa monographie Lebesgue fait aussi figurer la note suivante:

«Nous conviendrons de dire qu'une propriété a lieu presque partout dans un intervalle (a, b), ou sur un ensemble $\mathscr{E}$, si les points de (a, b) ou de $\mathscr{E}$ en lesquels elle n'a pas lieu ou bien n'existent pas, ou bien forment un ensemble de mesure nulle.

Cette locution, introduite dans la première édition de ce Livre, a été généralement adoptée» ([45] p. 179).

Répondant à une objection formulée par Denjoy, Lebesgue ajoute:

«Presque partout serait inadmissible si, dans la langue usuelle, cette expression avait un sens précis, mais cela n'est pas; de sorte que le lecteur [...] ne peut lui donner aucun sens précis sans se reporter à la définition posée pour presque partout. Aucune erreur n'est donc possible [...].

Et ce n'est pas diminuer l'inconvénient que de modifier, fût-ce même pour le perfectionner, un vocabulaire dont l'usage commence à se répandre» ([loc. cit.]).

Soit un ensemble mesurable E; le rapport $\frac{m(e)}{\beta-\alpha}$ de la mesure de la partie e de E située dans l'intervalle $[\alpha, \beta]$ à la mesure de cet intervalle est appelé par Lebesgue *densité moyenne de E dans* (α, β). Si ce rapport tend vers une limite quand la mesure de $[\alpha, \beta]$ tend vers zéro, elle constitue la *densité* en un point. Lebesgue ne se prive pas d'insérer aussi la réplique suivante:

«M. Denjoy remplace le mot densité par épaisseur; il est choqué par des phrases telles que celle-ci: 'Un ensemble partout non dense peut avoir presque partout une densité égale à un'. Il est certain qu'une telle phrase semble un calembour; mais des deux mots dense et densité, c'est le premier qui est mal choisi, car il éveille, me semble-t-il, une idée de mesure, une idée quantitative.

Quoi qu'il en soit, le lecteur qui m'aura suivi jusqu'ici ne se laissera pas troubler par ces questions puisqu'il aura consenti à s'occuper de la recherche de la définition de l'intégrale définie — ce qui peut paraître ridicule, — et de la définition de l'intégrale qui reste indéfinie même après qu'on l'a définie — ce qui est un peu humiliant» ([45] p. 195).

En 1930, Nikodým [57] met au point une généralisation des intégrales de Radon. Il envisage des mesures positives μ sur un *corps d'ensembles* $\mathcal{H}$, qui est stable pour les réunions ou intersections dénombrables et le passage au complémentaire. Il expose ses buts:

«Les intégrales $\int_E f\,d\mu$ que je vais étudier jouissent d'une propriété fondamentale [...] qui représente une généralisation de la propriété de la continuité absolue des intégrales de M. Lebesgue. La notion en question, qui fut introduite par M. Vitali pour les fonctions d'une variable réelle, se traduit pour les intégrales lebesguiennes par le fait suivant:

$$\text{si mes } E_n \longrightarrow 0, \text{ on a } \int_{E_n} f\,dx \longrightarrow 0;$$

et une propriété analogue subsiste pour les intégrales de M. Radon. Pour avoir une analogie avec la continuité ordinaire, nous allons introduire la notion de la 'μ-distance' de deux ensembles du corps considéré, en posant

$$||E, F||_\mu =_{\text{df}} \mu(E \backslash F) + \mu(F \backslash E)»$$

([57] p. 133).

Nikodým établit la propriété suivante pour la fonction d'ensembles $\mathcal{F}(E) = \int_E f\,d\mu$: Si $E_n \in \mathcal{H}$ et $\lim\limits_{n\to\infty} \mu(E_n) = 0$, on a $\lim\limits_{n\to\infty} \mathcal{F}(E_n) = 0$. La fonction $\mathcal{F}$ est dite μ-continue si $\lim\limits_{n\to\infty} \mathcal{F}(E_n) = \mathcal{F}(E)$ pourvu que $\lim\limits_{n\to\infty} ||E_n, E||_\mu = 0$. Le

O. Nikodým

théorème fondamental de Nikodým exprime l'équivalence des quatre conditions suivantes pour la fonction complètement additive $\mathcal{F}$: 1°) La fonction $\mathcal{F}$ est μ-continue. 2°) La relation $\mathcal{F}(E) \neq 0$ implique $\mu(E) > 0$. 3°) Si $\lim_{n\to\infty} \mu(E_n) = 0$, alors $\lim_{n\to\infty} \mathcal{F}(E_n) = 0$. 4°) Il existe une fonction μ-sommable f telle que $\mathcal{F}(E) = \int_E f\, d\mu$ pour tout $E \in \mathcal{K}$.

En 1931, Burkill [12] revient à l'idée de dérivée approximative, observant que cette notion a un sens même si la fonction n'est pas continue. Il introduit une intégrale approximativement continue. Alors qu'une telle démarche est déjà apparue chez Young, Burkill juge cette théorie insuffisante, vu que cette intégrale est dépourvue de quelques propriétés indispensables comme l'additivité. La méthode de Burkill suit la procédure de Perron.

Un sous-ensemble E de $\mathbb{R}$ est dit de densité 1 en x si quel que soit $k \in]0,1[$, il existe x_1 voisin de x appartenant à E et un ensemble de mesure $k|x_1 - x|$ dans $[x_1, x] \cap E$. Soit f une fonction mesurable. On appelle dérivée approximative supérieure $AD^* f(x)$ en x la limite inférieure de l'ensemble des $\alpha \in \mathbb{R}$ tels que l'ensemble des ζ pour lesquels

$$\frac{f(\zeta) - f(x)}{\zeta - x} < \alpha$$

soit de densité 1 en x. L'opérateur AD^* est sous-additif. On définit de même la dérivée approximative inférieure $AD_* f(x)$ en x. Si $AD_* f(x) = AD^* f(x)$, on parle de la dérivée approximative $ADf(x)$. La notion de continuité approximative en un point se définit d'une manière analogue. On appelle fonction majeure de f toute fonction M approximativement continue sur $[a, b]$ telle que $M(a) = 0$ et $AD_* M(x) > -\infty$, $AD_* M(x) \geq f(x)$ quel que soit $x \in [a, b]$. La définition d'une fonction mineure m est corrélative.

On pose $I(b) = \inf_M M(b)$, $J(b) = \sup_m m(b)$. Si $I(b) = J(b)$, cette valeur est dite intégrale de Perron approximative $(AP) \int_a^b f(x)dx$. Cette intégrale possède les propriétés habituelles, notamment la linéarité. Toute fonction intégrable au sens de Perron l'est au sens de Burkill. La fonction $F : x \mapsto (AP) \int_a^x f(t)dt$ est approximativement continue en x. En particulier, si G admet une dérivée approximative g, alors

$$(AP) \int_a^b g(x)dx = G(b) - G(a).$$

Burkill démontre que si $F(x) = (AP) \int_a^x f(t)dt$, alors $ADF = f$ presque partout.

Burkill [13] rappelle qu'Alexandrov et Looman ont montré que l'intégrale de Perron est équivalente à l'intégrale de Denjoy généralisée. Il définit une intégrale qui, à certains égards, peut être plus maniable; la convergence est considérée en moyenne de Cesàro.

Van der Pol et Niessen, travaillant dans un laboratoire industriel, publient une note [77] expliquant l'application pratique du calcul symbolique, qui pour eux est le calcul opérationnel de Heaviside dans la forme mise au point par Carson. Comme point de départ d'une longue collection de relations mathématiques ils prennent la transformée de Laplace qui via une intégrale convergente lie symboliquement à une fonction h donnée la fonction f:

$$f(p) = p \int_0^\infty e^{-px} h(x)dx.$$

S. Saks

En 1932, Riesz retrace le rôle prépondérant joué par la notion d'intégrale en théorie des fonctions réelles [65].

La monographie [67] de Saks constituera une référence au cours d'une longue période. L'auteur y développe systématiquement et en détail toutes les théories d'intégration. Il expose la notion de fonction d'ensemble, la mesure de Lebesgue, la notion de fonction à variation bornée, les intégrales de Lebesgue, de Perron, de Hake, de Denjoy, de Burkill. Il ajoute une annexe sur l'intégrale de Lebesgue dans les espaces abstraits, d'après les travaux de Radon et de Fréchet.

Bochner [3] inaugure la théorie de l'intégration des fonctions à valeurs vectorielles. Elle concerne des fonctions définies dans $\mathbb{R}^q$ ($q = 1, 2, \ldots$) et à valeurs dans un espace de Banach E. Soit A une partie mesurable de $\mathbb{R}^q$. La fonction $f : A \to E$ est dite *finie* s'il existe une partition mesurable finie de A telle que les restrictions de f aux différentes parties soient des fonctions constantes. Elle est dite *mesurable* s'il existe une suite (f_n) de fonctions *finies* telle que $f = \lim_{n\to\infty} f_n$ presque partout sur A; elle est appelée *sommable* si elle est mesurable et la fonction réelle $||f||$ est sommable au sens habituel. L'intégrale $I(f, A) \in E$ doit vérifier les propriétés suivantes : 1°) $||I(f, A)|| \leq \int_A ||f(x)||dx$. 2°) Si la fonction f prend la valeur constante a sur A, alors $I(f, A) = a$ mes(A). 3°) I satisfait à l'additivité. 4°) Si $A = \bigcup_{i=1}^{m} A_i$ et les parties mesurables A_i sont deux à deux disjointes, alors $I(f, A) = \sum_{i=1}^{m} I(f, A_i)$. 5°) Si la suite (f_n) de fonctions sommables sur A converge presque partout et $||f_n(x)|| \leq S(x)$ pour presque tout $x \in A$, S étant une fonction réelle sommable donnée, alors

$$I\big(\lim_{n\to\infty} f_n, A\big) = \lim_{n\to\infty} I(f_n, A).$$

La théorie de Bochner est une généralisation de celle de Lebesgue. Bochner établit dans ce cadre le théorème de Fubini, l'inégalité de Hölder; il étudie le problème de la différentiabilité et celui du développement en série de Fourier.

Haar [33] construit sur un groupe localement compact séparable une mesure invariante par translations à droite; sa procédure transcrit la méthode de Lebesgue. Il révèle que von Neumann et Riesz, ayant eu connaissance de ses résultats, lui ont fait savoir que la plupart des difficultés techniques de sa démonstration peuvent être éliminées si on recourt à la théorie de la mesure mise au point par Carathéodory.

Kolmogorov [42] inaugure l'époque moderne du calcul des probabilités, interprétant la probabilité comme étant une mesure abstraite positive normalisée.

La première édition de la monographie [75] de de la Vallée Poussin paraît en 1916. L'intégrale de Lebesgue y est exposée d'une façon claire et détaillée; il en est de même de la théorie de la mesure d'après Lebesgue, les contributions de Borel y comprises. C'est, d'autre part, une étude systématique de la théorie des fonctions additives d'ensemble ainsi que de la théorie des fonctions de Baire. Dans la préface à la seconde édition, paraissant en 1934, l'auteur écrit:

«J'ai voulu [...] proscrire tout recours, fût-il seulement apparent, à l'axiome du choix et ne fonder les démonstrations d'existence que sur des procédés de construction effectifs rigoureusement précisés» ([75] p. *xiii*).

Cette édition comporte un appendice sur l'intégrale de Stieltjes.

L'important travail [49] de Leray *Sur le mouvement d'un liquide visqueux emplissant l'espace*, paraissant en 1934, peut être considéré par certains aspects comme annonciateur de la théorie des distributions.

Dunford [25] se propose d'approcher la théorie de l'intégration pour des fonctions à valeurs dans un espace de Banach, en analogie avec la définition des nombres réels par Cantor. Il fait un large appel à la convergence de suites d'intégrales. L'exposé s'étend moins loin que celui de Bochner. A la fin, Dunford indique les possibilités d'extension de la méthode à des fonctions prenant leurs valeurs dans un espace métrique arbitraire.

Ward [80] observe que l'intégrale de Lebesgue-Stieltjes est associée à des fonctions à variation bornée, alors que l'intégrale de Stieltjes ordinaire n'exige pas ce type de restriction. Il réussit à faire usage des méthodes de Perron pour obtenir une intégrale englobant à la fois l'intégrale de Riemann-Stieltjes et l'intégrale de Lebesgue-Stieltjes. Il compare son travail à un manuscrit de Young appliquant l'intégrale de Denjoy à l'intégration de Stieltjes. Ward avance:

«In fact, the final theorems [...] are not far from a demonstration that the Perron-Stieltjes integral is equivalent to a Denjoy-Stieltjes integral» ([80] p. 579).

Kempisty [41] décrit des procédés d'intégration dans $\mathbb{R}^2$. Il considère des *divisions régulières* formées de rectangles *élémentaires* $[a, a+h] \times [b, b+k]$, où $h \leq 2k$, $k \leq 2h$. Le point de départ est constitué par des fonctions additives F, G définies sur ces rectangles. Si quand les dimensions des rectangles de la subdivision régulière de R tendent vers 0, $F(R)$ converge, la limite constitue l'intégrale de F; si quand R tend vers (x, y), $\frac{F(R)}{G(R)}$ converge, la limite est la *dérivée régulière de* F *par rapport à* G *au point* (x, y). Si $F(S)$ tend vers 0 en même temps que $G(S)$, on dit que la fonction F est absolument et régulièrement continue par rapport à S. Kempisty peut décrire ainsi une généralisation de l'intégrale de Stieltjes, au sens de Lebesgue et de Denjoy.

La méthode de dérivation, définie ultérieurement pour les distributions, est utilisée par Sobolev en 1936 pour la dérivation d'une forme linéaire continue dans son travail consacré au problème de Cauchy relatif à la résolution d'une équation aux dérivées partielles. La procédure *suit de l'intégration par parties* ([71] p. 62).

Temple notera:

«The concepts which were implicit in Dirac's bold, heuristic theory received explicit expression in the work of Sobolev, which seems to have been developed independently of Dirac's investigations, and which is primarly concerned with the existence of solutions of linear hyperbolic partial differential equations» ([72] p. 165).

Levi rappelle que l'intégrale d'une fonction est apparue sous deux aspects différents, comme primitive et comme limite de sommes. Il dit que l'*équivalence* des deux notions peut être visualisée grâce à quelques observations simples. Si $\int$ désigne indifféremment l'intégrale, l'intégrale supérieure ou l'intégrale inférieure, on doit avoir

$$\int_a^b + \int_b^c = \int_a^c, \int_a^b f(x)dx \geq \int_a^b g(x)dx \text{ pour } g \geq f, \tag{1}$$

$$\lim \int f_n dx = \limsup \int f_n dx \leq \int f dx \leq \liminf \int g_n dx = \lim \int g_n dx \tag{2}$$

pour une suite (f_n) croissante et une suite (g_n) décroissante de fonctions convergeant vers f. Il écrit:

«L'osservazione diviene più significativa se l'$\int$ si concreta nell'integrale superiore di Lebesgue; vale infatti allora la proposizione seguente che ci conviene enunciare considerando l'integrale di campo: Se una successione di funzioni $\{f_n\}$ definite in un campo $\mathcal{C}$ (di cui sia $d\omega$ il differenziale) tende crescendo ad una funzione limitata f,

$$\overline{\int_{\mathcal{C}}} f d\omega \leq \lim \overline{\int_{\mathcal{C}}} f_n d\omega = \limsup \overline{\int_{\mathcal{C}}} f_n d\omega, \tag{3}$$

perchè, supposte — come è sempre permesso — le f_n positive, e fissata arbitrariamente una successione di somme S_n superiori rispettivamente alle f_n e approssimanti convenientemente i rispettivi integrali superiori, se ne puó dedurre una summa S superiore ad f e $< \lim S_n + \varepsilon$, dove ε è un numero positivo arbitrario. Insiema con (2), ne segue che

$$\overline{\int_{\mathcal{C}}} f d\omega = \lim \overline{\int_{\mathcal{C}}} f_n d\omega»$$

([51] p. 516–517).

Levi transcrit les mêmes relations pour l'intégrale inférieure et envisage le cas des fonctions de plusieurs variables.

Weil recourt à la théorie de Radon dans sa monographie. Il peut écrire:

«Il y a donc correspondance biunivoque entre les mesures de Radon et les fonctionnelles $I(f)$» ([81] p. 32).

En fait, Weil n'explicite pas la démonstration. Il renvoie au fascicule de Bourbaki [9]; mais ce dernier sera publié seulement onze ans plus tard.

Pour un espace localement compact E soit $\mathcal{K}(E)$ l'espace vectoriel des fonctions continues sur E, à support compact. H. Cartan [19] appelle mesure de Radon positive ou distribution de masses positives une fonction additive μ sur $\mathcal{K}_+(E)$, à valeurs dans $\mathbb{R}_+$. Alors aussi $\alpha\mu(f) = \mu(\alpha f)$ quels que soient $f \in \mathcal{K}_+(E)$, $\alpha \in \mathbb{R}_+$. Cette mesure positive s'étend en une mesure sur $\mathcal{K}(E)$ puisque tout $f \in \mathcal{K}(E)$ s'écrit $f = f_1 - f_2$, avec $f_1, f_2 \in \mathcal{K}_+(E)$. Soient $\mathcal{I}$ [resp. $\mathcal{S}$] l'ensemble des fonctions positives, semi-continues inférieurement [resp. supérieurement] sur E. Si $g \in \mathcal{I}_+$, on pose

$$\mu_i(g) = \sup_{k \in \mathcal{K}_+(E), k \leq g} \mu(k);$$

si $h \in \mathcal{S}_+$, on pose

$$\mu_s(h) = \inf_{k \in \mathcal{K}_+(E), k \geq h} \mu(k).$$

Une fonction f à valeurs dans $\bar{\mathbb{R}}_+$ est dite sommable pour μ si

$$\inf_{\substack{g\in\mathcal{J}\\ g\geq f}} \mu_i(g) = \sup_{\substack{h\in\mathcal{S}\\ h\leq f}} \mu_s(h) \in \mathbb{R}_+.$$

La valeur commune constitue l'*intégrale de Lebesgue-Stieltjes* $\int f(x)d\mu(x)$. La mesure de Haar, construite par Weil sur le groupe localement compact, est un exemple d'une mesure de Radon.

H. Cartan introduit une topologie sur l'ensemble $\mathcal{M}$ des mesures de Radon positives. Il s'agit de la topologie la moins fine rendant continues toutes les fonctions $\mu \mapsto \mu(f)$, $f \in \mathcal{K}(E)$. H. Cartan dit que d'une manière vague, si $\int f(x)d\mu(x)$ varie d'une manière continue, alors μ varie d'une manière continue. Il considère aussi la composition des mesures de Radon. Pour $\mu, \nu \in \mathcal{M}$, il pose

$$[\mu,\nu](f) = \int\Big(\int f(xy)d\mu(x)\Big)d\nu(y).$$

Le crochet [.,.] est associatif, mais en général, non commutatif. Si E est compact et S est un sous-espace dense non trivial dans $\mathcal{C}(E)$, H. Cartan s'appuie sur un lemme de la théorie des opérateurs linéaires de Banach pour prouver l'existence d'une mesure de Radon non nulle μ telle que $\mu(f) = 0$ pour tout $f \in S$.

En 1941, Kakutani [40] rédige une théorie formelle pour des espaces de Banach abstraits constituant des treillis. Dans ce travail il vérifie que si Ω est un espace compact, toute forme linéaire positive F sur $\mathcal{C}(\Omega)$, de norme 1, est caractérisée par une relation

$$\langle f, F\rangle = \int_\Omega f(t)d\mu(t),$$

$f \in \mathcal{C}(\Omega)$, où μ est une fonction d'ensemble positive, complètement additive, définie sur les boréliens de Ω, telle que $\mu(\Omega) = 1$.

MacNeille [56] réalise une approche unifiée des intégrales de Riemann, Stieltjes, Lebesgue, Bochner, d'une part, des intégrales de Denjoy, d'autre part. Pour chaque type il formule une définition reposant sur une approximation. Une fonction s est dite sommable au sens de Lebesgue s'il existe une série $\sum_{n=1}^{\infty} u_n$ de fonctions étagées telle que $\sum_{n=1}^{\infty} u_n(x) = s(x)$ en tout point x pour lequel $\sum_{n=1}^{\infty} |u_n(x)| < +\infty$; $\int s(x)dx = \sum_{n=1}^{\infty} \int u_n(x)dx$. Une fonction s est dite L^0-intégrable sur l'intervalle I si elle est intégrable au sens de Riemann. On procède alors par induction. La fonction s est dite $L^{\zeta+1}$-intégrable sur I si a) quel que soit $\varepsilon \in \mathbb{R}_+^*$, il existe une

L. Schwartz

suite (I_n) d'intervalles dans I de longueur totale inférieure à ε, s n'est L^ζ-intégrable sur aucun I_n, mais L^ζ-intégrable sur $I\setminus\bigcup_{n=1}^{\infty} I_n$, b) $\lim_{\varepsilon\to 0}(L^\zeta)\int_{I\setminus\bigcup_{n=1}^{\infty} I_n} s(x)dx$ existe pour tout choix de (I_n). Cette limite constitue $(L^{\zeta+1})\int_I s(x)dx$. Les fonctions intégrables au sens de Lebesgue sont effectivement les fonctions L^1-intégrables. Les fonctions intégrables au sens de Denjoy sont les fonctions L^Ω, où Ω est le premier ordinal de classe 3. Le processus d'induction s'y arrête.

Le travail [52] de Lévy, qui se place juste avant la première présentation de la théorie des distributions par Schwartz, se propose d'illustrer l'importance du *calcul symbolique* pour la dérivation et l'intégration. Envisageant le problème de la convergence de développements, Lévy dresse ce constat:

«L'intégrale est un outil plus maniable que la dérivée : il s'applique à toutes les fonctions continues et les intégrations répétées conduisent souvent à des suites ou à des séries convergentes, ce qui n'est pas le cas pour les dérivations répétées, même si elles sont possibles» ([52] p. 45).

Il rappelle que Heaviside a désigné par p l'opération de dérivation par rapport à une variable qui se trouve être souvent le temps. Dans le cas e^{px}, *l'opération p équivaut à une multiplication par p.* A partir de

$$f(x) = \int_0^\infty e^{sx} d\Phi(s),$$

Lévy définit la dérivée d'ordre α par la formule

$$p^\alpha[f(x)] = \int_0^\infty s^\alpha e^{sx} d\Phi(s).$$

A propos de la relation entre l'équation de Laplace

$$\psi(t) = \int_0^\infty e^{-st} \varphi(s) ds$$

et le calcul symbolique, Lévy peut affirmer qu'elle

«a été bien mise en évidence par Carson (dans la revue technique The Bell System, 1923; éditée par l'American Telephon and Telegraph Company). Le livre de Heaviside, d'une lecture assez difficile, était peu connu, sauf en Angleterre, et ce sont surtout les travaux de Carson qui ont attiré l'attention sur le calcul symbolique d'Heaviside» ([52] p. 49).

Lévy estime enfin que le *calcul symbolique de Volterra*, relatif au produit de convolution $*$ généralise notablement celui de Heaviside. La transformation de Fourier

$$\hat{f}(z) = \int_{-\infty}^\infty e^{izx} f(x) dx$$

conduit à la relation $\widehat{f_1 * f_2} = \hat{f}_1 \hat{f}_2$ qui *permet de définir des puissances symboliques d'ordres non entiers.*

L'article de Schwartz inaugurant la théorie des distributions en 1945 débute ainsi:

«Depuis l'introduction du calcul symbolique, les physiciens se sont constamment servis de certaines notions ou de certaines formules dont le succès était incontestable, alors qu'elles n'étaient pas justifiées mathématiquement. C'est ainsi que la fonction $y(x)$ de la variable réelle x, égale à 0 pour $x \leq 0$, à 1 pour $x > 0$, est couramment considérée comme ayant pour dérivée la 'fonction de Dirac' $y'(x) = \delta(x)$, nulle pour $x \neq 0$, égale à $+\infty$ pour $x = 0$, et telle que, de plus $\int_{-\infty}^{+\infty} \delta(x) dx = +1$. Un tel 'abus de langage' est malgré tout incompatible avec la notion habituelle de fonction et de dérivation! Et que penser alors de la considération des dérivées successives de la fonction de Dirac! Et pourtant de telles expressions rendent de constants services en électricité et sont très adaptées à l'étude de la transformation de Laplace ou de Fourier et de la mécanique ondulatoire. Le but de cet article est de faire un très bref résumé (et sans démonstrations) d'un travail qui sera publié ultérieurement sous forme de mémoire ou de monographie et qui apportera une justification complète au langage précédent.

[...] Sans vouloir faire des affirmations prématurées, je pense qu'il y a intérêt à utiliser ce formalisme dans un grand nombre de domaines de l'analyse» ([68] p. 57–58).

Schwartz se propose de raisonner sur des objets plus généraux que les fonctions. Ainsi, on ne considère pas δ comme une fonction, mais une *masse* ou *distribution de masses*, qui comporte la masse $+1$ placée à l'origine. Une distribution de masses μ doit être finie et vérifier une condition d'additivité; on définit

$$\mu(A) = \int_A d\mu$$

et l'intégrale de Stieltjes

$$\mu(\varphi) = \int_{-\infty}^{\infty} \varphi(x)dx.$$

Par exemple, dans le cas de la *distribution de Dirac*,

$$\delta(\varphi) = \int_{-\infty}^{\infty} \varphi(x)d\delta = \varphi(0).$$

La formule générale *se trouve être la mieux adaptée à la définition même d'une distribution de masses*. La forme linéaire μ converge vers 0 si la fonction φ est nulle en dehors d'un intervalle borné et converge vers 0. Réciproquement, d'après le théorème de Riesz, toute forme linéaire possédant cette propriété est de ce type. Toute fonction f, sommable sur un intervalle borné, définie presque partout, détermine une *mesure de densité* μ; on écrit

$$\int_{-\infty}^{\infty} \varphi(x)d\mu(x) = \int_{-\infty}^{\infty} \varphi(x)f(x)dx.$$

Comme autres exemples de distributions interviennent des distributions correspondant aux *couches multiples* employées en théorie du potentiel. Un *doublet* de *moment* +1 placé à l'origine est formé des masses $\frac{1}{\varepsilon}$, $-\frac{1}{\varepsilon}$ placées respectivement aux points ε et 0. La forme linéaire ainsi définie vérifie

$$T_\varepsilon(\varphi) = \frac{\varphi(\varepsilon) - \varphi(0)}{\varepsilon};$$

elle n'admet de limite pour ε convergeant vers 0 que si φ est dérivable en 0; alors $T(\varphi) = \varphi'(0)$. On étend ces considérations aux dérivations d'ordre supérieur.

Schwartz introduit alors des concepts fondamentaux. Soit Φ l'ensemble des fonctions φ définies sur $\mathbb{R}^n$, indéfiniment dérivables et nulles en dehors d'ensembles bornés respectifs. Le noyau de φ est le complémentaire du plus grand ensemble ouvert sur lequel la fonction s'annule. Une distribution est toute forme linéaire T sur Φ vérifiant la condition de continuité suivante: Pour toute suite de fonctions (φ_m) admettant leurs noyaux dans un compact fixé, convergeant uniformément vers 0, ainsi que leurs dérivées, on a $\lim_{m\to\infty} T(\varphi_m) = 0$. Toute mesure est une distribution.

Si φ est une fonction indéfiniment dérivable sur $\mathbb{R}^n$, pour

$$f(\varphi) = \int \cdots \int_{\mathbb{R}^n} f(x_1, \ldots, x_n)\varphi(x_1, \ldots, x_n)dx_1 \ldots dx_n$$

on a

$$f'_{x_i}(\varphi) = \int \cdots \int_{\mathbb{R}^n} f'_{x_i}(x_1, \ldots, x_n)\varphi(x_1, \ldots, x_n)dx_1 \ldots dx_n.$$

Comme la fonction φ s'annule en dehors d'une partie bornée, on peut intégrer par parties; d'où

$$f'_{x_i}(\varphi) = f(-\varphi'_{x_i}).$$

Ainsi est suggérée la définition de la dérivée partielle d'une distribution T;

$$T'_{x_i}(\varphi) = T(-\varphi'_{x_i})$$

et, plus généralement,

$$\frac{\partial^{p_1+\ldots+p_n} T}{\partial x_1^{p_1} \ldots \partial x_n^{p_n}}(\varphi) = T\left((-1)^{p_1+\ldots+p_n} \frac{\partial^{p_1+\ldots+p_n} \varphi}{\partial x_1^{p_1} \ldots \partial x_n^{p_n}}\right).$$

L'intégration d'une distribution est toujours possible.

Par extension de la notion de mesure produit, pour des distributions S et T, on peut écrire

$$S \times T(\varphi) = S(\varphi_1)T(\varphi_2),$$

si $\varphi = \varphi_1\varphi_2$. Le produit de convolution passe aussi aux distributions; on étend la formule

$$\begin{aligned} h(\varphi) = \int_{-\infty}^{\infty} h(x)\varphi(x)dx &= \int\int f(x-y)g(y)\varphi(x)dx \\ &= \int\int f(u)g(v)\varphi(u+v)du\,dv \end{aligned}$$

où $h(x) = f * g(x) = \int_{-\infty}^{\infty} f(x-y)g(y)dy$. Alors $\delta * T = T$ et $\frac{\partial\delta}{\partial x_i} * T = \frac{\partial T}{\partial x_i}$.

Au terme de la première moitié de ce siècle, Schwartz publie sa monographie sur les distributions [69].

Au-delà de l'époque sous revue, la théorie de l'intégration, l'une des plus anciennes de l'histoire des mathématiques, continuera à être, sinon l'une des plus actuelles, du moins l'une des plus indispensables.

Bibliographie

[1] ALEXANDROV, PAUL. Über die Äquivalenz des Perronschen und des Denjoyschen Integralbegriffes. *Math. Z.* **20**, 213–222 (1924).

[2] BAUER, HANS. Der Perronsche Integralbegriff und seine Beziehung zum Lebesgueschen. *Monatsh. Math. Phys.* **26**, 153–198 (1915).

[3] BOCHNER, SALOMON. Integration von Funktionen, deren Werte die Elemente eines Vektorraumes sind. *Fund. Math.* **20**, 262–276 (1933).

[4] BOKS, T.J. Sur les rapports entre les méthodes d'intégration de Riemann et Lebesgue. *Rend. Circ. Mat. Palermo* **XLV**, 1–54 (1921).

[5] BOREL, EMILE. *Leçons sur la théorie des fonctions.* Gauthier-Villars, Paris, 1898.

[6] BOREL, EMILE. La théorie des ensembles et les progrès récents de la théorie des fonctions. *Rev. Gén. Sci.* **20**, 315–324 (1909). *Oeuvres* **3**, 1277–1307; CNRS, Paris, 1972.

[7] BOREL, EMILE. Le calcul des intégrales définies. *J. Math. Pures Appl.* **8**, 159–210 (1912).

[8] BOREL, EMILE. Sur l'intégration des fonctions non bornées et sur les définitions constructives. *Ann. Sci. Ecole Norm. Sup.* **36**, 71–91 (1919). *Oeuvres* **2**, 879–899; CNRS, Paris, 1972.

[9] BOURBAKI, NICOLAS. *Eléments de mathématique. Intégration.* Chapitres I–IV. Hermann, Paris, 1952.

[10] BOURBAKI, NICOLAS. *Eléments d'histoire des mathématiques.* Hermann, Paris, 1960.

[11] BOURBAKI, NICOLAS. *Eléments de mathématique. Fonctions d'une variable réelle.* Hermann, Paris, 1976.

[12] BURKILL, JOHN C. The approximately continuous Perron integral. *Math. Z.* **34**, 270–278 (1931).

[13] BURKILL, JOHN C. The Cesàro-Perron integral. *Proc. London Math. Soc.* **34**, 314–322 (1932).

[14] CANTOR, GEORG. Über endliche lineare Punktmannigfaltigkeiten. *Math. Ann.* **23**, 453–488 (1884).

[15] CANTOR, GEORG. *Gesammelte Abhandlungen.* Julius Springer, Berlin, 1932.

[16] CARATHÉODORY, CONSTANTIN. Über das lineare Maß von Punktmengen — eine Verallgemeinerung des Längenbegriffes. *Nachrichten von der Königlichen Gesellschaft der Wissenschaften zu Göttingen*, 404–426, 1914.

[17] CARATHÉODORY, CONSTANTIN. *Vorlesungen über reelle Funktionen.* Teubner, Leipzig-Berlin, 1918.

[18] CARSON, J.R. The Heaviside operational calculus. *Bull. Amer. Math. Soc.* **32**, 43–68 (1926).

[19] CARTAN, HENRI. Sur les fondements de la théorie du potentiel. *Bull. Soc. Math. France* **69**, 71–96 (1941).

[20] DANIELL, P.J. A general form of integral. *Ann. of Math.* **19**, 279–294 (1917).

[21] DENJOY, ARNAUD. Une extension de l'intégrale de M. Lebesgue. *C. R. Acad. Sci. Paris* **154**, 859–862 (1912).

[22] DENJOY, ARNAUD. Sur la dérivation et son calcul inverse. *C. R. Acad. Sci. Paris* **162**, 377–380 (1916).

[23] DENJOY, ARNAUD. Sur l'intégration riemannienne. *C. R. Acad. Sci. Paris* **169**, 219–221 (1919).

[24] DUGAC, PIERRE. Lettres d'Henri Lebesgue à Emile Borel. *Cah. Sém. Hist. Math.* **12** (1991).

[25] DUNFORD, NELSON. Integration in general analysis. *Trans. Amer. Math. Soc.* **37**, 441–453 (1935).

[26] EGOROV, D.-TH. Sur les suites de fonctions mesurables. *C. R. Acad. Sci. Paris* **158**, 244–246 (1911).

[27] FATOU, PIERRE. Séries trigonométriques et séries de Taylor. *Acta Math.* **30**, 335–400 (1906).

[28] FEJÉR, LEOPOLDO. Untersuchungen über Fouriersche Reihen. *Math. Ann.* **58**, 51–69 (1904).

[29] FÉLIX, LUCIENNE. *Message d'un mathématicien.* Albert Blanchard, Paris, 1974.

[30] FRÉCHET, MAURICE. Sur l'intégrale d'une fonctionnelle étendue à un ensemble abstrait. *Bull. Soc. Math. France* **XLIII**, 248–265 (1915).

[31] FUBINI, GUIDO. Sugli integrale multipli. *Atti Accad. Naz. Lincei* **16**, 608–614 (1907).

[32] GRAY, JEREMY. The shaping of the Riesz representation theorem: A chapter in the history of analysis. *Arch. Hist. Exact Sci.* **31**, 127–187 (1984).

[33] HAAR, ALFRED. Der Maßbegriff in der Theorie der kontinuierlichen Gruppen. *Ann. of Math.* **34**, 147–169 (1933).

[34] HADAMARD, JACQUES. Sur les opérations fonctionnelles. *C. R. Acad. Sci. Paris* **136**, 351–354 (1903).

[35] HAHN, HANS. Über eine Verallgemeinerung der Riemannschen Integraldefinition. *Monatsh. Math. Phys.* **26**, 3–18 (1915).

[36] HAKE, HEINRICH. Über de la Vallée Poussins Ober- und Unterfunktionen einfacher Integrale und die Integraldefinition von Perron. *Math. Ann.* **83**, 119–142 (1921).

[37] HELLINGER, ERNST. Neue Begründung der Theorie quadratischer Formen von unendlichvielen Veränderlichen. *J. Reine Angew. Math.* **136**, 210–271 (1909).

[38] HILBERT, DAVID. *Mathematische Probleme*. Akad. Wiss. Göttingen, 253–297 (1900).

[39] JORDAN, CAMILLE. *Cours d'analyse de l'Ecole polytechnique*; 3e édition; 3 tomes. Gauthier-Villars, Paris, 1909–1915.

[40] KAKUTANI, SHIZUO. Concrete representations of abstract (M)-spaces. *Ann. of Math.* **42**, 994–1024 (1941).

[41] KEMPISTY, STÉFAN. Intégrale Denjoy-Stieltjes d'une fonction de deux variables. *C. R. Acad. Sci. Paris* **202**, 1241–1244 (1936).

[42] KOLMOGOROV, ANDREÏ. *Grundbegriffe der Wahrscheinlichkeitsrechnung*. Springer, Berlin, 1933.

[43] LEBESGUE, HENRI. Sur une généralisation de l'intégrale définie. *C. R. Acad. Sci. Paris* **132**, 1025–1028 (1901).

[44] LEBESGUE, HENRI. Intégrale, longueur, aire. *Ann. Mat. Pura Appl.* **7**, 231–259 (1902).

[45] LEBESGUE, HENRI. *Leçons sur l'intégration et la recherche des fonctions primitives*. Gauthier-Villars, Paris, 1904; 2e édition: 1928. Réédition: Chelsea, Bronx.

[46] LEBESGUE, HENRI. Recherches sur la convergence des séries de Fourier. *Math. Ann.* **61**, 251–280 (1905).

[47] LEBESGUE, HENRI. Sur l'intégration des fonctions discontinues. *Ann. Ec. Norm. Sup.* **XXVII**, 361–450 (1910).

[48] LEBESGUE, HENRI. Notice sur la vie et les travaux de Camille Jordan. *Ac. Sci. Mém. Paris* **58**, $xxxix$–ℓxvi (1926).

[49] LERAY, JEAN. Sur le mouvement d'un liquide visqueux emplissant l'espace. *Acta Math.* **63**, 193–248 (1934).

[50] LEVI, BEPPO. Sul principio di Dirichlet. *Rend. Circ. Mat. Palermo* **22**, 293–360 (1906).

[51] LEVI, BEPPO. Sulla nozione dell'integrale. *Analisi matematica, algebraica ed infinitesimale*, 515–524. Zanichelli, Bologna, 1937.

[52] LÉVY, PAUL. Le calcul symbolique et ses principales applications. *Ann. Univ. Grenoble* **21**, 41–56 (1945).

[53] LOOMAN, H. Über die Perronsche Integraldefinition. *Math. Ann.* **93**, 153–156 (1925).

[54] LUSIN, NIKOLAI M. Sur les propriétés des fonctions mesurables. *C. R. Acad. Sci. Paris* **154**, 1688–1690 (1912).

[55] LUSIN, NIKOLAI M. Sur les propriétés de l'intégrale de M. Denjoy. *C. R. Acad. Sci. Paris* **155**, 1475–1478 (1912).

[56] MACNEILLE, H.M. A unified theory of integration. *Proc. Nat. Acad. Sci.* **27**, 71–76 (1941).

[57] NIKODÝM, OTTON M. Sur une généralisation des intégrales de M. J. Radon. *Fund. Math.* **15**, 131–179 (1930).

[58] PEANO, GIUSEPPE. Sulla integrabilità delle funzioni di primo ordine. *Atti Accad. Sci. Torino* **18**, 439–446 (1883).

[59] PERRON, OSKAR. Über den Integralbegriff. *Sitzb. Heidelb. Akad. Wiß. Math.-naturw. Kl.* **A14**, 1–16 (1914).

[60] PIER, JEAN-PAUL. Carathéodory's fundamental contribution to measure theory. *Constantin Carathéodory: An international tribute*, 1120–1145. World Scientific, Singapore, 1991.

[61] RADON, JOHANN. Theorie und Anwendungen der absolut additiven Mengenfunktionen. *Sitzb. Math.-naturw. Kl. Akad. Wiß. Wien* **CXXII**, 1295–1438 (1913). *Collected works* **1**, 45–188. Birkhäuser, Basel, 1987.

[62] RIESZ, FRIGYES. Sur les opérations fonctionnelles linéaires. *C. R. Acad. Sci. Paris* **149**, 974–977 (1909).

[63] RIESZ, FRIGYES. Untersuchungen über Systeme integrierbarer Funktionen. *Math. Ann.* **69**, 449–497 (1910).

[64] RIESZ, FRIGYES. Sur quelques points de la théorie des fonctions sommables. *C. R. Acad. Sci. Paris***154**, 641–643 (1912).

[65] RIESZ, FRIGYES. Sur l'existence de la dérivée des fonctions d'une variable réelle et des fonctions d'intervalle. Verh. Int. Kongr. Math. 258–269, 1932. *Oeuvres complètes* **I**, 264–275. Gauthier-Villars, Paris, 1960.

[66] RIESZ, FRIGYES. L'évolution de la notion d'intégrale depuis Lebesgue. *Ann. Inst. Fourier (Grenoble)* **1**, 29–42 (1949).

[67] SAKS, STANISŁAW. *Théorie de l'intégrale. Theory of the integral.* Monografie Matematyczne, VII. Varsovie-Lwow. Version française, 1933; version anglaise, 1937.

[68] SCHWARTZ, LAURENT. Généralisation de la notion de fonction, de dérivation, de transformation de Fourier et applications mathématiques et physiques. *Ann. Univ. Grenoble* **21**, 57–74 (1945).

[69] SCHWARTZ, LAURENT. *Théorie des distributions*. Hermann, Paris, 1950–1951. Troisième édition: Hermann, Paris, 1966.

[70] SIERPIŃSKI, WACŁAW et NIKOLAI LUSIN. Sur un problème concernant les ensembles mesurables superficiellement. *Fund. Math.* **1**, 112–115 (1920).

[71] SOBOLEV, SERGEI L. Méthode nouvelle à résoudre le problème de Cauchy pour les équations linéaires hyperboliques normales. *Recueil mathématique* **1**, 39–72 (1936).

[72] TEMPLE, GEORGE. *100 Years of Mathematics*. Duckworth, London, 1981.

[73] TONELLI, LEONIDA. Sulla nozione di integrale. *Annali di Mat.* **1**, 105–145 (1923).

[74] VALLÉE POUSSIN, CHARLES DE LA. *Cours d'analyse infinitésimale*. Louvain et Paris, 1903; 4e édition: Uystpruyst-Dieudonné, Louvain, 1922; 10e édition: Librairie universitaire, Louvain, 1947.

[75] VALLÉE POUSSIN, CHARLES DE LA. *Intégrales de Lebesgue, fonctions d'ensemble, classes de Baire*; 2e édition. Gauthier-Villars, Paris, 1934.

[76] VALLÉE POUSSIN, CHARLES DE LA. Sur l'intégrale de Lebesgue. *Trans. Amer. Math. Soc.* **16**, 435–501 (1915).

[77] VAN DER POL, BALTHASAR, et K.F. NIESSEN. Symbolic calculus. *Phil. Mag.* **13**, 537–577 (1932). *Balthasar Van der Pol. Selected Scientific Papers* **II**, 661–701. North-Holland, Amsterdam, 1960.

[78] VITALI, GIUSEPPE. *Sul problema della misura dei gruppi di punti di una retta*. Bologna, 1905.

[79] VITALI, GIUSEPPE. Sulle funzioni integrali. *Atti R. Accad. delle Sci. di Torino* **40**, 753–766 (1905).

[80] WARD, A.J. The Perron-Stieltjes integral. *Math. Z.* **41**, 578–604 (1936).

[81] WEIL, ANDRÉ. *L'intégration dans les groupes topologiques et ses applications*; 1940. Réédition: Hermann, Paris, 1953.

[82] WIENER, NORBERT. The operational calculus. *Math. Ann.* **95**, 557–584 (1926).

[83] YOUNG, W.H. On a new method in the theory of integration. *Proc. London Math. Soc.* **9**, 15–50 (1911).

JEAN-PAUL PIER
Séminaire de mathématique
Centre universitaire de Luxembourg
162A, Avenue de la Faïencerie
L-1511 Luxembourg

Some Remarks on the History of the Prime Number Theorem from 1896 to 1960

Wolfgang Schwarz
Johann Wolfgang Goethe-Universität, Frankfurt am Main

Abstract. The first proof of the *Prime Number Theorem* $\pi(x) \sim x \cdot (\log x)^{-1}$ was given in 1896 by JACQUES HADAMARD (Bull. Soc. Math. France 24, 199–220) and CHARLES-JEAN DE LA VALLÉE-POUSSIN (Ann. Soc. Sci. Brux. 20, 183-256), independently.

Starting with a short discussion of the mathematical achievements culminating in the proof of the Prime Number Theorem, in this article it is tried to sketch the impact of this important result on forthcoming investigations in the field of analytic number theory, in the first half of the twentieth century.

Some highlights, mentioned in this article, are the following:

- The development of sieve methods BRUN, SELBERG,
- The HARDY-LITTLEWOOD method in additive prime number theory,
- PAGE-SIEGEL-WALFISZ prime number theorem,
- Investigations on the zeros and order of the zeta-function in the critical strip LEVINSON,
- The ERDÖS-SELBERG elementary proof of the prime number theorem,
- The BOMBIERI-VINOGRADOV prime number theorem.

In some sections it was felt to be appropriate to mention also more recent developments after 1950.

1 Introduction

Prime numbers are an object of mathematical interest that fascinated mathematicians as well as laymen for at least two millennia. EUCLID knew that there are infinitely many primes, and ERATOSTHENES studied a kind of sieve-method, the so-called *Sieve of Eratosthenes* or better *Small Sieve of Eratosthenes*, which is a convenient way to produce all prime numbers below some rather large bound, say below 10^7 or 10^8.

We shall return to this subject later (see 3.3, 5.3).

Beginning with DIRICHLET, RIEMANN, and TCHEBYCHEFF, deep and important results were proved concerning the theory of the distribution of primes, and many sub-disciplines of number theory arose in the 19th and 20th century from investigations dealing with primes.[1])

In spite of the efforts of a large number of mathematicians there are still many apparently difficult open problems in analytic prime number theory, and therefore it seems worthwile to try to sketch the historical development of the theory of the distribution of prime numbers.

---

1) In Figure 1 the life-spans of important contributors to the theory of prime numbers are given graphically.

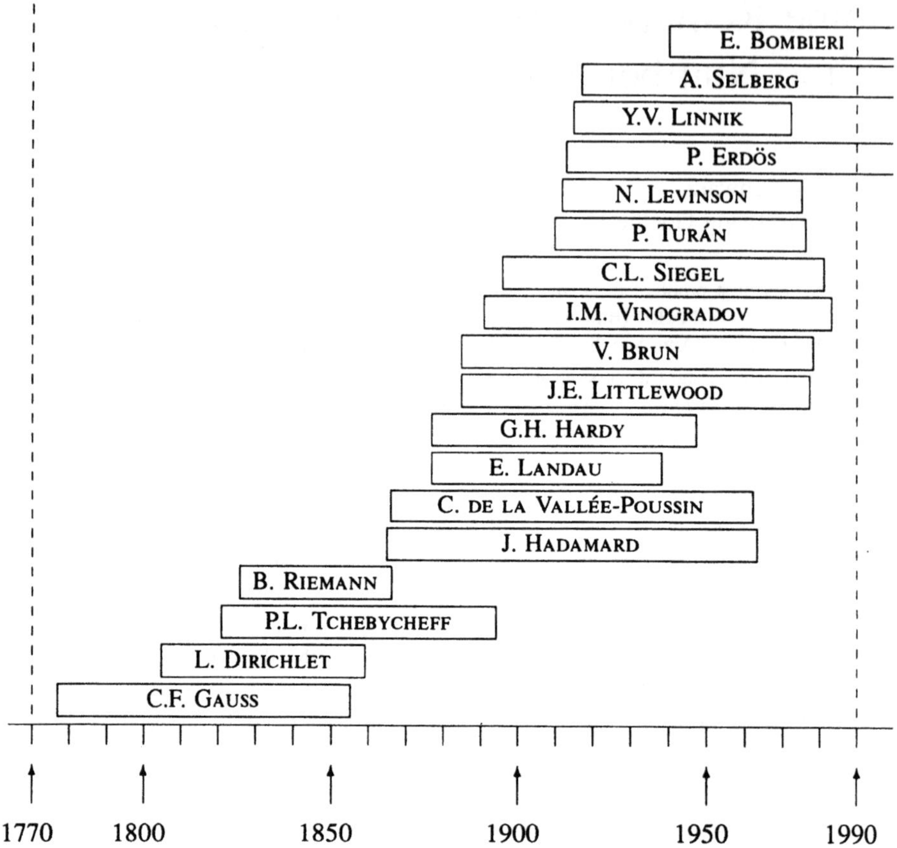

Fig. 1 Chronological Table

1.1 Monographs Dealing with Primes

The task of preparing this lecture on a wellknown, but important part of analytic number theory was comparatively easy; there are many survey-papers and monographs in number theory dealing in part with historical aspects of the theory of primes; some of these are mentioned in the bibliography. For example, let us quote the monographs by P. BUNDSCHUH, H. DAVENPORT, H. EDWARDS, G.H. HARDY and E.M. WRIGHT, L.K. HUA, A. IVIĆ, E. LANDAU, P. RIBENBOIM, and E.C. TITCHMARSH. There is R.K. GUY's book on «Unsolved Problems in Number Theory», 1981, there are surveys on results in analytic number theory by H.H. OSTMANN and LOO-KENG HUA, and there exist the three volumes of L.E. DICKSON's *History of the Theory of Numbers*, and *Reviews in Number Theory*, edited by W.J. LE VEQUE and R.K. GUY. Finally the *Festschrift zum hundertjährigen Jubiläum der Deutschen Mathematiker-Vereinigung*, «Hundert Jahre Mathematik 1890–1990» is to be mentioned.[2])

2) For important hints and valuable help I have to thank my assistant RAINER TSCHIERSCH.

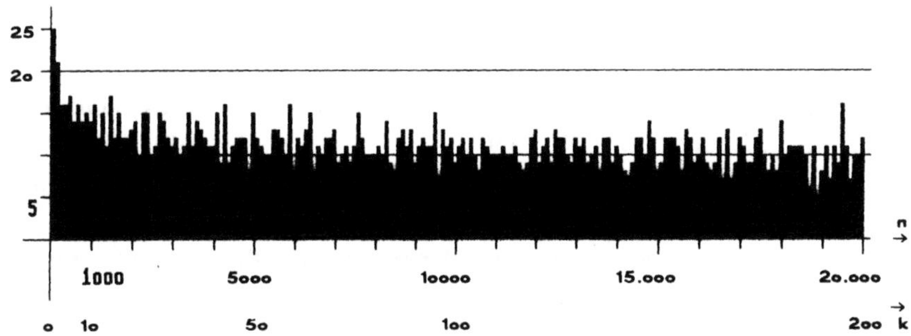

Fig. 2 Irregularity of Primes.

1.2 The Prime-Counting Function

It is easy to find many «small» primes by checking their divisibility properties. However, efforts to find a simple function, producing all the primes or at least infinitely many, were not very successful.[3])

So it is not surprising that, apart from a study of special primes, interest shifted to a study of the counting function

$$\boxed{\pi(x) = \sum_{p \leq x,\ p \text{ prime}} 1} \qquad (1)$$

of the primes. Figure 2 shows[4]) the irregular behaviour of this counting function. This figure gives the number of primes in 200 intervals $[100 \cdot k+1, 100 \cdot (k+1)]$ of length 100. In spite of this apparent irregularity, there is a rather smooth asymptotic formula for $\pi(x)$; see 2.1, 3.1.

1.3 Some Problems Connected with Primes

In spite of the fact that nobody knew how to deal with the problems mentioned below, *conjectures* like

- GOLDBACH's conjecture (1742) on the representation of even (resp. odd) numbers as a sum of two (resp. three) primes,[5])
- the twin prime conjecture on the existence of infinitely many pairs $p, p+2$, where both of these numbers are primes,

---

3) In connection with the negative solution of HILBERT's tenth problem by Y. MATIJASEVIČ in 1970, polynomials in (many) variables were found, whose positive values give exactly all the primes — See, for example, P. JONES, D. SATO, H. WADA and D. WIENS, Amer. Math. Monthly 83, 1975.

4) The graphical illustrations were produced on an ATARI 1040 ST, using the Pascal-SC-system by U. KULISCH and his coauthors from Karlsruhe.

5) This is an excellent example for problems arising when the multiplicative and additive structure of the integers are mixed up. For the analytic part of additive prime number theory we refer to I.M. VINOGRADOV's «The Method of Trigonometrical Sums in the Theory of Numbers». CHEN's famous result, $N = p + P_2$, is given in HALBERSTAM-RICHERT's monograph on Sieve Methods (1974).

- the so-called BERTRAND's postulate (1845): There is always a prime p in the interval $[N, 2N]$, if $N > 2$,
- the question of the infinitude of primes in arithmetical progressions $p \equiv a \bmod q$ (e.g. LEGENDRE 1808, 1830),
- the consideration of FERMAT numbers $F_n = 2^{2^n} + 1$ and MERSENNE primes $M_p = 2^p - 1$, which are of some interest on behalf of their connection with the question of constructing regular polygons with ruler and compass, respectively the problem of the existence of [infinitely many?] even perfect numbers,[6])

showed a permanent interest in questions centering around primes. However, only a few *results* were proved in this time. For example, P. FERMAT got results from the elementary theory of numbers. An important one was on the maximal order of elements in the residue class group $(\mathbf{Z}/q\mathbf{Z})^{\times}$.

The FERMAT conjecture concerning the diophantine equation $x^n + y^n = z^n$ induced countless investigations and helped Algebraic Number Theory coming into being. L. EULER seems to have been the first one to consider the RIEMANN Zeta-Function — but only in the real domain; he gave a new proof that there are infinitely many primes. In particular, he was able to show that the series $\sum_p p^{-1}$ is divergent (1737).

2 The Nineteenth Century

During the course of the nineteenth century *important contributions* to the theory of primes were achieved, which must be mentioned to know the state of knowledge at the end of the 19th century, which is the very beginning of the time considered in this Colloquium.

2.1 C.F. Gauss

Based on extensive, laborious numerical calculations, CARL FRIEDRICH GAUSS[7]) conjectured (see his letter from 1849) the asymptotic behaviour of the function $\pi(x)$ to be

$$\pi(x) \sim \mathrm{li}(x), \text{ as } x \to \infty. \tag{2}$$

This approximation is much better than LEGENDRE's [incorrect] conjecture

$$\pi(x) = x/(\log x - B(x)),$$

where $B(x)$ tends to $1.08\ldots$, as $x \to \infty$, or the asymptotics

$$\pi(x) \sim \frac{x}{\log x}. \tag{3}$$

---

6) P. FERMAT believed that all of «his» numbers are primes. But already L. EULER showed that 641 divides F_5.

7) *April 30, 1777 at Braunschweig, †February 23, 1855 at Göttingen. — See also GOLDSTEIN's article quoted in footnote on page 593.

C.F. GAUSS

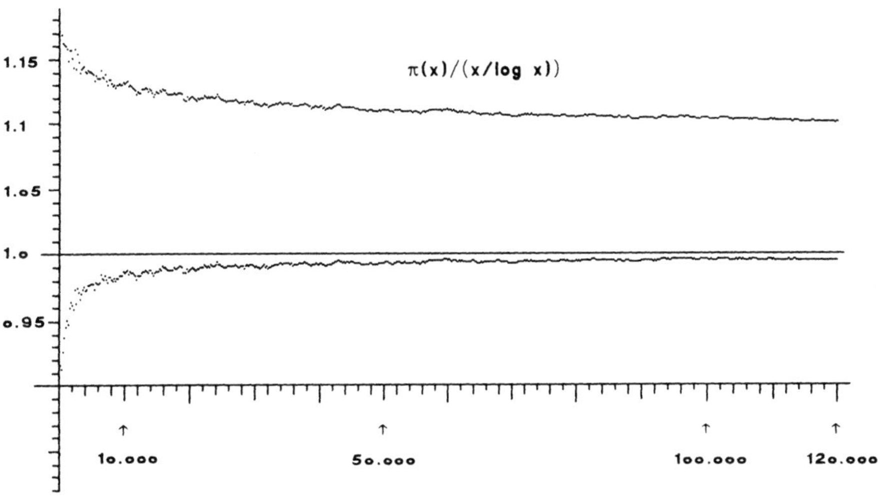

Fig. 3 Number of Primes up to 120000

This may be guessed from Figures 3 and 4, which compare the counting function $\pi(x)$ with the functions $x/\log(x)$ and

$$\mathrm{li}(x) = \mathrm{li}(e) + \int_e^x \frac{1}{\log u}\, du, \tag{4}$$

where $\mathrm{li}(e) = 1.895\ldots$, for

$$x = 200,\ 400,\ \ldots,\ 120000,\ \text{resp.}$$
$$x = 10000,\ 20000,\ \ldots,\ 6\cdot 10^6.$$

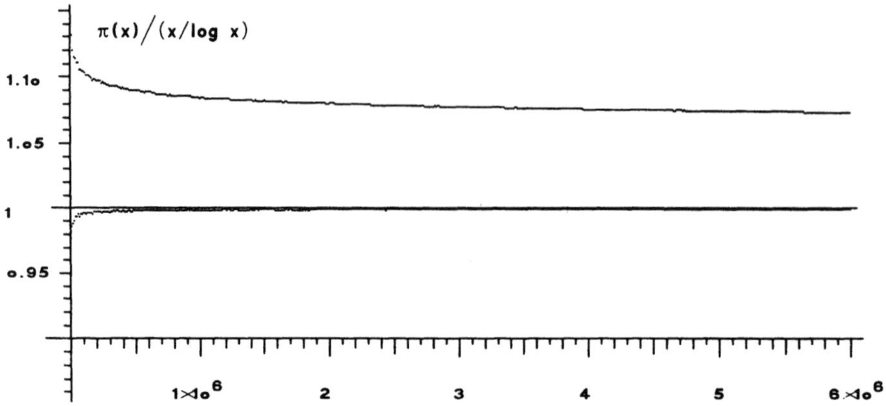

Fig. 4 Number of Primes up to Six Million

The lower graph in Figures 3 and 4 gives the quotient $\pi(x)/\mathrm{li}x$. «Obviously» the integral logarithm is a much better approximation to $\pi(x)$ than the function $x/\log x$, and the conjecture[8]) $\mathrm{li}(x) > \pi(x)$ is supported by Figure (7), giving the graph of the difference $\mathrm{li}(x) - \pi(x)$.

2.2 P.L. Tchebycheff

The Russian mathematician P.L. TCHEBYCHEFF[9]) used elementary estimates to prove BERTRAND's postulate, in fact more precisely[10]) he proved in 1851, 1852 the lower and upper bounds[11])

$$0.92129 \leq \liminf_{x\to\infty} \frac{\pi(x)}{x} \cdot \log x \leq \limsup_{x\to\infty} \frac{\pi(x)}{x} \cdot \log x \leq 1.1056, \tag{5}$$

and, concerning the possible asymptotics for $\pi(x)$, he showed:

$$\text{If} \quad \lim_{x\to\infty} \frac{\pi(x)}{x} \cdot \log x \quad \text{exists, it is equal to } 1. \tag{6}$$

8) This conjecture is wrong, as J.E. LITTLEWOOD showed in 1914.

9) *May 16, 1821, †December 8, 1894, Petersburg

10) It is possible to achieve this by using

$$T(N) = \log(N!) = \sum_{n\leq N} \Lambda(n) \cdot \left[\frac{x}{n}\right]$$

and considering the combination

$$T(N) - T\left(\frac{1}{2}N\right) - T\left(\frac{1}{3}N\right) - T\left(\frac{1}{5}N\right) + T\left(\frac{1}{30}N\right).$$

11) In 1891, SYLVESTER gave the slightly better bounds 0.95695 and 1.071584.

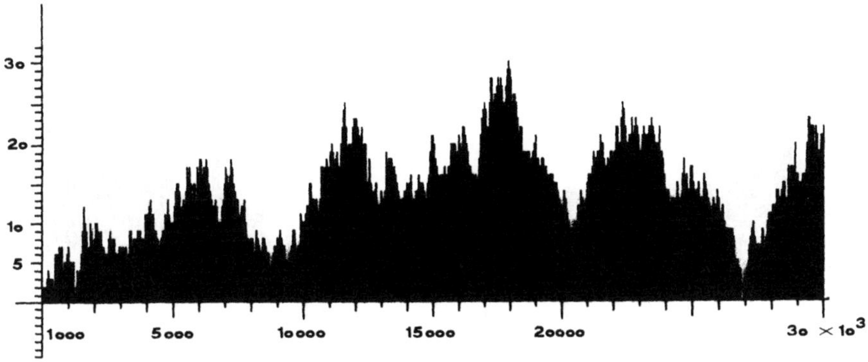

Fig. 5 The Difference $\pi(x;4,3) - \pi(x;4,1)$

The formula

$$\sum_{p\leq x} p^{-1} = \log\log x + A + \mathcal{O}\left(\frac{1}{\log x}\right)$$

is only due to F. MERTENS (1874).

The conjectures of R.D. VON STERNECK, 1897, 1901, and MERTENS, 1897, concerning the order of $\sum_{n\leq x}\mu(n)$, where $\mu(n)$ denotes the MÖBIUS function, will be dealt with in section 8.

There is a (conjectural) statement by TCHEBYCHEFF (1853),

$$\lim_{r\to 0+} \sum_{p>2} (-1)^{\frac{1}{2}\cdot(p-1)} \cdot e^{-pr} = -\infty,$$

which may be imprecisely interpreted as «there are more primes congruent to 3 mod 4 than congruent to 1 mod 4». This statement lies at the basis of PAUL TURÁN's «Comparative Prime Number Theory». TCHEBYCHEFF's statement is equivalent to the truth of the RIEMANN conjecture for the function $L(s,\chi)$, where χ is the non-principal character modulo 4 (HARDY and LITTLEWOOD, Acta Math. 41, and LANDAU, MZ 1, 1918). So it is very unlikely that TCHEBYCHEFF had a proof for his assertion.[12])

2.3 Dirichlet's L-Functions

J.P. LEJEUNE DIRICHLET[13]) used [Dirichlet] characters χ of the residue class group $(\mathbf{Z}/q\mathbf{Z})^\times$ to isolate special residue classes. Using his [Dirichlet] L-functions

$$L(s,\chi) = \sum_{n=1}^{\infty} \frac{1}{n^s} \cdot \chi(n), \tag{7}$$

12) The difference $\pi(x;4,3) - \pi(x;4,1)$ changes sign infinitely often. See also J. LEECH, 1957, and Figures 5 and 6.

13) *February 13, 1805, at Düren, †May 5, 1859, Göttingen.

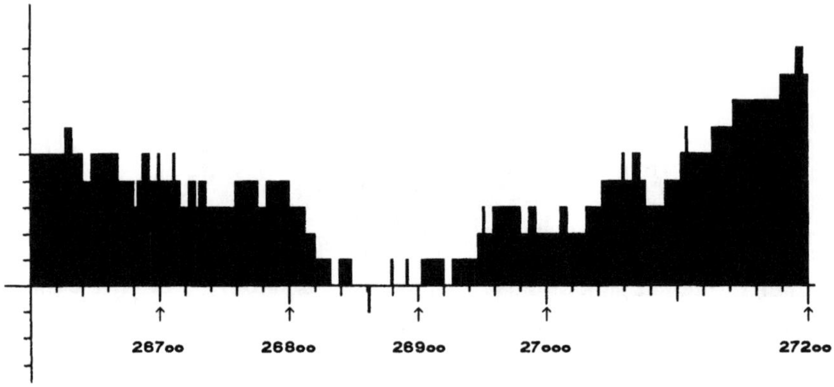

Fig. 6 $\pi(x;4,3) - \pi(x;4,1)$ near $x = 27000$

he introduced transcendental tools into number theory. The problem of the existence of infinitely many primes in residue-classes a mod q, where $\gcd(a,q) = 1$, that is the assertion

$$\pi(x;q,a) = \sum_{p \le x, p \equiv a \bmod q} 1 \to \infty, \quad \text{as} \quad x \to \infty, \tag{8}$$

is reduced to the question of non-vanishing of $L(s,\chi)$ at $s = 1$ (1837). The most difficult case of this assertion, when χ is a real-valued, quadratic character mod q, was dealt with the class-number-formula, proved in 1839.14)

This formula represents the class-number of cyclotomic fields as a product of values $L(1,\chi)$ with some other numbers, connected with the cyclotomic number-field under consideration,

$$h = w \cdot \frac{(|\text{Discriminant}|)^{\frac{1}{2}}}{2^{r_1+r_2} \cdot \pi^{r_2} \cdot \text{Regulator}} \cdot \prod_{\chi \ne \chi_0} L(1,\chi),$$

and so $L(1,\chi)$ cannot be zero.

2.4 B. Riemann's Memoir

BERNHARD RIEMANN,15) in his famous paper «*Über die Anzahl der Primzahlen unter einer gegebenen Größe*», 1859, noticed the importance of studying the zeta-function

$$\zeta(s) = \sum_{n=1}^{\infty} \frac{1}{n^s} = \prod_{p \text{ prime}} \left(1 - \frac{1}{p^s}\right)^{-1}, \quad Re(s) > 1, \tag{9}$$

14) Later, elementary proofs for the fact $L(1,\chi) \ne 0$, were given for example by H.N. SHAPIRO, and A. SELBERG, 1950

15) *September 17, 1826, Breselenz, †July 20, 1866, Selasca, Italy.

R. DEDEKIND

as a function of the complex variable $s = \sigma + it$. RIEMANN proved that $\zeta(s) - (s-1)^{-1}$ is an entire function, and he proved the *functional equation*

$$\pi^{-\frac{1}{2}s} \cdot \Gamma\left(\frac{1}{2}s\right) \cdot \zeta(s) = \pi^{-\frac{1}{2}(1-s)} \cdot \Gamma\left(\frac{1}{2}(1-s)\right) \cdot \zeta(1-s). \tag{10}$$

Showing extremely deep insight, he stated (without proofs) the **conjectures**

- $\zeta(s)$ has infinitely many zeros in the critical strip $0 \leq \sigma \leq 1$.

L. DIRICHLET

- The number $N(T)$ of zeros of $\zeta(s)$ in the critical strip, having imaginary part t within the interval $]0, T]$ satisfies the asymptotic relation

$$N(T) = \frac{T}{2\pi} \cdot \log\left(\frac{T}{2\pi}\right) - \left(\frac{T}{2\pi}\right) + \mathcal{O}(\log T). \tag{11}$$

- The entire function

$$\xi(s) = \frac{1}{2}s \cdot (s-1) \cdot \pi^{-\frac{1}{2}s} \cdot \Gamma\left(\frac{1}{2}s\right) \cdot \zeta(s) \tag{12}$$

is representable as a WEIERSTRASS-HADAMARD product

$$\xi(s) = e^{A+Bs} \cdot \prod_{\rho} \left(1 - \frac{s}{\rho}\right) \cdot \exp\left(\frac{s}{\rho}\right),$$

where ρ runs over the zeros of $\zeta(s)$ in the critical strip.

- There is an *explicit formula* for the difference $\pi(x) - \mathrm{li}(x)$ similar to the simpler one, given below for the related function

$$\psi(x) = \sum_{n \leq x} \Lambda(n) = x - \sum_{\rho} \frac{x^{\rho}}{\rho} - \frac{\zeta'(0)}{\zeta(0)} - \frac{1}{2} \cdot \log\left(1 - \frac{1}{x^2}\right), \tag{13}$$

where the H. VON MANGOLDT function Λ is defined by $\Lambda(n) = \log(p)$, if n is a power of the prime p, and where $\Lambda(n) = 0$ otherwise; it is assumed that $x > 1$ is not a power of a prime.
- All the [complex] zeros ρ of the zeta-function with real part $0 \leq Re(\rho) \leq 1$ have real part equal to $\frac{1}{2}$. This is the famous RIEMANN conjecture, which is unsettled up to now.

More than thirty years were to elapse, before any of the conjectures mentioned above was proved. H. DAVENPORT [1967], p. 63 wrote:

«There is very little indication of how RIEMANN was led to some of these conjectures. In 1932 SIEGEL (*Quellen und Studien zur Geschichte der Mathematik* vol. 2, 45–80 (1932)) published an asymptotic expansion for $\zeta(s)$, valid in the critical strip, which had its origin in notes of RIEMANN preserved in the Göttingen University Library. From SIEGEL's description of the notes, it is plain that RIEMANN had more knowledge about $\zeta(s)$ than is apparent from his published memoir; but there is no reason to think that he had proofs of any of his conjectures.»

From the just mentioned paper of SIEGEL we quote

> *«Die Legende,* RIEMANN *habe die Resultate seiner mathematischen Arbeit durch «große allgemeine» Ideen gefunden, ohne die formalen Hilfsmittel der Analysis zu benötigen, ist wohl jetzt nicht mehr so verbreitet wie zu* KLEIN*'s Lebzeiten. Wie stark* RIEMANNS *analytische Technik war, geht besonders deutlich aus seiner Ableitung und Umformung der semikonvergenten Reihe für* $\zeta(s)$ *hervor.»*

The state of knowledge, sketched in section 2, is the starting point for the developments during the next 50 or 60 years, the very period of time we are interested in.[16])

3 Hadamard and de la Vallée-Poussin

RIEMANN's memoir had shown the importance of a profound knowledge of the zeta-function as a function in the complex domain in order to obtain results in the theory of primes stronger than those already known. In 1892, 1893, JACQUES

16) In 1900, at the *International Congress of Mathematicians* at Paris, in his famous lecture on «Mathematical Problems», DAVID HILBERT dealt with «Problems of Prime Numbers» in his eighth problem.

B. RIEMANN

HADAMARD[17])[18]) began a systematic study of entire functions of finite order and proved «his» factorization theorem, which includes RIEMANN's product formula for $\xi(s)$. With a weaker error term, H. VON MANGOLDT proved the formula for the number $N(T)$ of zeros in the critical strip in 1894, 1895,[19]) and the «explicit

17) *December 8, 1865 at Versailles, †October 17, 1963 at Paris.

18) One section (p. 36–45) in M. VON RENTELN [1987] is devoted to the life and achievements of J. HADAMARD. He mentions the tragic fate of the three sons of J. HADAMARD: PIERRE and ETIENNE lost their lives in 1916 at the front-line during the first world-war, and the last son, MATHIEU, lost his life during the second world-war.

formula» in the same year. These investigations paved the way for a proof of the prime number theorem.

3.1 The Prime Number Theorem

The asymptotic behaviour of the prime-counting function $\pi(x)$, conjectured by C.F. GAUSS, the so-called *Prime Number Theorem*

$$\boxed{\pi(x) \sim \frac{x}{\log x}} \tag{14}$$

was finally proved in 1896 independently by JACQUES HADAMARD (Bull. Soc. Math. France 24, 199–220) and CHARLES-JEAN DE LA VALLÉE-POUSSIN[20]) (Ann. Soc. Sci. Brux. 20, 183–256).

The often used remainder term (15) is due to CH.-J. DE LA VALLÉE-POUSSIN (1899):

$$\boxed{\pi(x) = \mathrm{li}(x) + \mathcal{O}\left(x \cdot \exp\left(-c \cdot \sqrt{\log x}\right)\right).} \tag{15}$$

Consequences of the prime number theorem are, for example, asymptotic formulae for the number of square-free integers, for sums or products containing primes, like

$$\sum_{p \le x} \frac{\log p}{p}$$

or

$$\prod_{p \le x} \left(1 - p^{-1}\right),$$

or for the behaviour of the MÖBIUS function $\mu(n)$ in the mean, or the MERTENS formula mentioned in section 2.2.

Primes in arithmetical progressions were treated also, with the result

$$\pi(x; q, a) = \sum_{p \le x, p \equiv a \bmod q} 1 \sim \frac{1}{\varphi(q)} \cdot \frac{x}{\log x}, \quad \text{if } \gcd(a, q) = 1. \tag{16}$$

---

19) He proved the formula with the error term stated in (11) in 1905. A simpler proof is due to E. LANDAU, 1908. If the RIEMANN conjecture is true, then, according to H. CRAMÉR, 1918, the remainder term can be slightly improved to

$$\mathcal{O}\left(\log T \cdot \sqrt{\frac{\log\log\log T}{\log\log T}}\right).$$

20) *August 14, 1866, Louvain, Belgium, †March 2, 1962 at Brussels. There is an informative article *En marge d'un anniversaire et d'une inauguration: le mathématicien louvaniste Charles-Jean de la Vallée Poussin* on the life of DE LA VALLÉE-POUSSIN by JEAN MAWHIN, UCL, in *Louvain* **88**/2, 12–14, 1988.

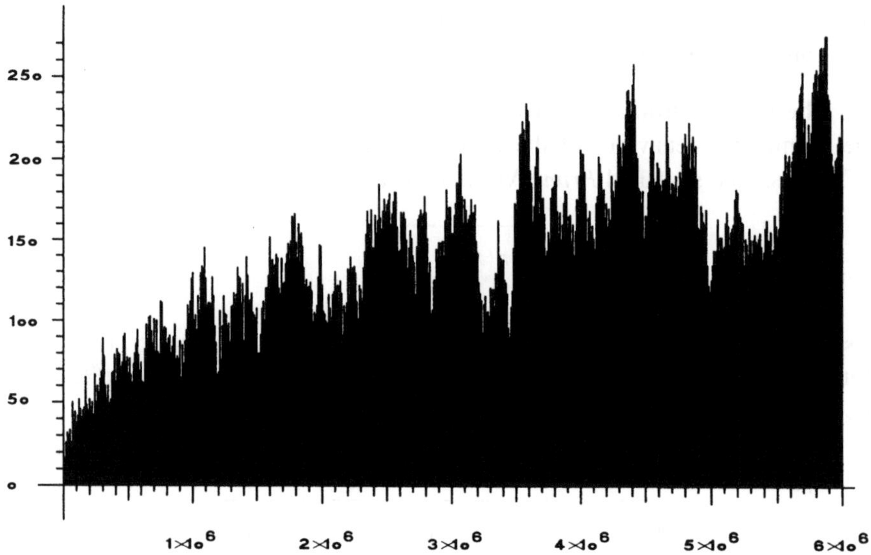

Fig. 7 Difference $\mathrm{li}(x) - \pi(x)$

The functional equation for DIRICHLET L-functions was first given by A. HURWITZ[21]) in 1882. The counting function for prime-ideals (with norm less than some bound) in algebraic number fields could not be dealt with by the methods, developed by J. HADAMARD, because sufficient information on the analytic behaviour of the DEDEKIND zeta-function in algebraic number fields was not available at that time,[22]) and so it was not possible to give a proof of the prime-ideal-theorem.

3.2 Further Research Problems Concerning the Distribution of Primes

Subsequently, the proof of the prime number theorem (abbreviated by PNT) having been achieved — what was left in the field of prime numbers for number theorists, working in the first half of the twentieth century? In general, the solution of an important problem (often) creates an abundance of new questions, and this was true for the prime number theorem, too. Let us state some problems, which were dealt with in the sequel.

- Simplification of the analytical proof, specification and explanation of the connection between zero-free regions for $\zeta(s)$ and the order of the remainder term in the PNT.
- Search for different methods of proving the PNT (use of TAUBERian Theorems (LANDAU-IKEHARA, WIENER), «elementary» proofs (P. ERDÖS, A. SELBERG)),[23]) use of Fourier analysis (HELSON).

---

21) *March 26, 1859, †November 18, 1919.

22) The necessary information came only in 1917 from results of E. HECKE. See also LANDAU's «Einführung in die elementare und analytische Theorie der algebraischen Zahlen und der Ideale», 1917, 1927.

- Improvements of the remainder term in the prime number theorem; by the connection of this question with the position of the zeros of the RIEMANN zeta-function in the complex plane, mentioned above, this means to give an enlargement of the zero-free region of the ζ-function, and investigations of this kind depend upon estimates of exponential sums

$$\sum_{N<n\leq N_1} n^{it},$$

where $N_1 \leq 2N$.
- So it became urgent to study the zeta-function for its own sake, in particular its zeros, and therefore an approximate functional equation, zero-free regions, zeros on the critical line $\mathrm{Re}(s) = \frac{1}{2}$, gaps between the zeros on the critical line, density theorems for the zeros, and estimates of the order of the zeta-function (in the critical strip $0 \leq \mathrm{Re} \leq 1$)[24]) were dealt with. See the monographs by TITCHMARSH [1951], EDWARDS [1974], and IVIĆ [1985].
- Upper and lower estimates for gaps between consecutive primes.
- Study of analytical properties, order estimates, and zeros of the DIRICHLET L-functions, including «Density-Theorems».
- Definition of more general zeta-functions[25])[26]) and exhibition of their properties (meromorphic behaviour, poles, zeros, functional equation, product representation, etc.).
- Characterization of the RIEMANN zeta-function as a meromorphic function of finite genus, which is representable as a DIRICHLET series, and possessing a functional equation (HAMBURGER, 1922).

---

23) The remarks in the popular book «Who got Einstein's office?» by ED REGIS, concerning a comparison between the proofs of the PNT by SELBERG on the one hand, and HADAMARD and DE LA VALLÉE-POUSSIN on the other, are misleading.

24) The question is to give upper bounds for $\zeta(\sigma + it)$ in terms of t (and σ).

25) Such zeta-functions are connected, for example, with the names of E. ARTIN, R. DEDEKIND, P. EPSTEIN, H. HASSE, E. HECKE, A. HURWITZ, A. SELBERG, C.L. SIEGEL, A. WEIL; others are known as the «congruence zeta-function» or «zeta-functions of elliptic curves», and there are the «p-adic L-functions». The BIRCH-SWINNERTON-DYER conjecture connects algebraic properties of the group of rational points on an elliptic curve with analytic properties (order of the zero at $s = 1$) of the corresponding L-function.

26) A famous example of a zeta-function defined by HECKE operators is the RAMANUJAN τ-function

$$\sum_{n=1}^{\infty} \tau(n) \cdot n^{-s} = \prod_{p} \left(1 - \tau(p) \cdot p^{-s} + p^{11-2s}\right)^{-1},$$

where $\tau(n)$ is defined by the FOURIER expansion

$$\Delta(z) = \exp(2\pi iz) \cdot \prod_{m=1}^{\infty} (1 - \exp(2\pi imz))^{24} = \sum_{n=1}^{\infty} \tau(n) \cdot \exp(2\pi inz).$$

- The RIEMANN conjecture and its implications, analogues of this conjecture, equivalent formulations, study of weaker conjectures like LINDELÖF's conjecture.
- Exhibition of irregularities in the distribution of primes, including the so-called «Comparative Prime Number theory» in the terminology of PAUL TURÁN.[27)]
- Study of sums over the MÖBIUS function, connected with the generating DIRICHLET series
$$\zeta^{-1}(s) = \sum_{n=1}^{\infty} \mu(n) \cdot n^{-s}.$$
For example, $\sum_{n=1}^{\infty} n^{-1} \cdot \mu(n) = 0$, and
$$\lim_{x \to \infty} \frac{1}{x} \cdot \sum_{n \leq x} \mu(n) = 0.$$
- Study of primes p in arithmetical progressions, $p \equiv a \bmod q$. **One** new aspect is the problem of obtaining results holding **uniformly** in a and q in some range. Another aspect is to obtain upper bounds for the *least prime* $p(q,a)$ in the arithmetical progression $p \equiv a \bmod q$ in terms of q:
$$p(q,a) \ll q^L$$
with some constant L (YU.V. LINNIK 1944, simplification of the proof by K.A. RODOSSKII, 1954). Values for the constant L are due to M. JUTILA (1977, $L = 80$), CHEN JING RUN (1979, $L = 17$).
In the opposite direction, in 1980, C. POMERANCE showed
$$\liminf_{q \to \infty} \max_{a,\ \gcd(a,q)=1} \frac{p(q,a)}{\varphi(q) \log q} \geq e^C = 1.781\ldots.$$
In this result, the constant 1.781 is important, the constant '1' would be trivial.
- Generalization of the Prime Number Theorem to algebraic number fields (Prime ideal theorem, TCHEBOTARËV's density theorem).
- Prime numbers in «arithmetical semigroups» (see A. BEURLING, J.E. NYMAN, P. MALLIAVIN, H. DIAMOND, J. KNOPFMACHER, and others). Of course, the results depend strongly on the asymptotic formula assumed for the number of elements in the arithmetical semigroup under consideration.
- Distribution of primes in sequences of the shape $\mathrm{INT}[n^\alpha]$, where α is a little bit larger than 1, more precisely, $1 \leq \alpha \leq \frac{10}{9}$, according to G. KOLESNIK,

27) We cannot touch this subject here, but mention the names of P. TURÁN, S. KNAPOWSKI, and J. PINTZ. Results of this kind are treated in TURÁN's book «A new method in analysis».

1967; the result is known as the prime number theorem of PJATECKII-SHAPIRO (1953). More recent [and better] results are due to LEITMANN, 1980, HEATH-BROWN, 1983, KOLESNIK, 1985; A. BALOG and G. DUFNER dealt with the problem of primes in the sequence $[p^c]$.

- Asymptotic results concerning the function $\pi_r(x)$, the number of integers $n \leq x$ composed of exactly r primes. The first results for this problem are due to E. LANDAU, 1908, and G.H. HARDY and S. RAMANUJAN, 1917. For the difficult question of obtaining results which are **uniform** in some range of r there are important theorems due to P. ERDÖS, L.G. SATHE, A. SELBERG, J.-L. NICOLAS, A. HILDEBRAND and G. TENENBAUM. We cannot go into details, but refer to the survey article «On the number of prime factors of an integer» by A. HILDEBRAND in «Ramanujan Revisited», 1988, pp. 167–185.
- Study of the number of prime twins or prime-k-tuples, and investigations on the number of primes in more general sequences, for example primes in the sequence $\{f(n)\}$, where f an integer-valued polynomial. A lot of interesting conjectures are given in a paper of G.H. HARDY and J.E. LITTLEWOOD in 1923. But there are only a few results concerning these conjectures — except for upper and lower estimates obtained with sieve methods.[28])
- Applications of the prime number theorem (in particular, the PNT for arithmetical progressions) to additive prime number theory (Theorem of GOLDBACH-VINOGRADOV,[29]) Theorem of GOLDBACH-WARING), and to a study of diophantine inequalities with primes. An important ingredient is the estimation of exponential sums with primes; estimates of this kind are also useful for the derivation of results on the uniform distribution of sequences dependent on the sequence of primes.[30])

3.3 New Branches of Number theory

It was to be expected, that the investigation of deep open problems like the RIEMANN conjecture concerning the non-trivial zeros of the zeta-function or the problem of the infinitude of prime twins would result in the appearance of new methods and techniques in number theory, but it was not foreseeable at the beginning of our century which new branches of number theory would emerge.

---

28) See the monograph of H. HALBERSTAM and H.-E. RICHERT. Lower estimates in general are obtainable only for «almost-primes»; these are integers without 'small' prime-divisors, and so they have only 'few' prime-factors.

29) Under the assumption of [parts of] the Great RIEMANN hypothesis, the *Three-Primes-Theorem* was proven by HARDY and LITTLEWOOD. Without any unproven conjecture, using the sieve method, L.G. SCHNIRELMANN was able to show that there is some constant c with the property that every integer greater than one is a sum of at most c primes (1930); explicit values for c were given by ROMANOFF, HEILBRONN, LANDAU & SCHERK, and others.

30) Concerning this subject we quote the Reviews by L.K. HUA, H.H. OSTMANN and monographs by L.K. HUA and I.M. VINOGRADOV. An excellent reference for questions concerning uniform distribution mod 1, is the monograph «Uniform Distribution of Sequences» by L. KUIPERS and H. NIEDERREITER, 1974.

- *Computational Number Theory* or *Algorithmic Number Theory* aimed at a design of algorithms for large-scale numerical calculations in number theory; at first, calculations were done using mechanical desk-calculators, however, many of the striking successes of Computational Number Theory depend on the development of programmable, fast electronic computers, and, of course, on the construction of adequate algorithms, reducing the 'complexity' of the necessary calculations. Problems dealt with in Algorithmic Number Theory are, for example
  - Prime Number Tests (A.E. WESTERN, J.C. MOREHEAD, D. SHANKS, J. POLLARD, G.L. MILLER, M.O. RABIN, V. STRASSEN, H.W. LENSTRA and others),[31])
  - Factorization of large integers (J.D. BRILLHART, J.D. DIXON, M.A. MORRISON, J.C.P. MILLER, C.-P. SCHNORR, and others),
  - Counting the number of primes (E.D.F. MEISSEL, D.H. LEHMER, J.C. LAGARIAS, V. MILLER, A.M. ODLYZKO, and others),
  - Investigation of MERSENNE primes[32]) and FERMAT numbers,
  - Calculations connected with the zeros of the RIEMANN zeta-function on the critical line (J.P. GRAM, R. BACKLUND, L.J. COMRIE, E.C. TITCHMARSH, R.P. BRENT, J. VAN DE LUNE, H.J.J. TE RIELE, D.T. WINTER). The first 1.5 billion zeros of the zeta-function are exactly **on** the critical line (1986). Sign-changes of the real-valued function

$$Z(t) = \zeta\left(\frac{1}{2} + it\right) \cdot \pi^{-\frac{1}{2}it} \cdot \frac{\Gamma\left(\frac{1}{4} + \frac{1}{2}it\right)}{\left|\Gamma\left(\frac{1}{4} + \frac{1}{2}it\right)\right|}$$

    correspond exactly to the zeros of the zeta-function on the critical line, and so the 'only' problem left is to calculate the exact number of zeros in the critical strip $0 \leq \sigma \leq 1$ with imaginary part $0 \leq \gamma \leq T$.
  - The possibility of providing many large primes and the extreme difficulty of factoring large integers into its (few) prime factors is exploited in the public key RSA-crypto-encoding method by R.L. RIVEST, A. SHAMIR, L. ADLEMAN, Comm. ACM 21, 120–126, 1978.[33])
- The *Development of Sieve Methods* started between 1915 and 1920 with VIGGO BRUN,[34]) with the surprising result, that the series

$$\sum_{p,p \text{ and } p+2 \text{ prime}} p^{-1}$$

---

31) See, for example, LENSTRA [1984], POMERANCE [1980].

32) The largest known MERSENNE primes (April 1992, Am. Math. Monthly) are $2^{110503} - 1, 2^{216092} - 1$, and $2^{756839} - 1$.

33) See KRANAKIS [1986], KOBLITZ [1987].

34) *October, 13, 1885, †Aug. 15, 1978 near Oslo.

is convergent (1919), in contrast to the divergence of $\sum_p p^{-1}$. Starting about 1920, applications were given to the problem of representing large even numbers N as a sum of two «almost-primes»,

$$N = P_r + P_s,$$

where P_r denotes an integer composed of at most r primes.[35]) A. SELBERG (1947) created a much simpler manageable version of the sieve method, which superseded BRUN's method and lead to more precise results, and in the mean-time this so-called «small sieve» is a standard device of number theory.[36])

Another kind of sieve method, the «Large Sieve», depending on ideas from probability theory, dates back to YU.V. LINNIK, 1941, and A. RÉNYI, 1949, 1950; this method was developed and applied to difficult questions from analytic number theory (representation of even numbers as a sum of a prime and an almost-prime) by RÉNYI himself, by K.F. ROTH, E. BOMBIERI, H.L. MONTGOMERY and others (See also M.B. BARBAN's survey [1966] and MONTGOMERY's monograph on «Multiplicative Number Theory» [1971]). A well-readable account of the ideas leading to the different sieve methods may be found in the monograph «Sequences» by H. HALBERSTAM and K.F. ROTH, Oxford 1966, Chapter IV.

In the next sections we are going to describe in short some of the developments (between about 1900 and 1950) for some of the topics of prime number theory mentioned above. At first we deal with the question of the possibility of simplifying the proof of the prime number theorem.

4 Other Proofs of the Prime Number Theorem

The proof of the prime number theorem known so far (about 1900) was difficult, a good deal of knowledge about entire functions of finite order was necessary. So, similar to the attempts to simplify the transcendence proof of F. LINDEMANN for π, it was considered desirable to find simpler proofs for the PNT, and to find proofs which could be generalized to algebraic number fields.

---

35) See papers of V. BRUN, H. RADEMACHER (1924), T. ESTERMANN (1932), A.A. BUCHSTAB 1938, and others.

36) «Indeed, Selberg's upper-bound method has a certain air of finality, ...» was written by HALBERSTAM and ROTH [1966], p. 190. For the development of sieve methods until 1975 and the countless important applications see the monograph «Sieve Methods» by H. HALBERSTAM and H.-E. RICHERT [1974].

4.1 E. Landau

I think one should start with the mathematical activities of EDMUND LANDAU,[37]) beginning with his dissertation in 1899; in this paper LANDAU gave a new proof of H. VON MANGOLDT's result

$$\sum_{n=1}^{\infty} n^{-1} \cdot \mu(n) = 0.$$

LANDAU introduced new techniques into number theory, and he had strong ambitions aiming at absolute correctness and simplicity of proofs, giving all details, but, unfortunately, often hiding important ideas in a detailed presentation; his style of writing was later described as the «Landau-style». Next, E. LANDAU tried to «elementarize» known results from prime number theory, and he deduced «elementary equivalences» between different, seemingly disconnected results in prime number theory.[38])

G.H. HARDY and H. HEILBRONN stated (1938, JLMS, p. 309), that no one was ever more passionately devoted to mathematics than LANDAU, and L. MIRSKY wrote in 1977:

«LANDAU's discoveries were legion, their originality is undoubted, and almost all his work (whether concerned with new results or with simplifications of results of other authors) is devoted to exceptionally difficult questions. ... Where previously there had been (broadly speaking) only isolated forays, he organized a systematic attack along the entire front of the analytic theory of numbers and of much of analysis ....»

In 1903, E. LANDAU published an important paper «Neuer Beweis des Primzahlsatzes und Beweis des Primidealsatzes»; in this paper he was able to avoid an application of HADAMARD's machinery from the theory of entire functions, and he worked with the zeta-function in a region very near to the straight line $\mathrm{Re}(s) = 1$.

37) *February 14, 1877, in Berlin, †February 19, 1938 Berlin. In 1933 the political situation forced him to resign his chair at Göttingen. His enforced retirement must have been a terrible blow to him (according to HARDY and HEILBRONN, 1938 — see the «Collected Works of E. Landau», Thales Verlag, Vol.1 to 10, with comments (main editor: H. WEFELSCHEID)). — See also M. VON RENTELN [1987], p. 73–96.

38) In an elementary way, E. LANDAU showed, for example, that $\vartheta(x) \sim x$ implies

$$\sum_{n=1}^{\infty} n^{-1} \cdot \mu(n) = 0 \text{ and } \sum_{n \le x} \mu(n) = o(x),$$

and this implies

$$\sum_{n \le x} \mu^2(n) = \frac{6}{\pi^2} \cdot x + o\left(\sqrt{x}\right).$$

See, for more details, «Collected works of E. Landau», Vol. 3, p. 46.

E. LANDAU

E. LANDAU showed in some small region to the left of the vertical line $\mathrm{Re}(s) = 1$

$$\zeta(s) \neq 0 \text{ in } \sigma > 1 - \frac{B}{\log(3 + |t|)^9},$$

and he obtained the estimate

$$\frac{\zeta'(s)}{\zeta(s)} \ll \log^9 |t|$$

in this region. Using the complex inversion formula

$$\sum_{p\leq x} \log p \cdot \log\left(\frac{x}{p}\right) + \mathcal{O}\left(\sqrt{x}\cdot\log^2 x\right) = -\frac{1}{2\pi i}\cdot\int_{2-i\infty}^{2+i\infty}\frac{x^s}{s^2}\cdot\frac{\zeta'(s)}{\zeta(s)}ds,$$

he got the Prime Number Theorem with the (weaker than (15)) remainder term

$$\mathcal{O}\left(x\cdot\exp(-\sqrt[13]{\log x})\right).$$

One of the advantages of this definitely simpler method was that it could be applied to the problem of counting prime ideals in algebraic number fields, and so it was possible for LANDAU to give a proof of the long searched-for prime ideal theorem for the first time. Many results proven by LANDAU up to 1909 are given in his famous monograph «Handbuch der Lehre von der Verteilung der Primzahlen», and therefore we omit these results. However, I would like to stress that E. LANDAU's continuing efforts were of great influence and importance for the development of analytic number theory. Later, beginning about 1912, strong impetus for progress in number theory came from researchers in Great Britain (G.H. HARDY, J.E. LITTLEWOOD, and S. RAMANUJAN, who spent the years 1914 until 1919 in England, invited by HARDY). In Russia, I.M. VINOGRADOV and A. WALFISZ dealt with difficult questions, and C.L. SIEGEL[39]) exerted strong influence on number theory in Germany.

4.2 Tauberian Theorems

In 1907 E. LANDAU proved a TAUBERian theorem, which allowed the deduction of the PNT from order conditions on the DIRICHLET series $-\zeta'(s)/\zeta(s)$ and from the fact that $\zeta(s) \neq 0$ on the vertical line $\text{Re}(s) = 1$. The order conditions were removed by S. IKEHARA in 1931, and his famous TAUBERian theorem gives a simple method to infer the prime number theorem from the non-vanishing of $\zeta(s)$ on the vertical line $\text{Re}(s) = 1$.

Assume the function $A : [0,\infty[\to \mathbf{R}$ is monotonely non-decreasing, $A(0) = 0$, and the LAPLACE*-integral*

$$f(s) = s\cdot\int_0^\infty A(u)\cdot e^{-su}\,du$$

is convergent in $Re(s) > 1$. Further assume the following TAUBER*ian condition: For any $\lambda > 0$, uniformly in $|t| \leq 2\lambda$, the function*

$$(\sigma+it)^{-1}\cdot f(\sigma+it) - a\cdot(\sigma+it-1)^{-1}$$

converges to some (continuous) function $h(t)$, as $\sigma\to 1+$. If $a \neq 0$ then

$$A(u)\cdot e^{-u}\to a, \qquad \text{as} \quad u\to\infty.$$

---

39) *December 31, 1896, Berlin, †April 4, 1981, Göttingen; Professor in Frankfurt 1922–1938, in Göttingen 1938–1959 (emigration 1940–1951).

Application of this «IKEHARA Theorem» to the DIRICHLET series $-\zeta'(s)/\zeta(s)$ immediately gives $\psi(x) \sim x$.

The *General* TAUBER*ian Theorem* of N. WIENER, 1933:

Assume

- *the functions* f_1, f_2 *are in* $L_1(]-\infty,\infty[)$,
- p *is bounded (and measurable), and*
- *the* FOURIER *transform* $\hat{f}_1$ *does not vanish.*

If

$$\lim_{x\to\infty} (f_1 * p)(x) = A \cdot \int_{-\infty}^{\infty} f_1(t)dt,$$

then

$$\lim_{x\to\infty} (f_2 * p)(x) = A \cdot \int_{-\infty}^{\infty} f_2(t)dt,$$

provides further possibilities of proving the prime number theorem.

There are also versions of TAUBERian theorems with rather weak remainder terms, leading to very weak estimates of the remainder term in the PNT; see T.H. GANELIUS, *Tauberian Remainder Terms*, Springer 1971, for further information. In 1973, H. HELSON (Conf. Montpellier 1972) used ideas from FOURIER analysis to prove the PNT with a (weak) remainder term. He used PLANCHEREL's theorem and the boundedness of the integral

$$\int_{-\infty}^{+\infty} |s \cdot \zeta(s+1)|^{-2} dt$$

for $\sigma \to 0+$, to show $\sum_{n=1}^{\infty} n^{-1} \cdot \mu(n) = 0$, and to estimate

$$\sum_{n\le x} \mu(n) = \mathcal{O}\left(x \cdot \log^{-k} x\right).$$

4.3 Elementary Proofs

In spite of the fact, that elementary proofs for difficult theorems of analytic number theory were found around 1940,[40]) it was totally unexpected and surprising that it was possible to give an *elementary* proof of the prime number theorem. The word *elementary* indicates that methods neither from the theory of complex-valued functions nor from FOURIER analysis are used.[41])

40) For example, in 1942 P. ERDÖS gave an elementary proof for the partition formula.

41) G.H. HARDY wrote in 1921: «You may ask me what an 'elementary' method is, and I must explain precisely what I understand by this expression. I do not mean an easy or trivial method; an elementary method may be quite desperately ingenious and subtle. I am using the word in a definite and technical sense, ... I mean, by an elementary method, a method which makes no use of the notion of an analytic function.»

P. ERDÖS

These elementary proofs were achieved independently in 1949 by A. SELBERG[42]) and P. ERDÖS, and these proofs were considered to be so important that SELBERG got a FIELDS medal, and ERDÖS later got the WOLF prize. In these elementary proofs, TAUBERian arguments play an important rôle.[43])

---

42) A photopraph of A. SELBERG may be found in SELBERG's *Collected Works*, for example.

43) Later, simpler variants of elementary proofs for the PNT were given, for example, by V. NEVANLINNA, and R. BREUSCH; E. WIRSING used estimates of STIELTJES integrals, and TH. BANG utilized the isoperimetric inequality.

SELBERG's method used a kind of mean-value, the SELBERG formula,

$$\psi(x) \cdot \log x + \sum_{p \le x} \psi\left(\frac{x}{p}\right) \cdot \log p = 2x \log x + \mathcal{O}(x),$$

together with the inequality

$$|\psi(x) - x| \cdot \log x \le \int_1^x \left|\psi\left(\frac{x}{u}\right) - \frac{x}{u}\right| du + \mathcal{O}(x \log\log x),$$

and an *oscillation condition*: In $1 \le y < x \le 2y$ one has

$$0 \le \psi(x) - \psi(y) \le 2 \cdot (x - y) + \mathcal{O}\left(\frac{x}{1 + \log x}\right).$$

Soon afterwards, several mathematicians (J.G. VAN DER CORPUT, P. KUHN, R. BREUSCH, and others) realized that it was possible to obtain a remainder term in the PNT by elementary methods — the improvement was by a [small] power of a logarithm only. Important progress was achieved independently by E. BOMBIERI and E. WIRSING in 1962, 1964; they were able to prove the prime number theorem with an error term of the form

$$\mathcal{O}_k\left(\frac{x}{(\log x)^k}\right), \quad \text{where } k \text{ is arbitrary.}$$

E. BOMBIERI used higher analogues of SELBERG's formula, and E. WIRSING worked with convolutions of arithmetical functions. Finally, by elementary methods, an even better remainder term

$$\mathcal{O}\left(x \cdot \exp\left(-\gamma \cdot \sqrt[7]{\log x}\right)\right)$$

was obtained by H. DIAMOND and J. STEINIG (Invent. Math. 11, 199–258, 1970), and afterwards this was slightly improved by A.F. LAVRIK and A.S. SOBIROV (Dokl. AN SSSR 211, 534–536, 1973). More recently new variants of elementary proofs were published. The proof by HÉDI DABOUSSI provided the possibility of giving an elementary proof of the mean-value-theorem of G. HALÁSZ, as was shown by H. DABOUSSI and K.-H. INDLEKOFER jointly in 1990.

E. WIRSING

5 Improvement of the Remainder Term

5.1 Remainder Term of the Prime Number Theorem

As the diagrams (Figures 3, 4 and 7) indicate, the integral logarithm $\mathrm{li}(x)$ seems to be a rather close approximation to $\pi(x)$; therefore, the remainder estimate $\mathcal{O}\left(x \cdot \exp(-\gamma \cdot \sqrt{\log x})\right)$ in (15) does not seem to be optimal. Taking into account that the RIEMANN conjecture implies a remainder term $\mathcal{O}\left(\sqrt{x} \cdot \log x\right)$ in the prime number theorem (H. VON KOCH, 1901), the remainder term given above is rather weak. But it took more than twenty years until DE LA VALLÉE-POUSSIN's remainder term was improved; in 1922, J.E. LITTLEWOOD proved that the PNT holds with the remainder term

$$\mathcal{O}\left(x \cdot \exp\left(-c \cdot \sqrt{\log x \cdot \log\log x}\right)\right),$$

in fact a rather weak improvement, but an important one: J.E. LITTLEWOOD paved the way for further improvements by showing that estimates of $\zeta(\sigma + it)$, obtained by utilizing non-trivial upper bounds for exponential sums

$$\sum \exp\left(2\pi i t \cdot \log\left(\frac{n}{2\pi}\right)\right)$$

lead to better remainder terms. Improvements concerning the techniques of estimating exponential sums gave better order estimates for the zeta-function (N.G. TCHUDAKOFF (1936), E.C. TITCHMARSH (1938)), and finally the zero-free region

$$\zeta(s) \neq 0 \text{ in } \mathrm{Re}(s) \geq 1 - \gamma \cdot \frac{\log^{\frac{2}{3}} t}{(\log\log t)^{\frac{1}{3}}}$$

was obtained.[44])

This zero-free region implies the remainder term

$$\mathcal{O}\left(x \cdot \exp\left(-\gamma \cdot \frac{(\log x)^{\frac{3}{5}}}{(\log\log x)^{\frac{1}{5}}}\right)\right). \tag{17}$$

As far as we know there is no improvement of this remainder term. The VINOGRADOV zero-free region was later deduced in an 'elementary' manner by Y. MOTOHASHI, using the sieve method (and [simple] estimates of exponential sums).

5.2 Irregularities of the Distribution of Primes

5.2.1 The Differences $\psi(x) - x$ and $\pi(x) - x$

Concerning the order of the remainder term of the PNT, in the opposite direction E. SCHMIDT showed already in 1903 that

$$\psi(x) - x = \Omega_{\pm}\left(x^{\frac{1}{2}-\delta}\right)$$

for any $\delta > 0$. So, the difference $\psi(x) - x$ is sometimes larger than $\gamma \cdot x^{\frac{1}{2}-\delta}$. In 1914, J.E. LITTLEWOOD improved this result to

$$\pi(x) - \mathrm{li}(x) = \Omega_{\pm}\left(x^{\frac{1}{2}} \cdot \frac{\log\log\log x}{\log x}\right). \tag{18}$$

These results are one of the starting points of the «Comparative Prime Number Theory» (see P. TURÁN's «A New Method in Analysis», in particular 41, and papers of P. TURÁN, S. KNAPOWSKI, J. PINTZ, and others for similar problems for primes in arithmetical progressions).

---

44) I.M. VINOGRADOV, N.M. KOROBOV 1958; see also the correction of the result by H.-E. RICHERT, 1960, mentioned in A. WALFISZ, [1963], p. 226. The theory of exponential sums is described, for example, in IVIĆ [1985], KRÄTZEL [1988], WALFISZ [1963].

5.2.2 The Sign of the difference $\mathrm{li}(x) - \pi(x)$
The inequality $\mathrm{li}(x) - \pi(x) > 0$ seems to be correct by numerical calculations extended up to 10^9 (see also Figure 7). However, in number theory numerical calculations often only scratch the surface.[45])

LITTLEWOOD showed in 1914 that the inequality $\mathrm{li}(x) - \pi(x) > 0$ is false for some value of x. In fact this difference changes sign infinitely often (and later E. LANDAU gave a simpler method of proving results of this kind, using the fact, that DIRICHLET series with positive coefficients, converging exactly in the half-plane $Re(s) > 1$ do have a singularity at $s = 1$). From a numerical point of view one would like to know where the first change of sign is to occur.[46])

S. SKEWES, in two papers (1933, 1955), made it his business to provide numerical constants for all the necessary estimates entering the proofs; in one paper (1933) the truth of the RIEMANN hypothesis is assumed, in the other one (1955) it is supposed that this conjecture is violated. So SKEWES was able to show that the first change of sign occurs below

$$10^{10^{10^{34}}}$$

if the RIEMANN conjecture is true (the largest number, that ever had a definite purpose, according to HARDY), and below

$$\exp\exp\exp\exp(7.705)$$

otherwise. This incredibly large bound, the «Skewes-Number», was reduced by R.S. LEHMANN in 1966, and then by H. TE RIELE in 1986, to 6.7×10^{370}; but even this number is far beyond the possible reach of any supercomputer.

Concerning the number of sign-changes of the difference

$$\mathrm{li}(x) - \pi(x),$$

in 1979 JANOS PINTZ showed that there are effectively computable constants c_1 and c_2 so that the number of sign-changes of the function

$$\mathrm{li}(x) - \pi(x)$$

in the interval $1 \leq x \leq Y$ is greater than

$$c_1 \cdot \sqrt{\log Y}, \quad \text{as soon as } Y \geq c_2.$$

In 1985, J. KACZOROWSKI got for the number of sign-changes the larger lower bound $\gamma \cdot \log Y$, but in an ineffective way.[47])

45) The same remark applies to the MERTENS conjecture or to the ERDÖS problem that almost every integer has two divisors d, d' satisfying the inequality $d < d' \leq 2d$; this deep problem was solved by H. MAIER and G. TENENBAUM in 1984.

46) The logician G. KREISEL showed that effectivization is possible.

47) This means, that it is not possible, even in principle, to obtain numerical values for some of the constants entering into the result.

x	$\pi(x)$	x	$\pi(x)$
1000	168	10^7	664 579
10 000	1 229	10^8	5 751 455
100 000	9 592	10^9	50847 534
0.5×10^6	41 538	10^{10}	455 052 511
1×10^6	78 498	10^{11}	4 118 054 813
1.5×10^6	114 155	10^{12}	37 607 912 018
2×10^6	148 933	10^{13}	346 065 536 839
2.5×10^6	183 072	10^{14}	3 204 941 750 802
3×10^6	216 816	10^{15}	29 844 570 422 669
5×10^6	348 513	10^{16}	279 238 341 033 925

Table 1 Values of the prime-counting function $\pi(x)$

5.3 Calculation of $\pi(x)$

At this point we would like to mention that the *calculation* of $\pi(x)$ for $x = 10^8$ and $x = 10^9$ was already performed by E. MEISSEL, by hand, in 1870 and 1885. He had considerably improved on the «Sieve of Eratosthenes», to reduce the necessary amount of calculations.48) D.H. LEHMER, 1959, D.C. MAPES, 1963 and J. BOHMANN, 1972, used electronic computers in order to calculate values of $\pi(x)$ up to $x = 10^{13}$. But a skilful improvement and refinement of the algorithm, reducing the complexity and the necessary amount of storage, was required to calculate values of $\pi(x)$ near 10^{16}. This was achieved by J.C. LAGARIAS, V.S. MILLER & A.M. ODLYZKO in 1984. The difficulties which had to be overcome, make it very unlikely that values of $\pi(x)$ around $x = 10^{370}$ can be calculated in the near future. It should be mentioned that the three authors also proposed a method for calculating values of $\pi(x)$ by utilizing the theory of the RIEMANN zeta-function.

5.4 Consecutive Primes

Of course, the prime number theorem also gives an answer to the question of estimating the *distance between consecutive primes* and more generally, it gives an asymptotic formula for the number of primes $\pi(x + h) - \pi(x)$ in «short» intervals; but «short» means here that h is larger than the function appearing in the remainder term of the PNT, and this seems to be a rather weak result. Assuming

---

48) Nowadays, for home computers or simple personal computers, it is no problem to produce all the primes up to 10^8 in some hours, using the (unaltered) sieve of ERATOSTHENES. The values up to 10^8 in Table 1 are produced in this way. Other values given are from a paper of LAGARIAS, MILLER & ODLYZKO. By the way, the values in the paper «A history of the prime number theorem», Amer. Math. Monthly 80, pp. 599–615 (in particular p. 614), by L.J. GOLDSTEIN, for 1.5, 2, 2.5 and 3 million, which were reproduced in the author's paper «Über einige Probleme aus der Theorie der Primzahlen», Sitzungsberichte der Wissenschaftlichen Gesellschaft Frankfurt, 1985, 61–118, are one unit too large. In GOLDSTEIN's paper the [incorrect] values, calculated by GAUSS, are given too.

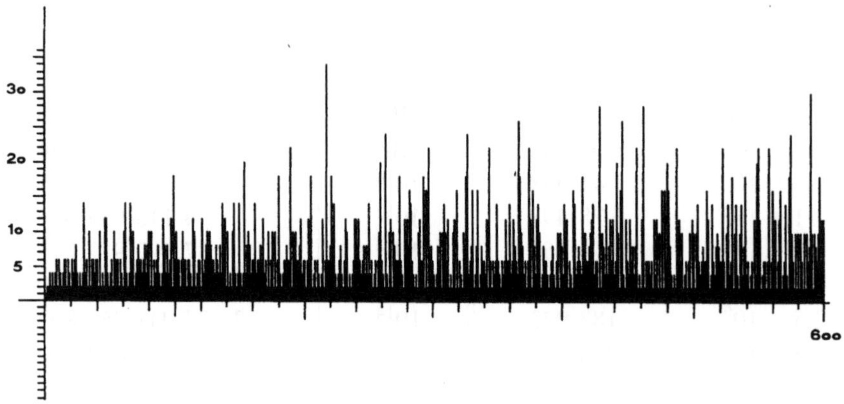

Fig. 8 The Difference between Consecutive Primes

the truth of the RIEMANN hypothesis, it follows that

$$\Delta p_n = p_{n+1} - p_n = \mathcal{O}\left(\sqrt{p_n} \cdot \log p_n\right)$$

(CRAMÉR 1921); he conjectured later (1936/37) that[49])

$$\Delta p_n = \mathcal{O}\left(\log^2 p_n\right).$$

The problem is of interest on behalf of the (unproven) conjecture, that there exists always a prime between consecutive squares $n^2, (n+1)^2$. In the opposite direction, P. ERDÖS (1935, Quart. J. 6) was able to prove that the inequality

$$\Delta p_n \geq C \cdot \left(\frac{\log p_n \cdot \log\log p_n}{(\log\log\log p_n)^2}\right), \tag{19}$$

holds infinitely often. In 1938, R.A. RANKIN was able to increase the right-hand side of (19) by a factor $\log\log\log\log p_n$ (and with $C = \frac{1}{3} - \varepsilon$). A. SCHÖNHAGE increased the value of C, and finally RANKIN again obtained

$$C = \exp(\mathcal{C} - \varepsilon).$$

Figure 8 gives the first 600 differences $p_{n+1} - p_n$ of consecutive primes.

---

49) Assuming the truth of the Riemann hypothesis A. SELBERG showed in 1943 that

$$\sum_{p_n \leq x} (p_{n+1} - p_n)^2 \ll x \log^3 x$$

in support of this conjecture. The (unconditional) result

$$\liminf_{n \to \infty} (p_{n+1} - p_n)/\log p_n \leq A$$

with some constant $A > 1$ is due to P. ERDÖS, 1940.

H. IWANIEC

Concerning upper estimates for the difference of consecutive primes, surprisingly it was possible to obtain better estimates than those coming from the prime number theorem. This breakthrough was achieved by G. HOHEISEL (1930), who proved the estimate

$$\Delta p_n = \mathcal{O}\left(p_n^{1-\delta}\right), \tag{20}$$

where $\delta = \frac{1}{33000}$. Of course, the RIEMANN conjecture implies a much stronger result. HOHEISEL's idea was that for deriving (20) one does not need to know that there are no zeros of the zeta-function in some strip $\mathrm{Re}(s) > 1 - \delta'$; it is sufficient that there are only «few» zeros. Upper estimates for the number $N(\alpha, T)$ of zeros ρ of $\zeta(s)$ (or more generally, for all L-functions mod q) satisfying

$$\mathrm{Re}(\rho) \geq \alpha, \quad 0 \leq \mathrm{Im}(\rho) \leq T$$

are known under the name «Density Theorems».

Later, better density theorems were proven, and so, by refined techniques, it was possible to improve considerably on HOHEISEL's exponent $\eta = 1 - \delta$. We mention results of[50]) H. HEILBRONN, ($\eta = \frac{1}{250} + \varepsilon$, 1933), A.E. INGHAM

50) For short, the results of MONTGOMERY, HUXLEY, IVIĆ, JUTILA & IWANIEC, HEATH-BROWN & IWANIEC are not mentioned.

E. WIRSING J. PINTZ A. SCHINZEL

($\eta = \frac{5}{8} + \varepsilon$, 1937), H. IWANIEC and J. PINTZ ($\eta = \frac{23}{42} = 0.54762\ldots$, 1984), and C.J. MOZZOCHI ($\eta = \frac{11}{20} - \frac{1}{384} + \varepsilon = 0.5474$), coming near the wanted-for value $\frac{1}{2}$. At the moment, a proof for CRAMÉR's conjecture seems hopeless.

6 Primes in Arithmetic Progressions

The prime number theorem for (individual) arithmetic progressions (see (16)),

$$\pi(x;q,a) = \sum_{p \le x, p \equiv a \bmod q} 1 \sim \frac{1}{\varphi(q)} \cdot \frac{x}{\log x}, \text{ if } \gcd(a,q) = 1,$$

was mentioned already. However, for applications in additive prime number theory for example, it turned out that results for $\pi(x;q,a)$ holding *uniformly* in some range were necessary. This was the case with I.M. VINOGRADOV's famous result:[51])

51) HARDY and LITTLEWOOD (1923) had already proved this result, however, assuming a weak form of the Great RIEMANN Hypothesis for L-series.

Every sufficiently large odd integer N is representable as a sum of three primes.

VINOGRADOV's proof represents the number $r(N)$ of solutions of the equation $N = p_1 + p_2 + p_3$ by the integral

$$r(N) = \int_0^1 S^3(\alpha) \cdot \exp(-2\pi i\alpha \cdot N)d\alpha, \text{ where}$$
$$S(\alpha) = \sum_{p \le N} \exp(2\pi i \cdot \alpha \cdot p).$$

The integral is split

- into «major arcs», small intervals (of length $N^{-1} \cdot \log^A N$ around rational numbers $\frac{a}{q}$ with small denominator $q \le \log^A N$; in these intervals a good approximation to $S(\alpha)$ by reasonably simple functions is possible, utilizing the prime number theorem in arithmetic progressions, **if** this theorem holds uniformly in $q \le \log^A N$, and
- into «minor arcs»; in these intervals VINOGRADOV was able to give a non-trivial estimate for the exponential sum $S(\alpha)$. This estimate shows that the contribution of the «minor arcs» to the integral is 'negligible'.

6.1 The Page-Siegel-Walfisz Theorem

A proof of the PNT in arithmetic progressions may be achieved by the method of complex integration. This method gives formulae for $\pi(x; q, a)$ (Theorem of PAGE and WALFISZ), but — compared to the explicit formula for $\pi(x)$ — with an extra term which comes from a possible contribution of one zero on the real axis near 1 of some function $L(s, \chi)$ mod q (with a real-valued character). It was of importance to know how near this zero might be to 1. In 1936 (Acta Arithm. 1), treating the problem of estimating the size of class numbers $h(d)$ (see (21)), CARL LUDWIG SIEGEL proved a lower bound for $L(1, \chi)$, and provided the result

For every $\varepsilon > 0$ *there exists a constant* $c(\varepsilon) > 0$ *with the property*

$$L(\sigma, \chi) \ne 0 \quad \text{in} \quad \sigma \ge 1 - c(\varepsilon) \cdot q^{-\varepsilon}$$

for every character χ *mod* q.

For the class-number $h(q)$ of a quadratic field with negative fundamental discriminant $-q$ this implies the estimate

$$h(q) > c(\varepsilon) \cdot q^{\frac{1}{2}-\varepsilon}. \tag{21}$$

All these results just mentioned were *ineffective*: Given ε (small enough), it is not possible to give the value of $c(\varepsilon)$. In 1948, SIEGEL's theorem was proved in a simpler way by T. ESTERMANN. His proof is non-effective again: Starting

C.L. SIEGEL

with a zero of $L(s,\chi)$ on the real axis near 1 (which probably does *not* exist), a consideration of the combination $L(s,\chi_1)\cdot L(s,\chi_2)\cdot L(s,\chi_1\cdot\chi_2)$ leads to a suitable lower bound for $L(1,\chi_2)$ and so to the result of SIEGEL.

Applied to the prime number theorem, one obtains (A. WALFISZ, 1936): *Given $A>0$ and $B>0$ arbitrarily,* **uniformly** *in a and q, $1\leq q\leq \log^A x, \gcd(a,q)=1$, the estimate*

$$\left|\pi(x;q,a)-\frac{li(x)}{\varphi(q)}\right| \ll_A x\cdot\exp(-\gamma\cdot\sqrt{\log x}) \ll_{A,B} x\cdot\log^{-B}x \qquad (22)$$

is true, with ineffective $\mathcal{O}$-constants.

Another uniform result concerning the PNT for arithmetic progressions is provided by the prime number theorem of T. TATUZAWA and K.A. RODOSSKII, 1949, 1950: The range of uniformity is considerably larger, but certain moduli q are excluded.

6.2 The Brun-Titchmarsh Theorem

For many applications intended the region of uniformity in the PAGE-SIEGEL-WALFISZ Theorem is *not* sufficient, but for larger moduli q only an upper estimate is required. Estimates of this kind follow from sieve methods (BRUN's or SELBERG's sieve or the «Large Sieve» can be used). Using BRUN's sieve, E.C. TITCHMARSH proved in 1930[52)] the nowadays so-called BRUN-TITCHMARSH inequality: If $\gcd(a, q) = 1$, then

$$\pi(x; q, a) \ll \frac{1}{\varphi(q)} \cdot \frac{x}{\log(x/q)}, \text{ if } 1 \leq q \leq x^{\beta}, \quad 0 < \beta < 1. \tag{23}$$

There are many improvements of inequality (23); finally H.L. MONTGOMERY and R.C. VAUGHAN proved in 1973: In $1 \leq q < y \leq x$, $\gcd(a, q) = 1$,

$$\pi(x; q, a) - \pi(x - y; q, a) \leq \frac{2}{\varphi(q)} \cdot \frac{y}{\log(y/q)}. \tag{24}$$

An improvement of the constant 2 to $2 - \delta$ would have consequences for SIEGEL's exceptional zero (of $L(s, \chi)$, near 1); so this improvement does not seem to be too likely. However, an improvement of the constant 2 is possible «on average».

6.3 The Bombieri-Vinogradov Theorem

Now let us leave the first half of the 20th century again, as we have done rather often already. For the treatment of several problems of analytic and additive number theory (for example, the TITCHMARSH divisor problem, the problem of representing numbers in the form $n = p + k^2 + m^2$ and $n = p + P_r$) the PNT for arithmetical progressions (and the result (23)) were not sufficient. It seemed to be necessary to assume the RIEMANN hypothesis to deal with these problems. However, it was realized that an approximation to $\pi(x; q, a)$ **in the mean**, but in a larger range, would suffice. This result was provided, beginning with RÉNYI (1948), M. BAR-

---

52) HARDY and LITTLEWOOD (1923) had already proved

$$\pi(x + y) - \pi(x) \ll y / \log y.$$

There is an old unsettled conjecture stating that

$$\pi(x) - \pi(x - y) \leq \pi(y)$$

in $1 \leq y \leq x$, which could be false; for HENSLEY and RICHARDS (1973) showed it to be incompatible with the (also unsettled) prime-k-tuples conjecture.

BAN (1961), PAN CHENG DONG (1962), finally by A.I. VINOGRADOV (1965, in the weaker range $Q = x^{\frac{1}{2}-\varepsilon}$), and E. BOMBIERI (1965, with the value of Q stated below):

Theorem of BOMBIERI-VINOGRADOV:

To any $A > 0$ there is a constant[53]) $B > 0$ so that the estimate

$$\sum_{q\leq Q} \max_{y\leq x} \max_{a,\ \gcd(a,q)=1} \left|\pi(y;q,a) - \frac{1}{\varphi(q)} \cdot \pi(y)\right| = \mathcal{O}\left(\frac{x}{\log^{-A} x}\right), \tag{25}$$

holds, where $Q = \sqrt{x} \cdot \log^{-B} x$.

The RIEMANN conjecture implies (25) with $B = A + 1$. So the BOMBIERI-VINOGRADOV theorem may be considered as a proof of the RIEMANN conjecture *in the mean.*

The proof of (25) depends on the observation that few zeros not lying on the critical line are not of great importance, and so one is lead again to «Density Theorems» concerning the number

$$N(\alpha, T, \chi) \text{ of zeros of } L(s,\chi) \text{ in } \quad \alpha \leq \mathrm{Re}(s) \leq 1, \quad 0 \leq \mathrm{Im}(s) \leq T.$$

BOMBIERI's result is: Denote by $\mathcal{A}$ a set of integers less than Q, put $D = \max_{q\in\mathcal{A}} \tau(\chi)$, where $\tau(\chi)$ stands for the GAUSS sum. Then

$$\begin{aligned}&\sum_{q\in\mathcal{A}} \frac{1}{\varphi(q)} \cdot \sum_{\chi \bmod q} |\tau(\chi)|^2 \cdot N(\alpha,T,\chi)\\ &\quad = \mathcal{O}\left(DT \cdot \left(Q^2 + QT\right)^{4(1-\alpha)/(3-2\alpha)} \cdot \log^{10}(Q+T)\right).\end{aligned} \tag{26}$$

The proof of (26) depends on the «Large Sieve». We state BOMBIERI's version:

$$\begin{aligned}&\sum_{q\leq Q} \sum_{1\leq a\leq q,\ \gcd(a,q)=1} \left|\sum_{M<n\leq M+N} \alpha_n \cdot e\left(\frac{a}{q}\cdot n\right)\right|^2\\ &\quad = \mathcal{O}\left((N+Q^2) \cdot \sum_{M<n\leq M+N} |\alpha_n|^2\right),\end{aligned} \tag{27}$$

---

53) It is possible to take $B = 3A+23$. P.X. GALLAGHER, 1968, gave the value $B = 16A+103$, and he simplified the proof of the BOMBIERI-VINOGRADOV Theorem. Another simple method of proof was provided by R.C. VAUGHAN, 1980, 1981. Analogues of the theorem are due to H. SIEBERT and D. WOLKE 1971. The critical exponent $1/2$ in Q ought to be replaced with $1-\delta$ (HALBERSTAM's conjecture); but only E. FOUVRY and H. IWANIEC gave an enlargement to $11/21$ under certain additional hypotheses. PAN CHENG DONG (1981) introduced weights into (25).

and this estimate (27) holds for the sum

$$\sum_{q\le Q}\frac{q}{\varphi(q)}\sum_{\chi \bmod q,\ \text{primitive}}\left|\sum_{M<n\le M+N}\alpha_n\cdot\chi(n)\right|^2 = \mathcal{O}\left((N+Q^2)\cdot\sum_{M<n\le M+N}|\alpha_n|^2\right), \tag{28}$$

too.

7 The Riemann Conjecture

7.1 Some Support for this Conjecture

The BOMBIERI-VINOGRADOV Theorem may be considered as a hint that the RIEMANN conjecture might be true. Of course, many mathematicians dealt with this challenging conjecture (some thought they were able to prove it); they obtained results stretching into different directions, and we would like to mention the following developments.

- **Analogy**. In an algebraic function field K over $k = \mathbf{F}_q$, the RIEMANN conjecture for the «congruence zeta-function»

$$\zeta_K(s) = \sum_{\mathbf{a}\text{ integral divisor}} q^{-\deg(\mathbf{a})\cdot s} = \prod_{\wp}\left(1-\frac{1}{q^{\deg(\wp)\cdot s}}\right)^{-1}$$

  was proved by ANDRÉ WEIL (1940, 1948). Later proofs of somewhat weaker results, using methods from the Theory of Diophantine Approximations, were given by S.A. STEPANOV 1969–1972, W.M. SCHMIDT 1973, 1974, and E. BOMBIERI 1973. These investigations do have important consequences for statements concerning the number of solutions of congruences in several variables.

- **Zeros on the critical line**. From a theoretical point of view it seems to be important that already in 1914 G.H. HARDY proved that there are infinitely many zeros of $\zeta(s)$ exactly on the critical line.[54]) Denoting their number in the interval [0,T] by $N_0(T)$, J.E. LITTLEWOOD showed the inequality

$$N_0 > \gamma\cdot T.$$

54) E. LANDAU, in 1915, began his paper *«Über die Hardysche Entdeckung ...»* with «Zu den bedeutendstem Fortschritten der Mathematik aus der letzten Zeit gehört die Note von Herrn G.H. HARDY *Sur les zéros de la fonction $\zeta(s)$ de Riemann ....*»

In 1942, A. SELBERG was able to show that, on behalf of the RIEMANN-MANGOLDT-formula

$$N(T) = \frac{T}{2\pi} \cdot \left(\log \left(\frac{T}{2\pi} \right) - 1 \right) + \mathcal{O}(\log T)$$

and his result

$$\boxed{N_0(T) > \gamma \cdot T \log T} \tag{29}$$

(with a small constant γ) a positive portion of all the zeros in the critical strip are **on** the critical line. Progress concerning this question of zeros on the critical line stopped until 1974, when N. LEVINSON proved, that in equation (29) one may take

$$\boxed{\gamma = \frac{1}{3}}$$

In the following years, some authors dealt with this problem again, studying the number of *simple* zeros of $\zeta(s)$ on the critical line (R. HEATH-BROWN), and improving the constant γ (PAN CHENG DONG, H. CONREY, P. BAUER).

- **Long-range Numerical Calculations**. As mentioned earlier (section 3.3), more than 1.5 billion of zeros of $\zeta(s)$ were shown to lie exactly on the critical line Re(s) = $\frac{1}{2}$. This may be considered as supporting the truth of the RIEMANN conjecture. However, as pointed out earlier in connection with the difference $\pi(x) - \mathrm{li}(x)$, or with the ERDÖS problem on divisors, numerical calculations in number theory are often misleading; in connection with his calculations of zeros of $\zeta(s)$ on the critical line, D.H. LEHMER wrote already in 1956:

  «A tendency for the behaviour of the zeta-function to become more and more capricious as t increases is evident from the values thus far computed ....»

- GRAM's law, exhibiting an interval on Re(s) = $\frac{1}{2}$ with exactly one t inside it, for which $\zeta(\frac{1}{2} + it)$ vanishes, and which was guessed from rather early numerical calculations, is wrong also.

7.2 Zeros of Partial Sums

P. TURÁN showed that RIEMANN's conjecture is true, **if** all the partial sums

$$\zeta_N(s) = \sum_{1 \le n \le N} n^{-s}$$

of the DIRICHLET series for the zeta-function with sufficiently large index N do not vanish in the half-plane Re(s) > 1. However, H.L. MONTGOMERY destroyed the idle hope of proving the RIEMANN hypothesis via this approach; in fact, he showed that for any $c < 4/\pi$ there is an integer $N_0(c)$ so that for all indices $N > N_0$ there are zeros of $\zeta_N(s)$ in

$$Re(s) > 1 + c \cdot \frac{\log \log N}{\log N}.$$

H.L. MONTGOMERY

7.3 Some Technical Tools

Very helpful in connection with investigations concerning the RIEMANN conjecture and calculations connected with it were

- The **Approximate Functional Equation**

$$\zeta(s) = \sum_{n \leq x} n^{-s} + \Xi \cdot \sum_{n \leq y} n^{s-1} + \mathcal{O}\left(x^{-\sigma} + t^{\frac{1}{2}-\sigma} \cdot y^{\sigma-1}\right),$$

where $0 < \sigma < 1$, $x, y > h > 0$, $2\pi xy = t$, and

$$\Xi(s) = 2^s \pi^{s-1} \Gamma(1-s) \cdot \sin\left(\frac{1}{2}\pi s\right)$$

(G.H. HARDY and J.E. LITTLEWOOD, 1922, 1929, C.L. SIEGEL 1932).

- **Order estimates** of $\zeta(\sigma + it)$, estimates of «moments»

$$\int_1^T |\zeta(\sigma + it)|^k \, dt$$

of the zeta-function (A. IVIĆ, R. HEATH-BROWN),

- TURÁN's **method.** PAUL TURÁN realized the importance of estimating sums of powers of complex numbers

$$\inf_z \max_\nu \left| \sum_{1 \le j \le n} b_j z_j^\nu \right| \quad \text{and} \quad \inf_z \max_\nu \operatorname{Re}\left(\sum_{1 \le j \le n} b_j z_j^\nu \right)$$

in order to obtain connections between estimates of exponential sums and (for example) density results on L-series.

8 The Möbius Function

8.1 Some Conjectures concerning $\mu(n)$

An equivalent formulation of the prime number theorem in the form of relation (14) is the fact that the mean-value of the MÖBIUS function is zero.[55)][56)]

If the RIEMANN conjecture is true, then

$$M(x) = \sum_{1 \le n \le x} \mu(n) = \mathcal{O}\left(x^{\frac{1}{2}+\varepsilon}\right)$$

---

55) There is an old conjecture (P. ERDÖS, A. WINTNER), stating that every multiplicative function, assuming only the values 0, +1, and −1, possesses a mean-value. This conjecture gave raise to a thorough study of mean-values of multiplicative functions (see, for example H. DELANGE 1961, E. WIRSING 1961). The ERDÖS-WINTNER conjecture was proved in 1967 by EDUARD WIRSING, and a more general result (with a simpler method of proof) was given by G. HALÁSZ in 1968. Recently, HEDI DABOUSSI and KARL-HEINZ INDLEKOFER provided two elementary proofs for the theorem of HALÁSZ (1990, 1992). An elementary proof of WIRSING's theorem was given in 1986 by A. HILDEBRAND.

56) The convergence of $\sum_1^\infty \mu(n)/\sqrt{n}$, conjectured by STIELTJES, was disproved by LANDAU (MZ, 1918). Figure (9) gives the values of the sum

$$-\sum_{n \le x} \frac{\mu(n)}{\sqrt{n}}$$

for $x = 25, 50, \dots 15000$.

P. TURÁN (The author is grateful to PAUL TURÁN for warm and continuing encouragement from about 1963 until TURÁN's early death in 1976.)

(LITTLEWOOD, 1912), and in 1924 LANDAU replaced the exponent in the $\mathcal{O}$-term by

$$\frac{1}{2} + \gamma \cdot \frac{\log\log\log x}{\log\log x}.$$

The **conjectures** of STERNECK and MERTENS state

$$\left| \sum_{n \le x} \mu(n) \cdot x^{-\frac{1}{2}} \right| \le \frac{1}{2}, \text{ resp. } \le 1, \text{ if } x > 200.$$

R. HEATH-BROWN

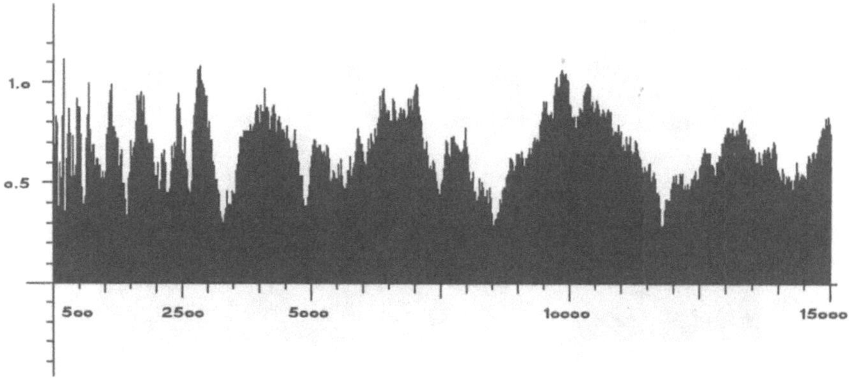

Fig. 9 Stieltjes Conjecture: Values of $-\sum_{n\leq x}\mu(n)\cdot n^{-1/2}$

These conjectures are supported by Figure (10). However, if the RIEMANN conjecture is false, these conjectures are «totally» false. If the RIEMANN conjecture is true and if the imaginary parts of the zeros on the critical line are **Q**-linearly independent, then

$$\mathcal{M} = \limsup_{x\to\infty}\left|\sum_{n\leq x}\mu(n)\cdot x^{-\frac{1}{2}}\right| = \infty,$$

A. HILDEBRAND

according to A.E. INGHAM, and so the truth of the conjectures in question is rather unlikely.

In fact, the first conjecture was refuted by numerical calculations (G. NEUBAUER, 1963): the STERNECK conjecture is wrong near 7.7×10^9. The MERTENS conjecture was harder to reject. In 1976, W. JURKAT and A. PEYERIMHOFF prepared the ground with a method, mixing up theoretical investigations and numerical calculations, to show that $\mathcal{M}$ is at least 0.75.

Figure 10 seems to be in favour of the conjectures, given in Section 8. This figure gives, for $x = 2, 4, \ldots, 1196$, the values of the sum

$$\sum_{n \leq x} \mu(n) \cdot x^{-1/2}.$$

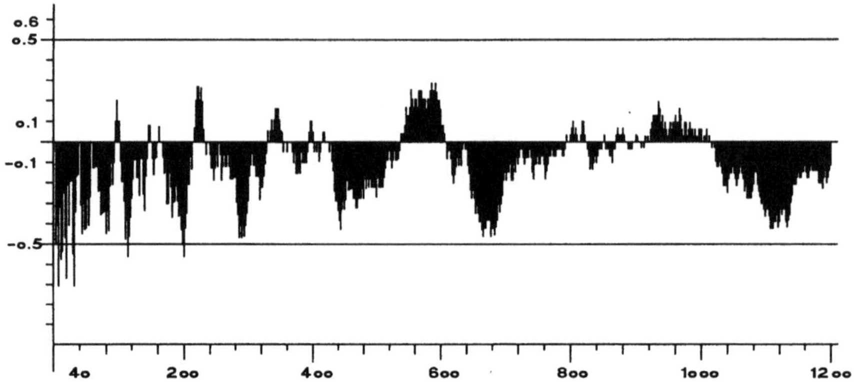

Fig. 10 Values of $\sum_{n\leq} \mu(n) \cdot x^{-1/2}$

8.2 The Mertens Conjecture Disproved

The final disprove of the conjeture of MERTENS was achieved by A. ODLYZKO and H. TE RIELE in 1983, 1985. A representation of the sum $\sum_{n\leq x} \mu(n)$ by a complex integral leads, via the method of complex integration, to the problem of investigating sums like

$$\sum_{\rho} (\rho \cdot \zeta'(\rho))^{-1} \cdot x^{\rho},$$

and it is required to find large values for these sums; in the usual notation, $\rho = \frac{1}{2} + i\gamma$ runs over zeros of $\zeta(s)$ on the critical line. Forming means, one is lead to find large values of

$$\sum_{0<|\gamma|\leq T} (\rho \cdot \zeta'(\rho))^{-1} e^{i\gamma w} \cdot \left(1 - \frac{|\gamma|}{T}\right).$$

This task was finished numerically, using more than 10 000 zeros of the zeta-function to 28 decimals, and using the LENSTRA-LENSTRA-LOVÁSZ-algorithm. So it was possible to disprove the MERTENS conjecture, a triumph for Computational Number theory.

9 Concluding Remarks

9.1 Fermat's Equation

Finally we emphasize that there are connections between the FERMAT equation

$$x^n + y^n = z^n \tag{30}$$

and the theory of primes. It is wellknown, that (30), for every $n > 2$, has at most finitely many (non-trivial) integer solutions, «primitive» in the sense that

A. ODLYZKO

gcd$(x, y, z) = 1$, by G. FALTINGS's proof of the MORDELL conjecture.[57]) In fact, there are many criterions for special values of n, showing insolubility of the FERMAT equation; using these criterions and a computer, H. WAGSTAFF was able to show that (30) does not have any (non-trivial) solutions for exponents $2 < n \leq 125000$.

9.2 The Result of Adleman, Fouvry and Heath-Brown

A generalization of an old criterion of SOPHIE GERMAIN reduces the question of solubility of the FERMAT equation (with exponent p, a prime greater than 2)

---

57) In the mean-time there are other proofs for this result, by PAUL VOJTA and ENRICO BOMBIERI. In 1992 P. VOJTA got the F.N. COLE-Prize (in algebra).

E. FOUVRY

in the «first case», when the possible solutions x, y, z are not divisible by p, to a problem in prime number theory. So, in 1985, L.M. ADLEMAN, E. FOUVRY and D.R. HEATH-BROWN were able to show that, in the «first case», the FERMAT equation does not have any non-trivial integer solutions for *infinitely* many prime exponents p; in fact, there are no solutions for approximately more than two thirds of all the primes.

The decisive result, needed from prime number theory, is the following:

Denote by $A(x)$ the number of primes p, $a+2 < p \leq x$, so that the maximal prime divisor of $(p-a)$ is greater than x^{δ}. Then, for any $a \neq 0$, if x is sufficiently large, and if $\delta \leq \Delta$, the inequality

$$A(x) > \gamma \cdot \frac{x}{\log x}$$

is true.

The admissible value of Δ is crucial. In 1973, C. HOOLEY got $\Delta = 0.625 - \varepsilon$, in 1981 J.M. DESHOUILLERS and H. IWANIEC obtained the value 0.6563. But it was necessary to have a value of Δ larger than $2/3$, a goal that could only be achieved by E. FOUVRY in 1985, $\Delta = 0.6687$. ROGER HEATH-BROWN wrote in 1985:

> *«The final value, the one due to Fouvry, incorporates five new estimates for $C(r)$, for different ranges of r. It is often the case that much effort is expended in improving some exponent or other, ... To outsiders this can appear to be a refuge for weaker researchers, as if the improvements could be obtained merely by more care, more pages of working, and more iterations of some established technique. Nonetheless, it is a fact that each such improvement, however small, requires a new idea. Fouvry's improvement from 0.6587 to 0.6687 needed five new ideas to achieve a 1.5% increase. It must be said, however, that the present example is the only instance to date where this process of improvement by small steps has taken us across a critical threshold (namely $\Delta = 2/3$ in this case) to produce a qualitatively new result.»*

9.3 Conclusion

Hopefully it was possible to show that the investigation of topics centering around the prime number theorem is still a vivid and active field of research, and that many difficult problems are left for later investigators.

Some photos of mathematicians, having worked or working in the field of prime number theory, are reproduced at different places in the text.

10 Bibliography

There are (nearly) countless papers and many monographs dealing with prime number theory. By no means, it was possible, to include all the articles and monographs referred to in the text, into this bibliography. Only a few of these can be mentioned.

[1] ANDREWS, G.E., et al., *Ramanujan Revisited*, Proc. Cent. Conf. Univ. of Illinois, 1988

[2] BARBAN, M.B., *The 'Large Sieve' Method and its Applications in the Theory of Numbers*, Russ. Math. Surveys **21**, 49–103, 1966

[3] BUNDSCHUH, P., *Einführung in die Zahlentheorie*, Springer 1988

[4] DAVENPORT, H., *Multiplicative Number Theory*, Chicago 1967

[5] DIAMOND, H.G., *Elementary methods in the study of the distribution of prime numbers*, Bull. AMS **7**, 553–589, 1982

[6] DIAMOND, H.G. & ERDÖS, P., *On sharp elementary prime number estimates*, L'Enseignement mathém. **26** (1980), 313–321, 1981

[7] EDWARDS, H.M., *Riemann's Zeta-Function*, New York 1974

[8] ERDÖS, P., *On a new method in elementary number theory which leads to an elementary proof of the prime number theorem*, Proc. Nat. Acad. Sci. USA **35**, 374–384, 1949

[9] ERDÖS, P., *On a Tauberian Theorem Connected with the New Proof of the Prime Number Theorem*, J. Indian Math. Soc. (N. S.) **13**, 131–144, 1949

[10] FALTINGS, G., *Endlichkeitssätze für abelsche Varietäten über Zahlkörpern*, Invent. Math. **73**, 349–366, 1983

[11] FOUVRY, E., *Théorème de Brun-Titchmarsh. Application au Théorème de Fermat*, Invent. Math. **79**, 383–407, 1985

[12] GELFOND, A.O., & LINNIK, YU.V., *Elementary Methods in the Analytic Theory of Numbers*, Cambridge (Mass.) 1966

[13] GUY, R.K., *Unsolved Problems in Number Theory*, Springer 1981

[14] HALBERSTAM, H. & RICHERT, H.E., *Sieve Methods*, London 1974

[15] HALBERSTAM, H. & ROTH, K.F., *Sequences I*, Oxford 1966

[16] HEATH-BROWN, D.R., *The First Case of Fermat's Last Theorem*, Math. Intelligencer, vol. **7**, 40–47, 1985

[17] HUA, LOO-KENG, *Additive Primzahltheorie*, Leipzig 1959

[18] HUA, LOO-KENG, *Die Abschätzung von Exponentialsummen und ihre Anwendung in der Zahlentheorie*, Leipzig 1959

[19] HUXLEY, M.N., *The Distribution of Prime Numbers*, Oxford 1972

[20] INDLEKOFER, K.-H., *Zahlentheorie*, Birkhäuser 1978

[21] INGHAM, A.E., *The distribution of prime numbers*, Cambridge 1932

[22] IVIĆ, A., *The Riemann Zeta-Function*, Wiley 1985

[23] KOLMOGOROV, A.N. & YUSHKEVICH, A.P., Editors, *Mathematics of the* 19th *Century*, Birkhäuser 1992

[24] KOBLITZ, N., *A Course in Number Theory and Cryptography*, Springer 1987

[25] KRÄTZEL, E., *Lattice Points*, Berlin 1988

[26] KRANAKIS, E., *Primality and Cryptography*, Wiley-Teubner 1986

[27] LANDAU, E., *Handbuch der Lehre von der Verteilung der Primzahlen, I, II*, Leipzig 1909

[28] LENSTRA JR., H.W., *Primality Testing*, Computational Methods in Number Theory, I, ed. by H.W. Lenstra jr. & R. Tijdeman, 55–87, 1987

[29] LEVINSON, N., *More than one third of the zeros of Riemann's zeta-function are on* $\sigma = \frac{1}{2}$, Adv. Math. **13**, 383–436, 1974

[30] LITTLEWOOD, J.E., *Sur la distribution des nombres premiers*, C. R. Paris **158**, 1869–1872, 1914

[31] MONTGOMERY, H.L., *Topics in Multiplicative Number Theory*, Springer 1971

[32] OSTMANN, H.H., *Additive Zahlentheorie I, II*, Springer 1956

[33] POMERANCE, C., *Recent Developments in Primality Testing*, Math. Intell. **3**, 97–105, 1980

[34] PRACHAR, K., *Primzahlverteilung*, Springer 1957

[35] VON RENTELN, M., *Aspekte zur Geschichte der Analysis im 20. Jahrhundert*, Skriptum zur Vorlesung Karlsruhe 1987

[36] RIBENBOIM, P., *The Book of Prime Number Records*, Sec. Ed., Springer 1989

[37] RIEMANN, B., *Über die Anzahl der Primzahlen unter einer gegebenen Größe*, Akad. Berlin 671–680, **1859**

[38] SCHMIDT, W.M., *Equations over Finite Fields, An Elementary Approach*, Springer 1976

[39] SCHWARZ, W., *Einführung in Methoden und Ergebnisse der Primzahltheorie*, BI Mannheim 1969

[40] SCHWARZ, W., *Über einige Probleme aus der Theorie der Primzahlen*, Sitzungsberichte der Wiss. Gesellschaft an der Univ. Frankfurt a. M., vol. **21**, 61–118, 1985

[41] SCHWARZ, W., *Einführung in die Zahlentheorie*, Sec. Ed., Wiss. Buchges. Darmstadt 1987

[42] SCHWARZ, W., *Geschichte der analytischen Zahlentheorie seit* 1890, Ein Jahrhundert Mathematik 1890–1990, Festschrift zum Jubiläum der DMV, 741–780, 1991

[43] SELBERG, A., *An elementary proof of the prime number theorem*, Ann. Math. (2) **50**, 305–313, 1949

[44] SELBERG, A., *The General Sieve Method in Prime Number Theory*, Proc. ICM Cambridge (Mass.) **1**, 286–292, 1950

[45] SPECHT, W., *Elementare Beweise der Primzahlsätze*, Berlin 1956

[46] TITCHMARSH, E.C., *The Theory of the Riemann Zeta-Function*, Oxford 1951

[47] TURÁN, P., *On a New Method of Analysis and its Applications*, Wiley 1984

[48] VAUGHAN, R.C., *An Elementary Method in Prime Number Theory*, Acta Arithm. **37**, 111–115, 1980

[49] VAUGHAN, R.C., *The Hardy-Littlewood Method*, Cambridge 1981

[50] VINOGRADOV, I.M., *The Method of Trigonometrical Sums in the Theory of Numbers*, New York 1954

[51] WALFISZ, A., *Weylsche Exponentialsummen in der neueren Zahlentheorie*, Berlin 1963

[52] WOLFART, J., *Primzahltests und Primfaktorzerlegung* , Überbl. Math. 161–188, BI Mannheim 1981

[53] WOLKE, D., *Ein Vierteljahrhundert Bombieri-Vinogradov'scher Primzahlsatz*, preprint [Lecture at Ulm, 1992]

11 Facsimiles

Within the first ten sections some photographs of mathematicians having given important contributions to the Distribution of Primes were reproduced.[58])

In this last section there are given

- the last page from a letter of C.F. GAUSS to ENCKE,
- a page from B. RIEMANN's manuscript *«Über die Anzahl der Primzahlen unterhalb einer gegebenen Größe»*.[59])[60])

WOLFGANG SCHWARZ
Fachbereich Mathematik
J.W. Goethe-Universität
Robert-Mayer-Strasse, 10
D-60084 Frankfurt am Main 1

58) For providing me with photographs I have to thank Prof. Dr. S.J. PATTERSON, Göttingen, Prof. Dr. C.J. MOZZOCHI, and Prof. Dr. K. JACOBS, Erlangen (Collection of Pictures of Mathematicians, 'Mathematisches Forschungsinstitut Oberwolfach').

59) For valuable help in connection with facsimiles of the letter of GAUSS and of RIEMANN's manuscript, and for photographs, I have to thank Dr. H. ROHLFING, Niedersächsische Staats- und Universitätsbibliothek Göttingen, Abteilung für Handschriften und seltene Drucke.

60) Last not least I would like to thank Prof. Dr. J.-P. PIER and his collaborators for the excellent organisation of the Colloquium «The Development of Mathematics from 1900 to 1950» in the ancient, uncommon, but homelike atmosphere of Bourglinster castle.

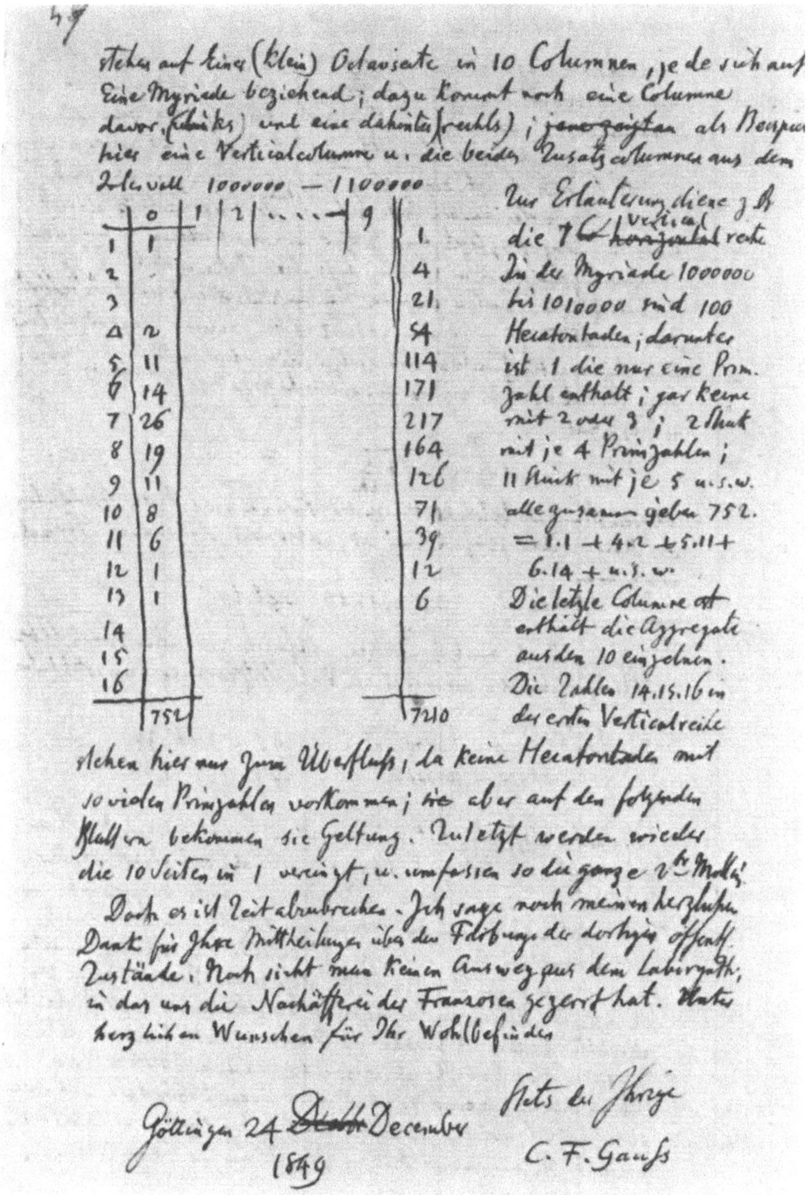

49

stehen auf einer (klein) Octavseite in 10 Columnen, jede sich auf eine Myriade beziehend; dazu kommt noch eine Columne davor (links) und eine dahinter (rechts); ~~jene zeigtan~~ als Beispiel hier eine Verticalcolumne u. die beiden Zusatzcolumnen aus dem Zehnmill 1000000 — 1100000

	0	1	2	…	9	
1	1					1
2						4
3						21
4	2					54
5	11					114
6	14					171
7	26					217
8	19					164
9	11					126
10	8					71
11	6					39
12	1					12
13	1					6
14						
15						
16						
	752					7210

Zur Erläuterung diene z. B. die 1te ~~horizontal~~ verticale Reihe. In der Myriade 1000000 bis 1010000 sind 100 Hecatontaden; darunter ist 1 die nur eine Prim.zahl enthält; gar keine mit 2 oder 3; 2 Stück mit je 4 Primzahlen; 11 Stück mit je 5 u.s.w. alle zusammen geben 752. = 1.1 + 4.2 + 5.11 + 6.14 + u.s.w. Die letzte Columne ist enthält die Aggregate aus den 10 einzelnen. Die Zahlen 14.15.16 in der ersten Verticalreihe stehen hier nur zum Überfluss, da keine Hecatontaden mit so vielen Primzahlen vorkommen; wie aber auf den folgenden Blättern bekommen sie Geltung. Zuletzt werden wieder die 10 Seiten in 1 vereinigt, u. umfassen so die ganze 1te Million.

Doch es ist Zeit abzubrechen. Ich sage noch meinen herzlichsten Dank für Ihre Mittheilungen über den Fortgang der dortigen öffentl. Zustände. Noch sieht man keinen Ausweg aus dem Labyrinth, in das uns die Nachäfferei der Franzosen gezerrt hat. Unter herzlichen Wünschen für Ihr Wohlbefinden

Stets der Ihrige

Göttingen 24 ~~Decemb~~ December 1849

C. F. Gauss

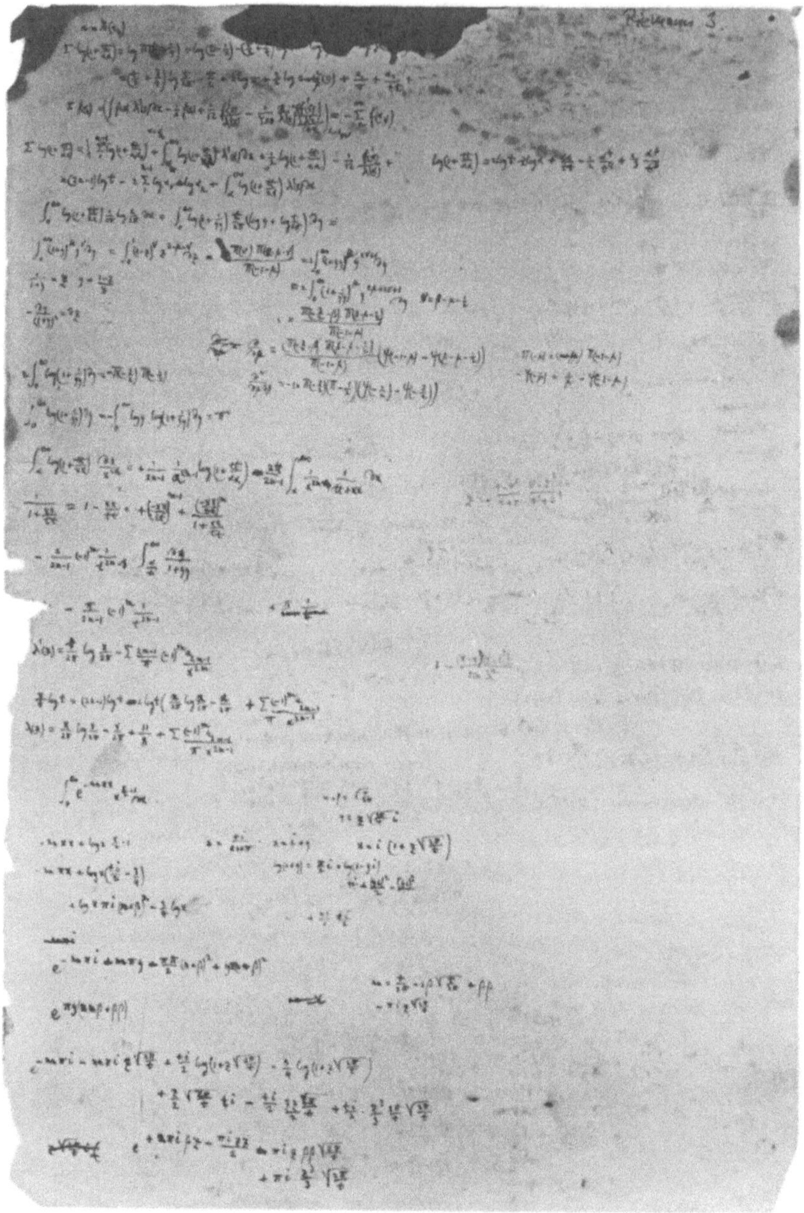

Bibliography

(coordinated by Pierre Dugac)

List of Journals

⟨1⟩ Abhandlungen aus dem Mathematischen Seminar der Universität Hamburg.

⟨2⟩ Abhandlungen der Königlichen Gesellschaft der Wissenschaften zu Göttingen, mathematisch-physikalische Klasse.

⟨3⟩ Abhandlungen der Preussischen Akademie der Wissenschaften zu Berlin, physikalisch-mathematische Klasse.

⟨4⟩ Acta Arithmetica.

⟨5⟩ Acta Eruditorum.

⟨6⟩ Acta Litterarum ac Scientiarum Regiae Universitatis Hungaricae Francisco-Josephinae, Szeged.

⟨7⟩ Acta Mathematica.

⟨8⟩ Acta Scientiarum Mathematicarum Szeged.

⟨9⟩ Acta Societatis Scientiarum Fennicae.

⟨10⟩ Akademie der Wissenschaften zu Wien, mathematisch-naturwissenschaftliche Klasse. Anzeiger.

⟨11⟩ Almanach der Österreichischen Akademie der Wissenschaften.

⟨12⟩ American Journal of Mathematics.

⟨13⟩ The American Mathematical Monthly.

⟨14⟩ American Mathematical Society Notices.

⟨15⟩ Analele Universitatii Bucaresti.

⟨16⟩ Anales de la Asociacio espanola para el Progreso de las Ciencias.

⟨17⟩ Annalen der Physik.

⟨18⟩ Annales Accademiae Scientiarum Fennicae, series A I, Mathematics.

⟨19⟩ Annales de la Société Polonaise de Mathématiques.

⟨20⟩ Annales de la Société Scientifique de Bruxelles.

⟨21⟩ Annales de l'Institut Fourier.

⟨22⟩ Annales de l'Institut Henri Poincaré.

⟨23⟩ Annales de l'Université de Grenoble, section sciences mathématiques et physiques.

⟨24⟩ Annales scientifiques de l'Ecole Normale Supérieure.

⟨25⟩ Annali della Reale Scuola Normale Superiore di Pisa, scienze fisiche e matematiche.

⟨26⟩ Annali di Matematica pura ed applicata.

⟨27⟩ Annals of Mathematical Statistics.

⟨28⟩ Annals of Mathematics.

⟨29⟩ Annals of Probability.

⟨30⟩ Annals of Science.

⟨31⟩ The Annals of Statistics.

⟨32⟩ Annuaire des Anciens Elèves de l'Ecole Normale Supérieure.

⟨33⟩ Archive for the History of Exact Sciences.

⟨34⟩ Arkiv för Mathematik, Astronomi och Fysik.

⟨35⟩ Arkiv for Mathematik og Naturvidenskab.

⟨36⟩ Atti della Reale Accademia dei Lincei. Rendiconti. Classe di scienze fisiche, matematiche e naturali.

⟨37⟩ Atti della Reale Accademia delle Scienze di Torino.

⟨38⟩ Atti della Reale Accademia Nazionale dei Lincei. Memorie della classe di scienze fisiche, matematiche e naturali.

⟨39⟩ Atti della Società Italiana per il Progresso delle Scienze.

⟨40⟩ Australia Mathematical Society Gazette.

⟨41⟩ Avtomatika i Telemehanika.

⟨42⟩ Bell System Technical Journal.

⟨43⟩ Berichte über die Verhandlungen der Königlichen Sächsischen Gesellschaft der Wissenschaften zu Leipzig, mathematisch-naturwissenschaftliche Klasse.

⟨44⟩ Boletin de la Facultad de Ingenieria de Montevideo.

⟨45⟩ Bolletino di Storia delle Scienze Matematiche.

⟨46⟩ Byulleten moskovskogo gosudarstvennogo Universiteta. Matematika.

⟨47⟩ Bulletin Astronomique.

⟨48⟩ Bulletin de la Société Mathématique de Belgique.

⟨49⟩ Bulletin de la Société Mathématique de France.

⟨50⟩ Bulletin de la Société Mathématique Roumaine.

⟨51⟩ Bulletin de l'Institut polytechnique de Jassy.

⟨52⟩ Bulletin de l'Université d'Etat de Moscou, série internationale. Section A. Mathématique et Mécanique.

⟨53⟩ Bulletin des Sciences Mathématiques.

⟨54⟩ Bulletin international de l'Académie de Cracovie, classe des sciences mathématiques et physiques.

⟨55⟩ Bulletin of the American Mathematical Society.

⟨56⟩ Bulletin of the London Mathematical Society.

⟨57⟩ Cahiers du Séminaire d'Histoire des Mathématiques.

⟨58⟩ Cahiers d'Histoire et de Philosophie des Sciences.

⟨59⟩ Cahiers Fundamenta Scientiae.
⟨60⟩ Canadian Journal of Mathematics.
⟨61⟩ Časopis po Pestovani Matematiky a Fysiky.
⟨62⟩ Časopis Praha.
⟨63⟩ Commentarii Mathematici Helvetici.
⟨64⟩ Communications on Pure and Applied Mathematics.
⟨65⟩ Compositio Mathematica.
⟨66⟩ Comptes Rendus de l'Académie des Sciences de Paris.
⟨67⟩ Comptes Rendus des Séances de la Société Mathématique de France.
⟨68⟩ Le Courrier du CNRS.
⟨69⟩ Doklady Akademii Nauk SSSR.
⟨70⟩ Duke Mathematical Journal.
⟨71⟩ Econometrica.
⟨72⟩ Elefteria.
⟨73⟩ Elemente der Mathematik.
⟨74⟩ L'Enseignement Mathématique.
⟨75⟩ Epistemologia.
⟨76⟩ Ergebnisse der Mathematik und ihrer Grenzgebiete.
⟨77⟩ Ericsson Technics.
⟨78⟩ Expositiones Mathematicae.
⟨79⟩ Facultad de Ingenieria Montevideo. Publicaciones del Instituto de Matematica y Estadistica.
⟨80⟩ Fundamenta Mathematicae.
⟨81⟩ Gazette des Mathématiciens.
⟨82⟩ Harkovski Matematičeskii Zapiski.
⟨83⟩ Historia Mathematica.
⟨84⟩ History and Philosophy of Logic.
⟨85⟩ Indigationes Mathematicae.
⟨86⟩ Institut des Hautes Etudes Scientifiques. Publications Mathématiques.
⟨87⟩ Interchange.
⟨88⟩ Istoriko-matematičeskie Issledovaniya.
⟨89⟩ Ivanov. Učeniye Zapiski Gosudarstvennogo pedagogičeskogo Instituta, fiziko-matematičeskogo Fakulteta.
⟨90⟩ Izvestiya Akademii Nauk SSSR, seriya matematičeskaya.
⟨91⟩ Izvestiya fiziko-matematičeskogo Obščestva pri Kazanskom Universitete.
⟨92⟩ Izvestiya politehničeskogo Instituta.
⟨93⟩ Jahrbuch der Akademie der Wissenschaften zu Göttingen.
⟨94⟩ Jahrbuch der Kurt Gödel Gesellschaft.

⟨95⟩ Jahrbuch Überblicke Mathematik.

⟨96⟩ Jahresbericht der Deutschen Mathematiker Vereinigung.

⟨97⟩ Journal de Mathématiques pures et appliquées.

⟨98⟩ Journal für die reine und angewandte Mathematik.

⟨99⟩ Journal of Mathematics and Physics.

⟨100⟩ Journal of Science of the Hiroshima University.

⟨101⟩ Journal of Symbolic Logic.

⟨102⟩ Journal of the College of Science of Tokyo.

⟨103⟩ Journal of the Faculty of Science Imperial University of Tokyo, section I: Mathematics, Astronomy, Physics, Chemistry.

⟨104⟩ The Journal of the London Mathematical Society.

⟨105⟩ Journal of the Mathematical Society of Japan.

⟨106⟩ Der Kongelige Norske Videnskabers Selskabs Skrifter.

⟨107⟩ Koninklijke Akademie van Wetenschappen te Amsterdam, Proceedings.

⟨108⟩ Koninklijke Akademie van Wetenschappen te Amsterdam, Verhandlingen 1^e sectie.

⟨109⟩ Koninklijke Nederlandsche Akademie van Wetenschappen, Verslagen van de Gewone Vergaderingen der Afdeeling Natuurkunde.

⟨110⟩ Mathematica.

⟨111⟩ The Mathematical Intelligencer.

⟨112⟩ Mathematical Journal of the Okahama University.

⟨113⟩ Mathematical Magazine.

⟨114⟩ Mathematical Medley.

⟨115⟩ Mathematical Spectrum.

⟨116⟩ Mathematičeskii Sbornik.

⟨117⟩ Mathematics and its Applications (Japanese series).

⟨118⟩ Mathématique, Informatique et Sciences Humaines.

⟨119⟩ Mathematische Annalen.

⟨120⟩ Mathematische Nachrichten.

⟨121⟩ Mathematische Semesterberichte.

⟨122⟩ Mathematische Zeitschrift.

⟨123⟩ Mathesis.

⟨124⟩ Meddelanden fram Lunds Universitets Matematiska Seminarium.

⟨125⟩ Menagement Science.

⟨126⟩ The Messenger of Mathematics.

⟨127⟩ Mitteilungen der Mathematischen Gesellschaft der Deutschen Demokratischen Republik.

⟨128⟩ Mlody Matematyk.

⟨129⟩ Monatshefte für Mathematik und Physik.

⟨130⟩ Nachrichten von der Königlichen Gesellschaft der Wissenschaften zu Göttingen, mathematisch-physikalische Klasse.

⟨131⟩ Nieuw Archief voor Wiskunde.

⟨132⟩ Norsk Mathematisk Forening, Skrifter series 2.

⟨133⟩ Notices of the American Mathematical Society.

⟨134⟩ Notre Dame Journal of Formal Logic.

⟨135⟩ Nouvelles Annales de Mathématiques.

⟨136⟩ Les Nouvelles de l'Académie des Sciences de Paris.

⟨137⟩ NTM-Schriftenreihe für Geschichte der Naturwissenschaften, Technik und Medizin.

⟨138⟩ Nyt Tidsskrift for Mathematik.

⟨139⟩ Öfversigt af Kungl. Vetenskaps-Akademiens Förhandligar, Stockholm.

⟨140⟩ Osiris.

⟨141⟩ Philosophia Mathematica.

⟨142⟩ The Philosophical Magazine and Journal of Science.

⟨143⟩ Philosophical Transactions of the Royal Society of London, series A.

⟨144⟩ Philosophy of Science.

⟨145⟩ Physical Review.

⟨146⟩ Pontificia Academia Scientiarum Relationes de Auctis Scientiis Tempore Belli, A.

⟨147⟩ Proceedings of the American Academy of Arts and Sciences.

⟨148⟩ Proceedings of the Cambridge Philosophical Society.

⟨149⟩ Proceedings of the Japan Academy.

⟨150⟩ Proceedings of the London Mathematical Society.

⟨151⟩ Proceedings of the National Academy of Sciences of the United States of America.

⟨152⟩ Proceedings of the physico-mathematical Society of Japan.

⟨153⟩ Proceedings of the Royal Irish Academy, section A.

⟨154⟩ Proceedings of the Royal Society of London, series A.

⟨155⟩ Progress in Mathematics.

⟨156⟩ The Quarterly Journal of Mathematics.

⟨157⟩ Reale Accademia Italiana, Memorie.

⟨158⟩ Rendiconti del Circolo Matematico di Palermo.

⟨159⟩ Review of Modern Physics.

⟨160⟩ Revista Matematica Hispano-Americana.

⟨161⟩ Revue d'Economie Politique.

⟨162⟩ Revue de Métaphysique et de Morale.

⟨163⟩ Revue d'Histoire des Sciences.
⟨164⟩ Rice University Studies.
⟨165⟩ Rivista di Matematica della Università di Parma.
⟨166⟩ Ruch Filozoficzny.
⟨167⟩ Russian Mathematical Surveys.
⟨168⟩ Science.
⟨169⟩ Sciences.
⟨170⟩ Scientia.
⟨171⟩ Scripta Mathematica.
⟨172⟩ Séminaire d'Analyse de l'Université Blaise Pascal.
⟨173⟩ Séminaire de Mathématiques de Gaston Julia.
⟨174⟩ SIAM Review.
⟨175⟩ Sitzungsberichte der Berliner Mathematischen Gesellschaft.
⟨176⟩ Sitzungsberichte der Heidelberger Akademie der Wissenschaften, mathematisch-naturwissenschaftliche Klasse.
⟨177⟩ Sitzungsberichte der Königlichen Preussischen Akademie der Wissenschaften zu Berlin, physikalisch-mathematische Klasse.
⟨178⟩ Sitzungsberichte der mathematisch-naturwissenschaftlichen Klasse der Kaiserlichen Akademie der Wissenschaften zu Wien, II, Abteilung.
⟨179⟩ Skrifter utgitt av det Norske Videnskaps-Akademi i Oslo, I: Mathematisk-naturvidenskapelig Klasse.
⟨180⟩ Southeast Asian Bulletin of Mathematics.
⟨181⟩ Studia Mathematica.
⟨182⟩ Studia Scientiarum Mathematicarum Hungarica.
⟨183⟩ Summa Brasiliensis Mathematicae.
⟨184⟩ The Tôhoku Mathematical Journal.
⟨185⟩ Transactions of the American Mathematical Society.
⟨186⟩ Travaux de la Société des Sciences et des Lettres de Varsovie.
⟨187⟩ Travaux de l'Institut physico-mathématique de Stekloff. Section mathématique.
⟨188⟩ Travaux mathématiques. Publications du Centre Universitaire de Luxembourg.
⟨189⟩ Učeniye Zapiski Moskovskogo Gosudarstvennogo Universiteta, Matematika.
⟨190⟩ Ukrainian Mathematical Journal.
⟨191⟩ Uspehi matematičeskih Nauk.
⟨192⟩ La Vie des Sciences.
⟨193⟩ Wissenschaftliche Zeitschrift. Ernst Moritz Arndt Universität, mathematisch-wissenschaftliche Reihe.

⟨194⟩ Wissenschaftliche Zeitschrift. Humboldt Universität Berlin, mathematisch-wissenschaftliche Reihe.

⟨195⟩ Zeitschrift der Kristallographie.

⟨196⟩ Zeitschrift Wahrscheinlichkeitstheorie und verwandte Gebiete.

⟨197⟩ Zenkoku Shijô Sûgaku Danwakai.

References

The symbol ⟨ ⟩ refers to the List of Journals.

ABBOTT J.C. (editor), *The Chauvenet papers. A collection of prizewinning expository papers in mathematics*, The Mathematical Association of America, 1978.

ACKERMANN W., *Zum Hilbertschen Aufbau der reellen Zahlen* (⟨120⟩, 99(1928), 118–133).

ACZEL J., *Léopold Fejér in memoriam* (⟨5⟩, 8(1961), 1–24).

D'ADHÉMAR R., *Sur une classe d'équations aux dérivées partielles du second ordre, de type hyperbolique, à 3 ou 4 variables indépendantes* (⟨97⟩, (5), 10(1904), 131–207).

ADO I.D., *Über die Darstellung von S. Lieschen Gruppen durch lineare Substitutionen* (⟨69⟩, 1936_{III}, 7–8).

AHLFORS L.V., *Sur le nombre des valeurs asymptotiques d'une fonction entière d'ordre fini* (⟨66⟩, 188(1929), 688–689) = p. 3–4, vol. I, *Collected papers*, Boston (Birkhäuser), 1982. (*Commentary*, p. 1).

AHLFORS L.V., *Untersuchungen zur Theorie der konformen Abbildung und der ganzen Funktionen* (⟨9⟩, A, (2), 1(1930), n° 9).

AHLFORS L.V., *Zur Theorie der Überlagerungsflächen* (⟨7⟩, 65(1935), 157–194) = p. 214–251, vol. I, *Collected papers*, Boston (Birkhäuser), 1982.

AHLFORS L.V., *An extension of Schwarz's lemma* (⟨185⟩, 43(1938), 359–364).

AHLFORS L.V. and BEURLING A., *Conformal invariants and function-theoretic null sets* (⟨7⟩, 83(1950), 101–129) = p. 406–434 (p. 394 *Commentary*), vol. I, L.V. AHLFORS, *Collected papers*, Boston (Birkhäuser), 1982 = p. 171–199, vol. I, A. BEURLING, *Collected papers*, Boston (Birkhäuser), 1989.

AHLFORS L.V. and GEHRING F.W., *Editors' preface*, O. TEICHMÜLLER, *Gesammelte Abhandlungen*, S. V–VI, Berlin (Springer-Verlag).

AIZERMAN M.A., *On the convergence of processes of automatic regulation after large initial deflections* (in Russian), (⟨41⟩, 7(1946), 148–167).

ALEXANDER J.W., *A proof of the invariance of certain constants of analysis situs* (⟨185⟩, 16(1915), 148–154).

ALEXANDER J.W., *On transformations with invariant points* (⟨185⟩, 23(1922), 89–95).

ALEXANDER J.W., *An example of a simply connected surface bounding a region which is not simply connected* (⟨151⟩, 10(1924), 8–10).

ALEXANDER J.W., *Combinatorial analysis situs* (⟨185⟩, 28(1926), 201–239).

ALEXANDER J.W., *On the chains of a complex and their duals* (⟨151⟩, 21(1935), 509–511).

ALEXANDER J.W. and BRIGGS G.B., *On types of knotted curves* (⟨28⟩, (2), 28(1927), 562–586).

ALEXANDROV P.S., *Über die Äquivalenz des Perronschen und des Denjoyschen Integralbegriffes* (⟨123⟩, 20(1924), 213–222).

ALEXANDROV P.S., *Über die Metrisation der im Kleinen kompakten topologischen Räume* (⟨120⟩, 92(1924), 294–301).

ALEXANDROV P.S., *Über kombinatorische Eigenschaften allgemeiner Kurven* (⟨120⟩, 96(1927), 512–554).

ALEXANDROV P.S., *Untersuchungen über Gestalt und Lage abgeschlossener Mengen beliebiger Dimension* (⟨28⟩, (2), 30(1929), 101–187).

ALEXANDROV P.S., *Dimensionentheorie. Ein Beitrag zur Geometrie der abgeschlossenen Mengen* (⟨120⟩, 106(1932), 160–238).

ALEXANDROV P.S., *Autobiographic pages* (in Russian) (⟨191⟩, 34(1979), N° 6, 219–249).

ALEXANDROV P.S. und HOPF H., *Topologie*, Berlin (Julius Springer), 1935.

ALEXANDROV P.S. und URYSOHN P.S., *Zur Theorie der topologischen Räume* (⟨120⟩, 92(1924), 258–266).

ALEXITS G., *Convergence problems of orthogonal series*, New York (Pergamon Press), 1961.

ALLENDOERFER C.B., *The Euler number of a Riemannian manifold* (⟨12⟩, 62 (1940), 243–248).

ALLENDOERFER C.B. and WEIL A., *The Gauss-Bonnet theorem for Riemannian polyhedra* (⟨185⟩, 53(1943), 101–129) = p. 299–327, vol. I, A. WEIL, *Oeuvres scientifiques*, New York (Springer-Verlag), 1980.

AMBROSE W., *Spectral resolution of groups of unitary operators*, (⟨70⟩, 11 (1944), 584–595).

AMBROSE W., *Structure theorems for a special class of Banach algebras*, (⟨185⟩, 57(1945), 364–386).

ANDRONOV A.A., *Les cycles limites de Poincaré et la théorie des oscillations autoentretenues* (⟨66⟩, 189(1929), 559–561).

ANDRONOV A.A. et PONTRYAGIN L.S., *Systèmes grossiers* (⟨69⟩, 14(1937), 247–250).

ANDRONOV A.A., VITT A.A. et KHAIKIN S.E., *Theory of oscillations* (in Russian), Moscow, 1937.

ANTOINE L., *Sur l'homéomorphie de deux figures et de leurs voisinages* (⟨97⟩, (8), 4(1921), 221–325).

ARBOLEDA L.C., *Les débuts de l'école topologique soviétique: notes sur les lettres de Paul S. Alexandroff et Paul S. Urysohn à Maurice Fréchet* (⟨33⟩, 20(1979), 73–89).

ARBOLEDA L.C., *Les recherches de M. Fréchet, P. Alexandrov, W. Sierpinski et K. Kuratowski sur la théorie des types de dimension et les débuts de la topologie générale* (⟨33⟩, 24(1981), 339–388).

ARTIN E., *Quadratische Körper im Gebiet der höheren Kongruenzen* (⟨123⟩, 19(1924), 153–246) = p. 1–94, *The collected papers*, Reading, Massachusetts (Addison-Wesley), 1965.

ARTIN E., *Theorie der Zöpfe* (⟨1⟩, 4(1925), 47–72).

ARTIN E., *Über die Zerlegung definiter Funktionen in Quadrate* (⟨1⟩, 5(1926), 100–115).

ARTIN E., *Beweis des allgemeinen Reziprozitätssatzes* (⟨1⟩, 5(1927), 353–363) = p. 131–141, *The collected Papers*, Reading, Massachusetts (Addison-Wesley), 1965.

ARTIN E., *Galois theory*, University of Notre Dame, Ind., 1942.

ARTIN E., *Theory of braids* (⟨28⟩, 48(1947), 101–126).

ARTIN E., *The influence of J.H.M. Wedderburn on the development of modern algebra* (⟨55⟩, 56(1950), 65–72) = p. 526–533, *The collected papers*, Reading, Massachusetts (Addison-Wesley), 1965.

ARTIN E., *Die Bedeutung Hilberts für die moderne Mathematik* (⟨93⟩, 1962) = p. 547–551, *The collected papers*, Reading, Massachusetts (Addison-Wesley), 1965.

ARTIN E. und SCHREIER O., *Algebraische Konstruktion reeller Körper* (⟨1⟩, 5 (1926), 85–99) = p. 258–272, E. ARTIN, *The collected papers*, Reading, Massachusetts (Addison-Wesley), 1965.

ARTIN M., *Introduction*, p. 3–10, vol. II, O. ZARISKI, *Collected papers*, Cambridge, Massachusetts (The MIT Press), 1973.

ASCOLI G., *Sugli spazi lineari e le loro varietà lineari* (⟨26⟩, (4), 10(1932), 33–81, 202–232).

ASPRAY W.F., *From mathematical constructivity to computer science: Alan Turing, John von Neumann, and the origins of computer science in mathematical logic*, Madison (University of Wisconsin), 1980.

ASPRAY W.F., *John von Neumann and the origins of modern computing*, Cambridge, Massachusetts (The MIT Press), 1990.

AUDIN M. et IGLESIAS P., *Questions à V.I. Arnold* (⟨81⟩, avril 1992, n° 52, 5–12).

BACHELIER L., *Théorie de la spéculation* (⟨24⟩, (3), 17(1900), 21–86).

BACHELIER L., *Calcul des probabilités*, Paris (Gauthier-Villars), 1912.

BAER R., *Homogeneity of projective planes* (⟨12⟩, 64(1942), 137–152).

BAIRE R., *Leçons sur les fonctions discontinues*, Paris (Gauthier-Villars), 1905 = p. 195–327, *Oeuvres scientifiques*, Paris (Gauthier-Villars, 1990).

BAIRE R. et SCHOENFLIES A., *Théorie des ensembles*, p. 489–531, tome I, vol. 1, *Encyclopédie des Sciences mathématiques*, Paris (Gauthier-Villars), 1909 = p. 478–520, R. BAIRE, *Oeuvres scientifiques*, Paris (Gauthier-Villars), 1990.

BAKER A., *Transcendental number theory*, Cambridge (University Press), 1990.

R. Baire

BANACH S., *Sur les opérations dans les ensembles abstraits et leur application aux équations intégrales* (⟨80⟩, 3(1922), 133–181) = p. 305–348, vol. II, *Oeuvres*, Warszawa (PWN — Editions scientifiques de Pologne), 1979.

BANACH S., *Sur les fonctionnelles linéaires* (⟨181⟩, 1(1929), 211–216, 223–239) = p. 375–395, vol. II, *Oeuvres*, Warszawa (PWN — Editions scientifiques de Pologne), 1979.

BANACH S., *Théorie des opérations linéaires*, Warszawa, 1932 = p. 13–219, vol. II, *Oeuvres*, Warszawa (PWN), 1979.

BANACH S. et STEINHAUS H., *Sur le principe de condensation des singularités* (⟨80⟩, 9(1927), 50–61) = p. 365–374, vol. II, S. BANACH, *Oeuvres*, Warszawa (PWN — Editions scientifiques de Pologne), 1979.

BANACH S. et TARSKI A., *Sur la décomposition des ensembles de points en parties respectivement congruentes* (⟨80⟩, 6(1924), 244–277) = p. 118–148, vol. I, S. BANACH, *Oeuvres*, Warszawa (PWN — Editions Scientifiques de Pologne), 1967.

BARBILIAN D., *Zur Bewegungstheorie der Septuoren* (⟨50⟩, 36(1934), 29–74).

BARONE J. and NOVIKOV A., *A history of the axiomatic formulation of probability from Borel to Kolmogorov* (⟨33⟩, 18(1977–1978), 123–190).

BARTLE R.G., *Nets and filters in topology* (⟨13⟩, 62(1955), 551–557).

BATT J., *Die Verallgemeinerungen des Darstellungssatzes von F. Riesz und ihre Anwendungen* (⟨96⟩, 74(1973), 147–181).

BAUER H., *Radons Arbeiten zur Mass- und Integrationtheorie sowie Potentialtheorie*, p. 29–44, vol. I, J. RADON *Collected works*, Basel (Birkhäuser) 1987.

BAUMGÄRTEL H., *John von Neumann, aus Leben und Werk* (Mitteilungen (⟨127⟩, 1982, N° 3–4, 87–98).

BEAULIEU L., *Bourbaki. Une histoire du groupe de mathématiciens français et ses travaux (1934–1944)*, Institut d'histoire et de sociopolitique des sciences, Faculté des arts et des sciences, Université de Montréal, 1989.

BEAULIEU L., *Proofs in expository writing — some examples from Bourbaki's early drafts* (⟨87⟩, 21(1990), n° 2, 35–45).

BECKER O., *Grundlagen der Mathematik in geschichtlicher Entwicklung*, Freiburg (Karl Alber), 1954.

BEHNKE H. und KÖTHE G., *Otto Toeplitz zum Gedächtnis* (⟨96⟩, 66 (1963), 1–16).

BEHNKE H. und THULLEN P., *Theorie der Funktionen mehrerer komplexer Veränderlichen*, Berlin (J. Springer), 1934.

BELL E.T., *Fifty years of algebra in America, 1888–1938*, p. 1–34, vol. II, *Semicentennial addresses of the American Mathematical Society*, New York, 1938.

BENDIXSON I., *Sur les courbes définies par des équations différentielles* (⟨7⟩, 24(1901), 1–88).

BENIS-SINACEUR H., *De D. Hilbert à E. Artin: les différents aspects du dix-septième problème et les filiations conceptuelles de la théorie des corps réels clos* (⟨33⟩, 29(1983–1984), 267–286).

BERGER M., *La géométrie de Riemann. Aperçu historique et résultats récents* (⟨68⟩, 1990, *Images de mathématiques*, Supplément au n° 76, 5–15).

BERNAYS P., *Axiomatische Untersuchungen des Aussagen-Kalküls der «Principia Mathematica»* (⟨123⟩, 25(1926), 305–320).

BERNAYS P., *A system of axiomatic set theory I* (⟨101⟩, 2(1937), 65–77).

BERNAYS P., *Axiomatic set theory.* With a historical introduction by A.A. FRAENKEL, Amsterdam (North-Holland), 1958.

BERNHARDT H., *Der Berliner Mathematiker R. v. Mises und sein wahrscheinlichkeitstheoretisches Konzept* (⟨194⟩, 39(1990), N° 3, 205–209).

BERNKOPF M., *The development of function spaces with particular reference to their origins in integral equation theory* (⟨33⟩, 3(1966–1967), 1–96).

BERNKOPF M., *A history of infinite matrices. A study of denumerably infinite linear systems as the first step in the history of operators defined on function spaces* (⟨33⟩, 4(1967–1968), 308–358).

BERNSTEIN F., *Untersuchungen aus der Mengenlehre*, Göttingen, 1901.

BERNSTEIN S.N., *Sur la nature analytique des solutions des équations aux dérivées partielles du second ordre* (⟨120⟩, 59(1904), 20–70).

BERNSTEIN S.N., *On the generalization of Dirichlet's problem* (⟨120⟩, 69(1910), 82–136) = (in Russian), p. 141–190, vol. III, *Collected works*, Moscow (Izdatel'stvo Akademii Nauk SSSR), 1960.

BERNSTEIN S.N., *Attempt of an axiomatic exposition of probability theory* (in Russian) (⟨82⟩, (2), 15(1917), 209–274).

BERS L., *Quasiconformal mappings, with applications to differential equations, function theory and topology* (⟨55⟩, 83(1977), 1083–1100).

BERTIN M.J., DECOMPS-GUILLOUX A., GRANDET-HUGOT M., PATHIAUX-DELEFOSSE M. and SCHREIBER J.P., *Pisot and Salem numbers*, Basel (Birkhäuser), 1992.

BERZOLARI L., *La matematica del siglo XX. II. Las correspondencias algebraicas* (⟨16⟩, 1934, 25–28).

BESICOVITCH A.S., *Almost periodic functions*, Cambridge (University Press), 1932.

BEURLING A., *Etudes sur un problème de majoration*, Upsal (Almqvist and Wiksell), 1933 = p. 1–107, vol. II, *The collected works*, Boston (Birkhäuser), 1989.

BEURLING A., *Sur les intégrales de Fourier absolument convergentes et leur application à une transformation fonctionnelle*, p. 345–366, *9ème Congrès Math. Scand. 1938*, 1939 = p. 39–60, vol. II, *The collected works*, Boston (Birkhäuser), 1989 (p. XII–XIII, commentaire de L. AHLFORS et L. CARLESON).

BEURLING A., *Un théorème sur les fonctions bornées et uniformément continues sur l'axe réel* (⟨7⟩, 77(1945), 127–136).

BEURLING A., *On two problems concerning linear transformations in Hilbert space* (⟨7⟩, 81(1949), 239–255).

BIDAL P. et DE RHAM G., *Les formes différentielles harmoniques* (⟨63⟩, 19(1946–1947), 1–49).

BIEBERBACH L., *Über die Bewegungsgruppen des n-dimensionalen euklidischen Raumes mit einem endlichen Fundamentalbereich* (⟨130⟩, 1910, 75–84).

BIEBERBACH L., *Über die Koeffizienten derjenigen Potenzreihen, welche eine schlichte Abbildung des Einheitskreises vermitteln* (⟨177⟩, 1916, 940–955).

BIERMANN K.-R., *David Hilbert und die Berliner Akademie* (⟨121⟩, 27(1963–1964), 377–384).

BIERMANN K.-R., *Die Mathematik und ihre Dozenten an der Berliner Universität 1810–1920*, Berlin (Akademie-Verlag), 1973.

BIGGS N.L., LLOYD E.K. and WILSON R.J., *Graph theory 1736–1936*, Oxford (Clarendon Press), 1976.

BILODEAU G.G., *The origin and early development of non-analytic infinitely differentiable functions* (⟨33⟩, 27(1982), 115–135).

BIRKHOFF G., *On the combination of subalgebras* (⟨148⟩, 29(1933), 441–464) = p. 9–33, *Selected papers on algebra and topology*, Boston (Birkhäuser), 1987. (*Lattices*, p. 1–2).

BIRKHOFF G., *On the structure of abstract algebras* (⟨148⟩, 31(1935), 433–454) = p. 117–138, *Selected papers on algebra and topology*, Boston (Birkhäuser), 1987. (*Universal algebra*, p. 111–112).

BIRKHOFF G., *Moore-Smith convergence in general topology* (⟨28⟩, (2), 38 (1937), 39–56).

BIRKHOFF G., *Lattice theory*, Providence, Rhode Island (American Mathematical Society), 1940.

BIRKHOFF G., *Lattice-ordered groups* (⟨28⟩, (2), 43(1942), 298–331).

BIRKHOFF G., *Current trends in algebra* (⟨13⟩, 80(1973), 760–782) = p. 490–512, *Selected papers on algebra and topology*, Boston (Birkhäuser), 1987.

BIRKHOFF G. (ed.), *Proceedings of the American Academy Workshop on the evolution of modern mathematics* (Historia Mathematica, 2(1975), 425–615).

BIRKHOFF G., *The rise of modern algebra*, p. 41–85, *Men and institutions in America* (Graduate Studies Texas Tech University, 13(1976)).

BIRKHOFF G. and KREYSZIG E., *The establishment of functional analysis* (⟨83⟩, 11(1984), 258–321).

BIRKHOFF G. and VON NEUMANN J., *The logic of quantum mechanics* (⟨28⟩, (2), 37(1936), 823–843).

BIRKHOFF G.D., *Proof of Poincaré's geometric theorem* (⟨185⟩, 14(1913), 14–22) = p. 673–681, vol. I, *Collected mathematical papers*, New York City (American Mathematical Society), 1950.

BIRKHOFF G.D., *The reducibility of maps* (⟨12⟩, 35(1913), 115–128).

BIRKHOFF G.D., *The generalized Riemann problem for linear differential equations and the allied problems for linear difference and q-difference equations* (⟨147⟩, 49(1913), 521–568).

BIRKHOFF G.D., *Dynamical systems with two degrees of freedom* (⟨185⟩, 18 (1917), 199–300) = p. 1–102, vol. II, *Collected mathematical papers*, New York City (American Mathematical Society), 1950.

BIRKHOFF G.D., *Recent advances in dynamics* (⟨168⟩, N.S., 51(1920), n° 1307, 51–55) = p. 106–110, vol. II, *Collected mathematical papers*, New York City (American Mathematical Society), 1950.

BIRKHOFF G.D., *Relativity and modern physics*, with the cooperation of R.E. Langer, Cambridge (Harvard University Press), 1923.

BIRKHOFF G.D., *Dynamical systems*, Providence, Rhode Island (American Mathematical Society), 1927.

BIRKHOFF G.D., *Proof of the ergodic theorem* (⟨151⟩, 17(1932), 656–660).

BIRKHOFF G.D., *Sur quelques courbes fermées remarquables* (⟨49⟩, 60 (1932), 1–26).

BIRKHOFF G.D. and KELLOGG O.D., *Invariant points in function space* (⟨185⟩, 23(1922), 96–115).

BIRMAN J.S., *Braids, links, and mapping class groups*, Princeton (University Press), 1974.

BLANC-LAPIERRE A. et FORTET R., *Résultats sur la décomposition spectrale des fonctions aléatoires stationnaires d'ordre 2* (⟨66⟩, 222(1946), 713–714).

BLANC-LAPIERRE A. et FORTET R., *Sur la décomposition spectrale des fonctions aléatoires d'ordre deux* (⟨66⟩, 222(1946), 467–468).

BLANC-LAPIERRE A. et FORTET R., *Sur la structure des fonctions aléatoires strictement stationnaires à spectre totalement discontinu* (⟨66⟩, 222(1946), 1155–1157).

BLASCHKE W., *Eine Frage über konvexe Körper* (⟨96⟩, 25(1916), 121–125).

BLASCHKE W., *Vorlesungen über Differentialgeometrie und geometrische Grundlagen von Einsteins Relativitätstheorie I*, Berlin (J. Springer), 1921.

BLASCHKE W., *Vorlesungen über Differentialgeometrie II: Affine Differentialgeometrie*, Berlin (J. Springer), 1923.

BLASCHKE W., *Vorlesungen über Integralgeometrie I*, Leipzig (B.G. Teubner), 1935.

BLASS P., *The influence of Oskar Zariski on modern algebraic geometry* (⟨40⟩, 16(1989), 157–168).

BLISS G.A., *The scientific work of E.M. Moore* (⟨55⟩, 40(1934), 501–514).

BLOCH S., *The proof of the Mordell conjecture* (⟨112⟩, 6(1984), n° 2, 41–47).

BOCHNER S., *Beiträge zur Theorie der fastperiodischen Funktionen, I. Funktionen einer Variablen* (⟨120⟩, 96(1927), 119–147).

BOCHNER S., *Darstellung reellvariabler und analytischer Funktionen durch verallgemeinerte Fourier- und Laplace-Integrale* (⟨120⟩, 97(1927), 635–662).

BOCHNER S., *Vorlesungen über Fouriersche Integrale*, Leipzig (Akademie Verlagsgesellschaft), 1932.

BOCHNER S., *Integration von Funktionen deren Werte die Elemente eines Vektorraumes sind* (⟨80⟩, 20(1933), 262–276).

BOCHNER S., *Remark on the theorem of Green* (⟨70⟩, 3(1937), 334–338).

BOCHNER S., *Vector fields and Ricci curvature* (⟨55⟩, 52(1946), 776–797).

BOCHNER S., *Curvature and Betti numbers* (⟨28⟩, (2), 49(1948), 379–390).

BOCHNER S., *Singularities and discontinuities*, p. 21–41, *Complex analysis*, vol. II, Houston (Rice University), 1973.

BOCHNER S. and CHANDRASEKHARAN K., *Fourier transforms*, Princeton, 1949.

BOERNER H., *Representations of groups with special consideration for the needs of modern physics*, Amsterdam (North-Holland), 1963.

BOGOLIOUBOV N.N. and KRYLOV N.M., *Fundamental problems in the non linear mechanics*, p. 270–272, vol. II, *Verhandlungen des internationalen Mathematiker-Kongresses, Zürich 1932*, Zürich, 1933.

BOHL P., *Über die Bewegung eines mechanischen Systems in der Nähe einer Gleichgewichtslage* (⟨98⟩, 127(1904), 179–276).

BOHL P., *Über die Differentialgleichung der Störungstheorie* (⟨98⟩, 131 (1906), 268–321).

BOHL P., *Über ein in der Theorie der säkularen Störungen vorkommendes Problem* (⟨98⟩, 135(1909), 189–283).

BOHR H., *Sur les fonctions presque périodiques* (⟨66⟩, 177(1923), 737–739) = C. 1, vol. II, *Collected mathematical papers*, København (Dansk Matematisk Forening), 1952.

BOHR H., *Fastperiodische Funktionen*, Berlin (J. Springer), 1932.

BOKS T.J., *Sur les rapports entre les méthodes d'intégration de Riemann et Lebesgue* (⟨158⟩, 45(1921), 1–54).

BOLLINGER M., *Geschichtliche Entwicklung des Homologiebegriffs* (⟨33⟩, 9 (1972–1973), 94–170).

BOLZA O., *Lectures on the calculus of variations*, Chicago (University Press), 1904.

BOLZA O., *Über den «Anormalen Fall» beim Lagrangeschen und Mayerschen Problem mit gemischten Bedingungen und variablen Endpunkten* (⟨120⟩, 74 (1913), 430–446).

BONNESEN T. und FENCHEL W., *Theorie der konvexen Körper*, Berlin (Springer), 1934.

BOREL A., *The school of mathematics at the Institute for Advanced Study*, p. 119–147, Part III, P. DUREN (edited by), *A century of mathematics in America*, Providence, Rhode Island (American Mathematical Society), 1989.

BOREL A., *The work of Chevalley in Lie groups and algebraic groups*, p. 1–22, *Proceedings of the Hyderabad conference on algebraic groups (Hyderabad 1989)*, Madras (Monoy Prakashan), 1991.

BOREL A. et SERRE J.-P., *Impossibilité de fibrer un espace euclidien par des fibres compactes* (⟨66⟩, 230(1950), 2258–2260) = p. 3–4, vol. I, J.-P. SERRE, *Collected papers*, Berlin (Springer-Verlag), 1986. (*Note*, p. 584).

BOREL E., *Les probabilités dénombrables et leurs applications arithmétiques* (⟨158⟩, 27(1909), 247–270) = p. 1055–1079, vol. II, *Oeuvres*, Paris (CNRS), 1972.

BOREL E., *Notice sur les travaux scientifiques*, Paris (Gauthier-Villars), 1912 = p. 119–201, vol. I, *Oeuvres*, Paris (C.N.R.S.), 1972.

BOREL E., *Les séries de fonctions analytiques et les fonctions quasi-analytiques* (⟨66⟩, 154(1912), 1491–1493) = p. 769–771, vol. II, *Oeuvres*, Paris (C.N.R.S.), 1972.

BOREL E., *La théorie du jeu et les équations intégrales à noyau symétrique* (⟨66⟩, 173(1921), 1304–1308) = p. 901–904, vol. II, *Oeuvres*, Paris (C.N.R.S.), 1972.

BORG G., *Eine Umkehrung der Sturm-Liouvilleschen Eigenwertaufgabe. Bestimmung der Differentialgleichung durch die Eigenwerte* (⟨7⟩, 78(1945), 1–96).

BORSUK K., *Sur les rétractes* (⟨80⟩, 17(1931), 152–170).

BORSUK K., *Drei Sätze über die n-dimensionale euklidische Sphäre* (⟨80⟩, 20 (1933), 177–190).

BORSUK K., *Sur les prolongements des transformations continues* (⟨80⟩, 28 (1937), 99–100).

BOS H.J.M., *Hans Freudenthal (1905–1990)* (⟨83⟩, 19(1992), 106–108).

BOSECK H., *Zur Entwicklung der Lie-Theorie* (⟨127⟩, N° 3–4, 1989, 3–18).

BOTT R., *Marston Morse and his mathematical works*, p. XI–XLIX, vol. I, *Collected papers*, Singapore (World Scientific), 1987.

BOTT R., *Georges de Rham: 1901–1990* (⟨14⟩, 38(1991), 114–115).

BOTTAZINI U., *Il flauto di Hilbert: storia della matematica moderna e contemporanea*, Torino (UTET Libreria), 1990.

BOULEAU N., *La jonction entre la théorie du potentiel et les probabilités* (⟨57⟩, 8(1987), 43–66).

BOULIGAND G., *Introduction à la géométrie infinitésimale directe*, Paris (Librairie Vuibert), 1932.

BOURBAKI N., *Sur un théorème de Carathéodory et la mesure dans les espaces topologiques* (⟨66⟩, 201(1935), 1309–1311).

BOURBAKI N., *Théorie des ensembles.* Fascicule de résultats, Paris (Hermann), 1939.

BOURBAKI N., *Topologie générale, I. Structures topologiques, II. Structures uniformes*, Paris (Hermann), 1940.

BOURBAKI N., *Eléments d'histoire des mathématiques*, Paris (Masson), 1984.

BOURGUIGNON J.-P. (propos recueillis par), *Entretien avec un optimiste, S.S. Chern* (⟨81⟩, avril 1991, n° 48, 5–10).

BOX J.F., *R.A. Fisher. The life of a scientist*, New York (Wiley), 1978.

BRAUER R., *Über Systeme hyperkomplexer Zahlen* (⟨123⟩, 30(1929), 79–107) = p. 40–68, vol. I, *Collected papers*, Cambridge, Massachusetts (The MIT Press), 1980.

BRAUER R., *Über die Konstruktion der Schiefkörper, die von endlichem Rang in Bezug auf ein gegebenes Zentrum sind* (⟨98⟩, 168(1932), 44–64) = p. 121–141, vol. I, *Collected papers*, Cambridge, Massachusetts (The MIT Press).

BRAUER R., *Über die Darstellung von Gruppen in Galoisschen Feldern*, Paris (Hermann), 1935 = p. 323–335, vol. I, *Collected papers*, Cambridge, Massachusetts (The MIT Press), 1980.

BRAUER R., *On Artin's L-series with general group characters* ((⟨28⟩, (2), 48 (1947), 502–514) = p. 539–551, vol. I, *Collected papers*, Cambridge, Massachusetts (The MIT Press), 1980.

BRAUER R., *On the representation of groups of finite order*, p. 33–36, vol. II, *Proceedings of the International Congress of Mathematicians, Cambridge, Massachusetts 1950*, Providence, R.I. (Amer. Math. Soc.), 1952.

BRAUER R., *Algebra der hyperkomplexen Zahlensysteme (Algebren)* (⟨113⟩, 21 (1979), 53–89) = p. 153–189, vol. I, *Collected papers*, Cambridge, Massachusetts (The MIT Press), 1980.

BRAUER R., *Symmetrische Funktionen. Invarianten von linearen Gruppen endlicher Ordnung* (⟨113⟩, 21(1979), 91–113) = p. 622–644, vol. III, *Collected papers*, Cambridge, Massachusetts (The MIT Press), 1980.

BRAUER R., *Preface*, p. XV–XIX, vol. I, *Collected papers*, Cambridge, Massachusetts (The MIT Press), 1980.

BRAUER R., HASSE H. und NOETHER E., *Beweis eines Hauptsatzes in der Theorie der Algebren* (⟨98⟩, 167(1932), 399–404).

BRAUER R. and WEYL H., *Spinors in n dimensions* (⟨12⟩, 57(1935), 425–449) = p. 418–442, vol. III, R. BRAUER, *Collected papers*, Cambridge, Massachusetts (The MIT Press), 1980 = S. 493–516, Band III, H. WEYL, *Gesammelte Abhandlungen*, Berlin (Springer-Verlag), 1968.

BRELOT M., *Sur un problème de Dirichlet généralisé* (⟨66⟩, 191(1930), 697–699).

BRELOT M., *Einige neuere Untersuchungen über das Dirichletsche Problem* (⟨96⟩, 42(1932–1933), 1–14).

BRELOT M., *Familles de Perron et problème de Dirichlet* (⟨8⟩, 9(1939), 133–153).

BRELOT M., *Sur la théorie autonome des fonctions sous-harmoniques* (⟨53⟩, (2), 65(1941), 72–98).

BRELOT M., *Sur les ensembles effilés* (⟨49⟩, 68(1944), 12–36).

BRELOT M., *Sur le point à l'infini dans la théorie des fonctions harmoniques* (⟨24⟩, (3), 61(1944), 301–332).

BRELOT M., *Minorantes sousharmoniques extrémales et capacité* (⟨97⟩, 24 (1945), 1–32).

BRELOT M., *La théorie moderne du potentiel* (⟨21⟩, 4(1952), 113–140).

BRELOT M., *Les étapes et les aspects multiples de la théorie du potentiel* (⟨74⟩, (2), 18(1972), 1–36).

BRENNER J.L., *Student days — 1930* (⟨13⟩, 86(1979), 350–356).

BREZINSKI C., *History of continued fractions and Padé approximants*, Berlin (Springer-Verlag), 1991.

BRIGAGLIA A., *Sur les relations des mathématiciens français et italiens au début du XXe siècle* (⟨57⟩, 5(1984), 21–48).

BRIGAGLIA A., *Introduction of modern algebra in Italy* (in Italian) (⟨57⟩, 10 (1989), 323–349; 11(1990), 129–132).

BROADBENT J.A.A., *Ramsey, Frank Plumpton*, p. 285–286, vol. XI, *Dictionary of scientific biography*, New York (Charles Scribner's Sons), 1975.

BROUWER L.E.J., *Over de Grondslagen der Wiskunde*, Amsterdam (Maas), 1907 = *On the foundations of mathematics*, p. 11–101, vol. I, *Collected works*, Amsterdam (North-Holland), 1975.

BROUWER L.E.J., *Über eineindeutige, stetige Transformationen von Flächen in sich* (⟨120⟩, 69(1910), 176–180) = p. 244–249, vol. II, *Collected works*, Amsterdam (North-Holland), 1976.

BROUWER L.E.J., *Beweis der Invarianz der Dimensionenzahl* (⟨120⟩, 70(1911), 161–165) = p. 430–434, vol. 2, *Collected works*, Amsterdam (North-Holland), 1976.

BROUWER L.E.J., *Über Abbildung von Mannigfaltigkeiten* (⟨120⟩, 71 (1911), 97–115) = p. 454–472, vol. II, *Collected works*, Amsterdam (North-Holland), 1976.

BROUWER L.E.J., *A. Schoenflies und H. Hahn, Die Entwicklung der Mengenlehre und ihrer Anwendungen. Erste Hälfte: Allgemeine Theorie der unendlichen Mengen und Theorie der Punktmengen von A. Schoenflies* (⟨96⟩, 23(1914), Literarisches, 78–83).

BROUWER L.E.J., *Begründung der Mengenlehre unabhängig vom logischen Satz vom ausgeschlossenen Dritten. Erster Teil: Allgemeine Mengenlehre* (⟨108⟩, 12(1918), n° 5) = p. 150–190, vol. I, *Collected works*, Amsterdam (North-Holland), 1975.

BROUWER L.E.J., *Begründung der Funktionenlehre unabhängig vom logischen Satz vom ausgeschlossenen Dritten. Erster Teil: Stetigkeit, Messbarkeit, Derivierbarkeit* (⟨108⟩, 13(1923), n° 2) = p. 246–267, vol. I, *Collected works*, Amsterdam (North-Holland), 1975.

BROWDER F.E., *The relation of functional analysis to concrete analysis in 20th century mathematics* (⟨83⟩, 2(1975), 577–590).

BROWDER F.E. (edited by), *The mathematical heritage of Henri Poincaré*, Providence, Rhode Island (American Mathematical Society), 1983.

BRUN V., *Über das Goldbachsche Gesetz und die Anzahl der Primzahlpaare* (⟨35⟩, 34(1915), 3–19).

BRUN V., *Le crible d'Eratosthène et le théorème de Goldbach* (⟨66⟩, 168(1919), 544–546).

BRUN V., *La série* $\frac{1}{5}+\frac{1}{7}+\frac{1}{11}+\frac{1}{13}+\frac{1}{17}+\frac{1}{19}+\frac{1}{29}+\frac{1}{31}+\frac{1}{41}+\frac{1}{43}+\frac{1}{59}+\frac{1}{61}+\dots$ *où les dénominateurs sont «nombres premiers jumeaux» est convergente ou finie* (⟨53⟩, (2), 43(1919), 100–104, 124–128).

BRUSH S.G., *Foundations of statistical mechanics 1845–1915* (⟨33⟩, 4 (1967–1968), 145–183).

BRUSH S.G., *A history of random processes* (⟨33⟩, 5(1968–1969), 1–36).

BRYLEVSKAYA L.I., *On the discussion on existence of nonmeasurable sets* (in Russian) (Istoriya Metod. Estestv. Nauk, 36(1989), 28–33).

BULLEN P.S., LEE P.Y., MAWHIN J.L., MULDOWNEY P. and PFEFFER W.F. (eds), *New integrals*, Berlin (Springer-Verlag), 1990.

BUREAU F., *Sur la question du Prix Bordin 1933* (⟨57⟩, 7(1986), 1–14).

BURKHARDT H., *Entwicklungen nach oscillierenden Functionen und Integration der Differentialgleichungen der mathematischen Physik* (⟨96⟩, 10-2(1901–1908) = Leipzig (Teubner), 1908.

BURKHARDT H. et VOGT H., *Sur les groupes finis discontinus*, p. 532–616, vol. 1, tome I, *Encyclopédie des sciences mathématiques*, Paris (Gauthier-Villars), 1909.

BURKS A.W., GOLDSTINE H.H. and VON NEUMANN J., *Preliminary discussion of the logical design of an electronic computing instrument*, Part I, vol. I. Report prepared for U.S. Army Ord. Dep. under contract W-36-034-ORD-7481 (June 28, 1946).

BURNSIDE W., *On some properties of groups of odd order* (⟨150⟩, 33 (1901), 162–185, 257–268).

BURNSIDE W., *On groups of order* $p^\alpha q^\beta$ (⟨150⟩, (2), 1(1904), 388–392).

BURNSIDE W., *On an unsettled question in the theory of discontinuous groups* (⟨156⟩, 33(1904), 230–238).

BURNSIDE W., *On the condition of reducibility for any group of linear substitutions* (⟨150⟩, 3(1905), 430–434).

BURNSIDE W., *Theory of groups of finite order*, second edition, Cambridge (University Press), 1911 = New York (Dover), 1955.

BUTZER P.L., GIESELER S., KAUFMANN F., NESSEL R.J. und STARK E.L., *Eduard Helly (1884–1943). Eine nachträgliche Würdigung* (⟨96⟩, 82(1980), 128–151).

BUTZER P.L., SCHMIDT M. and STARK E.L., *Observations on the history of central B-splines* (⟨33⟩, 39(1988), 137–156).

CACCIOPPOLI R., *Sulle equazioni ellitiche a derivate parziali con n variabili indipendenti* (⟨36⟩, (6), 19(1934), 83–89) = p. 98–105, vol. II, *Opere*, Roma (Edizioni Cremonese), 1963.

CAIRNS S.S., *The cellular division and approximation of regular spreads* (⟨151⟩, 16(1930), 488–491).

CAIRNS S.S., *On the triangulation of regular loci* (⟨28⟩, (2), 35(1934), 579–587).

CAIRNS S.S., *Triangulation of the manifold of class one* (⟨55⟩, 41(1935), 549–552).

CAJORI F., *History of mathematical notation*, Chicago (Open Court), 1928–1929.

CALDI D.G. and MOSTOW G.D. (edited by), *Proceedings of the Gibbs symposium*, Providence (American Mathematical Society), 1990.

CALLENS S., *Ensemble, mesure et probabilité selon Emile Borel* (⟨119⟩, 110 (1990), 27–45).

CAMPBELL J.E., *Introductory treatise on Lie's theory of finite continuous transformation groups*, Oxford (Clarendon Press), 1903 = New York (Chelsea), 1966.

CAMPBELL P.J., *The origins of «Zorn's lemma»* (⟨83⟩, 5(1978), 77–89).

CARATHÉODORY C., *Über die gegenseitige Beziehung der Ränder beider konformen Abbildungen des Inneren einer Jordanschen Kurve auf einen Kreis* (⟨120⟩, 73(1913), 305–320).

CARATHÉODORY C., *Über das lineare Mass von Punktmengen — eine Verallgemeinerung des Längenbegriffs* (⟨130⟩, 1914, 404–426) = S. 249–275, Band IV, *Gesammelte mathematische Schriften*, München (C.H. Beck), 1956.

CARATHÉODORY C., *Vorlesungen über reelle Funktionen*, Leipzig (Teubner), 1918.

CARATHÉODORY C., *Variationsrechnung und partielle Differentialgleichungen erster Ordnung*, Leipzig (B.G. Teubner), 1935.

CARATHÉODORY C., *Funktionentheorie*, II. Band, Basel (Birkhäuser), 1950.

CARATHÉODORY C. und FEJÉR L., *Über den Zusammenhang der Extremen von harmonischen Funktionen mit ihren Koeffizienten und über den Picard-Landauschen Satz* (⟨158⟩, 32(1911), 218–239) = S. 111–138, Band III, C. CARATHÉODORY, *Gesammelte mathematische Schriften*, München (C.H. Beck), 1955 = S. 693–715, Band I, L. FEJÉR, *Gesammelte Arbeiten*, Basel (Birkhäuser), 1970.

CARAZZA B. e GUIDETTI G.P., *La nascita dell'equazione di Klein-Gordon* (⟨33⟩, 22(1980), 373–383).

CARLEMAN T., *Sur les équations intégrales singulières à noyau réel et symétrique*, Uppsala (Lundequitsha Bokhandeln), 1923.

CARLEMAN T., *Les fonctions quasi analytiques*, Paris (Gauthier-Villars), 1926.

CARLEMAN T., *Sur les systèmes linéaires aux dérivées partielles du premier ordre à deux variables* (⟨66⟩, 197(1933), 471–474).

CARLEMAN T., *Sur la théorie mathématique de l'équation de Schrödinger* (⟨34⟩, 24(1934), B, N° 11) = p. 463–469, *Edition complète des articles*, Malmö (Litos Reprotryck), 1960.

CARLEMAN T., *Propriétés des fonctions fondamentales des membranes vibrantes. Comptes rendus du VIIIe Congrès des Mathématiciens Scandinaves à Stockholm, 1934.*

CARLEMAN T., *Sur un problème d'unicité pour les systèmes d'équations aux dérivées partielles à deux variables indépendantes* (⟨34⟩, 26(1938), B, N° 17) = p. 497–505, *Edition complète des articles*, Malmö (Litos Reprotryck), 1960.

CARLEMAN T., *L'intégrale de Fourier et questions qui s'y rattachent*, Uppsala (Almquist et Wiksells), 1944.

CARTAN E., *Sur l'équivalence des systèmes différentiels* (⟨66⟩, 135(1902), 781–783) = p. 563–565, partie II, vol. 2, *Oeuvres complètes*, Paris (Gauthier-Villars), 1953.

CARTAN E., *Les sous-groupes des groupes continus de transformations* (⟨24⟩, (3), 25(1908), 57–194) = p. 719–856, partie II, vol. 2, *Oeuvres complètes*, Paris (Gauthier-Villars), 1953.

CARTAN E., *La structure des groupes de transformations continus et la théorie du trièdre mobile* (⟨49⟩, 34(1910), 250–284) = p. 145–178, partie III, vol. 1, *Oeuvres complètes*, Paris (Gauthier-Villars), 1955.

CARTAN E., *Les groupes projectifs qui ne laissent invariante aucune multiplicité* (⟨49⟩, 41(1913), 53–96) = p. 355–398, partie I, vol. I, *Oeuvres complètes*, Paris (Gauthier-Villars), 1952.

CARTAN E., *Les groupes simples, finis et continus* (⟨24⟩, (3), 31(1914), 263–355) = p. 399–491, partie 1, vol. I, *Oeuvres complètes*, Paris (Gauthier-Villars), 1952.

CARTAN E., *Leçons sur les invariants intégraux*, Paris (Hermann), 1922.

CARTAN E., *Les espaces à connexion conforme* (⟨19⟩, 2(1923), 171–221) = p. 747–797, partie III, vol. 1, *Oeuvres complètes*, Paris (Gauthier-Villars), 1955.

CARTAN E., *Les récentes généralisations de la notion d'espace* (⟨53⟩, (2), 48 (1924), 294–320) = p. 863–889, partie III, vol. 1, *Oeuvres complètes*, Paris (Gauthier-Villars), 1955.

CARTAN E., *La géométrie des espaces de Riemann*, Mémorial Sc. math. IX, Paris (Gauthier-Villars), 1925.

CARTAN E., *Les groupes d'holonomie des espaces généralisés* (⟨7⟩, 48 (1926), 1–42) = p. 997–1038, partie III, vol. 2, *Oeuvres complètes*, Paris (Gauthier-Villars), 1955.

CARTAN E., *La théorie des groupes et les recherches récentes de géométrie différentielle*, p. 85–94, vol. I, *Proceedings of the international mathematical Congress hold in Toronto, August 11–16, 1924*, The University of Toronto Press, 1928 = p. 891–904, partie III, vol. I, *Oeuvres complètes*, Paris (Gauthier-Villars), 1955.

CARTAN E., *Leçons sur la géométrie des espaces de Riemann*, Paris (Gauthier-Villars), 1928.

CARTAN E., *Groupes simples clos et ouverts et géométrie riemannienne* (⟨97⟩, (9), 8(1929), 1–33) = p. 1011–1043, partie I, vol. 2, *Oeuvres complètes*, Paris (Gauthier-Villars), 1952.

CARTAN E., *La théorie des groupes finis et continus et l'analysis situs*, Paris (Gauthier-Villars), 1930. = p. 1165–1225, partie I, vol. II, *Oeuvres complètes*, Paris (Gauthier-Villars), 1952.

CARTAN E., *La topologie des espaces représentatifs des groupes de Lie* (⟨74⟩, 35(1936), 177–200) = p. 1307–1330, partie I, vol. 2, *Oeuvres complètes*, Paris (Gauthier-Villars), 1952.

CARTAN E., *La topologie des groupes de Lie*, Paris (Hermann), 1936.

CARTAN E., *La théorie des groupes finis et continus et la géométrie différentielle traitées par la méthode du repère mobile*, leçons rédigées par J. LERAY, Paris (Gauthier-Villars), 1937.

CARTAN E., *Leçons sur la théorie des spineurs*. Paris (Hermann), 1938.

CARTAN E., *Notice sur les travaux scientifiques*, p. 1–112, partie I, vol. I, *Oeuvres complètes*, Paris (Gauthier-Villars), 1952.

Elie Cartan 1869–1959, Bucarest (Editura Akad. R.S.R.), 1975.

CARTAN H., *Sur les domaines d'existence des fonctions de plusieurs variables complexes* (⟨49⟩, 59(1931), 46–59) = p. 303–326, vol. I, *Oeuvres*, Berlin (Springer-Verlag), 1979.

CARTAN H., *Sur les groupes de transformations analytiques*, Paris (Hermann), 1935 = p. 474–523, vol. I, *Oeuvres*, Berlin (Springer-Verlag), 1979.

CARTAN H., *Théorie des filtres* (⟨66⟩, 205(1937), 595–598) = p. 953–956, vol. III, *Oeuvres*, Berlin (Springer-Verlag), 1979.

CARTAN H., *Filtres et ultrafiltres* (⟨66⟩, 205(1937), 777–779) = p. 957–959, *Oeuvres*, Berlin (Springer-Verlag), 1979.

CARTAN H., *Sur la mesure de Haar* (⟨66⟩, 211(1940), 759–762) = p. 1020–1022, vol. III, *Oeuvres*, Berlin (Springer-Verlag), 1979.

CARTAN H., *Sur les fondements de la théorie de potentiel* (⟨49⟩, 69(1941), 71–96) = p. 1023–1048, vol. III, *Oeuvres*, Berlin (Springer-Verlag), 1979.

CARTAN H., *Idéaux de fonctions analytiques de n variables complexes* (⟨24⟩, 61(1944)) = p. 565–613, vol. II, *Oeuvres*, Berlin (Springer-Verlag), 1979.

CARTAN H., *Méthodes modernes et topologie algébrique* (⟨63⟩, 18(1945), 1–15) = p. 1164–1178, vol. III, *Oeuvres*, Berlin (Springer), 1979.

CARTAN H., *Théorie du potentiel newtonien: énergie, capacité, suites de potentiels* (⟨49⟩, 73(1945), 74–106).

CARTAN H., *Théorie générale du balayage en potentiel newtonien* (⟨23⟩, 22 (1946), 59–77).

CARTAN H., *Séminaire de topologie algébrique, 1e année: 1948–1949*, Ecole Normale Supérieure, Paris.

CARTAN H., *Idéaux et modules de fonctions analytiques de variables complexes* (⟨49⟩, 78(1950), 29–64) = p. 618–653, vol. II, *Oeuvres*, Berlin (Springer-Verlag), 1979.

CARTAN H., *Notion d'algèbre différentielle; application aux groupes de Lie et aux variétés où opère un groupe de Lie*, p. 15–27, *Colloque de topologie (espaces fibrés), tenu à Bruxelles du 5 au 8 juin 1950*, Liège (Georges Thone), 1951.

CARTAN H., *La transgression dans un groupe de Lie et dans un espace fibré principal*, p. 57–71, *Colloque de topologie (espaces fibrés), tenu à Bruxelles du 5 au 8 juin 1950*, Liège (Georges Thone), 1951.

CARTAN H., *Espaces fibrés analytiques complexes*, p. 30-01-10, *Séminaire Bourbaki, 3e année : 1950/1951. Textes des conférences, exposés 33 à 49*, 2e édition, Paris (Secrétariat mathématique), 1959.

CARTAN H., *Emil Artin* (⟨1⟩, 28(1965), 1–5).

CARTAN H., *Brève analyse des travaux*, p. IX–XIX, vol. I, *Oeuvres*, Berlin (Springer-Verlag), 1979.

CARTAN H., *Sur quelques progrès dans la théorie des fonctions analytiques de variables complexes entre 1930 et 1950*, p. 45–58, HILTON P., HIRZEBRUCH F. and REMMERT R. (edited by), *Miscellanea mathematica*, Berlin (Springer-Verlag), 1991.

Colloque analyse et topologie en l'honneur de Henri Cartan, Astérisque, 32–33, 1976, Société Mathématique de France.

CARTAN H. et FERRAND J., *Le cas André Bloch* (⟨57⟩, 9(1988), 211–220).

CARTAN H. et GODEMENT R., *Théorie de la dualité et analyse harmonique dans les groupes abéliens localement compacts* (⟨24⟩, (3), 64 (1947), 79–99) = p. 1203–1223, vol. III, H. CARTAN, *Oeuvres*, Berlin (Springer-Verlag), 1979.

CARTAN H. und THULLEN P., *Zur Theorie der Singularitäten der Funktionen mehrerer komplexen Veränderlichen* (⟨120⟩, 106(1932), 617–647) = p. 376–406, vol. I, H. CARTAN, *Oeuvres*, Berlin (Springer-Verlag), 1979.

CARTWRIGHT M.L. and LITTLEWOOD J.E., *On the differential equations of the second order. II. The equation* $y'' + kf(y)y' + g(y,k) = p_1(t) + kp_2(t)$*;* $k > 0$, $f(y) \geq 1$ (⟨28⟩, (2), 48(1947), 472–494; 49(1948), 1010; 50(1949), 504–505) = p. 95–119, vol. I, J.E. LITTLEWOOD, *Collected papers*, Oxford (Clarendon Press), 1982.

CASIMIR H.B.G., *Rotation of a rigid body in quantum mechanics*, Leiden (Acad. Proefschrift), 1931.

CASSELS J.W.S., *Mordell, Louis Joel*, p. 653–655, vol. 18, supplement II, *Dictionary of Scientific Biography*, New York (Charles Scribner's Sons), 1990.

CASSELS J.W.S., *Obituary of Kurt Mahler* (⟨4⟩, 58(1991), 215–228). (*List of publications*, 229–237).

CASSINET J. et GUILLEMOT M., *L'axiome du choix dans les mathématiques de Cauchy (1821) à Gödel (1940)*, Toulouse (Université Pierre Sabatier), 1983.

ČECH E., *Théorie générale de l'homologie dans un espace quelconque* (⟨80⟩, 19(1932), 149–183).

ČECH E., *On bicompact spaces* (⟨28⟩, (2), 38(1937), 823–844).

CESARI L., *Surface area*, Princeton (University Press), 1956.

CHANDRASEKHARAN K. (edited by), *Hermann Weyl 1885–1985*, Berlin (Springer-Verlag), 1986.

CHAZY J., *Pierre Fatou* (⟨47⟩, 8(1934), 379–384).

CHENTSOV N.N., *The unfathomable influence of Kolmogorov* (⟨31⟩, 18 (1990), 987–998).

CHERN S.-S., *A simple intrinsic proof of the Gauss-Bonnet formula for closed Riemannian manifolds* (⟨28⟩, (2), 45(1944), 747–752) = p. 83–88, *Selected papers*, New York (Springer-Verlag), 1978.

CHERN S.-S., *Some new characterizations of the euclidean sphere* (⟨70⟩, 12 (1945), 279–290).

CHERN S.-S., *Characteristic classes of Hermitian manifolds* (⟨28⟩, (2), 47(1946), 85–121).

CHERN S.-S., *Some new viewpoints in differential geometry in the large* (⟨55⟩, 52(1946), 1–30).

CHERN S.-S., *Differential geometry: its past and its future*, p. 41–53, vol. I, *Actes du Congrès international des mathématiciens 1970*, Paris (Gauthier-Villars), 1971.

CHERN S.-S., *What is geometry* (⟨13⟩, 97(1990), 679–686).

CHERN S.-S., *Historical remarks on Gauss-Bonnet*, p. 209–217, P.H. RABINOWITZ and E. ZEHNDER (edited by), *Analysis, et cetera*, Boston (Academic Press), 1990.

CHERN S.-S. and CHEVALLEY C., *Elie Cartan and his mathematical work* (⟨55⟩, 58(1952), 217–250).

CHEVALLEY C., *Sur la théorie des restes normiques* (⟨66⟩, 191(1930), 426–428).

CHEVALLEY C., *Sur la théorie du corps des classes dans les corps finis et les corps locaux* (⟨103⟩, 2(1933), 365–476).

CHEVALLEY C., *L'arithmétique dans les algèbres de matrices*, Paris (Hermann), 1936.

CHEVALLEY C., *Généralisation de la théorie du corps de classes pour les extensions infinies* (⟨97⟩, (9), 15(1936), 359–371).

CHEVALLEY C., *La théorie du corps de classes* (⟨28⟩, (2), 41(1940), 394–418).

CHEVALLEY C., *On the theory of local rings* (⟨28⟩, (2), 44(1943), 690–708).

CHEVALLEY C., *Intersections of algebraic and algebroid varieties* (⟨185⟩, 57 (1945), 1–85).

CHEVALLEY C., *Theory of Lie groups I*, Princeton (University Press), 1946.

CHEVALLEY C., *Sur la classification des algèbres de Lie simples et de leurs représentations* (⟨66⟩, 227(1948), 1136–1138).

CHEVALLEY C. et WEIL A., *Hermann Weyl (1885–1955)* (⟨74⟩, (2), 3(1957), 157–187) = S. 655–694, Band IV, H. WEYL, *Gesammelte Abhandlungen*, Berlin (Springer-Verlag), 1968.

CHOQUET G., *Etude de certains réseaux de routes* (⟨66⟩, 206(1938), 310–313).

CHOQUET G., *Application des propriétés de la fonction contingent à la théorie des fonctions de variable réelle et à la géométrie différentielle des variétés cartésiennes* (⟨97⟩, (9), 26(1947), 115–226).

CHOQUET G., *Convergences* (⟨23⟩, (N.S.), 23(1948), 57–112).

CHOQUET G., *Fonctions croissantes d'ensembles et capacités*, p. 376, vol. I, *Proceedings of the International Congress of Mathematicians, Cambridge, Massachusetts, August 30-September 6, 1950*, Providence, R.I. (American Mathematical Society), 1952.

CHOQUET G., *Notice sur les travaux scientifiques* (⟨83⟩, 2(1975), 153–160).

CHOQUET G., *La vie et l'oeuvre de Marcel Brelot (1903–1987)* (⟨57⟩, 11(1990), 1–32).

CHUNG K.L., *Contributions to the theory of Markov chains II* (⟨185⟩, 76(1954), 397–419).

CHUNG K.L., *The general theory of Markov process according to Doeblin* (⟨196⟩, 2(1964), 230–254).

CHURCH A., *A set of postulates for the foundation of logic* (⟨28⟩, (2), 33(1932), 346–366).

CHURCH A., *An unsolvable problem of elementary number theory* (⟨12⟩, 58 (1936), 345–363).

CHURCH A., *The calculus of lambda-conversion*, Princeton (University Press), 1941.

CIMMINO G., *Nuovo tipo di condizione al contorno e nuovo metodo di trattazione per il problema generalizzato di Dirichlet* (⟨158⟩, 61(1937), 177–221).

CIMMINO G., *Sulle equazioni lineari alle derivate parziali del secondo ordine di tipo ellittico sopra una superficie chiusa* (⟨25⟩, (2), 7(1937), 73–96).

COHEN I.S., *On the structure and ideal theory of complete local rings* (⟨185⟩, 59(1946), 54–106).

COLLONGWOOD E.F., *Emile Borel* (⟨104⟩, 34(1959), 488–512).

CONNES A., *Géométrie non commutative*, Paris (Inter Editions), 1990.

COOLIDGE J.L., *The geometry of the complex domain*, Oxford (Clarendon Press), 1924.

CORLESS R.M., *Continued fractions and chaos* (⟨13⟩, 99(1992), 203–215).

VAN DER CORPUT J.G., *Über Summen, die mit den elliptischen θ-Funktionen zusammenhängen* (⟨120⟩, 87(1922), 66–77).

COURANT R., *Reminiscences from Hilbert's Göttingen* (⟨112⟩, 3(1980–1981), N° 4, 154–164).

COURANT R., FRIEDRICHS K. und LEWY H., *Über die partiellen Differenzgleichungen der mathematischen Physik* (⟨120⟩, 100(1928), 32–74).

COURANT R. und HILBERT D., *Methoden der mathematischen Physik II*, Berlin (Springer), 1937.

COVER T.M., GACS P. and GRAY R.M., *Kolmogorov's contributions to information theory and algorithmic complexity* (⟨29⟩, 17(1989), 840–865).

COXETER H.S.M., *The complete enumeration of finite groups of the form* $R_i^2 = (R_iR_j)^{k_{ij}} = 1$ (⟨104⟩, 10(1935), 21–25).

CRAMER H., *On the mathematical theory of risk*, p. 7–84, vol. II, *Försäkringsaktiebolaget Skandia-festskrift 1855–1930*, 1930.

CRAMER H., *Random variables and probability distributions*, Cambridge (University Press), 1937.

CRAMER H., *Den Lundbergska riskteorien och teorien för stokastiska processer*, p. 25–31, *Försäkrings matematiska studier tillägnade Filip Lundberg*, Stockholm, 1946.

CROSSLEY J.N., *Fifty years of computability* (⟨180⟩, 11(1988), N° 2, 81–99).

CURRY H.B., *Grundlagen der kombinatorischen Logik* (⟨12⟩, 52(1930), 509–536, 789–834).

CURTIS C.W., *Representation theory of finite groups: from Frobenius to Brauer* (⟨112⟩, 14(1992), n° 4, 48–57).

DALE A.I., *A history of inverse probability: from Thomas Bayes to Karl Pearson*, New York (Springer-Verlag), 1991.

VAN DALEN D. and MONNA A.F., *Sets and integration. An outline of the development*, Groningen (Wolters-Noordhoof), 1972.

DANIELL P.J., *A general form of integral* (⟨28⟩, (2), 19(1918), 279–294).

DANIELL P.J., *Stieltjes-Volterra products*, p. 130–136, *Comptes Rendus du Congrès international des mathématiciens (Strasbourg), 22–30 septembre 1920*, Toulouse (Edouard Privat), 1921.

VAN DANTZIG D., *Zur topologischen Algebra I* (⟨120⟩, 107(1932), 587–626).

DAUBEN J.W., *The history of mathematics from antiquity to the present. A selective bibliography*, New York (Garland), 1985.

DAVENPORT H., *On the distribution of quadratic residues (mod p)* (⟨104⟩, 6(1931), 49–54).

DAVIS C., *A bibliographical survey of simple groups of finite order, 1900–1965*, Courant Institute of Mathematical Sciences, New York University, 1969.

DAVIS H.T., *The present status of integral equations*, Indiana University, 1926.

DAY M.M., *Normed linear spaces*, 3rd ed., Berlin (Springer-Verlag), 1973.

DEAKIN M.A.B., *The development of the Laplace transform, 1737–1937* (⟨33⟩, 25(1981), 343–390; 26(1982), 351–381).

DE BRUIJN N.G. and ERDÖS P., *On a combinatorial problem* (⟨85⟩, 10(1948), 421–423).

DEHN M., *Über den Rauminhalt* (⟨120⟩, 55(1901), 465–478).

DEHN M., *Über unendliche diskontinuierliche Gruppen* (⟨120⟩, 71(1912), 116–144).

DEHN M., *Die Gruppen der Abbildungsklassen* (⟨7⟩, 69(1938), 135–206).

DEHN M. und HEEGAARD P., *Analysis situs*, III 1 AB 3, *Enzykl. der Math.*, Leipzig (Teubner), 1907.

DELLACHERIE C. et MEYER P.-A., *Commentaires*, p. 265–270, chapitres I à IV, 1975; p. 443–448, chapitres V à VIII, 1980; p. 207–212, chapitres IX à XI, 1983; p. 339–351, chapitres XII à XVI, Paris (Hermann), 1987: *Probabilités et potentiel.*

DELL'AGLIO L. et ISRAEL G., *La théorie de la stabilité et l'analyse qualitative des équations différentielles ordinaires dans les mathématiques italiennes: le point de vue de Tulio Levi-Cività* (⟨57⟩, 10(1989), 283–322).

DE LONG H., *A profile of mathematical logic*, London (Addison-Wesley), 1970.

DELSARTE J., *Les fonctions moyenne-périodiques* (⟨97⟩, (9), 14(1935), 403–453) = p. 239–289, vol. I, *Oeuvres*, Paris (C.N.R.S.), 1971.

DENJOY A., *Une extension de l'intégrale de M. Lebesgue* (⟨66⟩, 154(1912), 859–862).

DENJOY A., *Notice sur les travaux scientifiques*, Toulouse (Edouard Privat), 1921) = p. 615–683, vol. II, *Articles et mémoires*, Paris (Gauthier-Villars), 1955.

DENJOY A., *Les courbes définies par les équations différentielles à la surface du tore* (⟨97⟩, (9), 11(1932), 333–375) = p. 881–923, vol. II, *Articles et mémoires*, Paris (Gauthier-Villars), 1955.

DENJOY A., *Leçons sur le calcul des coefficients d'une série trigonométrique I*, Paris (Gauthier-Villars), 1941.

DENJOY A., *Notice sur la vie et l'oeuvre de Henri Lebesgue*, p. 35–65, vol. I, H. LEBESGUE, *Oeuvres scientifiques*, Genève (L'Enseignement Mathématique), 1972.

Arnaud Denjoy, évocation de l'homme et de l'oeuvre, Astérisque, 28–29, 1975, Société Mathématique de France.

DENJOY A., FÉLIX L. et MONTEL P., *Henri Lebesgue: le savant, le professeur, l'homme*, p. 67–84, vol. I, *Oeuvres scientifiques*, Genève (L'Enseignement Mathématique), 1972.

DEURING M., *Algebraische Funktionenkörper und algebraische Geometrie*, S. 149–162, Band II, *Naturforschung und Medizin in Deutschland 1939–1946*, Wiesbaden (Dietrerich'sche Verlagsbuchhandlung), 1948.

DEURING M., *Algebraische Begründung der komplexen Multiplikation* (⟨1⟩, 16 (1949), 32–47).

DHOMBRES J., *Quelques aspects de l'histoire des équations fonctionnelles liés à l'évolution du concept de fonction* (⟨33⟩, 36(1986), 91–181).

DICKE A., *Emmy Noether 1882–1935*, Basel (Birkhäuser), 1970.

DICKSON L.E., *Linear groups with an exposition of the Galois field theory*, Leipzig (B.G. Teubner), 1901 = New York (Dover), 1958.

DICKSON L.E., *On the theory of equations in a modular field* (⟨55⟩, 13(1906), 8–10).

DICKSON L.E., *History of the theory of numbers*, Washington (Carnegie Institution), 1919–1920, New York (G.E. Steckert), 1934.

Dictionary of scientific biography, New York (Charles Scribner's Sons), 1970–1990.

DIEUDONNÉ J., *Sur les fonctions continues numériques définies sur un produit de deux espaces compacts* (⟨66⟩, 205(1937), 593–595) = p. 141–143, vol. I, *Choix d'oeuvres mathématiques*, Paris (Hermann), 1981.

DIEUDONNÉ J., *Une généralisation des espaces compacts* (⟨97⟩, (9), 23(1944), 65–76) = p. 165–176, vol. I, *Choix d'oeuvres mathématiques*, Paris (Hermann), 1981.

DIEUDONNÉ J., *L'école française moderne des mathématiques* (⟨141⟩, (1964), 97–106).

DIEUDONNÉ J., *Pure mathematics in France from 1949 to 1955* (⟨141⟩, 1(1964), 38–44).

DIEUDONNÉ J., *Recent developments in mathematics* (⟨13⟩, 71(1964), 239–248).

DIEUDONNÉ J., *Regards sur Bourbaki* (⟨15⟩, 18(1969), n° 2, 13–25).

DIEUDONNÉ J., *The work of Nicolas Bourbaki* (⟨13⟩, 77(1970), 134–145).

DIEUDONNÉ J., *Histoire de l'analyse harmonique*, Moscow (Nauka), 1971.

DIEUDONNÉ J., *Notice nécrologique sur Paul Lévy* (⟨66⟩, 274(1972), Vie Académique, 137–144).

DIEUDONNÉ J., *The historical development of algebraic geometry* (⟨13⟩, 79 (1972), 827–866).

DIEUDONNÉ J., *Cours de géométrie algébrique. 1. Aperçu historique du développement de la géométrie algébrique*, Paris (Presses Universitaires de France), 1974.

DIEUDONNÉ J., *Le développement historique de la notion de groupe* (⟨48⟩, 28 (1976), 267–296).

DIEUDONNÉ J., *Panorama des mathématiques pures. Le choix bourbachique*, Paris (Gauthier-Villars), 1977.

DIEUDONNÉ J. (sous la direction de), *Abrégé d'histoire des mathématiques. 1700–1900*, Paris (Hermann), 1978 = *Geschichte der Mathematik. 1700–1900*, Wiesbaden (Vieweg), 1985 = traduit en japonais, Tokyo (Iwanami Shoten), 1985.

DIEUDONNÉ J., *Les grandes lignes de l'évolution des mathématiques* (⟨59⟩, 94 (1980), 1–13).

DIEUDONNÉ J., *Bourbaki et la philosophie des mathématiques* (⟨75⟩, 4(1981), 173–188).

DIEUDONNÉ J., *History of functional analysis*, Amsterdam (North-Holland), 1981.

DIEUDONNÉ J., *Notice sur les travaux scientifiques*, p. 693–722, vol. II, *Choix d'oeuvres mathématiques*, Paris (Hermann), 1981.

DIEUDONNÉ J., *Carl Ludwig Siegel* (⟨66⟩, Vie Académique, 296(1983), 63–75).

DIEUDONNÉ J., *The beginnings of topology from 1850 to 1914*, p. 585–600, vol. II, *Atti degli incontri di logica matematica 1983–1984*, Dipartimento di Matematica, Università di Siena.

DIEUDONNÉ J., *Les débuts de la topologie algébrique* (⟨78⟩, 3(1985), 347–357).

DIEUDONNÉ J., *Claude Chevalley* (⟨32⟩, 1986).

DIEUDONNÉ J., *Marc Krasner* (⟨57⟩, 7(1986), 29–30).

DIEUDONNÉ J., *A history of algebraic and differential topology*, Boston (Birkhäuser), 1989.

DIEUDONNÉ J., *Siegel, Carl Ludwig*, p. 826–832, vol. 18, supplement II, *Dictionary of Scientific Biography*, New York (Charles Scribner's Sons), 1990.

DIEUDONNÉ J. et SCHWARTZ L., *La dualité dans les espaces* ($\mathscr{F}$) *et* ($\mathscr{LF}$) (⟨21⟩, 1(1949), 61–101) = p. 296–336, vol. I, J. DIEUDONNÉ, *Choix d'oeuvres mathématiques*, Paris (Hermann), 1981.

DIEUDONNÉ J. et TITS J., *La vie et l'oeuvre de Claude Chevalley* (⟨192⟩, 3, 1986, 559–565).

DIMITRIĆ R., *Anatoly Ivanovich Maltsev* (⟨112⟩, 14(1992), n° 2, 26–30).

DIRAC P.A.M., *The physical interpretation of the quantum dynamics* (⟨154⟩, 113 (1926), 621–641).

DIRAC P.A.M., *The principles of quantum mechanics*, London (H. Milford), 1930 = *Die Prinzipien der Quantenmechanik*, Leipzig (S. Hirzel), 1930.

DOBRUSHIN R.L., SUKHOV YU.M. and FRITZ J., *A.N. Kolmogorov — the founder of reversible Markov processes* (⟨167⟩, 43(1988), N° 6, 157–182).

DOEBLIN W., *Etude de l'ensemble de puissances d'une loi de probabilités* (⟨66⟩, 206(1938), 718–720).

DOEBLIN W., *Sur deux problèmes de M. Kolmogoroff concernant les chaînes dénombrables* (⟨49⟩, 60(1938), 210–220).

DOEBLIN W., *Sur les propriétés asymptotiques de mouvements régis par certains types de chaînes simples*, Bucureşti (Imprimeria centrala), 1938.

DOEBLIN W., *Eléments d'une théorie générale des chaînes de Markoff* (⟨24⟩, (3), 57(1940), 61–111).

DOEBLIN W., *Sur l'ensemble de puissances d'une loi de probabilité* (⟨181⟩, 9 (1940), 71–96).

DOEBLIN W. et FORTET R., *Sur des chaînes à liaisons complètes* (⟨49⟩, 65 (1937), 132–148).

DOOB J.L., *Stochastic processes depending on a continuous parameter* (⟨185⟩, 42(1937), 107–140).

DOOB J.L., *Regularity properties of certain families of chance variables* (⟨185⟩, 47(1940), 455–486).

DOOB J.L., *Topics in the theory of Markoff chains* (⟨185⟩, 52(1942), 37–64).

DOOB J.L., *Asymptotic properties of Markoff transition probabilities* (⟨185⟩, 63 (1948), 393–421).

DOOB J.L., *Feller, William*, p. 287–289, vol. 17, supplement II, *Dictionary of scientific biography*, New York (Charles Scribner's Sons), 1990.

DOUGLAS J., *Solution of the problem of Plateau when the contour is an arbitrary Jordan curve in n-dimensional euclidean space* (⟨55⟩, 36(1930), 189–190).

DOUGLAS J., *Solution of the problem of Plateau* (⟨185⟩, 33(1931), 263–321).

DOUGLAS J., *The most general form of the problem of Plateau* (⟨12⟩, 61(1939), 590–608).

DRACH J., *Sur l'intégration logique des équations différentielles ordinaires*, p. 488–497, vol. I, *Proceedings 5th International Congress of Mathematicians, 1912*, Cambridge (University Press), 1913.

DRESDEN A., *Some recent work in calculus of variations* (⟨55⟩, 32(1926), 475–521).

DROESBEKE J.-J. et TASSI P., *Histoire de la statistique*, Paris (Presses Universitaires de France), 1990.

DRUCKER T. (ed.), *Perspectives on the history of mathematical logic*, Basel (Birkhäuser), 1991.

DUBREIL P., *Apparition et premiers développements de la théorie des demi-groupes en France* (⟨57⟩, 2(1981), 59–66).

DUBREIL P., *L'algèbre, en France, de 1900 à 1935* (⟨57⟩, 3(1982), 69–82).

DUBREIL P., *Souvenirs d'un boursier Rockefeller 1929–1931* (⟨57⟩, 4 (1983), 61–74).

DUBREIL P., *Emmy Noether* (⟨57⟩, 7(1986), 15–28).

DUDLEY R.M., COOK S.L., LLOPIS J. and PENG N.P., *A.N. Kolmogorov and statistics: a citation bibliography* (⟨31⟩, 18(1990), 1017–1031).

DUFFING G., *Erzwungene Schwingungen bei veränderlicher Eigenfrequenz und ihre technische Bedeutung*, Braunschweig (Vieweg), 1918.

DUISTERMAAT H., *Peculiarities in the development of the theory of Lie groups*, p. 263–282, P.H. RABINOWITZ and E. ZEHNDER (edited by), *Analysis, et cetera*, Boston (Academic Press), 1990.

DULAC H., *Sur les cycles limites* (⟨49⟩, 51(1923), 45–188).

DUNFORD N. and SCHWARTZ J.T., *Linear operators. Part I: General theory*, New York (Interscience Publishers), 1958.

DURANTON H. (sous la direction de), *Les sciences en Alsace 1538–1988*, Strasbourg (Editions Oberlin), 1989.

DUREN N.L., *The development of sufficient conditions in the calculus of variations*, p. 245–349, vol. I, *University of Chicago contributions to the calculus of variations*, Chicago (University Press), 1930.

DUREN P. (edited by), *A century of mathematics in America*, Providence (American Mathematical Society), vol. 1, 1988, vol. 2 and 3, 1989.

DVORETZKY A. and ERDÖS P., *Some problems on a random walk in space*, p. 353–367, *Proceedings of the second Berkeley symposium in mathematical statistics and probability, 1950*, Berkeley (University of California Press), 1951.

DYNKIN E.B., *Calcul des coefficients de la formule de Campbell-Hausdorff* (in Russian) (⟨69⟩, (N.S.), 57(1947), 323–326).

EARMAN J. and GLYMOUR C., *Einstein and Hilbert: two months in the history of relativity* (⟨33⟩, 19(1978), 291–308).

EBERLEIN W.F., *Abstract ergodic theorems and weak almost periodic functions* (⟨185⟩, 67(1949), 217–240).

EBBINGHAUS H.-D., HERMES H., HIRZEBRUCH F., KOECHER M., MAINZER K., NEUKIRCH J., PRESTEL A. and REMMERT R., *Numbers*, with an introduction by K. LAMOTKE, New York (Springer-Verlag), 1990.

ECKMANN B., *Zur Homotopietheorie gefaserter Räume* (⟨63⟩, 14(1941–1942), 141–192) = p. 1–52, *Selecta*, Berlin (Springer-Verlag), 1987.

EDDINGTON A.S., *The mathematical theory of relativity*, Cambridge (University Press), 1923.

EDWARDS A.W.F., *Celebration of the centenary of the birth of Sir Ronald Aylmer Fischer, 1990* (⟨83⟩, 19(1992), 81–82).

EDWARDS H.M., *Fermat's last theorem: a genetic introduction to number theory*, New York (Springer-Verlag), 1977.

EGOROV D.F., *Sur les suites de fonctions mesurables* (⟨66⟩, 152(1911), 244–246).

EHRESMANN C., *Sur la topologie de certains espaces homogènes* (⟨28⟩, (2), 35 (1934), 396–443).

EHRESMANN C., *Sur les espaces localement homogènes* (⟨74⟩, 35(1936), 317–333).

EHRESMANN C., *Sur les applications continues d'un espace dans un espace fibré ou dans un revêtement* (⟨49⟩, 72(1944), 27–54).

EHRESMANN C., *Sur les espaces fibrés différentiables* (⟨66⟩, 224(1947), 1611–1612) = p. 326–328, parties I-1 et I-2, *Oeuvres complètes et commentées*, Amiens, 1984.

EHRESMANN C., *Les connexions infinitésimales dans un espace fibré différentiable*, p. 29–55, *Colloque de topologie (espaces fibrés), Bruxelles 1950*, Liège (C.R.B.M.), 1951.

EHRESMANN C., *Oeuvres complètes et commentées*, Amiens, 1984.

EHRESMANN C. et FELDBAU J., *Sur les propriétés d'homotopie des espaces fibrés* (⟨66⟩, 212(1941), 945–948) = p. 310–312, parties I-1 et I-2, C.

EHRESMANN C. et REEB G., *Sur les champs d'éléments de contact de dimension p complètement intégrables dans une variété continuement différentiable* $\mathcal{V}_n$ (⟨66⟩, 218(1944), 955–957).

EILENBERG S., *Cohomology and continuous mappings* (⟨28⟩, (2), 41 (1940), 231–251).

EILENBERG S., *Singular homology* (⟨28⟩, (2), 45(1944), 407–447).

EILENBERG S. and MAC LANE S., *Group extensions and homology* (⟨28⟩, (2), 43(1942), 757–831) = p. 183–257, S. MAC LANE, *Selected papers*, New York (Springer Verlag), 1979.

EILENBERG S. and MAC LANE S., *Natural isomorphisms in group theory* (⟨151⟩, 28(1942), 537–543).

EILENBERG S. and MAC LANE S., *Relations between homology and homotopy groups* (⟨151⟩, 29(1943), 155–158) = p. 269–272, S. MAC LANE, *Selected papers*, New York (Springer-Verlag), 1979.

EILENBERG S. and MAC LANE S., *General theory of natural equivalences* (⟨185⟩, 58(1945), 231–294) = p. 273–336, S. MAC LANE, *Selected papers*, New York (Springer Verlag), 1979.

EILENBERG S. and MAC LANE S., *Cohomology theory in abstract groups I* (⟨28⟩, (2), 48(1947), 51–78).

EILENBERG S. and STEENROD N.E., *Axiomatic approach to homology theory* (⟨151⟩, 31(1945), 117–120).

EINSTEIN A., *Zur Elektrodynamik bewegter Körper* (⟨17⟩, 17(1905), 891–921) = p. 275–306, vol. II, *The collected papers*, Princeton (University Press), 1989.

EINSTEIN A., *Die formale Grundlage der allgemeinen Relativitätstheorie* (⟨177⟩, 1914, 1030–1085).

EINSTEIN A., *Zur allgemeinen Relativitätstheorie* (⟨177⟩, 1915, 778–786, 799–801).

EINSTEIN A., *Die Grundlagen der allgemeinen Relativitätstheorie* (⟨17⟩, (1916), 49).

EINSTEIN A., *Riemann-Geometrie mit Aufrechterhaltung des Begriffes des Fernparallelismus* (⟨177⟩, 1928, 217–221).

EINSTEIN A., *The collected papers*, vol. II, Princeton (University Press), 1989.

EISENHART C., *Pearson, Karl*, p. 447–473, vol. X, *Dictionary of scientific biography*, New York (Charles Scribner's Sons), 1974.

EISENHART L.P., *A treatise of the differential geometry of curves and surfaces*, Boston (Ginn and Co.), 1909.

EISENHART L.P., *Riemannian geometry*, Princeton (University Press), 1926.

EISENHART L.P., *Non-Riemannian geometry*, New York (American Mathematical Society), 1927.

ENRIQUES F., *Le superficie algebriche*, Bologna (Zanichelli), 1949.

ERDÖS P., *On the density of some sequences of numbers II* (⟨104⟩, 12 (1937), 7–11).

ERDÖS P., *On a new method in elementary number theory which leads to an elementary proof of the prime number theory* (⟨151⟩, 35(1949), 374–384).

ERDÖS P. and KAC M., *On the Gaussian law of errors in the theory of additive functions* (⟨151⟩, 25(1939), 206–207).

ERLANG A.K., *Sandsynlighedsregning og Telefonsanitaler* (⟨138⟩, 20 B (1909), 33–39).

ESCLANGON E., *Sur une extension de la notion de périodicité* (⟨66⟩, 135(1902), 891–894).

EWING J.H. and GEHRING F.W. (eds.), *Paul Halmos: celebrating 50 years of mathematics*, New York (Springer-Verlag), 1991.

FANG J., *Hilbert. Towards a philosophy of modern mathematics*, New York (Paideia Press), 1970.

FANTAPPIÈ L., *I funzionali analitici* (⟨38⟩, (6), 3(1929), 453–683) = p. 19–247, vol. I, *Opere scelte*, Unione Matematica Italiana, 1973.

FANTAPPIÈ L., *Su alcuni indirizzi delle Scienze matematiche nel momento scientifico presente* (⟨39⟩, 1(1932), 581–594).

FATOU P., *Séries trigonométriques et séries de Taylor* (⟨7⟩, 30(1906), 335–400).

FATOU P., *Sur les équations fonctionnelles* (⟨49⟩, 47(1919), 161–271; 48(1920), 33–94, 208–314).

FATOU P., *Sur certains systèmes d'équations différentielles dépendant d'un paramètre* (⟨66⟩, 186(1928), 416–418).

FATOU P., *Notice sur les travaux scientifiques*, Paris (Téqui), 1929.

FATOU P., *Fonctions automorphes*, APPELL P. et GOURSAT E., *Théorie des fonctions algébriques d'une variable et des transcendantes qui s'y rattachent*, tome II, Paris (Gauthier-Villars), 1930.

FEIT W. and THOMPSON J.G., *A solvability criterion for finite groups and some consequences* (⟨151⟩, 48(1962), 968–970).

FEJÉR L., *Sur les fonctions bornées et intégrables* (⟨66⟩, 131(1900), 984–987) = p. 37–41, I. Band, *Gesammelte Arbeiten*, Basel (Birkhäuser), 1970.

FEJÉR L., *Untersuchungen über Fouriersche Reihen* (⟨120⟩, 58(1904), 51–69) = p. 142–160, Band I, *Gesammelte Arbeiten*, Basel (Birkhäuser), 1970.

FEL'DMAN N.I. and SHIDLOVSKII A.B., *The development and present state of the theory of transcendental numbers* (⟨167⟩, 22(1967), 1–79).

FELLER W., *Über den zentralen Grenzwertsatz in der Wahrscheinlichkeitsrechnung* (⟨123⟩, 40(1935), 521–559).

FELLER W., *An introduction to probability theory and its applications I*, New York (John Wiley), 1950.

FENCHEL W., *On total curvatures of Riemannian manifolds* (⟨104⟩, 15(1940), 15–22).

FENCHEL W., *On conjugate convex functions* (⟨60⟩, 1(1949), 73–77).

FEYNMAN R.P., *Space-time approach to non-relativistic quantum mechanics* (⟨159⟩, 20(1948), 367–387).

FICHERA G., *Ceni sui problemi di analisi matematica contemporanea, I. Produzione italiana nel campo dell'analisi matematica durante il periodo 1940–1945* (⟨51⟩, 4(1948), 63–107).

FIENBERG S.E. and HINKLEY D.V., *R.A. Fisher: an appreciation*, New York (Springer), 1980.

DE FINETTI B., *La prévision: ses lois biologiques, ses sources subjectives* (⟨22⟩, 7(1937), 1–68).

DE FINETTI B., *Compte rendu critique du colloque de Genève sur la théorie des probabilités*, Paris (Hermann), 1939.

FINSLER P., *Über Kurven und Flächen in allgemeinen Räumen*, Zürich (O. Füssli), 1918.

FISCHER E., *Applications d'un théorème sur la convergence en moyenne* (⟨66⟩, 144(1907), 1148–1151).

FISCHER E., *Sur la convergence en moyenne* (⟨66⟩, 144(1907), 1022–1024).

FISCHER G., HIRZEBRUCH F., SCHARLAU W. und TÖRNIG W. (herausgegeben von), *Ein Jahrhundert Mathematik 1890–1990, Festschrift zum Jubiläum der DMV*, Braunschweig (Friedr. Vieweg und Sohn), 1990.

FISHER R.A., *Statistical methods for research workers*, Edinburgh (Oliver and Boyd), 1925.

FISHER R.A., *The genetical theory of natural selection*, London (Oxford University Press), 1930.

FISHER R.A., *Statistical inference and analysis. Selected correspondence*, Oxford (Clarendon Press), 1990.

FOMENKO A.T., *The Plateau problem*, New York (Gordon and Breach), 1991.

FOMENKO O.M. and KUZMINA G.V., *The last 100 days of the Bieberbach conjecture* (⟨112⟩, 8(1986), n° 1, 40–47).

FONG P. and WONG W.J., *Introduction*, p. 319–321, vol. I, R. BRAUER, *Collected papers*, Cambridge, Massachusetts (The MIT Press), 1980.

FORSTER W., *J. Schauder — Fragments of a portrait*, p. 417–425, W. FORSTER (ed.), *Numerical solution of highly nonlinear problems*, Amsterdam (North-Holland), 1980.

FRAENKEL A.A., *Über die Teiler der Null und die Zerlegung von Ringen* (⟨98⟩, 145(1914), 139–176).

FRAENKEL A.A., *Zu den Grundlagen der Cantor-Zermeloschen Mengenlehre* (⟨120⟩, 86(1922), 230–237).

FRAENKEL A.A., *Der Begriff «definit» und die Unabhängigkeit des Auswahlaxioms* (⟨3⟩, 1922, 253–257).

FRAENKEL A.A., *Die neueren Ideen zur Grundlegung der Analysis und Mengenlehre* (⟨96⟩, 33(1925), 97–103).

FRAENKEL A.A., *Discontinu et continu*, p. 193–200, *Travaux du IXe Congrès international de philosophie, VI, Logique et mathématique*, Paris (Hermann), 1937.

FRÉCHET M., *Généralisation d'un théorème de Weierstrass* (⟨66⟩, 139 (1904), 848–849).

FRÉCHET M., *Sur quelques points du calcul fonctionnel* (⟨158⟩, 22(1906), 1–74).

FRÉCHET M., *Sur les opérations linéaires* (⟨185⟩, 8(1907), 433–446).

FRÉCHET M., *Sur la notion de différentielle totale* (⟨135⟩, (4), 12(1912), 385–403, 433–449).

FRÉCHET M., *Sur l'intégrale d'une fonctionnelle étendue à un ensemble abstrait* (⟨49⟩, 43(1915), 248–265).

FRÉCHET M., *Les espaces abstraits*, Paris (Gauthier-Villars), 1928.

FRÉCHET M., *Notice sur les travaux scientifiques*, Paris (Hermann), 1933.

FRÉCHET M., *Pages choisies d'analyse générale*, Paris (Gauthier-Villars), 1953.

FRÉCHET M., *Le rôle d'Emile Borel dans la théorie des jeux* (⟨161⟩, 1959, 139–167).

FRÉCHET M., *La vie et l'oeuvre d'Emile Borel* (⟨74⟩, (2), 11(1965), 1–97) = p. 5–115, vol. I, E. BOREL, *Oeuvres*, Paris (C.N.R.S.), 1972.

FREDHOLM I., *Sur une nouvelle méthode pour la résolution du problème de Dirichlet* (⟨139⟩, n° 1, 1900, 39–46) = p. 61–68, *Oeuvres complètes*, Malmö (Litos Reprotryck), 1955.

FREDHOLM I., *Sur une classe d'équations fonctionnelles* (⟨7⟩, 27(1903), 365–390) = p. 81–106, *Oeuvres complètes*, Malmö (Litos Reprotryck), 1955.

FREDHOLM I., *Sur l'intégrale fondamentale d'une équation différentielle elliptique à coefficients constants* (⟨158⟩, 25(1908), 346–351) = p. 117–122, *Oeuvres complètes*, Malmö (Litos Reprotryck), 1955.

FREI G., *Helmut Hasse (1898–1979)* (⟨78⟩, 3(1985), 55–69).

FREI G. (herausgegeben von), *Der Briefwechsel David Hilbert — Felix Klein (1886–1918)*, Göttingen (Vandenhoech und Ruprecht), 1985.

FREUDENTHAL H., *Über die Enden topologischer Räume und Gruppen* (⟨123⟩, 33(1931), 692–713).

FREUDENTHAL H., *Über die Klassen der Sphärenabbildungen I. Grosse Dimensionen* (⟨65⟩, 5(1937), 299–314).

FREUDENTHAL H., *History of mathematics — a problem and a paradigm* (⟨131⟩, (4), 8(1990), 217–234).

FREWER M., *Das wissenschaftliche Werk Felix Bernsteins*, Göttingen (Institut für mathematische Statistik), 1977.

FRICKE R. und KLEIN F., *Vorlesungen über die Theorie der automorphen Funktionen*, II. Band, Leipzig (B.G. Teubner), 1912.

FRIEDRICHS K.O., *Spektraltheorie halbbeschränkter Operatoren und Anwendung auf die Spektralzerlegung von Differentialoperatoren* (⟨120⟩, 109 (1934), 465–487, 685–713; 110(1935), 777–779) = p. 9–65, vol. II, *Selecta*, Boston (Birkhäuser), 1986).

FRIEDRICHS K.O., *Über die Spektralzerlegung eines Integraloperators* (⟨120⟩, 115 (1938), 249–272).

FRIEDRICHS K.O., *On differential operators in Hilbert space* (⟨12⟩, 61 (1939), 523–544).

FRIEDRICHS K.O., *The identity of weak and strong extensions of differential operators* (⟨185⟩, 55(1944), 132–151) = p. 117–137, vol. I, *Selecta*, Boston (Birkhäuser), 1986.

FRIEDRICHS K.O., *Formation and decay of shock waves* (⟨64⟩, 1(1948), 211–245).

FRIEDRICHS K.O., *von Neumann's Hilbert space and partial differential equations* (⟨174⟩, 22(1980), 486–493).

FROBENIUS F.G., *Über die Characktere der symmetrischen Gruppe* (⟨177⟩, 1900, 516–534).

FROBENIUS F.G., *Über auflösbare Gruppen III–V* (⟨177⟩, 1901, 849–875, 1216–1230, 1324–1329).

FROBENIUS F.G., *Theorie der hyperkomplexen Grössen* (⟨177⟩, 1903, 504–537, 634–645) = S. 284–329, Band III, *Gesammelte Abhandlungen*, Berlin (Springer), 1968.

FROBENIUS F.G., *Über einen Fundamentalsatz der Gruppentheorie* (⟨177⟩, 1903, 987–991).

FROBENIUS F.G., *Über Matrizen aus positiven Elementen* (⟨177⟩, 1908, 471–476).

FROBENIUS F.G., *Über Matrizen aus nicht negativen Elementen* (⟨177⟩, 1912, 456–477) = S. 546–567, Band III, *Gesammelte Abhandlungen*, Berlin (Springer-Verlag), 1968.

FROBENIUS F.G. und SCHUR I., *Über die reellen Darstellungen der endlichen Gruppen* (⟨177⟩, 1906, 186–208).

FROSTMAN O., *Potentiel d'équilibre et capacité des ensembles avec quelques applications à la théorie des fonctions* (⟨125⟩, 3(1935)).

FUBINI G., *Sugli integrali multipli* (⟨38⟩, (5), 16-1(1907), 608–614) = p. 243–249, *Opere scelte*, vol. II, Roma (Edizioni Cremonese), 1958.

FUBINI G. e ČECH E., *Geometria proiettiva differentiale*, Bologna (N. Zanichelli), 1926–1927.

FUCHS L., *Abelian groups*, Oxford (Pergamon Press), 1960.

FUCHS W.H.J., *The development of the theory of deficient values since Nevanlinna* (⟨18⟩, 7(1982), 33–48).

FUETER R., *Vorlesungen über die singulären Moduln und die komplexe Multiplikation der elliptischen Funktionen I*, Leipzig (Teubner), 1924.

FURTWÄNGLER P., *Über das Reziprozitätsgesetz der ℓ-ten Potenzreste in algebraischen Zahlkörper, wenn ℓ eine ungerade Primzahl bedeutet* (⟨2⟩, 1902, 3–82).

FURTWÄNGLER P., *Allgemeiner Existenzbeweis für den Klassenkörper eines beliebigen algebraischen Zahlkörpers* (⟨120⟩, 63(1907), 1–37).

FURTWÄNGLER P., *Beweis des Hauptidealsatzes für die Klassenkörper algebraischen Zahlkörper* (⟨1⟩, 7(1929), 14–36).

GALERKIN B.G., (⟨92⟩, 1915, 1916, 1918).

GAMBIER E., *Sur les équations différentielles du second ordre dont l'intégrale générale est uniforme* (⟨66⟩, 142(1906), 266–269).

GAMKRELIDZE R.V. (editor-in-chief), *Encyclopaedia of mathematical sciences*, Berlin (Springer-Verlag), 1988–1990.

GARDING L., *Le problème de Dirichlet pour les équations aux dérivées partielles elliptiques à coefficients constants* (⟨66⟩, 230(1950), 1030–1032).

GARDING L., *History of mathematics of double refraction* (⟨33⟩, 40(1989), 355–385).

GARIO P., *Histoire de la résolution des singularités des surfaces algébriques (une discussion entre C. Segre et P. Del Pezzo)* (⟨57⟩, 9(1988), 123–138).

GARIO P., *The classical theory of equisingularity for plane curves* (⟨45⟩, 10(1990), 77–97).

GÂTEAUX R., *Sur la notion d'intégrale dans le domaine fonctionnel et sur la théorie du potentiel. — Fonctions d'une infinité de variables indépendantes* (⟨49⟩, 47(1919), 47–96).

GELFAND I.M., *On normed rings* (⟨69⟩, 23(1939), 430–432) = p. 169–171, vol. I, *Collected papers*, Berlin (Springer-Verlag), 1987.

GELFAND I.M., *Normierte Ringe* (⟨117⟩, 9(1941), 3–23) = p. 181–201, vol. I, *Collected papers*, Berlin (Springer-Verlag), 1987.

GELFAND I.M. and NAIMARK M., *On the imbedding of normed rings into the ring of operators in Hilbert space* (⟨117⟩, 12(1943), 197–213) = p. 241–257, vol. I, I.M. GELFAND, *Collected papers*, Berlin (Springer-Verlag), 1987.

GELFAND I.M. and RAIKOV D.A., *Représentations unitaires irréductibles de groupes localement bicompacts* (in Russian), (⟨117⟩, 13(1943), 301–316) = *Irreducible unitary representations of locally bicompact groups*, p. 3–17, vol. II, I.M. GELFAND, *Collected papers*, Berlin (Springer-Verlag), 1988.

GELFOND A.O., *Sur les propriétés arithmétiques des fonctions entières* (⟨184⟩, 30(1929), 280–285).

GELFOND A.O., *Sur le septième problème de D. Hilbert* (⟨69⟩, 2(1934), 4–6).

GELFOND A.O., *On algebraic independence of transcendental numbers of certain classes* (in Russian) (⟨191⟩, (N.S.), 4(1949), N° 5, 14–48).

GENTZEN G., *Untersuchungen über das logische Schliessen* (⟨123⟩, 39 (1935), 176–210, 405–431).

GENTZEN G., *Die Widerspruchsfreiheit der reinen Zahlentheorie* (⟨120⟩, 112 (1936), 423–465).

GEVREY M., *Sur les équations aux dérivées partielles du type parabolique* (⟨97⟩, (6), 9(1913), 305–471; 10(1914), 105–148).

GEVREY M., *Sur la nature des solutions des équations aux dérivées partielles* (⟨24⟩, (3), 35(1918), 243–304).

GIBBS J.W., *Elementary principles in statistical mechanics*, Yale (University Press), 1902.

GISPERT H., *La France mathématique. La Société mathématique de France (1870–1914)*, suivi de cinq études par R. BKOUCHE, C. GILAIN, C. HOUZEL, J.-P. KAHANE, M. ZERNER (⟨58⟩, N° 34, 1991).

GIVANT S.R., *A portrait of Alfred Tarski* (⟨112⟩, 13(1991), N° 3, 16–32).

GLIMM J., IMPAGLIAZZO J. and SINGER I. (editors), *The legacy of John von Neumann*, Providence, Rhode Island (American Mathematical Society), 1990.

GNEDENKO B.V., *On the theory of attraction domains of stable laws* (in Russian) (⟨189⟩, 30(1939), 61–81).

GODBILLON C., *Feuilletages: études géométriques*, Basel (Birkhäuser), 1991.

GÖDEL K., *Einige metamathematische Resultate über Entscheidungsdefinitheit und Widerspruchsfreiheit* (⟨10⟩, 67(1930), 214–215) = p. 140–141, vol. I, *Collected works*, New York (Oxford University Press), 1986.

GÖDEL K., *Über formal unentscheidbare Sätze der Principia matematica und verwandter Systeme I* (⟨129⟩, 38(1931), 173–198) = (with an English translation) *On formally undecidable propositions of Principia matematica and related systems I*, p. 144–195 (S.C. KLEENE, *Introductory note*, p. 126–141), vol. I, *Collected works*, New York (Oxford University Press), 1986.

GÖDEL K., *The consistency of the axiom of choice and the generalized continum hypothesis* (⟨151⟩, 24(1938), 556–557) = p. 26–27, vol. II, *Collected works*, New York (Oxford University Press), 1990.

GÖDEL K., *The consistency of the continuum hypothesis, with the axioms of set theory*, Princeton University Press, 1940 = p. 33–101, vol. II, *Collected works*, New York (Oxford University Press), 1990.

GÖDEL K., *On undecidable propositions of formal mathematical systems*, p. 346–371, vol. I, *Collected works*, New York (Oxford University Press), 1986. (p. 338–345, S.C. KLEENE, *Introductory note to 1934*).

GÖDEL R., TAUSSKY-TODD O., KLEENE S.C. and KREISEL G., *Gödel remembered*, Napoli (Bibliopolis), 1987.

GODEMENT R., *Extension à un groupe abélien quelconque des théorèmes taubériens de N. Wiener et d'un théorème de A. Beurling* (⟨66⟩, 223(1946), 16–18).

GODEMENT R., *Les fonctions de type positif et la théorie des groupes* (⟨185⟩, 63(1948), 1–84).

GOE G., *Łukasiewicz, Jan*, p. 546–547, vol. VIII, *Dictionary of scientific biography*, New York (Charles Scribner's Sons), 1973.

GOHBERG I. (edited by), *Toeplitz centennial*, Basel (Birkhäuser), 1982.

GOLDSTINE H.H., *The computer from Pascal to von Neumann*, Princeton University Press, 1972.

GONSETH F., *Les fondements des mathématiques. De la géométrie d'Euclide à la relativité générale et à l'intuitionisme*, préface de J. HADAMARD, Paris (Librairie scientifique Albert Blanchard), 1926.

GONSKA H.H. and KNOOP H.-B., *On Hermite-Fejér interpolation: a bibliography (1914–1987)* (⟨182⟩, 25(1990), 147–198).

GOURSAT E., *Sur la définition générale des fonctions analytiques d'après Cauchy* (⟨185⟩, 1(1900), 14–16).

GOURSAT E., *Leçons sur le problème de Pfaff*, Paris (Hermann), 1922.

GOURSAT E., *Cours d'analyse mathématique*, tome I, Paris (Gauthier-Villars), 1902 = *A course in mathematical analysis I*, New York (Grinn), 1904 = New York (Dover), 1959.

GOURSAT E., *Sur un cas élémentaire de l'équation de Fredholm* (⟨49⟩, 35(1907), 163–173).

GOWER J.C., *Sir Ronald Aylmer Fisher, 1890–1962* (⟨116⟩, 23(3)(1990–1991), 76–86).

GRANDEMANGE P. et SCHWARTZ P., *Aspects classiques du troisième problème de Hilbert* (⟨81⟩, avril 1992, n° 52, 73–86).

GRATTAN-GUINNESS I., *Mathematical bibliography for W.C. and G.C. Young* (⟨83⟩, 2(1975), 43–58).

GRATTAN-GUINNESS I., *Wiener on the logics of Russell and Schröder* (⟨30⟩, 32 (1975), 103–132).

GRATTAN-GUINNESS I., *Dear Russell — dear Jourdain. A commentary on Russell's logic based on his correspondence with Philip Jourdain*, London (Duckworth), 1977.

GRATTAN-GUINNESS I., *How Bertrand Russell discovered his paradox* (⟨83⟩, 5 (1978), 127–137).

GRATTAN-GUINNESS I., *In memoriam Kurt Gödel: his 1931 correspondence with Zermelo on his incompletability theorem* (⟨83⟩, 6(1979), 294–304).

GRATTAN-GUINNESS I., *Georg Cantor's influence on Bertrand Russell* (⟨84⟩, 1 (1980), 61–93).

GRAUERT H., *The methods of the theory of functions of several variables*, p. 129–143, *Miscellanea mathematica*, Berlin (Springer), 1991.

GRAVES L.M., *A transformation of the problem of Lagrange in the calculus of variations* (⟨185⟩, 35(1933), 675–682).

GRAY J., *Ideas of space: euclidean, non-euclidean and relativistic*, Oxford (University Press), 1979.

GRAY J.D., *The shaping of the Riesz representation theorem: a chapter in the history of analysis* (⟨33⟩, 31(1984–1985), 126–187).

GRAY J.W., *Fragments of the history of sheaf theory*, p. 1–79, FOURMAN M.P., MULVEY C.J. and SCOTT D.S. (edited by), *Applications of sheaves*, Berlin (Springer-Verlag), 1979.

GRELL H., *Beziehungen zwischen den Idealen verschiedener Ringe* (⟨120⟩, 97 (1927), 490–523).

GREEN J.A., *Brauer, Richard Dagobert*, p. 98–102, vol. 17, supplement II, *Dictionary of Scientific Biography*, New York (Charles Scribner's Sons), 1990.

GRÖTZSCH H., *Über die Verzerrung bei schlichten nichtkonformen Abbildungen und über eine damit zusammenhängende Erweiterung des Picardschen Satzes* (⟨43⟩, 80(1928), 503–507).

HAAR A., *Der Massbegriff in der Theorie der kontinuierlichen Gruppen* (⟨28⟩, (2), 34(1933), 147–169) = S. 600–622, *Gesammelte Arbeiten*, Budapest (Verlag der Akademie der Wissenschaften), 1959.

HADAMARD J., *Sur l'itération et les solutions asymptotiques des équations différentielles* (⟨49⟩, 29(1901), 224–228) = p. 1023–1026, vol. III, *Oeuvres*, Paris (C.N.R.S.), 1968.

HADAMARD J., *Leçons sur la propagation des ondes et les équations de l'hydrodynamique*, Paris (Hermann), 1903.

HADAMARD J., *Sur les opérations fonctionnelles* (⟨66⟩, 136(1903), 351–353) = p. 405–408, *Oeuvres*, Paris (C.N.R.S.), 1968.

HADAMARD J., *Résolution d'un problème aux limites pour les équations linéaires de type hyperbolique* (⟨49⟩, 32(1904), 242–268) = p. 1133–1159, vol. III, *Oeuvres*, Paris (C.N.R.S.), 1968.

HADAMARD J., *Théorie des équations aux dérivées partielles linéaires hyperboliques et du problème de Cauchy* (⟨7⟩, 31(1908), 333–380) = p. 1249–1296, vol. III, *Oeuvres*, Paris (C.N.R.S.), 1968.

HADAMARD J., *Index de Kronecker et théorie du point fixe*, p. 437–477, t. II, J. TANNERY, *Introduction à la théorie des fonctions d'une variable*, Paris (Hermann), 1910.

HADAMARD J., *Leçons sur le calcul des variations I*, Paris (Hermann), 1910.

HADAMARD J., *Notice sur les travaux scientifiques*, Paris (Hermann), 1912.

HADAMARD J., *Lectures on Cauchy's problem in linear partial differential equations*, New Haven (Yale University Press), 1923 = *Le problème de Cauchy et les équations aux dérivées partielles linéaires hyperboliques*, édition revue et notablement augmentée, Paris (Hermann), 1932.

HADAMARD J., *Le développement et le rôle scientifique du calcul fonctionnel*, p. 143–161, vol. I, *Atti del Congresso internazionale dei Matematici*, Bologna

(Nicola Zanichelli), 1929 = p. 435–453, vol. I, *Oeuvres*, Paris (C.N.R.S), 1968.

HADAMARD J., *Le problème de Cauchy et les équations aux dérivées partielles linéaires hyperboliques*, Paris (Hermann), 1932.

HAHN H., *Bericht über die Theorie der linearen Integralgleichungen* (⟨96⟩, 20 (1911), 69–117).

HAHN H., *Über Folgen linearer Operationen* (⟨129⟩, 32(1922), 3–88).

HAHN H., *Über lineare Gleichungssysteme in linearen Räumen* (⟨98⟩, 157(1927), 214–229).

HAJOS G., *Über einfache und mehrfache Bedeckung des n-dimensionalen Raumes mit einem Würfelgitter* (⟨123⟩, 47(1942), 427–467).

HALL P., *A contribution to the theory of groups of prime-power orders* (⟨150⟩, (2), 36(1933), 29–95).

HALL P., *On a theorem of Frobenius* (⟨150⟩, 40(1936), 468–501).

HALMOS P.R., *The legend of John von Neumann* (⟨13⟩, 80(1973), 382–394).

HALMOS P.R., *I want to be a mathematician, an automathography*, New York (Springer-Verlag), 1978.

HALMOS P.R., *Has progress in mathematics slowed down ?* (⟨13⟩, 97 (1990), 561–588).

HAMEL G., *Eine Basis aller Zahlen und die unstetigen Lösungen der Funktionalgleichung $f(x+y) = f(x) + f(y)$* (⟨120⟩, 60(1904), 459–462).

HAMEL G., *Über die Grundlagen der Mechanik* (⟨120⟩, 66(1909), 350–397).

HAMMERSTEIN A., *Nichtlineare Integralgleichungen nebst Anwendungen* (⟨7⟩, 54 (1930), 117–176).

HARARY F., *On the history of the theory of graphs*, p. 1–17, F. HARARY (edited by), *New directions in the theory of graphs*, New York (Academic Press), 1973.

HARDY G.H., *Sur les zéros de la fonction $\zeta(s)$ de Riemann* (⟨66⟩, 158 (1914), 1012–1014).

HARDY G.H., *The mean value of the modulus of an analytic function* (⟨150⟩, (2), 14(1915), 269–277) = p. 549–557, vol. III, *Collected papers.* (*Comments*, p. 558), Oxford (Clarendon Press), 1969.

HARDY G.H., *Orders of infinity*, 2nd ed., Cambridge (University Press), 1924.

HARDY G.H., *Ramanujan. Twelve lectures on subjects suggested by his life and work*, Cambridge (University Press), 1940.

HARDY G.H., *A mathematician's apology*, Cambridge (University Press), 1967 = *L'apologie d'un mathématicien*, Paris (Belin), 1985.

HARDY G.H. and LITTLEWOOD J.E., *A new solution of Waring's problem* (⟨156⟩, 48(1920), 272–293) = p. 382–404, vol. I, G.H. HARDY, *Selected papers*, Oxford (Clarendon Press), 1966.

HARDY G.H. and LITTLEWOOD J.E., *Some new properties of Fourier constants* ($\langle 120\rangle$, 97(1926), 159–209) = p. 348–398, vol. III (p. 398–399: *Comments*), G.H. HARDY, *Collected papers*, Oxford (Clarendon Press), 1969.

HARDY G.H. and LITTLEWOOD J.E., *A maximal theorem with function-theoretic applications* ($\langle 7\rangle$, 54(1930), 81–116).

HARDY G.H., LITTLEWOOD J.E. and POLYA G., *Inequalities*, Cambridge (University Press), 1934.

HARDY G.H. et RAMANUJAN S., *Une formule asymptotique pour le nombre de partitions de n* ($\langle 66\rangle$, 164(1917), 35–38) = p. 274–276, vol. I, G.H. HARDY, *Collected papers*, Oxford (Clarendon Press), 1966 = p. 239–241, S. RAMANUJAN, *Collected papers*, Cambridge (Univ. Press), 1927.

HARDY G.H. and WRIGHT E.M., *An introduction to the theory of numbers*, Oxford (Clarendon Press), 1938.

HARRIS T.E., *The theory of branching processes*, Berlin (Springer-Verlag), 1963 = *Les processus de ramification*, Paris (Dunod), 1969.

HARTOGS F., *Zur Theorie der analytischen Funktionen mehrerer unabhängiger Veränderlichen, insbesondere über die Darstellung derselben durch Reihen, welche nach Potenzen einer Veränderlichen fortschreiten* ($\langle 120\rangle$, 62(1906), 1–88).

HASSE H., *Über die Darstellbarkeit von Zahlen durch quadratische Formen im Körper der rationalen Zahlen* ($\langle 98\rangle$, 152(1923), 129–148) = S. 3–22, Band I, *Mathematische Abhandlungen*, Berlin (Walter de Gruyter), 1975.

HASSE H., *Äquivalenz quadratischer Formen in einem beliebigen algebraischen Zahlkörper* ($\langle 98\rangle$, 153(1924), 158–162) = S. 93–97, Band I, *Mathematische Abhandlungen*, Berlin (Walter de Gruyter), 1975.

HASSE H., *Bericht über neuere Untersuchungen und Probleme aus der Theorie der algebraischen Zahlkörper* ($\langle 96\rangle$, 35(1926), 1–55; 36(1927), 233–311; Ergänzungsband 6(1936); 42(1932), 85–86).

HASSE H., *Höhere Algebra I*, Berlin (Walter de Gruyter), 1926.

HASSE H., *Die Normenresttheorie relativ Abelscher Zahlkörper als Klassenkörpertheorie im Kleinen* ($\langle 98\rangle$, 162(1930), 145–154).

HASSE H., *Beweis des Analogons der Riemannschen Vermutung für die Artinschen und F.K. Schmidtschen Kongruenzzetafunktionen in gewissen elliptischen Fällen* ($\langle 130\rangle$, 42(1933), 253–262) = S. 85–94, Band 2, *Mathematische Abhandlungen*, Berlin (Walter de Gruyter), 1975.

HASSE H., *Zur Theorie der abstrakten elliptischen Funktionenkörper III. Die Struktur des Meromorphismenrings. Die Riemannsche Vermutung* ($\langle 98\rangle$, 175 (1936), 193–208) = S. 251–266, Band II, *Mathematische Abhandlungen*, Berlin (Walter de Gruyter), 1975.

HASSE H., *Zahlentheorie*, Berlin (Akademie-Verlag), 1949.

HASSE H. und SCHMIDT F.K., *Die Struktur diskret bewerteter Körper* (⟨98⟩, 170 (1933), 4–63) = S. 460–519, Band III, H. HASSE, Mathematische Abhandlungen, Berlin (Walter de Gruyter), 1975.

HAUSDORFF F., *Die symbolische Exponentialformel in der Gruppentheorie* (⟨43⟩, 58(1906), 19–48).

HAUSDORFF F., *Grundzüge der Mengenlehre*, Leipzig (Veit und Comp.), 1914.

HAUSDORFF F., *Dimension und äusseres Mass* (⟨120⟩, 79(1919), 157–179).

HAUSDORFF F., *Eine Ausdehnung des Parsevalschen Satzes über Fourier-Reihen* (⟨123⟩, 16(1923), 163–169).

HAWKINS T., *Lebesgue's theory of integration. Its origins and development*, Madison, Milwaukee (The University of Wisconsin Press), 1970.

HAWKINS T., *Burnside, William*, p. 615–616, vol. II, *Dictionary of scientific biography*, New York (Charles Scribner's Sons), 1973.

HAZEWINKEL M. (managing editor), *Encyclopaedia of mathematics. An updated and annotated translation of the Soviet «Mathematical Encyclopaedia»*, Dordrecht (Reidel), 1988–1991.

HECKE E., *Über die Zetafunktion beliebiger algebraischer Zahlkörper* (⟨130⟩, 1917, 77–89) = S. 159–171, *Mathematische Werke*, Göttingen (Vanderhoeck und Ruprecht), 1959.

HECKE E., *Über eine neue Anwendung der Zetafunktionen auf die Arithmetik der Zahlkörper* (⟨130⟩, 1917, 90–95).

HECKE E., *Über die L-Funktionen und den Dirichletschen Primzahlsatz für einen beliebigen Zahlkörper* (⟨130⟩, 1917, 299–318) = S. 178–197, *Mathematische Werke*, Göttingen (Vanderhoeck und Ruprecht), 1959.

HECKE E., *Eine neue Art von Zetafunktionen und ihre Beziehungen zur Verteilung der Primzahlen* (⟨123⟩, 1(1918), 357–376; 6(1920), 11–51).

HECKE E., *Vorlesungen über die Theorie der algebraischen Zahlen*, Leipzig (Akad. Verlagsges.), 1923.

HECKE E., *Über Modulfunktionen und die Dirichletschen Reihen mit Eulerscher Produktentwicklung* (⟨120⟩, 114(1937), 1–28, 316–351) = S. 644–703, *Mathematische Werke*, Göttingen (Vanderhoeck und Ruprecht), 1959.

VAN HEIJENOORT J., *From Frege to Gödel. A source book in mathematical logic 1879–1931*, Cambridge, Massachusetts (Harvard University Press), 1967.

HEILBRONN H., *On the class-number in imaginary quadratic fields* (⟨156⟩, 5(1934), 150–160).

HEISENBERG W., *La partie et le tout. Le monde de la physique atomique*, Paris (Flammarion), 1990.

HELLINGER E., *Hilberts Arbeiten über Integralgleichungen und unendliche Gleichungssysteme*, S. 94–145, III. Band, D. HILBERT, *Gesammelte Abhandlungen*, Bronx, New York (Chelsea), 1965.

HELLINGER E. und TOEPLITZ O., *Integralgleichungen und Gleichungen mit unendlichvielen Unbekannten,* II C 13, *Encyklopädie der mathematischen Wissenschaften,* Leipzig (Teubner), 1927.

HELLY E., *Über Systeme linearer Gleichungen mit unendlich vielen Unbekannten* (⟨129⟩, 31(1921), 60–91).

HENKIN L., *Completeness in the theory of types* (⟨101⟩, 15(1950), 81–91).

HENSEL K., *Über die arithmetischen Eigenschaften der Zahlen* (⟨96⟩, 16(1907), 299–319, 388–393, 473–496).

HENSEL K., *Theorie der algebraischen Zahlen I,* Leipzig (B.G. Teubner), 1908.

HENSEL K., *Zahlentheorie,* Berlin (G.J. Göschen), 1913.

HENSEL K. und LANDSBERG G., *Theorie der algebraischen Funktionen einer Variablen und ihre Anwendung auf algebraische Kurven und abelsche Integrale,* Leipzig (Teubner), 1902.

HENSTOCK R., *A short history of integration theory* (⟨180⟩, 12(1988), 75–95).

HERBRAND J., *Recherches sur la théorie de la démonstration* (⟨186⟩, n° 33, 1930).

HERGLOTZ G., *Über Potenzreihen mit positivem, reellem Teil im Einheitskreis* (⟨43⟩, 63(1911), 501–511) = S. 247–257, *Gesammelte Schriften,* Göttingen (Vandenhoeck und Ruprecht), 1979. (H. WITTICH, S. XXXIX).

HERGLOTZ G., *Über die Integration linearer partieller Differentialgleichungen mit konstanten Koeffizienten. III* (Berichte über die Verhandlungen der Sächsischen Akademie der Wissenschaften zu Leipzig, 80(1928), 69–114).

HERMANN A., *Schrödinger, Erwin,* p. 217–223, vol. XII, *Dictionary of scientific biography,* New York (Charles Scribner's Sons), 1975.

HERVE M., *Commentaires sur les travaux rassemblés dans les volumes I et II,* p. 89–104, vol. I, G. JULIA, *Oeuvres,* Paris (Gauthier-Villars), 1968.

HERVE M., *L'oeuvre de Gaston Julia* (⟨57⟩, 2(1981), 1–8).

HESTENES M.R., *RAND Corporation Report,* 1950.

HEWITT E. and HEWITT R.E., *The Gibbs-Wilbraham phenomenon: an episode in Fourier analysis* (⟨33⟩, 21(1979–1980), 129–160).

HEYTING A., *Intuitionistische axiomatiek der projektieve meetkunde,* Groningen (P. Noordhoof), 1925.

HEYTING A., *Die formalen Regeln der intuitionistischen Logik* (⟨177⟩, 1930, 42–71, 158–169).

HILBERT D., *Über den Zahlbegriff* (⟨96⟩, 8(1900), 180–184).

HILBERT D., *Über das Dirichlet'sche Princip* (⟨96⟩, 8(1900), 184–188) = S. 10–14, III. Band, *Gesammelte Abhandlungen,* Berlin (Springer), 1935.

HILBERT D., *Sur les problèmes futurs des mathématiques,* p. 58–114, *Compte rendu du deuxième congrès international des mathématiciens,* Paris (Gauthier-Villars), 1902 = *Mathematische Probleme* (⟨130⟩, 1900, 253–297) = p. 290–329, III. Band, *Gesammelte Abhandlungen,* Berlin (J. Springer), 1935.

HILBERT D., *Über die Grundlagen der Geometrie* (⟨120⟩, 56(1902), 381–422) = (introduction) p. 234–241, *Les fondements de la géométrie*, Paris (Dunod), 1971.

HILBERT D., *Zur Variationsrechnung* (⟨130⟩, 1905, 159–180) = S. 38–55, III. Band, *Gesammelte Abhandlungen*, Berlin (Springer), 1935.

HILBERT D., *Beweis für die Darstellbarkeit der ganzen Zahlen durch eine feste Anzahl n-ter Potenzen (Waringsches Problem)* (⟨120⟩, 67(1909), 281–300) = S. 510–527, I. Band, *Gesammelte Abhandlungen*, Berlin (Springer), 1932.

HILBERT D., *Grundlagen der Geometrie. Dritte durch Zusätze und Literaturnachweise von Neuem vermehrte und mit sieben Anhägen versehene Auflage*, Leipzig (B.G. Teubner), 1909.

HILBERT D., *Grundzüge einer allgemeinen Theorie der linearen Integralgleichungen*, Leipzig (B.G. Teubner), 1912.

HILBERT D., *Axiomatisches Denken* (⟨120⟩, 78(1918), 405–415) = S. 146–156, III. Band, *Gesammelte Abhandlungen*, Berlin (J. Springer), 1935.

HILBERT D., *Neubegründung der Mathematik I* (⟨1⟩, 1(1922), 157–177) = S. 157–177, III. Band, *Gesammelte Abhandlungen*, Berlin (Springer), 1935.

HILBERT D. und ACKERMANN W., *Grundzüge der theoretischen Logik*, Berlin (J. Springer), 1928.

HILBERT D. und BERNAYS P., *Grundlagen der Mathematik I*, Berlin (Julius Springer), 1934.

HILDEBRANDT T.H., *On uniform limitedness of sets of functional operations* (⟨55⟩, 29(1923), 309–315).

HILDEBRANDT T.H., *Integration in abstract spaces* (⟨55⟩, 59(1953), 111–139).

HILDEBRANDT T.H. and GRAVES L.M, *Implicit functions and their differentials in general analysis* (⟨185⟩, 29(1927), 127–153).

HILLE E., *Functional analysis and semi-groups*, Amer. Math. Soc., 1948.

HILTON P., *Working with Alan Turing* (⟨112⟩, 13(1991), N° 4, 22–25).

HIRONAKA H., *Introduction*, p. 307–313, vol. I, O. ZARISKI, *Collected papers*, Cambridge, Massachusetts (The MIT Press), 1972.

HLAWKA E., *Zur Geometrie der Zahlen* (⟨123⟩, 49(1943), 285–312).

HLAWKA E. und BINDER CH., *Über die Entwicklung der Theorie der Gleichverteilung in den Jahren 1909–1916* (⟨33⟩, 36(1986), 197–249).

HODGE W.V.D., *A Dirichlet problem for harmonic functionals, with applications to analytic varieties* (⟨150⟩, (2), 36(1933), 257–303).

HODGE W.V.D., *The theory and applications of harmonic integrals*, Cambridge (University Press), 1941.

HODGES A., *Alan Turing: the enigma*, New York (Simon and Schuster), 1983.

HOHEISEL G., *Primzahlprobleme in der Analysis* (⟨177⟩, 1930, 580–588).

HOLMGREN E., *Über Systeme von linearen partiellen Differentialgleichungen* ((⟨139⟩, 58(1901), 91–103).

HOPF E., *Elementare Bemerkungen über die Lösungen partieller Differentialgleichungen zweiter Ordnung vom elliptischen Typus* ((⟨177⟩, 1927, 147–152).

HOPF E., *Beweis des Mischungscharakters der geodätischen Strömung auf Flächen der Krümmung minus Eins und endlicher Oberfläche* ((⟨177⟩, 1938, 333–344).

HOPF E., *Abzweigung einer periodischen Lösung von einer stationären Lösung eines Differentalsystems* ((⟨43⟩, 95(1943), 3–22).

HOPF E., *The partial differential equation* $u_t + uu_x = \mu u_{xx}$ ((⟨64⟩, 3(1950), 201–230).

HOPF H., *Abbildungsklassen n-dimensionaler Mannigfaltigkeiten* ((⟨120⟩, 96 (1926), 209–224).

HOPF H., *Vektorfelder in n-dimensionalen Mannigfaltigkeiten* ((⟨120⟩, 96(1926), 225–250).

HOPF H., *Eine Verallgemeinerung der Euler-Poincaréschen Formel* ((⟨130⟩, 1928, 127–136) = p. 5–13, *Selecta*, Berlin (Springer-Verlag), 1964.

HOPF H., *Über die Abbildungen der dreidimensionalen Sphäre auf die Kugelfläche* ((⟨120⟩, 104(1931), 637–665) = p. 38–63, *Selecta*, Berlin (Springer-Verlag), 1964.

HOPF H., *Differentialgeometrie und topologische Gestalt* ((⟨96⟩, 41(1932), 209–229).

HOPF H., *Über die Abbildungen von Sphären auf Sphären niedriger Dimension* ((⟨80⟩, 25(1935), 427–440) = p. 95–106, *Selecta*, Berlin (Springer-Verlag), 1964.

HOPF H., *Fundamentalgruppe und zweite Bettische Gruppe* ((⟨63⟩, 14 (1941), 257–309) = p. 185–206, *Selecta*, Berlin (Springer-Verlag), 1964.

HOPF H., *Über die Topologie der Gruppen-Mannigfaltigkeiten und ihrer Verallgemeinerungen* ((⟨28⟩, (2), 42(1941), 22–52) = S. 119–151, *Selecta*, Berlin (Springer-Verlag), 1964.

HOPF H., *Über die Bettischen Gruppen die einer beliebigen Gruppe gehören* ((⟨63⟩, 17(1944–1945), 39–79) = p. 211–234, *Selecta*, Berlin (Springer), 1964.

HOPF H., *Vom Bolzanoschen Nullstellensatz zur algebrischen Homotopietheorie der Sphären* ((⟨96⟩, 56(1953), 59–76).

HOPF H., *Einige persönliche Erinnerungen aus der Vorgeschichte der heutigen Topologie*, p. 9–20, *Colloque de topologie tenu à Bruxelles du 7 au 10 septembre 1964*, Louvain (Librairie Universitaire), 1966.

HOPF H. und RINOW W., *Über den Begriff der vollständigen differential-geometrischen Fläche* ((⟨63⟩, 3(1931), 209–225).

HORVATH J., *An introduction to distributions* ((⟨13⟩, 77(1970), 227–240).

HORVATH J., *L'oeuvre mathématique de Marcel Riesz* ((⟨57⟩, 3(1982), 83–122; 4(1983), 1–60).

HOSTINSKY B., *Sur les progrès récents de la théorie des probabilités* (⟨62⟩, 64 (1935), 94–106).

HULSBERGEN W.W.J., *Conjectures in arithmetic algebraic geometry. A survey*, Braunschweig (Vieweg), 1992.

HUNT J.C.R., PHILLIPS O.M. and WILLIAMS D. (edited by), *Turbulence and stochastic processes: Kolmogorov's ideas 50 years on*, London (Royal Society), 1991.

HUREWICZ W., *Beiträge zur Topologie der Deformationen* (⟨108⟩, 38 (1935), 521–528).

HUREWICZ W., *On duality theorems* (⟨55⟩, 47(1941), 562–563).

HUREWICZ W. and STEENROD N.E., *Homotopy relations in fibre spaces* (⟨151⟩, 27(1941), 60–64).

HUREWICZ W. and WALLMAN H., *Dimension theory*, Princeton (University Press), 1941.

IBSCH H., *L'oeuvre mathématique de Louis Antoine et son influence* (⟨78⟩, 9 (1991), 251–274).

IKEHARA S., *On Tauberian theorems of Hardy and Littlewood and a note on Wintner's paper* (⟨99⟩, 10(1931), 75–83).

INGHAM A.E., *The distribution of prime numbers*, Cambridge (University Press), 1990.

ITÔ K., *Equations différentielles déterminant les processus de Markov* (in Japanese) (⟨197⟩, 1077(1942), 1352–1400) = *Differential equations determining Markov processes*, p. 42–75, *Selected papers*, New York (Springer-Verlag), 1987.

ITÓ K., *On a stochastic integral equation* (⟨149⟩, 22(1946), 32–35) = p. 92–95, *Selected papers*, New York (Springer-Verlag), 1987.

ITÔ K. (edited by), *Encyclopedic dictionary of mathematics*, second edition, volumes I to IV, Cambridge, Massachusetts (The MIT Press), 1987.

IWASAWA K., *On some types of topological groups* (⟨28⟩, (2), 50(1949), 507–557).

JACOBSON N., *The fundamental theorem of the Galois theory for quasi-fields* (⟨28⟩, 41(1940), 1–7).

JACOBSON N., *Structure theory of simple rings without finiteness assumptions* (⟨185⟩, 57(1945), 228–245).

JÄGER W., MOSER J. and REMMERT R. (edited by), *Perspectives in mathematics. Anniversary of Oberwolfach 1984*, Basel (Birkhäuser), 1984.

JAHNKE H.N., *Hilbert, Weyl und die Philosophie der Mathematik* (⟨122⟩, 37 (1990), 157–179).

JEFFREYS H., *On certain approximate solutions of linear differential equations of the second order* (⟨150⟩, (2), 23(1924), 428–436).

JENSEN B., MARCINKIEWICZ J. and ZYGMUND A., *Note on the differentiability of multiple integrals* (⟨80⟩, 25(1935), 217–234).

JENSEN J.L.W.V., *Sur les fonctions convexes et les inégalités entre les valeurs moyennes* (⟨7⟩, 30(1906), 175–193).

JOHNSON D.M., *The problem of the invariance of dimension in the growth of modern topology* (⟨33⟩, 20(1979), 97–188; 25(1981), 85–267).

JONES B.W., *The arithmetic theory of quadratic forms*, Buffalo (Carus Monographs), 1950.

JORDAN P., VON NEUMANN J. and WIGNER E., *On an algebraic generalization of the quantum mechanical formalism* (⟨28⟩, (2), 35(1934), 29–64) = p. 409–444, vol. II, J. VON NEUMANN, *Collected works*, Oxford (Pergamon Press), 1961.

JULIA G., *Mémoire sur l'itération des fonctions rationnelles* (⟨97⟩, (8), 1(1918), 47–245) = p. 121–319, vol. I, *Oeuvres*, Paris (Gauthier-Villars), 1968).

JULIA G., *Sur les familles de fonctions de plusieurs variables* (⟨66⟩, 170(1920), 791–793) = p. 293–295, vol. II, *Oeuvres*, Paris (Gauthier-Villars), 1968.

JULIA G., *Essai sur le développement de la théorie des fonctions de variables complexes*, p. 102–127, I. Band, *Verhandlungen des Internationalen Mathematiker-Kongresses Zürich 1932*, Zürich (Orell Füssli).

JULIA G., *Notice sur les travaux scientifiques*, p. 1–87, vol. I, *Oeuvres*, Paris (Gauthier-Villars), 1968.

KAC M., *Probability methods in some problems of analysis and number theory* (⟨55⟩, 55(1949), 641–665).

KACZMARZ S. und STEINHAUS H., *Theorie der Orthogonalreihen*, Warsaw (Monograf. mat.), 1935.

KAHANE J.-P., *Séries de Fourier absolument convergentes*, Berlin (Springer-Verlag), 1970.

KAHANE J.-P., *A central object in mathematics: the brownian motion* (⟨83⟩, 2 (1975), 595–600).

KAHANE J.-P., *Leopold Fejér et l'analyse mathématique au début du XXe siècle* (⟨57⟩, 2(1981), 67–84).

KAHANE J.-P., *Le 13ème problème de Hilbert: un carrefour de l'algèbre, de l'analyse et de la géométrie* (⟨57⟩, 3(1982), 1–26).

KAHANE J.-P., *Some random series of functions*, second edition, Cambridge (University Press), 1985.

KAHANE J.-P., *Jacques Hadamard* (⟨112⟩, 13(1991), n° 1, 23–29).

KÄHLER E., *Forme differenziali e funzioni algebriche* (⟨157⟩, 3(1932), Nr 3, 1–19).

KÄHLER E., *Einführung in die Theorie der Systeme von Differenzialgleichungen*, Leipzig (B.G. Teubner), 1934.

KAKUTANI S., *Concrete representation of abstract ($\mathcal{L}$) spaces and the mean ergodic theorem* (⟨28⟩, (2), 42(1941), 523–537).

KAKUTANI S., *On brownian motion in n-space* (⟨149⟩, 20(1944), 648–652).

KAKUTANI S., *Two-dimensional Brownian motion and harmonic functions* (⟨149⟩, 20(1944), 706–714).

KAKUTANI S., *Ergodic theory*, p. 128–142, vol. II, *Proceedings of the international congress of mathematicians 1950*, Providence R.I. (American Mathematical Society), 1952.

KAMINSKI A. and SKORNIK K., *Facts about the life and the research of professor Jan Mikusinski* (⟨19⟩, ser. II, Wiad. Mat., 28(1988), 35–52).

VAN KAMPEN E.R., *Locally bicompact abelian groups and their character groups* (⟨28⟩, (2), 36(1935), 448–463).

KANIGEL R., *The man who knew infinity: a life of the genius Ramanujan*, New York (C. Scribner's), 1991.

KANTOROVIČ L.V., *On semi-ordered linear spaces and their applications to the theory of linear operators* (in Russian) (⟨69⟩, 4(1935), 13–16).

KANTOROVIČ L.V., *Méthodes mathématiques d'organisation et de planification de la production* (in Russian), Leningrad, 1939 = *Mathematical methods of organizing and planning production* (⟨110⟩, 6(1959–1960), 366–422).

KANTOROVIČ L.V., *Functional analysis and applied mathematics* (⟨191⟩, (N.S.), 3(1948), N° 6, 89–185).

KANTOROVOČ L.V. and KRYLOV V.I., *Approximate methods in higher analysis*, Groningen (P. Noordhoof), 1958.

KANTOROVIČ L.V., VULIKH B.Z. and A.G. PINSKER, *Functional analysis in partially ordered spaces* (in Russian), Moskva (Gosudarstv. Izdat. Tehn.-Teor. Lit.), 1950.

KARAMATA J., *Sur certains «Tauberian theorems» de MM. Hardy et Littlewood* (⟨111⟩, 3(1930), 33–48).

KARAMATA J., *Sur un mode de croissance régulière des fonctions* (⟨111⟩, 4 (1930), 38–53).

KARZEL H. und KROLL H.-J., *Geschichte der Geometrie seit Hilbert*, Darmstadt (Wissenschaftliche Buchgesellschaft), 1988.

KATZ V.J., *Differential forms — Cartan to de Rham* (⟨33⟩, 33(1985), 321–336).

KELLEY J. and PITCHER E., *Exact homomorphisms sequences in homology theory* (⟨28⟩, (2), 48(1947), 682–709).

KELLEY W.G. and ELIASON S.B. (eds), *Optimal control and differential equations*, Proc. Conf. Optimal control and differential equations, Oklahoma (Norman), 1977.

KELLOG O.D., *Foundations of potential theory*, Berlin (Springer), 1929.

KENDALL D.G., *Branching processes since 1873* (⟨104⟩, 41(1966), 385–406).

KENDALL D.G., *Obituary A.N. Kolmogorov (1903–1987)* (⟨56⟩, 22 (1990), 31–100).

KHINTCHINE A.J., *Über dyadische Brüche* (⟨123⟩, 18(1923), 109–116).

KHINTCHINE A.J., *Über einen Satz der Wahrscheinlichkeitsrechnung* (⟨80⟩, 6 (1924), 9–20).

KHINTCHINE A.J., *Asymptotische Gesetze der Wahrscheinlichkeitsrechnung* (⟨76⟩, 2(1933), Nr.4).

KHINTCHINE A.J., *Korrelationstheorie der stationären stochastischen Prozesse* (⟨120⟩, 109(1934), 604–615).

KHINTCHINE A.J. und KOLMOGOROV A.N., *Über Konvergenz von Reihen, deren Glieder durch den Zufall bestimmt werden* (⟨117⟩, 32(1925), 668–677).

KIERNAN B.M., *The development of Galois theory from Lagrange to Artin* (⟨33⟩, 8(1971–1972), 40–154).

KITCHER P., *Hilbert's epistemology* (⟨144⟩, 43(1976), 99–115).

KLEENE S.C., *General recursive functions of natural numbers* (⟨120⟩, 112(1936), 727–742).

KLEENE S.C., *λ-definability and recursiveness* (⟨70⟩, 2(1936), 340–353).

KLEIN F., *Vorlesungen über die Entwicklung der Mathematik im 19. Jahrhundert*, Berlin (Springer-Verlag), 1979.

KLINE M., *Mathematical thought from ancient to modern times*, New York (Oxford University Press), 1972.

KLOOSTERMANN H.D., *Over een in de additieve gestallenleer optredende som van eenheidswortels* (⟨131⟩, (2), 15(1925), 14–20).

KLUGE W., *Edmund Landau, sein Werk und sein Einfluss auf die Entwickelung der Mathematik*, Staatsexamensarbeit, Duisburg, 1983.

VON KOCH H., *Sur une courbe continue sans tangente, obtenue par une construction géométrique élémentaire* (⟨34⟩, 1(1904), 681–702).

KODAIRA K., *Harmonic fields in Riemannian manifolds, generalized potential theory* (⟨28⟩, (2), 50(1949), 587–665).

KÖHLER E., *Gödel and Carnap in Vienna* (⟨94⟩, 1991, 54–62).

KOLMOGOROV A.N., *Une série de Fourier-Lebesgue divergente presque partout* (⟨80⟩, 4(1923), 324–328).

KOLMOGOROV A.N., *On the principle tertium non datur* (⟨117⟩, 32 (1924–1925), 646–667).

KOLMOGOROV A.N., *Une série de Fourier-Lebesgue divergente partout* (⟨66⟩, 183(1926), 1327–1328).

KOLMOGOROV A.N., *Über das Gesetz des iterierten Logarithmus* (⟨120⟩, 101 (1929), 126–135).

KOLMOGOROV A.N., *Über die analytischen Methoden in der Wahrscheinlichkeitsrechnung* (⟨120⟩, 10(1931), 415–458).

KOLMOGOROV A.N., *Zur Deutung der intuitionistischen Logik* (⟨123⟩, 35 (1932), 58–65).

KOLMOGOROV A.N., *Grundbegriffe der Wahrscheinlichkeitsrechnung*, Berlin (J. Springer), 1933.

KOLMOGOROV A.N., *Zur Theorie der stetigen zufälligen Prozesse* (⟨120⟩, 108 (1933), 149–160).

KOLMOGOROV A.N., *Zur Normierbarkeit eines allgemeinen topologischen linearen Raumes* (⟨181⟩, 5(1934), 29–33).

KOLMOGOROV A.N., *Anfangsgründe der Theorie der Markoffschen Ketten mit unendlich vielen möglichen Zuständen* (⟨117⟩, 1(1936), 607–610).

KOLMOGOROV A.N., *Zur Umkehrbarkeit der statistischen Naturgesetze* (⟨120⟩, 113(1936), 766–772).

KOLMOGOROV A.N., *Wienersche Spiralen und einige andere interessante Kurven im Hilbertschen Raum* (⟨69⟩, 26(1940), 115–118).

KOLMOGOROV A.N., *Interpolation und Extrapolation von stationären zufälligen Folgen* (⟨90⟩, 5(1941), 3–14).

KOLMOGOROV A.N., *Wie ich Mathematiker geworden bin* (⟨127⟩, 1989, n° 1–2, 83–87).

KOLMOGOROV A.N., *Selected works*, vol. I, edited by V.M. TIKHOMOROV, *Commentary*, p. 410–527, 1992.

KOLMOGOROV A.N., PETROVSKY I.G. und PISKUNOV N.S., *Untersuchungen der Diffusionsgleichung bei zunehmender Stoffmenge und ihre Anwendung auf ein biologisches Problem* (⟨46⟩, Nr. 6, 2(41), 1937).

KOLMOGOROV A.N. et YOUSCHKEVITCH A.P., *Mathematics of the XIXth century* (in Russian), Moscow (Nauka), 1978, 1981, 1987.

KÖNIG D., *Theorie der endlichen und unendlichen Graphen. Kombinatorische Topologie der Streckenkomplexe*, Leipzig (Akademische Verlagsgesellschaft), 1936 = *Theory of finite and infinite graphs.* With a commentary by W.T. Tutte and a biographical sketch by T. Gallai, Boston (Birkhäuser), 1990.

KOSZUL J.-L., *Homologie et cohomologie des algèbres de Lie* (⟨49⟩, 78 (1950), 65–127).

KOSZUL J.-L., *Sur un type d'algèbres différentielles en rapport avec la transgression*, p. 73–81, *Colloque de topologie (espaces fibrés), Bruxelles 1950*, Liège (Georges Thone), 1951.

KÖTHE G., *Die Stufenräume, eine einfache Klasse linearer vollkommener Räume* (⟨123⟩, 51(1948), 317–345).

KÖTHE G., *Banach und die Lemberger Schule der Funktionsanalysis* (⟨122⟩, 36 (1989), 145–158).

KÖTHE G. und TOEPLITZ O., *Lineare Räume mit unendlichvielen Koordinaten und Ringe unendlicher Matrizen* (⟨98⟩, 171(1934), 193–226).

KRAGH H.S., *Dirac: A scientific biography*, Cambridge (University Press), 1990.

KRAMER E.E., *The nature and growth of modern mathematics*, Princeton (University Press), 1982.

KRASNER M., *Une généralisation de la notion de corps* (⟨97⟩, (9), 17 (1938), 367–385).

KRASNOSELSKII M.A. and RUTICKII YA.B., *Convex functions and Orlicz spaces*, Groningen (Noordhoff), 1961.

KRAZER A., *Lehrbuch der Thetafunktionen*, Leipzig (B.G. Teubner), 1903.

KRAZER A. und WIRTINGER W., *Abelsche Funktionen und allgemeine Thetafunktionen*, S. 604–873, II B 7, *Encyklopädie der mathematischen Wissenschaften*, Leipzig (B.G. Teubner), 1920.

KREIN M.G., *A generalization of some investigations of A.M. Lyapunov on linear differential equations with periodic coefficients* (in Russian) (⟨69⟩, (N.S.), 73(1950), 445–448).

KREIN M.G. and MILMAN D., *On extreme points of regularly convex sets* (⟨181⟩, 9(1940), 133–138).

KREIN M.G. and RUTMAN M.A., *Linear operators for which a cone remains invariant in a Banach space* (in Russian) (⟨191⟩, (N.S.), 3(1948), n° 1, 3–95).

KREISEL G., *K. Gödel, Collected works* (⟨134⟩, 29(1988), 160–181; 31 (1990), 602–641).

KREYSZIG E., *Zur Entwickelung der zentralen Ideen in der Funktionanalysis* (⟨73⟩, 41(1986), n° 2, 25–35; n° 3, 49–57).

KREYSZIG E., *Friedrich Riesz als Wegbereiter der Funktionalanalysis* (⟨73⟩, 45 (1990), n° 5, 117–130).

KRÜGER L., DASTON L.J. and HEIDELBERGER M., *The probabilistic revolution*, Cambridge, MA (The MIT Press), 1990.

KRULL W., *Über verallgemeinerte endliche Abelsche Gruppen* (⟨123⟩, 23(1925), 161–196).

KRULL W., *Galoissche Theorie der unendlichen algebraischen Erweiterungen* (⟨120⟩, 100(1928), 687–698).

KRULL W., *Allgemeine Bewertungstheorie* (⟨98⟩, 167(1932), 160–196).

KRULL W., *Idealtheorie*, Berlin (Springer), 1935.

KRULL W., *Beiträge zur Arithmetik kommutativer Integritätsbereiche III* (⟨123⟩, 42(1937), 745–766).

KRULL W., *Dimensionstheorie in Stellenringen* (⟨98⟩, 179(1938), 204–226).

KRYLOV N.M. et BOGOLYUBOV N.N., *La théorie générale de la mesure dans son application à l'étude des systèmes dynamiques de la mécanique non linéaire* (⟨28⟩, (2), 38(1937), 65–113).

KUHN H.W., *Extensive games* (⟨151⟩, 36(1950), 570–576).

KUIPERS L. and NIEDERREITER H., *Uniform distribution of sequences*, New York (Wiley), 1974.

KÜNNETH H., *Über die Bettischen Zahlen einer Produktmannigfaltigkeit* (⟨120⟩, 90(1923), 65–85).

KUPKA J., *Measure theory: the heart of the matter* (⟨112⟩, 8(1986), n° 4, 47–56).

KURATOWSKI C., *Sur l'opération $\bar{\mathscr{A}}$ de l'analysis situs* (⟨80⟩, 3(1922), 182–199).

KURATOWSKI C., *Une méthode d'élimination des nombres transfinis des raisonnements mathématiques* (⟨80⟩, 3(1922), 76–108).

KURATOWSKI C., *Topologie I, espaces métrisables, espaces complets*, Warszawa 1933.

KURATOWSKI C., *A half century of Polish mathematics: remembrances and reflections*, New York (Pergamon), 1980.

KÜRSCHAK J., *Über Limesbildung und allgemeine Körpertheorie* (⟨98⟩, 142 (1913), 211–253).

KUZAWA M.G., *Modern mathematics: the genesis of a school in Poland*, New Haven (College and University Press), 1968.

LALESCO T., *Introduction à la théorie des équations intégrales*, avec une préface de E. PICARD, Paris (Hermann), 1912.

LAMSON K.W., *A general implicit function theorem with an application to problems of relative minima* (⟨12⟩, 42(1920), 243–256).

LANCZOS C., *Space through the ages*, New York (Academic Press), 1970.

LANDAU E., *Über die zu einem algebraischen Zahlkörper gehörige Zetafunktion und die Ausdehnung der Tchebyschefschen Primzahltheorie auf das Problem der Vertheilung der Primideale* (⟨98⟩, 125(1902), 64–188) = p. 201–325, vol. 1, *Collected works*, Essen (Thales Verlag).

LANDAU E., *Über eine Verallgemeinerung des Picardschen Satzes* (⟨177⟩, 1904, 1118–1133) = p. 129–144, vol. II, *Collected works*, Essen (Thales Verlag).

LANDAU E., *Handbuch der Lehre von der Verteilung der Primzahlen*, Leipzig (B.G. Teubner), 1909.

LANDAU E., *Über die Anzahl der Gitterpunkte in gewissen Bereichen* (⟨130⟩, 1912, 687–770).

LANDAU E., *Darstellung und Begründung einiger neuerer Ergebnisse der Funktionentheorie*, Berlin (Springer), 1916.

LANDAU E., *Grundlagen der Analysis. (Rechnen mit ganzen, rationalen, irrationalen, komplexen Zahlen). Ergänzung zu den Lehrbüchern der Differential und Integralrechnung*, Leipzig (Akademische Verlagsgesellschaft), 1930.

LANDAU L.D., *On the problem of turbulence* (⟨69⟩, (N.S.), 44(1944), 311–314).

LANG S., *Transcendental numbers and diophantine approximations* (⟨55⟩, 77 (1971), 635–677).

LANG S. and TATE J., *Preface*, p. VII–X, E. ARTIN, *The collected papers*, Reading, Massachusetts (Addison-Wesley), 1965.

LANGER E.R., *Fourier's series. The genesis and the evolution of a theory* (⟨13⟩, 54(1947), n° 7, 1–82).

LANGER R.E., *On the asymptotic solutions of ordinary differential equations, with an application to the Bessel functions of large order* (⟨185⟩, 33(1931), 23–64).

LAPPO-DANILEVSKY I.A., *Mémoires sur la théorie des sytèmes des équations linéaires III* (⟨187⟩, 8(1936)).

LASKER E., *Zur Theorie der Moduln und Ideale* (⟨120⟩, 60(1905), 21–116).

DE LA VALLÉE POUSSIN CH.-J., *Intégrales de Lebesgue, fonctions d'ensemble, classes de Baire*, Paris (Gauthier-Villars), 1916.

DE LA VALLÉE POUSSIN CH.-J., *Critérium d'existence de la dérivée normale d'un potentiel de surface en un point de la surface active* (⟨20⟩, 50(1930) A, 132–139).

LAVRENTIEV M.A., *Sur une équation différentielle du premier ordre* (⟨123⟩, 23 (1925), 197–209).

LAVRENTIEV M.A., *Sur quelques propriétés des fonctions univalentes* (⟨69⟩, 1 (1935), 1–4).

LAVRENTIEV M.A. and BICADZE A.V., *On the problem of mixed-type equations* (in Russian) (⟨69⟩, (N.S.), 64(1949), 25–28).

LEBESGUE H., *Sur une généralisation de l'intégrale définie* (⟨66⟩, 132 (1901), 1025–1027) = p. 197–199, vol. I, *Oeuvres scientifiques*, Genève (L'Enseignement Mathématique), 1972.

LEBESGUE H., *Intégrale, longueur, aire* (⟨26⟩, (3), 7(1902), 231–359) = p. 201–331, vol. I, *Oeuvres scientifiques*, Genève (L'Enseignement Mathématique), 1972.

LEBESGUE H., *Sur les séries trigonométriques* (⟨24⟩, (3), 20(1903), 453–485) = p. 27–59, vol. III, *Oeuvres scientifiques*, Genève (L'Enseignement Mathématique), 1972.

LEBESGUE H., *Leçons sur l'intégration et la recherche des fonctions primitives*, Paris (Gauthier-Villars), 1904.

LEBESGUE H., *Sur les fonctions représentables analytiquement* (⟨66⟩, 139(1904), 29–31) = p. 79–81, vol. III, *Oeuvres scientifiques*, Genève (L'Enseignement Mathématique), 1972.

LEBESGUE H., *Leçons sur les séries trigonométriques*, Paris (Gauthier-Villars), 1906.

LEBESGUE H., *Sur le problème de Dirichlet* (⟨158⟩, 24(1907), 1–32) = p. 91–122, vol. IV, *Oeuvres scientifiques*, Genève (L'Enseignement Mathématique), 1973.

LEBESGUE H., *Sur l'intégration des fonctions discontinues* (⟨24⟩, (3), 27(1910), 361–450) = p. 185–274, vol. II, *Oeuvres scientifiques*, Genève (L'Enseignement Mathématique), 1972.

LEBESGUE H., *Sur des cas d'impossibilité du problème de Dirichlet ordinaire* (⟨67⟩, 41(1913), 17 = p. 131, vol. IV, *Oeuvres scientifiques*, Genève (L'Enseignement Mathématique), 1973.

LEBESGUE H., *Notice sur les travaux scientifiques*, Toulouse (Privat), 1922 = p. 97–175, vol. I, *Oeuvres scientifiques*, Genève (L'Enseignement Mathématique), 1972.

LEBESGUE H., *A propos de quelques travaux mathématiques récents* (⟨74⟩, 17 (1971), 1–48).

LEBESGUE H., *Lettres à Emile Borel* (⟨57⟩, 12(1991)).

LE CAM L., *Asymptotic methods in statistical decision theory*, New York (Springer-Verlag), 1986.

LEFSCHETZ S., *Continuous transformations on manifolds* (⟨151⟩, 9 (1923), 90–93).

LEFSCHETZ S., *L'analysis situs et la géométrie algébrique*, Paris (Gauthier-Villars), 1924.

LEFSCHETZ S., *Intersections and transformations of complexes and manifolds* (⟨185⟩, 23(1926), 1–49) = p. 199–247, *Selected papers*, Bronx, New York (Chelsea), 1971.

LEFSCHETZ S., *Manifolds with a boundary and their transformations* (⟨185⟩, 29(1927), 429–462, 848) = p. 248–281, *Selected papers*, Bronx, New York (Chelsea), 1971.

LEFSCHETZ S., *The residual set of a complex on a manifold and related questions* (⟨151⟩, 13(1927), 614–622, 805–807).

LEFSCHETZ S., *Closed point sets on a manifold* (⟨28⟩, (2), 29(1928), 232–254) = p. 545–568, *Selected papers*, New York (Chelsea), 1971.

LEFSCHETZ S., *Duality relations in topology* (⟨151⟩, 15(1929), 367–369).

LEFSCHETZ S., *Géométrie sur les surfaces et variétés algébriques*, Paris (Gauthier-Villars), 1929.

LEFSCHETZ S., *Topology*, New York (American Mathematical Society), 1930.

LEFSCHETZ S., *On singular chains and cycles* (⟨55⟩, 39(1933), 124–129).

LEFSCHETZ S., *Lectures in algebraic geometry*, Princeton University, 1937.

LEFSCHETZ S., *Algebraic topology*, New York City (American Mathematical Society), 1942.

LEFSCHETZ S., *Topics in topology*, Princeton (University Press), 1942.

LEFSCHETZ S., *Existence of periodic solutions for certain differential equations* (⟨151⟩, 29(1943), 29–32).

LEFSCHETZ S., *A page of mathematical autobiography* (⟨55⟩, 74(1967), 854–879) = p. 13–38, *Selected papers*, Bronx, New York (Chelsea), 1971.

LEJA F., *Sur les groupes abstraits continus* (⟨19⟩, 3(1925), 153).

LEJA F., *Sur la notion de groupe abstrait topologique* (⟨80⟩, 9(1927), 37–44).

LE LIONNAIS F. (présentés par), *Les grands courants de la pensée mathématique*, nouvelle édition, Paris (Librairie scientifique et technique Albert Blanchard), 1962.

LELONG P., *Définition des fonctions plurisousharmoniques* (⟨66⟩, 215 (1942), 398–400).

LELONG P., *Les objets souples de l'analyse complexe* (⟨78⟩, 3(1985), 149–164).

LERAY J., *Essai sur les mouvements plans d'un liquide visqueux que limitent les parois* (⟨97⟩, (9), 13(1934), 331–418).

LERAY J., *Sur le mouvement d'un liquide visqueux emplissant l'espace* (⟨7⟩, 63 (1934), 193–248).

LERAY J., *La théorie de T. Carleman* (⟨173⟩, 2(1934–1935), 1935).

LERAY J., *Les complexes d'un espace topologique* (⟨66⟩, 214(1942), 781–783).

LERAY J., *Sur la forme des espaces topologiques et sur les points fixes des représentations* (⟨97⟩, (9), 24(1945), 95–248).

LERAY J., *L'anneau d'homologie d'une représentation* (⟨66⟩, 222(1946), 1366–1368).

LERAY J., *L'homologie filtrée*, p. 61–82, *Colloque international de topologie algébrique, Paris 26 juin – 2 juillet 1947*, Paris (C.N.R.S.), 1949.

LERAY J., *La vie et l'oeuvre de Serge Sobolev* (⟨192⟩, 7(1990), N° 6, 467–471).

LERAY J. et SCHAUDER J., *Topologie et équations fonctionnelles* (⟨66⟩, 197 (1933), 115–117) = p. 317–319, J. SCHAUDER, *Oeuvres*, Warszawa (PWN — Editions scientifiques de Pologne), 1978.

LERAY J. et SCHAUDER J., *Topologie et équations fonctionnelles* (⟨24⟩, (3), 51 (1934), 43–78).

LETTA G., PAPINI P.L. e PEPE L., *Cesare Arzelà e l'analisi reale in Italia*, p. XIII–XXVII, vol. I, C. ARZELÀ, *Opere*, Bologna (Tecnoprint), 1992.

LEVI B., *Sul principio di Dirichlet* (⟨158⟩, 22(1906), 293–360).

LEVI E.E., *Sulla struttura dei gruppi finiti e continui* (⟨37⟩, 40(1905), 551–565) = p. 101–115, vol. I, *Opere*, Roma (Edizioni Cremonese), 1959.

LEVI E.E., *Sopra una proprietà caratteristica delle funzioni armoniche* (⟨36⟩, (5), 18–1(1909), 10–15) = p. 180–186, vol. I, *Opere*, Roma (Edizioni Cremonese).

LEVI E.E., *Studii sui punti singolari essenziali delle funzioni analitiche di due o più variabili complesse* (⟨26⟩, (3), 17(1910), 61–87) = p. 187–213, vol. I, *Opere*, Roma (Edizioni Cremonese), 1959.

LEVI-CIVITA T., *Nozione di parallelismo in una varietà qualunque e conseguente specificazione geometrica della curvatura riemanniana* (⟨158⟩, 42(1917), 173–215) = p. 1–39, vol. IV, *Opere matematiche*, Bologna (Nicola Zanichelli), 1960.

LEVINSON N., *The asymptotic behaviour of a system of linear differential equations* (⟨12⟩, 68(1946), 1–6).

LÉVY P. *Leçons d'analyse fonctionnelle*, Paris (Gauthier-Villars), 1922.

LÉVY P., *Calcul des probabilités*, Paris (Gauthier-Villars), 1925.

LÉVY P., *Sur la convergence absolue des séries de Fourier* (⟨65⟩, 1(1934), 1–14) = p. 119–132, vol. II, *Oeuvres*, Paris (Gauthier-Villars), 1974.

LÉVY P., *Sur les intégrales dont les éléments sont des variables aléatoires indépendantes* (⟨25⟩, (2), 3(1934), 332–366) = p. 9–40, vol. IV, *Oeuvres*, Paris (Gauthier-Villars), 1980.

LÉVY P., *Détermination générale des lois limites* (⟨66⟩, 203(1936), 698–700) = p. 265–266, vol. III, *Oeuvres*, Paris (Gauthier-Villars), 1976.

LÉVY P., *Théorie de l'addition des variables aléatoires*, Paris (Gauthier-Villars), 1937.

LÉVY P., *Le mouvement brownien plan* (⟨12⟩, 62(1940), 487–550).

LÉVY P., *Processus stochastiques et le mouvement brownien*, Paris (Gauthier-Villars), 1948.

LÉVY P. *W. Doeblin (1915–1940)* (⟨163⟩, 1955, 107–115).

LÉVY P., *Nouvelle notice sur les travaux scientifiques*, Paris, 1964.

LÉVY P. *Les mathématiques*, p. 143–212, Académie des Sciences, *Troisième centenaire 1666–1966 I*, Paris (Gauthier-Villars), 1967. (*Index des noms cités*, 59–62, 1(1980), ⟨57⟩).

LÉVY P., *Quelques aspects de la pensée d'un mathématicien*, Paris (Blanchard), 1970.

LÉVY P., MANDELBROJT S., MALGRANGE B. et MALLIAVIN P., *La vie et l'oeuvre de Jacques Hadamard*, Genève (L'Enseignement Mathématique), 1967.

LEWIS C.I., *A survey of symbolic logic*, Berkeley (University of California Press), 1918 = New York (Dover), 1960.

LIAPOUNOV A.M., *Une propriété générale du calcul des probabilités* (⟨66⟩, 132 (1901), 814–815).

LIAPOUNOV A.M., *Sur les figures d'équilibre peu différentes des ellipsoïdes d'une masse liquide homogène douée d'un mouvement de rotation. Quatrième partie: Nouvelles formules pour la recherche des figures d'équilibre*, St. Pétersbourg (Académie impériale des Sciences), 1914.

LICHNEROWICZ A., *Problèmes globaux en mécanique relativiste*, Paris (Hermann), 1939.

LICHNEROWICZ A., *Sur une proposition fondamentale de la théorie relativiste de la gravitation* (⟨66⟩, 221(1945), 652–654).

LICHNEROWICZ A., *Courbures et nombres de Betti d'une variété riemanienne compacte* (⟨66⟩, 226(1948), 1678–1680).

LICHNEROWICZ A., *Quelques théorèmes de géométrie différentielle globale* (⟨63⟩, 22(1949), 271–301).

LICHNEROWICZ A., *Courbures, nombres de Betti et espaces symétriques*, p. 216–223, vol. II, *International Congress of Mathematicians, Cambridge, Massachusetts, U.S.A., 1950*, Providence (American Mathematical Society), 1952.

LICHNEROWICZ A., *Avant-propos du choix d'oeuvres mathématiques*, Paris (Hermann), 1982.

Lie groups: history, frontiers and applications, Brookline, Massachusetts (Math. Sci. Press), 1975-1976-1977-1978.

LINDVALL T., *W. Doeblin, 1915–1940* (⟨29⟩, 19(1991), 929–934).

LINNIK YU.V., *The large sieve* (⟨69⟩, 30(1941), 292–294).

LINNIK YU.V., *An elementary solution of Waring's problem by Schnirelman's method* (in Russian) (⟨117⟩, N.S., 12(54)(1943), 225–230).

LITTLEWOOD D.E., *The theory of group characters and matrix representations of groups*, New York (Oxford University Press), 1940.

LITTLEWOOD J.E., *The converse of Abel's theorem on power series* (⟨150⟩, (2), 9(1911), 434–448).

LITTLEWOOD J.E., *Sur la distribution des nombres premiers* (⟨66⟩, 158(1914), 1869–1872).

LOÈVE M., *Paul Lévy, 1886–1971* (⟨29⟩, 1(1973), 1–18).

LORCH E.R., *Normed rings — the first decade*, p. 249–258, *Proceedings of the symposium on spectral theory and differential problems*, Stillwater, Oklahoma (The Mathematics Department Oklahoma Agricultural and Mechanical College), 1951.

LOWE V., *The first period of Whitehead's work* (in Spanish) (⟨124⟩, 5(1989), 1–31, 299–316).

LÖWENHEIM L., *Über Möglichkeiten im Relativkalkül* (⟨120⟩, 76(1916), 447–470).

LUCE R. and RAIFFA H., *Games and decisions: introduction and critical survey*, New York (John Wiley), 1957.

ŁUKASIEWICZ J., *O logice trojwartosciowej* (⟨166⟩, 5(1920), 170–171).

LÜTZEN J., *Heaviside's operational calculus and the attempts to rigorise it* (⟨33⟩, 21(1979–1980), 161–200).

LÜTZEN J., *The prehistory of the theory of distributions*, New York (Springer-Verlag), 1982.

LUZIN N.N., *Sur les propriétés des fonctions mesurables* (⟨66⟩, 154(1912), 1688–1690).

LUZIN N.N., *On convergence of trigonometric Fourier series* (⟨66⟩, 156 (1913), 1655–1658) (in Russian) = *Selected works* (in Russian), p. 45–47, vol. I, Moscow (Izdatel'stvo Akademii Nauk SSSR), 1953.

LUZIN N.N., *Integral and trigonometric series* (in Russian), Moscow (Tip. Lissnera i Sobko), 1915 = Moscow (GITTL), 1951.

LUZIN N.N., *Leçons sur les ensembles analytiques et leurs applications*, Paris (Gauthier-Villars), 1930.

LYUSTERNIK L.A. et SCHNIRELMANN L.G., *Méthodes topologiques dans les problèmes variationnels*, Paris (Hermann), 1934.

MA Z.M. and LEE P.Y., *An overview of classical integration theory* (⟨115⟩, 18 (1990), n° 2, 49–61).

MAASS H., *Über eine neue Art von nichtanalytischen automorphen Funktionen und die Bestimmung Dirichletscher Reihen durch Funktionalgleichungen* (⟨120⟩, 121(1949), 141–183).

MACAULAY F.S., *On the resolution of a given modular system into primary systems including some properties of Hilbert numbers* (⟨120⟩, 74 (1913), 66–121).

MACAULAY F.S., *The algebraic theory of modular systems*, Cambridge (University Press), 1916.

MAC DUFFEE C.C., *The theory of matrices*, Berlin (Springer), 1933 = New York (Chelsea), 1946.

MACKEY G.W., *On convex topological linear spaces* (⟨151⟩, 29(1943), 315–319, 30(1944), 30).

MACKEY G.W., *Harmonic analysis as the exploitation of symmetry. A historical survey* (⟨164⟩, 64(1978), 73–228).

MACLANE S., *Some recent advances in algebra* (⟨13⟩, 46(1939), 3–19) = p. 9–34, ALBERT A.A. (ed.), *Studies in modern algebra*, The Mathematical Association of America, 1963.

MACLANE S., *Some additional advances in algebra*, ALBERT A.A. (ed.), *Studies in modern algebra*, The Mathematical Association of America, 1963.

MACLANE S., *Topology and logic as a source of algebra* (⟨55⟩, 82(1976), 1–40).

MACLANE S., *Origins of the cohomology of groups* (⟨74⟩, (2), 24(1978), 1–29).

MAGNUS W., *A connection between the Baker-Hausdorff formula and a problem of Burnside* (⟨28⟩, (2), 52(1950), 111–126).

MAGNUS W., KARRASS A. and SOLITAR D., *Combinatorial group theory*, New York (Interscience), 1966.

MAHLER K., *Über das Mass der Menge aller S-Zahlen* (⟨120⟩, 106(1932), 131–139).

MAHLER K., *On lattice points in n-dimensional star bodies. I. Existence theorems* (⟨154⟩, 187(1946), 151–187).

MALCEV A.I., *Untersuchungen aus dem Gebiet der mathematischen Logik* (⟨117⟩, (2), 1(1936), 323–336).

MALCEV A.I., *On a general method for obtaining local theorems in group theory* (in Russian) (⟨89⟩, 1(1941), n° 1, 3–9).

MALLIAVIN P., *Olaf Arnfinn Laudal* (⟨136⟩, mars-avril 1991, 30–31).

MANDELBROJT S., *Séries de Fourier et classes quasi-analytiques de fonctions*, Paris (Gauthier-Villars), 1935.

MANDELBROJT S., *Notice sur les titres et travaux scientifiques*, Paris (Gauthier-Villars), 1964.

MANDELBROJT S., *Souvenirs à bâtons rompus, recueillis en 1970 et préparés par Benoît Mandelbrot* (⟨57⟩, 6(1985), 1–46).

MANDELKERN M., *Brouwerian counterexamples* (⟨114⟩, 62(1989), 3–27).

MANHEIM J.H., *The genesis of point set topology*, Oxford (Pergamon Press), 1964.

MANN H.B., *A proof of the fundamental theorem on the density of sums of sets of positive integers* (⟨28⟩, (2), 43(1942), 523–527).

MARCINKIEWICZ J., *Sur la convergence absolue des séries de Fourier* (⟨111⟩, 16(1940), 66–73).

MARDZHANISHVILI K.K., *Ivan Matveevič Vinogradov: a brief outline of his life and works*, p. 385–392, I.I. VINOGRADOV, *Selected works*, Berlin (Springer-Verlag), 1985.

MARKOV A.A., *Extension of the law of large numbers to mutually independent quantities* (in Russian) (⟨91⟩, (2), 15(1906), n° 4, 135–176) = p. 339–361, *Selected works* (in Russian), Moscow (Izdatel'stvo Akademii Nauk SSSR), 1951.

MARKOV A.A., *Calcul des probabilités*, 2e éd., Saint-Pétersbourg, 1908 = *Wahrscheinlichkeitsrechnung*, Leipzig (Teubner), 1912.

MARKOV A.A., *Sur les espaces vectoriels considérés comme groupes topologiques* (⟨66⟩, 197(1933), 610–612).

MARKOV A.A., *Über die freie Äquivalenz geschlossener Zöpfe* (⟨117⟩, 1(1936), 73–78).

MARKOV A.A., *On impossibility of certain algorithms in the theory of associative systems* (in Russian) (⟨69⟩, (N.S.), 55(1947), 583–586; 58 (1947), 553–556).

MARKOV A.A., *Selected works* (in Russian), Moscow (Izdatel'stvo Akademii Nauk SSSR), 1951.

MARTIN R.S., *Minimal positive harmonic functions* (⟨185⟩, 49(1941), 137–172).

MASANI P.R., *Norbert Wiener: 1894–1964*, Basel (Birkhäuser), 1990.

MASSERA J.L., *The existence of periodic solutions of systems of differential equations* (⟨70⟩, 17(1950), 457–475).

MASSERA J.L., *Remarks on the periodic solutions of differential equations* (in Spanish) (⟨44⟩, 4(Ano 14)(1950), 37–45) = (⟨79⟩, 2(1950), 43–53).

Mathematical developments arising from Hilbert problems, Providence, Rhode Island (American Mathematical Society), 1976.

20-volume author index of Mathematical Reviews 1940–1959, Providence (American Mathematical Society), 1961.

Mathematics in USSR during thirty years 1917–1947 (in Russian), Moskva (OGIZ), 1948.

Mathématiques, Congrès de la victoire, 20 au 26 octobre 1945, Association Française pour l'Avancement des Sciences, 64ème session, supplément au fascicule 9 (janvier 1947) de l'*Intermédiaire des Recherches Mathématiques.*

Les mathématiques, p. 13–127, vol. II, tome III, *Le XXe siècle*, R. TATON (publié sous la direction de), *Histoire générale des Sciences*, 2e édition, Paris (Presses Universitaires de France), 1983.

MATHIAS M., *Über positive Fourier-Integrale* (⟨123⟩, 16(1923), 103–125).

MAWHIN J., *Présences des sommes de Riemann dans l'évolution du calcul intégral* (⟨57⟩, 4(1983), 117–148).

MAWHIN J., *The forced pendulum: A paradigm for nonlinear analysis and dynamical systems* (⟨78⟩, 6(1988), 271–287).

MAWHIN J., *Problème de Cauchy pour les équations différentielles et théories de l'intégration: influences mutuelles* (⟨57⟩, 9(1988), 231–246).

MAWHIN J., *Analyse. Fondements, techniques, évolution*, Bruxelles (De Boeck), 1992.

MAYER J., *Le théorème des quatre couleurs: notice historique et aperçu technique* (⟨57⟩, 3(1982), 43–62).

MAZUR S., *Über die kleinste konvexe Menge die eine gegebene kompakte Menge enthält* (⟨181⟩, 2(1930), 7–9).

MAZUR S., *Sur les anneaux linéaires* (⟨66⟩, 207(1938), 1025–1027).

MCHUGH J.A.M., *A historical survey of ordinary linear differential equations with a large parameter and turning points* (⟨33⟩, 7(1970–1971), 277–324).

MCSHANE E.J., *Recent developments in the calculus of variations*, p. 69–97, *Semicentennial addresses of the American Mathematical Society*, New York, 1938.

MCSHANE E.J., *The calculus of variations from the beginning through optimal control theory*, p. 3–49, SCHWARZKOPF A.B., KELLEY W.G. and ELIASON S.B. (eds), *Optimal control and differential equations*, Proc. Conf. Optimal control and differential equations, Oklahoma (Norman), 1977.

MEDVEDEV F.A., *The development of the idea of Stieltjes integral* (in Russian) (⟨88⟩, 15(1961), 171–224).

MEDVEDEV F.A., *A.M. Liapunov's contribution to the Stieltjes integral theory* (in Russian) (⟨88⟩, 14(1961), 211–234).

MEDVEDEV F.A., *Development of the notion of integral* (in Russian), Moscow (Nauka), 1974.

MEDVEDEV F.A., *The French school of the theory of functions and sets at the turn of the twenthieth century* (in Russian), Moscow (Nauka), 1976.

MEDVEDEV F.A., *History of the origins of the axiom of choice* (in Russian), Moscow (Nauka), 1982.

MEDVEDEV F.A., *Scenes from the history of real functions*, Basel (Birkhäuser), 1991.

MEHRTENS H., *Die Entstehung der Verbandstheorie*, Hildesheim (Gerstenberg Verlag), 1979.

MEHRTENS H., *Felix Hausdorff: Ein Mathematiker in seiner Zeit*, Fachschaftsrat Mathematik und das Mathematische Institut der Universität Bonn, 1980.

MENCHOV D.E., *Sur l'unicité du développement trigonométrique* (⟨66⟩, 163 (1916), 433–436).

MENGER K., *Dimensionstheorie*, Leipzig (Teubner), 1928.

MENGER K., *Kurventheorie*, Leipzig (B.G. Teubner), 1932.

MENGER K. und NÖBELING G., *Über den n-Beinsatz in lokalzusammenhängenden Kontinua* (⟨10⟩, 67(1930), 86–88).

MÉTIVIER M., NEVEU J. et VARADHAN S.R.S. (éditeurs), *Colloque Paul Lévy sur les processus stochastiques*, Astérisque 157–158, 1988, Paris (Société Mathématique de France).

METROPOLIS N., HOWLETT J. and ROTA G.-C. (ed.), *A history of computing in the twentieth century*, New York (Academic Press), 1980.

MICHEL A., *Constitution de la théorie moderne de l'intégration*, Paris (Vrin), 1992.

MICHLIN S., *Singular integral equations with respect to two independent variables* (in Russian) (⟨117⟩, (N.S.), 1(43)(1936), 535–552).

MIKOLAS M., *Some historical aspects of the development of mathematical analysis in Hungary* (⟨83⟩, 2(1975), 304–308).

MIKUSINSKI J.G., *Sur la méthode de généralisation de Laurent Schwartz et sur la convergence faible* (⟨80⟩, 35(1948), 235–239).

MILLER G.A., *Note on the history of group theory during the period covering this volume*, p. 1–18, vol. II; p. 1–15, vol. III, *Collected works*, Urbana (University of Illinois), 1938, 1946.

MILLIONŚČIKOV M., *Decay of homogeneous isotropic turbulence in a viscous incompressible fluid* (⟨69⟩, (2), 22(1939), 231–235).

MINKOWSKI H., *Raum und Zeit* (Physikalische Zeitschrift, 10(1909), 104–111) = (⟨96⟩, 18(1909), 75–88) = Leipzig (B.G. Teubner), 1909 = S. 431–444, II. Band, *Gesammelte Abhandlungen*, Leipzig (B.G. Teubner), 1911.

MINKOWSKI H., *Geometrie der Zahlen*, Leipzig (Teubner), 1910.

MINKOWSKI H., *Theorie der konvexen Körper, insbesondere Begründung ihres Oberfächenbegriffs*, S. 131–229, II. Band, *Gesammelte Abhandlungen*, Leipzig (B.G. Teubner), 1911.

MINTS G., *Proof theory in the URSS 1925–1969* (⟨101⟩, 56(1991), 385–424).

MIRANDA C., *Renato Caccioppoli* (⟨26⟩, (4), 47(1959), V–VII).

MIRIMANOV D., *Les antinomies de Russel et de Burali-Forti et le problème fondamental de la théorie des ensembles* (⟨74⟩, 19(1917), 37–52).

VON MISES R., *Grundlagen der Wahrscheinlichkeitsrechnung* (⟨123⟩, 5 (1919), 52–99).

MITROPOL'SKIJ YU.A., *The ideas of Krylov and Bogolyubov in the theory of differential equations and mathematical physics and their development* (⟨190⟩, 42(1990), N° 3, 259–269).

MONNA A.F., *Sur une généralisation du problème de Dirichlet pour ensembles bornés mesurables (B) quelconques* (⟨107⟩, 42(1939), 745–752).

MONNA A.F., *On weak convergence and strong convergence in the p-adic Banach space* (in Dutch) (Nederl. Akad. Wetensch. Verslagen, Afd. Natuurkunde, 52(1943), 207–211).

MONNA A.F., *Remarques sur les métriques non-archimédiennes* (⟨85⟩, 12(1950), 122–133, 179–191).

MONNA A.F., *The concept of function in the 19th and 20th centuries, in particular with regard to the discussion between Baire, Borel and Lebesgue* (⟨33⟩, 9(1972–1973), 57–84).

MONNA A.F., *Functional analysis in historical perspective*, Utrecht (Oosthoek), 1973.

MONNA A.F., *Banach's Théorie des opérateurs linéaires* (⟨131⟩, (3), 23(1975), 67–71).

MONNA A.F., *Dirichlet's principle. A mathematical comedy of errors and its influence on the development of analysis*, Utrecht (Oosthoek, Scheltema and Holkema), 1975.

MONNA A.F., *Methods, concepts and ideas in mathematics*, Amsterdam (Math. Center), 1985.

MONNA A.F., *A la frontière de deux mondes. Expériences mathématiques dans une période de transition* (in Dutch), Amsterdam (Centrum voor Wiskunde en Informatica), 1989.

MONTEL P., *Intégration et dérivation*, p. 171–209, tome II, vol. 1, *Encyclopédie des Sciences mathématiques*, Paris (Gauthier-Villars), 1909.

MONTEL P., *Sur les fonctions analytiques qui admettent deux valeurs exceptionnelles dans un domaine* (⟨66⟩, 153(1911), 996–998).

MONTEL P., *Sur les familles de fonctions analytiques qui admettent des valeurs exceptionnelles dans un domaine* (⟨24⟩, (3), 29(1912), 487–535).

MONTEL P., *Leçons sur les familles normales de fonctions analytiques et leurs applications*, Paris (Gauthier-Villars), 1927.

MONTGOMERY D., *Oswald Veblen* (⟨55⟩, 69(1963), 26–36).

MONTGOMERY D. and ZIPPIN L., *Topological transformation groups I* (⟨28⟩, (2), 41(1940), 778–791).

MOORE E.H., *On a form of general analysis with application to linear differential and integral equations*, p. 98–114, vol. 2, *Atti del IV Congresso internazionale dei matematici (Roma, 6–11 aprile 1908)*, Roma (R. Accademia dei Lincei), 1909.

MOORE E.H., *Introduction to a form of general analysis*, New Haven (Yale University Press), 1910.

MOORE E.H., *Definition of limit in general integral analysis* (⟨151⟩, 1(1915), 628–632).

MOORE E.H. and SMITH H.L., *A general theory of limits* (⟨12⟩, 44(1922), 102–121).

MOORE G.H., *Beyond first-order logic: the historical interplay between mathematical logic and axiomatic set theory* (⟨84⟩, 1(1980), 95–137).

MOORE G.H., *Zermelo's axiom of choice. Its origins, development, and influence*, New York (Springer-Verlag), 1982.

MOORE G.H., *Bernays, Paul Isaac*, p. 75–78, vol. 17, supplement II, *Dictionary of scientific biography*, New York (Charles Scribner's Sons), 1990.

MOORE G.H., *Sixty years after Gödel* (⟨112⟩, 13(1991), N° 3, 6–10).

MORDELL L.J., *On Mr. Ramanujan's empirical expansion of modular functions* (⟨148⟩, 19(1917), 117–124).

MORDELL L.J., *On the rational solutions of the indeterminate equations of the third and fourth degrees* (⟨148⟩, 21(1922), 179–192).

MORGAN M.S., *The history of econometric ideas*, Cambridge (University Press), 1990.

MORREY C.B. JR., *On the solutions of quasilinear elliptic partial differential equations* (⟨185⟩, 43(1938), 126–166).

MORSE M., *Relations between the critical points of a real function of n independent variables* (⟨185⟩, 27(1925), 345–396) = p. 77–128, *Collected papers*, Singapore (World Scientific), 1987.

MORSE M., *The calculus of variations in the large*, New York (American Mathematical Society), 1934.

MORSE M., *George David Birkhoff and his mathematical work* (⟨55⟩, 52(1946), 357–391).

MOSTOWSKI A., *Thirty years of foundational studies*, New York (Branes and Noble), 1966.

MÜLLER G.H. (edited by), *Sets and classes. On the work by Paul Bernays*, Amsterdam (North-Holland), 1976.

MURRAY F.J. and VON NEUMANN J., *On rings of operators* (⟨28⟩, (2), 37(1936), 116–229) = p. 6–119, vol. III, J. VON NEUMANN, *Collected papers*, Oxford (Pergamon Press), 1961.

MYCIELSKI J., *Commentaire*, p. 325–327, vol. I, S. BANACH, *Oeuvres*, Warszawa (PWN — Editions Scientifiques de Pologne), 1967.

MYERS S.B., *Riemannian manifolds in the large* (⟨70⟩, 1(1935), 39–49).

NAGELL M.T., *L'analyse indéterminée de degré supérieur*, Paris (Gauthier-Villars), 1929.

NAGUMO M., *Über die Differentialgleichung* $y'' = f(x, y, y')$ (⟨152⟩, (3), 19 (1937), 861–866).

NAGUMO M., *Notice* (Kansu-Hoteiski, n° 15, 1939) (in Japanese).

NETTO E., *Lehrbuch der Kombinatorik*, Leipzig (Teubner), 1901.

VON NEUMANN J., *Eine Axiomatisierung der Mengenlehre* (⟨98⟩, 154 (1925), 219–240) = p. 35–56, vol. I, *Collected works*, Oxford (Pergamon Press), 1961.

VON NEUMANN J., *Die Axiomatisierung der Mengenlehre* (⟨123⟩, 27 (1928), 669–752) = p. 339–422, vol. I, *Collected works*, Oxford (Pergamon Press), 1961.

VON NEUMANN J., *Sur la théorie des jeux*, note présentée par E. Borel (⟨66⟩, 186(1928), 1689–1691).

VON NEUMANN J., *Zur Theorie der Gesellschaftsspiele* (⟨120⟩, 100(1928), 295–320) = p. 1–26, vol. VI, *Collected works*, Oxford (Pergamon Press), 1963.

VON NEUMANN J., *Zur allgemeinen Theorie des Masses* (⟨80⟩, 13(1929), 73–116) = p. 599–643, vol. I, *Collected works*, Oxford (Pergamon Press), 1961.

VON NEUMANN J., *Allgemeine Eigenwerttheorie Hermitischer Funktionaloperatoren* (⟨120⟩, 102(1929), 49–131) = p. 3–85, vol. II, *Collected works*, Oxford (Pergamon Press), 1961.

VON NEUMANN J., *Zur Algebra der Funktionaloperationen und Theorie der normalen Operatoren* (⟨120⟩, 102(1929), 370–427) = p. 86–143, vol. II, *Collected works*, Oxford (Pergamon Press), 1961.

VON NEUMANN J., *Zur Algebra der Funktionaloperationen und Theorie der normalen Operatoren* (⟨120⟩, 102(1930), 370–427) = p. 86–143, vol. II, *Collected works*, Oxford (Pergamon Press), 1961.

VON NEUMANN J., *Die Eindeutigkeit der Schrödingerschen Operatoren* (⟨120⟩, 104(1931), 570–578) = p. 220–229, vol. II, *Collected works*, Oxford (Pergamon Press), 1961.

VON NEUMANN J., *Mathematische Grundlagen der Quantenmechanik*, Berlin (J. Springer), 1932.

VON NEUMANN J., *Proof of the quasi-ergodic hypothesis* (⟨151⟩, 18(1932), 70–82) = p. 260–273, vol. II, *Collected works*, Oxford (Pergamon Press), 1961.

VON NEUMANN J., *Die Einführung analytischer Parameter in topologischen Gruppen* (⟨28⟩, 34(1933), 170–190) = p. 366–386, vol. II, *Collected works*, Oxford (Pergamon Press, 1961.

VON NEUMANN J., *Zum Haarschen Mass in topologischen Gruppen* (⟨65⟩, 1 (1934), 106–114) = p. 445–453, vol. II, *Collected works*, Oxford (Pergamon Press), 1961.

VON NEUMANN J., *Almost periodic functions on a group I* (⟨185⟩, 36 (1934), 445–492) = p. 454–601, vol. II, *Collected works*, Oxford (Pergamon Press, 1961.

von Neumann J., *On complete topological spaces* (⟨185⟩, 37(1935), 1–20) = p. 508–527, vol. II, *Collected works*, Oxford (Pergamon Press), 1961.

von Neumann J., *The uniqueness of Haar's measure* (⟨117⟩, 1(1936), 721–734) = p. 91–104, vol. IV, *Collected works*, Oxford (Pergamon Press, 1962.

von Neumann J., *Communication on the Borel notes* (⟨71⟩, 21(1953), 124–125) = p. 27–28, vol. VI, *Collected works*, Oxford (Pergamon Press), 1963.

von Neumann J. and Morgenstern O., *Theory of games and economic behavior*, Princeton University Press, 1944.

Nevanlinna F. und R., *Über die Eigenschaften analytischer Funktionen in der Umgebung einer singulären Stelle oder Linie* (⟨9⟩, 50(1922), Nr.5).

Nevanlinna R., *Sur les fonctions méromorphes* (⟨66⟩, 178(1924), 367–370).

Nevanlinna R., *Zur Theorie der meromorphen Funktionen* (⟨7⟩, 46 (1925), 1–99).

Nevanlinna R., *Le théorème de Picard-Borel et la théorie des fonctions méromorphes*, Paris (Gauthier-Villars), 1929.

Nevanlinna R., *Souvenirs* (in Finnish), Helsinki, 1976.

Neyman J. and Pearson E.S., *On the problem of the most efficient tests of statistical hypotheses* (⟨143⟩, 231(1933), 289–337).

Nikodým O., *Sur le principe de minimum dans le problème de Dirichlet* (⟨19⟩, 10(1912), 120–121).

Nikodým O., *Sur une généralisation des intégrales de M. J. Radon* (⟨80⟩, 15 (1930), 131–179).

Noether E., *Invariante Variationsprobleme* (⟨130⟩, 1918, 235–257).

Noether F., *Über eine Klasse singulärer Integralgleichungen* (⟨120⟩, 82(1921), 42–63).

Noether E., *Idealtheorie in Ringbereichen* (⟨120⟩, 83(1921), 24–66).

Noether E., *Abstrakter Aufbau der Idealtheorie in algebraischen Zahl- und Funktionenkörpern* (⟨120⟩, 96(1926), 26–61).

Noether E., *Hyperkomplexe Grösse und Darstellungstheorie* (⟨123⟩, 30(1929), 641–692).

Oelschlägel D., *Entwicklungstendenzen der numerischen Mathematik in unserem Jahrhundert* (⟨193⟩, 38(1989), N° 4, 68–69).

Oka K., *Domaines convexes par rapport aux fonctions rationnelles* (⟨100⟩, 6 (1936), 245–255) = *Rationaly convex domains*, p. 1–9, *Collected papers*, Berlin (Springer-Verlag), 1984.

Oka K., *Deuxième problème de Cousin* (⟨100⟩, 9(1939), 7–19) = *The second Cousin problem*, p. 24–34, *Collected papers*, Berlin (Springer-Verlag), 1984.

Oka K., *Domaines pseudoconvexes* (⟨184⟩, 49(1942), 15–52) = *Pseudoconvex domains*, p. 48–76, *Collected papers*, Berlin (Springer-Verlag), 1984.

OKA K., *Sur les fonctions analytiques de plusieurs variables (VII. Sur quelques notions arithmétiques)* (⟨49⟩, 78(1950), 1–27) = *On some arithmetical notions*, p. 80–106, *Collected papers*, Berlin (Springer-Verlag), 1984.

OKA K., *Collected papers*, Berlin (Springer-Verlag), 1984.

ONSAGER L., *Crystal statistics. I. A two-dimensional model with an order-disorder transition* (⟨145⟩, (2), 65(1944), 117–149).

ORE Ø., *Some recent developments in abstract algebra* (⟨55⟩, 37(1931), 537–548).

ORE Ø., *L'algèbre abstraite*, Paris (Hermann), 1936.

OSTROWSKI A., *Über einige Lösungen der Funktionalgleichung* $\varphi(x)\varphi(y) = \varphi(xy)$ (⟨7⟩, 41(1917), 271–284) = p. 322–335, vol. II, *Collected mathematical papers*, Basel (Birkhäuser), 1983.

OSTROWSKI A., *Untersuchungen zur arithmetischen Theorie der Körper (Die Theorie der Teilbarkeit in allgemeinen Körpern)* (⟨123⟩, 39(1934), 269–404) = p. 350–485, vol. II, *Collected mathematical papers*, Basel (Birkhäuser), 1983.

OWEN D.B. (ed.), *On the history of statistics and probability*, New York (Marcel Dekker), 1976.

OXTOBY J.C., PETTIS B.J. and PRICE G.B. (edited by), *John von Neumann 1903–1957* (⟨55⟩, 64(1958), n° 3, part 2) = Providence, RI (American Mathematical Society), 1988.

PAINLEVÉ P., *Analyse des travaux scientifiques jusqu'à 1900*, Paris (Gauthier-Villars), 1900 = p. 75–196, vol. I, *Oeuvres*, Paris (C.N.R.S.), 1972.

PAINLEÉ P., *Mémoire sur les équations différentielles dont l'intégrale est uniforme* (⟨49⟩, 28(1900), 201–261) = p. 123–183, vol. III, *Oeuvres*, Paris (C.N.R.S.), 1975.

PALEY R.E.A.C. and WIENER N., *Fourier transforms in the complex domain*, Providence, R.I. (American Mathematical Society), 1934.

PALEY R.E.A.C., WIENER N. and ZYGMUND A., *Notes on random functions* (⟨123⟩, 37(1932), 647–668).

PALEY R.E.A.C. and ZYGMUND A., *On some series of functions* (⟨148⟩, 26 (1930), 337–357, 458–474; 28(1932), 190–205).

PALM C., *Intensitätsschwankungen im Fernsprechverkehr* (⟨77⟩, n° 44, 1943).

PAPLAUSKAS A.B., *Trigonometric series from Euler to Lebesgue* (in Russian), Moscow (Nauka), 1966.

PARIKH C., *The unreal life of Oscar Zariski*, Boston (Academic Press), 1991.

PARSHAL B.J., *Nathan Jacobson: collected mathematical papers* (⟨83⟩, 18(1991), 378–380).

PARSHALL K.H., *Joseph H.M. Wedderburn and the structure theory of algebras* (⟨33⟩, 32(1985), 223–349).

PARSHALL K.V.H., *A study in group theory: Leonard Eugene Dickson's Linear groups* (⟨112⟩, 13(1991), n° 1, 7–11).

PARSHALL K.V.H., *A survey of modern algebra. The fiftieth anniversary of its publication. Garett Birkhoff and Saunders Mac Lane* (⟨112⟩, 14(1992), N° 1, 26–31).

PAUC C., *Introduction historique*, p. 1–14, *Les méthodes directes en calcul des variations et en géométrie différentielle*, Paris (Hermann), 1941.

PEARSON K., *On the criterion that a given system of deviations from the probable in the case of a correlated system of variables is such it can be reasonably supposed to have arisen from random sampling* (⟨142⟩, (5), 50(1900), 157–175).

PECKHAUS V., *Hilbertprogramm und kritische Philosophie. Das Göttinger Modell interdisziplinären Zusammenarbeit zwischen Mathematik und Philosophie*, Göttingen (Vandenhoeck und Ruprecht), 1990.

PELCZYNSKI A. and BESSAGA C., *Some aspects of the present theory of Banach spaces*, p. 221–302, vol. II, S. BANACH, *Oeuvres*, Warszawa (PWN — Editions scientifiques de Pologne), 1979.

PERES J., *Les sciences exactes*, Paris (E. de Boccard), 1930.

PERRON O., *Zur Theorie der Matrices* (⟨120⟩, 64(1907), 248–263.

PERRON O., *Die Lehre von Kettenbrüchen*, Leipzig (B.G. Teubner), 1912.

PERRON O., *Über den Integralbegriff* (⟨176⟩, Nr.14, S. 16, 1914).

PERRON O., *Ein neuer Existenzbeweis für die Integrale der Differentialgleichung* $y' = f(x,y)$ (⟨120⟩, 76(1915), 471–484).

PERRON O., *Eine neue Behandlung der ersten Randwertaufgabe für* $\Delta u = 0$ (⟨123⟩, 18(1923), 42–54).

PESIN I.N., *Classical and modern integration theories*, New York (Academic Press), 1970.

PETER F. und WEYL H., *Die Vollständigkeit der primitiven Darstellungen einer geschlossenen kontinuierlichen Gruppe* (⟨120⟩, 97(1927), 737–755) = S. 58–75, Band III, H. WEYL, *Gesammelte Abhandlungen*, Berlin (Springer-Verlag), 1968.

PETER R., *Rekursive Funktionen*, S. 336–337, II. Band, *Verhandlungen des Internationalen Mathematiker-Kongresses Zürich 1932*, Zürich (Orell Füssli).

PETERSSON H., *Das wissenschaftliche Werk von E. Hecke* (⟨1⟩, 16(1949), 7–31).

PETROVSKY I.G., *Sur la topologie des courbes réelles et algébriques* (⟨66⟩, 197 (1933), 1270–1273).

PETROVSKY I.G., *Über das Cauchysche Problem für Systeme von partiellen Differentialgleichungen* (⟨117⟩, (2), 2(1937), 815–870).

PETROVSKY I.G., *On the topology of real plane algebraic curves* (⟨28⟩, (2), 39 (1938), 189–209).

PETROVSKY I.G., *Über das Cauchysche Problem für ein System linearer partieller Differentialgleichungen im Gebiet der nicht analytischen Funktionen* (⟨52⟩, 1(1938), Nr 7, 1–74).

PETROVSKY I.G., *On the diffusion of waves and the lacunas for hyperbolic equations* (⟨117⟩, N.S., 17(1945), 289–370).

PETROVSKY I.G. et OLEINIK O.A., *Topology of real algebraic surfaces* (in Russian) (⟨90⟩, 13(1949), 389–402).

PHILLIPS E.R., *Nicolai Nicolaevich Luzin and the Moscow school of the theory of functions* (⟨83⟩, 5(1978), 275–305).

PHILLIPS E.R. (editor), *Studies in the history of mathematics*, The Mathematical Association of America, 1987.

PHRAGMÉN E. et LINDELÖF E., *Sur une extension d'un principe classique de l'analyse et sur quelques propriétés de fonctions monogènes dans le voisinage d'un point singulier* (⟨7⟩, 31(1908), 381–406).

PICARD E. et SIMART G., *Théorie des fonctions algébriques de deux variables indépendantes*, tome II, Paris (Gauthier-Villars), 1906.

PICONE M., *Sulla vita et sulle opere di Eugenio Elia Levi*, p. V–XVI, vol. I, E.E. LEVI, *Opere*, Roma (Edizioni Cremonese), 1959.

PIER J.-P., *Historique de la notion de compacité* (⟨83⟩, 7(1980), 425–443).

PIER J.-P., *Marc Krasner* (⟨57⟩, 7(1986), 43–44).

PIER J.-P., *L'apparition de la théorie des groupes topologiques* (⟨57⟩, 9(1988), 1–22).

PIER J.-P., *L'analyse harmonique. Son développement historique*, Paris (Masson), 1990.

PIER J.-P., *Aspects de l'évolution de l'analyse mathématique 1900–1950* (⟨188⟩, II(1990), 45–161).

PIER J.-P., *Repères de l'évolution de l'algèbre 1900–1950* (⟨188⟩, III(1991), 51–65).

PIER J.-P., *Repères de l'évolution du calcul des probabilités 1900–1950* (⟨188⟩, III(1991), 67–78).

PIER J.-P., *Aspects de l'évolution des mathématiques entre 1900 et 1950. A. Fondements des mathématiques. B. Théorie des nombres. C. Géométrie différentielle et topologie algébrique* (⟨188⟩, IV(1992), 183–236).

PIER J.-P. (published by), *L'émergence de l'analyse harmonique abstraite (1930–1950)* (⟨57⟩, (2), 2(1992)).

PIETSCH A., *Banach Räume* (⟨95⟩, 1992, 1–26).

PINCHERLE S., *La matematica e il futuro*, p. 61–84, *Annuario della Regia Università di Bologna per l'anno accademico 1915–1916*, Bologna, 1916.

PISOT C., *La répartition modulo 1 et les nombres algébriques* (⟨25⟩, (2), 7(1938), 205–248).

PLANCHEREL M., *Contribution à l'étude de la représentation d'une fonction arbitraire par des intégrales définies* (⟨158⟩, 30(1910), 289–335).

PLANCHEREL M., *Le développement de la théorie des séries trigonométriques dans le dernier quart de siècle* (⟨74⟩, 24(1924–1925), 19–58).

PLANCHEREL M., *Mathématiques et mathématiciens en Suisse (1850–1950)* (⟨74⟩, (2), 6(1960), 194–218).

PLEMELJ J., *Riemannsche Funktionenscharen mit gegebener Monodromiegruppe* (⟨129⟩, 19(1908), 211–246).

J.C. Poggendorff's biographisch-literarisches Handwörterbuch zur Geschichte der exacten Wissenschaften, Band IV, Leipzig (Johan Ambrosius Barth), 1904; Band V, Leipzig (Chemie), 1926; Band VI, Leipzig (Chemie), 1936–1940; Band VII A, Berlin (Akademie-Verlag), 1956–1961; Band VII B, Berlin (Akademie-Verlag), 1967–1989.

POINCARÉ H., *Second complément à l'analysis situs* (⟨150⟩, 32(1900), 277–308) = p. 338–370, vol. VI, *Oeuvres*, Paris (Gauthier-Villars), 1953.

POINCARÉ H., *Sur les propriétés arithmétiques des courbes algébriques* (⟨97⟩, (5), 7(1901), 161–233) = p. 483–548, vol. V, *Oeuvres*, Paris (Gauthier-Villars), 1950.

POINCARÉ H., *L'espace et ses trois dimensions* (⟨162⟩, 11(1903), 281–301, 407–429).

POINCARÉ H., *Cinquième complément à l'analysis situs* (⟨158⟩, 18 (1904), 45–110) = p. 435–498, vol. VI, *Oeuvres*, Paris (Gauthier-Villars), 1953.

POINCARÉ H., *Sur les lignes géodésiques des surfaces convexes* (⟨185⟩, 6(1905), 237–274) = p. 38–85, vol. VI, *Oeuvres*, Paris (Gauthier-Villars), 1953.

POINCARÉ H., *Les fonctions analytiques de deux variables et la représentation conforme* (⟨158⟩, 23(1907), 185–220) = p. 244–289, vol. IV, *Oeuvres*, Paris (Gauthier-Villars), 1950.

POINCARÉ H., *L'avenir des mathématiques*, p. 167–182, vol. I, *Atti del IV Congresso internazionale dei Matematici (Roma, 6–11 Aprile 1908)*, Roma (R. Accademia dei Lincei), 1909.

POINCARÉ H., *Calcul des probabilités*, 2e édition, Paris (Gauthier-Villars), 1912.

POINCARÉ H., *Sur un théorème de géométrie* (⟨158⟩, 33(1912), 375–407) = p. 499–538, vol. VI, *Oeuvres*, Paris (Gauthier-Villars), 1953.

VAN DER POL B., *De invloed van een geioniseerd gas op het voortschrijden van electromagnetische golven en toepassingen daarvan op het gebied der draadlooze telegraphie en bij metingen aan glimlichtontladingen*, Haarlem (Joh. Enschede en Zonen), 1920 = p. 129–219, *Selected scientific papers*, Amsterdam (North-Holland), 1960.

POLAK E., *A historical survey of computational methods in optimal control* (⟨174⟩, 15(1973), 553–584).

POLICHTCHUK E.M. et CHAPOCHNIKOVA J.O., *Jacques Hadamard 1865–1963* (in Russian), Leningrad (Nauka), 1990.

POLLACZEK F., *Über eine Aufgabe der Wahrscheinlichkeitstheorie* (⟨123⟩, 32 (1930), 64–100, 729–750, 796).

PÓLYA G., *Sur une question concernant les fonctions entières* (⟨66⟩, 158(1914), 330–333).

PÓLYA G., *Über die Analogie der Kristallsymetrie in der Ebene* (⟨195⟩, 60(1924), 278–282).

PÓLYA G. und SZEGÖ G., *Aufgaben und Lehrsätze aus der Analysis*, Berlin (Springer), 1925.

POMPEIU D., *Sur une classe de fonctions d'une variable complexe* (⟨158⟩, 33 (1912), 108–113).

PONTRYAGIN L.S., *Sur les groupes topologiques et le cinquième problème de Hilbert* (⟨66⟩, 198(1934), 238–240).

PONTRYAGIN L.S., *The general topological theorem of duality for closed sets* (⟨28⟩, (2), 35(1934), 904–914).

PONTRYAGIN L.S., *Sur les groupes abéliens continus* (⟨66⟩, 198(1934), 328–330).

PONTRYAGIN L.S., *The theory of topological commutative groups* (⟨28⟩, (2), 35 (1934), 361–388).

PONTRYAGIN L.S., *Continuous groups* (in Russian), Moscow (GITTL), 1938 = *Topological groups*, Princeton (University Press), 1939.

PONTRYAGIN L.S., *The homotopy group* $\pi^{n+1}(K_n)$ *(*$n \leq 2$*) of dimension* $n+1$ *of a connected finite polyhedron* K_n *of arbitrary dimension, whose fundamental group and Betti groups of dimension* $2, \dots, n-1$ *are trivial* (in Russian) (⟨69⟩, (N.S.), 65(1949), 797–800).

PONTRYAGIN L.S., *Homotopic classification of the mappings of the* $(n+2)$*-dimension sphere on the* n*-dimension sphere* (in Russian) (⟨69⟩, 70 (1950), 957–959).

PORTER T.I., *A history of the classical isoperimetric problem*, p. 475–517, vol. II, *University of Chicago contributions to the calculus of variations*, 1933.

POST E.L., *Introduction to a general theory of elementary propositions* (⟨12⟩, 43(1921), 163–185).

POST E.L., *Finite combinatory processes-formulation I* (⟨101⟩, 1(1936), 103–105).

POST E.L., *Recursively enumerable sets of positive integers and their decision problems* (⟨55⟩, 50(1944), 284–316).

POST E.L., *Recursive unsolvability of a problem of Thue* (⟨101⟩, 12(1947), 1–11).

POST E.L., *Recursively enumerable sets of positive integers and their decision problems* (⟨160⟩, (4), 7(1947), 187–229).

PRANDTL L., *Über Flüssigkeitsbewegung bei sehr kleiner Reibung*, S. 484–491, *Verhandlungen des dritten Internationalen Mathematiker-Kongresses in Heidelberg vom 8. bis 13. August 1904*, Leipzig (B.G. Teubner), 1905.

PREISSMANN A., *Quelques propriétés globales des espaces de Riemann* (⟨63⟩, 15(1942–1943), 175–216).

PRESBURGER M., *Über die Vollständigkeit eines gewissen Systems der Arithmetik ganzer Zahlen, in welchem die Addition als einzige Operation hervortritt*, p. 92–101, 395, *Sprawozdanie z I Kongresu matematykow krajow słowianskich, Warszawa 1929*, 1930.

PYENSON L., *Hermann Minkowski and Einstein's special theory of relativity* (⟨33⟩, 17(1977), 71–95).

PYENSON L., *Physics in the shadow of mathematics: the Göttingen electron-theory seminar of 1905* (⟨33⟩, 21(1979–1980), 55–89).

PYENSON L., *Relativity in late Wilhelmian Germany: the appeal to a preestablished harmony between mathematics and physics* (⟨33⟩, 27(1982), 137–155).

RADEMACHER H., *Einige Sätze über Reihen von allgemeinen Orthogonalfunktionen* (⟨120⟩, 87(1922), 112–138) = S. 231–258, vol. I, *Collected papers*, Cambridge, Massachusetts (The MIT Press), 1974.

RADO T., *Über den Begriff der Riemannschen Fläche* (⟨6⟩, 2(1925), 101–121).

RADO T., *On the problem of Plateau*, Berlin (Springer), 1930.

RADO T., *Subharmonic functions* (⟨76⟩, 5(1937), Nr.1).

RADON J., *Theorie und Anwendungen der absolut additiven Mengenfunktionen* (⟨178⟩, 122(1913), 1295–1438) = p. 45–188, vol. I, *Collected works*, Basel (Birkhäuser), 1987.

RADON J., *Über die Bestimmung von Funktionen durch ihre Integralwerte längst gewisser Mannigfaltigkeiten* (⟨43⟩, 69(1917), 262–277) = S. 11–26, Band 2, Gesammelte Abhandlungen, Wien (Verlag der Österreichischen Akademie der Wissenschaften), 1987. (S. 7–9, L. SCHMETTERER, *The Radon transform*).

RAMANUJAN S., *On certain arithmetical functions* (⟨148⟩, 22(1916), 159–184) = p. 136–162, *Collected papers*, Cambridge (University Press), 1927. (*Notes*, p. 340–341).

RAMANUJAN S., *On certain trigonometrical sums and their applications in the theory of numbers* (⟨148⟩, 22(1918), 259–276) = p. 179–189, *Collected papers*, Cambridge (University Press), 1927. (*Notes*, p. 343).

RAMANUJAN S., *Proof of certain identities in combinatory analysis* (⟨148⟩, 23 (1919), 214–216) = p. 214–215, *Collected papers*, Cambridge (University Press), 1927. (*Notes*, p. 344–346).

RAMSEY F.P., *On the problem of formal logic* (⟨150⟩, (2), 30(1929), 264–286).

RANG B. and THOMAS W., *Zermelo's discovery of the «Russel paradox»* (⟨83⟩, 8(1981), 15–22).

RASSIAS T.M. (ed.), *Constantin Carathéodory: an international tribute*, Singapore (World Scientific), 1991.

REEB G., *Variétés feuilletées, feuilles voisines* (⟨66⟩, 224(1947), 1613–1614).

REICH K., *Levi-Civitasche Parallelverschiebung, affiner Zusammenhang, Übertragungsprinzip: 1916/17–1922/23* (⟨33⟩, 44(1992), 77–105).

REID C., *Hilbert*, Berlin (Springer-Verlag), 1970.

REID C., *Courant in Göttingen and New York: the story of an improbable mathematician*, New York (Springer), 1976.

RELLICH F., *Ein Satz über mittlere Konvergenz* (⟨130⟩, 1930, 30–35).

RELLICH F., *Störungstheorie der Spektralzerlegung* (⟨120⟩, 113(1937), 600–619, 677–685).

REMMERT R., *Theory of complex functions*, New York (Springer-Verlag), 1991.

VON RENTELN M., *Aspekte zur Geschichte der Analysis im 20. Jahrhundert*, Skriptum zur Vorlesung an der Universität Karlsruhe, 1987.

VON RENTELN M., *Der Einfluss der Lebegueschen Integrationstheorie auf die komplexe Funktionentheorie zu Beginn dieses Jahrhunderts* (⟨95⟩, 1992, 75–96).

REY PASTOR J., *La matematica del siglo XX. I. La integral* (⟨16⟩, 1934, 13–24).

DE RHAM G., *Sur l'Analysis situs des variétés à n dimensions* (⟨97⟩, (9), 10(1931), 115–200) = p. 23–110, *Oeuvres mathématiques*, Genève (L'Enseignement mathématique), 1981.

DE RHAM G., *Sur la théorie des intersections et les intégrales multiples* (⟨63⟩, 4(1932), 151–157) = p. 117–123, *Oeuvres mathématiques*, Genève (L'Enseignement mathématique), 1981.

DE RHAM G., *Relations entre la topologie et la théorie des intégrales multiples* (⟨74⟩, 35(1936), 213–228) = p. 125–140, *Oeuvres mathématiques*, Genève (L'Enseignement mathématique), 1981.

DE RHAM G., *Complexes à automorphismes et homéomorphie différentielle* (⟨21⟩, 2(1950), 51–67) = p. 347–363, *Oeuvres mathématiques*, Genève (L'Enseignement mathématique), 1981.

DE RHAM G., *Quelques souvenirs des années 1925–1950* (⟨57⟩, 1(1980), 1–36) = p. 651–668, *Oeuvres mathématiques*, Genève (L'Enseignement mathématique), 1981.

DE RHAM G. and KODAIRA K., *Harmonic integrals*, Princeton (Institute for Advanced Study), 1950.

RIBENBOIM P., *13 lectures on Fermat's last theorem*, New York (Springer-Verlag), 1979.

RICCI M.G. et LEVI-CIVITA T., *Méthodes de calcul différentiel absolu et leurs applications* (⟨120⟩, 54(1901), 125–201) = G. RICCI, *Opere*, vol. II, p. 185–271.

RICKART C.E., *Banach algebras with an adjoint operation* (⟨28⟩, (2), 47(1946), 528–550).

RIESZ F., *Sur les systèmes orthogonaux de fonctions* (⟨66⟩, 144(1907), 615–619) = p. 378–381, vol. I, *Oeuvres complètes*, Paris (Gauthier-Villars), 1960.

RIESZ F., *Stetigkeitsbegriff und abstrakte Mengenlehre*, p. 18–24, vol. II, *Atti del IV Congresso internazionale dei Matematici (Roma, 6–11 Aprile 1908)*, Roma

(R. Accademia dei Lincei), 1909 = p. 155–161, vol. I, *Oeuvres complètes*, Paris (Gauthier-Villars), 1960.

RIESZ F., *Sur les opérations fonctionnelles linéaires* (⟨66⟩, 149(1909), 974–977) = p. 400–402, vol. I, *Oeuvres complètes*, Paris (Gauthier-Villars), 1960.

RIESZ F., *Untersuchungen über Systeme integrierbarer Funktionen* (⟨120⟩, 69 (1910), 449–497) = p. 441–489, vol. I, *Oeuvres complètes*, Paris (Gauthier-Villars), 1960.

RIESZ F., *Sur certains systèmes d'équations intégrales* (⟨24⟩, (3), 28 (1911), 33–62).

RIESZ F., *Les systèmes d'équations linéaires à une infinité d'inconnues*, Paris (Gauthier-Villars), 1913.

RIESZ F., *Über lineare Funktionalgleichungen* (⟨7⟩, 41(1918), 71–98) = p. 1053–1080, vol. I, *Oeuvres complètes*, Paris (Gauthier-Villars), 1960.

RIESZ F., *Sur les valeurs moyennes du module des fonctions harmoniques et des fonctions analytiques* (⟨8⟩, 1(1922–1923), 27–32) = p. 661–666, vol. I, *Oeuvres complètes*, Paris (Gauthier-Villars), 1960.

RIESZ F., *Über die Randwerte einer analytischen Funktion* (⟨123⟩, 18 (1923), 87–95) = p. 645–653, vol. I, *Oeuvres complètes*, Paris (Gauthier-Villars), 1960.

RIESZ F., *Über subharmonische Funktionen und ihre Rolle in der Funktionentheorie und in der Potentialtheorie* (⟨8⟩, 2(1925), 87–100) = p. 667–680, vol. I, *Oeuvres complètes*, Paris (Gauthier-Villars), 1960.

RIESZ F., *Sur les fonctions subharmoniques et leur rapport à la théorie du potentiel* (⟨7⟩, 48(1926), 329–343; 54(1930), 321–360).

RIESZ F., *Sur la décomposition des opérations fonctionnelles linéaires*, p. 143–148, t. III, *Atti del Congresso internazionale dei matematici, 1928*, Bologna (Nicola Zanichelli), 1930 = p. 1097–1102, t. II, *Oeuvres complètes*, Paris (Gauthier-Villars), 1960.

RIESZ F., *Sur les fonctions subharmoniques et leur rapport à la théorie du potentiel II* (⟨7⟩, 54(1930), 321–360).

RIESZ F., *Über Sätze von Stone und Bochner* (⟨8⟩, 6(1933), 184–198) = p. 1135–1149, vol. II, *Oeuvres complètes*, Paris (Gauthier-Villars), 1960.

RIESZ F., *Zur Theorie des Hilbertschen Raumes* (⟨8⟩, 7(1934), 34–38) = p. 1150–1154, vol. II, *Oeuvres complètes*, Paris (Gauthier-Villars), 1960.

RIESZ F., *Sur quelques notions fondamentales dans la théorie générale des opérations linéaires* (⟨28⟩, (2), 41(1940), 174–206) = p. 1224–1256, vol. II, *Oeuvres complètes*, Paris (Gauthier-Villars), 1960.

RIESZ F., *L'évolution de la notion d'intégrale depuis Lebesgue* (⟨21⟩, 1(1949), 29–42) = p. 327–340, vol. I, *Oeuvres complètes*, Paris (Gauthier-Villars), 1960.

RIESZ F. und M., *Über die Randwerte einer analytischen Funktion*, p. 27–44, *Quatrième Congrès des mathématiciens scandinaves*, Stockholm 1916, 1920 = p. 537–554, vol. I, F. RIESZ, *Oeuvres complètes*, Paris (Gauthier-Villars),

1960 = p. 195–212, M. RIESZ, *Collected papers*, Berlin (Springer-Verlag), 1988.

RIESZ M., *Les fonctions conjuguées et les séries de Fourier* (⟨66⟩, 178 (1924), 1464–1467) = p. 360–362, *Collected papers*, Berlin (Springer-Verlag), 1988.

RIESZ M., *Sur les maxima des formes bilinéaires et sur les fonctionnelles linéaires* (⟨7⟩, 49(1927), 465–497) = p. 377–409, *Collected papers*, Berlin (Springer-Verlag), 1988.

RIESZ M., *Intégrale de Riemann-Liouville et solution invariante du problème de Cauchy pour l'équation des ondes*, p. 44–45, vol. II, *Comptes Rendus du Congrès international des Mathématiciens Oslo 1936*, Oslo (A.W. Brøggers), 1937 = p. 477–478, *Collected papers*, Berlin (Springer-Verlag), 1988.

RIESZ M., *L'intégrale de Riemann-Liouville et le problème de Cauchy pour l'équation des ondes* (⟨49⟩, 66(1938), 1–18).

RITZ W., *Über eine neue Methode zur Lösung gewisser Variationsprobleme der mathematischen Physik* (⟨98⟩, 135(1908), 1–61).

RODRIGUES-CONSUEGRA F.A., *The mathematical philosophy of Bertrand Russell: origins and development*, Basel (Birkhäuser), 1991.

ROMANOVSKY J.V., *L.V. Kantorovich's works in mathematical programming*, Mathematical programming (Tokyo, 1988), 365–382 (⟨118⟩), 6 KTK, Scientific, Tokyo, 1989.

ROTA G.-C., *Les ambiguïtés de la pensée mathématique* (⟨81⟩, n° 45, juin 1990, 54–64).

ROUSSELOT T., *Sur le théorème de Stone-Weierstrass*, Centre des Sciences et des Techniques, Institut de Mathématiques, Université de Nantes, 1991.

ROWE D.E., *Klein, Hilbert, and the Göttingen mathematical tradition* (⟨140⟩, 5(1989), 186–213).

RUDIN W., *Functional analysis*, New York (McGraw-Hill), 1973.

RUELLE D., *Hasard et chaos*, Paris (Odile Jacob), 1991.

RUKHIN A.L., *Kolmogorov's contributions to mathematical statistics* (⟨31⟩, 18 (1990), 1011–1016).

RUSSELL B., *The principles of mathematics I*, Cambridge (University Press), 1903.

RUSSELL B., *Mathematical logic as based on the theory of types* (⟨12⟩, 30(1908), 222–262).

RÜTHING D., *Some definitions of the concept of function from Joh. Bernoulli to N. Bourbaki* (⟨112⟩, 6(1984), n° 4, 72–77).

RYBNIKOV K.A., *Esquisse de la théorie des graphes* (Istoriya Metod. Estestv. Nauk, 36(1989), 109–122).

SAATY T.L. and KAINEN P.C., *The four-color problem: assaults and conquest*, New York (McGraw-Hill), 1977.

I.R. Šafarevič

SACCOMAN J.J., *Extension theorems by Helly and Riesz revisited* (⟨165⟩, (4), 16(1990), 223–230).

SACCOMAN J.J., *On the origin of extrema points and the Krein-Milman theorem* (⟨40⟩, 17(1990), N° 1, 10–12).

ŠAFAREVIČ I.R., *A general reciprocity law* (in Russian) (⟨69⟩, (N.S.), 64(1949), 25–28).

SAKS S., *Théorie de l'intégrale*, Warszawa, 1933 = 2th ed., *Theory of the integral*, 1937.

SALEM R., *Essai sur les séries trigonométriques*, Paris (Hermann), 1940.

SALEM R., *Sets of uniqueness and sets of multiplicity* (⟨185⟩, 54(1943), 218–228).

SALEM R., *On a theorem of Bohr and Pal* (⟨55⟩, 50(1944), 579–580).

SALEM R., *Sets of uniqueness and sets of multiplicity II* (⟨185⟩, 56(1944), 32–49).

SAMUEL P., *Sur les anneaux locaux* (⟨66⟩, 225(1947), 1244–1245).

SARD A., *The measure of the critical values of differentiable maps* (⟨55⟩, 48 (1942), 883–890).

SAVAGE C.W. and ANDERSON C.A. (edited by), *Rereading Russell: essays in Bertrand Russell's metaphysics and epistemology*, Minneapolis (University of Minnesota Press), 1989.

SCHAPPACHER N., *Développement de la loi de groupe sur une cubique*, p. 159–184, *Séminaire de la théorie des nombres, Paris, 1988–1989* (⟨155⟩, 91), Boston (Birkhäuser), 1990.

SCHAPPACHER N., *Edmund Landau's Göttingen: from the life and death of a great mathematical center* (⟨112⟩, 13(1991), N° 4, 12–18).

SCHATZ J.A., *Some super-classics of mathematics* (⟨133⟩, 18(1971), 723–725).

SCHAUDER J., *Zur Theorie stetiger Abbildungen in Funktionalräumen* (⟨123⟩, 26(1927), 47–65, 417–431).

SCHAUDER J., *Der Fixpunktsatz in Funktionalräumen* (⟨181⟩, 2 (1930), 171–180) = p. 168–176, *Oeuvres*, Warszawa (PWN — Editions scientifiques de Pologne), 1978.

SCHAUDER J., *Sur les équations linéaires du type elliptique à coefficients continus* (⟨66⟩, 199(1934), 1366–1368).

SCHAUDER J., *Über lineare elliptische Differentialgleichungen zweiter Ordnung* (⟨122⟩, 38(1934), 257–282) = p. 354–379, *Oeuvres*, Warszawa (PWN — Editions scientifiques de Pologne), 1978.

SCHAUDER J., *Das Anfangsproblem einer quasilinearen hyperbolischen Differentialgleichung zweiter Ordnung in beliebiger Anzahl von unabhängigen Veränderlichen* (⟨80⟩, 24(1935), 213–246) = p. 380–409, *Oeuvres*, Warszawa (PWN — Editions scientifiques de Pologne), 1978.

SCHLESINGER L., *Bericht über die Entwicklung der Theorie der linearen Differentialgleichungen seit 1865* (⟨96⟩, 18(1909), 133–266).

SCHLISSEL A., *The development of asymptotic solutions of linear ordinary differential equations 1817–1920* (⟨33⟩, 16(1976–1977), 307–378).

SCHLISSEL A., *The initial development of the WHB solutions of linear second order ordinary differential equations and their use in the connection problem* (⟨83⟩, 4(1977), 183–204).

SCHMIDT E., *Entwicklung willkürlicher Funktionen nach System vorgeschriebener*, Diss. Göttingen.

SCHMIDT E., *Über die Anzahl der Primzahlen unter gegebener Grenze* (⟨120⟩, 57(1903), 195–204).

SCHMIDT E., *Über die Auflösung linearer Gleichungen mit unendlich vielen Unbekannten* (⟨158⟩, 25(1908), 53–77).

SCHMIDT E., *Zur Theorie der linearen und nicht linearen Integralgleichungen. III. Teil: über die Auflösung der nichtlinearen Integralgleichung und die Verzweigung ihrer Lösungen* (⟨120⟩, 65(1908), 370–399).

SCHMIDT M. (entretiens et photographies de), *Hommes de science*, Paris (Hermann), 1990.

SCHNEIDER T., *Transzendenzuntersuchungen periodischer Funktionen* (⟨98⟩, 172 (1934), 65–74).

SCHNEIDER T., *Zur Theorie der Abelschen Funktionen und Integrale* (⟨98⟩, 183 (1941), 110–128).

SCHNEIDER T., *Introduction aux nombres transcendants*, Paris (Gauthier-Villars), 1959.

SCHOENBERG I., *Metric spaces and positive definite functions* (⟨185⟩, 44(1938), 522–536).

SCHOENEBERG B., *Hecke, Erich*, p. 214–215, vol. VI, *Dictionary of scientific biography*, New York (Charles Scribner's Sons), 1972.

SCHOENFLIES A., *Die Entwicklung der Lehre von den Punktmannigfaltigkeiten*, II. Teil. Leipzig (B.G. Teubner), 1908.

SCHOENFLIES A., *Entwicklung der Mengenlehre und ihrer Anwendungen*, II. Auflage, Leipzig (Teubner), 1913.

SCHOLZ E. *et alii*, *Geschichte der Algebra: eine Einführung*, Mannheim (Bibliographisches Institut), 1990.

SCHOT S.H., *Eighty years of Sommerfeld's radiation condition* (⟨83⟩, 19(1992), 385–401).

SCHOTTKY F., *Über den Picardschen Satz und die Borelschen Ungleichungen* (⟨177⟩, 1904, 1244–1263).

SCHOUTE P.H., *Mehrdimensionale Geometrie I. Teil: Die linearen Räume*, Leipzig (G.J. Göschen), 1902.

SCHOUTEN J.A., *Über die verschiedenen Arten der Übertragung in einer n-dimensionalen Mannigfaltigkeit, die einer Differentialgeometrie zugrunde gelegt werden können* (⟨122⟩, 13(1922), 56–81; 15(1922), 168).

SCHOUTEN J.A., *Der Ricci-Kalkül*, Berlin (Julius Springer), 1924.

SCHREIER O., *Abstrakte kontinuierliche Gruppen* (⟨1⟩, 4(1925), 15–32).

SCHREIER O., *Die Verwandtschaft stetiger Gruppen im Grossen* (⟨1⟩, 5(1927), 233–244).

SCHRÖDER K., *Erhard Schmidt* (⟨121⟩, 25(1963), 1–3).

SCHRÖDINGER E., *Quantisierung als Eigenwertproblem I* (⟨17⟩, (4), 79(1926), 361–374) = S. 82–97, Band III, *Gesammelte Abhandlungen*, Wien (Österreich. Akademie der Wissenschaften), 1984.

SCHRÖDINGER E., *A method of determining quantum-mechanical eigenvalues and eigenfunctions* (⟨153⟩, 46(1940), 9–16).

SCHUR I., *Über eine Klasse von Matrizen, die sich einer gegebenen Matrix zuordnen lassen*, Dissertation, Berlin, 1901 = S. 1–72, Band I, *Gesammelte Abhandlungen*, Berlin (Springer-Verlag), 1973.

SCHUR I., *Neue Begründung der Theorie der Gruppencharaktere* (⟨177⟩, 1905, 406–432) = S. 143–169, Band I, *Gesammelte Abhandlungen*, Berlin (Springer-Verlag), 1973.

SCHWARTZ L., *Généralisation de la notion de fonction, de dérivation, de transformation de Fourier et applications mathématiques et physiques* (⟨23⟩, nouvelle série, 21(1945), 1946, 57–74).

SCHWARTZ L., *Théorie générale des fonctions moyenne-périodiques* (⟨28⟩, (2), 48(1947), 857–929).

SCHWARTZ L., *Sur une propriété de synthèse spectrale dans les groupes non compacts* (⟨66⟩, 227(1948), 424–426).

SCHWARTZ L., *Théorie des distributions*, tome I, Paris (Hermann), 1950.

SCHWARTZ L., *Les mathématiques en France pendant et après la guerre*, p. 49–67, *Proceedings of the second Canadian Mathematical Congress*, University of Toronto Press, 1951.

SCHWARTZ L., *Théorie des noyaux*, p. 220–230, vol. I, *Proceedings of the International Congress of Mathematicians, Cambridge, Massachusetts, August 30 – September 6, 1950*, American Mathematical Society, 1952.

SCHWARTZ L., *L'énumération transfinie et l'oeuvre de Denjoy* (⟨53⟩, (2), 79 (1955), 1e partie, 78–96).

SCHWARTZ L., *La vie et l'oeuvre d'Andréi Kolmogorov* (⟨192⟩, 6(1989), 573–781).

SCORZA DRAGONI G., *Renato Caccioppoli* (⟨38⟩, Appendice, 30 (1963), 85–93).

SEGAL I.E., *The group ring of a locally compact group I* (⟨151⟩, 27(1941), 348–352).

SEGRE B., *Geometria algebrica nei paesi anglo-sassoni (del 1939 al 1945)* (⟨146⟩, 11, 1946).

SEGRE B., *Lezioni di geometria moderna I. Fondamenti di geometria sopra un corpo qualsiasi*, Bologna (Nicola Zanichelli), 1948.

SEIFERT H., *Topologie dreidimensionaler geschlossener Räume* (⟨7⟩, 60 (1932), 147–238).

SEIFERT H. und THRELFALL W., *Lehrbuch der Topologie*, Leipzig (Teubner), 1934.

SELBERG A. *On an elementary method in the theory of primes* (Norske Vid. Selsk. Forh., 19(1947), 64–67).

SELBERG A., *An elementary proof of the prime-number theorem* (⟨28⟩, (2), 50 (1949), 305–313).

SELBERG A., *Lectures on sieves*, p. 65–247, vol. II, *Selected papers*, Berlin (Springer-Verlag), 1991.

SENETA E., *On the history of the strong law of large numbers and Boole's inequality* (⟨83⟩, 19(1992), 24–39).

SERGESCU P., *Les sciences mathématiques*, p. 5–182, *Tableau du XXe siècle (1900–1933)*, P. SERGESCU, J. ROSTAND et A. BOUTARIC, *Les sciences*, Paris (Denoël et Steele), 1933.

SEVERI F., *Sul principio della conservazione del numero* (⟨158⟩, 33(1912), 313–327) = p. 117–136, vol. I, *Memorie scelte*, Bologna (Cesare Zuffi), 1950.

SEVERI F., *Vorlesungen über algebraische Geometrie*, Leipzig (B.G. Teubner), 1921.

SEVERI F., *Moderni indirizzi nelle matematiche* (⟨39⟩, diciasettesima riunione, 15–22 settembre 1928, 1929, 85–115).

SEVERI F., *La matematica della prima meta del secolo XX* (⟨170⟩, 92 (1957), 20–26) = *Les mathématiques pendant la première moitié du vingtième siècle* (⟨170⟩, 92(1957), Supplément. Traductions françaises, 11–17).

SHADWICK W.F. and SLUIS W.M., *On E. Cartan's equivalence of differential systems* (⟨66⟩, 313(1991), Série I, 455–459).

SHANKER S.G. (edited by), *Gödel's theorem in focus*, London (Routledge), 1990.

SHANNON C.E., *A mathematical theory of communication* (⟨42⟩, 27 (1948), 379–423, 623–656).

SHEYNIN O.B., *H. Poincaré's work on probability* (⟨33⟩, 42(1991), 137–171).

SHIELDS A., *Luzin and Egorov* (⟨112⟩, 9(1987), n° 4, 24–27).

SHIELDS A., *Banach algebras 1939–1989* (⟨112⟩, 11(1989), n° 3, 15–17).

SIDON S., *Verallgemeinerung eines Satzes über absolute Konvergenz von Fourierreihen mit Lücken* (⟨120⟩, 97(1927), 675–676).

SIEGEL C.L., *Approximation algebraischer Zahlen* (⟨123⟩, 10(1921), 173–213) = S. 6–46, Band I, *Gesammelte Abhandlungen*, Berlin (Springer-Verlag), 1966.

SIEGEL C.L., *Über einige Anwendungen diophantischer Approximationen* (⟨3⟩, Nr.1, 1929) = S. 209–266, Band I, *Gesammelte Abhandlungen*, Berlin (Springer-Verlag), 1966.

SIEGEL C.L., *Über die analytische Theorie der quadratischen Formen* (⟨28⟩, (2), 36(1935), 527–606).

SIEGEL C.L., *Über die Classenzahl quadratischer Zahlkörper* (⟨4⟩, 1 (1935), 83–86).

SIEGEL C.L., *Einführung in die Theorie der Modulfunktionen n-ten Grades* (⟨120⟩, 116(1939), 617–657).

SIEGEL C.L., *Iteration of analytic functions* (⟨28⟩, (2), 43(1942), 607–612).

SIEGEL C.L., *Analytic functions of several complex variables*, Princeton (Institute for Advanced Study), 1949.

SIEGEL C.L., *Das wissenschaftliche Werk von E. Hecke*, S. 16–17, E. HECKE, *Mathematische Werke*, Göttingen (Vandenhoeck und Rupprecht), 1959.

SIEGMUND-SCHULTZE R., *Der Strukturwandel in der Mathematik um die Wende vom 19. zum 20. Jahrhundert, untersucht am Beispiel der Entstehung der ersten Begriffsbildungen der Funktionalanalysis* (⟨137⟩, 18(1981), 4–20).

SIEGMUND-SCHULTZE R., *Die Anfänge der Funktionalanalysis und ihr Platz im Umwälzungsprozess der Mathematik um 1900* (⟨33⟩, 26(1982), 13–71).

SIEGMUND-SCHULTZE R., *Der Beweis des Hilbert-Schmidt-Theorems* (⟨33⟩, 36 (1986), 251–270).

SIERPINSKI W., *Sur la valeur asymptotique d'une certaine somme* (⟨54⟩, A, 1910, 9–11).

SIERPINSKI W., *Sur une courbe dont tout point est un point de ramification* (⟨66⟩, 160(1915), 302–305).

SIERPINSKI W., *Sur l'hypothèse du continu* ($2^{\aleph_0} = \aleph_1$) (⟨80⟩, 5(1924), 117–187) = p. 527–536, vol. II, *Oeuvres choisies*, Warszawa (PWN — Editions scientifiques de Pologne), 1975.

SIERPINSKI W., *La notion de dérivé comme base d'une théorie des ensembles abstraits* (⟨120⟩, 97(1926), 321–337) = p. 598–615, vol. II, *Oeuvres choisies*, Warszawa (PWN — Editions scientifiques de Pologne), 1975.

SINACEUR H., *Corps et modèles: essai sur l'histoire de l'algèbre réelle*, Paris (Vrin), 1991.

SINACEUR H., *Préhistoire de la géométrie algébrique réelle: de Descartes à Tarski* (⟨57⟩, (2), 1(1991), 1–17).

SINAI YA.G., *Kolmogorov's work on ergodic theory* (⟨29⟩, 17(1989), 833–839).

SINGER I., *Basis in Banach spaces*, Berlin (Springer-Verlag), 1970–1981.

SKOLEM T., *Logisch-kombinatorische Untersuchungen über die Erfüllbarkeit und Beweisbarkeit mathematischer Sätze nebst einem Theorem über dichte Mengen* (⟨179⟩, Nr.4, 1920, 1–36).

SKOLEM T., *Begründung der elementären Arithmetik durch die rekurrierende Denkweise ohne Anwendung scheinbarer Veränderlichen mit unendlichem Ausdehnungsbereich* (Skrifter, Videnshabsakademiet i Kristiania, I, 1923, n° 6) = p. 153–188, *Selected works in logic*, Oslo (Universitetsforlaget), 1970.

SKOLEM T., *Einige Bemerkungen zur axiomatischen Begründung der Mengenlehre*, p. 217–232, *Conférences faites au Cinquième Congrès des Mathématiciens Scandinaves tenu à Helsingfors du 4 au 7 juillet 1922*, Helsingfors (Librairie Académique), 1923 = p. 137–152, *Selected works in logic*, Oslo (Universitetsforlaget), 1970.

SKOLEM T., *Über einige Grundlagenfragen der Mathematik* (⟨179⟩, 1929, Nr 4) = p. 227–273, *Selected works in logic*, Oslo (Universitetsforlaget), 1970.

SKOLEM T., *Über die Unmöglichkeit einer Charakterisierung der Zahlenreihe mittels eines endlichen Axiomensystems* (⟨132⟩, n° 1–12, 73–82, 1933) = p. 345–354, *Selected works in logic*, Oslo (Universitetsforlaget), 1970.

SKOLEM T., *Über die Nichtcharakterisierbarkeit der Zahlenreihe mittels endlich oder abzählbar unendlich vieler Aussagen mit ausschliesslich Zahlenvariablen* (⟨80⟩, 23(1934), 150–161) = p. 355–366, *Selected works in logic*, Oslo (Universitetsforlaget), 1970.

SKOLEM T., *Einige Sätze über p-adische Potenzreihen mit Anwendung auf gewisse exponentielle Gleichungen* (⟨120⟩, 111(1935), 399–424).

SKOLEM T., *Diophantische Gleichungen*, Berlin (Julius Springer), 1938.

SMAIL L.C., *History and synopsis of the theory of summable infinite processes*, Eugene (University of Oregon Press), 1925.

SOBOLEV S.L., *Méthode nouvelle à résoudre le problème de Cauchy pour les équations linéaires hyperboliques normales* (⟨117⟩, 1(1936), 39–71).

SOBOLEV S.L., *On a theorem of Functional Analysis* (in Russian) (⟨117⟩, 4(46) (1938), 471–497).

SOBOLEV S.L., *Applications of Functional Analysis in mathematical physics* (in Russian), Leningrad, 1950.

SOMMERFELD A., *Die Greensche Funktion der Schwingungsgleichung* (⟨96⟩, 21 (1912), 309–353).

SOUSLIN M.YA., *Sur une définition des ensembles analytiques B sans nombres transfinis* (⟨66⟩, 164(1917), 88–91).

SPERNER E., *Neuer Beweis für die Invarianz der Dimensionszahl und des Gebietes* (⟨1⟩, 6(1928), 265–272).

STEEN L.A., *Highlights in the history of spectral theory* (⟨13⟩, 80(1973), 359–381).

STEENROD N.E., *Regular cycles on compact metric spaces* (⟨28⟩, (2), 41(1940), 833–851).

STEENROD N.E., *Products of cocycles and extensions of mappings* (⟨28⟩, (2), 48(1947), 290–320).

STEENROD N.N., *Cohomology invariants of mappings* (⟨28⟩, (2), 50 (1949), 954–968).

STEINHAUS H., *Les propriétés dénombrables et leur rapport à la théorie de la mesure* (⟨80⟩, 4(1923), 286–310).

STEINHAUS H., *Über die Wahrscheinlichkeit dafür, dass der Konvergenzkreiss einer Potenzreihe ihre natürlich Grenze ist* (⟨123⟩, 31(1930), 408–416).

STEINHAUS H., *Stefan Banach, 1896–1945* (⟨171⟩, 26(1961), 93–100).

STEINITZ E., *Beiträge zur analysis situs* (⟨175⟩, 7(1908), 29–49).

STEINITZ E., *Algebraische Theorie der Körper* (⟨98⟩, 137(1910), 167–309).

STEINITZ E., *Algebraische Theorie der Körper mit Erläuterungen und einem Anhang von R. Baer und H. Hasse*, Berlin (Springer), 1930.

STEINITZ E., *Vorlesungen über die Theorie der Polyeder unter Einschluss der Elemente der Topologie. Aus dem Nachlass herausgegeben und ergänzt von* H. RADEMACHER, Berlin (J. Springer), 1934.

STEKLOV V.A., *Principal problems in mathematical physics II* (in Russian), Petersburg, 1923.

STERNBERG W., *Über die Gleichung der Wärmeleitung* (⟨120⟩, 101 (1929), 394–398).

STIEFEL E., *Richtungsfelder und Fernparallelismus in n-dimensionalen Mannigfaltigkeiten* (⟨63⟩, 8(1935–1936), 305–353).

VAN STIGT W.P., *Brouwer's intuitionism*, Amsterdam (North-Holland), 1990.

STOILOW S., *Sur les transformations continues et la topologie des fonctions analytiques* (⟨24⟩, (3), 45(1928), 347–382).

STONE M.H., *Linear transformations in Hilbert space. III: Operational methods and group theory* (⟨151⟩, 16(1930), 172–175).

STONE M.H., *Linear transformations in Hilbert space and their applications to analysis*, New York (American Mathematical Society), 1932.

STONE M.H., *On one-parameter unitary groups in Hilbert space* (⟨28⟩, (2), 33 (1932), 643–648).

STONE M.H., *Boolean algebras and their application to topology* (⟨151⟩, 20 (1934), 197–202).

STONE M.H., *Postulates for Boolean algebras and generalized Boolean algebras* (⟨12⟩, 57(1935), 703–732).

STONE M.H., *The theory of representation for Boolean algebras* (⟨185⟩, 40(1936), 37–111).

STONE M.H., *Applications of the theory of Boolean rings to general topology* (⟨185⟩, 41(1937), 375–481).

STONE M.H., *A general theory of spectra I* (⟨151⟩, 26(1940), 280–283).

STONE M.H., *The generalized Weierstrass approximation theorem* (⟨114⟩, 21 (1948), 167–184, 237–254).

STUDY E., *Gruppen zweiseitiger Kollineationen* (⟨130⟩, 1912, 453–479).

SZABADOS J. and TANDORI K. (edited by), *A. Haar memorial conference*, Amsterdam (North-Holland), 1987.

SZABO M.E., *Introduction*, p. 1–28, G. GENTZEN, *The collected papers*, Amsterdam (North-Holland), 1969.

SZEGÖ G., *Leopold Féjer: in memoriam* (⟨55⟩, 66(1960), 346–352).

SZMIELEW W., *Decision problem in group theory*, p. 763–766, vol. I, *Proceedings of the tenth International Congress of Philosophy, Amsterdam, August 11–18, 1948*, 1949.

SZYMANSKI W., *Who was Otto Nikodym* (⟨112⟩, 12(1990), n° 2, 27–31).

TAKAGI T., *Über eine Theorie des relativ Abelschen Zahlkörpers* (⟨102⟩, 41 (1920), Nr 9, 1–133).

TARSKI A., *Sur les ensembles finis* (⟨80⟩, 6(1924), 45–95).

TARSKI A., *Fundamentale Begriffe der Methodologie der deduktiven Wissenschaften I* (⟨129⟩, 37(1930), 361–404).

TARSKI A., *O pojęciu prawdy w odniesieniu do sformalizowanych nauk dedukcyjnych* (⟨166⟩, 12(1930), 210–311) = *Der Wahrheitsbegriff in den formalisierten Sprachen* (Studia Philosophica, 1(1936), 261–405) = *The concept of truth in formalized languages*, p. 152–278, *Logic, semantics, metamathematics. Papers from 1923 to 1938*, Oxford (Clarendon Press), 1956.

TARSKI A., *O stopniu rownowaznosci wielokatow* (⟨128⟩, 1(1931), 37–44).

TARSKI A., *Der Wahrheitsbegriff in den Sprachen der deduktiven Disziplinen* (⟨10⟩, 69(1932), 23–25).

TARSKI A., *Cancellation laws in the arithmetic of cardinals* (⟨80⟩, 36 (1949), 77–92).

TARWATER D. (ed.), *The bicentennial tribute to American mathematics: 1776–1976*, Washington (The Mathematical Association of America), 1977.

TATON R. (published under the direction of), *Histoire générale des Sciences*, tome III, vol. II, *Le XXe siècle*, 2e éd., Paris (Presses Universitaires de France), 1983.

TAYLOR A.E., *Notes on the history of the uses of analyticity in operator theory* (⟨13⟩, 78(1971), 331–342).

TAYLOR A.E., *The differential: nineteenth and twentieth century developments* (⟨33⟩, 12(1974), 355–383).

TAYLOR A.E., *A study of Maurice Fréchet* (⟨33⟩, 27(1982), 233–295; 34(1985), 279–380; 37(1987), 25–76).

TAYLOR J.G. (ed.), *Tributes to Paul Dirac*, Bristol (Adam Hilger), 1987.

TEICHMÜLLER O., *Extremale quasikonforme Abbildungen und quadratische Differentiale* (⟨3⟩, Nr. 22, 197(1939)) = S. 335–531, *Gesammelte Abhandlungen*, Berlin (Springer-Verlag), 1982.

TEMPLE G., *100 years of mathematics*, London (Duckworth), 1981.

TERPE F., *Andrej Nikolajevic Kolmogorov und die Hilbertschen Probleme* (⟨193⟩, 38(1989), N° 4, 54–58).

TESSIER B., *Quelques points de l'histoire des variétés polaires, de Poncelet à nos jours* (⟨172⟩, 1987–1988, N° 4, 1988).

THIEL C., *Löwenheim, Leopold*, p. 571–572, vol. 18, supplement II, *Dictionary of scientific biography*, New York (Charles Scribner's Sons), 1990.

THOM R., *Sur une partition en cellules associée à une fonction sur une variété* (⟨66⟩, 228(1949), 973–975).

THOM R., *Problèmes rencontrés dans mon parcours mathématique: un bilan* (⟨86⟩, N° 70(1989), 199–214).

THOM R., *La vie et l'oeuvre de Hassler Whitney* (⟨192⟩, 7(1990), N° 6, 473–476).

THOMAS T.Y., *The differential invariants of generalized spaces*, Cambridge (University Press), 1934.

THORIN G.O., *Convexity theorems generalizing those of M. Riesz and Hadamard with some applications* (⟨125⟩, 9(1948), 1–58).

TIETZE H., *Über die topologischen Invarianten mehrdimensionaler Mannigfaltigkeiten* (⟨129⟩, 19(1908), 1–118).

TIETZE H., *Beiträge zur allgemeinen Topologie. I. Axiome für verschiedene Fassungen des Umgebungsbegriffs* (⟨120⟩, 88(1923), 290–312).

TIKHOMIROV V.M., *A.N. Kolmogorov and approximation theory* (⟨167⟩, 44 (1989), 101–152).

TIKHOMIROV V.M., *A.N. Eduard Helly* (in Russian) (⟨88⟩, 32–33(1990), 137–145).

TIKHOMIROV V.M., *Stories about maxima and minima*, American Mathematical Society, 1990.

TITCHMARSH E.C., *The zeros of the Riemann zeta-function* (⟨154⟩, 151(1935), 234–255; 157(1936), 261–263).

TITCHMARSH E.C., *On expansions in eigenfunctions* $\mathscr{V}$ (⟨156⟩, 12(1941), 89–107).

TODD O.T., *Brouwer and Hilbert* (⟨112⟩, 13(1991), N° 3, 4).

TOEPLITZ O., *Über die Fourier'sche Entwickelung positiver Funktionen* (⟨158⟩, 32(1911), 191–192).

TONELLI L., *Fondamenti di calcolo delle variazioni*, vol. I, Bologna (Nicola Zanichelli), 1921.

TRAYLOR D.R. et al., *Creative teaching: heritage of R.L. Moore*, University of Houston, 1972.

TREVES F., *Applications of distributions to PDE theory* (⟨13⟩, 77(1970), 241–248).

TRICOMI F., *Sulle equazioni lineari alle derivati parziali di* 2^0 *ordine di tipo misto* (⟨36⟩, (5), 14(1923), 133–247).

TROPP H.S., *The origins and history of the Fields medal* (⟨83⟩, 3(1976), 167–181).

TUCCIARONE J., *The development of the theory of summable divergent series from 1880 to 1925* (⟨33⟩, 10(1973), 1–40).

TURAN P., *On a theorem of Hardy and Ramanujan* (⟨104⟩, 9(1934), 274–276).

TURAN P., *On a new method in the analysis with applications* (⟨61⟩, 74(1949), 123–131).

TURING A.M., *On computable numbers, with an application to the Entscheidungsproblem* (⟨150⟩, (2), 42(1936), 230–265).

TURING A.M., *Systems of logic based on ordinals* (⟨150⟩, (2), 45(1939), 161–228).

Two decades of mathematics in the Netherlands 1920–1940, Amsterdam (Wiskundig Genootschap), 1978.

TYCHONOV A.N., *Über die topologische Erweiterung von Räumen* (⟨120⟩, 102 (1930), 544–561).

ULAM S.M., *John von Neumann, 1903–1957* (⟨55⟩, 64(1958), 1–49).

ULAM S.M., *Adventures of a mathematician*, New York (Scribner's), 1976.

URYSOHN P.S., *Sur la ramification des lignes cantoriennes* (⟨66⟩, 175 (1922), 481–483) = (in Russian), p. 223–226, t. I, *Contributions on topology and other mathematical domains* (in Russian), Moscow (Gosudarstvennye izdatel'stvo tehniko-teoretičeskoi literaturi), 1951.

URYSOHN P.S., *Les multiplicités cantoriennes* (⟨66⟩, 175(1922), 440–442) = (in Russian), p. 219–222, t. I, *Contributions on topology and other mathematical domains* (in Russian), Moscow (Gosudarstvennye izdatel'stvo tehniko-teoretičeskoi literaturi), 1951.

URYSOHN P.S., *Über die Mächtigkeit zusammenhängender Mengen* (⟨120⟩, 94 (1925), 262–295).

VALIRON G., *Théorie des fonctions*, Paris (Masson), 1948.

VEBLEN O., *A system of axioms for geometry* (⟨185⟩, 5(1904), 343–384).

VEBLEN O., *Projective geometry II*, Boston (Ginn), 1918.

VEBLEN O., *Analysis situs*, New York (American Mathematical Society), 1922.

VEBLEN O., *Normal coordinates for the geometry of paths, Projective and affine geometry of paths* (⟨151⟩, 8(1922), 192–197, 347–350).

VEBLEN O. and WHITEHEAD J.H.C., *The foundations of differential geometry*, Cambridge (University Press), 1932.

VESSIOT E., *Sur la théorie des groupes continus* (⟨24⟩, (3), 20(1903), 411–451).

VESSIOT E., *Notice sur les travaux scientifiques*, Paris (Eyrolles), 1932.

VIETORIS L., *Über den höheren Zusammenhang kompakter Räume und eine Klasse von zusammenhangstreuen Abbildungen* (⟨120⟩, 97(1927), 454–472).

VIETORIS L., *Heinrich Tietze* (⟨11⟩, 114(1964), 360–377).

VILLAT H., *Notice sur la vie et l'oeuvre de Jules Drach (1871–1949)*, p. 209–223, vol. III, *Académie des Sciences, Notes et discours*, Paris (Gauthier-Villars), 1957.

VILLE J., *Etude critique de la notion de collectif*, Paris (Gauthier-Villars), 1939.

VINOGRADOV I.M., *Nouvelle solution du problème de Waring* (⟨69⟩, (2), 2 (1934), 337–341).

VINOGRADOV I.M., *Representation of an odd number as a sum of three primes* (⟨69⟩, 15(1937), 291–294) = p. 129–132, *Selected works*, Berlin (Springer-Verlag), 1985.

VINOGRADOV I.M., *Trigonometric sums method in number theory* (in Russian) (⟨187⟩, n° 23, 1947).

VITALI G., *Sul problema della misura dei gruppi di punti di una retta*, Bologna (Gamberini e Parmeggiani), 1905 = p. 231–235, *Opere*, Roma (Edizioni Cremonese), 1984.

VITALI G., *Sui gruppi di punti e sulle funzioni di variabili reali* (⟨37⟩, 43(1908), 229–246) = p. 257–276, *Opere*, Bologna (Ed. Cremonese), 1984.

VLADUT S.G., *Kronecker's Jugendtraum and modular funktions*, New York (Gordon and Breach), 1991.

VOLTERRA V., *Leçons sur les fonctions de lignes*, Paris (Gauthier-Villars), 1913.

VOLTERRA V., *Sur la théorie mathématique des phénomènes héréditaires* (⟨97⟩, (9), 7(1928), 249–298) = p. 130–169, vol. V, *Opere matematiche*, Roma (Accademia Nazionale dei Lincei), 1962.

VORONOÏ G.F., *Sur un problème du calcul des fonctions asymptotiques* (⟨98⟩, 126(1903), 241–282).

VAN DER WAERDEN B.L., *Nullstellentheorie der Polynomideale* (⟨120⟩, 96(1926), 183).

VAN DER WAERDEN B.L., *Beweis einer Baudetschen Vermutung* (⟨131⟩, 15(1927), 212–216).

VAN DER WAERDEN B.L., *Topologische Begründung des Kalküls der abzählenden Geometrie* (⟨120⟩, 102(1930), 337–362).

VAN DER WAERDEN B.L., *Moderne Algebra*, B. I–II, Berlin (J. Springer), 1930–1931.

VAN DER WAERDEN B.L., *Die Klassifikation der einfachen Lieschen Gruppen* (⟨123⟩, 37(1933), 446–462).

VAN DER WAERDEN B.L., *Zur algebraischen Geometrie I–III* (⟨120⟩, 108 (1933), 113–125; 253–259, 694–698).

VAN DER WAERDEN B.L., *Gruppen von linearen Transformationen*, Berlin (J. Springer), 1935.

VAN DER WAERDEN B.L., *Einführung in die algebraische Geometrie*, Berlin (J. Springer), 1939.

VAN DER WAERDEN B.L., *The foundation of algebraic geometry from Severi to André Weil* (⟨33⟩, 7(1970–1971), 171–180).

VAN DER WAERDEN B.L., *Ein Briefwechsel zwischen P. Finsler und B.L. van der Waerden über die Grundlagen der algebraischen Geometrie* (⟨33⟩, 9(1972–1973), 249–256).

VAN DER WAERDEN B.L., *Die Galois-Theorie von Heinrich Weber bis Emil Artin* (⟨33⟩, 9(1972–1973), 240–248).

VAN DER WAERDEN B.L., *On the sources of my book Moderne Algebra* (⟨83⟩, 2(1975), 31–40).

VAN DER WAERDEN B.L., *A history of Algebra from al-Khwārizmī to Emmy Noether*, Berlin (Springer-Verlag), 1985.

WALD A., *Statistical decision functions*, New York (John Wiley), 1950.

WALD A. and WOLFOWITZ J., *Optimum character of the sequential probability ratio test* (⟨27⟩, 19(1948), 326–339).

WALL H.S., *Analytic theory of continued fractions*, New York (D. Van Nostrand), 1948.

WALTER W., *Differential and integral inequalities*, Berlin (Springer-Verlag), 1970.

WANG H., *Reflections on Kurt Gödel*, Cambridge, Massachusetts (The MIT Press), 1987 = *Kurt Gödel*, Paris (Armand Colin), 1990.

WAŻEWSKI T., *Une méthode topologique de l'examen du phénomène asymptotique relativement aux équations différentielles ordinaires* (⟨36⟩, (8), 3(1947), 210–215).

VON WEBER E., *Vorlesungen über das Pfaff'sche Problem und die Theorie der partiellen Differentialgleichungen erster Ordnung*, Leipzig (B.G. Teubner), 1900.

WEBER H., *Lehrbuch der Algebra*, zweite Auflage, dritter Band: *Elliptische Funktionen und algebraische Zahlen*, Braunschweig (Vieweg), 1908.

WEDDENBURN J.H. MACLAGAN, *A theorem of finite algebras* (⟨185⟩, 6(1905), 349–352).

WEDDENBURN J.H. MACLAGAN, *On hypercomplex numbers* (⟨150⟩, (2), 6 (1908), 77–118).

WEIL A., *L'arithmétique sur les courbes algébriques* (⟨7⟩, 52(1928), 1–35) = p. 11–45, vol. I, *Oeuvres scientifiques*, New York (Springer-Verlag), 1980.

WEIL A., *Sur les espaces à structure uniforme et sur la topologie générale*, Paris (Hermann), 1937 = p. 147–183, vol. I, *Oeuvres scientifiques*, New York (Springer-Verlag), 1980.

WEIL A., *L'intégration dans les groupes topologiques et ses applications*, Paris (Hermann), 1940.

WEIL A., *Sur les fonctions algébriques à corps de constantes fini* (⟨66⟩, 210(1940), 592–594) = p. 257–259, vol. I, *Oeuvres scientifiques*, New York (Springer-Verlag), 1980.

WEIL A., *Foundations of algebraic geometry*, New York (American Mathematical Society), 1946.

WEIL A., *Sur les courbes algébriques et les variétés qui s'en déduisent*, Paris (Hermann), 1948.

WEIL A., *Variétés abéliennes et courbes algébriques*, Paris (Hermann), 1948.

WEIL A., *L'avenir des mathématiques*, p. 307–320, F. LE LIONNAIS (présentés par), *Les grands courants de la pensée mathématique*, Cahiers du Sud, 1948 = p. 359–372, vol. I, *Oeuvres scientifiques*, New York (Springer-Verlag), 1980.

WEIL A., *Numbers of solutions of equations in finite fields* (⟨55⟩, 55(1949), 497–508) = p. 399–410, vol. I, *Collected papers*, New York (Springer-Verlag), 1980.

WEIL A., *De la métaphysique aux mathématiques* (⟨169⟩, 1960, 52–56) = p. 408–412, vol. II, *Oeuvres scientifiques*, New York (Springer-Verlag), 1980.

WEIL A., *The collected papers of Emil Artin* (⟨171⟩, 28(1967), 237–238) = p. 173–174, vol. III, *Oeuvres scientifiques*, New York (Springer-Verlag), 1980.

WEIL A., *L'oeuvre mathématique de Delsarte*, p. 29–47, tome I, J. DELSARTE, *Oeuvres*, Paris (C.N.R.S.), 1971.

WEIL A., *S.S. Chern as geometer and friend*, p. IX–XII, S.-S. CHERN, *Selected papers*, New York (Springer-Verlag), 1978.

WEIL A., *Commentaire*, p. 522–578, vol. I, *Oeuvres scientifiques*, New York (Springer-Verlag), 1980.

WEIL A., *Souvenirs d'apprentissage*, Bâle (Birkhäuser), 1991.

WEITZENBÖCK R., *Invariantentheorie*, Groningen (P. Noordhoff), 1923.

WELLS R.O. JR. (editor), *The mathematical heritage of Hermann Weyl*, Providence, Rhode Island (American Mathematical Society), 1988.

WEYL H., *Über die gewöhnlichen Differenzialgleichungen mit Singularitäten und die zugehörigen Entwicklungen willkürlicher Funktionen* (⟨120⟩, 68(1910), 220–269) = S. 248–297, Band I, *Gesammelte Abhandlungen*, Berlin (Springer-Verlag), 1968.

WEYL H., *Die Idee der Riemannschen Fläche*, Leipzig (B.G. Teubner), 1913.

WEYL H., *Über die Gleichverteilung von Zahlen mod. Eins* (⟨120⟩, 77 (1916), 313–352).

WEYL H., *Reine Infinitesimalgeometrie* (⟨123⟩, 2(1918), 384–411).

WEYL H., *Das Kontinuum. Kritische Untersuchungen über die Grundlagen der Analysis*, Leipzig (Veit), 1918.

WEYL H., *Raum, Zeit, Materie. Vorlesungen über allgemeine Relativitätstheorie*, Berlin (J. Springer), 1918 = *Temps, espace, matière. Leçons de la relativité générale*, Paris (Blanchard), 1922.

WEYL H., *Theorie der Darstellung kontinuierlicher halbeinfacher Gruppen durch lineare Transformationen* (⟨123⟩, 23(1925), 271–309; 24(1926), 328–395, 789–791) = S. 543–647, Band II, *Gesammelte Abhandlungen*, Berlin (Springer-Verlag), 1968.

WEYL H., *Gruppentheorie und Quantenmechanik*, Leipzig (S. Hirzel), 1928.

WEYL H., *Emmy Noether* (⟨171⟩, 3(1935), 201–220) = S. 425–444, Band III, *Gesammelte Abhandlungen*, Berlin (Springer-Verlag), 1968.

WEYL H., *Cartan on groups and differential geometry* (⟨55⟩, 44(1938), 598–601) = S. 592–595, Band IV, *Gesammelte Abhandlungen*, Berlin (Springer-Verlag), 1968.

WEYL H., *On the volume of tubes* (⟨12⟩, 61(1939), 461–472).

WEYL H., *The classical groups, their invariants and representations*, Princeton (University Press), 1939.

WEYL H., *Algebraic theory of numbers* , Princeton University Press, 1940.

WEYL H., *The method of orthogonal projection in potential theory* (⟨70⟩, 7(1940), 411–444) = S. 758–791, Band III, *Gesammelte Abhandlungen*, Berlin (Springer-Verlag), 1968.

WEYL H., *David Hilbert and his mathematical work* (⟨55⟩, 50(1944), 612–654) = S. 130–172, Band IV, *Gesammelte Abhandlungen*, Berlin (Springer-Verlag), 1968.

WEYL H., *A half-century of mathematics* (⟨13⟩, 58(1951), 523–553) = p. 464–494, Band IV, *Gesammelte Abhandlungen*, Berlin (Springer-Verlag), 1968.

WHITEHEAD A.N. and RUSSELL B., *Principia mathematica*, vol. I, Cambridge (University Press), 1910.

WHITEHEAD G.W., *A generalization of the Hopf invariant* (⟨28⟩, (2), 51(1950), 192–237).

WHITEHEAD G.W., *The* $(n+2)^{nd}$ *homotopy group on the n-sphere* (⟨28⟩, (2), 52(1950), 245–247).

WHITEHEAD G.W., *Fifty years of homotopy theory* (⟨55⟩, new series, 8(1983), 1–29).

WHITEHEAD J.H.C., *Simplicial spaces, nuclei and m-groups* (⟨150⟩, (2), 45 (1938), 243–327).

WHITEMORE H., *Writing about Alan Turing* (⟨112⟩, 13(1991), N° 4, 26–30).

WHITNEY H., *Analytic extensions of differentiable functions in closed sets* (⟨185⟩, 36(1934), 63–89).

WHITNEY H., *Differentiable manifolds in euclidian space* (⟨151⟩, 21 (1935), 462–463).

WHITNEY H., *On regular closed curves in the plane* (⟨65⟩, 4(1935), 276–284).

WHITNEY H., *Sphere space* (⟨151⟩, 21(1935), 464–468).

WHITNEY H., *Differentiable manifolds* (⟨28⟩, (2), 37(1936), 645–680).

WHITNEY H., *The maps of an n-complex into an n-sphere* (⟨70⟩, 3(1937), 51–55).

WHITNEY H., *Topological properties of differentiable manifolds* (⟨55⟩, 43(1937), 785–805).

WHITNEY H., *On products in a complex* (⟨28⟩, (2), 39(1938), 397–432).

WHITNEY H., *On the theory of sphere-bundles* (⟨151⟩, 26(1940), 148–153).

WHITNEY H., *On the topology of differentiable manifolds*, p. 101–141, *Lecture in topology*, Ann Arbor (University of Michigan Press), 1941.

WHITNEY H., *On the extension of differentiable functions* (⟨55⟩, 50 (1944), 75–81).

WHITNEY H., *The singularities of a smooth n-manifold in* $(2n-1)$ *spaces* (⟨28⟩, (2), 45(1944), 247–293).

WHITNEY H., *On ideals of differentiable functions* (⟨12⟩, 70(1948), 635–658).

WHITTAKER E.T., *A course of modern analysis: an introduction of infinite series and of analytic functions; with an account of the principal transcendental functions*, Cambridge (University Press), 1902.

WIEFERICH A., *Zum letzten Fermatschen Theorem* (⟨98⟩, 136(1909), 293–302).

WIENER N., *On the theory of sets of points in terms of continuous transformations*, p. 312–315, *Comptes rendus du Congrès international des mathématiciens*

(Strasbourg, 22–30 septembre 1920), Toulouse (Edouard Privat), 1921 = p. 281–284, vol. I, *Collected works*, Cambridge, Massachusetts (The MIT Press), 1976.

WIENER N., *Limit in terms of continuous transformations* (⟨49⟩, 50 (1922), 119–134) = p. 303–318, vol. I, *Collected works*, Cambridge, Massachusetts (The MIT Press), 1976.

WIENER N., *Differential space* (⟨99⟩, 2(1923), 131–174) = p. 455–498, vol. I, *Collected works*, Cambridge, Massachusetts (The MIT Press), 1976.

WIENER N., *Certain notions in potential theory* (⟨99⟩, 3(1924), 24–51) = p. 364–391, vol. I, *Collected works*, Cambridge, Massachusetts (The MIT Press), 1976.

WIENER N., *The operational calculus* (⟨120⟩, 95(1926), 557–584) = p. 397–424, vol. II, *Collected works with commentaries*, Cambridge, Massachusetts (The MIT Press), 1979.

WIENER N., *Generalized harmonic analysis* (⟨7⟩, 55(1930), 117–258) = p. 183–260, vol. II, *Collected works*, Cambridge, Massachusetts (The MIT Press), 1979.

WIENER N., *Tauberian theorems* (⟨28⟩, 33(1932), 1–100) = p. 519–619, vol. II, *Collected works*, Cambridge, Massachusetts (The MIT Press), 1979.

WIENER N., *The Fourier integral and certain of its applications*, New York (Cambridge University Press), 1933.

WIENER N., *The historical background of harmonic analysis*, p. 56–68, vol. II, *Semicentennial addresses of the American Mathematical Society*, New York, 1938 = p. 794–806, vol. II, *Collected works*, Cambridge, Massachusetts (The MIT Press), 1979.

WIENER N., *The ergodic theorem* (⟨70⟩, 5(1939), 1–18) = p. 672–689, vol. I, *Collected works*, Cambridge, Massachusetts (The MIT Press), 1976.

WIENER N., *Cybernetics or control and communication in the animal and the machine*, Paris (Hermann), 1948.

WIENER N., *Extrapolation, interpolation, and smoothing of stationary time series*, New York (John Wiley), 1949.

WIENER N., *I am a mathematician*, New York (Doubleday), 1956.

Norbert Wiener (1894–1964) (⟨55⟩, 72(1966), n° 1, part II) = Providence, RI (American Mathematical Society), 1988.

WIENER N., *Collected works*, volume I, Cambridge, Massachusetts (The MIT Press), 1976.

WIENER N., *Collected works*, vol. II, Cambridge, Massachusetts (The MIT Press), 1979.

WIENER N. und HOPF E., *Über eine Klasse singulärer Integralgleichungen* (⟨177⟩, 1931, 696–706) = p. 33–43, vol. III, N. WIENER, *Collected works with com-*

mentaries, Cambridge, Massachusetts (The MIT Press), 1981. (p. 44–53, J. PINCUS, *Commentary*).

WIGNER E., *On unitary representations of the inhomogeneous Lorentz group* (⟨28⟩, (2), 40(1929), 149–204).

WILCZYNSKI E.J., *Projective differential geometry of curves and ruled surfaces*, Leipzig (B.G. Teubner), 1906.

WILDER R.L., *Evolution of the topological concept of «connected»* (⟨13⟩, 85 (1978), 720–726).

WILDER R.L., *The mathematical work of R.L. Moore: its background nature and influence* (⟨33⟩, 26(1982), 73–97).

WILLARD S., *General topology*, London (Addison-Wesley), 1970.

WILSON E.B., *Vector analysis founded upon the lectures of J. Willard Gibbs*, New York, 1901.

WITT E., *Theorie der quadratischen Formen in beliebigen Körpern* (⟨98⟩, 176 (1937), 31–44).

WUSSING H., *Die Genesis des abstrakten Gruppenbegriffes*, Berlin (VEB Deutscher Verlag der Wissenschaften), 1969.

YAGLOM I.M., *Combinatorial geometry and discrete geometry — their places in science and in the contemporary world* (in Russian) (⟨182⟩, 22(1987), 139–148).

YOSIDA K., *On the differentiability and the representation of one-parameter semigroup of linear operators* (⟨105⟩, 1(1948), 15–21).

YOUNG L.C., *On approximation by polygons in the calculus of variations* (⟨154⟩, 141(1933), 325–341).

YOUNG W.H., *A note in trigonometrical series* (⟨126⟩, (2), 38(1908), 44–48).

YOUNG W.H., *On a new method in the theory of integration* (⟨150⟩, (2), 9(1910), 15–50).

YOUNG W.H., *On the determination of the summability of a function by means of its Fourier constants* (⟨150⟩, (2), 12(1913), 71–88).

YOUNG W.H., *Some characteristic features of twentieth century pure mathematical research*, p. 157–169, vol. I, *Proceedings of the international mathematical congress*, 1924, Toronto (The University of Toronto Press), 1928.

YOUNG W.H. and G.C., *On the theorem of Riesz-Fischer* (⟨156⟩, 44 (1913), 49–88).

YOURGRAU P., *The disappearance of time. Kurt Gödel and the idealistic tradition in philosophy*, Cambridge (University Press), 1991.

YOUSCHKEVITCH A.A., *Khinchin, Aleksandr Yakovlevich*, p. 351–352, vol. VII, *Dictionary of scientific biography*, New York (Charles Scribner's Sons), 1973.

YOUSCHKEVITCH A.P., *History of mathematics in Russia up to 1917* (in Russian), Moscow (Nauka), 1968.

ZAREMBA S., *Sur le principe du minimum* (⟨54⟩, 7(1909), 197–264).

ZARISKI O., *Algebraic surfaces*, Berlin (Springer), 1935.

ZARISKI O., *The reduction of the singularities of an algebraic surface* (⟨28⟩, (2), 40(1939), 639–689) = p. 325–375, vol. I, *Collected papers*, Cambridge, Massachusetts (The MIT Press), 1972.

ZARISKI O., *Reduction of the singularities of algebraic three dimensional varieties* (⟨185⟩, 55(1944), 132–151).

ZARISKI O., *The compactness of the Riemann manifold of an abstract field of algebraic functions* (⟨55⟩, 50(1944), 683–691).

ZARISKI O., *Generalized semi-local rings* (⟨183⟩, 1(1946), 169–195) = p. 13–19, vol. II, *Collected papers*, Cambridge, Massachusetts (The MIT Press), 1973.

ZEILON N., *Ivan Fredholm*, p. I–XVI, I. FREDHOLM, *Oeuvres complètes*, Malmö (Litos Reprotryck), 1955.

Zentralblatt für Mathematik, Kumulatives Autorenregister zu den Bänden 1–25, 26–41 (62(1974); 63(1976)).

ZERMELO E., *Beweis, dass jede Menge wohlgeordnet werden kann* (⟨120⟩, 59 (1904), 514–516).

ZERMELO E., *Untersuchungen über die Grundlagen der Mengenlehre I* (⟨120⟩, 65(1908), 261–281).

ZERVOS S.P., *Marc Krasner (1912–1985)* (⟨72⟩, 1985, 9–37).

ZORN M., *Theorie der alternativen Ringe* (⟨1⟩, 8(1930), 123–147).

ZORN M., *A remark on method in transfinite algebra* (⟨55⟩, 41(1935), 667–670).

ZYGMUND A., *Trigonometrical series*, Warszawa, 1935.

ZYGMUND A., *Notes on the history of Fourier series*, p. 1–19, J.M. ASH (editor), *Studies in harmonic analysis*, The American Association of America, 1976.

I. Schur

O. Toeplitz

Scientific council — Faculty of mathematics and mechanics, Moscow, 1967. 1: A.O. Gel'fond, N.V. Efimov, A.B. Shidlovskiy; 2: *, *, I.S. Berezin, A.N. Tihonov, P.S. Alexandrov; 3: P.K. Rashevskiy, A.G. Kurosh, V.V. Nemickiy, A.A. Markov, S.A. Gal'iern; 4: Ishlinskiy, D.E. Men'shov, Slezkin, Kosmodamianskiy, G.G. Chern'iy, Sretenskiy; 5: B.V. Gnedenko, Roytenberg, B.P. Kishkin; 6: Komarov, *

Mathematical Society of Moscow, 1975

Index of Names

We acknowledge contributions of pictures by Z. Ciesielski, S. Demidov, J. Mawhin, F. Pescatore, M. von Renteln.

MIX
Papier aus verantwortungsvollen Quellen
Paper from responsible sources
FSC® C105338

If you have any concerns about our products,
you can contact us on
ProductSafety@springernature.com

In case Publisher is established outside the EU,
the EU authorized representative is:
Springer Nature Customer Service Center GmbH
Europaplatz 3, 69115 Heidelberg, Germany

Printed by Libri Plureos GmbH
in Hamburg, Germany